D0912964

CHEMICAL HAZARDS OF THE WORKPLACE

Also by the Authors

The Pocket Guide to Chemical Hazards of the Workplace

CHEMICAL HAZARDS OF THE WORKPLACE

Second Edition

Nick H. Proctor, Ph.D.
ASSISTANT ADJUNCT PROFESSOR OF OCCUPATIONAL MEDICINE
UNIVERSITY OF CALIFORNIA SAN FRANCISCO

James P. Hughes, M.D.
ASSOCIATE CLINICAL PROFESSOR OF OCCUPATIONAL MEDICINE
UNIVERSITY OF CALIFORNIA SAN FRANCISCO

Michael L. Fischman, M.D., M.P.H.
ASSISTANT CLINICAL PROFESSOR OF OCCUPATIONAL MEDICINE
UNIVERSITY OF CALIFORNIA SAN FRANCISCO
MEDICAL DIRECTOR, INTEL CORPORATION
CONSULTING PHYSICIAN IN OCCUPATIONAL MEDICINE AND TOXICOLOGY
OAKLAND, CALIFORNIA

J. B. Lippincott Company
PHILADELPHIA
London Mexico City New York
St. Louis São Paulo Sydney

Science Information Resource Center

Publisher: Eugene M. Falken
Sponsoring Editor: Delois Patterson
Manuscript Editor: Ann Blum
Design Coordinator: Caren Erlichman
Interior Design: Rita Naughton
Cover Design: Paul Autodore
Production Manager: Carol A. Florence
Production Editor: Rosanne Hallowell
Production Coordinator: Kathryn Rule
Composition: Maryland Composition Company, Inc.
Text Printer/Binder: R. R. Donnelly & Sons Company
Cover Printer: Phoenix Studios

Second Edition

1 3 5 6 4 2

Library of Congress Cataloging-in-Publication Data

Chemical hazards of the workplace.
Includes bibliographies and index.
1. Industrial toxicology. 2. Industrial hygiene.
I. Proctor, Nick H. II. Hughes, James P., 1920– .
III. Fischman, Michael L., 1954– . [DNLM: 1. Occupational Diseases—chemically induced. 2. Poisoning.
3. Poisons. WA 465 C517]
RA1229.C47 1988 363.1′7 87-30982
ISBN 0-397-53025-0

The authors and publishers have exerted every effort to ensure that drug selection and dosage set forth in this text are in accord with current recommendations and practice at the time of publication. However, in view of ongoing research, changes in government regulations, and the constant flow of information relating to drug therapy and drug reactions, the reader is urged to check the package insert for each drug for any change in indications and dosage and for added warnings and precautions. This is particularly important when the recommended agent is a new or infrequently employed drug.

Contributors

Gideon Letz, M.D., M.P.H.
Medical Director
State Compensation Insurance Fund
San Francisco, California

Gloria J. Hathaway, Ph.D.
Consultant to Hughes, Lewis, Fischman and Associates
Oakland, California

Foreword

In the western world, man's curiosity about the origins of occupational disease and his attempts to develop methods of prevention and treatment date back more than 300 years. Although some notable attempts were made in the last 100 years to develop public policies and strategies to control occupational hazards in industrial nations, it has required the impetus of rapid technological and industrial growth after World War II, with its associated new hazards, to generate new levels of concern about the health hazards of the workplace.

With the enactment of the Occuptional Safety and Health Act (OSHA) in 1970, these concerns were translated into law and regulatory strategies, which delineate responsibility and accountability for the control of occupational hazards. In order for these strategies to fulfill their objectives, the quality of occupational health practice requires our attention. It must be based in sound education, for knowledge, understanding, and skill, if properly applied, become wisdom, good judgment, and effective problem solving.

For some time it has been apparent to the occupational health professional that up-to-date references and educational materials, especially those providing essential information and guidelines for occupational health practice, are limited in number and scope. Source books of basic information and manuals of occupational medicine are urgently needed.

The Standards Completion Project, which was sponsored by the National Institute for Occupational Safety and Health and OSHA, provided Drs. Nick H. Proctor and James P. Hughes with an unusual opportunity to develop monographs for nearly 400 potentially hazardous chemical materials. These monographs assembled basic information about chemical, physical, and toxicologic characteristics, as well as diagnostic criteria, including special tests. This material has been updated and additional substances have been added by the authors and their associates in this revised edition of their 1978 handbook.

The introductory chapters provide guidelines for the recognition and evaluation of chemical hazards, and the diagnosis of occupational disease. It should also be noted that one of the unheralded benefits of the Standards Completion Project, had the program been fully implemented, was that it may have been the basis for an effective national occupational health surveillance system.

Everyone concerned with the recognition and control of occupational hazards, and the management of clinical problems arising from chemical exposures, should welcome this very useful volume.

Raymond R. Suskind, M.D.
Director Emeritus
Institute of Environmental Health
University of Cinicinnati Medical Center

Preface

Chemical Hazards of the Workplace, Second Edition, is intended primarily for the health professional seeking a brief introductory statement on the toxicology of over 400 chemicals most likely to be encountered at work.

There are introductory chapters on diagnostic principles, which are supplemented by lists of chemicals according to their major effects, such as respiratory irritation, central nervous system depression, or skin irritation. These lists are not all-inclusive; only the more commonly encountered chemical agents are given.

Chapters on industrial hygiene principles, treatment of occupational disease, and chemically induced pulmonary disease are omitted from this edition since that information is now readily available in other sources.

The main body of material consists of monographs on chemical substances arranged in alphabetical order. All of the 386 monographs from the 1974 NIOSH/OSHA Standards Completion Project have been revised and updated in this work. Monographs on other chemicals of special current interest have been added. Finally, additional information on carcinogenic factors and on current NIOSH recommendations on OSHA health standards are given in appendices.

In reviewing the monographs, the reader might keep in mind a few details: The

Threshold Limit Values (TLVs) listed are those given by the American Conference of Governmental Industrial Hygienists (ACGIH) for 1987. As explained in Chapter 1, the numerical values of the TLVs do not take into account the toxicity that might result from skin absorption. However, in instances where this may occur, skin absorption is listed as a route of exposure. Obviously, all substances existing in liquid or solid form could conceivably be ingested, but this is rare in the workplace and of minor importance except for certain extremely toxic substances such as arsenic and lead. Only in those cases have we listed ingestion as a route of exposure, since contamination of food taken at work may be a factor.

In each monograph we have indicated the chief signs and symptoms caused by overexposure to the chemical and the clinical effects in humans that are related, where data are available, to exposure levels. The reader may consider these clinical effects in the manner of a dose–response relationship.

If the literature does not provide sufficient data to illustrate human effects, data from animal tests are given to point out target organs that should be monitored closely in exposed workers. Sometimes no human data exist, and animal results must be relied upon entirely. In such cases, the experimental data from the literature are summarized. Furthermore, in many instances we have designated the odor threshold or other warning properties.

Treatment of overexposure is considered only for those substances such as methemoglobin formers, anticholinesterases, and certain heavy metals for which specific therapeutic measures are generally accepted. For most of these substances, diagnostic features and special tests are included.

We hope that the information presented here will be of value to physicians and other health professionals in the effective application of some basic toxicologic and clinical principles aimed at protecting the health of working people.

Nick H. Proctor, Ph.D.
James P. Hughes, M.D.
Michael L. Fischman, M.D., M.P.H.

Acknowledgments

We wish to thank the California Department of Health Services for access to the excellent Occupational and Environmental Health Library at Berkeley which greatly facilitated our literature review. We are particularly grateful to librarians Sharon Brunzel and Joyce Johnson for their warm and expert assistance.

We frequently called upon Dr. Gideon Letz for opinions on the text during his service as Public Health Medical Officer for the California Hazard Evaluation System and Information Service (HESIS) at Berkeley, and for contributions to the text (Chapters 2 and 3).

We are also grateful to Adelaide M. Hughes for editorial support and to Thelma Valentino for her valuable clerical assistance.

Some of the work upon which this publication was based initially was performed pursuant to Contract No CDC-99-74-43 with the National Institute for Occupational Safety and Health, now of the U.S. Department of Health and Human Services.

Contents

List of Tables

CHEMICAL HAZARDS OF THE WORKPLACE

Part One

GUIDELINES IN OCCUPATIONAL HEALTH PRACTICE

1

Setting Threshold Limit Values: Toxicologic Concepts

Nick H. Proctor

DEFINITIONS

In occupational health practice, the following terms describe the states of matter in which chemical contaminants may occur[1]:

Gas: A formless fluid that completely occupies the space of an enclosure at 25°C and 760 torr pressure.

Vapor: The gaseous phase of a material that is liquid or solid at 25°C and 760 torr pressure.

Aerosol: A dispersion of particles of microscopic size in a gaseous medium. These may be solid particles (dust, fume, smoke) or liquid particles (mist, fog).

Dust: Airborne solid particles (an aerosol) that range in size from 0.1 to 50 μ and larger in diameter. A person with normal eyesight can detect dust particles as small as 50 μ in diameter. Smaller airborne particles cannot be detected by unaided eyes unless strong light is reflected from the particles. Dust of respirable size (below 10 μ) cannot be seen without the aid of a microscope.

Fume: An aerosol of solid particles generated by condensation from the gaseous state, generally after volatilization from molten metals. The solid particles that make up a fume are extremely fine, usually less than 1.0 μ in diameter. In most cases, the volatilized solid reacts with oxygen in the air to form an oxide. A common example is cadmium oxide fume.

Smoke: An aerosol of carbon or soot particles less than 0.1 μ in diameter that results from the incomplete combustion of carbonaceous materials such as coal or oil. Smoke generally contains droplets as well as dry particles.

Mist: An aerosol of suspended liquid droplets generated by condensation from the gaseous to the liquid state or by the breaking up of a liquid into a dispersed state, such as by splashing, foaming, or atomizing. Mist is formed when a finely divided liquid is suspended in the atmosphere. Examples are the oil mist during cutting and grinding operations, acid mists from electroplating, acid or alkali mists from pickling operations, and paint spray mist from spraying procedures.

Fog: A visible aerosol of a liquid, formed by condensation.

The following terms of measurement are commonly used:

ppm: Parts of vapor or gas per million parts of contaminated air by volume.

mg/m³: Milligrams of a substance per cubic meter of air.

mppcf: Millions of particles of a particulate per cubic foot of air (mainly historical use).

TOXICOLOGIC CONCEPTS

Routes of Entry of Chemicals Into the Body

In the industrial setting, inhalation is the most important route of entry of chemical agents into the body. Next is contact with the skin. In either case, there may be irritation of contacted tissue and/or absorption into the blood with possible systemic intoxication. Although the gastrointestinal tract is a potential site of absorption, the ingestion of significant amounts of chemicals is rare in the industrial situation.

Inhalation. The water solubility of a gas or vapor is an important factor in determining the amount of inhaled material that reaches the lung.[2] Highly water-soluble gases, such as ammonia, hydrogen chloride, and hydrogen fluoride, dissolve readily in the moisture associated with the mucous membrane of the nose and upper respiratory tract. This causes irritation at these sites. At low airborne concentrations, relatively little of these substances will reach the lungs owing to this "scrubbing" effect. At high atmospheric concentrations, however, some of the gas or vapor will not be absorbed at these upper respiratory sites, and amounts sufficient to cause severe irritation and pulmonary edema can reach the alveoli.

Comparatively insoluble gases, such as nitrogen dioxide and phosgene, are not removed by the moisture in the nose and upper respiratory tract and can easily reach the terminal recesses of the lung. Substances of intermediate solubility, such as ozone and chlorine, cause irritation of both the upper respiratory tract and the lung.

Gases such as carbon monoxide do not irritate the respiratory tract but are rapidly absorbed into the blood, resulting in systemic intoxication.

The particle size of aerosols determines the extent of their accessibility to small airways. If the diameter of the particles is larger than 10 μ, impaction occurs on the mucous coat of the pharynx or nasal cavity, and the particles do not reach the alveoli. For this reason, particles of 10 μ or less in diameter are termed *respirable*. Particles between 1 μ and 5 μ often sediment within the bronchioles, whereas particles less than 1 μ can diffuse within the alveoli.[3]

Particulate matter may deposit on the ciliary epithelium, which covers the upper respiratory tract down to the level of the terminal bronchioles. These particles, either in a free state or after phagocytosis, are moved by cilia and the mucous blanket toward the glottis, where they are swallowed or expectorated. Particulate matter that deposits beyond the ciliary epithelium may be absorbed through the alveolar lining into the blood or, as free and phagocytized particles, may enter the lymphatic system. Several industrially important substances, such as crystalline silica, beryllium, and asbestos, resist solubilization by the blood or removal by phagocytes, and remain in the alveoli indefinitely. Irritation, inflammation, fibrosis, allergic sensitization, or malignant change may occur. However, substances such as iron oxide can be present for extended periods of time with no apparent ill effects.

Skin Contact. When a substance contacts the skin, four actions are possible.

- The skin and its associated film of lipid and sweat may act as an effective barrier that the substance cannot penetrate.
- The substance may react with the skin surface and cause primary irritation (acids, alkalis, many organic solvents).
- The substance may penetrate the skin and cause sensitization (formaldehyde, nickel, phthalic anhydride).
- The agent may penetrate the skin, enter the blood, and act systemically (aniline, parathion, tetraethyl lead).[4]

Only a small number of substances are absorbed through the skin in hazardous amounts. In order to pass into the skin, the substance must enter through one or more of the following four routes: the epidermal cells, the sweat glands, the sebaceous glands, or the hair follicles. The pathway through the epidermal cells and the overlying stratum corneum into the blood is probably the main avenue of penetration, since this tissue constitutes the majority of the surface area of the skin.[5] The stratum corneum plays a critical role in determining cutaneous permeability. Absorption is faster through skin that is abraded or inflamed. For this reason, chemicals that are not normally considered hazardous may be dangerous to persons suffering from active inflammatory dermatoses.

Ingestion. Ingestion occurs as a route of entry through eating or smoking with contaminated hands or in contaminated work areas. Ingestion of inhaled material also occurs, as previously described. The amounts ingested are usually of little significance. Exceptions are highly toxic substances, such as lead, arsenic, and mercury.

Dose and Response

Toxicology is the study of the noxious effects of chemical and physical agents. The most fundamental concept in toxicology states that a relationship exists between the dose of an agent and the response that is produced in a biological system. Certain assumptions are made.[6] The first assumption is that the response is due to the chemical administered. The second assumption is that the response is related to the dose, and (a) the chemical interacts with a molecular or receptor site to produce the response; (b) the response is related to the concentration of the chemical at the reactive site; and (c) the concentration at the site is related to the dose. The third assumption is that there is both a quantifiable method of measuring and a precise means of expressing the toxicity.

The first toxicity data on an uncharacterized agent are usually obtained by using mice or rats and by either oral or intraperitoneal administration. A range of doses is found that results in some deaths and some survivals. When the data are plotted, the usual form is a sigmoid curve, as depicted in Figure 1-1.

Statistical analysis of the data allows the calculation of the LD_{50} (*i.e.*, the dose expected to be lethal in 50% of the animals in a group of the same species when administered by the stated route). The agent may also be administered by inhalation at different atmospheric levels for a specific time period. The term then calculated is the LC_{50} (*i.e.*, the atmospheric concentration expected to be lethal to 50% of a group of animals of that species exposed for the specified time period). The LD_{50} or LC_{50} for an agent provides an initial index of com-

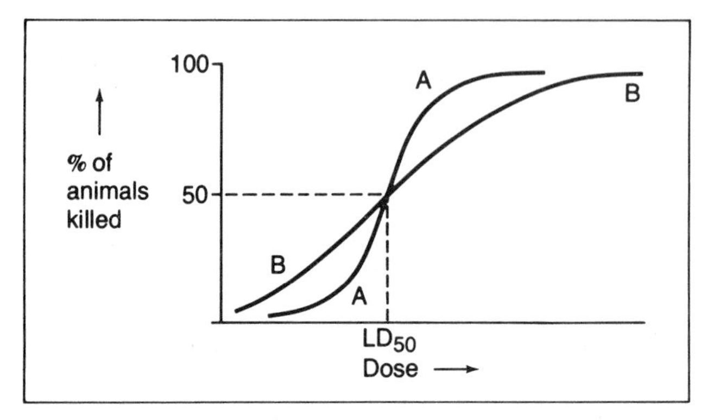

Fig. 1-1. Dose-response curve.

parative toxicity. Table 1-1 is a scheme of classification of toxic substances suggested by Hodge and Sterner.[7]

During the determination of the LD_{50} or LC_{50}, observation of the animals often provides valuable information about the effects that may occur in humans. Autopsy of the animals shows which organs were affected. For many substances, these are the only data available. In such cases, until proven otherwise, it is prudent to suspect that the same effects will occur in humans if exposed to a sufficiently high atmospheric concentration for a sufficient period of time.

The slope of the dose–response curve also provides useful information. It suggests the magnitude of the range between a no-effect level and a lethal dose, a concept known loosely as the margin of safety. If the dose–response curve is especially steep, the margin of safety is small. In Figure 1-1, substances A and B both have the same LD_{50}, but substance B has a wider margin of safety than substance A.

Toxicity and Hazard

Toxicity is the ability of a substance to cause injury to biologic tissue. The *hazard* associated with the substance is the likelihood that it will cause injury in a given environment or situation.

The hazard of a substance depends

TABLE 1-1. Toxicity Classes[7]

Toxicity Rating	*Descriptive Term*	*LD_{50}—wt/kg Single Oral Dose Rats*	*LC_{50}—ppm 4-hr Inhalation Rats*
1	Extremely toxic	1 mg or less	<10
2	Highly toxic	1–50 mg	10–100
3	Moderately toxic	50–500 mg	100–1000
4	Slightly toxic	0.5–5 g	1000–10,000
5	Practically nontoxic	5–15 g	10,000–100,000
6	Relatively harmless	15 g or more	>100,000

first on its toxicity, how it is absorbed, metabolized, and excreted, how rapidly it acts, and its warning properties. A hypothetical substance of extreme hazard would be an agent that is in the "extremely toxic" class (Table 1-1); rapidly absorbed by inhalation, skin contact, or ingestion; slowly metabolized and/or slowly excreted, thus resulting in accumulation in the body; capable of rapidly causing irreversible effects or death; and without warning properties. Clearly, such a substance would be hazardous to handle.

In a second group are other factors that determine the hazard of a material: its physical characteristics and the manner in which the substance will be encountered in the workplace. For example, a liquid with a high vapor pressure will reach a higher airborne concentration and will be more hazardous than an equally toxic liquid with a low vapor pressure. Factors that determine the potential for fire and explosion, such as flash point and lower explosive limit, are also important. Any work practices that will cause high ambient levels or frequent or prolonged exposure will contribute to the hazard associated with a substance.

EXPOSURE

Frequency and duration of exposure to chemicals in toxicologic tests of animals are rather arbitrarily divided into four types:

- *Acute exposure* is defined as exposure for less than 24 hours.
- *Subacute exposure* is repeated exposure to a chemical for 1 month or less.
- *Subchronic exposure* is exposure for 1 to 3 months.
- *Chronic exposure* is exposure for more than 3 months.

In the occupational setting, however, acute human exposure generally means exposure that causes an immediate effect, whereas chronic exposure is applied to any repeated exposures over time.

THE STANDARDS SETTING PROCESS

Threshold Limit Value (TLV)

The American Conference of Governmental Industrial Hygienists (ACGIH) has prepared a list of the threshold limit values (TLVs) for approximately 800 substances. The following three categories of TLVs are specified in the list for 1986–1987[8]:

- *Threshold Limit Value—Time-Weighted Average (TLV—TWA).* This is the time-weighted average concentration for a normal 8-hour workday and 40-hour workweek to which nearly all workers may be repeatedly exposed, day after day, without adverse effect.
- *Threshold Limit Value—Short-Term Exposure Limit (TLV—STEL).* This is the highest concentration to which workers can be exposed for short periods of time without suffering from (a) irritation, (b) chronic or irreversible tissue change, or (c) narcosis of sufficient degree to increase accident-proneness, impair self-rescue, or materially reduce work efficiency. This is provided that no more than four excursions per day are permitted within this limit, with at least 60 minutes between exposure periods at the STEL, and provided that daily TLV—TWA is also not exceeded. STELs are recommended only where toxic effects have been reported from high short-term exposures in either humans or animals.
- *Threshold Limit Value—Ceiling (TLV—C).* This is the concentration that should not be exceeded during any part of the working day.

It cannot be overemphasized that TLVs are guidelines and are not intended as absolute boundaries between safe and dangerous concentrations. (Exceptions are those substances that have a "C" or Ceiling Limit.) Time-weighted averages permit limited excursions above the TLV, provided that they are compensated for by equivalent excursions below the limit during the workday. TLVs for some substances bear the notation "skin." This designation calls attention to the fact that absorption of the substance through the skin, mucous membranes, or eyes is possible. The TLV does not take into account absorption by these routes.

TLVs range from 0.0002 ppm for osmium tetroxide to 5,000 ppm for carbon dioxide. TLVs for respirable dusts, such as crystalline silica and nonasbestiform talc, are expressed in terms of mg/m^3. TLVs are periodically revised as new information becomes available. Each year, new substances are added to the TLV list. Certain compounds that are proven or suspected carcinogens in humans, such as benzidine, β-naphthylamine, 4-aminodiphenyl, and 4-nitrodiphenyl, have no TLV value, and human exposure to these agents should be avoided.

Every occupational health professional should have a copy of the current *Threshold Limit Values and Biological Exposure Indices.* It may be obtained from the American Conference of Governmental Industrial Hygienists, 6500 Glenway Ave., Bldg. D-7, Cincinnati, Ohio 45211–4438, at a modest cost.

OSHA Standards

It is important to understand the relationship between the TLVs recommended by the ACGIH and the numerical standards (permissible exposure limits) promulgated by the Occupational Safety and Health Administration (OSHA). The genesis of the OSHA standards occurred when, with only minor changes, the 1968 ACGIH list of nearly 400 TLVs, as well as certain standards of the American National Standards Institute (ANSI), were incorporated into the Walsh-Healey Public Contracts Act. They thereby became limits of exposure for employees of government contractors. Subsequently, under the authority of the Occupational Safety and Health Act of 1970, these same 1968 TLVs and ANSI standards were promulgated by OSHA as the start-up permissible exposure limits for all workers.

It is much more difficult for OSHA to change a permissible exposure limit than it is for the ACGIH to change a TLV. Either a lengthy rule-making procedure or an emergency temporary standard is required before an OSHA permissible exposure limit can be changed or before a value for a newly added chemical substance can be promulgated. Examples of emergency temporary standards have been those for vinyl chloride and benzene. As mentioned previously, the ACGIH adds new substances to the TLV list every year and revises the existing values as necessary. Thus, there are substances for which the current TLVs are lower than the OSHA standards, and there are substances on the current TLV list for which there are no OSHA standards.

REFERENCES

1. Olishifski JB: Fundamental concepts of industrial hygiene. In Olishifski JB, McElroy FE (eds): Fundamentals of Industrial Hygiene, pp 1–49. Chicago, National Safety Council, 1971
2. Amdur MO: Industrial toxicology. In National Institute for Occupational Safety and Health: The Industrial Environment—Its Evaluation and Control, pp 61–73. Washington, DC, US Government Printing Office, 1973
3. Morrow PE, et al: The clearance of dust from the lower respiratory tract in man: An experimental study. In Davies CN (ed): Inhaled Particles and Vapours, pp 351–359. New York, Pergamon Press, 1965
4. Stokinger HE: Routes of entry and modes of action. In Key MM et al (eds): Occupational Diseases—A Guide to Their Recognition, rev ed,

pp 11–21. Washington, DC, US Government Printing Office, 1977
5. Loomis TA: Essentials of Toxicology, 2nd ed, pp 63–75. Philadelphia, Lea & Febiger, 1974
6. Klaassen CD: Principles of toxicology. In Klaassen CD, Amdur MO, Doull J (eds): Casarett and Doull's Toxicology, 3rd ed, pp 11–32. New York, Macmillan, 1986
7. Hodge HC, Sterner JH: Tabulation of toxicity classes. Am Ind Hyg Q 10:93, 1949
8. TLVs—Threshold Limit Values and Biological Exposure Indices for 1986–87. Cincinnati, American Conference of Governmental Industrial Hygienists (ACGIH) 1986

2

Clinical Manifestations of Occupational Chemical Exposure Seen in Emergency Medical Situations

James P. Hughes and Gideon Letz

The patient whose illness results from exposure to a toxic agent encountered at work may present to the physician either in acute distress or with evidence of a chronic condition. The acute intoxications likely to be seen in the emergency facility must be suspected in order to be identified.

WHEN TO SUSPECT OCCUPATIONAL DISEASE

Typically, the acutely ill person is brought in directly from the workplace, where an unusual episode of chemical exposure has occurred. However, exposure to irritant gases initially produces a mild upper respiratory reaction; more serious pulmonary effects develop within 6 to 12 hours following exposure.

It is essential that an accurate history be obtained from those persons accompanying the victim as well as from the patient. More than one worker may have similar complaints of varying severity but may have been taken to another treatment facility. The work supervisor can often provide the most useful information on working conditions and the chemicals in use. The patient may have only a vague idea of these chemicals, referring to them in colloquial terms used only in that shop (*e.g.*, "dope," or "slush,"). In any event, the supervisor or a fellow worker may often be pressed into service to obtain additional information quickly from the worksite, such as the chemical identification of substances in use. Access to Material Safety Data Sheets (MSDS) for substances suspected of exposure should be requested.

A first impression of possible occupational illness may be gained from a chemical odor about the patient, for example, from the breath, the hair and skin, or the clothing and shoes. It is useful to record the impression and to characterize the odor if possible. Contaminated clothing, including undergarments, should be removed promptly, because they may possibly be a source of further absorption by skin or through inhalation.

SOURCES OF EXPOSURE

Chemical exposures may result from an accident in the workplace, an explosion, a fire, or a spill. A less dramatic event, such as a leaking valve, vessel, or pipe, may also contaminate the atmosphere. An exhaust fan may have failed, or the cover of a tank may be damaged or carelessly replaced. Clothing may be contaminated from a dripping overhead line, and walking through a puddle of a spilled chemical may contaminate shoes. Exposures often occur during maintenance work on equipment or when an unfamiliar chemical is introduced into the laboratory or workshop.

The first warning sign of toxic effects from a hazardous exposure is often observed by a fellow worker. Aberrant behavior may be noted if the chemical has narcotic properties. Pallor or cyanosis may appear before the victim is aware of overexposure; "blue lip" is recognized by chemical workers where methemoglobin-forming agents such as aniline and nitrobenzene are encountered. This is an emergency indication for immediate removal from the site and thorough showering with soap and water.

If the primary victim has collapsed at the site of exposure, the rescuers should also be observed for signs of possible overexposure to the contaminated atmosphere.

PRESENTING COMPLAINTS

The patient's complaint may be of a quite sudden and unmistakable onset, as with an irritant gas such as chlorine. More commonly, the onset of symptoms is vague and insidious, as occurs from exposure to a cumulative poison such as lead or mercury, or to a neurotoxic solvent as methyl butyl ketone.

The delayed onset of acute respiratory distress, which often leads to pulmonary edema following exposure to an irritant gas, was mentioned previously. Except for mild upper respiratory irritation at the time of exposure, a symptom-free latent period of about 12 hours may occur. The patient typically awakens from sleep short of breath, coughing, and requiring prompt relief.

Also of delayed onset is the skin irritation from sunshine following exposure to a photosensitizing chemical such as coal tar pitch.

Another condition characterized by a latent period of several hours is the painful eye irritation termed *welder's flash*. Although resulting from exposure to the ultraviolet light emanating from an electric arc, rather than from a chemical exposure, this condition is encountered frequently among workers near a welding site as well as among welders themselves.

DIAGNOSTIC LEADS

The differential diagnosis of certain syndromes encountered in the emergency room should include consideration of possible occupational factors. It is always useful to inquire, "What is your work?" Consider occupations of the recent past as well as current occupation if the condition is acute, and of the remote past if chronic, such as pulmonary fibrosis. For example, think of possible chemical exposure in the following patients:

The Patient in Respiratory Distress

Upper respiratory—mild irritants (Table 2-1) or mild exposures to severe irritants (Table 2-2)
Tracheobronchitis, pulmonary edema, pneumonitis—severe irritants (Table 2-2)

Asthma—pulmonary sensitizers (Table 2-3)
Chronic respiratory failure—beryllium (granulomatosis), fibrogenic dusts (Table 2-4) to be differentiated from agents causing benign pneumoconiosis (Table 2-5), pulmonary carcinogens (Table 2-6)

The Cyanotic Patient

Cholinesterase inhibitors (Table 2-7)
Methemoglobin formers (Table 2-8)
Asphyxiants (Table 2-9)

The Unconscious Patient

Central nervous system depressants (Table 2-10)
Cholinesterase inhibitors (Table 2-7)
Asphyxiants (Table 2-9)

The Mildly Intoxicated Patient

A less critically ill person may be suspected of having a chemical intoxication arising from the work environment when certain clinical signs are present. The most frequent presenting complaints of milder intoxication are unusual fatigue, lassitude, irritability, headache, nausea, and light-headedness. Respiratory irritants produce lacrimation, nasal irritation, sneezing, cough, and tightness in the chest. Skin irritation due to the work environment is quite common.

The diagnosis of occupational poisoning is based upon a precise history of exposure to a specific toxic substance, a determination of the nature and severity of the exposure (preferably confirmed by an industrial hygienist's evaluation of the worksite), clinical signs and symptoms characteristic of the suspected toxic state, and supporting laboratory evidence of absorption of the toxic substance in hazardous amounts.

PHYSICAL CLUES TO OCCUPATIONAL POISONING

General Appearance

Odor. A few chemicals have such a characteristically obnoxious odor as to be diagnostic—namely, the alkyl mercaptans (usually encountered around petroleum refineries) and hydrogen sulfide.

Fever. Suspect the following: metal fumes (especially zinc oxide, copper, magnesium, cadmium); dinitro-*o*-cresol, dinitro-phenol, and pentachlorophenol; polymer fume (polytetrafluoroethylene); and dust of cotton (mill fever), bagasse, or moldy hay.

Sweating. This may be caused by organophosphates, dinitro-*o*-cresol, or pentachlorophenol.

Eye

Dr. W. Morton Grant of Harvard, in his monumental work *Toxicology of the Eye,* presents a detailed survey of eye or vision disturbances produced by either direct contact with chemicals or systemic routes. The following brief list of eye disturbances is selected from his work.*

A. Corneal and Conjunctival Disturbances

Chemical burns—caustic alkalis and acids
Solvent splashes—most organic solvents
Surfactant and detergent injuries—liquid splashes or dusts
Delayed or latent action injuries from splashes or dusts—dimethyl sulfate, formaldehyde, methyl bromide, methyl dichloroproprionate, osmic acid, sulfur dioxide
Delayed corneal epithelial edema (producing haloes around lights) from vapor exposure—allyl alcohol, amines, diethyldiglycolate, diisopropylamine, 3-dimethylaminopropylamine, ethylenediamine, tetramethylbutanediamine, triethylenediamine
Corneal epithelial vacuoles after chronic vapor exposure—*n*-butanol, xylene

* Grant WM: Toxicology of the Eye, 3rd ed, 1986. Courtesy of Charles C Thomas, Publisher, Springfield, Illinois.

Delayed corneal epithelial injury (keratitis epithelialis) with blurring and pain several hours after exposure to gases, vapors, or dusts—allyl alcohol, diazomethane, dichlorobutenes, dimethyl sulfate, ethylene oxide, ethylenimine, hydrogen sulfide, methyl bromide, osmic acid

Corneal and conjunctival discoloration with late corneal scarring and distortion—benzoquinone

Conjunctival discoloration from internal substances—aniline, arsine, nitrobenzene, silver

B. Lens, Iris, and Anterior Chamber Disturbances

Cataracts from systemic chemicals—dinitrol-*o*-cresol, dinitrophenol

Lens deposits and discoloration—copper, iron, mercury, phenylmercuric salts, silver

Myopia with miosis—anticholinesterase agents

C. Posterior Segment and Optic Nerve Disturbances

Retinal function; dark adaptation altered—carbon dioxide, carbon disulfide, carbon monoxide

Retinal edema—cyanide, methanol

Retinal hemorrhages—acetylphenylhydrazine, benzene, lead, methyl bromide, triethyl tin, warfarin

Papilledema—ethylene glycol, lead, phosphorus, triethyl tin

Optic neuritis—dinitrobenzene, dinitrotoluene, lead, methanol, thallium, triethyl tin

Peripheral visual field constriction—carbon dioxide, carbon monoxide, methyl mercury, methanol, naphthalene

D. Central Nervous System Effects

Cortical blindness or hemianopsia—carbon monoxide, lead

E. Intraocular Pressure Disturbances

Pressure elevation from chemical burns—ammonia, calcium hydroxide, formaldehyde

F. Eye and Lid Disturbances

Nystagmus—carbon disulfide, dieldrin, ethyl alcohol, ethylene glycol, methyl bromide, methyl chloride, methyl iodide

Diplopia—carbon disulfide, carbon monoxide, ethyl alcohol, ethylene glycol, lead, methyl bromide, methyl chloride, methyl iodide, triethyl tin

Paralysis of extraocular muscles—ethyl alcohol, lead, thallium, triethyl tin

Ptosis of eye lids—thallium

Lid edema from systemic substances—arsine, hydralazine

G. Pupillary Status

Dilated pupils—CNS (central nervous system) depressants (Table 2-10), organophosphates

Constriction of pupils—organophosphates

Nervous System

Severe headache—carbon monoxide, nitrites, alcohols, organic and inorganic lead compounds, methemoglobin formers (Table 2-8)

Drowsiness, disorientation, vertigo—CNS depressants (Table 2-10)

Peripheral neuropathy—neurotoxins (Table 2-11)

Behavioral changes—CNS depressants (Table 2-10), convulsants (Table 2-12), carbon disulfide, carbon monoxide, organic lead, mercury, methyl chloride, manganese

Irritability, hyperactivity, convulsions—convulsants (Table 2-12)

Cerebrospinal effects, tremor, spasticity, muscle weakness, ataxia, hyperreflexia, micrographia—manganese, carbon disulfide, organic lead, organic tin, mercury, DDT, methylene chloride

Nose

Rhinitis—irritants (Tables 2-1, 2-2)

Epistaxis—severe irritants (Table 2-2)

Perforated septum—severe irritants (Table 2-2), especially chromic acid mist, soluble chromate dusts, acid gases

Persistent sneezing—chlorine; riot control agents, especially *o*-chlorobenzylidene malononitrile

Carcinoma of the paranasal sinuses—nickel carbonyl, isopropyl oil (extremely rare)

Mouth and Throat

Green tongue—vanadium
Salivation—mercury, arsenic, organophosphates
Gingivitis, with or without punctate pigmentation—mercury, lead, bismuth
Mottled dental enamel—*not* a sign of occupational exposure to fluorides but may suggest childhood exposure through ingestion
Pharyngitis—severe irritants (Table 2-2)
Carcinoma of larynx—isopropyl oil (extremely rare)

Pulmonary System

Acute Conditions

Tracheobronchitis—severe irritants (Table 2-2)
Pneumonitis—special emphasis on ammonia, chlorine, oxides of nitrogen, ozone, phosgene, sulfur dioxide, vanadium pentoxide, mercury, manganese, cadmium dust and fume
Asthma—pulmonary sensitizers (Table 2-3)
Acute bronchiolar constriction (nonsensitization)—phthalic anhydride, maleic anhydride

Chronic Conditions

Fibrosis—fibrogenic dusts (Table 2-4)
Bronchiolitis obliterans—nitrogen dioxide
Granulomatosis—beryllium
Radiopaque dust deposits without fibrosis (benign pneumoconiosis)—iron, tin, barium (Table 2-5)
Bronchogenic carcinoma—pulmonary carcinogens (Table 2-6)
Mesothelioma—asbestos

Cardiovascular

Hypotension—CNS depressants (Table 2-10) nitroglycerin, nitrated glycols, nitrites
Hypertension—diphenyl, diphenyl oxide
Conduction impairment (EKG)—barium sulfate
Arrhythmias—trichloroethylene, carbon tetrachloride
Tachycardia—alcohol, arsenic, dinitro-*o*-cresol, dinitrophenol, pentachlorophenol
Myocardial degeneration—carbon disulfide, antimony, arsine
Sensitization of myocardium to epinephrine effects—fluorinated hydrocarbons

Breasts

Gynecomastia—estrogens (pharmaceutical manufacture)

Digestive System

Nausea, vomiting—ubiquitous symptoms of intoxication by many chemicals: CNS depressants (Table 2-10), cholinesterase inhibitors (Table 2-7), methemoglobin formers (Table 2-8), and irritants (Tables 2-1 and 2-2)
Abdominal colic—inorganic lead, arsenic, thallium, organophosphates
Diarrhea—arsenic, radioactive substances, barium, phosphorus
Constipation—lead, barium sulfate, thallium
Hepatomegaly—hepatotoxins (Table 2-15)
Jaundice—hepatotoxins (Table 2-15), hemolytic agents (Table 2-16)
Angiosarcoma—vinyl chloride

Genitourinary

Toxic nephrosis—nephrotoxins (Table 2-18)
Bladder tumors—β-naphthylamine, benzidine, 4-aminodiphenyl

Musculoskeletal

Osteonecrosis—phosphorus
Osteosarcoma—radium, other bone-seeking radioactive substances
Osteomalacia—cadmium
Osteosclerosis—fluorides
Calcification of ligaments—fluorides
Acroosteolysis—vinyl chloride

Hematopoietic System

Anemia—inorganic lead: hemolytic agents (Table 2-16), marrow suppressants (Table 2-17)

Leukopenia—bone marrow suppressants (Table 2-17)
Leukemia—benzene, radioactive substances
Hypoprothrombinemia—warfarin

Skin

The site of eruption is important in identifying dermatitis of occupational origin. It usually begins on the exposed parts: on the hands and forearms if a solid or liquid, on the face and neck if a vapor. The covered parts of the body may also be affected if a liquid or vapor penetrates the clothing. Points of friction (wrists, ankles, collar band, hat band, or beltline) are frequent sites. Generalized dermatitis may also occur in a highly sensitized person, usually a few days following a localized eruption or burn.

Stains

Orange—tetryl, chlorine gas, nitric acid
Yellow—picric acid, epoxy resins and hardeners (amine type)
Green—copper salts
Blue—oxalic acid
Brown—*p*-phenylenediamine
Black—silver salts, osmium trioxide

Eruptions and Other Reactions

Acute eczematous dermatitis, contact type (erythema, edema, papules, vesicles, bullae, crusts, desquamation) irritants (Table 2-13); sensitizers (Table 2-14)
Chronic eczematous dermatitis, contact type (erythema, lichenification, fissuring)—solvents, detergents
Punctate ulcers or scars ("chrome holes")—chromic acid
Painful burns—hydrofluoric acid (burns characterized by excruciating deep pain of delayed onset, often around fingernails, out of keeping with the apparently superficial nature of the lesion)
Purpura—hepatotoxins (Table 2-15)
Folliculitis and boils—oil and grease
Acne—chlorinated naphthalenes, chlorinated diphenyls (PCBs), dioxin, chlorobenzene, paraffin, and coal tar
Photosensitization—coal tar pitch, asphalt, anthracene, creosote, fluorescein and phenanthrane
Keratosis and cancer—coal tar, soot, pitch, anthracene, and some arsenic compounds
Granuloma—beryllium
Hyperpigmentation—coal tar, some petroleum oils
Depigmentation (leukoderma)—monobenzyl ether of hydroquinone
Hair damage—chlorobutadiene
Alopecia—boric acid, thallium

Fetal Abnormalities

These may be caused by methyl mercury and some other substances (see Chapter 3).

(Tables follow, pp 16–25)

TABLE 2-1. Mild Irritants (Eye, Mucous Membranes)

Acetaldehyde	Dibutyl phthalate
Acetic acid	*o*-Dichlorobenzene
Acetic anhydride	*p*-Dichlorobenzene
Acetone	1,3,-Dichloro-5,5-dimethylhydantoin
Acetylene dichloride	Diethylamine
Allyl alcohol	Diethylaminoethanol
Allyl chloride	Difluorodibromomethane
Allyl propyl disulfide	Diglycidyl ether
n-Amyl acetate	Diisobutyl ketone
sec-Amyl acetate	Dimethyl phthalate
Arsenic and compounds	Dioctyl phthalate
Benzoyl peroxide	Dioxane
Benzyl chloride	Diphenyl
Bromoform	Dipropylene glycol methyl ether
1,3-Butadiene	Epichlorhydrin
n-Butyl acetate	Epoxy resins
sec-Butyl acetate	2-Ethoxyethanol
tert-Butyl acetate	2-Ethoxyethyl acetate
sec-Butyl alcohol	Ethyl acetate
n-Butylamine	Ethyl acrylate
Butyl cellosolve	Ethyl alcohol
n-Butyl glycidyl ether	Ethylamine
p-*tert*-Butyltoluene	Ethyl amyl ketone
Calcium oxide	Ethylbenzene
Chloroacetaldehyde	Ethyl bromide
α-Chloroacetophenone	Ethyl butyl ketone
Chlorobenzene	Ethyl chloride
o-Chlorobenzylidene malononitrile	Ethylenediamine
Chlorobromomethane	Ethylene dibromide
Chloroprene	Ethyl ether
Copper dusts and mists	Ethyl formate
Copper fume	*n*-Ethylmorpholine
Crotonaldehyde	Ethyl silicate
Cumene	Ferbam
Cyclohexane	Ferrovanadium dust
Cyclohexanol	Fluoride
Cyclohexanone	Formaldehyde
Cyclohexene	Formic acid
Cyclopentadiene	Furfural
Diacetone alcohol	Furfural alcohol
Dibutyl phosphate	Glycidol

(continued)

TABLE 2-1. *(continued)*

n-Hexane
Hexachloroethane
sec-Hexyl acetate
Hydrazine
Hydrogen bromide
Hydrogen peroxide, 90%
Hydrogen selenide
Isoamyl acetate
Isoamyl alcohol
Isobutyl acetate
Isophorone
Isopropyl acetate
Isopropyl alcohol
Isopropylamine
Isopropyl ether
Isopropyl glycidyl ether
Lithium hydride
Magnesium oxide fume
Mesityl oxide
Methyl acetate
Methyl acrylate
Methylal
Methylamine
Methyl amyl ketone
Methyl butyl ketone
Methyl cellosolve
Methyl cellosolve acetate
Methylcyclohexanol
o-Methylcyclohexanone
Methyl ethyl ketone
Methyl formate
Methyl isobutyl carbinol
Methyl isobutyl ketone
Methyl methacrylate
Methyl propyl ketone
α-Methyl styrene
Morpholine
Naphtha, coal tar
Naphtha, petroleum distillates
Nitroethane
1-Nitropropane
Octane
Osmium tetroxide
Oxalic acid
Pentachlorophenol
Pentane
Phenol
Phenyl ether-biphenyl mixture
Phenyl glycidyl ether
Phosphoric acid
Phosphorus (yellow)
Phosphorus pentachloride
Phosphorus pentasulfide
Phosphorus trichloride
Propyl acetate
n-Propyl alcohol
Propylene dichloride
Propylene imine
Propylene oxide
Pyridine
Selenium compounds
Silver, metal and soluble compounds
Sodium hydroxide
Stoddard solvent
Styrene
Sulfur monochloride
Sulfuryl fluoride
Terphenyls
Tetrachloroethylene
Tetrahydrofuran
Tetranitromethane
Thiram
Tin, inorganic compounds
1,1,2-Trichloroethane
Trichloroethylene
1,2,3-Trichloropropane
1,1,2-Trichloro-1,2,2-trifluoroethane
Triethylamine
Turpentine
Vinyltoluene
Xylene

TABLE 2-2. Severe Pulmonary Irritants

Acrolein	Ketene
Ammonia	Maleic anhydride
Antimony	Methyl bromide
ANTU	Methylene bisphenyl isocyanate
Beryllium and beryllium compounds	Methyl iodide
Boron trifluoride	Methyl isocyanate
Bromine	Methyl mercaptan
Butyl mercaptan	Nickel carbonyl
Cadmium dust/fume	Nitric acid
Chlorine	Nitroethane
Chlorine dioxide	Nitrogen dioxide
Chlorine trifluoride	2-Nitropropane
1-Chloro-1-nitropropane	Oxygen difluoride
Chloropicrin	Ozone
Chromic acid and chromates	Paraquat
Chromium, metal and insoluble salts	Perchloromethyl mercaptan
Cotton dust, raw	Perchloryl fluoride
Diazomethane	Phosgene
Diborane	Phosphine
1,1-Dichloro-1-nitroethane	Phosphorus trichloride
Dichloroethyl ether	Phthalic anhydride
Diisopropylamine	Selenium hexafluoride
Dimethylamine	Silicone tetrafluoride
Dimethyl sulfate	Sulfur dioxide
Ethanolamine	Sulfuric acid
Ethylene chlorohydrin	Sulfur pentafluoride
Ethyleneimine	Tellurium hexafluoride
Ethylene oxide	Toluene-2,4-diisocyanate
Ethyl mercaptan	Tributyl phosphate
Fluorine	Uranium (natural), soluble and insoluble compounds
Hydrogen chloride	Vanadium pentoxide
Hydrogen fluoride	Zinc chloride fume
Hydrogen sulfide	
Iodine	

TABLE 2-3. Pulmonary Sensitizers

Castor bean pomace	Phthalic anhydride
Cobalt, metal fume and dust	Platinum salts
Enzymatic detergents	Polyvinyl chloride (fume from heated film: meat wrapper's asthma)
Grain dusts	Toluene 2,4-diisocyanate
Maleic anhydride	Tungsten carbide
Methylene bisphenyl isocyanate	Western red cedar dust
Methyl isocyanate	Wood pulp dust
Nickel, metal	
p-Phenylenediamine	

TABLE 2-4. Fibrogenic Dusts

Asbestos
Coal dust
Cobalt, metal fume and dust
Graphite, natural
Hematite
Kaolin
Silica, amorphous, including diatomaceous earth (when contaminated with crystalline silica)
Silica, crystalline
Soapstone
Talc, fibrous
Yttrium

TABLE 2-5. Agents Causing Benign Pneumoconiosis

Barium and compounds
Graphite, natural
Hematite
Iron oxide fume
Kaolin
Mica
Silica, amorphous
Soapstone
Stannic oxide
Talc, nonasbestos form

TABLE 2-6. Pulmonary Carcinogens

Arsenic and compounds
Asbestos
Bis (chloromethyl) ether
Chloromethyl methyl ether
Chromates (see Chromic acid and Chromates)

TABLE 2-7. Cholinesterase Inhibitors

Azinphosmethyl	Mevinphos
Carbaryl	Naled
Demeton	Parathion
Dichlorvos	Ronnel
EPN	Tetraethyl dithionopyrophosphate
Malathion	Tetraethyl pyrophosphate
Methyl parathion	

TABLE 2-8. Methemoglobin Formers

Aniline
Anisidine, ortho- and para-isomers
Dimethylaniline
Dinitrobenzene, all isomers
Dinitrotoluene
Monomethylaniline
p-Nitroaniline
Nitrobenzene
o-Nitrochlorobenzene
p-Nitrochlorobenzene
Nitrogen trifluoride
Nitrotoluene
Perchloryl fluoride
n-Propyl nitrate
Tetranitromethane
o-Toluidine
Xylidine

TABLE 2-9. Asphyxiants

CHEMICAL ASPHYXIANTS

Acetonitrile
Acrylonitrile
Carbon monoxide
Cyanide (alkali)
Hydrogen cyanide

SIMPLE ASPHYXIANTS

Acetylene
Argon, neon and helium
Carbon dioxide
Dichloromonofluoromethane
Dichlorotetrafluoroethane
Ethane
Ethylene
Hydrogen
Liquid petroleum gas
Methane
Nitrogen
Propane
Propylene

TABLE 2-10. Central Nervous System Depressants

Acetaldehyde	Decaborane
Acetone	Diacetone alcohol
Acetylene dichloride	Dichlorodifluoromethane
Acrylamide	Dichloroethyl ether
Allyl glycidyl ether	Difluorodibromomethane
n-Amyl acetate	Diglycidyl ether
sec-Amyl acetate	Diisobutyl ketone
Benzene	Dipropylene glycol methyl ether
Bromoform	2-Ethoxyethyl acetate
1,3-Butadiene	Ethyl acetate
n-Butyl acetate	Ethyl alcohol
sec-Butyl acetate	Ethyl amyl ketone
tert-Butyl acetate	Ethylbenzene
n-Butyl alcohol	Ethyl bromide
sec-Butyl alcohol	Ethyl butyl ketone
tert-Butyl alcohol	Ethyl chloride
n-Butyl glycidyl ether	Ethylene dibromide
Butyl mercaptan	Ethylene dichloride
Carbon disulfide	Ethylene oxide
Carbon tetrachloride	Ethyl ether
Chlorobenzene	Ethyl formate
Chlorobromomethane	Ethyl mercaptan
Chloroform	Ethylidene chloride
Cresol, all isomers	Furfuryl alcohol
Cumene	Glycidol
Cyclohexane	*n*-Heptane
Cyclohexanol	Hexachloroethane
Cyclohexanone	*n*-Hexane
Cyclohexene	*sec*-Hexyl acetate

(continued)

TABLE 2-11. Neurotoxins Producing Peripheral Neuropathy

Acrylamide	Lead arsenate
Allyl chloride	Mercury
Arsenic and compounds	Methyl bromide
Calcium arsenate	Methyl butyl ketone
Carbon disulfide	Thallium, soluble compounds
2,4-Dichlorophenoxyacetic acid(2,4 D)	2,4,6-Trinitrotoluene
n-Hexane	Tri-*o*-cresyl phosphate
Lead and inorganic lead compounds	

TABLE 2-10. (*continued*)

Isoamyl acetate	Naphtha, petroleum distillates
Isoamyl alcohol	Nitroethane
Isobutyl acetate	Octane
Isobutyl alcohol	Pentaborane
Isopropyl acetate	Pentane
Isopropyl alcohol	Phenyl glycidyl ether
Isopropyl ether	Propyl acetate
Mesityl oxide	*n*-Propyl alcohol
Methyl acetate	Propylene dichloride
Methyl acetylene	Propylene oxide
Methyl acetylene, propadiene mixture	Pyridine
Methylal	Stoddard solvent
Methyl amyl ketone	Styrene
Methyl butyl ketone	Sulfuryl fluoride
Methyl cellosolve acetate	1,1,1,2-Tetrachloro-2,2-difluoroethane
Methylcyclohexane	1,1,2,2-Tetrachloro-1,2-difluoroethane
Methylcyclohexanol	Tetrachloroethane
o-Methylcyclohexanone	Tetrachloroethylene
Methylene chloride	Tetrahydrofuran
Methyl ethyl ketone	Toluene
Methyl formate	1,1,1-Trichloroethane
Methyl iodide	1,1,2-Trichloroethane
Methyl isobutyl carbinol	Trichloroethylene
Methyl isobutyl ketone	Trichlorofluoromethane
Methyl mercaptan	1,2,3-Trichloropropane
Methyl propyl ketone	1,1,2-Trichloro-1,2,2-trifluoroethane
α-Methyl styrene	Turpentine
Naphtha, coal tar	Vinyltoluene
	Xylene

TABLE 2-12. Convulsants

Aldrin	Methyl iodide
2-Aminopyridine	Methyl mercaptan
Camphor	Monomethylhydrazine
Chlordane	Nicotine
Crag herbicide	Nitromethane
DDT	Oxalic acid
Decaborane	Pentaborane
2,4-Dichlorophenoxyacetic acid	Phenol
Dieldrin	Rotenone
1,1-Dimethylhydrazine	Sodium fluoroacetate
Endrin	Strychnine
Heptachlor	Tetraethyl lead
Hydrazine	Tetramethyl lead
Lindane	Tetramethylsuccinonitrile
Methoxychlor	Thallium, soluble compounds
Methyl bromide	Toxaphene
Methyl chloride	

TABLE 2-13. Primary Chemical Irritants of the Skin

Acetaldehyde
Acetic acid
Acetic anhydride
Acetone
Acrolein
Acrylamide
Acrylonitrile
Allyl alcohol
Allyl chloride
Allyl glycidyl ether
Ammonia
n-Amyl acetate
sec-Amyl acetate
Antimony
Arsenic and compounds
Barium and compounds
Benzene
Benzoyl peroxide
Benzyl chloride
Beryllium and beryllium compounds
Boron oxide
Boron trifluoride
Bromine
n-Butyl acetate
sec-Butyl acetate
n-Butyl alcohol
sec-Butyl alcohol
tert-Butyl alcohol
n-Butylamine
n-Butyl glycidyl ether
Calcium arsenate
Calcium oxide
Carbaryl
Carbon disulfide
Carbon tetrachloride
Chlorinated diphenyl oxide
Chlorine
Chlorine trifluoride
Chloroacetaldehyde
α-Chloroacetophenone
Chlorobenzene
o-Chlorobenzylidene malononitrile
Chlorobromomethane
Chlorodiphenyl, 42% chlorine
Chlorodiphenyl, 54% chlorine
Chloroform
Chloropicrin
Chloroprene
Chromic acid and chromates
Chromium, soluble chromic and chromous salts
Coal tar pitch volatiles
Copper dusts and mists
Crag herbicide
Cresol, all isomers
Crotonaldehyde
Cumene
Cyanides (alkali)
Cyclohexane
Cyclohexanol
Cyclohexanone
Cyclohexene
DDT
Diacetone alcohol
Dibutyl phosphate
o-Dichlorobenzene
p-Dichlorobenzene
1,1-Dichloro-1-nitroethane
Diethylamine
Diethylaminoethanol
Diglycidyl ether
Diisobutyl ketone
Dimethylamine
Dimethylformamide
1,1-Dimethylhydrazine
Dimethyl sulfate
Dioxane
Epichlorhydrin
Epoxy resins
Ethanolamine
2-Ethoxyethanol
2-Ethoxyethyl acetate
Ethyl acetate
Ethyl acrylate
Ethylamine
Ethyl amyl ketone
Ethylbenzene
Ethyl bromide
Ethyl butyl ketone
Ethylenediamine
Ethylene dibromide
Ethylene dichloride
Ethyleneimine
Ethylene oxide
Ethyl ether
Ethyl formate
Ethylidene chloride
Ethyl silicate
Fluoride

(continued)

TABLE 2-13. (*continued*)

Fluorine
Formaldehyde
Formic acid
Furfural
Glycidol
n-Heptane
Hexachloronaphthalene
n-Hexane
Hydrazine
Hydrogen bromide
Hydrogen chloride
Hydrogen fluoride
Hydrogen peroxide, 90%
Iodine
Isoamyl acetate
Isobutyl alcohol
Isophorone
Isopropylamine
Isopropyl ether
Isopropyl glycidyl ether
Ketene
Lead arsenate
Lithium hydride
Maleic anhydride
Mercury
Mercury, alkyl compounds
Mesityl oxide
Methyl acetate
Methyl acrylate
Methylal
Methyl alcohol
Methylamine
Methyl bromide
Methyl butyl ketone
Methyl chloride
Methylcyclohexane
Methylcyclohexanol
o-Methylcyclohexanone
Methylene bisphenyl isocyanate
Methylene chloride
Methyl ethyl ketone
Methyl iodide
Methyl isobutyl ketone
Methyl isocyanate
Methyl methacrylate
Methyl propyl ketone
α-Methyl styrene
Monomethylhydrazine
Morpholine
Naled
Naphtha, coal tar
Naphtha, petroleum distillates
Nickel, metal
Nitric acid
Nitrobenzene
Nitroethane
Nitromethane
Octachloronaphthalene
Octane
Osmium tetroxide
Oxalic acid
Pentaborane
Pentachloronaphthalene
Pentachlorophenol
Pentane
Perchloromethyl mercaptan
Phenol
p-Phenylenediamine
Phenyl ether
Phenyl ether-biphenyl mixture
Phenyl glycidyl ether
Phenylhydrazine
Phosgene
Phosphoric acid
Phosphorus (yellow)
Phosphorus pentachloride
Phosphorus pentasulfide
Phosphorus trichloride
Phthalic anhydride
Picric acid
Platinum, soluble salts
Portland cement
Propyl acetate
n-Propyl alcohol
Propylene dichloride
Propylene imine
Propylene oxide
Pyrethrum
Pyridine
Quinone
Rotenone
Selenium compounds
Silver, metal and soluble compounds
Sodium hydroxide
Stoddard solvent
Styrene
Sulfur dioxide
Sulfur monochloride
Sulfuric acid

(*continued*)

TABLE 2-13. (*continued*)

Tellurium	1,1,1-Trichloroethane
Terphenyls	Trichloroethylene
Tetrachloroethane	Trichloronaphthalene
Tetrachloroethylene	2,4,5-Trichlorophenoxyacetic acid
Tetrachloronaphthalene	1,2,3-Trichloropropane
Tetranitromethane	1,1,2-Trichloro-1,2,2–trifluoroethane
Tetryl	Triethylamine
Thiram	2,4,6-Trinitrotoluene
Tin, organic and inorganic compounds	Turpentine
Toluene	Uranium (natural), soluble and insoluble compounds
Toluene 2,4-diisocyanate	Vanadium pentoxide
o-Toluidine	Vinyltoluene
Toxaphene	Xylene
Tributyl phosphate	Zinc chloride fume

TABLE 2-14. Skin Sensitizers

Acetic anhydride	Isopropyl glycidyl ether
Allyl glycidyl ether	Maleic anhydride
Benzoyl peroxide	Mercury
Beryllium and beryllium compounds (granuloma)	Naphthalene
n-Butyl glycidyl ether	Nickel, metal
Chromic acid and chromates	*p*-Phenylenediamine
Cobalt, metal fume and dust	Phenyl glycidyl ether
Cresol, all isomers	Phenylhydrazine
o-Dichlorobenzene	Phthalic anhydride
Dinitrochlorobenzene	Picric acid
Epoxy resins	Platinum, soluble salts
Ethyl acrylate	Tetryl
Ethylenediamine	Thiram
Formaldehyde	Toluene 2,4-diisocyanate
Iodine	2,4,6-Trinitrotoluene
	Vanadium pentoxide

TABLE 2-15. Hepatotoxins

Acetylene tetrabromide	Kepone
Carbon disulfide	Nitroethane
Carbon tetrachloride	Octachloronaphthalene
Chlorodiphenyl, 42% chlorine	Pentachloronaphthalene
Chlorodiphenyl, 54% chlorine	Phosphorus
Chloroform	Picric Acid
p-Dichlorobenzene	TCDD
Dimethylacetamide	Tetrachloroethane
Dimethylformamide	Tetrachloroethylene
Dioxane	Tetrachloronaphthalene
Ethylene chlorohydrin	Tetryl
Ethylene dibromide	Trichloronaphthalene
Ethylene dichloride	2,4,6-Trinitrotoluene
Hexachloronaphthalene	Vinyl chloride
Hydrazine	

TABLE 2-16. Hemolytic Agents

Arsine
Butyl cellosolve
Naphthalene
Phenylhydrazine
Stibine

TABLE 2-17. Bone Marrow Suppressants

Benzene
Dinitrophenol
Tetryl
2,4,6-Trinitrotoluene

TABLE 2-18. Nephrotoxins

4-Aminodiphenyl
Anisidine
Carbon disulfide
Carbon tetrachloride
Chloroform
Dioxane
Ethylene chlorohydrin
Ethylene dibromide
Mercury
Oxalic acid
Picric acid
Tetrachloroethane
2,3,4-Trinitrotoluene
Turpentine
Uranium (natural), soluble and insoluble compounds

3

The Diagnosis of Occupational Disease

Gideon Letz

The identification of previously unrecognized occupational illness has been largely dependent on the alert clinician who suspects an association between work and a patient's ill health. Isolated cases of rare diseases or clusters of common illness associated with a specific occupation have sometimes been the first clue that a work exposure may be causally related to a specific disease. Percival Pott initiated the concept of occupational cancer when he observed in 18th century England that work as a chimney sweep was associated with death at an early age from scrotal cancer. In fact, some of the best known examples of chemically induced occupational disease were first identified by clinical observation:

- Tetraethyl lead—psychosis
- Bis (chloromethyl) ether (BCME)—oat-cell carcinoma of the lung
- Vinyl chloride—angiosarcoma of the liver
- Aniline dyes—bladder cancer
- Asbestos—mesothelioma
- Dimethylaminoproprionitrile (DMAPN)—urinary retention
- Dibromochloropropane(DBCP)—male infertility

Additionally, case reports of certain occupationally associated illnesses have generated hypotheses for animal studies or epidemiologic surveys that have eventually established an etiologic relationship between a specific chemical exposure and the a suspect illness.

The identification of new occupational diseases has also had broad impact on the general public health. For example, Jenner noted that dairy maids who had contracted the occupational disease, cowpox, from milking infected cows, did not develop smallpox. This observation led to the development of smallpox vaccination. More recently, an attending plant physician observed that workers exposed to certain organosulfides used as rubber accelerators could not tolerate even small amounts of alcohol. This led to the development of the drug disulfiram (chemically related to the rubber accelerators), which is a valuable adjunct in the control of chronic alcoholism.

Given the accelerating changés in industrial processes, particularly in the so-called high-tech industries, and the fact that many chemicals are newly introduced for industrial use each year, it is likely that cases of previously unrecognized occupational illness will continue to be identified. The first indication of a problem will often be the alert clinician, nurse, or industrial hygienist who notices an association between illness and workplace exposure. This may occur despite the increasing use of animal toxicity testing to screen potential industrial chemicals. Animal tests do not monitor all possible adverse effects, and rodent response is not a perfect predictor of human response. In addition, there will always be a small percentage of the work force who will demonstrate an immune-mediated hypersensitivity or idiosyncratic response to certain chemical agents at exposure levels that have no perceptible effect on the majority of exposed workers.

Several approaches are considered to be of potential value in the surveillance of occupational disease. The possible use of organized data systems such as insurance company records and Bureau of Labor Statistics data is sometimes attractive. But these data systems are not particularly useful for detecting new occupational diseases, because they generally reflect only those diseases that have been recognized previously as being causally related to occupation. The possibility of identifying new trends in disease rates associated with specific occupations exists in the thoughtful review of organized data bases, particularly the Cancer Registries, but there are no instances in which such data have suggested the presence of a previously unrecognized occupational disease.

Occupational disease is always preventable. Thus, the early recognition of occupationally related illness has ramifications that go beyond the individual patient being evaluated. For the index case, an accurate diagnosis may ensure that further exposure can be reduced or eliminated. Early diagnosis also implies that appropriate medical treatment is more likely to be successful. In addition, identification of a single patient with occupational disease may trigger further actions with significant public health impact, such as the following:

- The systematic clinical surveillance of other workers with similar exposures (case finding)
- Modifications in environmental conditions or work practices to keep exposures within established safe limits
- Revisions in accepted safe limits of exposure (Threshold Limit Values, TLVs)

GUIDELINES FOR HAZARD EVALUATION

Occupational health practitioners are sometimes called upon to determine whether or not a hazard exists based upon limited information on a single patient. A systematic approach to the initial evaluation results in the most effective utilization of sometimes limited health and safety resources in the workplace, as well as avoids unnecessary anxiety by offering reassurance when no work hazard exists. More elaborate investigation may be warranted when there are indications of a problem, and appropriate studies may be launched to identify chemical or physical hazards that may affect other workers.

The following suggestions convey a general sense of how the various disciplines of toxicology, epidemiology, and industrial hygiene are integrated to supplement the clinical evaluation of an individual patient.

Importance of the Occupational History

The key to accurate diagnosis of occupational disease is that aspect of the medical history that deals with the workplace. A thorough history alone sometimes establishes that an illness is work-related or non-work-related. The essential additions to the clinician's usual medical history include the following questions:

- Are the symptoms associated with the hours of work? Is there improvement during weekends or vacations?
- Do symptoms seem to be caused or aggravated by specific activities in the workplace?
- Are other workers similarly affected?
- Is there currently, or has there been in the past, obvious exposure to dust, fumes, or chemical vapors?

The diagnosis of work related disease, as with other types of illness, depends on the presence of specific physical findings and the results of selected clinical laboratory and radiographic studies. The results of environmental monitoring by industrial hygiene techniques, when available, are invaluable. All results should be consistent with the tentative diagnosis, and non-work-related causes should be ruled out. The initial identification of occupational disease does not necessarily require extensive knowledge of the health effects of industrial toxins, but rather a systematic clinical approach. As with other human disorders, the diagnosis of occupational illness cannot always be made with certainty. Informed clinical judgment is essential.

Many factors tend to obscure the diagnosis of occupational illness:

- Although a few toxins produce a characteristic pattern of illness or a specific disease entity, most occupational disease is clinically and pathologically indistinguishable from common disorders of nonoccupational origin. With the exception of the pneumoconioses such as silicosis and asbestosis, there are only a few conditions, such as the two rare cancers, mesothelioma (related to asbestos exposure) and angiosarcoma of the liver (from vinyl chloride exposure), that are specifically linked with an occupational factor.
- For many occupational diseases, there is a latent interval between the time of exposure and the initial clinical manifestations of disease. This is particularly significant for occupational cancers that typically have latent periods of 15 to 30 years. Some specific health effects may go unnoticed for years until an irreversible loss of function has occurred, such as renal failure. There may be progression of tissue damage long after exposure ceases, as occurs with asbestosis. The clinician should be aware that illness of recent onset may be due to toxic exposure that occurred years previously. Conversely, many toxins do not cause delayed reactions; thus, if no effects are observed at the time of exposure, there is no increased risk of illness occurring months or years later. For example, asthma of recent onset is unlikely to be related to an exposure that occurred uneventfully many years before.
- The risk of development of occupational illness in an individual may depend on the interaction of multiple risk factors, but very little is known about the relative importance of specific variables such as age, sex, genetic background, and pre-existing disease. All are assumed to be potentially important variables. Not much is understood, for example, of the synergistic effects of multiple chemical exposures. The best documented example of such a phenomenon is the multiplicative risk of lung cancer in those exposed to the two recognized carcinogens, asbestos dust and tobacco smoke.

Identification of Toxic Agents

The obvious first step in the process of hazard evaluation is the identification of potentially hazardous materials. This can be a difficult and time-consuming task. Despite right-to-know regulations, workers often are unaware of the chemical composition of materials they are handling. Under the OSHA hazard communication system, employers are required to have a Material Safety Data Sheet (MSDS) at the worksite for each potentially toxic material that may be encountered. An MSDS listing hazardous ingredients should be obtained if there is any question as to the chemical composition of a suspect substance that is in use. The MSDS customarily includes toxicity information, although some listings may be unreliable in this respect and require close scrutiny.

Review of Available Toxicologic Literature

Once a potentially harmful material has been identified, a next step is the development of a toxicologic profile of each ingredient. Reference to the alphabetized section of this volume may be useful; it is designed specifically for quick reference to basic toxicologic information on many commonly used industrial chemicals. Additional resources may be consulted if a less commonly used chemical is not discussed herein or if information on a specific toxic effect is lacking. When such an effect on a given organ system is not mentioned in this volume, it is only because there was no reliable published information about such an effect when the manuscript was prepared. It does not necessarily imply that studies have been made resulting in negative findings. Many commonly used chemicals have not been completely characterized from a toxicologic standpoint. This is especially true for carcinogenic potential, for reproductive system toxicity, and for some other delayed chronic effects.

Estimation of Exposure Dose

A clinician dealing with a case of possible occupational disease often must look beyond the clinical findings in the individual patient. The diagnosis of work-related illness involves the synthesis of information in addition to the usual clinical approach of medical history, physical examination, and laboratory studies. A vigorous effort must be made to characterize the work environment in terms of a dose–response profile for each potentially hazardous agent that the patient may have encountered at work.

For many volatile organic chemicals, for dusts, and for fumes, data derived from timed air sampling with qualitative and quantitative analysis of collected material may provide the best evidence of probable exposure dose. If such reliable industrial hygiene data are not available, there may still be clues in the medical history that are useful in estimating dosage levels. The following factors should be documented: duration of exposure (length of time on this job), hours per week at specific tasks, the effectiveness of available ventilation systems, the use and effectiveness of personal protective equipment, the proximity of the worker to the source of toxic exposure, general work practices such as rest periods and the prevalence of eating or smoking in the workroom, the presence of visible dust or fumes in the atmosphere, and the presence of odors. In addition, the history of acute symptoms such as mucous membrane irritation or central nervous system depression may be suggestive indicators of probable exposure levels, particularly when evaluating illness that is encountered by the clinician months or years after the onset of exposure. This aspect of the medical history may be the only information available for estimating exposure levels that occurred in the past, especially if the suspect work operation has been shut down or is not accessible.

For some chemical agents with well characterized acute effects, such as those organic solvents that produce central nervous system depression, a cause–effect relationship may be readily apparent from the history. But for those toxic agents that produce only delayed or chronic effects, such as asbestos dust and certain polychlorinated biphenyls, the absence of a history of acute symptoms is not very helpful in ruling out hazardous exposures in the past.

The physical properties of a substance such as vapor pressure may be useful in estimating probable airborne concentrations and for predicting routes of absorption. For example, materials that are lipid soluble are likely to be absorbed through intact skin.

Development of Dose–Response Model

The most critical phase of the hazard evaluation process is the synthesis of available clinical, industrial hygiene, and toxicologic data into a dose–response model.This model can then be used to determine whether a given clinical condition is related to work exposure and whether the suspect condition of work constitutes an unnecessary health hazard. Simply stated, this process involves evaluating all available information to determine whether (1) the probable dosage level of the suspect toxic agent was sufficient to explain observed clinical effects and (2) the exposure level that likely existed in the work environment presented an unacceptable risk to health.

The available information may be ambiguous or incomplete; for example, there may not be enough data to reasonably estimate the exposure level, or the observed clinical state may not be consistent with available toxicity data. The experienced observer will not neglect these key steps in the process of hazard evaluation. Whether investigating the cause of a possibly work-related disorder in an individual or evaluating the safety of a worksite, consideration of a dose–response relationship is essential: "All substances are poisons; there is none which is not a poison. The right dose differentiates a poison and a remedy." (Paracelsus 1493–1541)

Misunderstandings, erroneous judgments, and inappropriate actions in reference to a suspected toxic exposure are often traceable to disregard of the dose–response requirement. Highly toxic materials can be used safely if there are fully effective engineering controls and protective work practices, while exposure to materials of low toxicity can be hazardous under unusual conditions such as use in confined spaces without adequate ventilation.

CHEMICALLY INDUCED ILLNESS: ORGAN SYSTEM SPECIFICITY

Although many industrial chemicals have some potential to cause toxic effects, there are only a limited number of well-defined clinical syndromes documented by observations on workers or human volunteers. Many of the documented examples of industrial poisoning are case reports involving chemical exposures at levels well above safety limits (TLVs). In general, such hazardous exposures are much less likely to occur in the usual industrial setting now than in decades past, except as occasional isolated episodes such as explosions or major equipment failures. Other examples of hazard in which health effects were less obvious have been documented by systematic epidemiologic studies. A comparison between the morbidity and mortality experience of exposed and unexposed work groups may be invaluable in delineating such effects. We will now summarize by organ system the well-documented examples of chemically induced occupational disease.

In the tables presented in Chapter 2, chemicals are grouped according to target organ toxicity so that the most sensitive endpoints are emphasized. Although some organic chemicals have been shown to affect the central nervous system, the liver, or the kidneys of experimental animals given very high doses, such information may not be relevant in evaluating the possible effects of exposure to much lower concentrations in the work setting. However, some agents may be suspected of causing specific biochemical or tissue damage in humans based upon data from experimental animal studies. Although such information is sometimes useful in the development of occupational health standards, the aforementioned tables include chiefly those agents that produce effects confirmed by case reports or epidemiologic studies.

Respiratory Disorders

Since most environmental toxins are absorbed by way of the respiratory tract, it is not surprising that respiratory disorders occupy a prominent role in occupational medicine. The association of respiratory disease with workplace exposures may be complicated by the compounding effects of smoking, nonspecific allergic reactions, or viral infections; the pathologic response is only rarely specific or pathognomonic of occupational exposure. Moreover, many pulmonary disorders may be aggravated by occupational factors. Thus, the diagnosis of a work related respiratory illness depends on careful integration of clinical, epidemiologic, and industrial hygiene data.

Although the clinical spectrum of occupational respiratory disease is markedly variable, there are three basic mechanisms by which inhaled substances affect the respiratory tract:

- Deposition in the upper respiratory tract, lung parenchyma, or lymphatics with subsequent inflammatory or fibrotic reaction
- Irritation of the lining of the tracheobronchial tree with inflammation and/or edema
- Allergic or immunologic response

A complete cataloging of respiratory disorders is beyond the scope of this text, but representative examples are presented according to pathophysiologic category in the tables.

Pneumoconioses. Silica and asbestos are classic examples of agents that cause chronic, interstitial fibrosis. Many other particles, both fibrous and nonfibrous, are associated with pathologic reactions following deposition at the level of the terminal bronchioles and alveoli (see Table 2-4). The association of occupational exposure with disease of this type can usually be differentiated from the effects of smoking, because the pathology, clinical findings, and pulmonary function deficit are indicative of restrictive disease, which is readily differentiated from the obstructive disease pattern induced by smoking.

Some dusts are deposited in the distal respiratory tract without evidence of resulting functional impairment, the so-called benign pneumoconioses (see Table 2-5). These agents are retained in the lung and, if radiopaque, cause a characteristic radiographic pattern; but fibrogenic effects are minimal and there is no functional deficit.

Occupational Asthma. Numerous workplace agents have been implicated in causing asthma, and the list is steadily growing. Many of these agents are large molecular organic compounds such as animal or plant proteins, enzymes, and vegetable gums, which are not considered here. Certain inorganic and organic compounds of small molecular weight, including metals, anhydrides, isocyanates, and fluxes, may induce an allergic response. Some common examples are listed in Table 2-3. Although effects are primarily immunologically mediated, other mechanisms have been recognized in the pathogenesis of occupational asthma, including IgE-mediated hypersensitivity, IgG-delayed hypersensitivity, direct pharmacologic effect (cotton dust), or even simple irritant effects (toluene diisocyanate). In addition, single high dose exposures to airway irritants have been associated with subsequent, nonspecific bronchial reactivity, termed reactive airway dysfunction syndrome (RADS).

Hypersensitivity Pneumonitis (Extrinsic Alveolitis). Repeated inhalation of various organic antigens such as molds, fungi, bacteria, or animal protein can cause acute or chronic lung disease characterized by systemic symptoms, restrictive breathing defect, and presence in the serum of IgG precipitans to causal agents. Farmer's lung, related to thermophilic actinomycetes in moldy hay, is the classic example. There are very few examples of industrial chemicals associated with this disorder. The only documented example is methylene bisphenyl isocyanate (MDI), which is also associated with occupational asthma. There is some evidence that chronic berylliosis may be due to a form of hypersensitivity reaction.

Respiratory Irritation. Respiratory tract irritation has been associated with exposure to a diverse group of gases, metal dusts or fumes, and certain reactive chemicals. All produce an inflammatory reaction of the mucous membranes; depending upon particle size, any portion of the respiratory tract may be affected. For a given chemical agent, the extent and severity of injury to the respiratory tract is primarily dependent upon the airborne concentration of the substance and the duration of exposure. Many of these agents are water-soluble irritant gases, such as ammonia, with good warning properties. For these agents, severe reactions are unlikely except in unusual situations such as explosions, fires, and spills. The relatively insoluble irritants, such as nitrogen dioxide, cause fewer immediate symptoms but may result in severe injury to the lower respiratory tract with delayed pulmonary edema occurring up to 72 hours after exposure. Some gases and metals that act as severe pulmonary irritants are listed in Table 2-2, with particular note made of the delayed onset of symptoms, because this information may be critical to the proper management of an exposed person.

Metal/Polymer Fume Fever Syndromes. Metal fume generated by welding, smelting, or foundry operations may cause a unique clinical picture. Several hours after high exposure to metal oxide fume, a flu-like illness may occur with fever, myalgia, and headache; leukocytosis is often present. Chest x-ray and arterial blood gases are normal, and the illness is self-limited to 24 to 48 hours. Apparently tolerance is induced by daily exposure, and symptoms usually occur in newly assigned workers or after a few days away from work. Severe or permanent impairment has never been described, even after repeated episodes of this syndrome. Zinc and copper fume are the most commonly responsible agents, but other metals, such as magnesium and antimony, have also been implicated. A similar condition has been reported from exposure to combustion products of polytetrofluorethylene (Teflon), with symptoms typically developing after smoking contaminated cigarettes.

Renal Disease

A number of chemical agents have been reported to cause acute renal failure after high-dose exposure, which occurs only rarely in industry today, typically from spills, leaks, or heavy exposure in confined spaces. However, relatively few chemical exposures have been systematically evaluated for more subtle renal effects, and only a handful have been positively associated with chronic renal disease (Tables 2-18 and 3-1). This is puzzling, given the number of known suspect nephrotoxins in widespread industrial use and given the many theoretic reasons why the kidney should be highly susceptible to the presence of en-

TABLE 3-1. Occupational Exposures Associated with Chronic Renal Failure

Agent	*Type of Study*	*Effects Observed*	*Comments*
Lead	Cross-sectional survey	Asymptomatic ↓ in GFR; correlated with body burden Pb (EDTA chelation); interstitial nephritis on bx	Effects observed without other clinical signs or symptoms
Cadmium	Cross-sectional	Tubular proteinuria (low MW proteins); renal stones	Proteinuria observed in absence of clinical nephropathy or other organ effects
Mercury	Case reports; cross-sectional surveys	Nephrotic syndrome; glomerulonephritis; proteinuria (high MW) correlated with UHg; β-galactocidase (blood/urine)	Proteinuria, increased enzyme activities observed without other clinical signs of Hg toxicity
Beryllium	Registry data	Renal stones; hypercalciuria	Renal effects not observed in absence of pulmonary disease
Carbon disulfide	Cross-sectional survey	↓ PAH clearance; ↓ renal blood flow	Renovascular effects may be secondary to generalized effect on sympathetic nervous system and hypertension
Silica	Case reports	Immune-mediated glomerulonephritis; interstitial nephritis	Renal effects only observed in generalized systemic disease (vasculitis/hypertension) after heavy exposure
Solvents	Case control (retrospective)	Glomerulonephritis (all types)	Specific agents not identified; retrospective data only; common exposure, but rare disease; idiosyncratic immune response?

vironmental toxins in the body. A number of factors may explain this apparent paradox:

- The kidney has a great reserve capacity and can function adequately despite slowly developing damage.
- Currently used screening tests for renal insufficiency are relatively insensitive.
- Histopathologic changes may not be distinctive by the time clinically apparent renal impairment is recognized.
- Since renal disease is relatively rare, case-control studies are the only practical epidemiologic method to identify nephrotoxic agents. Such studies are hampered, however, by the lack of information in the typical medical record concerning occupation and past exposures.

The development of a new array of noninvasive tests of kidney function offers the prospect of earlier recognition of renal dysfunction and may allow identification of previously unrecognized occupational nephrotoxins. The measurement of urinary enzymes and other urinary proteins is particularly promising in this regard, but not enough is known about the correlation of these urinary indicators with the risk of irreversible kidney dysfunction.

Hepatotoxicity

The liver is the critical organ for the detoxification and elimination by metabolism of many chemicals that may be absorbed. As such, it is a likely target organ for toxic effects. The list of industrial chemicals with confirmed or suspected hepatotoxic potential in animals or humans is growing steadily. Of these agents, only a relatively few have been confirmed as hepatotoxic in occupationally exposed persons (see Table 2-15). Occupational toxicity has most frequently been observed as part of a toxic syndrome dominated by extrahepatic manifestations, and only rarely does liver damage occur as an isolated clinical finding.

Despite the large number of known or suspect hepatotoxins in widespread industrial use and the few well-documented examples of occupational exposure associated with acute or chronic liver injury, the extent of clinically significant liver dysfunction related to occupational exposure is poorly understood. The diagnosis of occupational liver disease is difficult because the incidence of nonoccupational liver disease owing to ethanol ingestion or viral infections is high, and the commonly used laboratory tests of liver function are of rather low specificity. In addition, the usual clinical and histologic features of occupational liver disease are not unique. Workplace chemicals can cause a number of the morphologic lesions seen in hepatopathology, and in the individual worker with suspected exposure to a hepatotoxic agent, a cause–effect relationship is often difficult to establish. One of the best clues to causation is a definite temporal relationship between exposure and the onset of clinical symptoms in a previously healthy person. For acute liver injury, the latency period is often a few days, but for subacute liver damage, it may be measured in months, and for chronic liver injury, years may have elapsed since the onset of toxic exposure. Other factors to look for include the following:

- Remission after withdrawal from exposure
- Similar findings in co-workers
- Symptoms or signs resulting from effects on other organ systems.

Although histologic examination of biopsy specimens is sometimes of value for prognostic purposes, histologic findings are usually nonspecific, and it may be impossible to determine whether the abnormality is related to ethanol ingestion or to workplace exposure. Because it is an invasive procedure, liver biopsy is not often done in asymptomatic or mildly symptomatic workers.

Hematotoxicity

Fifty years ago, examples of occupational hematotoxicity were more common from hazards related to coal tar derivatives, aromatic amines, organic solvents, and heavy metals. More recently, preventive and regulatory controls have markedly reduced the risk. However, work-related effects on the hematologic system due to lead, arsenic, benzene, and several other agents still occur sporadically, particularly in smaller industrial plants. A wide range of disorders has been reported, including the following:

- Inhibition of hemoglobin synthesis
- Bone marrow suppression
- Hemolysis of circulating red blood cells
- Methemoglobin production

Some of the recognized examples that may still be encountered are listed in Tables 2-16 and 2-17. Because the clinical findings are indistinguishable from those caused by exposure to certain drugs or to underlying systemic diseases, the correct diagnosis depends on detailed knowledge of the patient's exposure history.

Certain industrial chemicals have the potential to induce methemoglobin by oxidation of the ferrous iron (Fe^{++}) normally present in reduced hemoglobin to the ferric (Fe^{+++}) state. Methemoglobin is incapable of binding to either oxygen or carbon dioxide, with resulting decrease in oxygen carrying capacity of the blood and a left shift in the oxygen dissociation curve. Although extreme elevations of methemoglobin can be life threatening, this is extremely rare in an otherwise healthy person. Most reports of occupational methemoglobinemia involve only minor symptoms with reversible cyanosis. Chemicals responsible include nitrates, nitrites, and analine derivatives such as amino and nitro compounds of chlorobenzene, benzene, and its homologues (see Table 2-8).

Nervous System Disorders

Neurotoxicology is an emerging subspeciality based upon improved diagnostic procedures for evaluating nervous system dysfunction. The number of chemicals associated with neurotoxic effects in the workplace is growing. Although outbreaks of obvious neurologic disease of toxic origin are uncommon, the true prevalence of occupational neurotoxicity in its subtler aspects is undetermined. There is increasing concern about the occurrence of neurobehavioral disorders related to workplace exposure to organic solvents. But a cause–effect association with specific chemicals is often difficult to establish in an individual case, and epidemiologic studies have produced conflicting results. As more sensitive methods for evaluating subtle losses in cognitive function are refined, validated, and applied to the evaluation of chemically exposed persons, improved understanding of this subtopic of industrial toxicology may be expected.

Most neurotoxic agents affect both the central nervous system (CNS) and the peripheral nervous system (PNS). Because the symptoms and signs of peripheral neuropathies tend to be more specific and the applied diagnostic tools are more precise, the PNS effects of occupational exposure are more readily recognizable than CNS effects. However, several chemical agents are definitely linked with CNS effects, including behavioral manifestations, in exposed workers. Examples are presented in Table 2-10.

The more common neurologic dysfunction definitely associated with specific chemical exposure occurs as peripheral neuropathy. These toxic neuropathies usually occur as polyneuropathy, that is, symmetrical involvement of multiple peripheral nerves. Rather than segmental demyelinization, the neurologic lesion is axonal degeneration related to metabolic derangement of the entire neuron with degeneration beginning distally. Nerve conduction velocity is usually within normal limits until the condition is far advanced. Distal muscles develop denervation change. Recovery is by axonal regeneration but is usually very slow and incomplete. Clinical features vary with different agents and severity of exposure and depend upon whether there is involvement of sensory, motor, or mixed nerve fibers. Autonomic function is usually not affected. However, abdominal colic is historically associated with lead poisoning, and urinary bladder dysfunction has been reported from exposure to dimethylaminoproprionitrile (DAPN). Findings are rarely specific enough to be regarded as pathognomonic of occupational exposure, and the etiologic agent may be overlooked unless a detailed history of chemical exposure is obtained. Alcoholism, diabetes mellitus, uremia, and malignant disease, among other disorders, may be associated with peripheral neuropathy, but in nearly half the cases, specific etiology is not established. For this reason, the possible role of toxic exposure in the workplace should not be overlooked. Some commonly encountered industrial chemicals known to cause peripheral neuropathy are presented in Table 2-11.

Skin Disease

Skin constitutes almost 50% of all reported cases of occupational disease. This is not surprising because the skin with its large surface area is so often in direct contact with chemical agents in the work environment. Perhaps one of every four workers is afflicted with an occupational skin condition at some time, losing an average of 10 to 12 days from work for each episode. Because occupational dermatitis is often traceable to some minor chemical constituent of workplace materials such as preservatives, catalysts, or accelerators, the evaluation of exposure history must be detailed and precise. A product label or an MSDS may omit mention of a trace chemical component. If present in a concentration of less than 0.1%, an ingredient does not require mention on the MSDS under the OSHA Hazard Communication Standard, and rarely does the worker or the employer know of its presence. A call to the manufacturer is often necessary to rule out trace materials as a potential cause of a case of occupational dermatitis.

More than 80% of occupational skin diseases are attributable to chemical exposure. A variety of clinical responses have been observed, but the most important category is contact dermatitis. Contact dermatitis is any inflammatory condition precipitated by direct skin exposure to the offending agent, in contrast to those dermal conditions that are secondary to systemic absorption of toxic chemicals or drugs.

Skin Irritation. Most contact dermatitis is due to simple irritants. Classifying a chemical as an irritant is difficult because so many different substances produce a wide range of reactions. Almost any substance can be an irritant depending on its concentration in the environment and the duration of skin contact. A given chemical may cause irritation in only a small percentage of exposed persons who react adversely because of individual, local skin factors present at the exposure site at the time of contact. Examples are certain oils, alcohols, and glycols, especially propylene glycol. At the opposite extreme are the potent irritants such as sodium hydroxide and hydrofluoric acid, which, in sometimes low concentrations, will immediately produce a chemical burn on anyone's skin. A partial list of primary chemical irritants of the skin is given in Table 2-13.

The etiologic agent in irritant dermatitis may be difficult to identify because most industrial irritants are relatively weak and require frequent and prolonged contact before clinical effects are observed. Dermatitis resulting from cumulative chemical insults is frequently multifactorial with many different weak irritants such as soaps, detergents, or solvents playing a role, as well as physical conditions of humidity, friction, or heat. Pre-existing skin conditions due to trauma, infection, hyperhydrosis, or use of occlusive covering such as impervious gloves or clothing may also affect reaction to irritant chemicals.

Susceptibility to skin irritants probably increases with age, because the skin becomes thinner, drier, and less elastic. Persons with a history of pre-existing skin disease such as atopic eczema are much more likely to develop irritant dermatitis.

Allergic Contact Dermatitis. Allergic contact dermatitis is an immune-mediated type of reaction to a chemical agent. Fewer than 20% of cases of contact dermatitis are classed as allergic, although this figure varies with the industrial setting. The growing number of recently introduced organic chemicals has increased the potential for allergic skin reactions to occur. The development of allergic skin disease is critically dependent on host factors, and a recognized allergen will usually affect only a small percentage of exposed workers. The time between first exposure and onset of symptoms may be relatively short (a minimum of 5 days for potent sensitizers such as dinitrochlorobenzene), but most reactions occur only after several months or even years of exposure. Morphologically, allergic contact dermatitis cannot be distinguished from simple irritant dermatitis. The diagnosis can often be confirmed by patch testing with specific agents or mixtures, but false-negative results are not uncommon. Allergic contact dermatitis should be strongly suspected when there is a history of exposure to a known sensitizer (see Table 2-14) or when the patterns of skin reactivity occurring among a group of workers is consistent with an immune-mediated response.

Photo Contact Dermatitis. Any chemical on the skin surface that increases the absorption of ultraviolet radiation is a candidate for the production of photochemical reactions and photocontact dermatitis. Clinically, the appearance is that of an acute sunburn on sun-exposed areas of the body. Phototoxic reactions are analogous to irritant contact dermatitis and do not involve immunologic mechanisms. Polycyclic aromatic hydrocarbons such as creosote are well-known causes of photo contact dermatitis. A number of phototoxic agents are in commercial use. However, most photoallergic reactions are due to nonoccupational exposure to drugs. True photo allergic reactions are rarely occupationally related.

Cardiovascular System Disorders

There have been few studies of disorders of the heart or blood vessels in relation to occupational factors except those concerned with the cardiovascular risks of sedentary versus physically active work. Atherosclerotic heart disease (ASHD) has long been considered a degenerative disease of aging, but the list of definable risk factors is growing steadily and includes dietary agents and tobacco smoke. The role of chemical exposure as a factor in cardiovascular disease is difficult to assess, because an endpoint such as ASHD is so common in the population, and because there are multiple known risk factors that may confound the specific cause–effect association under study. However, a few chemical exposures of occupational origin have been definitely linked with a variety of adverse cardiovascular effects (Table 3-2).

It should be emphasized that the inconclusive nature of much of the existing data attempting to relate chemical exposure to cardiovascular disease emphasizes the need for further research rather than the adoption of an attitude rejecting the possible role of chemical exposure.

Because the prevalence of cardiovascular disease is so high in the general population, identifying and preventing occupational exposures that result in even a small increase in the relative risk of cardiovascular disease could potentially benefit large numbers of people.

Immune System Disorders

There are many documented examples of immune-mediated respiratory (see Table 2-3) and skin (see Table 2-14) disorders secondary to industrial chemical exposures. But despite emerging literature on the immunotoxic effects of chemicals observed in experimental animals, the more subtle subclinical effects on the human immune system in relation to chemical exposure have not been evaluated systematically.

Reproductive System Dysfunction

The potential hazard to human reproduction associated with chemical exposure has received relatively little attention. The birth defects resulting from the use of thalidomide in the 1960s and the occurrence in 1970 of an epidemic of methyl mercury poisoning in Minamata, Japan, demonstrated dramatically the teratogenic potential of certain chemical exposures. These events, along with the growing number of women in the industrial workforce, have focused greater attention on potential chemical hazard to the female reproductive function. In 1977 a high rate of sterility was recognized among male pesticide workers exposed to DBCP. This episode raised concern about the potential for occupational chemical hazards to the male reproductive function as well.

Occupational exposures are now recognized as producing a wide range of effects on human reproductive function. Hazardous exposures to either parent prior to conception or to the mother after fertilization must be considered. Human reproduction is a cycle of multiple interdependent events and physiologic functions. Virtually all of its phases have been shown to be affected by one or more environmental agents (Fig. 3-1). The recent development of new methods for detecting genetic damage has created an increased interest in the interrelationship of genetic, reproductive, and carcinogenic effects of chemical exposures. Unfortunately, human data on the subject are scarce, and there are enormous data gaps between interesting hypotheses and available scientific information.

In Table 3-3, some agents that have been shown to be associated with adverse reproductive outcome in humans are presented. Although data currently available indicate that animal and in vitro studies may be highly predictive of the adverse reproductive potential in humans, it should be emphasized that most chemical substances in commercial use have not been adequately studied for the potential to cause reproductive damage.

Occupational Cancer

The extent to which occupational exposures contribute to the etiology of

TABLE 3-2. Industrial Agents Associated with Adverse Effects on the Cardiovascular System

Agent	Effect	Comments
METALS		
Cadmium	? Hypertension	Hypertension observed in animals; renal toxicity in humans; occupational studies of hypertension have been negative
Arsenic	ECG abnormalities	Most studies of occupational cohorts show no increased CVD mortality.
Cobalt	Cardiomyopathy	"Beer drinkers" cardiomyopathy may involve other factors such as nutritional state, subclinical alcoholic heart disease.
Lead	? Hypertension	Relation to hypertension in humans is controversial; it may be secondary to renal effects.
SOLVENTS		
Methylene chloride	Cardiac ischemia	Metabolism to carbon monoxide; MI post exposure in case reports; chronic low level exposures not associated with excess mortality in one study
Carbon disulfide	ASHD	Multiple mortality studies show increased mortality from ischemic heart disease; first effect may be increased blood pressure or increased serum cholesterol with ASHD as secondary event
Chlorinated organic solvents (especially TCE)	Arrhythmias	Sudden death without pathologic findings in glue sniffers, but such high exposures not likely in industrial settings
Fluorocarbons	Arrhythmias	One study found various arrhythmias in pathology residents exposed during specimen preparation; sudden death reported from "sniffing" of aerosols
OTHER AGENTS		
Carbon monoxide	Cardiac ischemia ? ASHD	May explain strong association between smoking and ASHD; aggravates symptoms of ischemic heart disease but increased mortality has not been documented in occupationally exposed cohorts
Organonitrates	Cardiac ischemia secondary to vasospasm	Case reports of "Monday deaths" in explosives industry; coronary artery vasospasm probable mechanism; increased cardiovascular mortality after exposure greater than 20 years reported in one study

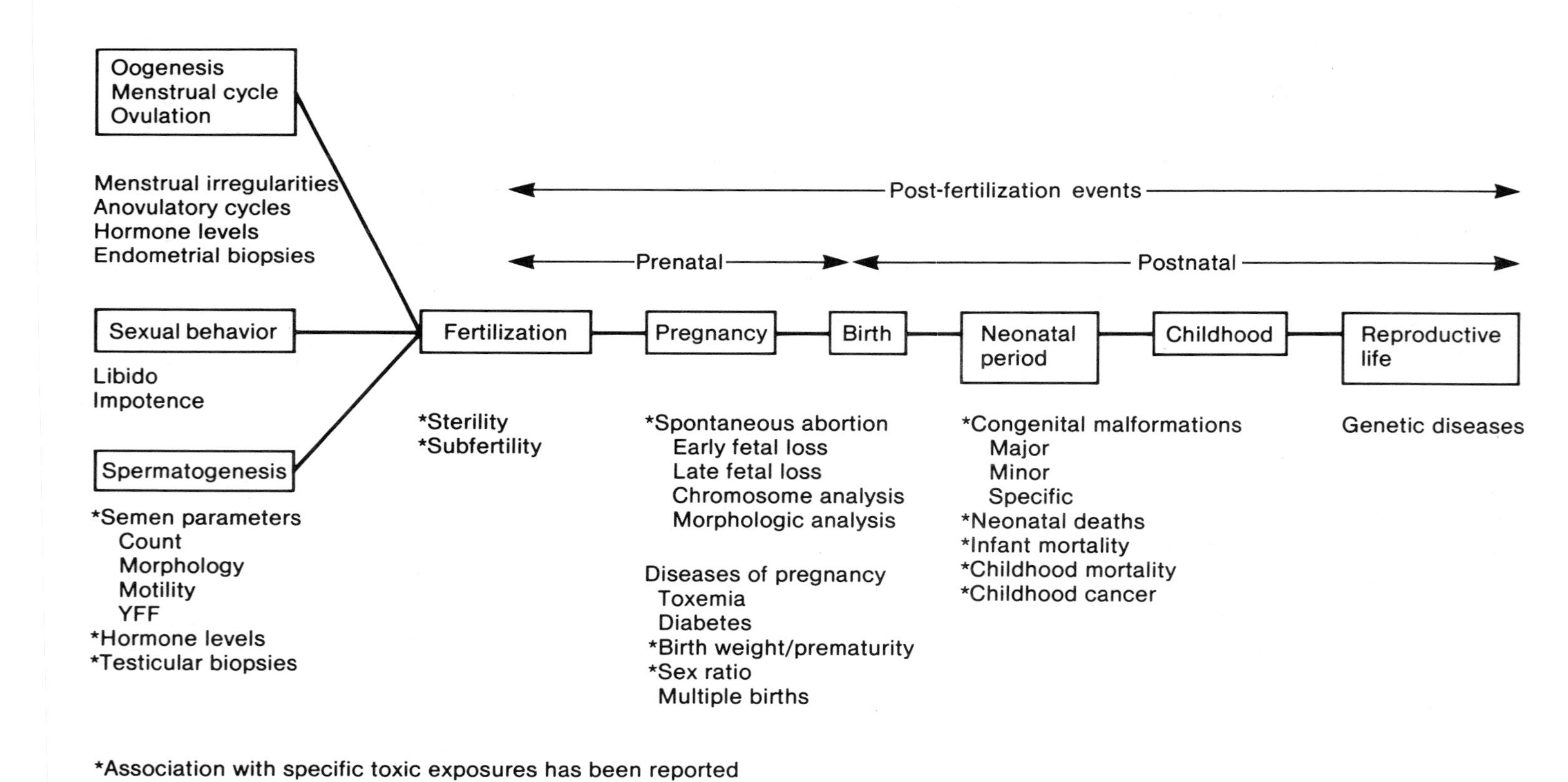

Fig. 3-1. Stages of the reproductive cycle: Endpoints available for monitoring in relation to toxic exposures.

cancer is a topic of considerable scientific and public debate. Several factors make such estimates difficult:

- Neoplastic disease is usually not evident clinically until long after the onset of exposure to a recognized carcinogen (latency period). This interval may vary from 10 to 40 years for some substances such as asbestos.
- Industrial workers are typically exposed to multiple chemicals, some of which may interact to increase the risk of cancer.
- There are many potentially confounding nonoccupational variables, such as smoking, alcohol intake, dietary habits, and genetic background, that may influence results of epidemiologic studies.
- Precise data on the levels and duration of exposure to specific chemical agents is generally not available for the worklife of the individual.
- There are sometimes significant errors in the diagnosis and classification of cancer, especially for unusual neoplasms such as mesothelioma.
- Cases of rare cancers such as hemangiosarcoma and mesothelioma can be more readily assigned to specific chemical exposures than is possible for more common cancers such as those of the lung and urinary bladder.

Despite the difficulty in estimating the overall portion of cancers that may be related to occupation, there can be little doubt that certain groups of workers exposed to specific chemical agents are at increased risk of developing cancer. Various scientific and governmental agencies have reviewed the available epidemiologic and animal data and have classified workplace chemicals and processes as to their potential carcinogenicity (see Table 3-3). Many industrial

TABLE 3-3. Industrial Agents/Processes Associated with Cancer in Humans*

Agent/Process	*Site*
4-Aminobiphenyl	Bladder
Arsenic and Trioxide compounds	Skin; lung; angiosarcoma; liver
Asbestos	Lung, mesothelioma, larynx, G.I.
Auramine (basic yellow 2)	Bladder
Benzene	Leukemia (acute nonlymphocytic)
Benzidine	Bladder
Bis (Chloromethyl) ether (BCME) and chloromethyl ether	Lung
Chromium and compounds	Lung; nasal sinuses
Coke oven emissions	Skin; lung; urinary tract
Hematite underground mining	Lung
Isopropyl alcohol manufacture	Lung; larynx; paranasal sinuses
Mustard gas	Lung
β-Naphthylamine	Bladder
Nickel and compounds; nickel refining	Lung; nasal passages; larynx
Soots, tars, and mineral oils	Skin; lung; bladder, G.I.
Vinyl chloride	Liver (angiosarcoma); lung; brain
Boot and shoe manufacture	Nasal sinuses; lung; bladder; leukemia
Furniture manufacture	Nasal sinuses
Rubber manufacture	Multiple sites

* Based on classification by International Agency for Research on Cancer (IARC) as carcinogenic to humans with sufficient epidemiologic evidence (Group 1) and by National Toxicology Program (NTP) as known to be carcinogenic with evidence from human studies (Group a).

chemicals cause tumors in experimental studies on rodents. Although almost all the chemicals identified as human carcinogens also cause cancer in at least one rodent species, it is not known how many of the large number of rodent carcinogens have significant carcinogenic effects on humans. Regulatory action has increasingly been based on the results of animal or in vitro tests that do not meet current criteria for acceptance of a chemical agent as a human carcinogen.

LABORATORY CONFIRMATION OF CHEMICAL INTOXICATION

Advances in the development of sensitive and precise analytic methods to measure chemicals and drugs in blood, urine, and other body fluids have provided useful tools for the recognition of occupational disease. Available laboratory procedures for toxicologic analysis vary in their accuracy, sensitivity, specificity, and cost. However, for a number of chemical agents, the laboratory ability to measure trace amounts of chemicals in body tissues or fluids has outstripped our ability to interpret the human health implications of the measured values. Particularly in regard to persistent organochlorines such as PCBs, DDT, and other halogenated compounds, there has been a tendency to overemphasize the clinical implications of the presence of often small concentrations of the chemical in human tissues.

In selected circumstances, toxicologic analyses can be an important adjunct to the occupational history, physical examination, and routine clinical laboratory evaluation in the diagnosis of disease secondary to chemical exposure. The toxicology laboratory can also be used as an effective tool in the periodic surveillance of persons exposed to potentially hazardous materials. Such biologic monitoring can be used along with environmental monitoring to estimate the dose of a toxic agent and the risk of adverse health effects. The primary goal of biologic monitoring is preventive rather than diagnostic in that it is done to ensure that current or past exposures are "safe", that is, they either do not result in the unacceptable risk of adverse effect, or they aid in the detection of excessive exposure before the onset of clinically evident disease. The advantages of biological monitoring include the following:

- Exposure by all routes of absorption, dermal and gastrointestinal as well as respiratory, is evaluated.
- Individual variability in factors such as personal activity, biological characteristics, and life-styles are integrated in the assessment.

The appropriate role of biological monitoring is to complement environmental monitoring. Both techniques are valuable and provide independent data that contribute to the hazard evaluation process.

Whether toxicologic analyses are done as part of a diagnostic workup on a person who may be ill or in the course of a preventive surveillance survey of well persons, the following multiple theoretic and practical limitations should be considered:

- Rational interpretation of results is possible only when sufficient information is available on the mechanism of toxicity and the pharmacokinetics of the substance being measured. At present, the main obstacle limiting the application of chemical analyses is not analytic sensitivity, but rather a lack of precise information on the health implications of the results.
- Levels of specific chemicals in body fluids or tissue can be interpreted only by comparing them with reference values in nonexposed or asymptomatic populations. For a few chemicals, such as lead and mer-

cury, information is available correlating levels of a chemical in body fluids or tissue with risk of adverse effect. However, for the majority of chemicals, no health-based reference values are available. This is of particular concern for those organochlorine compounds such as PCB and DDT that persist in tissue for long periods. These substances have received much attention because trace amounts are being detected in humans and other species worldwide. For these compounds, the acute toxic effects are generally minimal; however, there are many unanswered questions regarding potential long-term or delayed effects secondary to chronic low level exposure. Although laboratory analyses may be important in epidemiologic studies designed to address these public health concerns, the measured levels of such compounds in the tissues of a given person are not particularly useful in medical diagnosis and are rarely helpful in documenting past exposure levels.

TABLE 3-4. Methodologic Sources of Error in the Interpretation of Toxicologic Analyses (Nonanalytic Errors)

I. Physiologic sources of variation
   A. Hydration
   B. Posture
   C. Activity level
   D. Pregnancy
   E. Pre-existing disease
   F. Use of tourniquet for venipuncture
   G. Time since last meal
II. Kinetic sources of variation
   A. Timing of sample in reference to last exposure
   B. Tissue sampled (blood, urine, alveolar air)
   C. Skin absorption affecting local venous blood levels
III. Variation associated with specimen storage and collection
   A. Factors that cause spuriously *high* results
      1. External contamination from workplace air, skin, or clothing
      2. Contamination from specimen container or additives
      3. Contamination during laboratory analysis
   B. Factors that cause spuriously *low* results
      1. Chemical decomposition (influenced by temperature, *p*H, time to analysis)
      2. Adsorption to container
      3. Evaporative loss
IV. Nonoccupational environmental exposures
      1. Smoking
      2. Alcohol
      3. Diet
      4. Leisure-time activities (hobbies, home maintenance)

- Ideally, it is considered that the measured level of a chemical in tissue or body fluid is a function of its rate of absorption, distribution, metabolism, and elimination. Unfortunately, there are a number of methodologic factors that can affect the measured level apart from analytic errors. Some of these factors are summarized in Table 3-4.

Despite the pitfalls that must be considered, sufficient toxicologic, kinetic, and reference value information exist to render biologic monitoring feasible for a number of common industrial chemials.

MEDICAL MANAGEMENT OF OVEREXPOSURE TO CHEMICALS

While thousands of chemicals are used routinely in industry, the medical management of acute overexposure is, with few exceptions nonspecific and includes four basic steps:

- *Removal from exposure.* Prompt and safe removal is the first consideration. If rescue of a disabled victim is required, rescuers must protect themselves in the contaminated atmosphere. Multiple casualties are common when unprepared would-be rescuers enter a tank or other confined space. Supplied air respirators and lifelines are mandatory precautions, and chemically impervious suits are sometimes required.

- *Resuscitation.* If the victim is apneic, resuscitation must begin as soon as he is out of the hazardous area, further supportive care should be provided as with any other medical emergency.
- *Decontamination.* A victim whose skin or clothing has been contaminated requires immediate removal of garments and shoes and vigorous showering with soap and water, including attention to the fingernails and scalp. Unnecessary delays in thorough decontamination have resulted in worsening symptoms or even fatalities from exposure to substances such as organophosphate pesticides and aromatic amines.
- *Symptomatic treatment.* Acute overexposures may result in a variety of signs and symptoms that require general supportive medical management regardless of the specific toxic agent involved. Examples include the control of convulsive seizures and the treatment of bronchospasm, dehydration, or arrhythmias.

There are a few situations in which specific antidotes or management strategies are available; in those rare cases, a knowledge of effective treatment protocols may be lifesaving. The details of certain treatment protocols are provided in the discussion of specific chemical substances elsewhere in this text.

SUGGESTED READINGS

Parkes WR: Occupational Lung Disorders, 2nd ed. London, Butterworths, 1982

Zimmerman HJ: Hepatotoxicity: Adverse Effects of Drugs and Other Chemicals on the Liver. New York, Appleton-Century-Crofts, 1978

Spencer PS, Schaumburg HH: Experimental and Clinical Neurotoxicology. Baltimore, Williams & Wilkins, 1980

NIOSH Current Intelligence Bulletin #48, Organic Solvent Neurotoxicity. DHHS (NIOSH) Pub No 87-104. Washington, DC, US Government Printing Office, March 31, 1987

Maibach HI: Occupational and Industrial Dermatology, 2nd ed. Chicago, Yearbook Medical Publishers, 1986

Barlow SM, Sullivan FM: Reproductive Hazards of Industrial Chemicals. London, Academic Press, 1982

Alderson M: Occupational Cancer. London, Butterworths, 1985

Identification and Classification of Carcinogens: Guidelines for the Chemical Substances Threshold Limit Values Committee. Documentation of the TLV's and BEI's, 5th ed, pp A-3(86)–A-11(86). Cincinnati, American Conference of Governmental Industrial Hygienists (ACGIH), 1986

Lauwerys RR: Industrial Chemical Exposure: Guidelines for Biological Monitoring. Davis, California, Biomed Publications, 1982

Part Two

THE CHEMICAL HAZARDS

Nick H. Proctor
James P. Hughes
Michael L. Fischman
Gloria W. Hathaway

ACETALDEHYDE

CAS: 75-07-0

CH_3CHO 1987 TLV = 100 ppm

Synonyms: Ethanal; acetic aldehyde; ethylaldehyde

Physical Form. Liquid; boils at 20.6°C

Uses. In the manufacture of paraldehyde, acetic acid, plastics, synthetic rubber, synthetic resins, aniline dyes, mirrors, and disinfectants

Exposure. Inhalation

Toxicology. Acetaldehyde is an irritant of the eyes and mucous membranes; at high concentrations it causes narcosis and respiratory tract irritation in animals, and it is expected that severe exposure will produce the same effect in humans.

Human volunteers exposed for 15 minutes to 50 ppm experienced mild eye irritation.[1] Sensitive subjects have noted eye irritation following 15-minute exposures at 25 ppm.[2] An exposure to 134 ppm for 30 minutes produced mild upper respiratory irritation, whereas at 200 ppm for 15 minutes, all subjects had transient conjunctivitis.[1,3]

The irritant effects of the vapor at lower concentrations, such as cough and a burning sensation in the nose, throat, and eyes, usually prevent exposure sufficient to cause central nervous system depression or chronic effects.[3]

Splashed in the eyes, the liquid causes a burning sensation, lacrimation, blurred vision, and corneal injury. On the skin for a prolonged period of time, the liquid causes erythema and burns; repeated contact may result in dermatitis, due to primary irritation.[3] Exposure to high concentrations in rats and hamsters (750 ppm or more chronically) induced squamous cell carcinomas and adenocarcinomas of the nasal cavity.[4] Associated changes included degenerative changes of olfactory epithelium and squamous metaplasia. Acetaldehyde has demonstrated genotoxicity in a variety of cell culture systems.[5]

The fruity odor of acetaldehyde may be recognized by some persons at 25 ppm.

REFERENCES

1. Silverman L, Schulte HF, First MW: Further studies on sensory response to certain industrial solvent vapors. J Ind Hyg Toxicol 28:262–266, 1946
2. Chemical Hazard Information Profile: Acetaldehyde. Washington, DC, US Environmental Protection Agency, 1983
3. Acetaldehyde. Documentation of the TLVs for Substances in Workroom Air, 5th ed, p 3. Cincinnati, American Conference of Governmental Industrial Hygienists (ACGIH), 1986
4. Woutersen R et al: Inhalation toxicity of acetaldehyde in rats. II. Carcinogenicity study: Interim results after 15 months. Toxicology 31:123–133, 1984
5. Heck H: Mechanisms of aldehyde toxicity: Structure activity studies. CIIT Activities 5(10):1–6, 1985

ACETIC ACID

CAS: 64-19-7

CH_3COOH 1987 TLV = 10 ppm

Synonyms: Ethanoic acid; ethylic acid; methane carboxylic acid

Physical Form. Liquid

Uses. Chemical syntheses; in cellulose and vinyl resins, esters; manufacture of acetic anhydride; solvent

Exposure. Inhalation

Toxicology. Acetic acid vapor is a severe irritant of the eyes, mucous membranes, and skin.

Exposure to 50 ppm or more is intolerable to most persons and results in intensive lacrimation and irritation of the eyes, nose, and throat, with pharyngeal edema and chronic bronchitis.[1] Unacclimatized humans experience extreme eye and nasal irritation at concentrations in excess of 25 ppm; conjunctivitis from concentrations below 10 ppm has been reported.[1]

In a study of 5 workers exposed for 7 to 12 years to concentrations of 80 to 200 ppm at peaks, the principal findings were blackening and hyperkeratosis of the skin of the hands, conjunctivitis (but no corneal damage), bronchitis and pharyngitis, and erosion of the exposed teeth (incisors and canines).[2]

Glacial (100%) acetic acid caused severe injury when applied to the eyes of rabbits; in humans, it has caused permanent corneal opacification.[3] A splash of vinegar (4% to 10% acetic acid solution) in the human eye causes immediate pain and conjunctival hyperemia, sometimes with injury of the corneal epithelium.[3]

On the guinea pig skin, the liquid in concentrations in excess of 80% produced severe burns; concentrations of 50% to 80% produced moderate to severe burns; solutions below 50% caused relatively mild injury; no injury was produced by 5% to 10% solutions.[2]

Although ingestion is unlikely to occur in industrial use, as little as 1.0 ml of glacial acetic acid has resulted in perforation of the esophagus.[1]

REFERENCES

1. AIHA Hygienic Guide Series: Acetic Acid. Akron, Ohio, American Industrial Hygiene Association, 1978
2. Guest D et al: Aliphatic carboxylic acids. In Clayton GD, Clayton FE (eds): Patty's Industrial Hygiene and Toxicology, 3rd ed, rev, Vol 2C, Toxicology, pp 4909–4911. New York, Wiley–Interscience, 1982
3. Grant WM: Toxicology of the Eye, 2nd ed, pp 80–82. Springfield, Illinois, Charles C Thomas, 1974

ACETIC ANHYDRIDE

CAS: 108-24-7

$CH_3C(O)OC(O)CH_3$ 1987 TLV = C 5 ppm

Synonyms: Acetic oxide; acetyl oxide; ethanoic anhydrate

Physical Form. Colorless liquid

Uses. Acetylating agent in production of cellulose acetate; solvent

Exposure. Inhalation

Toxicology. Acetic anhydride vapor is a severe irritant of the eyes, mucous membranes, and skin.

Humans exposed to undetermined but high vapor concentrations complained immediately of severe conjunctival and nasopharyngeal irritation, harsh cough, and dyspnea.[1] Workmen exposed to vapors from a boiling mixture complained of severe eye irritation and lacrimation.[1] The immediate effect of exposure to vapor concentrations above 5 ppm is acute irritation of the eyes and upper respiratory tract; inhalation of high vapor concentrations may produce ulceration of the nasal mucosa and, in some instances, bronchospasm.[2]

Rats exposed to 2000 ppm for 4 hours died, but 1000 ppm for 4 hours was not lethal.[3]

Both the liquid and the vapor can cause severe damage to the human eye; this is characterized by immediate burning, followed some hours later by an increasing severity of reaction with corneal and conjunctival edema.[1] Interstitial corneal opacity may develop over a period of several days from progression of tissue infiltration; in mild cases, this condition is reversible, but permanent opacification with loss of vision may also occur. Workmen exposed to acetic anhydride vapor may show evidence of conjunctivitis with associated photophobia.[1]

Severe burns and vesiculation of human skin have been reported from liquid splashes; concentrated vapor produces primary irritation. Generalized skin reactions in guinea pigs sensitized to acetic anhydride have been demonstrated, and skin sensitization in humans occasionally occurs.[2] Although ingestion of the liquid is unlikely in ordinary industrial use, the highly corrosive nature of the substance may be expected to produce serious burns of the mouth and esophagus.

Acetic anhydride has good warning properties.

REFERENCES

1. AIHA Hygienic Guide Series: Acetic Anhydride. Akron, Ohio, American Industrial Hygiene Association, 1978
2. Fassett DW: Organic acids and related compounds. In Fassett DW, Irish DD (eds): Toxicology, Vol 2. In Patty FA (ed): Industrial Hygiene and Toxicology, 2nd ed, pp 1817–1818. New York, Interscience, 1963
3. Smyth HF Jr: Hygienic standards for daily inhalation. Am Ind Hyg Assoc J 17:129–185, 1956

ACETONE

CAS: 67-64-1
$(CH_3)_2CO$ 1987 TLV = 750 ppm

Synonyms: Dimethyl ketone; 2–propanone; pyroacetic ether

Physical Form. Liquid (highly flammable)

Uses. Solvent; in the production of lubricating oils; in the dyeing and celluloid industries; as an intermediate in the production of chloroform

Exposure. Inhalation

Toxicology. Acetone is an irritant of the eyes and mucous membranes; at very high concentration, it causes narcosis in animals, and severe exposure is expected to produce the same effect in humans.

Acetone is considered to be of low risk to health, because few adverse effects have been reported despite widespread use for many years.[1] One early study, often quoted, reports eye, nose, and throat irritation in volunteers exposed to 500 ppm.[2]

In more recent studies, subjects exposed to 500 ppm were aware of odor but exhibited no effects.[3] Mild eye irritation occurred around 1000 ppm.[4] Higher concentrations produced headache, light-headedness, and nose and throat irritation.[4] Concentrations above 12,000 ppm depress the central nervous system, causing dizziness, weakness, and loss of consciousness.[5] Topical application of 1 ml of acetone for 90 minutes produced reversible skin damage in humans.[6]

Acetone may be recognized by a sweet, pungent odor.[7]

REFERENCES

1. National Institute for Occupational Safety and Health, US Department of Health, Education and Welfare: Criteria for a Recommended Standard . . . Occupational Exposure to Ketones. DHEW (NIOSH) 78–173. Washington, DC, US Government Printing Office, 1978
2. Nelson KW et al: Sensory response to certain industrial solvent vapors. Am Ind Hyg Assoc J 25:282–285, 1943
3. DiVincenzo GO, Yanno FJ, Astill BD: Exposure of man and dog to low concentrations of acetone vapor. Am Ind Hyg Assoc J 34:329–336, 1973
4. Raleigh RL, McGee WA: Effects of short, high-concentration exposures to acetone as determined by observation in the work area. J Occup Med 14:607–610, 1972
5. Ross DS: Short communications—acute acetone intoxication involving eight male workers, Ann Occup Hyg 16:73–75, 1973
6. Lupulescu AP, Birmingham DJ: Effect of protective agent against lipid-solvent-induced damages—ultrastructural and scanning elec-

tron microscopical study of human epidermis. Arch Environ Health 31:33–36, 1976
7. Chemical Safety Data Sheet SD–87, Acetone, pp 5, 13. Washington, DC, MCA, Inc, 1962

ACETONITRILE

CAS: 75-05-8

CH_3CN 1987 TLV = 40 ppm; skin

Synonyms: Methyl cyanide; cyanomethane; ethanenitrile

Physical Form. Colorless liquid

Uses. Chemical intermediate; solvent; extractant for animal and vegetable oils

Exposure. Inhalation; skin absorption

Toxicology. Acetonitrile causes headache, dizziness, and nausea; at extremely high concentrations, it can cause convulsions, coma, and death.

Of 15 painters exposed to the vapor of a mixture containing 30% to 40% acetonitrile for 2 consecutive workdays, 10 developed symptoms ranging in severity from nausea, headache, and lassitude among the lesser exposed, to vomiting, respiratory depression, extreme weakness, and stupor in the more heavily exposed. Five of the painters required hospitalization, and one died; the worker who died experienced the onset of chest pain 4 hours after leaving the job on the second day of exposure, followed shortly by massive hematemesis, convulsions, shock, and coma, with death occurring 14 hours after cessation of exposure.[1] At autopsy, cyanide ion concentrations (in micrograms per cent) were as follows: blood, 796; urine, 215; kidney, 204; spleen, 318; and lung, 128. Cyanide ion was not detected in the liver.[1]

Two human subjects inhaled 160 ppm for 4 hours; one of them experienced a slight flushing of the face 2 hours later and a slight feeling of bronchial tightness 5 hours later. A week before this, the same two subjects had inhaled 80 ppm with no effects.[2] Blood cyanide and urine thiocyanate levels did not correlate with exposure and, therefore, are not reliable indicators of brief exposure to low concentrations.

In male rats, the LC_{50} was 7500 ppm for a single 8-hour exposure; there was prostration followed by convulsive seizures; at autopsy there was pulmonary hemorrhage.[2] Rats exposed 6 hours/day, 5 days/week for 4 weeks to concentrations greater than 600 ppm had respiratory and ocular irritation and anemia.[3] In another study, rats repeatedly exposed to 665 ppm for 7 hours daily developed pulmonary inflammation, and there were minor changes in the liver and kidneys in some animals.[2] In the rabbit eye, a drop of the liquid caused superficial injury.[4] The liquid on the belly of a rabbit caused a faint erythema of short duration.[5]

No malformations related to acetonitrile exposure were observed in the offspring of rats orally exposed at maternally toxic levels.[6] Inhalation of 5000 or 8000 ppm for 60 minutes by pregnant hamsters on day 8 of gestation was associated with production of severe axial skeletal disorders; maternal toxicity, including irritation, respiratory difficulty, lethargy, ataxia, hypothermia, and increased mortality were noted.[7] At lower doses, there were no signs of maternal toxicity and offspring were normal.[7]

The toxic effects of acetonitrile are attributed to the metabolic release of cyanide; cyanide in turn acts by inhibiting cytochrome oxidase and thus impairs cellular respiration.[8] Evidence of the cyanide effect is supported by the reported effectiveness of specific cyanide antidotes in acetonitrile poisonings.[8]

Diagnosis. *Signs and Symptoms:* Asphyxia and death can occur from severe exposure. Signs and symptoms of exposure include nausea and vomiting, chest pain, weakness, stupor, and convulsions. Acetonitrile may also cause eye and skin irritation.

The diagnosis is usually self-evident from history, but even if it is not completely established and acetonitrile poisoning is suspected, the recommended therapy should be considered.

Special Tests: A blood level of cyanide in excess of 0.2 μg/ml suggests a toxic reaction, but because of tight binding of cyanide to cytochrome oxidase, serious poisoning may occur with only modest blood levels, especially several hours after poisoning. In view of the likelihood of lactic acidosis, there should be measurement of blood *p*H, plasma bicarbonate, and blood lactic acid.[11]

Treatment. Preparedness and speed of action are prerequisites for successful treatment of overexposure.[9,10] The following treatment is recommended but is said not to be as effective for acetonitrile poisoning as for inorganic cyanide poisoning. All persons working with or around acetonitrile should be given specific and detailed instruction on the use of antidote kits containing amyl nitrite ampules for inhalation. Kits for use by medical personnel should contain the following:

2 boxes of amyl nitrite ampules, 0.3 ml/ampule
2 ampules of sterile sodium nitrite, 10 ml of 3% solution each
2 ampules of sterile sodium thiosulfate, 50 ml of 25% solution each
2 sterile 10-ml syringes with intravenous needles
1 sterile 50-ml syringe with intravenous needles
1 tourniquet
1 gastric tube (rubber)
1 nonsterile 100-ml syringe

A complete cyanide antidote package can be purchased from the Eli Lilly Co., stock number M76.

The exposed person should be removed as rapidly as possible to an uncontaminated area by someone with adequate respiratory protective equipment.

Resuscitate: If breathing has stopped, immediately institute resuscitation and continue until normal breathing is established.

Decontaminate: If the liquid has contaminated the skin or clothing, the clothing should be removed and the skin flushed with copious amounts of water. If the liquid has been taken by mouth, gastric lavage should be performed but should be postposed until after specific therapy has been initiated.

Administer Amyl Nitrite: In cases of severe exposure, the following should be considered: a pearl (ampule), if not provided with a fabric sleeve, should be wrapped lightly in a handkerchief or gauze pad, broken, and held about 1 inch from the patient's mouth and nostrils for 15 to 30 seconds of every minute while the sodium nitrite solution is being prepared. Repeat five times at 15-second intervals. Use a fresh pearl every 5 minutes.

Administer Sodium Nitrite and Sodium Thiosulfate: If the patient does not respond to the amyl nitrite administration or if severe exposure is suspected, administer intravenously (for adults) 0.3 g sodium nitrite (10 ml of a 3% solution) at the rate of 2.5 to 5 ml/min followed by an injection of 12.5 g sodium thiosulfate (50 ml of a 25% solution) over about 10 minutes by the same needle and vein.

Observe Patient: The patient should be kept under observation for 24 to 48 hours. The blood levels of methemoglobin should be monitored and not allowed to exceed 40%.[11] The desired methemoglobin level is approximately 25%. If signs of intoxication persist or reappear, the injection of nitrite and thiosulfate should be repeated in one-half the above doses. Even if the patient appears well, a second injection may be given 2 hours after the first injection for prophylactic purposes. *Note:* The use of oxygen should not be ignored, because the methemoglobinemia induced by this treatment reduces the ability of blood to carry oxygen to the brain.

Note: The therapeutic regimen (amyl nitrite/sodium nitrite plus sodium thiosulfate) just discussed is the "classical" one that has long been recommended for cyanide. However, there is growing support for a new regimen of 100% oxygen together with intravenous hydroxocobalamin and/or sodium thiosulfate.[11,12] Hydroxocobalamin reverses cyanide toxicity by combining with cyanide to form cyanocobalamin (vitamin B_{12}); one advantage is that hydroxocobalamin is apparently of low toxicity. The only side effects have been occasional urticaria and a brown-red discoloration of the urine. Hydroxocobalamin is neither currently available nor approved for treatment of cyanide intoxication in the United States.[13] Nonspecific supportive therapy is an essential part of the treatment; in some cases, patients have recovered from massive cyanide poisoning with supportive therapy only.

REFERENCES

1. Amdur MO: Accidental group exposure to acetonitrile. J Occup Med 1:627–633, 1959
2. Pozzani UC et al: An investigation of the mammalian toxicity of acetonitrile. J Occup Med 1:634–642, 1959
3. Roloff V et al: Comparison of subchronic inhalation toxicity of five aliphatic nitriles in rats. Toxicologist 5:30, 1985
4. Grant WM: Toxicology of the Eye, 2nd ed, p 84. Springfield, Illinois, Charles C Thomas, 1974
5. Toxicology Studies, Acetonitrile. New York, Union Carbide Corporation, 1965
6. Berteau PE et al: Teratogenic evaluation of aliphatic nitriles in rats. Toxicologist 2:118, 1982
7. Willhite CC: Developmental toxicology of acetonitrile in the Syrian golden hamster. Teratology 27:313–325, 1983
8. National Institute for Occupational Safety and Health, US Department of Health, Education, and Welfare: Criteria for a Recommended Standard . . . Occupational Exposure to Nitriles. DHEW (NIOSH) Pub No 78–212, p 155. Washington, DC, US Government Printing Office, 1978
9. Wolfsie JH: Treatment of cyanide poisoning in industry. AMA Arch Ind Hyg Occup Med 4:417–425, 1951
10. Chen KK, Rose CL: Nitrite and thiosulfate therapy in cyanide poisoning. JAMA 149:113–119, 1952
11. Graham DL, Laman D, Theodore J, Robin ED: Acute cyanide poisoning complicated by lactic acidosis and pulmonary edema. Arch Intern Med 137:1051–1055, 1977
12. Berlin C: Cyanide poisoning—a challenge. Arch Intern Med 137:993–994, 1977
13. Becker CE: The role of cyanide in fires. Vet Hum Toxicol 27:487–490, 1985

2-ACETYLAMINOFLUORENE

CAS: 53-96-3

$C_{15}H_{13}NO$ 1987 TLV = none established

Synonyms: *N*–2–Fluorenylacetamide; 2–fluorenamine; acetylaminofluorine

Physical Form. Crystals

Uses. Promising pesticide until carcinogenic activity was discovered in rodents

Exposure. Inhalation

Toxicology. 2-Acetylaminofluorene (AAF) is a potent carcinogen in dogs, hamsters, and rats.

There is no toxicity information on humans.[1]

Four of five dogs developed tumors of the liver and urinary bladder after ingestion of 0.6 to 1.2 g AAF/kg diet for up to 91 months.[2] Animals developing tumors received a total of 90 to 198 g AAF, whereas the animal with no tumor formation ingested 45 g; another group of four dogs receiving 32 to 37 g during 2.25 years did not develop tumors.[2] The extent of tumor formation was directly related to the amount of AAF consumed, being most marked in those animals that received nearly 200 g during the feeding period.[2] Liver tumors of varied types were observed. Multiple papillomas were produced in the urinary bladder, and in one dog, there was invasion of the submucosa and muscle by the tumor cells.

Intratracheal administration of 5 to 15 mg AAF one to two times per week for 17 months in hamsters (total dose 1100 mg) caused bladder tumors in 10 of 23 animals; all tumors were transitional cell carcinomas with or without focal squamous cell carcinomas.[3]

In rats, AAF had no demonstrable acute toxicity in quantities up to 50 mg/kg subcutaneously and 1 g/kg gastrically; however, AAF was very toxic when incorporated in the diet.[4] Incorporation of 0.031% AAF or higher for at least 95 days led to epithelial hyperplasia of the bladder, renal pelvis, liver, pancreas, and lung; 19 of 39 rats developed malignant tumors, 16 of which were carcinomas.[4]

Animal studies have indicated that *N*-hydroxy-2-acetylaminofluorene (*N*-hydroxy-AAF) is a proximate carcinogenic metabolite of AAF.[5] AAF is not carcinogenic in the guinea pig, and no *N*-hydroxylation of AAF has been detected in vivo or in vitro in this species; however, administration of *N*-hydroxy-AAF causes tumors in guinea pigs.[5] In addition, *N*-hydroxy-AAF has proved to be a carcinogen of much greater potency than AAF in rats, mice, hamsters, and rabbits at sites of local application.[5]

AAF is classified as a cytotoxic teratogen.[1] Because of demonstrated carcinogenicity in animals, contact by all routes should be avoided.[6] However, in recent years, this compound has been used only in laboratories as a model of tumorigenic activity in animals.[6] It is of little occupational health importance.

REFERENCES

1. Doull J, Klaassen CD, Amdur MO (eds): Toxicology—The Basic Science of Poisons, 2nd ed, p 163. New York, Macmillan, 1980
2. Morris HP, Eyestone WH: Tumors of the liver and urinary bladder of the dog after ingestion of 2-Acetylaminofluorene. J Natl Cancer Inst 13:1139–1165, 1953
3. Oyasu R, Kitajima T, Hoop ML, Sumie H: Induction of bladder cancer in hamsters by repeated intratracheal administrations of 2-acetylaminofluorene. J Natl Cancer Inst 50:503–506, 1973
4. Wilson RH, DeEds, F, Cox AJ Jr: The toxicity and carcinogenic activity of 2-Acetaminofluorene. Cancer Res 1:595–608, 1941
5. Miller EC, Miller JA, Enomoto M: The comparative carcinogenicities of 1-Acetylaminofluorene and its *N*-Hydroxy metabolite in mice, hamsters, and guinea pigs. Cancer Res 24:2018–2031, 1964
6. Stokinger HE: In Patty's Industrial Hygiene and Toxicology, 3rd ed, rev, p 2893. New York, John Wiley & Sons, 1981

ACETYLENE TETRABROMIDE

CAS: 79-27-6

$CHBr_2CHBr_2$ 1987 TLV = 1 ppm

Synonyms: Tetrabromoethane; Muthmann's liquid; 1,1,2,2–tetrabromoethane

Physical Form. Colorless to yellow liquid

Uses. Gauge fluid; solvent; refractive index liquid in microscopy

Exposure. Inhalation

Toxicology. Acetylene tetrabromide is a central nervous system depressant and hepatotoxin.

A chemist working with the substance for 7.5 hours with no local exhaust ventilation developed severe, nearly fatal, liver damage and was hospitalized for 9 weeks; his estimated exposure during most of the work shift prior to the onset of symptoms was 1 to 2 ppm, although he had a single 10-minute period exposure to approximately 16 ppm.[1] He complained first of headache, anorexia, and nausea within hours of the exposure, and within 5 days he developed abdominal pain with bilirubinuria and a monocytosis of 17%. In this case, exposure to higher concentrations or significant skin absorption might have occurred. The similarity of the symptoms to viral hepatitis is also noted. Other workers in the same laboratory complained only of slight eye and nose irritation, with headache and lassitude.[1]

Rats exposed to a saturated atmosphere for 7 hours exhibited slight eye and nose irritation.[2] Guinea pigs exposed for 90 minutes to a saturated vapor became comatose, seemed to recover, but died after several days; the same exposure for up to 3 hours was not lethal to rats and rabbits.[3] No mortality was observed in rats, guinea pigs, rabbits, mice, and a monkey exposed 7 hours/day to 14 ppm for 100 days; findings at 14 ppm did include edema of the lungs and slight fatty degeneration of the liver in all species except guinea pigs, which showed only growth depression.[2] Repeated exposure to 4 ppm for 180 days caused slight histopathologic changes in the liver and lungs of some animals, but no effects were observed at 1 ppm.[2]

Repeated application of 15 mg to the skin of mice caused a statistically significant increased incidence of forestomach papillomas.[4]

The liquid instilled in the rabbit eye caused slight to moderate pain, conjunctival irritation, and corneal injury, which disappeared after 24 hours.[1] When bandaged onto the shaved abdomen of the rabbit for 72 hours, moderate redness, edema, and blistering were observed.[1]

Acetylene tetrabromide has a sweetish unpleasant odor that is readily apparent and objectionable to most persons at concentrations greater than 1 to 2 ppm.[1,2]

REFERENCES

1. Van Haaften AB: Acute tetrabromoethane (acetylene tetrabromide) intoxication in man. Am Ind Hyg Assoc J 30:251–256, 1969
2. Hollingsworth RL, Rowe VK, Oyen F: Toxicity of acetylene tetrabromide determined on experimental animals. Am Ind Hyg Assoc J 24:28–35, 1963
3. Gray MG: Effect of exposure to the vapors of tetrabromoethane (acetylene tetrabromide). Arch Ind Hyg Occup Med 2:407–419, 1950
4. Van Duuren BL et al: Carcinogenicity of halogenated olefinic and aliphatic hydrocarbons in mice. J Natl Cancer Inst 63:1433–1439, 1979

ACROLEIN

CAS: 107-02-8

$CH_2{=}CHCHO$ 1987 TLV = 0.1 ppm

Synonyms: Acrylaldehyde; propenal; allyl aldehyde; ethylene aldehyde; aqualin

Physical Form. Liquid; boils at 52.7°C

Uses. In manufacture of pharmaceuticals, perfumes, food supplements, and resins; warning agent in methyl chloride refrigerating systems

Exposure. Inhalation

Toxicology. Acrolein is an intense irritant of the eyes and upper respiratory tract.

Exposure to high concentrations may cause tracheobronchitis and pulmonary edema.[1] The irritation threshold in humans is 0.25 ppm to all mucous membranes within 5 minutes.[1] Fatalities have been reported at levels as low as 10 ppm, and 150 ppm is lethal after 10 minutes.[2,3] The violent irritant effect usually prevents chronic toxicity in humans.[1] Prolonged or repeated contact produces skin irritation, burns, and sometimes sensitization. Eye splashes cause corneal damage, palpebral edema, blepharoconjunctivitis, and fibrinous or purulent discharge.[4]

Of 57 male rats, 32 died following exposure to 4 ppm for 6 hours/day for up to 62 days.[5] Bronchiolar necrosis and focal pulmonary edema were noted. Sublethal acrolein exposure in mice at 3 and 6 ppm suppressed pulmonary antibacterial defense mechanisms.[6]

Intra-amniotic administration of acrolein in rats induced a significant number of fetal malformations, whereas intravenous administration was embryolethal.[7]

The carcinogenic potential of acrolein has not been adequately determined, but one of its potential metabolites, glycidaldehyde, is considered to be carcinogenic.[1,8]

REFERENCES

1. Beauchamp RO et al: A critical review of the literature on acrolein toxicity. Crit Rev Toxicol 14:309–380, 1985
2. Henderson Y, Haggard HW: Noxious Gases, p 138. New York, Reinhold Publishing Co, 1943
3. Prentiss AM: Chemicals in War. A Treatise of Chemical Warfare, pp 139–140. New York, McGraw-Hill, 1937
4. Grant WM: Toxicology of the Eye, 3rd ed, pp 49–50. Springfield, Illinois, Charles C Thomas, 1986
5. Kutzman RS et al: Changes in rat lung structure and composition as a result of subchronic exposure to acrolein. Toxicology 34:139–151, 1985
6. Astry CL, Jakab GJ: The effects of acrolein exposure on pulmonary antibacterial defenses. Toxicol Appl Pharmacol 67:49–54, 1983
7. Slott VL, Hales BF: Teratogenicity and embryolethality of acrolein and structurally related compounds in rats. Teratology 32:65–72, 1985
8. IARC Monographs on the Evaluation of the Carcinogenic Risk of Chemicals to Man, Vol 19, Some Monomers, Plastics and Synthetic Elastomers, and Acrolein, pp 579–594. Lyon, International Agency for Research on Cancer, 1979

ACRYLAMIDE

CAS: 79-06-1

$CH_2{=}CHCONH_2$ 1987 TLV = 0.03 mg/m^3 (proposed change); skin

Synonyms: Acrylic amide; propenamide

Physical Form. White crystalline powder

Uses. Waterproofing; soil stabilizer; production of polymers used in paper manufacturing and flocculating processes

Exposure. Inhalation; skin absorption; ingestion

Toxicology. Acrylamide causes neurologic effects from systemic exposure and irritation of skin and mucous membrane from local exposure. It is carcinogenic in laboratory animals and is suspected of carcinogenic potential for humans.

A variety of signs and symptoms have been described in cases of acrylamide poisoning that suggest involvement of the central, peripheral, and autonomic nervous systems.[1] Effects on the central nervous system are characterized by abnormal fatigue, memory difficulties, and dizziness.[1] With severe poisoning, confusion, disorientation, and hallucinations occur. Truncal ataxia, nystagmus, and slurred speech have also been observed. Peripheral neuropathy symptoms can include muscular weakness, paresthesia, numbness in hands, feet, lower legs and lower arms, unsteadiness and difficulties in walking and standing. Clinical signs are loss of peripheral tendon reflexes, impairment of vibration sense, and muscular wasting in the extremities. Nerve biopsy shows loss of large diameter nerve fibers as well as regenerating fibers. Autonomic nervous system involvement is indicated by excessive sweating, peripheral vasodilation, and difficulties in micturation and defecation. Central nervous system effects appear to predominate in acute cases, whereas peripheral neuropathy is more common with repeated low-dose exposures or following a latency period of up to several weeks after acute exposure.

After cessation of exposure to acrylamide, most patients recover, although the course of improvement can extend over months to years.[1]

Because most cases of human poisoning have included skin absorption, the dose–response relationship has not been determined. On the skin, acrylamide causes local irritation characterized by blistering and desquamation of the palms and soles, combined with blueness of the hands and feet.[1]

For a number of species, the oral LD_{50} was approximately 150 to 180 mg/kg body weight. In cats, a total cumulative dose of 70 to 130 mg/kg was characterized by delayed onset of ataxia.[2] Cats fed 10 mg/kg diet per day developed definite hind limb weakness after 26 days; at 3 mg/kg/day, there was twitching in the hindquarters after 26 days and signs of hind limb weakness after 68 days.[3]

The underlying lesion involves distal retrograde degeneration of long and large-diameter axons.[4]

Teratogenic effects were not observed in the offspring of rats given up to 50 mg/kg diet for 2 weeks prior to mating and for 19 days during gestation.[1]

A statistically significant increase in mesothelioma of the scrotal cavity was observed in rats given drinking water formulated to provide 0.5 mg/kg of body weight per day for 2 years; in females, there were significant increases in the number of neoplasms of the central nervous system, thyroid, mammary gland, oral cavity, clitoral gland, and uterus.[5]

Acrylamide has also been reported to act as a skin tumor initiator in mice by three exposure routes and to increase the yield of lung adenomas in another strain of mice.[6]

In a human mortality study of 371 workers, no increase in total malignant neoplasms or any specific cancers attributable to acrylamide exposure were found.[7] Exposure levels reached 1.0 mg/m^3 prior to 1957 and were between 0.1 and 0.6 mg/m^3 after 1970. The American Conference of Governmental Industrial Hygienists (ACGIH) has determined that acrylamide should be treated as a suspected human carcinogen.[8]

REFERENCES

1. The International Programme on Chemical Safety: Environmental Health Criteria 49 Acrylamide, pp 1–121. Geneva, World Health Organization, 1985
2. Kuperman AS: Effects of acrylamide on the central nervous system of the cat. J Pharmacol Exp Ther 123:180–192, 1958
3. McCollister DD et al: Toxicology of acrylamide. Toxicol Appl Pharmacol 6:172–181, 1964
4. Miller MS, Spencer PS: The mechanisms of acrylamide axonopathy. Ann Rev Pharmacol Toxicol 25:643–666, 1985
5. Johnson KA et al: Chronic toxicity and onco-

genicity study on acrylamide incorporated in the drinking water of Fischer 344 rats. Toxicol Appl Pharmacol 85:154–168, 1986
6. Bull RJ et al: Carcinogenic effect of acrylamide in Sencar and A/J mice. Cancer Res 44:107–111, 1984
7. Sobel W et al: Acrylamide cohort mortality study. Br J Ind Med 43:785–788, 1986
8. Acrylamide. Documentation of TLVs and BEIs, 5th ed, pp 12–13. Cincinnati, American Conference of Governmental Industrial Hygienists (ACGIH), 1986

ACRYLIC ACID

CAS: 79-10-7
$CH_2{=}CHCOOH$ 1987 TLV = 10 ppm

Synonyms: 2-Propenoic acid; acroleic acid; ethylenecarboxylic acid; vinylformic acid

Physical Form. Colorless, fuming liquid

Uses. Starting material for acrylates and polyacrylates used in plastics, water purification, paper and cloth coatings, and medical and dental materials

Exposure. Inhalation

Toxicology. Acrylic acid is a severe irritant of the eyes, nose, and skin.

Medical reports of acute human exposures (concentration unspecified) include moderate and severe skin burns, moderate eye burns, and mild inhalation effects.[1]

Rats exposed to 1500 ppm for four 6-hour periods exhibited nasal discharge, weight loss, lethargy, and kidney congestion.[2] At 300 ppm, twenty 6-hour exposures produced all but the latter effect. No toxic signs resulted from exposure to 80 ppm for twenty 6-hour periods. In another study, rats and mice were exposed to 0, 5, 25, or 75 ppm 6 hours/day, 5 days/week, for 13 weeks.[3] At 75 ppm in rats, there was slight degenerative lesions of the nasal mucosa, but none at 25 ppm. In contrast, lesions of the nasal mucosa appeared in at least some of the mice at all dose levels, but not in the control group.

There is great variability in the reported values for the oral LD_{50} in rats, ranging from 350 to 3200 mg/kg.[4,5] The dermal LD_{50} in rabbits was 750 mg/kg.[6]

Application of 500 mg to the skin of rabbits produced severe irritation.[7] A 1% solution in the eye of a rabbit caused significant injury.[6]

Intraperitoneal injection of female rats on days 5, 10, and 15 of gestation with 4.7 or 8 mg/kg produced a significant increase in the number of gross abnormalities, including skeletal abnormalities at the higher dose level.[8] In another teratogenicity and embryolethality study in rats, acrylic acid was injected intra-amniotically on day 13 of gestation at doses of 10, 100, or 1000 μg per fetus.[9] The highest dose level was significantly embryotoxic. One fetus at the 100 μg level was malformed, but there was no dose–response relationship. Neither of these studies is useful in assessing teratogenic potential in the workplace because very small numbers of animals were used, the route of exposure bears no resemblance to normal exposure of mothers and embryos, and the injection process may have physically traumatized the embryo.

In an inhalation teratogenicity study, pregnant rats were exposed from day 6 to day 15 to 0, 40, 120, or 360 ppm.[10] Marked effects were observed in the dams at 360 ppm, including reduced weight gain, decreased food intake, and clinical signs of an irritant effect on mucous membranes. There were no signs of embryotoxicity or teratogenicity.

In 2-week studies preliminary to a lifetime dermal carcinogenicity study in mice, a concentration of 5% in acetone caused peeling and flaking of the skin.[11] At 1% in acetone, there were no adverse effects, and this level was chosen to apply a dose of 0.25 mg acrylic acid to the skin of 40 C3H/HEJ male mice 3 days/week for 1.5 years. No treatment-related tumors or effects on mortality were produced.

Acrylic acid was not mutagenic in five strains of *Salmonella typhimurium* with or without metabolic activation by liver microsomes.[12]

Alpha, beta-Diacryloxypropionic acid has been found to be a sensitizing impurity in commercial acrylic acid.[13]

REFERENCES

1. Sittig M: Handbook of Toxic and Hazardous Chemicals and Carcinogens, 2nd ed, p 43. New Jersey, Noyes Publishing, 1985
2. Gage JC: The subacute inhalation toxicity of 109 industrial chemicals. Br J Ind Med 27:1, 1970
3. Miller RR, Ayres JA, Jersey GC, McKenna MJ: Inhalation toxicity of acrylic acid. Fund Appl Toxicol 1:271, 1981
4. Carpenter CP, Weil CF, Smyth HF Jr: Range-finding toxicity data: List VIII. Toxicol Appl Pharmacol 28:313, 1974
5. Miller ML: Acrylic acid polymers. In Bikales NM (ed): Encyclopedia of Polymer Science and Technology, Plastics, Resins, Rubbers, Fibers, Vol 1, p 197. New York, Interscience, 1964
6. Toxicology Studies—Acrylic Acid, Glacial, Union Carbide Corporation, New York, Industrial Medicine and Toxicology Department, 1977
7. Unpublished data, Union Carbide Data Sheet, 1965
8. Singh AR, Lawrence WH, Autian J: Embryonic-fetal toxicity and teratogenic effects of a group of methacrylate esters in rats. J Dent Res 51:1632, 1972
9. Slott VL, Hales BF: Teratogenicity and embryolethality of acrolein and structurally related compounds in rats. Teratology 32:65, 1985
10. Klimisch HJ: Acrylic acid: Prenatal toxicity after inhalation in Sprague-Dawley rats. BASF, Ludwigshaven, unpublished data, December 30, 1983
11. DePass LR et al: Dermal oncogenicity bioassays of acrylic acid, ethyl acrylate and butyl acrylate. J Toxicol Environ Health 14:115, 1984
12. Lijinsky W, Andrews AW: Mutagenicity of vinyl compounds in *Salmonella typhimurium*. Ter Carcin Mut 1:259, 1980
13. Waegemaekers THJ, van der Walle HB: Alpha, beta-diacryloxypropionic acid, a sensitizing impurity in commercial acrylic acid. Dermatosen Beruj Umwelt 32:55, 1984

ACRYLONITRILE

CAS: 107-13-1

$CH_2{=}CHCN$ 1987 TLV = 2.0 ppm; skin

Synonyms: ACN; propenenitrile; vinyl cyanide

Physical Form. Explosive, flammable liquid

Uses. Manufacture of acrylic fibers; synthesis of rubber-like materials; pesticide fumigant

Exposure. Inhalation; skin absorption

Toxicology. Acrylonitrile is an eye, skin, and upper respiratory tract irritant; systemic effects are nonspecific but may include the central nervous system and hepatic, renal, cardiovascular and gastrointestinal systems. It is carcinogenic in experimental animals and is a suspected human carcinogen.

Two human fatalities from accidental poisoning have been reported; one was caused by inhalation of an unknown concentration of the vapor, and the other was thought to be caused by skin absorption or inhalation.[1] Most cases of intoxication from industrial exposure have been mild, with rapid onset of eye irritation, headache, sneezing, and nausea; weakness, light-headedness, and vomiting may also occur.[2] Acute exposure to higher concentrations may produce profound weakness, asphyxia, and death.[2] Acrylonitrile is metabolized to cyanide by hepatic microsomal reactions. Deaths from acute poisoning result from inhibition of mitochondrial cytochrome oxidase activity by metabolically liberated cyanide.[2]

Prolonged skin contact with the liquid results in both systemic toxicity and the formation of large vesicles after a latent period of several hours.[2] The affected skin may resemble a second degree thermal burn.

Administration of 65 mg/kg/day by gavage to rats on days 6 to 15 of gestation produced significant maternal toxicity and an increased incidence of malformation in the offspring.[3] Inhalation of 80 ppm 6 hours/day by the dams did not result in a statistically significant increase in the occurrence of any single malformation.[3]

In a number of chronic bioassays in rats, administration of acrylonitrile by gavage, by inhalation, and in the drinking water produced tumors of the mammary gland, the gastrointestinal tract, the zymbal glands, and the central nervous system.[4,5] In another study, administration of 500 ppm in drinking water caused a statistically significant increase in microscopically detectable primary brain tumors.[6] Neurologic signs were observed in 29 of 400 rats within 18 months, and brain tumors occurred in 49 of 215 animals that died or were sacrificed in the first 18 months.[6]

A study of 1345 workers potentially exposed to acrylonitrile and followed for 10 or more years showed a greater than expected incidence of lung cancer (8 obs. vs. 4.4 exp.).[7] A trend toward increased risks of cancer of all sites was also observed with increased duration of exposure and with higher severity of exposure.[7] In a follow-up of this cohort through 1983, the only statistically significant excess was for prostate cancer (6 obs. vs. 1.8 exp.).[8] An excess number of lung cancer cases remained (10 obs. vs. 7.2 exp.) but was not as marked.[8] Other epidemiologic studies have reported excess lung cancer deaths but lacked statistical significance because of small cohort size.[9]

Based on studies available in 1982, the International Agency for Research on Cancer (IARC) determined that there was sufficient evidence of carcinogenicity to animals and limited evidence for carcinogenicity to humans.[5]

Diagnosis. *Signs and Symptoms:* Asphyxia and death can occur from high exposure. Eye irritation, headache, sneezing, nausea and vomiting, weakness and light-headedness, and vesicles from prolonged skin contact may occur. Repeated skin contact may produce scaling dermatitis.

Special Tests: A blood level of cyanide in excess of 0.2 μg/ml suggests a toxic reaction, but because of tight binding of cyanide to cytochrome oxidase, serious poisoning may occur with only modest blood levels, especially several hours after poisoning. In view of the likelihood of lactic acidosis, there should be measurement of blood *p*H, plasma bicarbonate, and blood lactic acid.[10]

Treatment. Preparedness and speed of action are prerequisites for successful treatment for overexposure.[12–14] All persons working with or around acrylonitrile should be given specific and detailed instructions on the use of antidote kits containing amyl nitrite ampules for inhalation. Kits for use by medical personnel should contain the following:

2 boxes of amyl nitrite ampules, 0.3 ml/ampule
2 ampules of sterile sodium nitrite, 10 ml of 3% solution each
2 ampules of sterile sodium thiosulfate, 50 ml of 25% solution each
2 sterile 10-ml syringes with intravenous needles
1 sterile 50-ml syringe with intravenous needle
1 tourniquet
1 gastric tube (rubber)
1 nonsterile 100-ml syringe

A complete cyanide antidote package can be purchased from the Eli Lilly Co., stock number M76.

The exposed person should be removed as rapidly as possible to an uncontaminated area by someone with adequate respiratory protective equipment.

Resuscitate: If breathing has stopped, immediately institute resuscitation and continue until normal breathing is established.

Decontaminate: If liquid has contaminated the skin or clothing, the clothing should be removed and the skin flushed with copious amounts of water. If taken by mouth, gastric lavage should be performed but should be postponed until after specific therapy has been initiated.

Administer Amyl Nitrite: In cases of severe exposure, the following should be considered: A pearl (ampule), if not provided with a fabric sleeve, should be wrapped lightly in a handkerchief or gauze pad, broken, and held about 1 inch from the patient's mouth and nostrils for 15 to 30 seconds of every minute, while the sodium nitrite solution is being prepared. Repeat five times at 15-second intervals. Use a fresh pearl every 5 minutes.

Administer Sodium Nitrite and Sodium Thiosulfate: If the patient does not respond to the amyl nitrite administration or if severe exposure is suspected, administer intravenously for adults 0.3 g sodium nitrite (10 ml of a 3% solution) at the rate of 2.5 to 5 ml/minute followed by injection of 12.5 g sodium thiosulfate (50 ml of a 25% solution) over about 10 minutes by the same needle and vein.

Observe Patient: The patient should be kept under observation for 24 to 48 hours. The blood levels of methemoglobin should be monitored and not allowed to exceed 40%. The desired methemoglobin level is approximately 25%. If signs of intoxication persist or reappear, the injection of nitrite and thiosulfate should be repeated in one-half the aforementioned doses. Even if the patient appears well, a second injection may be given 2 hours after the first for prophylactic purposes. *Note:* The use of oxygen should not be ignored, because the methemoglobinemia induced by this treatment reduces the ability of blood to carry oxygen to the brain.

Note: The therapeutic regimen (amyl nitrite/sodium nitrite plus sodium thiosulfate) just discussed is the "classic" one that has long been recommended for cyanide. However, there is growing support for a new regimen of 100% oxygen together with intravenous hydroxocobalamin and/or sodium thiosulfate.[10,11] Hydroxocobalamin reverses cyanide toxicity by combining with cyanide to form cyanocobalamin (vitamin B_{12}); one advantage is that hydroxocobalamin is apparently of low toxicity. The only side effects have been occasional urticaria and brown-red discoloration of the urine. Hydroxocobalamin is not currently available or approved for treatment of cyanide intoxication in the United States.[15] Nonspecific supportive therapy is an essential part of the treatment; in two cases, patients have recovered from massive cyanide poisoning with supportive therapy only.

REFERENCES

1. Brieger H, Riders F, Hodes WA: Acrylonitrile: Spectrophotometric determination, acute toxicity, and mechanism of action. AMA Arch Ind Hyg Occup Med 6:128–140, 1952
2. Willhite CC: Toxicology updates. Acrylonitrile. J Appl Toxicol 2:54–56, 1982
3. Murray FJ et al: Teratogenicity of acrylonitrile given to rats by gavage or by inhalation. Food Cosmet Toxicol 16:547–551, 1979
4. Maltoni C et al: Carcinogenicity bioassays on rats of acrylonitrile administered by inhalation and ingestion. Med Lav 68:401–411, 1977
5. IARC Monographs on the Evaluation of the Carcinogenic Risk of Chemicals to Man. Suppl. 4, pp 25–27. Lyon, International Agency for Research on Cancer, 1982

6. Bigner DD et al: Primary brain tumors in Fischer 344 rats chronically exposed to acrylonitrile in their drinking water. Fed Chem Toxicol 24:129–137, 1986
7. O'Berg MT: Epidemiologic study of workers exposed to acrylonitrile. J Occup Med 22:245–252, 1980
8. O'Berg MT et al: Epidemiologic study of workers exposed to acrylonitrile: An update. J Occup Med 27:835–840, 1985
9. Koerselman W, van der Graaf M: Acrylonitrile: A suspected human carcinogen. Int Arch Occup Environ Health 54:317–324, 1984
10. Graham D, Laman D, Theodore J, Robin ED: Acute cyanide poisoning complicated by lactic acidosis and pulmonary edema. Arch Intern Med 137:1051–1055, 1977
11. Berlin C: Cyanide poisoning—A challenge. Arch Intern Med 137:993–994, 1977
12. Chen KK, Rose CL: Nitrite and thiosulfate therapy in cyanide poisoning. JAMA 149:113–119, 1952
13. Wolfsie JH: Treatment of cyanide poisoning in industry. AMA Arch Ind Hyg Occup Med 4:417–425, 1951
14. Gosselin RE, Smith RP, Hodge HC: Clinical Toxicology of Commercial Products, Section III, 5th ed, pp 123–130. Baltimore, Williams & Wilkins, 1984
15. Becker CE: The role of cyanide in fires. Vet Hum Toxicol 27:487–490, 1985

ALDRIN

CAS: 309-00-2

$C_{12}H_8Cl_6$ 1987 TLV = 0.25 mg/m^3; skin

Synonyms: 1,2,3,4,10,10–Hexachloro–1,4,4a,5,8,8a–hexahydro–1,4,5,8,–dimethanonnaphthalene; HHDN(ISO); aldrine

Physical Form. White, crystalline, odorless solid

Uses. Insecticide

Exposure. Inhalation; skin absorption; ingestion

Toxicology. Aldrin is a convulsant; its metabolite, dieldrin, causes liver cancer in mice.

In humans, early symptoms of intoxication may include headache, dizziness, nausea, vomiting, malaise, and myoclonic jerks of the limbs; clonic and tonic convulsions and sometimes coma follow and may occur without the premonitory symptoms.[1,2] A suicidal person who ingested 25.6 mg/kg developed convulsions within 20 minutes, which persisted recurrently until large amounts of barbiturates had been administered. Hematuria and azotemia occurred the day after ingestion and continued for 18 days. Liver function studies were within normal limits except for an elevated icterus index; an electroencephalogram revealed generalized cerebral dysrhythmia, which returned to normal after 5 months.[3]

Once aldrin is absorbed, it is rapidly metabolized to dieldrin.[4] In a study of five workers exposed to concentrations of aldrin of up to 8.5 mg/m^3 who had suffered convulsive seizures or myoclonic limb movements, the probable concentration of dieldrin in the blood during intoxication ranged from 16 to 62 μg/100 g of blood; in healthy workers, the concentration of dieldrin ranged up to 22 μg/100 g of blood.[4]

Aldrin is reported to have caused erythematobullous dermatitis in a single case. Minor erythema may be observed from skin contact, but dermatitis associated with aldrin is unusual.[5]

The hepatocarcinogenicity of dieldrin in mice has been confirmed in several experiments, and, in some cases, the liver-cell tumors metastasized.[6] Single doses of aldrin of 50 mg/kg administered orally to golden hamsters during the period of organogenesis causes a high incidence of fetal deaths, congenital anomalies, and growth retardation.[7] The oral LD_{50} in animals ranged from 39 to 99 mg/kg. Effects were renal tubular degeneration and hepatic necrosis.[5]

REFERENCES

1. Kazantzis G, McLaughlin AIG, Prior PF: Poisoning in industrial workers by the insecticide aldrin. Br J Ind Med 21:46–51, 1964
2. Hoogendam I, Versteeg JPJ, de Vlieger M: Nine years toxicity control in insecticide plants. Arch Environ Health 10:441–448, 1965
3. Spiotta EJ: Aldrin poisoning in man. AMA Arch Ind Hyg Occup Med 4:560–566, 1951
4. Brown VKH, Hunter CG, Richardson A: A blood test diagnostic of exposure to aldrin and dieldrin. Br J Ind Med 21:283–286, 1964
5. Hayes WJ Jr: Pesticides Studied in Man, pp 234–237. Baltimore, Williams & Wilkins, 1982
6. IARC Monographs on the Evaluation of the Carcinogenic Risk of Chemicals to Man. Vol 5, Some Organochlorine Pesticides, pp 25–38, 125–156. Lyon, International Agency for Research on Cancer, 1974
7. Ottolenghi AD, Haseman JK, Suggs F: Teratogenic effects of aldrin, dieldrin, and endrin in hamsters and mice. Teratology 9:11–16, 1974

ALLYL ALCOHOL

CAS: 107-18-6

$CH_2{=}CHCH_2OH$ — 1987 TLV = 2 ppm; skin

Synonyms: 2–Propen–1–ol; 1–propenol–3; vinyl carbinol

Physical Form. Colorless liquid

Uses. In manufacture of allyl compounds, resins, plasticizers; fungicide and herbicide

Exposure. Inhalation; skin absorption

Toxicology. Allyl alcohol is a potent lacrimator and is an irritant of the mucous membranes and skin.

In humans, severe eye irritation occurs at 25 ppm, and irritation of the nose is moderate at 12.5 ppm.[1] In workers exposed to a "moderate" vapor level, there was a syndrome of lacrimation, retrobulbar pain, photophobia, and blurring of vision.[1] The symptoms persisted for up to 48 hours. Skin contact with the liquid has a delayed effect, causing aching, which begins several hours after contact, followed by the formation of vesicles. Splashes of the liquid in human eyes have caused moderately severe reactions.[2]

In rats, the 1-, 4-, and 8-hour LC_{50}s are 1060, 165, and 76 ppm, respectively.[1] Signs of toxicity included apathy, excitability, tremors, convulsions, diarrhea, coma, pulmonary and visceral congestion, and varying degrees of liver injury. Repeated 7 hour/day exposure at 60 ppm caused gasping during the first few exposures, persistent eye irritation, and death of 1 of 10 rats.[1] In several species of animals exposed to 7 ppm for 7 hours/day for 6 months, observed effects were minimal, but, at autopsy, findings were focal necrosis of the liver and necrosis of the convoluted tubules of the kidneys.[3]

The warning properties are thought to be adequate to prevent voluntary exposure to acutely dangerous concentrations but inadequate for chronic exposure.

REFERENCES

1. Dunlap MK et al: The toxicity of allyl alcohol. Arch Ind Health 18:303–311, 1958
2. Grant WM: Toxicology of the Eye, 2nd ed, pp 105–106. Springfield, Illinois, Charles C Thomas, 1974
3. Torkelson TR, Wolf MA, Oyen F, Rowe VK: Vapor toxicity of allyl alcohol as determined on laboratory animals. Am Ind Hyg Assoc J 20:224–229, 1959

ALLYL CHLORIDE

CAS: 107-05-1

$CH_2{=}CHCH_2Cl$ 1987 TLV = 1 ppm

Synonyms: Chlorallylene; 3–chloroprene; 1–chloro–2–propene; 3–chloropropylene

Physical Form. Liquid

Uses. Synthesis of allyl compounds

Exposure. Inhalation; skin absorption

Toxicology. Allyl chloride is an irritant of the eyes, mucous membranes, and skin. In animals, it causes renal, hepatic, and pulmonary damage and, at high concentrations, central nervous system depression.

The most frequent effects in humans following overexposure have been conjunctival irritation and eye pain with photophobia; eye irritation occurs between 50 and 100 ppm.[1] Irritation of the nose occurs at levels below 25 ppm.

In one report from China, 26 factory workers exposed to allyl chloride ranging from 0.8 ppm to 2100 ppm complained of lacrimation and sneezing, which gradually diminished.[2] After 2.5 months to 5 years exposure, most had developed weakness, paresthesia, cramping pain, and numbness in the extremities, with sensory impairment in the glove–stocking distribution, as well as diminished ankle reflexes. Electroneuromyography showed neurogenic abnormalities in 10 of 19 subjects. Similar but much milder symptoms appeared in other workers exposed at 0.06 to 8 ppm for 1 to 4.5 years. Diagnostic findings suggested mild neuropathy in 13 of 27 of these subjects.

The liquid is a skin irritant and may be absorbed through the skin, causing deep-seated pain.[1] If splashed in the eye, severe irritation would be expected.

Rats survived 15 minutes at 32,000 ppm, 1 hour at 3,200 ppm, or 3 hours at 320 ppm, but 0.5-, 3- and 8-hour exposures, respectively, were lethal to all within the following 24 hours.[3] Exposure to 16,000 ppm for up to 2 hours in rats or 1 hour in guinea pigs caused eye and nose irritation, drowsiness, weakness, instability, labored breathing, and, ultimately, death.[3] Postmortem findings were severe kidney injury, alveolar hemorrhage in the lungs, and slight liver damage. A recent study found no significant effects in rats exposed at 200 ppm for 6 hours; renal toxicity appeared at 300 ppm, but mortality was not affected until 1,000 ppm was reached.[4] Several species exposed to 8 ppm for 7 hours daily for 1 month showed no apparent ill effects, but histopathologic examination revealed focal necrosis in the liver and necrosis of the convoluted tubules of the kidneys. At 3 ppm for 6 months, rats showed slight centrilobular degeneration in the liver.[5]

In other reports, rats and mice showed no effects at 20 ppm 7 hours/day for 90 days, but adverse effects were found following the 50-ppm regime.[4]

Allyl chloride was fetotoxic to rats exposed during gestation to 300 ppm, which also caused considerable maternal toxicity in the form of kidney and liver injury.[4]

Administered by gavage for 1.5 years, allyl chloride was not carcinogenic to rats but caused a low incidence of squamous cell carcinomas of the forestomach in mice.[6]

Although allyl chloride is detectable below 3 ppm, the warning properties are insufficient to prevent exposure to concentrations that may be hazardous with chronic exposure.[4]

REFERENCES

1. National Institute for Occupational Safety and Health, US Department of Health, Education, and Welfare: Criteria for a Recommended Standard . . . Occupational Exposure to Allyl Chloride. DHEW (NIOSH) Pub No 76–204, pp 19–38. Washington, DC, US Government Printing Office, 1976
2. He F, Zhang SL: Effects of allyl chloride on occupationally exposed subjects. Scand J Work Environ Health 11(suppl 4):43–45, 1985
3. Adams EM et al: The acute vapor toxicity of allyl chloride. J Ind Hyg Toxicol 22:79–86, 1940
4. Torkelson TR, Rowe VK: Halogenated aliphatic hydrocarbons. In Clayton GD, Clayton FE (eds): Patty's Industrial Hygiene and Toxicology, Vol 2B, Toxicology, pp 3568–3572. New York, Wiley–Interscience, 1981
5. Torkelson TR, Wolf MA, Oyen F, Rowe VK: Vapor toxicity of allyl chloride as determined on laboratory animals. Am Ind Hyg Assoc J 20:217–223, 1959
6. National Cancer Institute: Carcinogenesis Technical Report Series. Bioassay of Allyl Chloride for Possible Carcinogenicity. DHEW (NIH) Pub No 78–1323, p 53. Washington, DC, US Government Printing Office, 1978

ALLYL GLYCIDYL ETHER

CAS: 106-92-3

$C_6H_{10}O_2$ — 1987 TLV = 5 ppm; skin

Synonyms: AGE; allyl 2,3–epoxypropyl ether

Physical Form. Liquid

Uses. Reactive diluent in epoxy resin systems; stabilizer of chlorinated compounds, manufacture of rubber

Exposure. Inhalation; skin absorption

Toxicology. Allyl glycidyl ether causes dermatitis and eye irritation; in animals, at high concentrations, it causes eye and nose irritation, pulmonary edema, and narcosis.

Workers exposed to the vapor and/or liquid complained of dermatitis with itching, swelling, and blister formation.[1] Skin sensitization has occurred; cross-sensitization probably can occur with other epoxy agents.[2]

In rats, the LC_{50} for 8 hours was 670 ppm; effects were lacrimation, nasal discharge, dyspnea, and narcosis.[1] In rats repeatedly exposed to 600 ppm for 8 hours daily, effects were pronounced irritation of the eyes and respiratory tract; more than half of the rats developed corneal opacity; at necropsy, after 25 exposures, pulmonary findings were inflammation, bronchiectasis, and bronchopneumonia.[1] Percutaneous absorption has been documented in rabbits.[2] The liquid dropped into the eye of a rabbit caused severe but reversible conjunctivitis, iritis, and corneal opacity.[1] Cytotoxic effects on rat bone marrow cells, with reduction in leukocyte counts, and testicular degeneration were observed after intramuscular injections at 400 mg/kg/day. AGE has shown mutagenic activity in bacteria.[3]

REFERENCES

1. Hine CH et al: The toxicology of glycidol and some glycidyl ethers. AMA Arch Ind Health 13:250–264, 1956
2. Allyl glycidyl ether (AGE). Documentation of the TLVs and BEIs, 5th ed, p 20. Cincinnati, American Conference of Governmental Industrial Hygienists (ACGIH), 1986
3. National Institute for Occupational Safety and Health, US Department of Health, Education and Welfare: Criteria for a Recommended Standard . . . Occupational Exposure to Glycidyl Ethers. DHEW (NIOSH) Pub No 78–166. Washington, DC, US Government Printing Office, 1978

ALLYL PROPYL DISULFIDE

CAS: 2179-59-1

$CH_2{=}CHCH_2S_2C_3H_7$ 1987 TLV = 2 ppm

Synonyms: Disulfide, allyl propyl; onion oil

Physical Form. Pale, yellow oil

Source. Onions

Exposure. Inhalation

Toxicology. Allyl propyl disulfide vapor is a mucous membrane irritant.

No systemic effects have been reported from industrial exposure. At an average concentration of 3.4 ppm in an onion dehydrating plant, there was irritation of eyes, nose, and throat in some workers.[1]

REFERENCES

1. Feiner B, Burke WJ, Baliff J: An industrial hygiene survey of an onion dehydrating plant. J Ind Hyg Toxicol 28:278–279, 1946

α-ALUMINA

CAS: 1344-28-1

Al_2O_3 containing <1% quartz

1987 TLV = 10 mg/m^3

Synonyms: Aluminum oxide; corundum

Physical Form. White powder. There are six common crystalline forms of aluminum oxide:[1]

Gibbsite (α-$Al_2O_3 \cdot 3H_2O$), the prominent constituent of many bauxites and is also formed in the precipitation step of the Bayer process for producing alumina

Bayerite (β-$Al_2O_3 \cdot 3H_2O$), does not occur in nature

Boehmite (α-$Al_2O_3 \cdot H_2O$), found in many bauxites

Diaspore (β-$Al_2O_3 \cdot H_2$), found in some bauxites

Gamma alumina (γ-Al_2O_3), a name applied to numerous anhydrous aluminas with ill-defined structures. One can readily obtain this form of alumina, which is the main constituent of "activated" alumina, by dehydrating boehmite at about 400°C

Corundum (α-Al_2O_3), found in nature and can also be produced synthetically by heating γ-alumina for a few hours at 1200°C

Uses. Production of aluminum; synthetic abrasives; refractory material

Exposure. Inhalation

Toxicology. Alumina is a nuisance dust and has not been shown to injure the pulmonary system in humans.

In animals, experiments with alumina have shown that the type of reaction in lung tissue is dependent upon the form of alumina and its particle size, as well as the species of animal used. For example, intratracheal administration into rats of γ-alumina of 2 μ average size caused only a mild fibrous reaction (loose reticulin).[2] However, intratracheal administration of γ-alumina of 0.02 to 0.04 μ size into rats produced reticulin nodules that later developed into areas of dense collagenous fibrosis.[3] The latter alumina by the same route in mice and guinea pigs caused development of a reticulin network with occasional collagen, and, in rabbits, only a slight reticulin network was observed.[2] Intratracheal administration of another form of alumina in rats, corundum of particle size less than 1 μ, caused the development of compact nodules of reticulin.[2]

In rats, inhalation of massive levels of γ-alumina with an average particle size of 0.0005 to 0.04 μ for up to 285 days caused heavy desquamation of alveolar cells and secondary inflammation, but only slight evidence of fibrosis. The dust concentration in the exposure chamber was described as so high that visibility was reduced; a few breaths of the atmosphere by the investigators caused bronchial irritation and persistent cough.

See also Aluminum, Bauxite.

REFERENCES

1. Godard HP, Jepson WB, Bothwell MR, Kane RL: The Corrosion of Light Metals, pp 4–5. New York, John Wiley & Sons, 1967
2. Stacy BD et al: Tissue changes in rats' lungs caused by hydroxides, oxides and phosphates of aluminum and iron. J Pathol Bact 77:417–426, 1959
3. King EJ, Harrison CV, Mohanty GP, Nagelschmidt G: The effect of various forms of alumina on the lungs of rats. J Pathol Bact 69:81–93, 1955
4. Klosterkotter W: Effects of ultramicroscopic *gamma*-aluminum oxide on rats and mice. AMA Arch Ind Health 21:458, 1960

ALUMINUM

CAS: 7429-90-5

Al

1987 TLV = 10 mg/m^3—metal dust, as Al
5 mg/m^3—pyro powders, as Al
5 mg/m^3—welding fumes, as Al
2 mg/m^3—soluble salts, as Al
2 mg/m^3—alkyls, all types, as Al

Physical Forms. Metal dusts, pyro powders, welding fumes. When exposed to air, an aluminum surface becomes oxidized to form a thin coating of Al_2O_3, which protects against ordinary corrosion. Powder and flake aluminum are flammable and can form explosive mixtures in air, especially when treated to reduce surface oxidation (pyro powders).

Uses. Pyro powders are used in fireworks and aluminum paint

Exposure. Inhalation

Toxicology. The inhalation of very fine aluminum powder (pyropowder) in massive concentrations may rarely cause pneumoconiosis in some persons. The metallic dust produced by grinding aluminum products is regarded only as inert dust.

In humans, the symptoms of long-term overexposure to only some fine powders of aluminum may include dyspnea, cough, and weakness. Typically, there may be radiographic evidence of fibrosis and occasional pneumothorax. At autopsy, there is generalized interstitial fibrosis, predominantly in the upper lobes, with pleural thickening and adhesions. Particles of aluminum are found in the fibrotic tissue. A rare fatal case of pulmonary fibrosis from inhalation of a heavy concentration of fine aluminum dust was reported in a 22-year-old British worker; autopsy revealed a generalized non-nodular fibrosis and interstitial emphysema with right ventricular hypertrophy. There had been work exposure to varying concentrations of a wide range of particle sizes, but the quantity of dust in the atmosphere below 5 μ was of the order of 19 mg/m^3.[1]

Of 27 workmen with heavy exposures to aluminum powder in the same plant as the above-mentioned case, 6 were found to have evidence of pulmonary fibrosis. The finer dust was more dangerous than the coarse dust; of the 12 men exposed to fine aluminum powder, 2 died and 2 others were affected; and, of 15 men who worked exclusively with coarser powder, 2 had radiologic changes but no symptoms.[2]

A 49-year-old British workman who had spent over 13 years in the ball-mill room of an aluminum powder factory died of rapidly progressive encephalopathy.[3] At autopsy the brain showed no histologic evidence of abnormality. There was a moderate degree of pulmonary fibrosis and focal emphysema, although there had been no respiratory symptoms or measurable impairment of pulmonary function prior to the occurrence of a pulmonary embolus and terminal bronchopneumonia. The brain, lungs, and liver were reported as containing many times the average normal concentration of aluminum. No other cases of encephalopathy with elevated levels of aluminum in the brain of aluminum workers have been reported over the past 80 years. In view of recent reports of increased aluminum concentration in the brain of some patients with seriously impaired renal function, suggesting a disturbance in permeability of the blood–brain barrier in these individuals,[4,5] it is of interest that renal function and the condition of the kidneys at autopsy is not mentioned in this singular case report of encephalopathy in an aluminum worker.[3] Several mortality studies of aluminum reduction plant workers, in which the study cohorts totalled nearly 28,000 long-term employees, recorded no excess deaths due to organic brain disorders of dementia type,[6–7] and an analysis of the occupational mortality experience of nearly 430,000 men who died in Washington state during the years 1950 through 1979 showed no excess deaths from this cause among the 1238 former aluminum workers included in the study.[8]

Fine metallic aluminum powders inhaled by hamsters and guinea pigs caused no pulmonary fibrosis; in rats that inhaled the dust, small scars resulted from foci of lipid pneumonitis. Alveolar proteinosis developed in all three species; it resolved spontaneously, and the accumulated dust deposits cleared rapidly from the lungs after cessation of the exposure. The failure of inhaled aluminum powder to cause pulmonary fibrosis in experimental animals parallels the clinical experience in the United States, where pulmonary fibrosis has not been observed in aluminum workers.[9]

It has been suggested that the explanation of pulmonary disease among powder workers in other countries may lie in the duration of exposure, the size of the particles, the density of the dust, and especially the fact that all reported cases have been associated with exposure to a submicron-sized aluminum pyrotechnic flake (powder), which has been lubricated with a nonpolar aliphatic oil rather than with the usually employed stearic acid.[2,10]

Evidence of the relatively benign nature of aluminum dust in measured concentrations lies in the 27-year experience of administration of freshly milled metal particles to workers exposed to silica as a suggested means of inhibiting the development of silicosis. Disregarding the questionable rationale of "treating" the effects of exposure to a fibrogenic dust by adding exposure to another, however innocuous dust, the results are reassuring in assessing the effects of aluminum dust on the lung. Inhalation of aluminum powder of particle size of 1.2 μm (96%), given over 10- or 20-minute periods several times weekly, resulted in no adverse health effects among thousands of workers over several years. This experience has led to setting the TLV for aluminum metal dust at 10 mg/m^3, the standard for an inert dust.

In another rare case, a 35-year-old aluminum arc welder, who usually worked in confined spaces during small boat construction, developed shortness of breath and radiologic evidence of ill-defined, nodular infiltrative lesions of the lungs.[11] Pulmonary function tests revealed a mild to moderate mixed restrictive and obstructive deficit. Because the etiology of the pulmonary lesions was obscure and a malignant tumor could not be ruled out, the left upper lobe was removed surgically. The lung tissue exhibited a metallic sheen. The cut surfaces of the parenchyma were similarly discolored and disclosed fibrotic nodules with seemingly necrotic centers. There were several fibrous pleural adhesions. The hilar lymph nodes were gray. Histologically, there was diffuse and focal fibrosis accompanied by localized infiltrates of lymphocytes. The nodular lesions were composed of dense, partially hyalized, collagenous tissue and exhibited irregular areas of acellular, coagulative necrosis. Dense accumulations of macrophages containing a granular, gray-brown, metallic material were found in the fibrotic tissue. Air spaces also contained variable numbers of these cells. Examination of tissue secretions by scanning electron microscopy, using x-ray spectrometry and x-ray mapping, showed that the pigmented material in the macrophages was composed of aluminum.

This man, working in confined quarters, was exposed to an aerosol of oxidized aluminum. Under these conditions, concentrations of welding fumes may be increased appreciably over the amounts detectable in the air around welders working in open spaces. The role of irritant gases such as ozone and oxides of nitrogen in welding fume must be considered. The patient also was a heavy smoker (60 pack years). In this regard, experimental studies have demonstrated the retention of inhaled particulates and strikingly reduced lung clearance in smokers. One might conclude that this set of circumstances predisposed to aluminum particle accumulation in the lungs. No other cases of pulmonary disease in welders of aluminum products have been confirmed.

Aerosols of the soluble salts of aluminum, such as the chloride and sulfate, are irritants of little occupational importance. Although the aluminum alkyls may also be irritants, there is inadequate toxicity information on these compounds.[12]

REFERENCES

1. Mitchell J: Pulmonary fibrosis in an aluminum worker. Br J Ind Med 16:123–125, 1959
2. Mitchell J, Manning GB, Molyneux M, Lane RE: Pulmonary fibrosis in workers exposed to finely powdered aluminum. Br J Ind Med 18:10–20, 1961
3. McLaughlin AIG et al: Pulmonary fibrosis and encephalopathy associated with the inhalation of aluminium dust. Br J Ind Med 19:253–263, 1962
4. Alfrey AC, LeGendre GR, Koehny WD: The dialysis encephalopathy syndrome: Possible aluminum intoxication. N Engl J Med 294:184–188, 1976
5. Wisniewski HM: Neurotoxity of aluminium. In Hughes JP (ed): Health Protection in Primary Aluminium Production, Vol 2, pp 217–221. London, International Primary Aluminium Institute, 1981
6. Gibbs GW: Mortality experience in eastern Canada. In Hughes JP (ed): Health Protection in Primary Aluminum Production, Vol 2, pp 56–69. London, International Primary Aluminium Institute, 1981
7. Rockette HE, Arena VC: Mortality studies of aluminum reduction plant workers: Potroom and carbon department. J Occup Med 25:549–557, 1983
8. Milham S: Occupational Mortality in Washington State, 1950–1979. DHHS (NIOSH) Pub No 83–116, p 38. Washington, DC, US Government Printing Office, 1983
9. Gross P, Harley RA Jr, deTreville RTP: Pulmonary reaction to metallic aluminum powders. Arch Environ Health 26:227–236, 1973
10. Dinman BD: Aluminum in the lung: The pyropowder conundrum. J Occup Med 29:869–876, 1987
11. Vallyathan V et al: Pulmonary fibrosis in an aluminum arc welder. Chest 81:372–374, 1982

12. Aluminum. Documentation of TLVs and BEIs, 5th ed, p 22. Cincinnati, American Conference of Governmental Industrial Hygienists (ACGIH), 1986

4–AMINODIPHENYL

CAS: 92-67-1

$C_{12}H_{11}N$ 1987 TLV = none; recognized human carcinogen; skin

Synonym: *p*-Xenylamine

Physical Form. Colorless, crystalline compound; darkens on oxidation

Uses. Previously used as a rubber antioxidant; no longer produced on a commercial scale

Exposure. Inhalation, skin absorption

Toxicology. 4–Aminodiphenyl exposure is associated with a high incidence of bladder cancer in humans; in animals, it has produced bladder and liver tumors.

Of 171 workers exposed to 4–aminodiphenyl for 1.5 to 19 years, 11% had bladder tumors; the tumors appeared 5 to 19 years after initial exposure.[1]

In another study of 503 exposed workers, there were 35 histologically confirmed bladder carcinomas and an additional 24 men with positive cytology.[2]

Two bladder papillomas and three bladder carcinomas were observed in six dogs fed a total of 5.5 g to 7 g (1.0 mg/kg, 5 days/week for life).[3] In another study, each of four dogs developed urinary bladder carcinomas with predominantly squamous differentiation in 21 to 34 months after ingestion of 0.3 g of 4–aminodiphenyl three times per week (total dose 87.5 g to 144 g/dog); hematuria, salivation, loss of body weight, and vomiting were also noted, and all animals died within 13 months of the first appearance of a tumor.[4]

Rats injected subcutaneously with a total dose of 3.6 to 5.8 g/kg had an abnormally high incidence of mammary gland and intestinal tumors.[5] Nineteen of 20 newborn male mice and 6 of 23 newborn female mice developed hepatomas in 48 to 52 weeks after three subcutaneous injections of 200 μg of 4–aminodiphenyl; in control animals, 5 of 41 males and 2 of 47 females had hepatomas.[6]

The accumulated experimental and epidemiologic evidence has demonstrated that 4–aminodiphenyl may be the most hazardous of the aromatic amines regarding carcinogenic potential.[7] Because of demonstrated carcinogenicity, contact by all routes should be avoided.[8]

REFERENCES

1. Melick WF et al: The first reported cases of human bladder tumors due to a new carcinogen—xenylamine. J Urol 74:760–766, 1955
2. Koss LG, Myron R, Melamed MR, Kelly RE: Further cytologic and histologic studies of bladder lesions in workers exposed to para-aminodiphenyl: progress report. J Natl Cancer Inst 43:233, 1969
3. Deichmann WB et al: Synergism among oral carcinogens. Simultaneous feeding of four bladder carcinogens to dogs. Ind Med Surg 34:640, 1965
4. Deichmann WB et al: The carcinogenic action of *p*-aminodiphenyl in the dog. Ind Med Surg 27:25–26, 1958
5. Walpole AL, Williams MHC, Roberts DC: The carcinogenic action of 4–aminodiphenyl and 3:2′–dimethyl–4–aminodiphenyl. Br J Ind Med 9:255–261, 1952
6. Gorrod JW, Carter RL, Roe FJC: Induction of hepatomas by 4–aminodiphenyl and three of its hydroxylated derivatives administered to newborn mice. J Natl Cancer Inst 41:403–410, 1968
7. Department of Labor: Occupational Safety and Health Standards—Carcinogens. Federal Register 39:3756, 3781–3784, 1974

8. 4–Aminodiphenyl. Documentation of TLVs and BEIs, 5th ed, p 23. Cincinnati, American Conference of Governmental Industrial Hygienists (ACGIH), 1986

p–AMINOPHENOL

CAS: 123-30-8

$NH_2C_6H_4OH$ 1987 TLV = none established

Synonyms: Activol; 4–amino–1–hydroxybenzene; 4–hydroxyaniline; PAP

Physical Form. White or reddish-yellow crystals; discolors to lavender when exposed to air.

Uses. Oxidative dye; developing agent for photographic processes; precursor for pharmaceuticals; used in hair dyes

Exposure. Inhalation

Toxicology. *p*–Aminophenol is of moderately low toxicity but has caused kidney injury and dermal sensitization in animals; the potential for producing methemoglobin is of relatively minor importance. There are no reports of adverse health effects for human exposure.

The oral LD_{50} in rats was 671 mg/kg.[1] Effects included central nervous system depression. *p*–Aminophenol caused dermal sensitization in guinea pigs.[2] Solutions of 2.5% applied to abraded skin of rabbits was a mild irritant.[1] The dermal LD_{50} in rabbits was greater than 8 g/kg, which strongly suggests that absorption through the skin is minimal.[3] Single nonlethal acute doses in rats produced proximal renal tubular necrosis.[4]

Studies of the teratogenic effects of *p*–aminophenol indicated both positive and negative effects, depending upon the route of administration. Hamsters given intravenous or intraperitoneal injections of *p*–aminophenol at 100 to 250 mg/kg showed significant increases in malformed fetuses and resorptions in a dose-dependent manner.[5] However, oral studies using hamsters and topical application of hair dyes containing *p*–aminophenol on rats showed no teratogenic effects.[5,6]

REFERENCES

1. Lloyd GK et al: Assessment of the acute toxicity and potential irritancy of hair dye constituents. Fd Cosmet Toxicol 15:607, 1977
2. Kleniewska D, Maibach H: Allergenicity of aminobenzene compounds: Structure–function relationships. Derm Beruf Umwelt 28:11, 1980
3. Mallinckrodt, Inc. For your information (FYI) Submission FYI–OTS–1083–0272 Supp. Seq. C. Bio/Tox data on *p*–aminophenol from 1980. Washington, DC, Office of Toxic Substances, US Environmental Protection Agency, 1983
4. Briggs D, Calder I, Woods R, Tange J: The influence of metabolic variation on analgesic nephrotoxicity. Experiments with the Gunn rat. Pathology 14:349, 1982
5. Rutkowski JV, Fermn VH: Comparison of the teratogenic effects of isomeric forms of aminophenol in the Syrian golden hamster. Toxicol Appl Pharmacol 63:264, 1982
6. Burnett C et al: Teratology and percutaneous toxicity studies on hair dyes. Toxicol Environ Health 1:1027, 1976

2–AMINOPYRIDINE

CAS: 504-29-0

$NH_2C_4H_4N$ 1987 TLV = 0.5 ppm

Synonyms: α-aminopyridine; α-pyridylamine

Physical Form. Crystalline solid

Uses. Manufacture of pharmaceuticals, especially antihistamines

Exposure. Inhalation; skin absorption

Toxicology. 2–Aminopyridine is a convulsant.

In industrial experience, intoxication

has occurred from inhalation of the dust or vapor, or by skin absorption following direct contact.[1] Fatal intoxication occurred in a chemical worker who spilled a solution of 2–aminopyridine on his clothing during a distillation; he continued to work in contaminated clothing for 1.5 hours. Two hours later, he developed dizziness, headache, respiratory distress, and convulsions that progressed to respiratory failure and death; it is probable that skin absorption was a major factor in this case.

A nonfatal intoxication from exposure to an undetermined concentration of 2–aminopyridine in air resulted in severe headache, weakness, convulsions, and a stuporous state that lasted several days.[1] A chemical worker exposed to an estimated air concentration of 20 mg/m^3 (5.2 ppm) for approximately 5 hours developed severe pounding headache, nausea, flushing of the extremities, and elevated blood pressure, but he recovered fully within 24 hours.[1]

The LD_{50} in mice by intraperitoneal injection was 35 mg/kg; lethal doses in animals produced excitement, tremors, convulsions, tetany, and death.[1] Fatal doses were readily absorbed through the skin. A 0.2-M aqueous solution dropped in a rabbit's eye was only mildly irritating.[2]

REFERENCES

1. Reinhardt CF, Brittelli MR: Heterocyclic and miscellaneous nitrogen compounds. In Clayton GD, Clayton FE (eds): Patty's Industrial Hygiene and Toxicology, 3rd ed, rev, pp 2731–2732. New York, Wiley–Interscience, 1981
2. Grant WM: Toxicology of the Eye, 3rd ed, p 383 Springfield, Illinois, Charles C Thomas, 1986

AMMONIA

CAS: 7664-41-7

NH_3 1987 TLV = 25 ppm

Synonyms: Ammonia gas

Physical Form. Colorless gas

Uses. Refrigeration; petroleum refining; blueprint machines; manufacture of fertilizers, nitric acid, explosives, plastics, and other chemicals

Exposure. Inhalation

Toxicology. Ammonia is a severe irritant of the eyes, respiratory tract, and skin.

Exposure to and inhalation of concentrations of 2500 to 6500 ppm, as might result from accidents with liquid anhydrous ammonia, causes severe corneal irritation, dyspnea, bronchospasm, chest pain, and pulmonary edema, which may be fatal. Upper airway obstruction from laryngopharyngeal edema and desquamation of mucous membranes may occur early in the course and require endotracheal intubation or tracheostomy.[1–3] Case reports have documented chronic airway hyperreactivity and asthma, with associated obstructive pulmonary function changes following massive ammonia exposures.[3,4]

In a human experimental study that exposed 10 subjects to various vapor concentrations for 5 minutes, 134 ppm caused irritation of the eyes, nose, and throat in most subjects and 1 person complained of chest irritation; at 72 ppm, several subjects reported the same symptoms; at 50 ppm, 2 reported nasal dryness; and, at 32 ppm, only 1 person reported nasal dryness.[2] Surveys of workers have generally found that the maximal concentration not resulting in significant complaints is 20 to 25 ppm.[2]

Tolerance to usually irritating concentrations of ammonia may be acquired by adaptation, a phenomenon frequently observed among workers who became inured to the effects of exposure; no data are available on concentrations that are irritating to workers who are regularly exposed to ammonia and who presumably have a higher irritation threshold.

Liquid anhydrous ammonia in contact with the eyes may cause serious injury to the cornea and deeper structures and sometimes blindness; on the skin, it causes first- and second-degree burns, which are often severe and, if extensive, may be fatal. Vapor concentrations of 10,000 ppm are mildly irritating to the moist skin, whereas 30,000 ppm or greater cause a stinging sensation and may produce skin burns and vesiculation.[2] With skin and mucous membrane contact, burns occur in three ways: (1) cryogenic (from the liquid ammonia), (2) thermal (from the exothermic dissociation of ammonium hydroxide), and (3) chemical (alkaline).[3]

REFERENCES

1. Department of Labor: Exposure to ammonia, proposed standard. Federal Register 40:54684–54693, 1975
2. National Institute for Occupational Safety and Health, US Department of Health, Education and Welfare: Criteria for a Recommended Standard . . . Occupational Exposure to Ammonia. DHEW (NIOSH) Pub No 74–136. Washington, DC, US Government Printing Office, 1974
3. Arwood R, Hammond J, Ward G: Ammonia inhalation. Trauma 25(5):444–447, 1985
4. Flury K, Dines D, Rodarto J, Rodgers R: Airway obstruction due to inhalation of ammonia. Mayo Clin Proc 58:389–393, 1983

AMMONIUM SULFAMATE

CAS: 7773-06-0

$NH_4SO_3NH_2$ 1987 TLV = 10 mg/m^3

Synonym: Ammate

Physical Form. Hygroscopic crystals

Uses. Manufacture of weed-killing compounds and fire-retardant compositions

Exposure. Inhalation

Toxicology. Ammonium sulfamate is of low toxicity; there are no reports of systemic effects in humans.

Repeated application of a 4% solution to the anterior surface of one arm of each of five human subjects for 5 days caused no skin irritation.[1]

The oral LD_{50} values were 3900 mg/kg for rats and 5760 mg/kg for mice.[2]

In rats, the intraperitoneal injection of 0.8 g/kg caused the death of 6 of 10 animals; effects were stimulation of respiration and then prostration.[1]

Continuous feeding of 1% (10,000 ppm) in the diet of rats for 105 days caused no effect; 2% in the diet caused growth inhibition, but no histologic effects were observed.[2]

REFERENCES

1. Ambrose AM: Studies on the physiological effects of sulfamic acid and ammonium sulfamate. J Ind Hyg Toxicol 25:26–28, 1943
2. Ammonium Sulfamate. Documentation of the TLVs and BEIs, 5th ed, p 28. Cincinnati, American Conference of Governmental Industrial Hygienists (ACGIH), 1986

n-AMYL ACETATE

CAS: 628-63-7

$CH_3COOC_5H_{11}$ 1987 TLV = 100 ppm

Synonym: Amyl acetic ether

Physical Form. Liquid

Uses. Manufacture of lacquers, artificial leather, photographic film, artificial glass, celluloid, artificial silk, and furniture polish

Exposure. Inhalation; minor skin absorption

Toxicology. *n*-Amyl acetate is an irritant of mucous membranes; at high concentrations, it causes narcosis in animals, and it is expected that severe exposure will produce the same effect in humans.

Several grades of technical amyl acetate are known; isoamyl acetate is the major component of some grades, whereas *n*-amyl acetate predominates in others.[1]

Exposure at 200 ppm for 30 minutes is the lowest irritation dose reported for humans.[2]

Air saturated with 5200 ppm of technical amyl acetate (*n*-amyl acetate the principal component) was fatal to six of six rats in 9 hours but caused no deaths in 4 hours.[3]

In standardized testing on rabbit eyes, isoamyl acetate was graded as only slightly injurious.[4]

Amyl acetates may be recognized at concentrations of 7 ppm by the fruit-like odor characteristic of esters.[1]

REFERENCES

1. Hygienic Guide Series: Amyl acetate. Am Ind Hyg Assoc J 26:199–202, 1965
2. Sandmeyer EE, Kirwin CJ Jr: Esters. In Clayton GD, Clayton FE (eds): Patty's Industrial Hygiene and Toxicology, 3rd ed, rev, Vol 2A, Toxicology, p 2274. New York, Wiley–Interscience, 1981
3. *n*-Amyl acetate. Documentation of the TLVs and BEIs, 5th ed, p 29. Cincinnati, American Conference of Governmental Industrial Hygienists (ACGIH), 1986
4. Grant WM: Toxicology of the Eye, 3rd ed, pp 97–98 Springfield, Illinois, Charles C Thomas, 1986

sec-AMYL ACETATE

CAS: 626-38-0

$C_7H_{14}O_2$ 1987 TLV = 125 ppm

Synonyms: α-Methyl butyl acetate; banana oil

Physical Form. Liquid

Uses. Manufacture of lacquers, artificial leather, photographic film, artificial glass, celluloid, artificial silk, and furniture polish

Exposure. Inhalation

Toxicology. *sec*-Amyl acetate is an irritant of the eyes, mucous membranes, and skin; high concentrations cause narcosis in animals, and severe exposure is expected to produce the same effect in humans.

In humans, exposure to 5,000 to 10,000 ppm for short periods of time caused irritation of the eyes and nasal passages.[1] Exposure to 1,000 ppm for 1 hour may be expected to produce serious toxic effects.[1]

In guinea pigs, 2,000 ppm for 13.5 hours produced no abnormal signs except irritation of the eyes and nose; at 5,000 ppm, there was lacrimation after 5 minutes, incoordination in 90 minutes, and narcosis within 9 hours, from which animals recovered.[1] A concentration of 10,000 ppm was fatal after 5 hours.[1]

The *sec*-amyl acetates are more volatile than the primary isomers and appear to be somewhat less toxic.[2]

REFERENCES

1. von Oettingen WF: The aliphatic acids and their esters: Toxicity and potential dangers. AMA Arch Ind Health 21:28–64, 1960
2. *sec*-Amyl acetate. Documentation of the TLVs and BEIs, 5th ed, p 29. Cincinnati, American Conference of Governmental Industrial Hygienists (ACGIH), 1986

ANILINE

CAS: 62-53-3

$C_6H_5NH_2$ 1987 TLV = 2 ppm; skin

Synonyms: Aminobenzene; phenylamine

Physical Form. Colorless to light yellow liquid, which tends to darken on exposure to air and light

Uses. Intermediate in chemical synthesis; manufacture of synthetic dyestuffs

Exposure. Inhalation; skin absorption

Toxicology. Aniline absorption causes anoxia from the formation of methemoglobin.

Human exposure to vapor concentrations of 7 to 53 ppm has been observed to cause slight symptoms, whereas concentrations in excess of 100 to 160 ppm may cause serious disturbances if inhaled for 1 hour.[1] Rapid absorption through the intact skin is frequently the main route of entry; a small amount absorbed from contaminated clothing or shoes may cause intoxication, characterized by cyanosis.[1,2] The formation of methemoglobinemia is often insidious; following skin absorption, the onset of symptoms may be delayed for up to 4 hours.[2] Headache is commonly the first symptom and may become quite intense as the severity of methemoglobinemia progresses. Cyanosis occurs when the methemoglobin concentration is 15% or more. Blueness develops first in the lips, nose, and earlobes and is usually recognized by fellow workers. The person usually feels well, has no complaints, and is insistent that nothing is wrong until the methemoglobin concentration approaches approximately 40%. At methemoglobin concentrations of over 40%, there is usually weakness and dizziness; up to 70% concentration, there may be ataxia, dyspnea on mild exertion, and tachycardia. Coma may ensue with methemoglobin levels above 70%, and the lethal level is estimated to be 85% to 90%.[3] In general, higher ambient temperatures increase susceptibility to cyanosis from exposure to methemoglobin-forming agents.[4]

The development of intravascular hemolysis and anemia due to aniline-induced methemoglobinemia has been postulated, but neither is observed often in industrial practice, despite careful and prolonged study of numerous cases. Occasional deaths from asphyxiation caused by severe aniline intoxication are said to occur.[1,2] The existence of chronic aniline poisoning is controversial, but some investigators have suggested that continuous exposure to small doses of aniline may produce anemia, loss of energy, digestive disturbance, and headache.[1,5] The mean lethal dose by ingestion in humans has been estimated to be between 15 and 30 g, although death has been reported after as little as 1 g.[6] A significant elevation in methemoglobin levels was reported in adult volunteers given 25 mg orally.[6] Peak methemoglobin levels may occur several hours after exposure, and it has been postulated that metabolic transformation of aniline to phenylhydroxylamine is necessary for the production of methemoglobin.[6] Liquid aniline is mildly irritating to the eyes and may cause corneal damage.[7]

No evidence of embryolethal or teratogenic effect was observed in the offspring of rats dosed with aniline hydrochloride during gestation.[8] Signs of maternal toxicity included methemoglobinemia, increased relative spleen weight, decreased red blood cell count, and hematologic changes indicative of increased hematopoietic activity. Transient signs of toxicity were observed postnatally in the offspring through postnatal day 30.

Aniline hydrochloride was not carcinogenic to mice when administered orally.[9] In one experiment, it produced fibrosarcomas, sarcomas, and hemangiosarcomas of the spleen and body cavities in rats fed diets containing 3000 mg/kg or 6000 mg/kg for 103 weeks.

The high risk of bladder cancer observed originally in workers in the aniline dye industry was attributed to exposure to chemicals other than aniline.[9] Epidemiologic studies of workers exposed to aniline but to no other known bladder carcinogen have shown little evidence of increased risk; one study revealed one death from bladder cancer vs. 0.83 deaths expected in a population of 1223 men producing or using aniline.[9]

The IARC has determined that evidence for carcinogenicity is limited in animals and inadequate in humans.[9]

Diagnosis. Headache, signs of anoxia, including cyanosis of lips, nose, and earlobes; eye irritation; anemia; and hematuria

Differential Diagnosis: Other causes of cyanosis must be differentiated from methemoglobinemia owing to aniline exposure. These causes include hypoxia from lung disease, hypoventilation, and decreased cardiac output. Lung disease may be suspected from results of pulmonary function tests and arterial blood gas analysis. The arterial P_{O_2} may be normal in methemoglobinemia but tends to be decreased in cyanosis resulting from lung disease. Hypoventilation will cause elevation of arterial P_{CO_2}, which is not seen in aniline exposure. Decreased cardiac output states will cause cyanosis only when accompanied by arterial hypotension. If blood withdrawn from the vein shows the characteristic chocolate-brown coloration, the diagnosis of an abnormal pigment is almost certain, especially if the color remains after shaking the blood in air.[10]

Special Tests: These include examination of urine for blood, determination of methemoglobin concentration in the blood when aniline intoxication is suspected and at regular intervals until the methemoglobin has been fully reduced to normal hemoglobin.[11] Methemoglobin may be differentiated from sulfhemoglobin by the addition of a few drops of 10% potassium cyanide, which results in the rapid production of bright red cyanomethemoglobin but has no effect on the color of sulfhemoglobin. Spectrophotometry is required for the precise identification of the pigment and its quantitation. Normal acid methemoglobin has a characteristic absorption spectrum with peaks at 502 and 632 nm, which disappear with the addition of cyanide, whereas sulfhemoglobin has a peak at 620 nm, which does not disappear with cyanide.[10]

Treatment. All aniline on the body must be removed. Immediately remove all clothing, and wash the entire body from head to foot with soap and water. Pay special attention to the hair and scalp, finger and toenails, nostrils, and ear canals. Administer oxygen to alleviate the headache and general sense of weakness; confine the person to bed.

Determine the methemoglobin concentration in the blood, and repeat every 3 to 6 hours for 18 to 24 hours. Repeat skin cleansing if the methemoglobin concentration appears to rise after 3 to 4 hours. In general, patients will return to normal within 24 hours, provided that all sources of further absorption are completely eliminated. The methemoglobin will be reduced spontaneously to ferrous hemoglobin in 2 to 3 days.[10] Such therapy is not effective in subjects with glucose-6-phosphate dehydrogenase deficiency.

The only justifiable use of methylene blue would be in cases of coma or stupor, usually at methemoglobin levels over 60%; in those patients in whom therapy is necessary, methylene blue may be given intravenously, 1 to 2 mg/kg, of a 1% solution in saline over a 10-minute period; if cyanosis has not disappeared within an hour, a second dose of 2 mg/kg should be administered.[10,11] The total dose should not exceed 7 mg/kg, because methylene blue may cause toxic effects such as dyspnea, precordial pain, restlessness, apprehension, red cell hemolysis, and changes in the electrocardiogram (reduction in the height or even reversal of the T wave, frequently with lowering of the R wave).[11]

REFERENCES

1. Aniline and homologues. Documentation of TLVs and BEIs, 5th ed, p 30. Cincinnati, American Conference of Governmental Industrial Hygienists (ACGIH), 1986
2. Hamblin DO: Aromatic nitro and amino compounds. In Fassett DW, Irish DD (eds): Toxicology, Vol 2. In Patty FA (ed): Industrial Hygiene and Toxicology, 2nd ed, pp 2105–2133, 2242. New York, Wiley–Interscience, 1963
3. Chemical Safety Data Sheet, SD–21, Nitrobenzene, pp 56, 1214. Washington, DC, MCA, Inc, 1967
4. Linch AL: Biological monitoring for industrial exposure to cyanogenic aromatic nitro and amino compounds. Am Ind Hyg Assoc 35:426–432, 1974
5. Hazard Data Sheet: Aniline. Sheet Number 78, pp 44–45. The Safety Practitioner, June 3, 1986
6. Kearney TE et al: Chemically induced methemoglobinemia from aniline poisoning. West J Med 140:282–286, 1982
7. Chemical Safety Data Sheet SD–17, Aniline, pp 45, 1214. Washington, DC, MCA, Inc, 1967
8. Price CJ et al: Teratologic and postnatal evaluation of aniline hydrochloride in the Fischer 344 rat. Toxicol Appl Pharmacol 77:465–489, 1985
9. IARC Monographs on the Evaluation of the Carcinogenic Risk of Chemicals to Man. Suppl. 4, pp 49–50. Lyon, International Agency for Research on Cancer, 1982
10. Rieder RF: Methemoglobinemia and sulfhemoglobinemia. In Wyngaarden JB, Smith LH (eds): Cecil Textbook of Medicine, 16th ed, p 894. Philadelphia, WB Saunders, 1982
11. Mangelsdorff AF: Treatment of methemoglobinemia. AMA Arch Ind Health 14:148–153, 1956

ANISIDINE

CAS: 29191-52-4

$NH_2C_6H_4OCH_3$ 1987 TLV = 0.1 ppm; skin

Synonyms: Methoxyaniline; aminoanisole

Physical Form. *o*-Anisidine is a yellowish liquid that darkens on exposure to air; *p*-anisidine is a white solid.

Uses. In the preparation of azo dyes

Exposure. Inhalation; skin absorption

Toxicology. Anisidine, *o*- and *p*-isomers, cause anoxia owing to the formation of methemoglobin. *o*-Anisidine was carcinogenic in experimental animals.

Workers exposed to 0.4 ppm for 3.5 hours for 6 months did not develop anemia, but there were some cases of headache and vertigo, which may have been related to the increased levels of methemoglobin and sulfhemoglobin; erythrocytic inclusion bodies (Heinz bodies) were observed; absorption through the skin may have contributed.[1] Anisidine is a mild skin sensitizer, and local contact may cause dermatitis.

Mice exposed 2 hours/day at 2 to 6 ppm for a year developed anemia and reticulocytosis.[2]

The oral LD_{50} of *o*-anisidine is reported to be 2000 mg/kg in rats, 1400 mg/kg in mice, and 870 mg/kg in rabbits. The oral LD_{50} of *p*-anisidine is 1400 mg/kg body weight in rats, 1300 mg/kg in mice, and 2900 mg/kg in rabbits.[3] For both isomers, subacute effects included hematologic changes, anemia, and nephrotoxicity.

A significant increase in transitional cell carcinomas of the urinary bladder was found in mice and rats fed diets containing 5000 mg/kg *o*-anisidine hydrochloride for 103 weeks.[4]

The IARC has determined that there is sufficient evidence for the carcinogenicity of *o*-anisidine hydrochloride in experimental animals, and, in the absence of human data, it should be regarded as though it presented a carcinogenic risk to humans.[3]

Diagnosis. Headache; signs of anoxia including cyanosis of lips, nose, and earlobes; eye burns; anemia; and hematuria

Differential Diagnosis: Other causes of cyanosis must be differentiated from methemoglobinemia resulting from chemical exposure. These include hypoxia due to lung disease, hypoventilation, and decreased cardiac output. Lung disease may be suspected from results of pulmonary function tests and arterial blood gas analysis. The arterial Po_2 may be normal in methemoglobinemia but tends to be decreased in cyanosis due to lung disease. Hypoventilation will cause elevation of arterial Pco_2, which is not seen in chemical exposure. Decreased cardiac output states will cause cyanosis only when accompanied by arterial hypotension. If blood withdrawn from the vein shows the characteristic chocolate-brown coloration, the diagnosis of an abnormal pigment is almost certain, especially if the color remains after shaking the blood in air.[5]

Special Tests: Special tests include examination of urine for blood, determination of methemoglobin concentration in the blood when chemical intoxication is suspected and at regular intervals until the methemoglobin has been reduced to normal hemoglobin.[7] Methemoglobin may be differentiated from sulfhemoglobin by the addition of a few drops of 10% potassium cyanide, which results in the rapid production of bright red cyanomethemoglobin but has no effect on the color of sulfhemoglobin.[6] Spectrophotometry is required for the precise identification of the pigment and its quantitation. Normal acid methemoglobin has a characteristic absorption spectrum with peaks at 502 and 632 nm, which disappear with the addition of cyanide, whereas sulfhemoglobin has a peak at 620 nm, which does not disappear with cyanide.[6]

Treatment. All the contaminant on the body must be removed. Immediately remove all clothing, and wash the entire body from head to foot with soap and water. Pay special attention to the hair and scalp, finger and toenails, nostrils, and ear canals. Administer oxygen to alleviate the headache and general sense of weakness; confine the patient to bed. Determine the methemoglobin concentration in the blood, and repeat every 3 to 6 hours for 18 to 24 hours. Repeat skin cleansing if the methemoglobin concentration appears to rise after 3 to 4 hours. In general, patients will return to normal within 24 hours, provided that all sources of further absorption are completely eliminated. The methemoglobin will be reduced spontaneously to ferrous hemoglobin in 2–3 days.[5] Such therapy is not effective in subjects with glucose-6-phosphate dehydrogenase deficiency.

The only justifiable use of methylene blue would be in cases of coma or stupor, usually at methemoglobin levels over 60%. In those patients in whom therapy is necessary, methylene blue may be given intravenously, 1 to 2 mg/kg of a 1% solution in saline over a 10-minute period. If cyanosis has not disappeared within 1 hour, a second dose of 2 mg/kg may be administered.[5,6] The total dose should not exceed 7 mg/kg, because methylene blue may cause toxic effects such as dyspnea, precordial pain, restlessness, apprehension, red cell hemolysis, and changes in the electrocardiogram (reduction in the height or even reversal of the T wave, frequently with lowering of the R wave).[6]

REFERENCES

1. Pacseri I, Magos L, Batskor IA: Threshold and toxic limits of some amino and nitro compounds. AMA Arch Ind Health 18:1–8, 1958
2. Anisidine. Documentation of the TLVs and BEIs, 5th ed, p 31. Cincinnati, American Conference of Governmental Industrial Hygienists (ACGIH), 1986
3. IARC Monographs on the Evaluation of the Carcinogenic Risk of Chemicals to Humans, Vol 27, pp 63–77. Lyon, International Agency for Research on Cancer, 1982
4. National Cancer Institute: Bioassay of *o*-Anisidine Hydrochloride for Possible Carcinogenicity, TR-89. DHEW (NIH) Pub No 78-1339. Washington, DC, US Government Printing Office, 1978
5. Rider RF: Methemoglobinemia and sulfmethomoglobinemia. In Wyngaarden JB, Smith LH (eds): Cecil Textbook of Medicine, 16th ed, p 896. Philadelphia, WB Saunders, 1982
6. Mangelsdorff AF: Treatment of methemoglobinemia. AMA Arch Ind Health 14:148–153, 1956

ANTIMONY (and compounds)
CAS: 7440-36-0
Sb 1987 TLV = 0.5 mg/m^3

Compounds: Antimony salts including the chlorides, sulfides, and fluorides; antimony oxide

Physical Form. Silvery-white soft metal

Uses. Constituent of alloy with other metals (tin, lead, copper); sulfides used in compounding of rubber and manufacture of pyrotechnics; chlorides used as coloring agents and as catalysts; fluorides used in organic synthesis and pottery manufacture

Exposure. Inhalation

Toxicology. Antimony is an irritant of the mucous membranes, eyes, and skin.

Antimony poisoning was reported in 69 of 78 smelter workers during a 5-month period when antimony concentrations of breathing zone samples in the smelter building averaged 10.07 to 11.81 mg/m^3 of air (range 0.92 to 70.7 mg/m^3); dermatitis and rhinitis were reported most frequently, but other symptoms included irritation of eyes, sore throat, headache, pain or tightness in the chest, shortness of breath, metallic taste, nausea, vomiting, diarrhea, weight loss, and dysosmia.[1]

Symptomless radiographic lung changes resembling the simple pneumoconiosis of coal workers were found in 44 of 262 men exposed to antimony oxide at up to 10 times the TLV.[2,3] In another roentgenographic study of 51 workers exposed 9 or more years to antimony oxides, there were numerous small opacities densely distributed in the middle and lower lung fields.[4] There were no characteristic pulmonary function abnormalities, but chronic cough was a common symptom.[4] Brief exposures to antimony trichloride, approximately 73 mg Sb/m^3, caused gastrointestinal symptoms as well as irritation of the skin and respiratory tract; urinary antimony ranged up to 5 mg/liter.[5]

Six sudden deaths and two deaths due to chronic heart disease occurred among 125 abrasive wheel workers exposed to antimony trisulfide for 8 to 24 months.[6] At air concentrations averaging over 3.0 mg/m^3, 37 of 75 workers had electrocardiogram changes, and 38 had abnormalities in blood pressure.[6] The lack of electrocardiographic changes in the oxide exposures would seem to indicate a special effect of the sulfide.

Contact of antimony compounds with the skin causes papules and pustules around sweat and sebaceous glands.[2]

Female rats exposed to 4.2 and 3.2 mg/m^3 antimony trioxide 6 hours/day, 5 days/week for 1 year had lung tumors after an additional year of observation.[7]

Current evidence in humans is inconclusive regarding an increased risk of lung cancer and reproductive disorders from antimony exposure.[8]

REFERENCES

1. Renes LE: Antimony poisoning in industry. AMA Arch Ind Hyg Occup Med 7:99–108, 1953
2. McCallum RI: The work of an occupational hygiene service in environmental control. Ann Occup Hyg 6:55–64, 1963
3. McCallum RI: Detection of antimony in process workers lungs by x-radiation. Trans Soc Occup Med 17:134–138, 1967
4. Potkonjak V, Pavlovich M: Antimoniosis: A particular form of pneumoconiosis I. Etiology, clinical and x-ray findings. Int Arch Occup Environ Health 51:199–207, 1983
5. Taylor PJ: Acute intoxication from antimony trichloride. Br J Ind Med 23:318–321, 1966
6. Brieger H, Semisch CW, Stasney J, Piatnek DA: Industrial antimony poisoning. Ind Med Surg 23:521–523, 1954
7. Department of Labor: Antimony metal; antimony trioxide; and antimony sulfide response to the interagency testing committee. Federal Register 48:717–724, 1983
8. National Institute for Occupational Safety and Health, US Department of Health, Education and Welfare: Criteria for a Recommended Standard. Occupational Exposure to Antimony, DHEW (NIOSH) 78-216. Washington, DC, US Government Printing Office, 1978

ANTU (α-NAPHTHYL-THIOUREA)

CAS: 86-88-4

$C_{11}H_{10}N_2S$ 1987 TLV = 0.3 mg/m^3

Synonyms: α-Naphthylthiourea; α-naphthylthiocarbamide

Physical Form. Blue to gray powder

Uses. Rodenticide

Exposure. Inhalation; ingestion

Toxicology. ANTU dust causes pulmonary edema and pleural effusion in animals.

ANTU is probably not toxic to humans except in large amounts; the lethal dose by ingestion is estimated to be approximately 4 g/kg.[1] In an instance of human intoxication by ANTU, 80 g of a rat poison containing 30% ANTU was ingested along with a considerable amount of ethanol; signs attributable to ANTU were prompt vomiting, dyspnea, cyanosis, and coarse pulmonary rales; no pleural effusion occurred, and the pulmonary signs gradually cleared.[1]

Oral administration to rats of 35 mg/kg was fatal to 60% of the animals; effects were labored respiration and muscular weakness; autopsy revealed pleural and pericardial effusion as well as mild liver damage.[2] Tachyphylaxis or tolerance to the acute toxicity of ANTU has been observed following repeated administrations; intraperitoneal injection of 2.5 mg/kg produced moderate pulmonary edema and large pleural effusions, but two additional 2.5 mg/kg doses at 2-day intervals resulted in lesser degrees of edema and minimal pleural fluid.[3] Daily doses of 200 mg/kg (20% of the median lethal dose) in rabbits were cumulative, causing death in 5 to 6 days without pleural effusions.[2]

REFERENCES

1. Gosselin RE, Smith RP, Hodge HC: Clinical Toxicology of Commercial Products, 5th ed, Section III, pp 40–42. Baltimore, Williams & Wilkins, 1984
2. McClosky WT, Smith MI: Studies on the pharmacologic action and the pathology of alphanaphthylthiourea (ANTU). I. Pharmacology. Public Health Rep 60:1101–1113, 1945
3. Sobonya RE, Kleinerman J: Recurrent pulmonary edema induced by alpha naphthyl thiourea. Am Rev Respir Dis 108:926–932, 1973

ARSENIC (and compounds)

CAS: 7440-38-2

As — 1987 TLV = 0.2 mg/m^3, as As; none as As_2O_3

Synonyms and Compounds: Grey arsenic; metallic arsenic; arsenic trichloride; arsenic trioxide; arsenic salts

Physical Form. Metallic arsenic is a steel-grey, brittle metal; arsenic trichloride is an oily liquid; arsenic trioxide is a crystalline solid.

Uses and Sources. In metallurgy for hardening copper, lead, alloys; in pigment production; in the manufacture of certain types of glass; insecticides and fungicides, rodent poison; a by-product in the smelting of copper ores; dopant material in semiconductor manufacture

Exposure. Inhalation; skin absorption; ingestion

Toxicology. Arsenic compounds are irritants of the skin, mucous membranes, and eyes; arsenical dermatoses and epidermal carcinoma are reported risks of exposure to arsenic compounds, as are other forms of cancer.[1]

Acute arsenic poisoning is rare in the occupational setting and results primarily from ingestion of contaminated food and drink.[2] Initial symptoms include burning lips, constriction of the throat, and dysphagia, followed by excruciating abdominal pain, severe nausea, projectile vomiting, and profuse diarrhea.[3] Other toxic effects on the liver, blood-forming organs, central and peripheral nervous systems, and cardiovascular system may appear.[4] Convulsions, coma, and death follow within 24 hours in severe cases.[3] Levels of exposure associated with acute arsenic toxicity vary with the valency form of the element; trivalent arsenic compounds are the most toxic, presumably because of their avid binding to sulfhydryl groups. For arsenic trioxide, the reported estimated lethal dose ranges from 70 to 300 mg.[3,4]

Acute inhalation exposures have resulted in irritation of the upper respiratory tract, even leading to nasal perforations.[4] Occupational exposure to arsenic compounds results in hyperpigmentation of the skin and hyperkeratoses of palmar and plantar surfaces, as well as dermatitis of primary irritation type.[1] Impairment of peripheral circulation resulting in gangrene of the fingers and toes has been reported.

Chronic arsenic intoxication by ingestion is characterized by weakness, anorexia, gastrointestinal disturbances, peripheral neuropathy, and skin disorders; with the exception of skin disorders, these signs and symptoms are not common in industrial practice. Liver damage has been observed in animals after both ingestion and inhalation of arsenic compounds, but this has not been observed with occupational exposure.[1]

Arsenic trichloride is a vesicant and can cause severe damage to the respiratory system upon inhalation; it is rapidly absorbed through the skin, and a fatal case following a spill on the skin has been reported.[5] The vapor of arsenic trichloride is highly irritating to the eyes. Some organic arsenicals, such as arsanilates, have a selective effect on the optic nerve and can cause blindness.

Teratogenic effects, including exencephaly, skeletal defects, and genitourinary system defects, of arsenic compounds administered intravenously or intraperitoneally at high doses have been demonstrated in hamsters, rats, and mice.[4] Only minimal fetal effects have been observed in studies of pregnant rats or mice exposed to lower levels by means of drinking water.[4]

In a large number of studies, exposure to inorganic arsenic compounds in drugs, food, and water as well as in an occupational setting have been causally associated with the development of cancer, primarily of the skin and lungs.[1–4] An excess mortality from respiratory cancer has been found among smelter workers and workers engaged in the production and use of arsenical pesticides. It should be noted, however, that in a number of these studies, levels of exposure are uncertain, and there is simultaneous exposure to other agents. In the most recent follow-up of 8045 smelter workers, those with the highest estimated exposure and the longest follow-up had a ninefold increase in respiratory cancer mortality.[6]

Information on the association of arsenic with skin cancer has primarily involved nonoccupational populations exposed to contaminated drinking water.[4] Chronic ingestion of trivalent arsenic in medicinal preparations was also associated with an increased incidence of hyperkeratosis and skin cancer.[4]

Because there is no convincing evidence of the carcinogenicity of arsenic compounds in animals, it has been suggested that the compounds are not direct carcinogens but act in some other way. Despite the absence of suitable animal models, IARC has determined that there is sufficient evidence for carcinogenicity to humans.[7]

Diagnosis. *Signs and Symptoms:* Signs and symptoms include conjunctivitis and visual disturbances; ulceration and perforation of nasal septum; pharyngitis and pulmonary irritation; peripheral neuropathy; and hyperpigmentation of skin, palmar and plantar hyperkeratoses, dermatitis, and skin cancer. Arsenic may cause cancer of the lung, larynx, lymphoid system, or viscera.

Special Tests: Urinary levels of arsenic above 0.7 to 1.0 mg/liter in exposed persons may be indicative of harmful exposure, but dietary factors must be ruled out before significance is attached to any increase in the arsenic content of urine.[5] The biological half-life of arsenic in urine in subjects with normal renal function is 1 to 2 days.[8] Seafood, especially shellfish, is a rich source of organic arsenic compounds, which are essentially nontoxic but affect the total urinary arsenic determination. Levels of 1.35 mg/liter of urine have occurred after a meal of lobster tails.[5] In persons without known exposure to arsenic who have not ingested a seafood meal for 2 days before sampling, the urinary total arsenic is usually below 100 μg/g creatinine. Inorganic arsenic is excreted in the urine unchanged or as monomethylarsonic or dimethylarsonic (cacodylic) acid; these compounds can be measured by hydride-generation atomic absorption spectrophotometry, yielding an index of exposure to inorganic arsenic. By this method, levels of urinary arsenic are generally below 20 μg/g urinary creatinine in unexposed subjects.[8] One estimate is that a time-weighted average exposure to 50 $\mu g/m^3$ of inorganic arsenic over several days would yield an average urinary excretion of 220 μg of arsenic per gram creatinine by the latter method.[8] A determination of arsenic in hair and nails may be useful, although its value has been questioned in industrial exposures because of the difficulty in removing all external contamination.

Treatment. Severe acute arsenic poisoning from occupational exposure is unlikely; if it should occur, administer BAL (dimercaprol) 10% in oil, intramuscularly (gluteal), 3 mg/kg for each injection; first and second days, one injection every 4 hours day and night; third day, one injection every 6 hours; fourth to 14th day, one injection twice a day until recovery is complete.[9,10] Side effects of BAL (nausea, vomiting, pain, injection site sterile abscess formation, hypertension, tachycardia, headache, anxiety, diaphoresis, muscle cramps, seizures, coma, and urticaria) usually respond to supportive measures and decreased dosing and are typically transient.[11] With severe acute intoxication, d-penicillamine may also be given orally in a dose of 25 mg/kg (to maximum of 500 mg) every 6 hours for the first 10 days, in addition to BAL.[11] BAL may also be used to treat chronic arsenic intoxication.[12]

REFERENCES

1. National Institute for Occupational Safety and Health, US Department of Health, Education, and Welfare: Criteria for a Recommended Standard . . . Occupational Exposure to Inorganic Arsenic, New Criteria—1975. DHEW (NIOSH) Pub No 75-149, pp 14–71. Washington, DC, US Government Printing Office, 1975
2. Landrigan PJ: Arsenic—state of the art. Am J Ind Med 2:5–14, 1981
3. Winship KA: Toxicity of inorganic arsenic salts. Adv Drug React Ac Pois Rev 3:129–160, 1984
4. Health Assessment Document for Inorganic Arsenic. Final Report. Research Triangle Park, North Carolina, US Environmental Protection Agency, March 1984
5. Hygienic Guide Series: Arsenic and its compounds (except arsine). Am Ind Hyg Assoc J 25:610–613, 1964
6. Lee-Feldstein A: Cumulative exposure to arsenic and its relationship to respiratory cancer among copper smelter employees. J Occup Med 28:296–302, 1986
7. IARC Monographs on the Evaluation of the Carcinogenic Risk of Chemicals to Humans. Chemicals, Industrial Processes and Industries Associated with Cancer in Humans, Suppl 4, pp 50–51. Lyon, International Agency for Research on Cancer, 1982
8. Lauwerys RR: Industrial Chemical Exposure: Guidelines for Biological Monitoring, pp 12–15. Davis, California, Biomedical Publications, 1983
9. Martin DW Jr, Woeber KA: Arsenic poisoning. Calif Med 118:13, 1973
10. Arena JM: Poisoning, 3rd ed, pp 25–26. Springfield, Illinois, Charles C Thomas, 1974
11. Linden CH: Antidotes in poisoning. In Callaham ML (ed): Current Therapy in Emergency Medicine, pp 951–952. Toronto, BC Decker, Inc, 1987
12. Gilman AG, Goodman LS (eds): Goodman and Gilman's The Pharmacological Basis of Therapeutics, 6th ed, pp 1629–1632. New York, Macmillan, 1980

ARSINE

CAS: 7784-42-1

AsH_3 1987 TLV = 0.05 ppm

Synonyms: Arsenic hydride, arseniuretted hydrogen; arsenous hydride; hydrogen arsenide

Physical Form. Colorless, heavier-than-air gas

Uses. Produced accidentally as a result

of generation of nascent hydrogen in the presence of arsenic or by the action of water on a metallic arsenide; used as a dopant in semiconductor manufacture

Exposure. Inhalation

Toxicology. Arsine is a severe hemolytic agent; abdominal pain and hematuria are cardinal features of arsine poisoning and are frequently accompanied by jaundice.

Arsine is the most acutely toxic form of arsenic.[1] It binds with oxidized hemoglobin, causing profound hemolysis of sudden onset.[2] Inhalation of 250 ppm may be fatal within 30 minutes, whereas 10 to 50 ppm may cause anemia and death with more prolonged exposure. Human experience has indicated that there is usually a delay of 2 to 24 hours after exposure before the onset of headache, malaise, weakness, dizziness, and dyspnea, with abdominal pain, nausea, and vomiting.[4–7] Dark red urine is frequently noted 4 to 6 hours after exposure. This often progresses to brown urine, with jaundice appearing at 24 to 48 hours after exposure.

An unusual bronze skin color has been noted in some patients; pigmentation of the skin and mucous membranes is more often described as ordinary jaundice and is seen in most poisoning patients. Oliguria or anuria, the most serious manifestation, may become manifest before the third day. In fatal cases, death may result from renal shutdown. Kidney failure occurs as extensive lysis by-products precipitate in the tubules and/or from hypoxic damage resulting from the reduced oxygen carrying capacity of blood.[3] Other tissues at risk from hemolysis, anemia, and sludging of red cell debris within the microcirculation are the myocardium, liver, marrow, lungs, and skeletal muscles.[3] Reticulocytosis and leukocytosis are expected.[4–7] Normal red cell fragility and a negative Coombs' test are observed. Plasma hemoglobin values of greater than 2 g/100 ml are reported. Symptoms of arsenic poisoning, in addition to those of arsine, may be present. In two reported cases, arsenic encephalopathy with extreme restlessness, memory loss, agitation, and disorientation occurred several days after an acute exposure and lasted 10 days.[6] Peripheral neuropathy appeared within a few weeks, and symptoms included numbness of the hands and feet, severe muscle weakness, and photophobia.[6]

In a report of chronic arsine poisoning in workers engaged in the cyanide extraction of gold, there was severe anemia in the absence of other signs and symptoms.[8] Hemoglobin values ranged as low as 3.2 g/100 ml; marked basophilic stippling was observed. Previous exposure to trace amounts of arsine for a period of 8 months was documented. It appears that in very small concentrations, arsine exerts a cumulative effect.[4]

Inhaled arsine is oxidized to form elemental trivalent arsenic (AS^{+3}) and arsenous oxide (As_2O_3), two human carcinogens.[9] Excess cancers from trivalent arsenic and arsenic trioxide have been associated with cumulative lifetime arsenic exposure. Exposure to arsine above 0.004 ppm is associated with increased urinary arsenic excretion, indicating exposure to arsenic. Current exposure limits may not prevent potential chronic toxicity.[9]

Arsine is nonirritating, with a garlic-like odor. Warning properties of exposure to hazardous concentrations are inadequate.[1]

Diagnosis. *Signs and Symptoms:* Signs and symptoms include headache, malaise, weakness, dizziness; dyspnea; abdominal pain, nausea, vomiting; hematuria; jaundice; and oliguria and anuria.

Differential Diagnosis: Distinguish hemolysis due to arsine poisoning from paroxysmal nocturnal hemoglobinuria, cold agglutinin disease, thalassemia syndromes, sickle cell anemia, congenital hemolytic icterus, and poisoning by other hemolytic agents such as stibine.[1]

Special Tests: Special tests include plasma hemoglobin; white cell count; urinalysis, and urinary concentration of hemoglobin.[2] Changes in the hematocrit value may not be of sufficient magnitude to be useful in diagnosis.[2] Analysis of blood and urine for arsenic may be done, but, in a medical emergency, this is of less value than the aforementioned tests. Normal blood levels of arsenic are usually below 20 μg/100 ml.[2] In arsine poisoning, the urinary arsenic level usually remains elevated for several days or until normal renal function is restored.[2]

Treatment. The treatment of choice for acute and severe arsine poisoning is exchange transfusion and, if renal failure develops, hemodialysis.[1,2] As a rough guide when there is a history of arsine exposure, replacement transfusion should be done if the serum hemoglobin reaches 1.5 g/dl or higher or if oliguria develops.[2] Renal function should be closely monitored. Alkaline diuresis should be instituted in an attempt to avoid hemoglobin precipitation in the renal tubules. Dimercaprol (BAL) therapy is of no use in arsine poisoning, because it affords no protection against hemolysis.[2]

REFERENCES

1. NIOSH Current Intelligence Bulletin 32, Arsine (Arsenic Hydride) Poisoning in the Workplace. DHEW (NIOSH) Pub No 79–142. Cincinnati, National Institute for Occupational Safety and Health, 1979
2. Hesdorffer CS et al: Arsine gas poisoning: The importance of exchange transfusions in severe cases. Br J Ind Med 43:353–355, 1986
3. Luckey TD, Venugopal B: Metal Toxicity in Mammals, Vol 2, p 209. New York, Plenum Press, 1977
4. Fowler BA, Weissberg JB: Arsine poisoning. N Engl J Med 291:1171–1174, 1974
5. Pinto SS: Arsine poisoning: Evaluation of the acute phase. J Occup Med 18:633–635, 1976
6. Levinsky WJ, Smalley RV, Hillyer PN, Shindler RL: Arsine hemolysis. Arch Environ Health 20:436–440, 1970
7. Teitelbaum DT, Kier LC: Arsine poisoning. Arch Environ Health 19:133–143, 1969
8. Bulmer FMR et al: Chronic arsine poisoning among workers employed in the cyanide extraction of gold: A report of fourteen cases. J Ind Hyg Toxicol 22:111–124, 1940
9. Landrigan PJ et al: Occupational exposure to arsine. An epidemiologic reappraisal of current standards. Scand J Work Environ Health 8:169–177, 1982

ASBESTOS

Recognized human carcinogen

CAS: 1332-21-4
Amosite—CAS: 12172-73-5;
1987 TLV = 0.5 fiber/ml
Chrysotile—CAS: 12001-29-5;
1987 TLV = 2 fibers/ml
Crocidolite—CAS: 12001-28-4;
1987 TLV = 0.2 fiber/ml
Other forms: 1987 TLV = 2 fibers/ml

Synonyms: Asbestos is a generic term applied to a number of hydrated mineral silicates.

Physical Form. Fibers of various sizes, colors, and textures

Uses. Thermal and electrical insulation; fireproofing; cement products

Exposure. Inhalation

Toxicology. Asbestos causes chronic lung disease (asbestosis), inflammation of the pleura, and certain cancers of the lungs and digestive tract.

Asbestosis is a disorder characterized by a diffuse interstitial pulmonary fibrosis, at times including pleural changes of fibrosis and calcification.[1] Chest x-ray reveals a granular change chiefly in the lower lung fields; as the condition progresses, the heart outline becomes shaggy, and irregular patches of mottled shadowing may be seen. Typically, the patient exhibits restrictive pulmonary function. Accompanying clinical changes may include fine rales, finger clubbing, dyspnea, dry cough, and cyanosis.

The onset of asbestosis probably depends upon asbestos dust concentration, fiber morphology, and the length of exposure. There may also be some factors of individual susceptibility, although it is not possible to identify persons who may have undue susceptibility or resistance to developing asbestosis.[2] It is a progressive disease that may develop fully in 7 to 9 years and may cause death as early as 13 years after first exposure. Usually, pneumoconiosis becomes evident 20 to 40 years after the first exposure to asbestos. Once established, asbestosis progresses even after exposure has ceased.[1]

There is often thickening of the visceral pleura from extension of the parenchymal inflammation. The parietal pleura may show patches of severe thickening, particularly over the diaphragm and the lower portions of the chest wall, resulting in the so-called pleural hyaline plaques. These may be seen by x-ray, especially if calcified. Pleural plaques, which produce no symptoms, may develop from asbestos exposure in the absence of asbestosis.[3]

Bronchogenic carcinoma and mesothelioma of the pleura and peritoneum are causally associated with asbestos exposures; excesses of cancer of the stomach, colon, and rectum have also been observed.[4] Among 632 asbestos workers observed from 1943 to 1967, there were 99 excess deaths (above that expected on the basis of the US white male population) for three types of malignancies: (1) bronchogenic (63), (2) gastrointestinal, (26) and (3) all other sites combined (10).[1]

Mesothelioma, a relatively rare and rapidly fatal neoplasm seen chiefly in crocidolite workers, may occur without radiologic evidence of asbestosis at exposure levels lower than those required for prevention of radiologically evident asbestosis.[1] Mesothelioma can occur after a short intensive exposure; cases in patients younger than 19 years of age indicate that the latent time period for development may be shorter than first estimated, although the disease may occur following a very limited exposure 20 to 30 years earlier.

Cigarette smoking is strongly implicated as a cocarcinogen among asbestos workers.[5] The incidence of lung carcinoma among nonsmoking asbestos workers is not significantly greater than that of nonasbestos workers, whereas asbestos workers who smoke have a much higher incidence. Cigarette-smoking asbestos workers have approximately 15 times the risk of developing lung cancer compared with nonsmoking asbestos workers.[6]

REFERENCES

1. National Institute for Occupational Safety and Health, US Department of Health, Education, and Welfare: Criteria for a Recommended Standard . . . Occupational Exposure to Asbestos. DHEW (HSM) 72–10267. Washington, DC, US Government Printing Office, 1972
2. Parkes WR: Occupational Lung Disorders, 2nd ed, p 255. London, Butterworths, 1982
3. Asbestos. Documentation of TLVs and BEIs, 5th ed, pp 40–42. Cincinnati, American Conference of Governmental Industrial Hygienists (ACGIH), 1986
4. Selikoff IJ, Churg J, Hammond EC: Asbestos exposure and neoplasia. JAMA 188:22–28, 1968
5. Selikoff IJ, Hammond EC, Churg J: Asbestos

exposure, smoking and neoplasia. JAMA 204:106–112, 1968
6. Selikoff IJ, Lee DHK: Asbestos and Disease, p 327. New York, Academic Press, 1978

AZINPHOSMETHYL

CAS: 86-50-0

$C_{10}H_{12}N_3O_3PS_2$ TLV = 0.2 mg/m³; skin

Synonyms: O,O-Dimethyl S (4–oxo–1, 2, 3–benzotriazin–3(4H)–ylmethyl) phosphorodithioate; Guthion

Physical Form. Solid

Uses. Acaricide; insecticide

Exposure. Inhalation; skin absorption; ingestion

Toxicology. Azinphosmethyl is an indirect inhibitor of cholinesterase.

Dosages given to volunteers for approximately 30 days ranged from 4 to 20 mg/man/day, which did not produce clinical effects or a significant change in cholinesterase levels. Apparently, a level high enough to inhibit cholinesterase remains to be studied. It is regarded as being of only moderate toxicity to humans.[1]

In a study of eight workers engaged in the formulation of a Guthion wettable powder and exposed to concentrations of up to 9.6 mg/m³, the lowest activity of cholinesterase in blood serum was 78% of the value before exposure, and there were no signs or symptoms of illness.[2]

In animals, azinphosmethyl has an acute oral toxicity similar to that of parathion, although the acute dermal toxicity is less than that of parathion.[3] There were no significant toxicologic effects in rats at an equivalent human oral dose of 70 mg/man/day.[3]

Rats that inhaled azinphosmethyl at 4.72 mg/m³, 6 hours/day, 5 days/week for 12 weeks showed significant depression of red cell and plasma cholinesterases; concentrations of 0.195 and 1.24 mg/m³ were without effect.[1]

Diagnosis. *Signs and Symptoms:* By analogy to other anticholinesterases, the following signs and symptoms would be expected in extremely severe poisonings: *Initial intoxication:* headache, blurred vision, pallor, weakness, sweating, abdominal pain, nausea, vomiting, and diarrhea. *Moderate to severe intoxication:* miosis, lacrimation, excessive salivation, muscle fasciculations, dyspnea, cyanosis, convulsions, shock, cardiac arrhythmias, and coma.

Differential Diagnosis: Diagnosis is based primarily on a history of exposure and clinical evidence of diffuse parasympathetic stimulation. Careful observation of the effects of atropine and pralidoxime may be valuable. Patients with organophosphate poisoning are resistant to the action of atropine at moderate dosages. Failure of 1 to 2 mg of atropine administered parenterally to produce signs of atropinization (flushing, mydriasis, tachycardia, or dryness of mouth) indicates organophosphate poisoning. Intravenous injection of 1 g of pralidoxime generally causes some recovery from signs and symptoms.

Special Tests: Two types of cholinesterase are clinically significant: (1) true acetylcholinesterase, found principally in the nervous system and the red blood cell; and (2) pseudo- or butyrylcholinesterase, found in the plasma, liver, and nervous system. Whereas the action of both types is inhibited by organophosphates, the level of depression of red blood cell cholinesterase is a better indicator of clinically significant reduction of cholinesterase activity in the nervous system.

Laboratory evidence of depression of red blood cell cholinesterase to a level substantially below pre-exposure levels (at least 50% and usually much lower) is verification of organophosphate poisoning. There is an imperfect correlation between the degree of depression of cholinesterase enzymes and the occurrence of symptoms. With a rapid drop in cholinesterase activity, generally reflecting an acute heavy exposure, there may be symptoms with only a 30% depression, whereas with slower drops to 70% depression, reflecting chronic low level exposure, there may be no symptoms.[4]

If no pre-exposure baseline has been performed but symptoms are not sufficient to justify treatment with atropine, repeated testing during the recovery period demonstrating progressively increasing plasma and red blood cell cholinesterase levels over several days and weeks, respectively, suggests the diagnosis of anticholinesterase poisoning.

There are many different methods for estimation of cholinesterase content of blood, and associated with each method is a different set of normal values and a different set of reporting units. The laboratory report of a cholinesterase determination should state the units involved along with the appropriate normal range. Based on the Michel method, the normal range of red blood cell cholinesterase activity (delta *p*H/hour) is 0.39 to 1.02 for men and 0.34 to 1.10 for women.[5] The normal range of the enzyme activity (delta *p*H/hour) of plasma is 0.44 to 1.63 for men and 0.24 to 1.54 for women.

Treatment. Treatment of organophosphate poisoning ranges from simple removal from exposure in very mild cases to the provision of very rigorous supportive and antidotal measures in severe cases.[1,6–8] In the moderate to severe case, because of pulmonary involvement, there may be need for artificial respiration using a positive-pressure method. Careful attention must be paid to removal of secretions and to maintenance of a patent airway. Anticonvulsants such as thiopental sodium may be necessary. Maintenance of respiration is critical, because death usually results from weakness of the muscles of respiration and accumulation of excessive secretions in the respiratory tract.[6]

As soon as cyanosis has been overcome, 2 to 4 mg of atropine should be given intravenously. (Atropine may induce ventricular fibrillation in the presence of cyanosis.) *This dose of atropine is approximately 10 times the amount that is administered for other conditions in which atropine is considered therapeutic.* This dose should be repeated at 5- to 10-minute intervals until signs of atropinization appear (dry, flushed skin, tachycardia as high as 140 beats/minute, and pupillary dilatation). A mild degree of atropinization should be maintained for at least 48 hours.[6]

Pralidoxime (2–PAM, Protopam) chloride is a cholinesterase reactivator that complements the action of atropine. It has its greatest effect in reversing the nicotinic action of anticholinesterase agents at skeletal neuromuscular junctions but virtually no effect on central nervous system manifestations. In the moderate-to-severe case, the dose for adults is 1 to 2 g injected intravenously at a rate not in excess of 500 mg/minute. After an hour, a second dose of 1 g is indicated if muscle weakness has not been relieved. Treatment with pralidoxime chloride will be most effective if given within 24 hours after poisoning.[6] Morphine, aminophylline, and phenothiazines are contraindicated because of documented experience of adverse reactions in cases of organophosphate poisoning.[7]

It is of great importance to decontaminate the patient. Contaminated clothing should be removed at once, and the skin should be washed with generous amounts of soap or detergent and a flood of water; this is best accomplished under a shower or by submersion in a pond or other body of water if the exposure occurred in the field. Careful attention should be paid to cleansing of the skin and hair.

The patient should be attended and monitored continuously for not less than 24 hours, because serious and sometimes fatal relapses have occurred as a result of continuing absorption of the toxin or dissipation of the effects of the antidote.

Regeneration of cholinesterase is primarily by synthesis of new enzyme and takes place at the rate of approximately 1%/day.[7] A patient who has recovered from the acute phase of poisoning remains hypersusceptible to anticholinesterases for up to several weeks.

Medical Control. Medical control involves preplacement and annual physical examination with determination of pre-exposure red blood cell cholinesterase activity. A person whose red blood cell cholinesterase falls to or below 40% of the pre-exposure baseline should be removed from further exposure until the activity returns to within 80% of the pre-exposure baseline.

REFERENCES

1. Hayes WJ Jr: Organic phosphorus pesticides. In Pesticides Studied in Man, pp 358–359. Baltimore, Williams & Wilkins, 1982
2. Jegier Z: Exposure to Guthion during spraying and formulating. Arch Environ Health 8:565–569, 1964
3. Azinphos-methyl. Documentation of the TLVs and BEIs, 5th ed, p 46. Cincinnati, American Conference of Governmental Industrial Hygienists (ACGIH), 1986
4. Coye MJ, Lowe JA, Maddy KT: Biological monitoring of agricultural workers exposed to pesticides. I. Cholinesterase activity determinations. J Occup Med 28:619–627, 1986
5. Michel HO: Electrometric method for determination of red blood cell and plasma cholinesterase activity. J Lab Clin Med 34:1564–1568, 1949
6. Taylor P: Anticholinesterase agents. In Gilman AG et al (eds): Goodman and Gilman's The Pharmacological Basis of Therapeutics, 7th ed, pp 110–129. New York, Macmillan, 1985
7. Milby TH: Prevention and management of organophosphate poisoning. JAMA 216:2131–2133, 1971
8. Namba T, Nolte CT, Jackrel J, Grob D: Poisoning due to organophosphate insecticides. Am J Med 50:475–492, 1971

BARIUM (and compounds)

CAS: Ba = 7440-39-3; $BaSO_4$ = 7727-43-7

$BaCO_3$; $Ba(OH)_2$; $BaSO_4$

1987 TLV = 0.5 mg/m³, soluble compounds as Ba; 10 mg/m³ $BaSO_4$, total dust

---

Synonyms: Soluble—barium nitrate, barium sulfide, barium carbonate, barium chloride

Insoluble—barium sulfate ($BaSO_4$)

Physical Form. Powders

Uses. Catalyst for organic reactions; lubricating oil additive; rat poison; manufacture of paper electrodes; in fireworks; in electroplating

Exposure. Inhalation; ingestion

Toxicology. Certain compounds of barium are irritants of the eyes, mucous membranes, and skin; inhalation of $BaSO_4$ is innocuous.

The toxicity of barium compounds depends upon their solubility, with the more soluble forms being the most toxic.[1] Inhalation of insoluble barium-containing dusts may produce a benign pneumoconiosis termed *baritosis*. The condition is without clinical significance. Characteristic x-ray changes are those of small, extremely dense circumscribed nodules evenly distributed throughout the lung fields, reflecting the radio-opacity of the barium dust. Exposure of workers to concentrations ranging to 92 mg/m^3 of barium sulfate caused no abnormal signs or symptoms, including no interference with lung function or liability to develop pulmonary or bronchial infection.[2] Ingestion of insoluble barium compounds also presents no problems to health—barium sulfate being widely used as a contrast agent in radiography.[3]

Barium ion is a muscle poison causing stimulation and then paralysis. Initial symptoms are gastrointestinal, including nausea, vomiting, colic, and diarrhea, followed by myocardial and general muscular stimulation with tingling in the extremities.[1] Severe cases continue to loss of tendon reflexes, general muscular paralysis, and death from respiratory arrest or ventricular fibrillation. Threshold of a toxic dose in humans is reported to be about 0.2 to 0.5 g of barium absorbed from the gut; the lethal dose is 3 to 4 g of barium. Administration of a soluble sulfate causes precipitation of barium sulfate in the alimentary tract and inhibits intestinal absorption.[1]

The barium ion is a physical antagonist of potassium, and it appears that the symptoms of barium poisoning are attributable to Ba^{2+}-induced hypokalemia.[1] The effect is probably due to a transfer of potassium from extracellular to intracellular compartments rather than to urinary or gastrointestinal losses. Signs and symptoms are relieved by intravenous infusion of K^+.[1]

Barium hydroxide and barium oxide are strongly alkaline in aqueous solution, causing severe burns of the eye and irritation of the skin.[4]

REFERENCES

1. Reeves AL: Barium. In Friberg L et al (eds): Handbook on the Toxicology of Metals, pp 321–328. New York, Elsevier North-Holland, 1979
2. Barium sulfate. Documentation of the TLVs and BEIs, 5th ed, p 48. Cincinnati, American Conference of Governmental Industrial Hygienists (ACGIH), 1986
3. Dare PRM et al: Short communication. Barium in welding fume. Ann Occup Hyg 2:445–448, 1984
4. Grant WM: Toxicology of the Eye, 3rd ed, p 134. Springfield, Illinois, Charles C Thomas, 1986

BAUXITE

$Al_2O_3 \cdot 2H_2O$ 1987 TLV = 10 mg/m^3
(see nuisance particulates)

Synonym: Beauxite

Physical Form. Dust (red, brown, or yellow)

Uses. Ore for production of alumina; adsorbent in oil refining

Exposure. Inhalation

Toxicology. Bauxite can be considered to be a nuisance particulate; long experience with mining and refining of bauxite has not resulted in adverse health effects.

Nuisance particulates have little adverse effect on lungs and do not produce significant organic disease or toxic effect when exposures are kept under reasonable control.[1] The nuisance dusts have also been called (biologically) "inert" dusts, but this term is inappropriate to the extent that there is no dust that does not evoke some cellular response in the lung when inhaled in sufficient amount. However, the lung–tissue reaction caused by inhalation of nuisance particulates has the following characteristics: (1) The architecture of the air spaces remains intact. (2) Collagen (scar tissue) is not formed to a significant extent. (3) The tissue reaction is potentially reversible.

The nuisance dust aspect of bauxite is in sharp contrast to the limited industrial situation where lung injury was reported in the 1940s in Canadian workers engaged in the manufacture of alumina abrasives in the virtual absence of fume control.[2,3] Fusing of bauxite at 2000°C gave rise to a fume composed of freshly formed particles of amorphous silica and aluminum oxide. In spite of the poor choice of the term *bauxite fume pneumoconiosis* sometimes used to describe the disease, scientific opinion favors the silica component as the probable toxic agent. It should be emphasized that bauxite from some sources may contain small amounts of silica.

REFERENCES

1. Nuisance particulates. Documentation of TLVs and BEIs, 5th ed, p 445. Cincinnati, American Conference of Governmental Industrial Hygienists (ACGIH), 1986
2. Hatch TF: Summary. In Vorwald AJ (ed): Pneumoconiosis, Beryllium, Bauxite Fumes, pp 498–501. New York, Harper & Brothers, 1950
3. Shaver CG, Riddel AR: Lung changes associated with the manufacture of alumina abrasives. J Ind Hyg Toxicol 29:145–157, 1947

BENZENE

CAS: 71-43-2

C_6H_6

OSHA amended Standard 1987:
PEL/TWA = 1 ppm
STEL = 5 ppm

Synonyms: Benzol; cyclohexatriene

Physical Form. Colorless liquid

Uses. Intermediate in the production of styrene, phenol, cyclohexane, and other organic chemicals; manufacture of detergents, pesticides, solvents, and paint removers; found in gasoline

Exposure. Inhalation; skin absorption

Toxicology. Acute benzene exposure causes central nervous system depression; chronic exposure results in depression of the hematopoietic system and is associated with an increased incidence of leukemia and possibly of multiple myeloma.

Human exposure to very high concentrations, approximately 20,000 ppm, is fatal in 5 to 10 minutes.[1] Convulsive movements and paralysis followed by unconsciousness follow severe exposures. Brief exposure to concentrations in excess of 3000 ppm is irritating to the eyes and respiratory tract; continued exposure may cause euphoria, nausea, a staggering gait, and coma.[2] Inhalation of lower concentrations (250 to 500 ppm) produces vertigo, drowsiness, headache, and nausea.[3]

The most significant toxic effect of benzene exposure is an insidious and often irreversible injury to the bone marrow. Long-term exposures to low concentrations have been observed to have an initial stimulant effect on the bone marrow, followed by aplasia and fatty degeneration.[3] Clinically, an initial increase followed by a decrease in the erythrocytes, leukocytes, or platelets is observed, with progression to leukopenia, thrombocytopenia, pancytopenia, and/or aplastic anemia.[3] The hypocellularity varies greatly from conditions in which the marrow is completely devoid of recognizable hematopoietic precursors to those in which the precursors of only one cell line are absent or arrested in their development.[4] Typical symptoms may include lightheadedness, headache, loss of appetite, and abdominal discomfort. With more severe intoxication, there may be weakness, blurring of vision, and dyspnea on exertion; the mucous membranes and skin may appear pale, and a hemorrhagic tendency may result in petechiae, easy bruising, epistaxis, bleeding from the gums, or menorrhagia.[5] The most serious cases of aplastic anemia succumb within 3 months of diagnosis owing to infection or hemorrhage.[6]

Accumulated case reports and epidemiologic studies suggest a leukemogenic action of benzene in humans—the leukemia tending to be acute and myeloblastic in type, often following aplastic changes in the bone marrow. Benzene may also induce chronic types of leukemia.[7]

One study indicated a fivefold excess of all leukemias and a tenfold excess of myelomonocytic leukemia among benzene-exposed workers as compared with the U.S. Caucasian male population.[8] Among shoemakers chronically exposed to benzene, the annual incidence of leukemia was 13.5 per 100,000 population, whereas the incidence in the general population was 6 per 100,000 population.[9] Four cases of acute leukemia were reported in shoemakers exposed to concentrations of benzene up to 210 ppm for 6 to 14 years; two of the four shoemakers had aplastic anemia prior to leukemia; three of the cases were acute myeloblastic leukemia; the fourth patient developed thrombocythemia in the second year after an episode of aplastic anemia, and acute monocytic leukemia developed later.[10]

A recent retrospective cohort study in China of 28,460 benzene-exposed workers found a leukemia mortality rate of 14 per 100,000 person-years in the benzene cohort and 2 per 100,000 in the control cohort.[11] The standardized mortality ratio (SMR) was 574, and the mean latency period of benzene leukemia was 11.4 years. Concentrations in the workplace where the patients had been employed were reported to range from 3 to 300 ppm but were mostly in the range of 16 to 160 ppm. The SMR in this study was similar to that in a study of two pliofilm manufacturing plants with 748 workers and exposures ranging from 16 to 100 ppm (SMR = 560).[12] In another report, a mortality update through 1982 for 956 employees exposed to benzene, there was a nonsignificant excess of total death from leukemia based on four observed cases; however, all four cases involved myelogenous leukemias, which represented a significant excess in this subcategory.[13]

Persons with aplastic anemia as a result of benzene exposure have been found to be at a much greater risk for developing leukemias. A follow-up of 51 benzene-exposed workers with pancytopenia revealed 13 cases of leukemia.[14] The cumulative incidence of leukemia among people with clinically ascertained benzene hemopathy has ranged from 10% to 17% in various studies.[15]

The IARC has concluded that epidemiologic studies have established the relationship between benzene exposure and the development of acute myelogenous leukemia and that there is sufficient evidence that benzene is carcinogenic to humans.[15] Although a benzene–leukemia association has been made, the exact shape of the dose–response curve and/or the existence of a threshold for the response is unknown and has been the source of speculation and controversy.[16–18] A recent risk assessment of the mortality experience of rubber workers suggests exponential increases in relative risk (of leukemias) with increasing cumulative exposure to benzene.[17] There was also a statistically significant increase in deaths from multiple myeloma, although the numbers were small. This study was cited by OSHA in the 1987 amended standard which lowered the PEL/TWA from 10 ppm to 1 ppm.[18a] In addition to cumulative dose, other factors such as multiple solvent exposure, familial connection, and individual susceptibility may play a role in the development of leukemia.[14]

An increased incidence of neoplasms at multiple sites has been found in chronic inhalation and gavage studies in rodents. Anemia, lymphocytopenia, bone marrow hyperplasia, and an increased incidence of lymphoid tumors occurred in male mice exposed at 300 ppm for life.[19] Gavage administration to rats in one study and rats and mice in another study caused an increase in tumors; especially significant was an increase in zymbal gland tumors (tumors of the auditory sebaceous glands) in both reports.[20,21]

Although consistent findings of chromosomal aberrations (stable and unstable) in the nuclei of lymphocytes have been reported in exposed workers, the implications with respect to leukemia are not clear.[22] Data on exposure levels are limited but are said to range from 10 to a few hundred ppm.[15] In controlled rat studies, exposure to 1, 10, 100, or 1000 ppm for 6 hours caused a dose–response relationship in the percentage of cells with abnormalities and aberrations at the two highest dose levels.[23] An increase in polychromatic erythrocytes with micronuclei (thought to be broken fragments of chromosomes that are left behind) have also been observed in benzene-treated animals.[15] Mice exposed at 10 ppm for 6 hours had a significantly increased incidence of micronuclei compared with controls.[24]

In addition to tumor induction and cytogenic damage, inhaled benzene in mice can cause immunodepressive effects at 100 ppm as manifested by reduced host resistance to a transplantable syngeneic tumor.[25]

Exposure to benzene vapor produces fetotoxicity in mice and rats; in one study, a teratogenic potential as evidenced by exencephaly, angulated ribs, and dilated brain ventricles was observed in rats exposed to 500 ppm.[26]

Tests for phenol levels in urine have been used as an index of benzene exposure; urinary phenol concentrations of 200 mg/liter are indicative of exposure to approximately 25 ppm of benzene in air.[27]

Direct contact with the liquid may cause erythema and vesiculation; prolonged or repeated contact has been associated with the development of a dry scaly dermatitis or with secondary infections.[3] Some skin absorption can occur with lengthy exposure to solvents containing benzene and may contribute more to toxicity than originally believed.[28]

REFERENCES

1. Flury F: Moderne gewerbliche Vergiftungen in pharmakologischtoxikologistche Hinsicht. Arch Exp Pathol Pharmacol 183:65–82, 1928
2. Herarde HW: Toxicology and Biochemistry of Aromatic Hydrocarbons, pp 97–108. New York, Elsevier, 1960
3. Department of Labor: Occupational exposure to benzene. Federal Register 42:22516–22529, 1977
4. Goldstein BD: Hematotoxicity in humans. In Laskin S, Goldstein BD: Benzene toxicity. A critical evaluation. J Toxicol Environ Health (Suppl) 2:69–105, 1977
5. Committee on Toxicology of the National Research Council: Health Effects of Benzene—A Review. US Department of Commerce, National Technical Information Service PB–254 388, pp 1–23. Washington, DC, National Academy of Sciences, 1976
6. Rappaport JM, Nathan DG: Acquired aplastic anemias: Pathophysiology and treatment. Adv Intern Med 27:547–590, 1982
7. Vigliani EC: Leukemia associated with benzene exposure. Ann NY Acad Sci 271:143–151, 1976
8. Infante PF, Rinsky RA, Wagoner JK, Young RJ: Leukemia Among Workers Exposed to Benzene. National Institute for Occupational Safety and Health (NIOSH), US Department

of Health, Education and Welfare. Washington, DC, US Government Printing Office, April 26, 1977
9. Aksoy M et al: Leukemia in shoe-workers exposed chronically to benzene. Blood 44:837–841, 1974
10. Aksoy M, Dincol K, Erden S, Dincol G: Acute leukemia due to chronic exposure to benzene. Am J Med 52:160–165, 1972
11. Yin SN et al: Leukaemia in benzene workers: A retrospective cohort study. Br J Ind Med 44:124–128, 1987
12. Rinsky RA et al: Leukemia in benzene workers. Am J Ind Med 2:217–245, 1981
13. Bond GG et al: An update of mortality among chemical workers exposed to benzene. Br J Ind Med 43:685–691, 1986
14. Aksoy M: Malignancies due to occupational exposure to benzene. Am J Ind Med 7:395–402, 1985
15. IARC Monographs on the Evaluation of the Carcinogenic Risk of Chemicals to Humans, Vol 29, Some Industrial Chemicals and Dyestuffs, pp 93–148. Lyon, International Agency for Research on Cancer, May 1982
16. Marcus WL: Chemical of current interest—benzene. Toxicol Ind Health 3:205–266, 1987
17. Rinsky RA et al: Benzene and leukemia. An epidemiologic risk assessment. N Engl J Med 316:1044–1050, 1987
18. Infante PF, White MC: Benzene: Epidemiologic observations of leukemia by cell type and adverse health effects associated with low-level exposure. Environ Health Perspect 52:75–82, 1983
18a. Occupational Safety and Health Administration: Occupational exposure to benzene; final rule, 29 CFR Part 1910. Federal Register 52(176):34460, 1987
19. Snyder CA et al: The inhalation toxicology of benzene: Incidence of hematopoietic neoplasms and hematotoxicity in AKR/J and C57BL/6J mice. Toxicol Appl Pharmacol 54:323–331, 1980
20. Maltoni C et al: Benzene: A multipotential carcinogen. Results of long-term bioassays performed at the Bologna Institute of Oncology. Am J Ind Med 4:589–630, 1983
21. National Toxicology Program: Toxicology and Carcinogenesis Studies of Benzene in F344N Rats and B6C3F1 Mice (Gavage Studies), pp 1–277. National Cancer Institute, NTP Technical Report 289, 1986
22. Tough IM, Brown WM: Chromosome aberrations and exposure to ambient benzene. Lancet 1:684, 1965
23. Styles JA, Richardson CR: Cytogenetic effects of benzene: Dosimetric studies on rats exposed to benzene vapor. Mutat Res 135:203–209, 1984
24. Erixson GL et al: Induction of sister chromatid exchanges and micronuclei in male DBA/2 mice by the inhalation of benzene (abstr). Environ Mutagen 6:408, 1984
25. Rosenthal GJ, Synder CA: Inhaled benzene reduces aspects of cell-mediated tumor surveillance in mice. Toxicol Appl Pharmacol 88:35–43, 1987
26. Kuna RA, Kapp RW Jr: The embryotoxic/teratogenic potential of benzene vapor in rats. Toxicol Appl Pharmacol 57:1–7, 1981
27. Walkley JE, Pagnotto LD, Elkins HB: The measurement of phenol in urine as an index of benzene exposure. Am Ind Hyg Assoc J 22:362–367, 1961
28. Susten AS et al: Percutaneous penetration of benzene in hairless mice: An estimate of dermal absorption during tire building operations. Am J Ind Med 7:323–335, 1985

BENZIDINE

CAS: 92-87-5

$NH_2C_6H_4C_6H_4NH_2$ 1987 TLV = none; recognized human carcinogen; skin

Synonyms: 4,4′-Biphenyldiamine; 4,4′-diaminobiphenyl

Physical Form. Colorless, crystalline compound; darkens on oxidation

Uses. Manufacture of dyestuffs; hardener for rubber; laboratory reagent

Exposure. Skin absorption; inhalation

Toxicology. Benzidine exposure is associated with a high incidence of bladder cancer in humans.

Of 25 workers involved in benzidine manufacture, 13 developed urinary bladder tumors, and four renal tumors also occurred. The average duration of exposure was 13.6 years, and the average induction time from first exposure to detection of the first tumor was 16.6 years. Initial tumors made their appearance as late as 9 years after cessation of exposure. It is not known whether the cancers were influenced by concurrent exposure to other chemicals in the occupational environment.[1]

In a plant that manufactured β-naphthylamine and benzidine, a cohort of 639 male employees with exposure from 1938 or 1939 to 1965 was studied; concentration of initial exposure, duration of exposure, and years of survival after the exposure are factors that affect the incidence of tumor formation.[5] Thirty-five per cent of all malignant neoplasms were of the bladder and kidney.[2] The observed mortality rate for cancer of the bladder was 78 per 100,000 in the cohort, compared with 4.4 per 100,000 expected for men of the same age (78 ÷ 4.4 = 17.7).[2] Of 42 bladder and kidney neoplasms, 16 were attributed to benzidine exposure, and 18 were attributed to combined exposure.[2]

During a 17-year period, 83 workers in a benzidine department were examined cystoscopically; 34 workers had congestive lesions, 3 had pedunculated papillomas, 4 had sessile tumors, and carcinoma was found in 13 of the workers.[3]

The onset of occupational bladder tumors is insidious, and, occasionally, the disease may be in an advanced stage before any signs or symptoms appear. Recurrences are frequent, and tumors may recur as papillomas or carcinomas irrespective of the nature of the original lesion.[4]

Because of the high incidence of bladder tumors among workers with benzidine exposure, contact by any route should be avoided.[6,7]

REFERENCES

1. Zavon MR, Hoegg U, Bingham E: Benzidine exposure as a cause of bladder tumors. Arch Environ Health 27:1–7, 1973
2. Mancuso TF, El-Attar AA: Cohort study of workers exposed to *beta*-naphthylamine and benzidine. J Occup Med 9:277–285, 1967
3. Barsotti M, Vigliani EC: Bladder lesions from aromatic amines. AMA Arch Ind Hyg Occup Med 5:234–241, 1952
4. Scott TS, Williams MHC: The control of industrial bladder tumours. Br J Ind Med 14:150–163, 1957
5. Rye WA, Woolrich PF, Zanes RP: Facts and myths concerning aromatic diamine curing agents. J Occup Med 12:211–215, 1970
6. Haley TJ: Benzidine revisited: A review of the literature and problems associated with the use of benzidine and its congeners. Clin Toxicol 8:13–42, 1975
7. Benzidine. Documentation of TLVs and BEIs, 5th ed, p 53. Cincinnati, American Conference of Governmental Industrial Hygienists (ACGIH), 1986

BENZOYL PEROXIDE

CAS: 94-36-0

$(C_6H_5CO)_2O_2$ 1987 TLV = 5 mg/m^3

Synonyms: Benzoyl superoxide; dibenzoyl peroxide; lucidol; oxylite

Physical Form. Granular solid

Uses. Bleaching flour and edible oils; additive in self-curing of plastics

Exposure. Inhalation

Toxicology. Benzoyl peroxide is an irritant of mucous membranes and causes both primary irritation and sensitization dermatitis.

Exposure of workers to levels of 12.2 mg/m^3 and higher has caused pronounced irritation of the nose and throat.[1]

Application to the face as lotion for acne treatment in two persons caused facial erythema and edema; patch tests with benzoyl peroxide were positive.[2] In contact with the eyes, it may produce irritation, and, if allowed to remain on the skin, it may produce inflammation.[3] No systemic effects have been reported in humans. The major hazards of benzoyl peroxide are fires and explosions, which have caused serious injuries and death.[4]

Rats exposed at an atmospheric concentration of 24.3 mg/liter of 78% benzoyl peroxide showed the following signs during a 4-hour exposure period: eye squint, difficulty in breathing, salivation, lacrimation, erythema, and an increase followed by a decrease in motor activity.[4] All rats appeared normal at 24 and 48 hours postexposure.

Benzoyl peroxide has been tested for carcinogenicity in mice and rats by oral administration in the diet and by subcutaneous administration, and in mice by skin application.[5] Although no significant increases in tumor incidences were found, the IARC has determined that all the studies were inadequate for a complete evaluation of carcinogenicity. Two studies indicated that benzoyl peroxide may act as a cancer promoter on mouse skin.[6,7]

Among a small factory population, two cases of lung cancer were found in men primarily involved in the production of benzoyl peroxide, but they were also exposed to benzoyl chloride and benzotrichloride.[8] No evaluation of the carcinogenicity of benzoyl peroxide to humans can be made according to IARC.[5]

REFERENCES

1. Benzoyl peroxide. Documentation of the TLVs and BEIs, 5th ed, p 54. Cincinnati, American Conference of Govermental Industrial Hygienists (ACGIH), 1986
2. Eaglstein WH: Allergic contact dermatitis to benzoyl peroxide—report of cases. Arch Dermatol 97:527, 1968
3. Chemical Safety Data Sheet SD–81, Benzoyl Peroxide, pp 3–4, 10. Washington, DC, MCA, Inc, 1960
4. National Institute for Occupational Safety and Health: Criteria for a Recommended Standard . . . Occupational Exposure to Benzoyl Peroxide. DHEW (NIOSH) Pub No 77–166, p 117. Washington, DC, US Government Printing Office, June 1977
5. IARC Monographs on the Evaluation of the Carcinogenic Risks of Chemicals to Humans, Vol 36, Allyl Compounds, Aldehydes, Epoxides and Peroxides, pp 267–283. Lyon, International Agency for Research on Cancer, 1985
6. Reiners JJ Jr et al: Murine susceptibility to two-stage skin carcinogenesis is influenced by the agent used for promotion. Carcinogenesis 5:301–307, 1984
7. Slaga TJ et al: Skin tumor—promoting activity of benzoyl peroxide, a widely used free radical-generating compound. Science 213:1023–1025, 1981
8. Sakabe H, Fukuda K: An updating report on cancer among benzoyl chloride manufacturing workers. Ind Health 15:173–174, 1977

BENZYL CHLORIDE

CAS: 100-44-7

$C_6H_5CH_2Cl$ 1987 TLV = 1 ppm

Synonyms: α-Chlorotoluene; ω-chlorotoluene; (chloromethyl) benzene

Physical Form. Colorless liquid

Uses. In manufacture of benzyl compounds, cosmetics, dyes, resins

Exposure. Inhalation

Toxicology. Benzyl chloride is a severe irritant of the eyes, mucous membranes, and skin.

Benzyl chloride is a powerful lacrimator (an immediate warning sign), and, at 31 ppm, it is unbearably irritating to the eyes and nose.[1] At 16 ppm, it is intolerable after 1 minute. Workers exposed to 2 ppm complained of weakness, irritability, and persistent headache.[2]

One author reported disturbances of liver function and mild leukopenia in some exposed workers, but this has not been confirmed.[2]

Splashes of the liquid in the eye will

produce severe irritation and will lead to corneal injury; skin contact may produce dermatitis. Skin sensitization has been reported in guinea pigs.[3]

The LC_{50}s in mice and rats for a 2-hour inhalation exposure are 80 and 150 ppm, respectively.[4] In another investigation, it was found that all mice and rats survived 400 ppm for 1 hour.[5] Cats exposed to 100 ppm 8 hours/day for 6 days exhibited eye and respiratory tract irritation, which appeared sooner and with increasing severity each exposure day.[6]

While repeated high-dose subcutaneous injections produced local tumors in some animals, the carcinogenic potential has not been determined in humans.[6,7] Evidence of efficient detoxification mechanisms suggests that the risk from chronic low-level exposure is small.[6]

REFERENCES

1. Smyth HF Jr: Hygienic standards for daily inhalation. Am Ind Hyg Assoc Q 17:147, 1956
2. Mikhailova TV: Benzyl chloride. In International Labour Office: Encyclopaedia of Occupational Health and Safety, Vol I, A–K, pp 169–170. New York, McGraw-Hill, 1971
3. Landsteiner K, Jacobs J: Studies on the sensitization of animals with simple chemical compounds—II. J Exp Med 64:625–639, 1936
4. Mikhailova TV: Comparative toxicity of chloride derivatives of toluene—benzyl chloride, benzal chloride and benzotrichloride. Fed Proc 24:T877–880, 1965
5. Back KC et al: Reclassification of Materials Listed as Transportation Health Hazards, Report TSA-20-72-3, pp 24-25, A-264 to A-265. Washington, DC, Department of Transportation, Office of Hazardous Materials, Office of Assistant Secretary for Safety and Consumer Affairs, 1972
6. National Institute for Occupational Safety and Health: Criteria for a Recommended Standard . . . Occupational Exposure to Benzyl Chloride. DHEW (NIOSH) Pub No 78-182, p 92. Washington, DC, US Government Printing Office, 1978
7. Preussman R: Direct alkylating agents as carcinogens. Food Cosmet Toxicol 6:576–577, 1968

BERYLLIUM (and beryllium compounds)

CAS: 7440-41-7

Be 1987 TLV = 0.002 mg/m^3

Synonyms: Glucinium; beryllium oxide; beryllium salts

Physical Form. Hard, noncorrosible grey metal

Uses. Beryllium metal sheet or wire; ceramics; hardening agent in alloys used especially in the electronics field

Exposure. Inhalation

Toxicology. Exposure to compounds of beryllium may cause dermatitis, acute pneumonitis, and chronic pulmonary granulomatosis (berylliosis) in humans. The compounds are carcinogenic in experimental animals.

Acute lung disease, now chiefly of historical importance owing to improved working conditions, has resulted from brief exposures to high concentrations of the oxide, or phosphor mixtures, or the acid salts.[1] All segments of the respiratory tract may be involved, with rhinitis, pharyngitis, tracheobronchitis, and pneumonitis.[2] The pneumonitis may be fulminating following high exposure levels, or it may be less severe, with gradual onset, from lesser exposures.[3] In the majority of cases with acute beryllium pneumonitis, recovery occurs within 1 to 6 months; however, fatalities as a result of pulmonary edema or spontaneous pneumothorax have been reported.The human threshold of an injurious concentration by inhalation is approximately 30 mg Be/m^3 for the high fired oxide, 1 to 3 mg Be/m^3 for the low fired oxide, and 0.1 to 0.5 mg Be/m^3 for the sulfate.[2]

Beryllium disease is regarded as chronic if it persists for a year or more and is usually due to granulomas in the lungs.[1,4] The onset of berylliosis may be insidious, with only slight cough and fatigue, which can occur as early as 1 year or as late as 25 years following exposure.[2] Progressive pulmonary insufficiency, anorexia, weight loss, weakness, chest pain, and constant hacking cough characterize the advanced disease. Cyanosis and clubbing of fingers may be seen in approximately one third of cases, and cor pulmonale is another frequent sequela.[2]

Early x-rays show a fine diffuse granularity in the lungs, whereas a diffuse reticular pattern is observed in the second stage; finally, in the third stage, distinct nodules appear.[5]

There are many similarities between berylliosis and sarcoidosis, but, in sarcoidosis, the systemic effects are much more pronounced.[6]

An immunologic basis for chronic beryllium disease has been postulated, and a hypersensitivity phenomenon has been demonstrated.[4,6] Consistent with the concept of chronic berylliosis as a form of hypersensitivity disease are the following factors: Persons with berylliosis also show delayed cutaneous hypersensitivity reactions to beryllium compounds; their peripheral blood lymphocytes undergo blast transformation and release of macrophage inhibition factor after exposure to beryllium in vitro; and there is lack of a dose–response relationship in chronic beryllium cases.[2,4] Hypersensitization may lead to berylliosis in people with relatively low exposures, whereas nonsensitized persons with higher exposures may have no effects.

Skin contact with soluble beryllium salts may produce either primary irritation or sensitization dermatitis characterized by pruritis with an eruption of an erythematous, a papular, or a papulovesicular nature; the eruption usually subsides within 2 weeks after cessation of exposure.[1] Accidental implantation of beryllium or its compounds beneath the skin may cause necrosis of adjacent tissue and formation of an ulcer; implantation of comparatively insoluble compounds may produce a localized granuloma, as has occurred from lacerations with old fluorescent tubes containing the phosphor.[3] Healing of ulcers and granulomas requires the surgical removal of the beryllium substance.[3] Conjunctivitis may accompany contact dermatitis resulting from exposure to soluble beryllium compounds; angioneurotic edema may be striking.[1,3]

Beryllium metal, beryllium–aluminum alloy, beryl ore, beryllium chloride, beryllium fluoride, beryllium hydroxide, beryllium sulfate, and beryllium oxide all produce lung tumors in rats exposed by inhalation or intratracheally.[7] The oxide and the sulfate produce lung tumors in monkeys after intrabronchial implantation or inhalation. A number of compounds produce osteosarcomas in rabbits following their intravenous or intramedullary administration.[7]

Although a number of epidemiologic studies have reported an increased risk of lung cancer among occupationally exposed beryllium workers, deficiencies in the studies limit any unequivocal conclusion.[7–9] Specific criticisms include no consideration of latent effects, of smoking history, or of exposure to other potential carcinogens and underestimation of expected lung cancer deaths in comparison populations.[10]

The IARC has determined that there is sufficient evidence of carcinogenicity in animals and limited evidence of carcinogenicity in humans.[7]

Diagnosis. Signs and symptoms include conjunctivitis; periorbital edema; nasopharyngitis, tracheobronchitis;

pneumonitis; pulmonary granulomatosis with cough, dyspnea, weakness and marked weight loss; hepatomegaly; contact dermatitis; ulcers; and granulomata of skin.

Differential Diagnosis: Chronic beryllium disease must be differentiated from sarcoidosis; in beryllium disease, there is neither lymph node nor ocular involvement.

Special Tests: Diagnostic studies should include electrocardiogram, sputum gram stain and culture, differential white blood count, and arterial blood gas analysis. Urinalysis for beryllium has shown little quantitative correlation with either exposure to beryllium or with clinical findings; the positive identification of beryllium in urine indicates only that exposure to beryllium has occurred; the test is seldom used. Beryllium assay of tissue, the seat of typical histologic evidence of beryllium effects, may be undertaken; the presence of the element indicates exposure, although the concentration often is not in direct proportion to the severity of the disease.

Treatment. If a solution of a salt is splashed on the skin, immediately flush with water. Cuts or puncture wounds, where beryllium or its compounds may be embedded under the skin, should be thoroughly cleansed immediately by a physician. Any implanted beryllium must be excised. Steroid therapy should be considered in the case of either acute or chronic beryllium disease.

If inhaled in a massive exposure, immediate hospitalization and observation for 72 hours for delayed onset of pneumonitis is advisable.

REFERENCES

1. Tepper LB, Hardy HL, Chamberlin RI: Toxicity of Beryllium Compounds, pp 31–80. New York, Elsevier, 1961
2. Reeves AL: Beryllium. In Friberg L et al (eds): Handbook on the Toxicology of Metals, 2nd ed, Vol II, pp 95–116. New York, Elsevier/ North Holland Biomedical Press, 1986
3. Hygienic Guide Series: Beryllium and its compounds. Am Ind Hyg Assoc J 25:614–617, 1964
4. Deodhar SD, Barna B, Van Ordstrand HS: A study of the immunologic aspects of chronic berylliosis. Chest 63:309–313, 1973
5. Hardy HL, Tabershaw IR: Delayed chemical pneumonitis occurring in workers exposed to beryllium compounds. J Ind Hyg Toxicol 28:197–211, 1946
6. National Institute for Occupational Safety and Health: Criteria for a Recommended Standard . . . Occupational Exposure to Beryllium. DHEW (NIOSH) Pub No 72–10268. Washington, DC, US Government Printing Office, 1972
7. IARC Monographs on the Evaluation of the Carcinogenic Risk of Chemicals to Humans. Chemicals, Industrial Processes and Industries Associated with Cancer in Humans, Suppl 4, pp 60–62. Lyon, International Agency for Research on Cancer, 1982
8. Mancuso TF: Mortality study of beryllium industry workers' occupational lung cancer. Environ Res 21:48–55, 1980
9. Wagoner JK et al: Beryllium: An etiologic agent in the induction of lung cancer, nonneoplastic respiratory disease, heart disease among industrially exposed workers. Environ Res 21:15–34, 1980
10. Smith RJ: Beryllium Report Disputed by Listed Author: A standing controversy over an NIOSH study is revived by one of the scientists involved. Science 211:556–557, 1981

BIPHENYL

CAS: 92-52-4

$C_6H_5C_6H_5$ 1987 TLV = 0.2 ppm

Synonyms: Diphenyl; phenylbenzene; bibenzene; 1,1–biphenyl; PhPh

Physical Form. Colorless to yellow solid

Uses. Heat transfer agent; fungistat for citrus fruits; in organic synthesis

Exposure. Inhalation; skin absorption

Toxicology. Biphenyl is an irritant of the eyes and mucous membranes.

In a study of 33 workers in one plant with prolonged exposure to concentrations ranging up to 123 mg/m^3, the most common complaints were headache, gastrointestinal symptoms (diffuse pain, nausea, indigestion), numbness and aching of limbs, and general fatigue.[1] Neurophysiologic examination of 22 of these workers showed that 19 had changes consistent with central and/or peripheral nervous system damage. In one fatal case in this plant, exposure was high for 11 years, symptoms were as just described, and, at autopsy, there was widespread liver necrosis with some cirrhotic areas, nephrotic changes, heart muscle degeneration, and edematous brain tissue.[1]

Exposure of rats to biphenyl dust impregnated on diatomaceous earth at a concentration of 300 mg/m^3 for 7 hours/day for 64 days caused irritation of the nasal mucosa, bronchopulmonary lesions, and slight injury to the liver and kidneys.[2]

REFERENCES

1. Hakkinen I et al: Diphenyl poisoning in fruit paper production. Arch Environ Health 26:70–74, 1973
2. Deichmann WB, Kitzmiller KV, Dierker M, Witherup S: Observations on the effects of diphenyl, *o*- and *p*-Aminodiphenyl, *o*- and *p*-nitrodiphenyl and dihydroxyoctachlorodiphenyl upon experimental animals. J Ind Hyg Toxicol 29:1–3, 1947

BISMUTH TELLURIDE

CAS: 1304-82-1

Bi_2Te_3

1987 TLV = 10 mg/m^3—undoped
5 mg/m^3 as Bi_2Te_3—doped

Synonyms: None

Physical Form. Gray solid

Uses. Semiconductors; thermoelectric cooling; power generation application; for commercial use, Bi_2Te_3 is "doped" with selenium sulfide to alter its conductivity.

Exposure. Inhalation

Toxicology. Bismuth telluride, either alone or doped with selenium sulfide, is apparently of very low toxicity.

In limited industrial experimental work with Bi_2Te_3 under controlled conditions (vacuum hoods), no adverse health effects were encountered other than tellurium breath.[1]

In a multispecies study, dogs, rabbits, and rats were exposed to 15 mg/m^3 of Bi_2Te_3 doped with Bi_2Se_3 and SnTe for 6 hours/day, 5 days/week, for 1 year.[1] Small granulomatous lesions without fibrosis occurred in the lungs of dogs at 6 months. In dogs autopsied 4 months after an 8-month exposure, the lesions had regressed, indicating a reversible process. Rabbits showed a similar reaction, but with a decreased number of pulmonary macrophages, no fibrous tissue activity, and no cellular or fibrous tissue reaction around the dust deposits in the lymph nodes. The rats exhibited no fibrosis and no lymph node reactions. The pulmonary lesions seen in the study were present in all three exposed species but were interpreted as mild and reversible and not of serious physiologic consequence.

In a similar 11-month study in which animals were exposed to undoped bismuth telluride dust of 0.4 μm diameter at 15 mg/m^3, no adverse responses of any type were observed other than the pulmonary responses to the inhalation of an inert dust.

REFERENCES

1. Stokinger HE: The metals. In Clayton GD, Clayton FE (eds): Patty's Industrial Hygiene and Toxicology, 3rd ed, rev, Vol 2A, Toxicology, pp 1558–1563. New York, Wiley–Interscience, 1981

BORON OXIDE

CAS: 1303-86-2

B_2O_3 1987 TLV = 10 mg/m^3

Synonyms: Boric anhydride; boron sesquioxide; boron trioxide; fused boric acid

Physical Form. Vitreous colorless crystals

Uses. In preparation of fluxes; component of enamels and glass; catalyst in organic reaction

Exposure. Inhalation

Toxicology. Boron oxide is an eye and respiratory mucosa irritant.

In 113 workers exposed to boron oxide and boric acid dusts, there were statistically significant increases in symptoms of eye irritation; dryness of the mouth, nose, and throat; sore throat; and productive cough as compared with controls.[1] The mean exposure level was 4.1 mg/m^3, with a range of 1.2 to 8.5 mg/m^3. Exposures may occasionally have exceeded 10 mg/m^3. Because of mixed exposures, the study does not indicate whether boron oxide or boric acid dust is more important in causing symptoms, nor does it indicate the minimal duration of exposure necessary to produce symptoms.

Repeated exposure of rats to an aerosol at a concentration of 470 mg/m^3 for 10 weeks caused only mild nasal irritation; repeated exposure of rats to 77 mg/m^3 for 23 weeks resulted in elevated creatinine and boron content of the urine in addition to increased urinary volume.[2] Conjunctivitis resulted when the dust was applied to the eyes of rabbits, probably the result of the exothermic reaction of boron oxide with water to form boric acid; topical application of boron oxide dust to the clipped backs of rabbits produced erythema that persisted for 2 to 3 days.[2]

REFERENCES

1. Garabrant DH et al: Respiratory and eye irritation from boron oxide and boric acid dusts. J Occup Med 26:584–586, 1984
2. Wilding JL et al: The toxicity of boron oxide. Am Ind Hyg Assoc J 20:284–289, 1959

BORON TRIFLUORIDE

CAS: 7637-07-2

BF_3 1987 Ceiling Limit = 1 ppm

Synonyms: Boron fluoride

Physical Form. Colorless gas; forms dense white fume in moist air

Uses. In catalysis with and without promoting agents; fumigant; flux for soldering magnesium

Exposure. Inhalation

Toxicology. Boron trifluoride gas is a severe irritant of the lungs, eyes, and skin.

Of 13 workers with present or past occupational exposure, examination revealed that 8 had abnormalities of pulmonary function; chest x-rays were negative, and preshift urinary fluoride concentrations did not exceed 4 mg/liter.[1] Air sampling showed concentrations ranging from 0.1 to 1.8 ppm. Dryness of the nasal mucosa and epistaxis were attributed to boron trifluoride exposure in workers exposed to high concentrations for 10 to 15 years.[2]

Cotton soaked with boron trifluoride in water and placed on the skin for a day or so resulted in a typical acid burn; there was no evidence of the more severe hydrogen fluoride burn occurring, because boron trifluoride has the ability to complex the fluoride ion effectively.[1]

In rats, the 4-hour LC_{50} was 436 ppm for boron trifluoride dihydrate, which is formed when boron trifluoride gas reacts with moisture. Clinical signs included gasping, excessive oral and nasal discharge, and lacrimation.[3] In a 2-week study, all animals exposed at 67 ppm, 6 hours/day, died prior to the sixth exposure, and histopathology showed necrosis and pyknosis of the proximal tubular epithelium of the kidneys; at 24 ppm and 9 ppm, signs of respiratory irritation, depression of body weight, increased lung weights, and depressed liver weights were observed. Repeated exposure for 13 weeks at 6 ppm, 6 hours/day, 5 days/week, resulted in renal toxicity for 2 of 40 rats; although clinical signs of respiratory irritation were seen, morphologic examination showed no evidence of damage. The same 13-week exposure regime at 2 ppm caused elevation of urinary, serum, and bone fluoride levels but did not result in a toxic response.[3] Guinea pigs and rats showed pneumonitis and congestion in the lungs after a 6-month exposure to a calculated concentration of 3.0 ppm (1.5 ppm by analysis), and a 4-month exposure at 1.0 ppm caused reversible tracheitis and bronchitis.[1,4]

Boron trifluoride combines with atmospheric moisture to form a white mist containing hydration and hydrolysis products.[1] The odor is detectable at 3.0 ppm, but this does not serve as an adequate warning.[1]

REFERENCES

1. National Institute for Occupational Safety and Health: Criteria for a Recommended Standard . . . Occupational Exposure to Boron Trifluoride. DHEW (NIOSH) Pub No 77–122, p 83. Washington, DC, US Government Printing Office, 1976
2. Kasparov AA: Boron trifluoride. In International Labor Office: Encyclopaedia of Occupational Health and Safety, Vol I, pp 204–205. New York, McGraw-Hill, 1974
3. Rusch GM et al: Inhalation toxicity studies with boron trifluoride. Toxicol Appl Pharmacol 83:69–78, 1986
4. Torkelson TR, Sadek SE, Rowe VK: The toxicity of boron trifluoride when inhaled by laboratory animals. Am Ind Hyg Assoc J 22:263, 1961

BROMINE

CAS: 7726-95-6

Br_2 — 1987 TLV = 0.1 ppm

Synonyms: None

Physical Form. Dark reddish-brown, fuming, volatile liquid

Uses. Antiknock compounds for gasoline; fire retardants; sanitation preparations

Exposure. Inhalation

Toxicology. Bromine is a severe irritant of the eyes, mucous membranes, lungs, and skin.

In humans, 10 ppm is intolerable, causing severe irritation of the upper respiratory tract; lacrimation occurs at levels below 1 ppm.[1] Symptoms and signs in humans also include dizziness, headache, epistaxis, and cough, followed some hours later by abdominal pain, diarrhea, and, sometimes, a measles-like eruption on the face, trunk, and extremities.[2] Exposure at 40 to 60 ppm is thought to be dangerous for brief periods, and 1000 ppm may be rapidly fatal because of choking caused by edema of the glottis and pulmonary edema.[3] Pneumonia may be a late complication of severe exposures.[4] The liquid or concentrated vapor in contact with the eye will cause severe and painful burns.[4] Liquid bromine spilled on the skin causes a mild, cooling sensation on first contact, followed by a burning sensation.[4] If bromine is not removed from the skin immediately, deep surface burns result; a brown discoloration appears, leading to the development of deep-seated ulcers, which heal slowly.[4]

Nearly 50% of mice exposed at 240 ppm for 2 hours died within 30 days; at 750 ppm, a 7-minute exposure was lethal to 40% during the same follow-up period.[5]

REFERENCES

1. AIHA Hygienic Guide Series: Bromine. Akron, Ohio, American Industrial Hygiene Association, 1978
2. Stokinger HE: The halogens and the nonmetals boron and silicon. In Clayton GD, Clayton FE (eds): Patty's Industrial Hygiene and Toxicology, 3rd ed, rev, Vol 2B, Toxicology, pp 2965–2968. New York, Wiley–Interscience, 1981
3. Henderson Y, Haggard HW: Noxious Gases. New York, Reinhold, 1943
4. Chemical Safety Data Sheet SD–49, Bromine, pp 5, 16–18. Washington, DC, MCA, Inc, 1968
5. Bitron MD, Aharonson EF: Delayed mortality of mice following inhalation of acute doses of CH_2O, SO_2, CL_2 and Br_2. Am Ind Hyg Assoc J 39:129–138, 1978

BROMOFORM

CAS: 75-25-2

$CHBr_3$ 1987 TLV = 0.5 ppm; skin

Synonyms: Methenyl tribromide; tribromomethane

Physical Form. Colorless liquid

Uses. Synthesis of pharmaceuticals; used in shipbuilding and aircraft industries

Exposure. Inhalation; skin absorption; ingestion

Toxicology. Bromoform is a respiratory irritant and causes central nervous system depression.

Human exposure to the vapor causes irritation of the eyes and throat.[1] Accidental ingestion of the liquid has produced central nervous system depression with coma and loss of reflexes; smaller doses have led to listlessness, headache, and vertigo.[1] Chronic effects have not been reported from industrial exposure.

A saturated atmosphere of 7000 ppm or more produced anesthesia in dogs in 8 minutes and death in 60 minutes.[2]

Chronic exposure of rats at 25 ppm 4 hours/day for 2 months caused unspecified adverse effects in the liver and kidneys.[3]

The undiluted liquid was moderately irritating to rabbit eyes, but healing was complete in 1 to 2 days. Repeated skin contact caused moderate irritation to rabbit skin.[3]

Bromoform has a chloroform-like odor.[1]

REFERENCES

1. von Oettingen WF: The Halogenated Aliphatic, Olefinic, Cyclic, Aromatic, and Aliphatic-Aromatic Hydrocarbons, Including the Halogenated Insecticides, Their Toxicity and Potential Dangers. US Public Health Service Pub No 414, pp 65–67. Washington, DC, US Government Printing Office, 1955
2. Irish DD: Aliphatic halogenated hydrocarbons. In Fassett DW, Irish DD (eds): In Patty FA (ed): Industrial Hygiene and Toxicology, 2nd ed, Vol 2, Toxicology, pp 1262–1263. New York, Interscience, 1963
3. Torkelson TR, Rowe VK: Halogenated aliphatic hydrocarbons. In Clayton GD, Clayton FE (eds): Patty's Industrial Hygiene and Toxicology, 3rd ed, rev, Vol 2B, Toxicology, pp 3469–3470. New York, Wiley–Interscience, 1981

1,3–BUTADIENE

CAS: 106-99-0

C_4H_6 1987 TLV = 10 ppm

Synonyms: Butadiene; divinyl; biethylene

Physical Form. Colorless gas

Uses. Manufacture of synthetic rubber

Exposure. Inhalation

Toxicology. Butadiene is an irritant of the eyes and mucous membranes; at extremely high concentration, it causes narcosis in animals, and severe exposure is expected to produce the same effect in humans. It is carcinogenic in experimental animals.

Human subjects tolerated 4000 ppm for 6 hours without apparent effect other than slight irritation of the eyes; tolerance to higher exposures appears to develop following a single exposure of 1,3–butadiene.[1] Exposure of two human volunteers to 8000 ppm for 8 hours caused eye and upper respiratory tract irritation.[1]

Dermatitis and frostbite may result from exposure to liquid and evaporating 1,3–butadiene.[2]

Deep anesthesia in rabbits was induced after 8 to 10 minutes at 200,000 to 250,000 ppm, and death occurred in 23 minutes at 250,000 ppm.[1] Recovery from brief periods of anesthesia occurred within 2 minutes of terminating exposure; no tissue changes were detectable microscopically after daily induction of anesthesia for as many as 34 times.[1] Daily exposure of various small animal species at 6700 ppm over 8 months resulted in no significant chronic effects.[1]

Exposure of mice to 625 or 1250 ppm 6 hours/day, 5 days/week, caused early deaths primarily from malignant neoplasms involving multiple organs.[3] At the end of 61 weeks, there were tumors in 20% of control males and in 12% of control females, compared with 80% and 94%, respectively, of the exposed mice. The most common tumors were malignant lymphomas, heart hemangiosarcomas, and alveolar–bronchiolar neoplasms. Non-neoplastic effects associated with these exposures included testicular and ovarian atrophy and nasal cavity lesions.

Chronic exposure of rats for 2 years 6 hours/day, 5 days/week, to 1000 or 8000 ppm caused a significant increase in neoplasms of the mammary gland, thyroid, uterus, and zymbal glands of exposed females and in neoplasms of the testes and pancreas (8000 ppm only) in exposed males.[4]

Pregnant rats exposed to 200, 1000, or 8000 ppm during days 6 to 15 of gestation had depressed body weight gain.[5] At the 8000 ppm exposure level, fetal growth was significantly retarded, and a significant increase in major skeletal abnormalities was observed.

No statistically significant excess in total or cause-specific mortality was found in 2756 styrene–butadiene rubber workers.[6] The average length of employment was 10 years, and butadiene exposures ranged from 0.11 to 174 ppm. In one plant group, there was a small excess of lymphatic and hematopoietic cancers, which was not statistically significant. In another study of nearly 14,000 workers, there was a low overall mortality rate, with a slight but statistically insignificant increase in cancers of the digestive system, kidney, lymph nodes, and larynx.[7]

NIOSH recommends that 1,3–butadiene be considered a potential occupational carcinogen and a possible reproductive hazard.[8]

REFERENCES

1. Carpenter CP, Shaffer CB, Weil CS, Smyth WJ Jr: Studies on the inhalation of 1,3–butadiene; with a comparison of its narcotic effect with benzol, toluol and styrene, and a note on the elimination of styrene by the human. J Ind Hyg Toxicol 26:69–78, 1943
2. Weaver NK: Encyclopedia of Occupational Health and Safety, 3rd ed, rev, Vol 1, A–K, 1,3–Butadiene, pp 347–348. Geneva, International Labor Office, 1982

3. Huff JE et al: Multiple organ carcinogenicity of 1,3–butadiene in B6C3F_1 mice after 60 weeks of inhalation exposure. Science 227:548–549, 1985
4. Owen PE: The Toxicity and Carcinogenicity of Butadiene Gas Administered to Rats by Inhalation for Approximately 24 Months. Unpublished report submitted to The International Institute of Synthetic Rubber Producers, Inc, by Hazleton Laboratories Ltd, Harrogate, England, Nov 1981 (cited by NIOSH Current Intelligence Bulletin 41, Feb 9, 1984)
5. Owen PE, Irvine LFH: 1,3–Butadiene: Inhalation Teratogenicity Study in the Rat. Final Report. Unpublished report submitted to The International Institute of Synthetic Rubber Producers, Inc, by Hazleton Laboratories Ltd, Harrogate, England, Nov 1981 (cited by NIOSH Current Intelligence Bulletin 41, Feb 9, 1984)
6. Meinhardt TJ et al: Environmental epidemiologic investigation of the styrene–butadiene rubber industry. Mortality patterns with discussion of the hematopoietic and lymphatic malignancies. Scand J Work Environ Health 8:250–259, 1982
7. Matanoski GM et al: Mortality of Workers in the Styrene–Butadiene Rubber Polymer Manufacturing Industry. Unpublished Report submitted to The International Institute of Synthetic Rubber Producers, Inc, by John Hopkins University School of Hygiene and Public Health, Baltimore, Maryland, June 1982 (cited by NIOSH Current Intelligence Bulletin 41, Feb 9, 1984)
8. NIOSH Current Intelligence Bulletin 41, 1,3–Butadiene. DHHS (NIOSH) Pub No 84–105, pp 1–18. National Institute for Occupational Safety and Health, US Department of Health and Human Services. Washington, DC, US Government Printing Office, Feb 9, 1984

---

n-BUTYL ACETATE

CAS: 123-86-4

$CH_3(CH_2)_2C(O)OC_2H_5$

1987 TLV = 150 ppm

Synonyms: Butyl ethanoate; acetic acid, butyl ester

Physical Form. Colorless liquid

Uses. Solvent for nitrocellulose, oils, fats, resins, waxes, and camphor; manufacture of lacquer and plastics

Exposure. Inhalation

Toxicology. *n*-Butyl acetate causes irritation of mucous membranes and the eyes; at high concentrations, it causes narcosis in animals. It is expected that severe exposure will cause the same effect in humans.

In humans, *n*-butyl acetate affected the throat at 200 ppm; severe throat irritation occurred at 300 ppm, and the majority of the subjects also complained of eye and nose irritation.[1]

Guinea pigs exhibited signs of eye irritation at 3300 ppm; at 7000 ppm, there was narcosis within 700 minutes but no deaths following exposure for 810 minutes; 14,000 ppm was lethal after 4 hours.[2]

Cats exposed to 4200 ppm for 6 hours for 6 days showed weakness, loss of weight, and minor blood changes.[3]

REFERENCES

1. Nelson KW, Ege JF, Ross M, Woodman LE, Silverman L: Sensory response to certain industrial solvent vapors. J Ind Hyg Toxicol 25:282–285, 1943
2. Sayers RR et al: Acute response of guinea pigs to vapors of some new commercial organic compounds, XII: Normal butyl acetate. Public Health Rep 51:1229–1236, 1936
3. Sandmeyer EE, Kirwin CJ: Esters. In Clayton GD, Clayton FE (eds): Patty's Industrial Hygiene and Toxicology, 3rd ed, p 2273. New York, Wiley–Interscience, 1981

---

sec-BUTYL ACETATE

CAS: 105-46-4

$CH_3COOCH(CH_3)C_2H_5$

1987 TLV = 200 ppm

Synonyms: 2-Butanol acetate; acetic acid, secondary butyl ester

Physical Form. Liquid

Uses. In solvents, especially lacquer solvents; textile sizes, and paper coatings

Exposure. Inhalation

Toxicology. *sec*-Butyl acetate causes irritation of the eyes and respiratory tract. By analogy with chemically similar substances, it is considered a central nervous system depressant at very high concentrations.[1]

Skin irritation may occur. *sec*-Butyl acetate has not been studied regarding its toxicity, nor are there any reports concerning harmful effects on humans.[2]

The odor of *sec*-butyl acetate is milder than *n*-butyl acetate, and it appears less irritative to the eyes and respiratory tract.[1]

REFERENCES

1. *sec*-Butyl acetate. Documentation of the TLVs for Substances in Workroom Air, 4th ed, pp 52–53. Cincinnati, American Conference of Governmental Industrial Hygienists (ACGIH), 1984
2. von Oettingen WF: The aliphatic acids and their esters: Toxicity and potential dangers. AMA Arch Ind Health 21:28–65, 1960

tert-BUTYL ACETATE

CAS: 540-88-5

$CH_3COOC(CH_3)_3$ 1987 TLV = 200 ppm

Synonyms: Acetic acid, *tert*-butyl ester

Physical Form. Colorless liquid

Uses. Gasoline additive; lacquer solvent

Exposure. Inhalation

Toxicology. By analogy to other acetate esters, *tert*-butyl acetate is expected to cause irritation of the eyes and throat. It is considered a central nervous system depressant at very high concentrations.[1]

The toxicity of *tert*-butyl acetate has not been studied, nor are there any published reports concerning harmful effects on humans.[2] Comparative tests indicate it to be definitely less irritating to the throat than *n*-butyl acetate.[1]

REFERENCES

1. *tert*-Butyl acetate. Documentation of the TLVs and BEIs, 4th ed, p 53. Cincinnati, American Conference of Governmental Industrial Hygienists (ACGIH), 1984
2. von Oettingen WF: The aliphatic acids and their esters: Toxicity and potential dangers. AMA Arch Ind Health 21:28–65, 1960

n-BUTYL ACRYLATE

CAS: 141-32-2

$H_2C{=}CHC(O)O\text{-}(CH_2)_3\text{-}CH_3$

1987 TLV = 10 ppm

Synonyms. Acrylic acid butyl ester; 2–propenoic acid butyl ester

Physical Form. Colorless liquid with acrid odor; commercial form contains hydroquinone (1000 ppm) or hydroquinone methyl ether (15 or 200 ppm) to prevent polymerization

Uses. Manufacture of polymers and resins for textiles, paints, and leather finishes

Exposure. Inhalation; skin contact/absorption

Toxicology. Butyl acrylate is an irritant of the eyes and skin.

In an early range-finding study, exposure of rats to 1000 ppm for 4 hours was lethal to five of six animals.[1] In a more recent study, the LC_{50} for 4 hours in rats was 2730 ppm.[2] Behavior of the animals suggested irritation of the eyes, nose, and respiratory tract, with labored breathing. At necropsy, there were no discernible gross abnormalities of the major organs.

The rabbit dermal LD_{50} was on the order of 1800 mg/kg[3]. On the skin of rabbits, butyl acrylate was moderately irritating.[1] In the rabbit eye, the liquid produced corneal necrosis.

A woman with dermatitis from the plastic nose pads of her spectacle frames was found on patch testing to react to butyl acrylate 1% but not to ethyl or methyl acrylate.[4] The sensitization was attributed to butyl acrylate, which might have been present in the plastic nose pads.

Hamsters and rats were exposed to an average concentration of 817 and 820 ppm, respectively, for 4 days.[5] In both animal species, there were distinct signs of toxicity; 4 of 10 hamsters died during the exposure. The chromosome analysis carried out in the bone marrow after the exposure indicated no chromosome-damaging effects.

In a dermal carcinogenesis study, 25 μl of 1% butyl acrylate in acetone was applied three times weekly to mice for their lifetime.[6] No epidermal tumors were observed, and there was, therefore, no evidence of local carcinogenic activity.

In a reproductive study, inseminated rats were exposed to butyl acrylate at 0, 25, 135, and 250 ppm 6 hours/day from the 6th to the 15th day postcoitum.[6] During the inhalation period, the two high doses led to maternal toxicity, including signs of mucous membrane irritation. The same levels induced embryolethality, measured as an increased number of dead implantations. The 25 ppm level did not cause maternal toxicity or embryolethality. A teratogenic effect was not seen at any of the exposure levels.

REFERENCES

1. Smyth HF Jr, Carpenter CP, Weil CS: Range-finding toxicity data: List IV. AMA Arch Ind Hyg 4:119, 1951
2. Oberly R, Tansy MF: LC_{50} values for rats acutely exposed to vapors of acrylic and methacrylic acid esters. J Toxicol Environ Health 16:811, 1985
3. Carpenter CP et al: Range-finding toxicity data: List VIII. Toxicol Appl Pharmacol 28:313, 1974
4. Hambly EM, Wilkinson DS: Contact dermatitis to butyl acrylate in spectacle frames. Contact Dermatitis 4:115, 1978
5. Englehardt G, Klimisch JJ: *n*-Butyl acrylate monomer: Cytogenetic investigations in the bone marrow of Chinese hamsters and rats after 4-day inhalation. Fund Appl Toxicol 3:640, 1983
6. De Pass LR et al: Acrylic acid, ethyl acrylate, and butyl acrylate: Dermal oncogenicity bioassays of acrylic acid, ethyl acrylate, and butyl acrylate. J Toxicol Environ Health 14:115, 1984
7. Merkle J, Klimisch HJ: *n*-Butyl acrylate: Prenatal inhalation toxicity in the rat. Fund Appl Toxicol 3:443, 1983

---

n-BUTYL ALCOHOL

CAS: 71-36-3

C_4H_9OH — 1987 C = 50 ppm; skin

Synonyms: *n*-Butanol; butyric alcohol; propyl carbinol; butyl hydroxide, 1–butanol

Physical Form. Colorless liquid

Uses. Lacquer solvent; manufacture of plastics and rubber cements

Exposure. Inhalation; skin absorption

Toxicology. *n*-Butyl alcohol is an irritant of the eyes and mucous membranes and may cause central nervous system depression at very high concentrations.

Chronic exposure of humans to concentrations above 50 to 200 ppm causes irritation of the eyes with lacrimation, blurring of vision, and photophobia.[1,2]

In a 10-year study of workers exposed to average concentrations of 100 ppm, no systemic effects were observed.[1] Two studies have suggested mild hearing impairment in exposed workers compared with a control group, but the possible contribution of noise-induced hearing loss was not adequately investigated.[3] Contact dermatitis of the hands may occur owing to a defatting action of the liquid, and toxic amounts can be absorbed through the skin.[4] Direct contact of the hands with *n*-butyl alcohol for 1 hour results in an absorbed dose that is four times that of inhalation of 50 ppm for 1 hour.[4]

No effects were observed in mice exposed to 3300 ppm for 7 hours, whereas exposure to 6600 ppm produced prostration within 2 hours, narcosis after 3 hours, and some deaths.[4]

n-Butyl alcohol may be recognized by its pungent odor, resembling fuel oil. The odor threshold is approximately 15 ppm, but following adaptation, the threshold can increase to 10,000 ppm.[4]

REFERENCES

1. Sterner JI, Crouch HC, Brockmyre HF, Cusack M: A ten-year old study of butyl alcohol exposure. Am Ind Hyg Assoc J 10:53–59, 1949
2. Tabershaw IR, Fahy JP, Skinner JB: Industrial exposure to butanol. J Ind Hyg Toxicol 26:328–330, 1944
3. *n*-Butyl alcohol. Documentation of the TLVs for Substances in Workroom Air, 4th ed, p 54. Cincinnati, American Conference of Governmental Industrial Hygienists (ACGIH), 1984
4. Rowe VK, McCollister SB: Alcohols. In Clayton GD, Clayton FE (eds): Patty's Industrial Hygiene and Toxicology, 3rd ed, Vol 2C, Toxicology, pp 4571–4578. New York, Wiley–Interscience, 1982

sec-BUTYL ALCOHOL

CAS: 78-92-2

$CH_3CH_2CHOHCH_3$ 1987 TLV = 100 ppm

Synonyms: 2–Butanol; ethylmethyl carbinol; butylene hydrate; 2–hydroxybutane

Physical Form. Colorless liquid

Uses. Polishes, cleaning materials, paint removers, fruit essences, perfumes, and dyestuffs; synthesis of methyl ethyl ketone; lacquer solvent

Exposure. Inhalation

Toxicology. At high concentrations, *sec*-butyl alcohol causes narcosis in animals, and it is expected that severe exposure will produce the same effect in humans.

By analogy to *n*-butyl alcohol, it is regarded as an eye irritant, although no symptoms or signs were observed in workers exposed regularly to 100 ppm.[1] Heavy exposure reportedly causes eye, nose, and throat irritation; headache; nausea; fatigue; and dizziness.[2] Mild skin irritation may occur owing to defatting action, although it was not irritating when applied to rabbit skin.[3]

In mice, ataxia, prostration, and narcosis occurred at various times after exposure to concentrations ranging from 3,300 ppm to 19,800 ppm.[2] Exposure to 16,000 ppm for 4 hours was lethal to five of six rats.[3] When instilled directly into a rabbit eye, the liquid causes severe corneal injury.[3]

sec-Butyl alcohol has an odor similar to but less pungent than *n*-butyl alcohol. The malodorous and irritating properties probably prevent exposure to toxic levels.

REFERENCES

1. *sec*-Butyl alcohol. Documentation of the TLVs and BEIs, 5th ed, p 77. Cincinnati, American Conference of Governmental Industrial Hygienists (ACGIH), 1986
2. Rowe VK, McCollister SB: Alcohols. In Clayton GD, Clayton FE (eds): Patty's Industrial Hygiene and Toxicology, 3rd ed, Vol 2C, Toxicology, pp 4582–4585. New York, Wiley–Interscience, 1982
3. Smyth HF Jr et al: Range-finding toxicity data: List V. AMA Arch Ind Hyg Occup Med 20:61–68, 1954

tert-BUTYL ALCOHOL

CAS: 75-65-0

$(CH_3)_3COH$ 1987 TLV = 100 ppm

Synonyms: 2–Methyl–2–propanol; trimethyl carbinol

Physical Form. Colorless liquid

Uses. Plastics, lacquers, cellulose esters, fruit essences, perfumes, and chemical intermediates

Exposure. Inhalation

Toxicology. At high concentrations, *tert*-butyl alcohol causes narcosis in animals, and it is expected that severe exposure in humans will result in the same effect.

Heavy exposure may cause irritation of the eyes, nose, and throat; headache; nausea; fatigue; and dizziness.[1] Systemic effects have not been reported. Application of *tert*-butyl alcohol to human skin causes slight erythema and hyperemia.[1]

Signs of intoxication in rats were ataxia and narcosis; the oral LD_{50} was 3.5 g/kg.[2]

The malodorous quality and irritant effects of *tert*-butyl alcohol may prevent inadvertent exposure to toxic levels.

REFERENCES

1. Rowe VK, McCollister SB: Alcohols. In Clayton GD, Clayton FE (eds): Patty's Industrial Hygiene and Toxicology, 3rd ed, Vol 2C, Toxicology, pp 4585–4588. New York, Wiley–Interscience, 1982
2. Schaffarzick RW, Brown BJ: The anticonvulsant activity and toxicity of methyl parafynol (Vormison) and some other alcohols. Science 116:663–665, 1952

BUTYLAMINE

CAS: 109-73-9

$CH_3(CH_2)_3NH_2$

1987 C = 5 ppm; skin

Synonyms: 1–Aminobutane; *n*-butylamine

Physical Form. Colorless liquid

Uses. Intermediate for pharmaceuticals, dyestuffs, rubber chemicals, emulsifying agents, insecticides, synthetic tanning agents

Exposure. Inhalation; skin absorption

Toxicology. Butylamine is an irritant of the eyes, mucous membranes, and skin.

In humans, the liquid on the skin causes severe primary irritation and second-degree burns with vesiculation.[1] Workers exposed daily to 5 to 10 ppm complained of irritation of the nose, throat, and eyes and, in some instances, headache and flushing of the skin of the face.[1,2] Concentrations of 10 to 25 ppm are unpleasant and even intolerable to some subjects for exposure of more than a few minutes duration; daily exposures of workers to less than 5 ppm (usually 1 to 2 ppm) resulted in no symptoms.[1]

In rats exposed to 3000 to 5000 ppm, there was an immediate irritant response, followed by labored breathing, pulmonary edema, and death within minutes to hours.[1]

The oral LD_{50} for *n*-butylamine in rats was 372 mg/kg versus 228, 152, and 80 mg/kg for isobutylamine, *sec*-butylamine, and *tert*-butylamine, respectively.[3] Signs of toxicity included sedation, ataxia, nasal discharge, gasping, salivation, and death. Pathologic examination showed pulmonary edema.[3]

The liquid produced severe eye damage and skin burns in animals.[1]

REFERENCES

1. Beard RR, Noe JT: Aliphatic and alicyclic amines. In Clayton GD, Clayton FE (eds): Patty's Industrial Hygiene and Toxicology, 3rd ed, rev, Vol 2B, Toxicology, pp 3135–3155. New York, Wiley–Interscience, 1981
2. Hygienic Guide Series: *n*-Butylamine. Am Ind Hyg Assoc J 21:532–533, 1960
3. Cheever KL et al: Short communication. The acute oral toxicity of isomeric monobutylamines in the adult male and female rat. Toxicol Appl Pharmacol 63:150–152, 1982

n-BUTYL GLYCIDYL ETHER

CAS: 2426-08-6

$C_7H_{14}O_2$ 1987 TLV = 25 ppm

Synonyms: BGE; 1–*n*-butoxy–2,3–epoxypropane

Physical Form. Colorless liquid

Uses. Viscosity-reducing agent, acid acceptor for solvents, chemical intermediate

Exposure. Inhalation

Toxicology. *n*-Butyl glycidyl ether (BGE) causes central nervous system depression and is a mild irritant of the eyes and skin in animals; it is expected that severe exposure will cause the same effects in humans.

No chronic systemic effects have been reported in humans. However, sensitization dermatitis may occur with repeated skin contact.[1]

Intragastric and intraperitoneal injection of BGE in animals produced incoordination and ataxia followed by coma.[1] In rats exposed to graded vapor concentrations of BGE, effects were lacrimation, nasal irritation, and labored breathing. The LC_{50} for an 8-hour exposure in rats was 1030 ppm and greater than 3500 ppm for 4 hours in mice. At autopsy, pneumonitis was frequently observed. Three intramuscular injections of 400 mg/kg produced minimal toxic effects and a slight increase in leukocyte counts.[2] In male mice topically treated with 1.5 g/kg for 8 weeks and then mated, there was a significant increase in the number of fetal deaths as compared with controls.[3] BGE produced widely disparate degrees of skin irritation, ranging from very mild to severe, in tests by different investigators using similar methodology.[3] After a series of intracutaneous injections, 16 of 17 guinea pigs became sensitized.[4] The undiluted liquid in rabbit eyes caused mild eye irritation.[1]

REFERENCES

1. Hine CH et al: The toxicology of glycidol and some glycidyl ethers. AMA Arch Ind Health 14:250–264, 1956
2. Kodama JK et al: Some effects of epoxy compounds on the blood. Arch Environ Health 2:56–67, 1961
3. National Institute for Occupational Safety and Health: Criteria for a Recommended Standard . . . Occupational Exposure to Glycidyl Ethers. DHEW (NIOSH) Pub No 78–166. Washington, DC, US Government Printing Office, 1978
4. Weil CS et al: Experimental carcinogenicity and acute toxicity of representative epoxides. Am Ind Hyg Assoc J 24:305–325, 1963

n-BUTYL MERCAPTAN

CAS: 109-79-5

C_4H_9SH 1987 TLV = 0.5 ppm

Synonyms: 1–Butanethiol; thiobutyl alcohol; butyl sulfhydrate

Physical Form. Colorless liquid

Uses. Solvent; intermediate in the production of insecticides and herbicides; gas odorant

Exposure. Inhalation

Toxicology. *n*-Butyl mercaptan is a central nervous system depressant.

Accidental exposure of seven workers to concentrations estimated at between 50 and 500 ppm for 1 hour caused muscular weakness and malaise; six of the workers experienced sweating, nausea, vomiting, and headache; three experienced confusion, and one lapsed into a coma for 20 minutes.[1] On admission to the hospital, all of the workers had flushing of the face, increased rate of breathing, and obvious mydriasis. Six recovered within a day, but the most seriously affected patient experienced profound weakness, dizziness, vomiting, drowsiness, and depression.

In rats, the LC_{50} for 4 hours was 4020 ppm; effects were irritation of mucous membranes, increased respiration, incoordination, staggering gait, weakness, partial skeletal muscle paralysis beginning in the hind limbs, light to severe cyanosis, and mild to heavy sedation.[2] Animals that survived single near-lethal doses by the intraperitoneal and oral routes frequently had liver and kidney damage at autopsy up to 20 days post-treatment.[2] The liquid dropped in the eyes of rabbits caused slight to moderate irritation.[2] No dermal changes were observed when 0.2 ml of a 20% solution was applied to the clipped skin of guinea pigs for 10 days.[1]

The disagreeable, skunk-like odor is detectable at about 0.0001 to 0.001 ppm.[1]

REFERENCES

1. National Institute for Occupational Safety and Health: Criteria for a Recommended Standard . . . Occupational Exposure to n-Alkane Mono Thiols, Cyclohexanethiol, Benzenethiol. DHEW (NIOSH) Pub No 78–213, p 129. Washington, DC, US Government Printing Office, 1978
2. Fairchild EJ, Stokinger HE: Toxicologic studies on organic sulfur compounds. 1. Acute toxicity of some aliphatic and aromatic thiols (mercaptans). Am Ind Hyg Assoc J 19:171–189, 1958

---

p-tert-BUTYLTOLUENE

CAS: 98-51-1

$(CH_3)_3C—C_6H_4CH_3$ 1987 TLV = 10 ppm

Synonyms: p-Methyl-*tert*-butylbenzene; TBT; 1–methyl–4–tertiary-butylbenzene

Physical Form. Clear colorless liquid

Uses. Solvent for resins; intermediate in organic synthesis

Exposure. Inhalation

Toxicology. *p-tert*-Butyltoluene is an irritant of the mucous membranes.

Exposure of human volunteers for 5 minutes to concentrations of 5 to 160 ppm caused complaints of irritation of the nose and throat, nausea, and metallic taste; moderate eye irritation occurred at 80 ppm.[1] Exposed workers have complained of nasal irritation, nausea, headache, malaise, and weakness.[1] Signs and symptoms included decreased blood pressure, increased pulse rate, tremor, anxiety, and evidence of chemical irritation from skin contact. Laboratory findings suggested slight bone marrow depression.

The LD_{50} in female rats ranged from 934 ppm for 1 hour to 165 ppm for 8 hours.[1] Principal effects were irregular gait, paralysis, narcosis, and dyspnea as well as eye irritation. In some animals there was pulmonary edema and severe hemorrhage. Repeated exposures of rats to 50 ppm produced liver and kidney changes and lesions in the spinal cord and brain. The liquid on the rabbit skin was only a mild irritant.

The odor is recognized by most people at 5 ppm, but tolerance may be readily acquired. The irritating property may not be sufficient to protect one from hazardous concentrations.

REFERENCES

1. Hine CH et al: Toxicological studies on *p*-tertiary-butyltoluene. Arch Ind Hyg Occup Health 9:227–244, 1954

CADMIUM (and compounds)

CAS: 7440-43-9
Cd — 1987 TLV = 0.05 mg/m³ —dusts, salts

CADMIUM OXIDE
CAS: 1306-19-0
CdO — = 0.05 mg/m³ —cadmium oxide production
1987 Ceiling Limit = 0.05 mg/m³ —cadmium oxide fume

Synonyms: None

Physical Form. The metal is soft, ductile, silver-white, electropositive; cadmium oxide may be in the form of a colorless amorphous powder or as red or brown crystals

Uses. The metal is used in electroplating, solder for aluminum, constituent of easily fusible alloys, deoxidizer in nickel plating, process engraving, in cadmium–nickel batteries, and in reactor control rods. Cadmium compounds are used as TV phosphors; pigments in glazes and enamels; dyeing and printing; semiconductors; and rectifiers.

Exposure. Inhalation

Toxicology. Cadmium oxide fume is a severe pulmonary irritant; cadmium dust is also a pulmonary irritant, but it is less potent than cadmium fume because it has a larger particle size. The dust, at high concentrations, could be expected to cause the same physiologic effects as those arising from fume exposure. Several inorganic cadmium compounds cause malignant tumors in animals.

Most acute intoxications have been caused by inhalation of cadmium oxide fume at concentrations that did not provide warning symptoms of irritation. Concentrations of fume responsible for fatalities have been 40 to 50 mg/m³ for 1 hour, or 9 mg/m³ for 5 hours.[1] Nonfatal pneumonitis has been reported from concentrations of 0.5 to 2.5 mg/m³, while relatively mild cases have been attributed to even lower concentrations. Following an asymptomatic latent period of 4 to 10 hours, there is characteristically nasopharyngeal irritation followed by a feeling of chest constriction or substernal pain, with cough and dyspnea. The patient may also experience headache, chills, muscle aches, nausea, vomiting, and diarrhea.[2,3] Pulmonary edema may then develop rapidly, with decreased vital capacity and markedly reduced carbon monoxide diffusing capacity.[3] In approximately 20% of cases, dyspnea is progressive, accompanied by wheezing or hemoptysis, and may result in death within 7 to 10 days after exposure; at autopsy, the lungs are markedly congested, and there is an intra-alveolar fibrinous exudate, as well as alveolar cell metaplasia.[2,3] Among survivors, the subsequent course is unpredictable; most cases resolve slowly, but respiratory symptoms may linger for several weeks; impairment of pulmonary function may persist for months.[3]

In experimental animals, cadmium exposure has caused pulmonary fibrosis, but this has not been documented in humans.[3]

Inhalation overexposure may also result in renal calculi and irreversible renal tubular injury. In one fatal human case, there was renal cortical necrosis.[3]

Repeated exposure to lower levels of

cadmium in air has resulted in chronic poisoning characterized by irreversible lung injury of emphysematous type. This is accompanied by abnormal lung function and renal tubular damage, with urinary excretion of a specific low molecular weight protein.[4] Clinical evidence of the cumulative effects of cadmium may appear after exposure has terminated. The disease then tends to be progressive.[1] The frequency of occurrence of proteinuria increases with length of exposure. In one study, those persons exposed to cadmium compounds for less than 2 years had no proteinuria, whereas most of those exposed for 12 years or more had proteinuria with little other evidence of renal damage.[3] The urinary excretion of cadmium bears no known relationship to the severity or duration of exposure and is only a confirmation of absorption.[2] Absorbed cadmium is retained to a large extent by the body, and excretion is very slow.[5]

Other consequences of cadmium exposure are anemia, eosinophilia, yellow discoloration of the teeth, rhinitis, occasional ulceration of the nasal septum, damage to the olfactory nerve, and anosmia.[3,5,7] The long-term ingestion of water, beans, and rice contaminated with cadmium has been proposed as the probable cause of a crippling condition ("itai-itai" or "ouch-ouch" disease) among Japanese women who have had multiple pregnancies; characteristics of the disorder include pain in the back and joints, a waddly gait, osteomalacia, bone fractures, and, occasionally, fatal renal failure.[8]

Occupational exposure to cadmium has been implicated in a significant increase in prostate and respiratory tract cancer.[9] IARC classified the metal as probably carcinogenic to humans, but lacking human evidence, whereas cadmium chloride is classified only as showing sufficient evidence of carcinogenicity in experimental animals. NIOSH found the evidence of cadmium's ability to cause tumors to be contradictory.[10] However, in reconsidering this position in 1984, NIOSH cited recent epidemiologic evidence of a significant excess of respiratory cancer deaths among a cohort of cadmium production workers and concluded that cadmium and its compounds are potential carcinogens.[11]

Rats injected subcutaneously or intramuscularly developed rhabdomyosarcomas and fibrosarcomas; with cadmium sulfate or cadmium sulfide, local sarcomas; and with cadmium chloride, local pleomorphic sarcomas and testicular interstitial cell tumors.[9,12–14] Cadmium sulfate injected into the lingual vein of female hamsters on day 8 of pregnancy caused a high incidence of resorption and malformed offspring.[15] Acute necrosis of rat testes follows large doses orally or parenterally, but testicular effects have not been reported thus far in humans.[13]

REFERENCES

1. Cadmium and compounds. Documentation of TLVs and BEIs, 5th ed, pp 87–89. Cincinnati, American Conference of Governmental Industrial Hygienists (ACGIH), 1986
2. Dunphy B: Acute occupational cadmium poisoning: A critical review of the literature. J Occup Med 9:22–26, 1967
3. Louria DB, Joselow MM, Browder AA: The human toxicity of certain trace elements. Ann Intern Med 76:307–319, 1972
4. Tsuchiya K: Proteinuria of workers exposed to cadmium fume. Arch Environ Health 14:875–890, 1967
5. Stokinger HE: The metals. In Clayton GD, Clayton FE (eds): Patty's Industrial Hygiene and Toxicology, 3rd ed, Vol 2, Toxicology, pp 1563–1583. New York, Wiley–Interscience, 1981
6. Bonnell JA: Cadmium poisoning. Ann Occup Hyg 8:45–49, 1965
7. Fassett DW: Cadmium: Biological effects and occurrence in the environment. Ann Rev Pharmacol 15:425–435, 1975

8. Emerson BT: "Ouch-Ouch" disease: The osteomalacia of cadmium nephropathy. Ann Intern Med 73:854–855, 1970
9. IARC Monographs on the Evaluation of the Carcinogenic Risk of Chemicals to Man, Vol 11, Cadmium, Nickel, Some Epoxides, Miscellaneous Industrial Chemicals and General Considerations on Volatile Anaesthetics, pp 39–74. Lyon, International Agency for Research on Cancer, 1976
10. National Institute for Occupational Safety and Health: Criteria for a Recommended Standard . . . Occupational Exposure to Cadmium. DHEW (NIOSH) Pub No 76-192. Washington, DC, US Government Printing Office, 1976
11. NIOSH Current Intelligence Bulletin 42. DHHS (NIOSH) Pub No 84-116. National Institute for Occupational Safety and Health, US Department of Health and Human Services. Washington, DC, US Government Printing Office, 1984
12. Heath JC, Daniel MR, Dingle JT, Webb M: Cadmium as a carcinogen. Nature 193:592–593, 1962
13. Haddow A, Roe FJC, Dukes CE, Mitchley BCV: Cadmium neoplasia: Sarcomata at the site of injection of cadmium sulphate in rats and mice. Br J Cancer 18:667–673, 1964
14. Gunn SA, Gould TC, Anderson WAD: Specific response of mesenchymal tissue to cancerigenesis by cadmium. Arch Pathol 83:483–499, 1967
15. Holmberg RE Jr, Ferm VH: Interrelationships of selenium, cadmium, and arsenic in mammalian teratogenesis. Arch Environ Health 18:873–877, 1969

CALCIUM HYDROXIDE

CAS: 1305-62-0

$Ca(OH)_2$ 1987 TLV = 5 mg/m^3

Synonyms: Slaked lime; hydrated lime; calcium hydrate

Physical Form. White, microcrystalline powder

Uses. Manufacture of mortar, plaster, whitewash; in lubricants; in drilling fluids

Exposure. Inhalation; skin or eye contact

Toxicology. Calcium hydroxide is a relatively strong base and, therefore, a caustic irritant of all exposed surfaces of the body, including the respiratory tract.

The oral LD_{50} for rats is between 4.8 and 11.1 g/kg.[1] Rats given tap water containing 50 and 350 mg/liter had reduced food intake and were restless and aggressive at 2 months; at 3 months, there was a loss in body weight, decreased counts for erythrocytes and phagocytes, and decreased hemoglobin.[2] Autopsy showed inflammation of the small intestine and dystrophic changes in the stomach, kidneys, and liver.[2]

It is one of the most common causes of severe chemical eye burns.[3] In almost all cases, there is a semisolid particulate paste in contact with the cornea and conjunctiva, tending to adhere and to dissolve slowly. Strongly alkaline calcium hydroxide solution is formed and causes severe injury if not removed promptly.

REFERENCES

1. Smyth HF Jr et al: Range-finding toxicity data: List VII. Am Ind Hyg Assoc J 30:470, 1969
2. Wands RC: Alkaline materials. In Clayton GD, Clayton FE (eds): Patty's Industrial Hygiene and Toxicology, 3rd ed, rev, Vol 2B, Toxicology, pp 3052–3053. New York, Wiley–Interscience, 1981
3. Grant WM: Toxicology of the Eye, 3rd ed, pp 167–172. Springfield, Illinois, Charles C Thomas, 1986

CALCIUM OXIDE

CAS: 1305-78-8

CaO 1987 TLV = 2 mg/m^3

Synonyms: Burnt lime; calx; lime; quicklime

Physical Form. Crystals, white or greyish-white lumps, or granular powder

Uses. In construction materials; manufacture of steel, aluminum, and magnesium; manufacture of glass, paper, industrial chemicals; in fungicides, insecticides, lubricants

Exposure. Inhalation

Toxicology. Calcium oxide is an irritant of the eyes, mucous membranes, and skin.

The irritant effects are probably caused primarily by its alkalinity, but dehydrating and thermal effects may also be contributing factors. Strong nasal irritation was observed from exposure to a mixture of dusts containing calcium oxide in the range of 25 mg/m^3, but levels of 9 to 10 mg/m^3 produced no observable irritation.[2] Inflammation of the respiratory tract, ulceration and perforation of the nasal septum, and pneumonia have been attributed to inhalation of calcium oxide dust; severe irritation of the upper respiratory tract ordinarily causes persons to avoid serious inhalation exposure.[1,2]

Particles of calcium oxide can cause severe burns of the eyes.[3] Calcium oxide can produce skin burns and fissuring and brittleness of the nails.[4]

REFERENCES

1. Calcium oxide. Documentation of TLVs and BEIs, 5th ed, p 92. Cincinnati, American Conference of Governmental Industrial Hygienists (ACGIH), 1986
2. Wands RC: Alkaline materials. In Clayton GD, Clayton FE (eds): Patty's Industrial Hygiene and Toxicology, 3rd ed, rev, Vol 2B, Toxicology, pp 3053–3054. New York, Wiley–Interscience, 1981
3. Grant WM: Toxicology of the Eye, 3rd ed, p 173. Springfield, Illinois, Charles C Thomas, 1986
4. Fisher AA: Contact Dermatitis, p 17. Philadelphia, Lea & Febiger, 1973

CAMPHOR

CAS: 76-22-2

$C_{10}H_{16}O$ 1987 TLV = 2 ppm

Synonyms: 2–Bornanone; 2–camphanone; 2–keto–1,7,7–trimethylnorcamphane

Physical Form. Translucent crystals with characteristic odor

Uses. Plasticizer for cellulose esters and ethers; manufacture of plastics; in lacquers and varnishes; in explosives; in pyrotechnics; as moth repellent; as preservative in pharmaceuticals and cosmetics

Exposure. Inhalation; skin absorption

Toxicology. Camphor is an irritant of the eyes and nose; at high concentrations it is a convulsant.

Camphor is readily absorbed from all sites of administration, producing a feeling of coolness on the skin, whereas oral doses cause a sensation of warmth in the stomach.[1]

Symptoms of vapor exposure in humans include irritation of the eyes and nose and anosmia; these symptoms occur at concentrations above 2 ppm.[2] Heavy exposures cause nausea, anxiety, confusion, headache, dizziness, twitching of facial muscles, spasticity, convulsions, and coma.[1,3,4]

Most poisonings in humans are due to accidental ingestion.[5] With mild poisoning, gastrointestinal tract symptoms are more common than neurologic symptoms and include irritation of the mouth, throat, and stomach.[5] Severe poisoning is characterized by convulsions.[5]

Ingestion of 6 to 10 g of camphor by two men resulted in psychomotor agitation and hallucinations.[6] The probable lethal dose for humans is in the 50 to

500 mg/kg range.[5] It may be expected to be somewhat irritating on contact with the eye, but no serious injuries have been reported.[7]

REFERENCES

1. Gosselin RE, Smith RP, Hodge HC: Clinical Toxicology of Commercial Products, Section III, 5th ed, pp 84–86. Baltimore, Williams & Wilkins, 1984
2. Gronka PA, Bobkoskie RL, Tomchick GJ, Rakow AB: Camphor exposures in a packaging plant. Am Ind Hyg Assoc J 30:276–279, 1969
3. Arnow R: Camphor poisoning. JAMA 235:1260, 1976
4. Ginn HE et al: Camphor intoxication treated by lipid dialysis. JAMA 203:164–165, 1968
5. Segal S: Camphor: Who needs it? Pediatrics 62:404–406, 1978
6. Köppel C et al: Camphor poisoning. Abuse of camphor as a stimulant. Arch Toxicol 51:101–106, 1982
7. Grant WM: Toxicology of the Eye, 3rd ed, p 173. Springfield, Illinois, Charles C Thomas, 1986

CARBARYL

CAS: 63-25-2

$C_{12}H_{11}NO_2$ 1987 TLV = 5 mg/m^3

Synonyms: 1–Naphthyl methylcarbamate; SEVIN

Physical Form. Crystals of a white or greyish odorless solid

Uses. Insecticide

Exposure. Inhalation; skin absorption; ingestion

Toxicology. Carbaryl is a short-acting anticholinesterase agent, with the important characteristic of rapid reversibility of inhibition of the enzyme.

Carbaryl inactivates cholinesterase, resulting in the accumulation of acetylcholine at synapses in the nervous system, skeletal and smooth muscle, and secretory glands.[1–4] Signs and symptoms of overexposure may include miosis, blurred vision, lacrimation, excessive nasal discharge or salivation, sweating, abdominal cramps, nausea, vomiting, diarrhea, tremor, cyanosis, and convulsions. The rapid reversibility of cholinesterase inhibition is of significance in monitoring exposure; measurements of this enzyme are likely to give values in the normal range of activity, which will not reflect the true magnitude of exposure and will thus be clinically misleading; carbaryl is also rapidly metabolized, which further diminishes the severity of its effect.[1]

A single dose of 250 mg (approximately 2.8 mg/kg) ingested by an adult resulted in moderate poisoning; after 20 minutes, there was sudden onset of abdominal pain followed by profuse sweating, lassitude, and vomiting; 1 hour after ingestion, and following administration of a total of 3 mg atropine sulfate, the person felt better and was completely recovered after another hour.[1]

In a study of 59 workers exposed to concentrations ranging from 0.23 to 31 mg/m^3 during a 19-month period, there were no signs or symptoms of anticholinesterase activity.[5] In the most heavily exposed workers, relatively large amounts of 1–naphthol (a metabolite of carbaryl) were excreted in the urine, and the blood cholinesterase activity was slightly depressed. It was concluded that an excretion level of total (free plus conjugated) 1–naphthol significantly above 400 μg/100 ml of urine indicates absorption and metabolism of carbaryl.

On the skin, concentrated solutions may cause irritation and systemic intoxication.[1]

Men exposed in error to 85% water-wettable powder as a dust complained of burning and irritation of the skin but recovered in a few hours without any treatment except bathing. Their blood cholinesterase levels were only slightly depressed.[2]

The possible effect of carbaryl on reproduction and/or teratogenesis has been explored in rats, mice, guinea pigs, rabbits, dogs, and other species, sometimes at dosages so high that the mother was endangered. Although there is no doubt that massive doses can interfere with reproduction, there is doubt that anything but fetal toxicity has been observed.[2]

Diagnosis. Signs and symptoms include miosis, blurred vision, tearing; excessive nasal discharge or salivation; sweating; abdominal cramps, nausea, vomiting, diarrhea; tremor; cyanosis; convulsions; skin irritation from contact with concentrated solutions.

Differential Diagnosis: Diagnosis is based primarily on a history of exposure and clinical evidence of diffuse parasympathetic stimulation.[2] Careful observation of the effects of atropine is valuable. Patients with severe poisoning are resistant to the action of atropine; failure of 1 to 2 mg of atropine administered parenterally to produce signs of atropinization (flushing, mydriasis, tachycardia, or dryness of mouth) indicates poisoning.

Special Tests: Two types of cholinesterase are clinically significant: (1) true acetylcholinesterase, found principally in the nervous system and the red blood cell; and (2) pseudo- or butyrylcholinesterase, found in the plasma, liver, and the nervous system. Whereas the action of both types is inhibited by carbamates, the level of depression of red blood cell cholinesterase is a better indicator of clinically significant reduction of cholinesterase activity in the nervous system.

Laboratory evidence of depression of red blood cell cholinesterase to a level substantially below pre-exposure levels (at least 50% and usually much lower) is verification of poisoning. There is an imperfect correlation between the degree of depression of cholinesterase enzymes and the occurrence of symptoms. With a rapid drop in cholinesterase activity, generally reflecting an acute heavy exposure, there may be symptoms with only a 30% depression, whereas with slower drops to 70% depression, reflecting chronic low-level exposure, there may be no symptoms.[6]

If no pre-exposure baseline has been performed but symptoms are not sufficient to justify treatment with atropine, repeated testing during the recovery period demonstrating progressively increasing plasma and red blood cell cholinesterase levels over several days and weeks, respectively, suggests the diagnosis of anticholinesterase poisoning.[6]

There are many different methods for estimation of cholinesterase content of blood, and associated with each method is a different set of normal values and a different set of reporting units. The laboratory report of a cholinesterase determination should state the units involved along with the appropriate normal range. Based on the Michel method, the normal range of red blood cell cholinesterase activity (delta *p*H/hour) is 0.39 to 1.02 for men and 0.34 to 1.10 for women.[7] The normal range of the enzyme activity (delta *p*H/hour) of plasma is 0.44 to 1.63 for men and 0.24 to 1.54 for women.

The urinary concentration of 1–naphthol, one of the principal metabolites of carbaryl, which is excreted in the urine free or as the sulfate or glucuronide, has been used as an index of exposure. No quantitative relationship has been established between exposure to airborne carbaryl and total urinary 1–naphthol excretion.[4]

Treatment. Treatment of carbaryl poisoning ranges from simple removal from exposure in very mild cases to the provision of rigorous supportive and antidotal measures in severe cases.[2] In the moderate to severe case, because of pulmonary involvement, there may be need for artificial respiration using a positive-pressure method. Careful attention must be paid to removal of secretions and to maintenance of a patent airway. Anticonvulsants such as diazepam or thiopental sodium may be necessary. Maintenance of respiration is critical, because death usually results from weakness of the muscles of respiration and accumulation of excessive secretions in the respiratory tract.

As soon as cyanosis has been overcome, 2 to 4 mg of atropine should be given intravenously. (Atropine may induce ventricular fibrillation in the presence of cyanosis.) *This dose of atropine is approximately 10 times the amount that is administered for other conditions in which atropine is considered therapeutic.* This dose should be repeated at 5- to 10-minute intervals until signs of atropinization appear (dry, flushed skin, tachycardia as high as 140 beats/minute, and pupillary dilatation). A mild degree of atropinization should be maintained for at least 48 hours.

It is of great importance to decontaminate the patient. Contaminated clothing should be removed at once, and the skin should be washed with generous amounts of soap or detergent and a flood of water; this is best accomplished under a shower or by submersion in a pond or other body of water if the exposure occurred in the field. Careful attention should be paid to cleansing of the skin and hair.

The patient should be attended and monitored continuously for not less than 24 hours, because serious and sometimes fatal relapses have occurred owing to continuing absorption of the toxin or dissipation of the effects of the antidote.

Note: Pralidoxime (2–PAM, Protopam) chloride and other oxime reactivators are not used because of their inherent toxicity and the rapidly reversible inhibition of cholinesterase by carbaryl.[4]

Medical Control. Medical control includes preplacement and annual physical examination with determination of pre-exposure red blood cell and plasma cholinesterase activity. A person whose red blood cell cholinesterase falls to or below 40% of the pre-exposure baseline should be removed from further exposure until the activity returns to within 80% of the pre-exposure baseline.[6]

REFERENCES

1. Hayes WJ Jr: Clinical Handbook on Economic Poisons. Emergency Information for Treating Poisoning, US Public Health Service Pub No 476, pp 44–46. Washington, DC, US Government Printing Office, 1963
2. Hayes WJ Jr: Pesticides Studied in Man, pp 438–447. Baltimore, Williams & Wilkins, 1982
3. Koelle GB: Anticholinesterase agents. In Goodman LS, Gilman A (eds): The Pharmacological Basis of Therapeutics, 5th ed, p 456. New York, Macmillan, 1975
4. National Institute for Occupational Safety and Health: Criteria for a Recommended Standard . . . Occupational Exposure to Carbaryl. DHEW (NIOSH) Pub No 77–107, pp 17–96, 109–117. Washington, DC, US Government Printing Office, 1976
5. Best EM Jr, Murray BL: Observations on workers exposed to Sevin insecticide: A preliminary report. J Occup Med 10:507–517, 1962
6. Coye MJ, Lowe JA, Maddy KT: Biological monitoring of agricultural workers exposed to pesticides. I. Cholinesterase activity determinations. J Occup Med 28:619–627, 1986
7. Michel HO: Electrometric method for determination of red blood cell and plasma cholinesterase activity. J Lab Clin Med 34:1564–1568, 1949

CARBON BLACK

CAS: 1333-86-4 1987 TLV = 3.5 mg/m^3

Synonyms: Carbon; activated carbon; decolorizing carbon; actibon; animal charcoal; carboraffin; channel black; furnace black; thermal black; gas black; lamp black, norite; opocarbyl; ultracarbon

Physical Form. Black crystal, powder that varies in particle size and degree of aggregation

Uses. Clarifying; deodorizing; decolorizing; rubber pigment; ink constitutent; electrical insulating apparatus

Exposure. Inhalation

Toxicology. There are no well-demonstrated health hazards to humans from acute exposure to carbon black.

Commercial carbon black is a spherical colloidal form of nearly pure carbon particles and aggregates with trace amounts of organic and inorganic impurities adsorbed on the surface. Potential health effects are usually attributed to these impurities rather than to the carbon itself. (Soots, by contrast, contain mixtures of particulate carbon, resins, tars, and so on in a nonadsorbed state.)[1]

A significant loss in pulmonary function was reported in a group of 125 Nigerian carbon black workers exposed to levels of up to 34 mg/m^3.[2] The most common respiratory symptom was cough with phlegm, but radiograms were normal. Significant annual declines in FEV_1 and FVC and radiologic lung changes were reported in another group of 35 workers exposed to concentrations less than 10 mg/m^3.[1] In contrast, a survey of more than 500 carbon black workers in the United States and in the United Kingdom found no statistical difference in spirometry, chest radiograph, physical examination, or reported symptoms.[3]

A retrospective cohort study of 1200 men employed at four carbon black plants from 1935 to 1974 found no significant increase in total mortality, mortality from heart disease, or mortality owing to malignant neoplasms.[4]

Repeated inhalation by monkeys caused deposition of the dust in the lungs with minimal or no fibrous tissue proliferation.[6] Repeated exposure of the monkeys to 56.5 mg/m^3 of furnace black for a total of 10,000 hours produced marked electrocardiographic changes interpreted as right atrial and right ventricular strain, probably a result of massive deposition of the dust in the lungs.[1,6] (In the original paper, the concentration was incorrectly reported as 1.5 mg/m^3 for furnace black.)

Carbon black itself does not appear to be a carcinogen. The major concern is the simultaneous exposure to polycyclic aromatic hydrocarbons, which are strongly adsorbed on the respirable carbon black particles and from which PAHs are elutriated *in vivo* under conditions of human exposure.[7] In a number of studies, attempts to elutriate PAH with biological fluids have been largely unsuccessful, and prolonged extraction with boiling aromatic solvents is required for quantitative desorption.[7]

The IARC has determined that there is sufficient evidence that solvent extracts of carbon black are carcinogenic, but there is inadequate evidence to evaluate the carcinogenicity of carbon black to humans.[1]

There are no reports on humans or animals suggesting a mutagenic or teratogenic potential for carbon black.[7]

REFERENCES

1. IARC Monographs on the Evaluation of the Carcinogenic Risk of Chemicals to Humans. Polynuclear Aromatic Compounds, Part 2, Car-

bon Blacks, Mineral Oils and Some Nitroarsines. Vol 33, pp 35–85. Lyon, International Agency for Research on Cancer, 1984
2. Oleru UG et al: Pulmonary function and symptoms of Nigerian workers exposed to carbon black in dry cell battery and tire factories. Environ Res 39:161–168, 1983
3. Crosbie WA et al: Survey of respiratory disease in carbon black workers in the UK and USA (abstr). Am Rev Respir Dis 119:209, 1979
4. Robertson JM, Ingalls TH: A mortality study of carbon black workers in the United States from 1935 to 1974. Arch Environ Health 35:181–186, 1980
5. Hodgson JT, Jones RD: A mortality study of carbon black workers employed at five United Kingdom factories between 1947 and 1980. Arch Environ Health 40:261–268, 1985
6. Nau CA, Neal J, Stembridge VA, Cooley RN: Physiological effects of carbon black. IV. Inhalation. Arch Environ Health 4:45–61, 1962
7. National Institute for Occupational Safety and Health: Criteria for a Recommended Standard . . . Occupational Exposure to Carbon Black. DHEW (NIOSH) Pub No 78–204, pp 1–99. Washington, DC, US Government Printing Office, 1978

CARBON DIOXIDE

CAS: 124-38-9

CO_2 1987 TLV = 5000 ppm

Synonym: Carbonic acid gas

Physical Form. Colorless gas (solid is "dry ice")

Uses and Sources. By-product of ammonia production, lime kiln operations, and fermentation; used in carbonation of beverages; propellant in aerosols; dry ice for refrigeration

Exposure. Inhalation

Toxicology. Carbon dioxide is usually considered a simple asphyxiant, although it is also a potent stimulus to respiration and both a depressant and excitant of the central nervous system.

Numerous human fatalities have occurred after entering fermentation vats, wells, and silos where the air had been replaced largely by carbon dioxide.[1,2] The most immediate and significant effects of acute exposure at high concentrations are those on the central nervous system.[1] Concentrations of 20% to 30% (200,000 to 300,000 ppm) result in unconsciousness and convulsions within 1 minute of exposure. At concentrations of approximately 120,000 ppm, unconsciousness may be produced with longer exposures of 8 to 23 minutes. Neurologic symptoms, including psychomotor agitation, myoclonic twitches, and eye flickering appeared after 1.5 minutes at 100,000 to 150,000 ppm.[1] Inhalation of concentrations from 60,000 to 100,000 ppm may produce dyspnea, headache, dizziness, sweating, restlessness, paresthesias, and a general feeling of discomfort; at 50,000 ppm, there may be a sensation of increased respiration, but subjects rarely experience dyspnea.[3] After several hours of exposure to 2% (20,000 ppm) carbon dioxide, subjects develop headache and dyspnea on mild exertion.[4] Circulatory effects in humans exposed to carbon dioxide include an increase in heart rate and cardiac output.[5]

Adaptation to low levels, 1.5% to 3.0% carbon dioxide, has occurred with chronic exposure.[1] Carbon dioxide at room temperature will not injure the skin, but frostbite may result from contact with dry ice or from the gas at low temperatures.

It is important to note that since carbon dioxide is heavier than air, pockets of the gas may persist in areas such as pits for some time unless ventilation is provided.

REFERENCES

1. National Institute for Occupational Safety and Health: Criteria for a Recommended Standard . . . Occupational Exposure to Carbon Dioxide. DHEW (NIOSH) Pub No 76–194, pp 17–105,

114–126. Washington, DC, US Government Printing Office, 1976
2. Williams HI: Carbon dioxide poisoning—report of eight cases, with two deaths. Br Med J 2:1012–1014, 1958
3. Smith TC et al: The therapeutic gases. In Gilman A et al (eds): Goodman and Gilman's The Pharmacological Basis of Therapeutics, 7th ed, pp 333–335. New York, Macmillan, 1985
4. Schulte JH: Sealed environments in relation to health and disease. Arch Environ Health 8:438–452, 1964
5. Cullen DJ, Eger EI: Cardiovascular effects of carbon dioxide in man. Anesthesiology 41:345–349, 1974

CARBON DISULFIDE

CAS: 75-15-0

CS_2 1987 TLV = 10 ppm; skin

Synonyms: Carbon bisulfide; carbon disulphide

Physical Form. Colorless liquid

Uses. Solvent for lipids, sulfur, rubber, phosphorus, oils, resins, and waxes; insecticide; preparation of rayon viscose fibers

Exposure. Inhalation; skin absorption

Toxicology. Carbon disulfide causes damage to the central and peripheral nervous systems and may accelerate the development of, or worsen, coronary heart disease.

Exposure of humans to 4800 ppm for 30 minutes causes coma and may be fatal.[1] Carbon disulfide intoxication can involve all parts of the central and peripheral nervous systems, including damage to the cranial nerves and development of peripheral neuropathy with paresthesias and muscle weakness in the extremities, unsteady gait, and dysphagia.[2] A follow-up of workers with clinical and electromyographic evidence of neuropathy attributed to carbon disulfide exposure showed no significant improvement 10 years after exposure was discontinued, suggesting a permanent axonal neuropathy.[3]

In extreme cases of intoxication, a Parkinsonism-like syndrome may result, characterized by speech disturbances, muscle spasticity, tremor, memory loss, mental depression, and marked psychic symptoms; permanent disability is likely.[2] Psychosis and suicide are established risks of overexposure to carbon disulfide.[4]

Other reported effects of exposure to carbon disulfide are ocular changes (blind spot enlargement, contraction of peripheral field, corneal anesthesia, diminished pupillary reflexes, nystagmus, and microscopic aneurysms in the retina), gastrointestinal disturbances (chronic gastritis and achlorhydria), renal impairment (albuminuria, microhematuria, elevated blood urea nitrogen, and diastolic hypertension), and liver damage.[2,5] Hearing loss to high-frequency tones has also been reported.[6]

Effects commonly caused by repeated exposure to carbon disulfide vapor are exemplified by a group of workers with a time-weighted average (TWA) exposure of 11.2 ppm (range 0.9 to 127 ppm) who complained of headaches and dizziness; in other workers with a TWA of 186 ppm (range 23 to 389 ppm), complaints also included sleep disturbances, fatigue, nervousness, anorexia, and weight loss; the end-of-the-day exposure coefficient of the iodine azide test on urine was a good indicator of workers who were or had been symptomatic.[7]

Overexposure to carbon disulfide has been associated with an increase in coronary heart disease. In a mortality study of viscose rayon workers, 42% of deaths were certified to coronary heart disease versus 17% in unexposed workers.[8] A recent follow-up of this cohort showed a similar pattern, with an SMR for ischemic heart disease of 172 in spinning operatives.[9] In this study, it was also found that the risk declined after exposure ceased, suggesting a direct cardiotoxic or thrombotoxic effect of carbon disulfide rather than an atherogenic effect. In a Finnish cohort, removal from exposure of workers with coronary risk factors and reduction of levels to 10 ppm caused a dramatic decrease in cardiovascular mortality and return to background levels.[10] Other cardiovascular effects observed in workers repeatedly exposed to carbon disulfide are bradycardia, tachycardia, other arrhythmias, and electrocardiographic changes consistent with both nonspecific and ischemic wave changes.[5]

Conflicting studies have appeared regarding the ability of carbon disulfide to affect reproductive function.[6] Hypospermia, abnormal sperm morphology, menstrual cycle irregularities, increased menstrual flow and pain, and a slight increase in miscarriages have been reported in some studies, whereas other studies have not found adverse effects. Pregnant rats and rabbits exposed at 20 and 40 ppm 7 hours/day showed no evidence of embryotoxicity or teratogenicity. In another report, hydrocephalia was observed in rats exposed to 32 and 64 ppm 8 hours/day throughout gestation.[6]

Chronic exposure of animals for periods less than 1 year has not shown a carcinogenic potential for carbon disulfide.[6] Furthermore, epidemiologic studies do not support a carcinogenic risk under moderate exposure conditions.[11]

Splashes of the liquid in the eye cause immediate and severe irritation; dermatitis and vesiculation may result from skin contact with the vapor or the liquid.[12] Although ingestion is unlikely to occur, it may cause coma and convulsions.[12]

Carbon disulfide has a foul odor, but it is not sufficient to give adequate warning of hazardous concentrations.

REFERENCES

1. Teisinger J: Carbon disulphide. In International Labour Office: Encyclopaedia of Occupational Health and Safety, Vol 1, pp 252–253. New York, McGraw-Hill, 1974
2. Tolonen M: Vascular effects of carbon disulfide: A review. Scand J Work Environ Health 1:63, 1975
3. Corsi G et al: Chronic peripheral neuropathy in workers with previous exposure to carbon disulphide. Br J Ind Med 40:209–211, 1983
4. Mancuso TF, Locke BZ: Carbon disulfide as a cause of suicide—epidemiological study of viscose rayon workers. J Occup Med 14:595, 1972
5. Davidson M, Feinleib M: Carbon disulfide poisoning: A review. Am Heart J 83:100, 1972
6. Beauchamp RO et al: A critical review of the literature on carbon disulfide toxicity. CRC Crit Rev Toxicol 11:169–278, 1983
7. Rosensteel RE, Shama SK, Flesch JP: Occupational health case report—No 1. J Occup Med 16:22, 1974
8. Tiller JR et al: Occupational toxic factor in mortality from coronary heart disease. Br Med J 4:407–411, 1968
9. Sweetnam PM et al: Exposure to carbon disulphide and ischaemic heart disease in a viscose rayon factory. Br J Ind Med 44:220–227, 1987
10. Nurminen M, Hernberg S: Effects of intervention on the cardiovascular mortality of workers exposed to carbon disulphide: A 15-year follow up. Br J Ind Med 42:32–35, 1985
11. Nurminen M, Hernberg S: Cancer mortality among carbon disulfide-exposed workers. J Occup Med 26:341, 1984
12. Chemical Safety Data Sheet SD-12, Carbon Disulfide, pp 5–6, 14–15. Washington, DC, MCA, Inc, 1967

CARBON MONOXIDE

CAS: 630-08-0

CO 1987 TLV = 50 ppm

Synonyms: Carbonic oxide; exhaust gas; flue gas

Physical Form. Odorless, colorless, tasteless gas

Sources. Incomplete combustion of organic fuels; vehicle exhaust; space heaters; gas and kerosene lanterns

Exposure. Inhalation

Toxicology. Carbon monoxide (CO) causes tissue hypoxia by preventing the blood from carrying sufficient oxygen.

Carbon monoxide combines reversibly with the oxygen-carrying sites on the hemoglobin molecule with an affinity ranging from 210 to 240 times greater than that of oxygen; the carboxyhemoglobin thus formed is unavailable to carry oxygen.[1] In addition, partial saturation of each hemoglobin molecule with CO results in tighter binding of oxygen to hemoglobin. This shifts the oxygen–hemoglobin dissociation curve, further reducing oxygen delivery to the tissues.[2] Carbon monoxide may also exert a direct toxic effect by binding to myoglobin and cellular cytochromes, such as those contained in respiratory enzymes.

Although CO poisoning represents a multisystem insult, the cardiac and central nervous systems are particularly sensitive to the effects of hypoxia.[1,2] Most clinical manifestations are referable to the central nervous system, but it is likely that myocardial ischemia is responsible for many carbon monoxide–induced deaths.[3]

With exposure to high concentrations such as 4000 ppm and above, transient weakness and dizziness may be the only premonitory warnings before coma supervenes; the most common early aftermath of severe intoxication is cerebral edema.[4–5] Exposure to concentrations of 500 to 1000 ppm causes the development of headache, tachypnea, nausea, weakness, dizziness, mental confusion, and, in some instances, hallucinations; the person is commonly cyanotic.[1–4] Because carboxyhemoglobin has a bright red color, occasionally someone will exhibit the unusual combination of hypoxia together with a bright red color of the fingernails, mucous membranes, and skin; however, this "cherry-red cyanosis" is usually seen only at autopsy.[4]

Exposure to 50 ppm for 90 minutes may cause aggravation of angina pectoris; exposed anginal patients may show a negative inotropic effect (weakened force of myocardial contraction); 50 ppm for 120 minutes may cause aggravation of intermittent claudication.[6] The clinical effects of CO exposure are aggravated by heavy labor, high ambient temperature, and altitudes above 2000 feet; pregnant women are more susceptible to the effects of CO.[1]

The reaction to a given blood level of carboxyhemoglobin is extremely variable; some persons may be in a coma with a carboxyhemoglobin level of 38%, whereas others may maintain an apparently clear sensorium with levels as high as 55%. Levels of carboxyhemoglobin over 60% are usually fatal; 40% is associated with collapse and syncope; above 25% there may be electrocardiographic evidence of a depression of the S–T segment; between 15% and 25%, there may be headache and nausea; levels below 15% rarely produce symptoms. The blood of cigarette smokers usually contains 2% to 10% and sometimes as high as 18% carboxyhemoglobin, and nonexposed persons have an average level of 1%; heme metabolism is an endogenous source of CO.[1]

Exposure of nonsmokers to 50 ppm (the TLV) for 6 to 8 hours results in carboxyhemoglobin levels of 8% to 10%.[1–3] Several investigators have suggested that the results of behavioral tests such as time discrimination, visual vigilance, choice response tests, visual evoked responses, and visual discrimination threshold may be altered at levels of carboxyhemoglobin below 5%.[1]

Transient central nervous system symptoms or rapid death are not the only results of CO poisoning.[3] The occurrence of late, fatal demyelination is a rare but dreaded complication.[3] Further, it is inappropriate to assume that because a patient with CO poisoning shows improvement, residual mental damage may not occur.[3] A report of 63 patients studied 3 years after CO poisoning indicated that 13% showed gross neuropsychiatic damage directly attributable to their CO intoxication; 33% showed a "deterioration of personality" after poisoning; and 43% reported memory impairment.[8]

Animal experiments suggest that prenatal exposure at maternally nontoxic levels may damage the developing central nervous system.[9,10] Exposure of pregnant rats to 150 ppm produced only minor reductions in pup birthweights, but evaluation of learning and memory processes suggested a functional deficit in the central nervous system that persisted into adulthood of exposed offspring.[9,10]

REFERENCES

1. National Institute for Occupational Safety and Health: Criteria for a Recommended Standard . . . Occupational Exposure to Carbon Monoxide. (HSM) 73–11000. Washington, DC, US Government Printing Office, 1972
2. Olson KR: Carbon monoxide poisoning: Mechanisms, presentation, and controversies in management. J Emerg Med 1:233–243, 1984
3. Winter PM, Miller JN: Carbon monoxide poisoning. JAMA 236:1502–1504, 1976
4. Swinyard EA: Noxious gases and vapors. In Goodman LS, Gilman A (eds): The Pharmacological Basis of Therapeutics, 5th ed, pp 900–904, 910–911. New York, Macmillan, 1975
5. Beard RR: Inorganic compounds of O, N, and C. In Clayton GD, Clayton FE (eds): Patty's Industrial Hygiene and Toxicology, Vol 2C, Toxicology, pp 4114–4124. New York, Interscience, 1982
6. Goldsmith JR, Aronow WS: Carbon monoxide and coronary heart disease: A review. Environ Res 10:236–248, 1975
7. Haldane JBS: Carbon monoxide as a tissue poison. Biochem J 21:1068–1075, 1927
8. Smith J, Brandon S: Morbidity from acute carbon monoxide poisoning at a three-year follow-up. Br Med J 1:318–321, 1973
9. Mactutus CF, Fechter LD: Prenatal exposure to carbon monoxide: Learning and memory deficits. Science 223:409–411, 1984
10. Mactutus CF, Fechter LD: Moderate prenatal carbon monoxide exposure produces persistent, and apparently permanent, memory deficits in rats. Teratology 31:1–12, 1985

CARBON TETRACHLORIDE

CAS: 56-23-5

CCl_4 1987 TLV = 5 ppm; skin

Synonym: Tetrachloromethane

Physical Form. Colorless liquid

Uses. Manufacture of fluorocarbon propellants; solvent for oils, fats, lacquers, varnishes, rubber waxes, and resins; used as a degreasing and cleaning agent; grain fumigant

Exposure. Inhalation; skin absorption; ingestion

Toxicology. Carbon tetrachloride causes central nervous system depression and severe damage to the liver and kidneys;

it is carcinogenic in experimental animals and has been classified as a potential human carcinogen.

In animals, the primary damage from intoxication is to the liver, but, in humans, most fatalities have been the result of renal injury with secondary cardiac failure.[1,2] Human autopsy results have included renal tubular necrosis. In humans, liver damage occurs more often after ingestion of the liquid than after inhalation of the vapor.

Human fatalities from acute renal damage have occurred after exposure for ½ to 1 hour to concentrations of 1000 to 2000 ppm; occasional sudden deaths may have been due to ventricular fibrillation.[1] Exposure to high concentrations results in symptoms of central nervous system depression, including dizziness, vertigo, incoordination, and mental confusion; abdominal pain, nausea, vomiting, and diarrhea are frequent.[1–4] Cardiac arrhythmias and convulsions have also occurred. Polycythemia followed by anemia and hemodilution may occur. Within a few days, jaundice may appear, and liver injury may progress to toxic necrosis. At the same time, acute nephritis may occur with albumin, red and white blood cells, and casts in the urine; there may be oliguria, anuria, and increased nitrogen retention, resulting in the development of uremia. The no adverse effect level for acute human exposure is 10 ppm for a 3-hour exposure.[5]

There are several reports of adverse effects in workmen who were repeatedly exposed to concentrations between 25 and 30 ppm; nausea, vomiting, dizziness, drowsiness, and headache were frequently noted.[1] Chronic exposure has caused cases of various abnormalities of the eyes, such as reduced visual field.[1]

The liquid splashed in the eye causes pain and minimal injury to the conjunctiva. Prolonged or repeated skin contact with the liquid may result in skin irritation and blistering.[1,4]

A number of substances, including ethanol, polybrominated biphenyls, phenobarbital, and benzo(a)pyrene, have been shown to synergistically affect CCl_4 toxicity.[4] Alcohol has been a concomitant factor in many of the human cases of poisoning, especially in cases in which severe liver and kidney damage have occurred.[1] Some substances such as chlordecone greatly potentiate the toxicity of CCl_4 at doses at which both substances are not considered toxic; effects include extensive hepatotoxicity characterized by total hepatic failure and greatly potentiated lethality.[6]

Animal studies demonstrate that carbon tetrachloride produces hepatocellular carcinomas in the mouse, rat, and hamster.[4] Mice administered 1250 or 2500 mg/kg by gavage 5 times weekly for 78 weeks approached nearly a 100% incidence of hepatocellular carcinomas vs. 6% or less in various controls.[4] Hamsters receiving 30 weekly doses of 190 and 380 mg/kg by gavage had a 100% liver-cell carcinoma incidence for those animals surviving past the 43rd week.[7]

Sensitivity to carbon tetrachloride-induced hepatocellular carcinomas varied widely among five strains of rats receiving twice weekly subcutaneous injections of 2080 mg/kg as a 50% solution in corn oil.[8]

A number of animal studies suggest that hepatomas occur only after liver necrosis and fibrosis have occurred and, therefore, that CCl_4 is not a direct liver carcinogen.[4] One early study, however, found that liver necrosis and its associated chronic regenerative state probably were not necessary for tumor induction, although a correlation was found between the degree of liver necrosis and the incidence of hepatomas.[9]

In humans cases of hepatomas that have appeared years following acute exposure to CCl_4, however, none of the cases could establish a causal link between the exposure and development of neoplasms.[4]

Epidemiologic studies have also given inconclusive results.[4] A cancer mortality study of a population of rubber workers reported significantly elevated odds ratios relating CCl_4 with lymphatic leukemia, lymphosarcoma, and reticulum cell carcinoma.[10,11] However, several solvents were used simultaneously, and effects should not be attributed solely to CCl_4.

The IARC evaluations for CCl_4 conclude that there is sufficient evidence for carcinogenicity in animals, inadequate evidence for carcinogenicity in humans, and an overall conclusion that CCl_4 is probably carcinogenic to humans.[12]

Dosing of pregnant rats on days 6 to 15 of gestation at 300 ppm 7 hours/day or 1000 ppm 7 hours/day was fetotoxic, causing an increase in skeletal anomalies owing to delayed development.[13] Maternal toxicity as evidenced by weight loss and hepatotoxicity was also observed.[13]

The mechanism of CCl_4 hepatotoxicity is generally viewed as an example of lethal cleavage, where the CCl_3–Cl bond is split in the mixed function oxidase system of hepatocytes.[4] Following this cleavage, damage may occur directly from the free radicals ($\cdot CCl_3$ and $\cdot Cl$) and/or from the formation of toxic metabolites such as phosgene.[4]

The sweetish odor is not a satisfactory warning of excessive exposure.

REFERENCES

1. Criteria for a Recommended Standard . . . Occupational Exposure to Carbon Tetrachloride. DHEW (NIOSH) Pub No 76–133, pp 15–68, 84–112. National Institute for Occupational Safety and Health, US Department of Health, Education and Welfare. Washington, DC, US Government Printing Office, 1975
2. Fassett DW: Toxicology of organic compounds: A review of current problems. Ann Rev Pharmacol 3:267–274, 1963
3. von Oettingen WF: The Halogenated Aliphatic, Olefinic, Cyclic, Aromatic, and Aliphatic-Aromatic Hydrocarbons including the Halogenated Insecticides, Their Toxicity and Potential Dangers. US Public Health Service Pub No 414, pp 75–112. Washington, DC, US Government Printing Office, 1955
4. Health Assessment Document for Carbon Tetrachloride. Cincinnati, US Environmental Protection Agency, Environmental Criteria and Assessment Office, 1984
5. Stewart RD et al: Human exposure to carbon tetrachloride vapor—relationship of expired air concentration to exposure and toxicity. J Occup Med 3:586–590, 1961
6. Mehendale HM: Potentiation of halomethane hepatoxicity: Chlordecone and carbon tetrachloride. Fund Appl Toxicol 4:295–308, 1984
7. Della Porta G et al: Induction with carbon tetrachloride of liver cell carcinomas in hamsters. J Natl Cancer Inst 26:855–863, 1961
8. Reuber MD, Glover EL: Cirrhosis and carcinoma of the liver in male rats given subcutaneous carbon tetrachloride. J Natl Cancer Inst 44:419–427, 1970
9. Eschenbrenner AB, Miller E: Studies on hepatomas—size and spacing of multiple doses in the induction of carbon tetrachloride hepatomas. J Natl Cancer Inst 4:385–388, 1943
10. Wilcosky TC et al: Cancer mortality and solvent exposures in the rubber industry. Am Ind Hyg Assoc J 45:809–811, 1984
11. Checkoway H et al: An evaluation of the associations of leukemia and rubber industry solvent exposures. Am J Ind Med 5:239–249, 1984
12. IARC Monographs on the Evaluation of the Carcinogenic Risk of Chemicals to Man, Vol 20, Some Halogenated Hydrocarbons, pp 371–399. Lyon, International Agency for Research on Cancer, 1979
13. Schwetz DW et al: Embryo- and fetotoxicity of inhaled carbon tetrachloride, 1,1–dichloroethane and methyl ethyl ketone in rats. Toxicol Appl Pharmacol 28:452–464, 1974

CHLORDANE

CAS: 57-74-9

$C_{10}H_8Cl_8$ 1987 TLV = 0.5 mg/m^3; skin

Synonyms: Chlordan; Velsicol 1068; CD–68; Toxichlor, Octa-Klor

Physical Form. Viscous amber liquid

Uses. Insecticide

Exposure. Skin absorption, ingestion, inhalation

Toxicology. Chlordane is a convulsant.

Established cases of chlordane poisoning have been associated with gross exposure either by ingestion or skin contact.[1] Typically, the poisoning is characterized by onset of violent convulsions within ½ to 3 hours and either death or recovery within a few hours to a day. Following ingestion, nausea and vomiting may precede signs of central nervous system overactivity. Convulsions may be accompanied by confusion, incoordination, excitability, or coma. In one instance, accidental ingestion of approximately 300 ml of a 75% chlordane solution (215 g chlordane) was survived despite rapid onset of respiratory, gastrointestinal, and neurologic effects.[2] In this case, the chlordane level in whole blood was 5 mg/liter at 3.5 hours postingestion. Kinetic analysis of blood chlordane levels with time suggested a half-life of 7 hours for distribution in the body and 34 days for elimination.[2]

Skin absorption of chlordane is rapid; a worker who spilled a 25% suspension of chlordane on the clothing, which was not removed, began having convulsions within 40 minutes and died shortly thereafter.[3]

Technical grade chlordane is stated to be irritating to the skin and mucous membranes, but this may be more true of earlier chlordane formulations with significant hexachlorocyclopentadiene contamination.[1,3]

Mice kept in saturated vapor of technical chlordane without hexachlorocyclopentadiene for 25 days showed no symptomatic effects.[1] The oral LD_{50} values for rats range from 200 to 590 mg/kg.[3]

In experimental animals, prolonged exposure to dietary levels exceeding 3 to 5 mg/kg resulted in the induction of hepatic microsomal enzymes and, at a later stage, liver hypertrophy with histologic changes.

At dosages above 30 mg/kg in the diet, chlordane interfered with reproduction in rats and mice, but this was reversible after exposure ceased.[3]

A dose-related increase in the incidence of hepatocellular carcinomas was found in male and female mice fed chlordane for 80 weeks.[4] IARC determined that there was limited evidence for the carcinogenicity of chlordane in experimental animals.[5]

Human studies have not shown an association between chlordane exposure and cancer mortality. The most recent study of 800 workers employed at a chlordane production plant for 3 months or more during the period 1946 to 1985 showed a slightly less than expected overall death rate and an inverse relationship of cancer mortality to length of employment.[6]

REFERENCES

1. Hayes WJ Jr: Pesticides Studied in Man, pp 229–233. Baltimore, Williams & Wilkins, 1982
2. Olanoff LS et al: Acute chlordane intoxication. J Toxicol Clin Toxicol 20:291–306, 1983
3. Environmental Health Criteria 34, Chlordane, p 82. Geneva, World Health Organization, 1984
4. National Cancer Institute: Bioassay of Chlordane for Possible Carcinogenicity, Technical Report Series No 8. DHEW (NIH) Pub No 77–808. Washington, DC, US Government Printing Office, 1977
5. IARC Monographs on the Evaluation of the Carcinogenic Risk of Chemicals to Man, Suppl

4, pp 25–27. Lyon, International Agency for Research on Cancer, 1982

6. Shindell S, Ulrich S: Mortality of workers employed in the manufacture of chlordane: An update. J Occup Med 28:497–501, 1986

CHLORDECONE

CAS: 143-50-0

$C_{10}Cl_{10}O$ 1987 TLV = none established

Synonym: Kepone

Physical Form. Powder

Uses. Pesticide (leaf-eating insects and fly larvae); production ceased in 1975 as a result of toxicity to workers.

Exposure. Inhalation; possible skin absorption

Toxicology. Kepone affects the brain, the peripheral nerves, muscle, liver, and testes.

The first reports of effects in humans concerned the cases of intoxication of workers of the Life Science Products Company in Hopewell, Virginia.[1,2] This was a small improvised unit lacking in control of dust generated by a process that was operated over a period of 16 months. On initial assessment, at least 9 of the current 33 employees of the company were severely affected and showed memory impairment, slurred speech, tremor, opsiclonus (eye twitching), and liver damage; blood levels of the compound of up to 25 ppm were found.[2]

Later examination of 117 of the 149 current or previous employees of this plant showed that 57 had present or past symptoms of intoxication, including weight loss, tremor of the upper extremities, ataxia, incoordination, arthralgia, skin rash, and abnormality of liver function tests. The incidence of illness was 67% in production workers and 16% for other employees of the plant. The wives of two workers had objective tremor; they both had washed their husband's work clothes.[3] Laboratory findings in this group are summarized by Hayes[3] as follows:

"Among 32 of the men who had manufactured chlordecone and who were suffering persistent poisoning involving the nervous system, testes, and liver, whole blood levels of the compound ranged from 0.6 to 32.0 ppm, with an average of 5.8 ppm. The concentration in subcutaneous fat of 16 of them was only 2.2 to 62.0 ppm with a mean of 21.5 ppm—remarkably low in view of the blood levels. In the ten samples that were taken, the highest concentrations were in the liver and ranged from 13.3 to 173.0 ppm with a mean of 75.9 ppm. No metabolites were identified in tissues, urine, or stool. Excretion of chlordecone was negligible in urine and sweat, but occurred mainly in the feces, where it amounted to 0.019 to 0.279 percent/day (mean 0.075 percent/day) of what was estimated to be the total amount stored in the body."

Sural nerve biopsies from affected workers showed significant histologic damage to nonmyelinated and smaller myelinated fibers, with relative sparing of larger myelinated fibers.[4]

In early animal studies, the compound caused tremor, the severity of which depended upon the dosage level and duration of exposure; this persisted for a week or more following single exposures and cumulatively developed from daily repeated, individually ineffective doses. In male rats, the oral LD_{50} was 132 mg/kg.[5]

Reproductive studies showed that 14 pairs of mice that received 40 ppm chlordecone in the diet for 2 months before mating and during the test produced no litters, whereas 14 control pairs produced 14 first litters and 14 second litters. The results of studies of carcinogenicity in animals were inconclusive.[3]

REFERENCES

1. Bureau of National Affairs, Inc: OSHA Cites Chemical Manufacturer, Labels Kepone Exposure "Catastrophe." Occup Safety Health Rep 5:379–380, 1975
2. Bureau of National Affairs, Inc: Allied, Hooker Chemical Firms Named as Defendants in Worker Suit. Occup Safety Health Rep 5:516–517, 1975
3. Hayes WJ Jr: Pesticides Studied in Man, pp 256–260. Baltimore, Williams & Wilkins, 1982
4. Martinez AJ et al: Chlordecone intoxication in man. II. Ultrastructure of peripheral nerves and skeletal muscle. Neurology 28:631–635, 1978
5. Toxicological Studies on Decachlorooctahydro–1,3,4–metheno–2H–cyclobuta[cd] pentalen–2–one (Compound No. 1189) (Kepone). New York, Allied Chemical Corporation, March 1960

CHLORINATED DIPHENYL OXIDE

CAS: 55720-99-5

$C_{12}H_9ClO$,
$C_{12}H_6Cl_4O$
$C_{12}H_8Cl_2O$,
$C_{12}H_5Cl_5O$
$C_{12}H_7Cl_3O$,
$C_{12}H_4Cl_6O$ 1987 TLV = 0.5 mg/m^3

Synonyms: Chlorinated phenyl ethers; monochlorodiphenyl oxide; dichlorodiphenyl oxide, etc., through hexachlorodiphenyl oxide

Physical Form. Varies from colorless oily liquids to yellowish waxy semisolids as the equivalents of chlorine increase from 1 to 6

Uses. Chemical intermediates; in the electrical industry

Exposure. Inhalation, skin absorption

Toxicology. Chlorinated diphenyl oxide causes an acneform dermatitis (chloracne).

Limited experience with humans has shown that exposure to even small amounts of the higher chlorinated derivatives, particularly hexachlorodiphenyl oxide, may result in appreciable acneform dermatitis.[1] No cases of systemic toxicity have been reported in humans.[1]

Chloracne is usually persistent and affects the face, ears, neck, shoulders, arms, chest, and abdomen (especially around the umbilicus and on the scrotum). The most sensitive areas are below and to the outer side of the eye (malar crescent) and behind the ear.[2] The skin is frequently dry with noninflammatory comedones, and pale yellow cysts containing sebaceous matter and keratin.

In laboratory animals, cumulative liver damage has resulted from repeated intake, and, in general, the toxicity increases with the degree of chlorination. Liver injury is characterized by congestion and varying degrees of fatty degeneration. In animals, these compounds cause severe skin irritation on topical application.[3] Animal experiments suggest that absorption from dermal application can result in systemic toxicity, including liver injury and weight loss. In guinea pigs, a single oral dose of 0.05 to 0.1 g/kg of material containing four or more equivalents of chlorine resulted in death 30 days after administration.[1]

REFERENCES

1. Sandmeyer EE, Kirwin CJ Jr: Ethers. In Clayton GD, Clayton FE (eds): Patty's Industrial Hygiene and Toxicology, 3rd ed, rev, Vol 2A, Toxicology, pp 2546–2551. New York, Wiley–Interscience, 1981
2. Crow KD: Chloracne (halogen acne). In Marzulli FN, Maibach HI (eds): Dermatotoxicology, 2nd ed, pp 462–470. New York, Hemisphere Publishing, 1983
3. von Oettingen WF: The Halogenated Aliphatic, Olefinic, Cyclic, Aromatic, and Aliphatic-Aromatic Hydrocarbons Including the Halogenated Insecticides, Their Toxicity and Potential Dangers. US Public Health Service Pub No 414, pp 311–313. Washington, DC, US Government Printing Office, 1955

CHLORINE
CAS: 7782-50-5
Cl_2 1987 TLV-TWA = 1 ppm
TLV-STEL = 3 ppm

Synonyms: None

Physical Form. Greenish-yellow gas with an irritating odor

Uses. Metal fluxing; sterilization of water supplies and swimming pools; bleaching agent; synthesis of chlorinated organic chemicals and plastics; pulp and paper manufacturing; detinning and dezincing iron

Exposure. Inhalation

Toxicology. Chlorine is a potent irritant of the eyes, mucous membranes, and skin, and exposure causes pulmonary irritation.

Mild mucous membrane irritation may occur at 0.2 to 16 ppm; eye irritation occurs at 7 to 8 ppm, throat irritation at 15 ppm, and cough at 30 ppm.[1] A level of 1000 ppm is fatal after a few deep breaths.[1] Other studies have shown that at least some persons develop eye irritation, headache, and cough in concentrations as low as 1 to 2 ppm.

The location and severity of respiratory tract involvement are functions of both the concentration and duration of exposure. With significant exposures, laryngeal edema with stridor, acute tracheobronchitis, chemical pneumonitis, and noncardiogenic pulmonary edema have been described.[2] Accidental exposure of humans to unmeasured but high concentrations for a brief period of time caused burning of the eyes with lacrimation, burning of the nose and mouth with rhinorrhea, cough, choking sensation, and substernal pain.[3] These symptoms were frequently accompanied by nausea, vomiting, headache, dizziness, and sometimes syncope. Of 33 of the victims who were hospitalized, all suffered tracheobronchitis, 23 progressed to pulmonary edema, and, of those, 14 progressed to pneumonitis.[2] Respiratory distress and substernal pain generally subsided within the first 72 hours; cough increased in frequency and severity after 2 to 3 days and became productive of thick mucopurulent sputum; cough disappeared by the end of 14 days.

In another accidental exposure of five chlorine-plant workers and 13 nonworkers, rales, dyspnea, and cyanosis were observed in the most heavily exposed, and cough was present in nearly all the patients. Pulmonary function tests 24 to 48 hours postexposure showed airway obstruction and hypoxemia; these cleared within 3 months except in four of the chlorine workers, who still showed reduced airway flow and mild hypoxemia after 12 to 14 months.[4]

Following acute exposures to chlorine gas, both obstructive and restrictive abnormalities on pulmonary function tests have been observed. Eighteen healthy subjects exposed after a leak from a liquid storage tank had diminished FEV 1.0, FEF 25% to 75%, and other flow rates within 18 hours of exposure. Follow-up studies at 1 and 2 weeks demonstrated resolution of these abnormalities in the 12 subjects with an initial chief complaint of cough, while the 6 subjects with a chief complaint of dyspnea had persistently reduced flow rates. Repeat studies in 5 months were normal in all patients studied except for mildly reduced flow rates in two patients who were smokers.[5]

Of 19 healthy persons exposed in an accident at a pulp mill and tested within 24 hours, 10 (53%) had a reduced FEV 1% (less than 75%), and 13 (68%) had increased residual volumes (greater than 120%), suggesting obstruction with air trapping. Periodic follow-up testing over the next 700 days demonstrated gradual resolution of these abnormalities in all but three subjects tested, who had persistently reduced FEV 1%. Two of these three patients were smokers.[6] In contrast, a study of four healthy patients exposed in a leak at a swimming pool showed acute mild reductions in forced vital capacity, total lung capacity, and diffusing capacity, presumably related to mild interstitial edema. These decrements were not evident until repeat testing in 1 month showed significant improvements above 100% predicted in all these parameters.[7]

In all these studies, some subjects acutely exhibited mild arterial hypoxemia, increases in alveolar-arterial oxygen tension difference, and respiratory alkalosis. Mild transient hyperchloremic metabolic acidosis, with a normal anion gap, has been described in a patient following chlorine inhalation, presumably related to systemic absorption of hydrochloric acid.[8]

Some studies of survivors of massive chlorine exposures have shown either persistent obstructive or restrictive deficits; but pre-exposure data on these patients was not available. Currently, there is insufficient evidence to conclude whether there is a potential for chronic impairment of pulmonary function following acute or chronic chlorine exposure.[9] In several cases, prolonged symptoms following chlorine exposure may be due to aggravation of pre-existing conditions such as tuberculosis, asthma, chronic obstructive pulmonary disease, or heart disease.[9]

In high concentrations, chlorine irritates the skin and causes sensations of burning and pricking, inflammation, and vesicle formation.[10] Liquid chlorine causes eye and skin burns on contact.[11]

The range of reported odor thresholds of chlorine is 0.03 to 3.5 ppm; however, because of olfactory fatigue, odor does not always serve as an adequate warning of exposure.[1]

REFERENCES

1. Committee on Medical and Biological Effects of Environmental Pollutants, National Research Council: Chlorine and Hydrogen Chloride, pp 116–123. Washington, DC, National Academy of Sciences, 1976
2. Chlorine poisoning (editorial). Lancet I: 321–322, 1984
3. Chasis H et al: Chlorine accident in Brooklyn. J Occup Med 4:152–176, 1947
4. Kaufman J, Burkons D: Clinical, roentgenologic and physiologic effects of acute chlorine exposure. Arch Environ Health 23:29–34, 1971
5. Hasan F et al: Resolution of pulmonary dysfunction following acute chlorine poisoning. Arch Environ Health 38:76–80, 1983
6. Charan N et al: Effects of accidental chlorine inhalation on pulmonary function. West J Med 143:333–336, 1985
7. Ploysonsang Y et al: Pulmonary function changes after acute inhalation of chlorine gas. South Med J 75:23–26, 1982
8. Szerlip H, Singer I: Hyperchloremic metabolic acidosis after chlorine inhalation. Am J Med 77:581–582, 1984
9. National Institute for Occupational Safety and Health: Criteria for a Recommended Standard . . . Occupational Exposure to Chlorine. DHEW (NIOSH) Pub No 76-170, pp 29, 36, 56, 84, 101. Washington, DC, US Government Printing Office, 1976
10. Heyroth FF: Halogens. In Patty FA (ed): Industrial Hygiene and Toxicology, 2nd ed, Vol 2, Toxicology, pp 845–849. New York, Interscience, 1963
11. Chemical Safety Data Sheet SD-80, Chlorine, pp 23–26. Washington, DC, MCA, Inc, 1970

CHLORINE DIOXIDE

CAS: 10049-04-4

ClO_2 1987 TLV = 0.1 ppm

Synonyms: Chlorine oxide; chlorine peroxide

Physical Form. Yellow to reddish-yellow gas

Uses. Bleaching cellulose, paper pulp, flour; purification, taste and odor control of water; oxidizing agent; bactericide and antiseptic

Exposure. Inhalation

Toxicology. Chlorine dioxide gas is a severe respiratory and eye irritant.

Exposure of a worker to 19 ppm for an unspecified period of time was fatal.[1] Repeated acute exposure of workers to undetermined concentrations caused eye and throat irritation, nasal discharge, cough, wheezing, bronchitis, and delayed onset of pulmonary edema.[2] Repeated exposure may cause chronic bronchitis.[2]

Delayed deaths occurred in animals after exposure to 150 to 200 ppm for less than 1 hour.[4] Rats exposed daily to 10 ppm died after 10 to 13 days of exposure; effects were nasal and ocular discharge and dyspnea; autopsy revealed purulent bronchitis. No adverse reactions were observed in rats exposed to about 0.1 ppm for 5 hours/day for 10 weeks.[3]

REFERENCES

1. Elkins HB: The Chemistry of Industrial Toxicology, 2nd ed. New York, John Wiley & Sons, 1959
2. Gloemme J, Lundgren KD: Health hazards from chlorine dioxide. AMA Arch Ind Health 16:169, 1957
3. Dalhamn T: Chlorine dioxide toxicity in animal experiments and industrial risks. AMA Arch Ind Health 15:101, 1957

CHLORINE TRIFLUORIDE

CAS: 7790-91-2

ClF_3 1987 Ceiling Limit = 0.1 ppm

Synonym: Chlorine fluoride

Physical Form. Colorless gas, liquefying at 11.2°C

Uses. Fluorinating agent; incendiary; igniter and propellant for rockets; in nuclear reactor fuel processing; pyrolysis inhibitor for fluorocarbon polymers

Exposure. Inhalation

Toxicology. Chlorine trifluoride gas is an extremely severe irritant of the eyes, respiratory tract, and skin in animals.

The injury caused by chlorine trifluoride is in part attributed to its hydrolysis products, including chlorine, hydrogen fluoride, and chlorine dioxide.[1,2] Effects in humans have not been reported but may be very severe; inhalation may be expected to cause pulmonary edema, and contact with eyes or skin may cause severe burns.[1-3]

Exposure of rats to 800 ppm for 15 minutes was fatal, but nearly all survived when exposed for 13 minutes. There was severe inflammation of all exposed mucosal surfaces, resulting in lacrimation, corneal ulceration, and burning of exposed areas of skin.[4] In another study, exposure of rats to 480 ppm for 40 minutes or to 96 ppm for 3.7 hours was fatal; in the latter group, effects were pulmonary edema and marked irritation of the bronchial mucosa.[4] Chronic exposure of dogs and rats to about 1 ppm 6 hours/day for up to 6 months caused severe pulmonary irritation and some deaths.[2]

REFERENCES

1. Horn HJ, Weir RJ: Inhalation toxicology of chlorine trifluoride. I. Acute and subacute toxicity. AMA Arch Ind Health 12:515–521, 1955
2. Horn HJ, Weir RJ: Inhalation toxicology of chlorine trifluoride. II. Chronic toxicity. AMA Arch Ind Health 13:340–345, 1956
3. Boysen JE: Health hazards of selected rocket propellants. Arch Environ Health 7:71–75, 1963
4. Dost FN, Reed DJ, Smith VN, Wang CH: Toxic properties of chlorine trifluoride. Toxicol Appl Pharmacol 27:527–536, 1974

CHLOROACETALDEHYDE

CAS: 107-20-0

$ClCH_2CHO$ 1987 Ceiling Limit = 1 ppm

Synonyms: Monochloroacetaldehyde; 2–chloroacetaldehyde

Physical Form. Liquid

Uses. In the manufacture of 2–aminothiazole; to facilitate bark removal from tree trunks

Exposure. Inhalation, skin absorption

Toxicology. Chloroacetaldehyde is a severe irritant of the eyes, mucous membranes, and skin.

A manufacturer of chloroacetaldehyde has commented on the irritant nature of the substance, but no details were given.[1] Contact of a strong solution with the eye will likely result in permanent impairment of vision. Contact with the skin will cause burns.

Inhalation of 5 ppm by rats caused eye and nasal irritation.[1] In rabbits, the LD_{50} for skin absorption was 0.22 ml/kg for 30% chloroacetaldehyde in water solution. This solution on the skin or in the eyes of rabbits produced severe damage.[2]

Chloroacetaldehyde was determined to be mutagenic in the Ames assay.[1]

REFERENCES

1. Chloroacetaldehyde. Documentation of the TLVs and BEIs, 5th ed, p 120. Cincinnati, American Conference of Governmental Industrial Hygienists (ACGIH), 1986
2. Lawrence WH, Dillingham EO, Turner JE, Autian J: Toxicity profile of chloroacetaldehyde. J Pharm Sci 61:19–25, 1972

α-CHLOROACETOPHENONE

CAS: 532-27-4

$C_6H_5COCH_2Cl$ 1987 TLV = 0.05 ppm

Synonyms: 2–Chloro–1–phenylethanone; phenacyl chloride; 2–chloroacetophenone; phenyl chloromethyl ketone

Physical Form. Crystals

Uses. Chemical warfare agent (CN); principal constituent of riot control agent "Mace"

Exposure. Inhalation

Toxicology. α-Chloroacetophenone is a potent lacrimating agent and an irritant of mucous membranes; it causes primary irritation of the skin.

In one fatal case of exposure, death occurred as a result of pulmonary edema; exposure occurred under unusual circumstances that caused inhalation of high concentrations.[1] Human volunteers exposed to levels of 200 to 340 mg/m^3 could not tolerate exposure for longer than 30 seconds.[2] Effects were lacrimation, burning of eyes, blurred vision, tingling of the nose, rhinorrhea, and burning of the throat.[2] Less frequent symptoms included burning in the chest, dyspnea, and nausea.

Sporadic cases of dermatitis from primary irritation by α-chloroacetophenone have been reported.[3,4] Allergic contact dermatitis to this substance in Chemical Mace has been documented by patch test evaluation, and it is said to be a potent skin sensitizer.[3,4]

Eye splashes cause marked conjunctivitis and may result in permanent corneal damage.[5] The lacrimation threshold ranges from 0.3 to 0.4 mg/m^3, and the odor threshold is 0.1 mg/m^3.[5]

REFERENCES

1. Stein AA, Kirwan WE: Chloroacetophenone (tear gas) poisoning: Clinico-pathologic report. J Forensic Sci 9:374–382, 1964
2. Punte CL, Gutentag PJ, Owens EJ, Gongwer LE: Inhalation studies with chloroacetophenone, diphenylaminoarsine and pelargonic morpholide. II. Human exposure. J Am Ind Hyg Assoc 23:199–202, 1962
3. Penneys NS: Contact dermatitis to chloroacetophenone. Fed Proc 30:96–99, 1971
4. Penneys NS, Israel RM, Indgin SM: Contact dermatitis due to 1-chloroacetophenone and Chemical Mace. N Engl J Med 281:413–415, 1969
5. Mackison FW et al: Occupational Health Guideline for *alpha*-chloroacetophenone. In NIOSH/OSHA Occupational Health Guidelines for Chemical Hazards. DHHS(NIOSH) Pub No 81-123. Washington, DC, US Government Printing Office, 1981

CHLOROBENZENE

CAS: 108-90-7

C_6H_5Cl 1987 TLV = 75 ppm

Synonyms: Phenyl chloride; monochlorobenzene; chlorobenzol; benzene chloride

Physical Form. Colorless liquid

Uses. Manufacture of phenol, aniline, DDT; solvent for paint; color printing; dry-cleaning industry

Exposure. Inhalation

Toxicology. Chlorobenzene is irritating to the skin and mucous membranes and causes central nervous system depression.

In humans, eye and nasal irritation occur at 200 ppm, and, at that level, the odor is pronounced and unpleasant; industrial experience indicates that occasional short exposures are not likely to result in more than minor skin irritation, but prolonged or frequently repeated contact may result in skin burns.[1] In one case of accidental poisoning from ingestion of the liquid by a child, there was pallor, cyanosis, and coma, followed by complete recovery.[2]

Cats exposed to 8000 ppm showed severe narcosis after ½ hour and died 2 hours after removal from exposure, but 660 ppm for 1 hour was tolerated.[3] Exposed animals showed eye and nose irritation, drowsiness, incoordination, and coma, followed by death from the most severe exposures. Several species of animals exposed daily to 1000 ppm for 44 days showed injury to the lungs, liver, and kidneys, but, at 475 ppm, there was only slight liver damage in guinea pigs.

Leukopenia and depressed bone marrow activity were found in mice exposed at 544 ppm 7 hours/day for 3 weeks or at 22 ppm 7 hours/day for 3 months.[4] Only slight transient hematologic effects were found in rats and rabbits exposed at 250 ppm 7 hours/day for 24 weeks.[5] Administered to dogs in capsule form, 272.5 mg/kg/day for up to 92 days caused an increase in immature leukocytes and some deaths.[6] Postmortem findings included gross and/or microscopic pathology in liver, kidneys, gastrointestinal mucosa, and hematopoietic tissue. No consistent effects were observed at 54.5 mg/kg/day.[6] Gastric intubation of 120 mg/kg/day for 2 years produced a slight but statistically significant increase in neoplastic nodules of the liver in male rats.[7]

Although the odor of chlorobenzene is pronounced and unpleasant, it is not sufficient to give warning of hazardous concentrations.[1]

REFERENCES

1. Hygienic Guide Series: Chlorobenzene. Am Ind Hyg Assoc J 25:97–99, 1964

2. von Oettingen WF: The Halogenated Aliphatic, Olefinic, Cyclic, Aromatic, and Aliphatic-Aromatic Hydrocarbons Including the Halogenated Insecticides, Their Toxicity and Potential Dangers. US Public Health Service Pub No 414, pp 283–285. Washington, DC, US Government Printing Office, 1955
3. Deichmann WB: Halogenated cyclic hydrocarbons. In Clayton GD, Clayton FE (eds): Patty's Industrial Hygiene and Toxicology, 3rd ed, rev, Vol 2B, Toxicology, pp 3604–3611. New York, Wiley–Interscience, 1981
4. Zub M: Reactivity of the white blood cell system to toxic actions of benzene and its derivatives. Acta Biol Cracoviensia 21:163–174, 1978
5. Toxicology Evaluation of Inhaled Chlorobenzene (Monochlorobenzene). NTIS PB–276–623. Cincinnati, National Institute for Occupational Safety and Health (NIOSH), Division of Biomedical and Behavioral Sciences, 1977
6. Knapp WK et al: Subacute oral toxicity of monochlorobenzene in dogs and rats. Toxicol Appl Pharmacol 19:393, 1971
7. National Toxicology Program: Toxicology and Carcinogenesis Studies on Chlorobenzene (CAS No 108–90–7) in F344/N Rats and B6C3F Mice (Gavage Studies). Technical Report Series 261, NIH Pub No 86–2517, pp 220. Washington, DC, US Department of Health and Human Services, October 1985

o-CHLOROBENZYLIDENE MALONONITRILE

CAS: 2698-41-1

$ClC_6H_4CH{=}C(CN)_2$

1987 C = 0.05 ppm; skin

Synonyms: CS; OCBM; chlorobenzylidene malononitrile; 2–chlorobenzylidene malononitrile

Physical Form. White, crystalline solid

Uses. Riot-control agent

Exposure. Inhalation; skin absorption

Toxicology. *o*-Chlorobenzylidene malononitrile (CS) aerosol is a potent lacrimator and upper respiratory irritant.

In human experiments, concentrations ranging from 4.3 to 6.7 mg/m³ were barely tolerated when reached gradually over a period of 30 minutes.[1] Following cessation of exposure, a burning sensation and deep pain in the eyes persisted for 2 to 5 minutes. Severe conjunctival injection lasted for 25 to 30 minutes, and erythema of the eyelids with some blepharospasm was present for 1 hour. There was a burning sensation in the throat with cough, followed by a constricting sensation in the chest; no therapy other than removal from exposure was necessary.[1]

At a concentration of 1.5 mg/m³, three of four men developed headache during a 90-minute exposure; one subject developed slight eye and nose irritation.[1,2] On the skin, the powder caused a burning sensation, which was greatly aggravated by moisture; erythema and vesiculation resembling second-degree burns were produced.

In animals, the manifestation of lethal toxicity is different following intravenous, intraperitoneal, oral and inhalation routes. After intravenous administration, there is rapid onset of signs characteristic of effects on the nervous system owing to the alkylating properties of CS.[2] High doses of CS intraperitoneally result in expression of the cyanogenic potential of the malononitrile radical. By the oral route, local inflammation in the gastrointestinal tract contributes to toxicity. Lethal toxicity from inhalation is from lung damage leading to asphyxia or, in the case of delayed deaths, bronchopneumonia secondary to respiratory tract damage. Rats survived a 10-minute exposure at 1800 mg/m³, but 20 of 20 succumbed following 60 minutes at 2700 mg/m³.

REFERENCES

1. Punte CL, Owens EJ, Gutentag PJ: Exposure to ortho-chlorobenzylidene malononitrile. Arch Environ Health 6:366–374, 1963

2. Ballantyne B, Swanston DW: The comparative acute mammalian toxicity of 1–chloroacetophenone (CN) and 2–chlorobenzylidene malononitrile (CS). Arch Toxicol 40:75–95, 1978

CHLOROBROMOMETHANE

CAS: 74-97-5

CH_2BrCl 1987 TLV = 200 ppm

Synonyms: Mono-chloro-mono-bromomethane; bromochloromethane; methylene chlorobromide; monobromomonochloromethane; chloromethyl bromide

Physical Form. Colorless liquid

Uses. Fire-fighting agent

Exposure. Inhalation

Toxicology. Chlorobromomethane is a mild irritant of the eyes and mucous membranes; at high concentrations, it causes central nervous system depression.

Exposure of three firefighters to unknown but very high vapor concentrations was characterized by disorientation, headache, nausea, and irritation of the eyes and throat. Two of the three firefighters became comatose; one had convulsive seizures, and the other had respiratory arrest from which he was resuscitated.[1] Recovery was slow but complete. Some effects may have been due to the inhalation of thermal decomposition products.

Prolonged skin contact may cause dermatitis.[1] The liquid in the eye causes an immediate burning sensation followed by corneal epithelial injury and conjunctival edema.[2]

Concentrations near 30,000 ppm were lethal to rats within 15 minutes; toxic signs included loss of coordination and narcosis; this level of exposure produced pulmonary edema and, in delayed deaths, interstitial pneumonitis.[3] Concentrations as low as 3000 ppm for 15 minutes produced light narcosis in rats.[3] No toxic effects were observed in rats, rabbits, and dogs exposed 7 hours/day, 5 days/week for 14 weeks to 1000 ppm.[4]

Metabolic studies of inhaled chlorobromomethane in rats have shown production of carbon monoxide, halide ions, and other reactive intermediates.[5]

Chlorobromomethane has a distinctive odor at 400 ppm; however, the odor is not disagreeable and does not provide sufficient warning properties.

REFERENCES

1. Rutstein HR: Acute chlorobromomethane toxicity. Arch Environ Health 7:440–444, 1963
2. Grant WM: Toxicology of the Eye, 3rd ed, pp 210–211. Springfield, Illinois, Charles C Thomas, 1986
3. Comstock CC, Fogleman RW, Oberst FW: Acute narcotic effects of monochloromonobromomethane vapor in rats. AMA Arch Ind Hyg Occup Med 7:526–528, 1953
4. Svirbely JL, Highman B, Alford WC, von Oettingen WF: The toxicity and narcotic action of monochloromonobromomethane with special reference to inorganic and volatile bromide in blood, urine and brain. J Ind Hyg Toxicol 29:382–389, 1947
5. Gargas ML et al: Metabolism of inhaled dihalomethanes in vivo: Differentiation of kinetic constants for two independent pathways. Toxicol Appl Pharmacol 82:211–223, 1986

CHLORODIFLUOROMETHANE

CAS: 75-45-6

$CHClF_2$ 1987 TLV = 1000 ppm

Synonyms: Freon 22; monochlorodifluoromethane; di-fluoromonochloromethane

Physical Form. Colorless, nearly odorless gas, nonflammable

Uses. Aerosol propellant; refrigerant; low-temperature solvent

Exposure. Inhalation

Toxicology. Chlorodifluoromethane gas causes central nervous system depression at very high concentrations.

Acute effects in humans have not been reported.[1]

The incidence of cardiac palpitations was compared in two employee groups.[2] One group of 118 employees was exposed to an average concentration of 300 ppm chlorodifluoromethane during its use as a tissue preservative. The control group of 85 employees were from a different department and had no chemical exposure. The number of employees exhibiting palpitations was significantly higher in the exposed group than in the control group.

An epidemiologic study involving workers exposed to chlorofluorocarbons, including chlorodifluoromethane, showed no increased mortality owing to heart, circulatory, or malignant disorders.[1]

Exposure of rats and guinea pigs for 2 hours to levels of 75,000 to 100,000 ppm caused excitation and/or dysfunction in equilibrium.[3] Narcosis occurred at 200,000 ppm, and animals died at 300,000 to 400,000 ppm.

Studies in the dog and other species show that high concentrations (above 50,000 ppm) in association with injected adrenalin are required to produce cardiac arrhythmias.[1] This is a relatively low order of effect by comparison with other chlorofluorocarbons.[1]

Pregnant rats exposed to 50,000 ppm 6 hours/day on days 6–15 of gestation had decreased bodyweight gain, and their offspring had an increased incidence of anophthalmia. At this dose, chlorodifluoromethane did not affect the pregnant rabbit or her offspring, nor was there any effect on male fertility in the rat or mouse.

Evaluation of tumor data from lifetime studies showed an increased incidence of fibrosarcomas, some involving the salivary gland, in male rats chronically exposed to 50,000 ppm. Lifetime rodent studies established a clear no-effect level of 10,000 ppm.

REFERENCES

1. Litchfield MH, Longstaff E: Summaries of toxicological data. The toxicological evaluation of chlorofluorocarbon 22 (CFC22). Fd Chem Toxic 22:465–475, 1984
2. Speizer FE, Wegman DH, Ramirez A: Palpitation rate associated with fluorocarbon exposure in a hospital setting. N Engl J Med 292:624, 1975
3. Chlorodifluoromethane. Documentation of the TLVs and BEIs, 5th ed, p 127. Cincinnati, American Conference of Governmental Industrial Hygienists (ACGIH), 1986

CHLORODIPHENYL, 42% CHLORINE

CAS: 53469-21-9

$C_{12}H_7Cl_3$ 1987 TLV = 1 mg/m^3; skin

Synonyms: Arochlor 1242; polychlorinated biphenyl; PCB

Physical Form. Straw-colored liquid

Uses. Dielectric in capacitors and transformers; investment casting processes; heat exchange fluid; hydraulic fluid

Exposure. Skin absorption, ingestion; inhalation

Toxicology. Chlorodiphenyl, 42% chlorine (a polychlorinated biphenyl or PCB), is an irritant of the eyes and mucous membranes, is toxic to the liver, and causes an acneform dermatitis (chloracne). It is a liver carcinogen in animals.

In humans, systemic effects include anorexia, nausea, edema of the face and hands, and abdominal pain.[1] In a survey of 34 workers exposed to concentrations of up to 2.2 mg/m^3, complaints were of a burning sensation of the face and hands, nausea, and a persistent (uncharacterized) body odor.[1] One worker had chloracne, and five exhibited an eczematous rash on the legs and hands.[1] Although hepatic function tests were normal, the mean blood level of chlorodiphenyl in the exposed group was approximately 400 ppb, whereas none was detected in the control group.[1]

Cases of mild to moderate skin irritation and chloracne have been reported in workers exposed to 0.1 mg/m^3 for several months. Levels of 10 mg/m^3 were unbearably irritating, presumably to mucous membranes and skin.[2] Chloracne does not appear to occur at concentrations below 0.1 mg/m^3.

Chloracne is usually persistent and affects the face, ears, neck, shoulders, arms, chest, and abdomen (especially around the umbilicus and on the scrotum). The most sensitive areas are below and to the outer side of the eye (malar crescent) and behind the ear. The skin is frequently dry with noninflammatory comedones, and pale yellow cysts containing sebaceous matter and keratin. Some evidence of liver disease is often seen in association with PCB-induced chloracne.[3]

Some studies of occupationally exposed groups have revealed evidence of liver injury by serum enzyme studies or other liver function tests. Adverse effects and dose–effect relationships have not been consistent within and between studies, raising the possibility that other factors (*e.g.*, alcohol intake, other exposures) could be responsible.[2] Review of these studies indicates that some liver effects may have occurred with repeated exposures at concentrations below 0.1 mg/m^3, assuming PCBs were responsible. Several deaths from toxic hepatitis have been reported among workers exposed to mixtures of PCBs with chlorinated naphthalenes; such effects have not been observed with PCB exposure alone.[2]

A cross-sectional survey of 205 capacitor manufacturing workers with a geometric mean serum PCB level of 18.2 ppb (SD 2.88) found no statistically significant correlations between PCB levels and clinical chemistry results, including SGOT, GGTP, and LDH levels.[4] The primary dielectric used in the plant was Arochlor 1242. However, another cross-sectional survey of 120 railroad transformer workers with mean plasma PCB levels of 33.4 ppb did reveal statistically significant correlations of PCB level with serum triglyceride and SGOT (but not SGPT or GGTP) levels.[5] There was a significant correlation between self-reported direct dermal contact with PCBs and plasma PCB level. In a survey of 80 heavily exposed capacitor or transformer manufacturing workers in Italy with mean blood PCB levels of about 340 ppb, there was a correlation between blood PCB levels and abnormal liver findings (including hepatomegaly and increased GGTP, SGOT, and SGPT levels).[6] Even in this latter group, except for a few cases of chloracne, no other symptoms or findings referable to PCB exposure were present. The biological significance of these generally mild elevations in serum enzymes is also unclear.

Industrial hygiene studies support the notion that the dermal and dermal/oral, rather than the respiratory, route of exposure are the predominant contributors to body burden among workers occupationally exposed to PCBs.[7]

The toxic effects of PCBs in humans are further illustrated by a 1968 outbreak of poisoning in Japan that involved more than 1000 people who ingested

PCB-contaminated rice bran oil for a period of several months.[8,9] The contamination of the oil (estimated 1500 to 2000 ppm) occurred when heat transfer pipes immersed in the oil during processing developed pin-sized holes. The clinical aspects of the poisoning included chloracne, brown pigmentation of the skin and nails, distinctive hair follicles, increased eye discharge, swelling of eyelids, transient visual disturbance, and systemic gastrointestinal symptoms with jaundice. In some patients, symptoms persisted 3 years after PCB exposure was discontinued. Infants born to poisoned mothers had decreased birthweights and showed skin discoloration. Chemical analysis of the contaminated rice oil has revealed significant amounts of polychlorinated dibenzofurans (PCDFs) as well as PCBs.[10] High concentrations of PCDFs were found in blood and adipose tissue of these so-called *Yusho* victims. In contrast, in a group of workers occupationally exposed to PCBs, PCB levels were higher than in the *Yusho* victims, but PCDFs were not generally detected.[10,11] Animal experiments have reproduced some findings seen in Yusho victims with administration of PCDFs but not PCBs. Thus, it appears that PCDFs were the main causative agents in the induction of *Yusho* disease.[10,11]

PCBs are poorly metabolized and tend to accumulate in animal tissues, including humans.[8] The accumulation, particularly in tissues and organs rich in lipids, appears to be higher in the case of penta- and more highly chlorinated biphenyls.[8] Studies have revealed PCBs in human fat tissue and blood plasma. PCBs, in amounts greater than 2 ppm, were reported in 198 of 637 (31%) samples of human fat tissue taken from the general population of 18 states and the District of Columbia.[8] PCB residues ranging up to 29 ppb were also found in 43% of 616 plasma samples collected from volunteers in a southeastern U.S. county.[8] A recent study of nonoccupationally exposed adults in Southern California found mean plasma PCB levels of 5 ± 4 ppb (±SD), with a range of 1 to 37 ppb.[12] Other studies of the general U.S. population have revealed mean plasma or serum levels of PCBs between 2.1 and 24.4 ppb.[12]

A cohort study of 544 male and 1557 female workers employed in a capacitor manufacturing plant using PCBs in Italy between 1946 and 1978 found statistically significant excesses of total cancer deaths in males (14 obs. vs. 7.6 exp.) and females (12 obs. vs. 5.3 exp.), cancer of the gastrointestinal tract in males (6 obs. vs. 2.2 exp.), and hematologic neoplasms in females (4 obs. vs. 1.1 exp.).[13] Of the six gastrointestinal tract malignancies in males, the primary sites were stomach (two), pancreas (two), liver (one), and biliary tract (one). There was an excess of hematologic neoplasms in males (3 obs. vs. 1.1 exp.), but this excess was not statistically significant. The authors qualified their conclusions regarding excess malignancies because of the small number of deaths in the cohort, the occurrence of some tumors in workers with minimal exposure or short latency intervals, and the disparate sites and types of tumors.[13]

An update of a retrospective cohort mortality study of 2588 workers exposed to PCBs in two capacitor manufacturing plants revealed no excess of all cancers (SMR 78), stomach cancer (SMR 36), pancreatic cancer (SMR 54), or lymphatic and hematopoietic cancer (SMR 68). The only statistically significant excess was for cancer of the liver and biliary passages (5 obs. vs. 1.9 exp., SMR 263). Both Aroclor 1254 (54% chlorine) and 1242 (42% chlorine) had been used at different times in both plants. Although the workers studied had positions involving greater exposure to PCBs than other workers in the plants, historical levels of exposure were unknown. Four of the five cases of liver and biliary tract cancer occurred in women in plant 2. Although all five workers were first employed in the 1940s and early 1950s, when exposures were presumed to be the highest, analysis did not reveal that risk was associated with time since first employment or length of employment in "PCB-exposed" jobs. Attempts to confirm the site of origin of the cancer as liver by review of records were unsuccessful in two of the cases; in one case, no records were available, and, in the other case, records indicated that the primary site was unknown. The author tempered his conclusions because of these limitations and the small number of cases identified.[14]

Single inhalation exposures of brief duration to trivial concentrations of PCBs in air, or to surface residues, are generally of no clinical consequence.

Guinea pigs died at intervals up to 21 days after the first of 11 daily applications of 34.5 mg to the skin; at necropsy, the liver showed fatty degeneration and central atrophy; rats, however, survived 25 daily applications, and only slight changes in the liver were observed.[15]

After application of radiolabeled PCB, 42% chlorine, to the skin of guinea pigs and rhesus monkeys, 33% and 15% to 34%, respectively, of the applied doses were absorbed.[16]

All PCB mixtures adequately tested in mice and rats have shown carcinogenic activity.[2] For example, of 20 rats fed Arochlor 1242 at 100 ppm in the diet for 24 months, 11 developed liver tumors, three of which were hepatomas. Evidence from bioassays suggests that the less highly chlorinated PCBs (*e.g.*, Arochlor 1242) have less carcinogenic potential than the more highly chlorinated mixtures (*e.g.*, Arochlor 1254).[2]

The IARC concluded in 1982 (prior to publication of the two aforementioned cohort studies) that there was inadequate evidence for carcinogenicity of PCBs to humans, but that there was sufficient evidence for carcinogenicity to animals, based on a number of studies in mice and rats in which oral administration resulted in benign and malignant liver neoplasms.[17]

Adverse reproductive effects, including fetotoxicity and teratogenicity, have been observed in animals fed PCBs in the diet. PCBs have been observed in human cord blood and in tissues of newborn humans and animals.[2]

REFERENCES

1. Ouw HK, Simpson GR, Siyali DS: Use and health effects of Arochlor 1242, a polychlorinated biphenyl, in an electrical industry. Arch Environ Health 31:189–194, 1976
2. National Institute for Occupational Safety and Health: Criteria for a Recommended Standard . . . Occupational Exposure to Polychlorinated Biphenyls. DHEW (NIOSH) Pub No 77–225. Washington, DC, US Government Printing Office, 1977
3. von Oettingen WF: The Halogenated Aliphatic, Olefinic, Cyclic, Aromatic, and Aliphatic-Aromatic Hydrocarbons Including the Halogenated Insecticides, Their Toxicity and Potential Dangers. US Public Health Service Pub No 414, pp 311–313. Washington, DC, US Government Printing Office, 1955
4. Acquevella JF et al: Assessment of clinical, metabolic, dietary, and occupational correlations with serum polychlorinated biphenyl levels among employees at an electric capacitor manufacturing plant. J Occup Med 28:1177–1180, 1986
5. Chase KH et al: Clinical and metabolic abnormalities associated with occupational exposure to polychlorinated biphenyls (PCBs). J Occup Med 24:109–114, 1982
6. Maroni M et al: Occupational exposure to polychlorinated biphenyls in electrical workers. II. Health effects. Br J Ind Med 38:55–60, 1981
7. Lees PSJ, Corn M, Breysse P: Evidence for dermal absorption as the major route of body entry during exposure of transformer main-

tenance and repairmen to PCBs. Am Ind Hyg Assoc J 48:257–264, 1987
8. Lloyd JW, Moore RM Jr, Woolf BS, Stein HP: Polychlorinated biphenyls. J Occup Med 18:109–113, 1976
9. Kuratsune M, Yoshimura T, Matsuzaka J, Yamasuchi A: Yusho, a poisoning caused by rice oil contaminated with polychlorinated biphenyls. HSMHA Health Rep 36:1083–1091, 1971
10. Masuda Y, Yoshimura H: Chemical analysis and toxicity of polychlorinated biphenyls and dibenzofurans in relation to Yusho. J Toxicol Sci 7:161–175, 1982
11. Kunita N et al: Causal agents of Yusho. Am J Ind Med 5:45–58, 1984
12. Sahl JD et al: Polychlorinated biphenyl concentrations in the blood plasma of a selected sample of non-occupationally exposed Southern California working adults. Sci Total Environ 46:9–18, 1985
13. Bertazzi PA et al: Cancer mortality of capacitator manufacturing workers. Am J Ind Med 11:165–176, 1987
14. Brown DP: Mortality of workers exposed to polychlorinated biphenyls—an update. Cincinnati, National Institute for Occupational Safety and Health (NIOSH), US Department of Health and Human Services, 1986
15. Miller JW: Pathologic changes in animals exposed to a commercial chlorinated diphenyl. Public Health Rep 59:1085–1093, 1944
16. Werter RC et al: Polychlorinated biphenyls (PCB): Dermal absorption, systemic elimination, and dermal wash efficiency. J Toxicol Environ Health 12:511–519, 1983
17. IARC Monographs on the Evaluation of the Carcinogenic Risk of Chemicals to Humans: Chemicals, Industrial Processes, and Industries, Associated with Cancer in Humans, Suppl 4, pp 217–219. Lyon, International Agency for Research on Cancer, 1982

CHLORODIPHENYL, 54% CHLORINE

CAS: 11097-69-1

$C_{12}H_5Cl_5$ 1987 TLV = 0.5 mg/m^3; skin

Synonyms: Arochlor 1254; polychlorinated biphenyl; PCB

Physical Form. Viscous liquid

Uses. Dielectric in capacitors and transformers; investment casting processes; heat exchange fluid; hydraulic fluid

Exposure. Inhalation; skin absorption

Toxicology. Chlorodiphenyl, 54% chlorine (a polychlorinated biphenyl or PCB) is toxic to the liver of animals, and severe exposure may produce a similar effect in humans. It is a liver carcinogen in animals.

Cases of mild to moderate skin irritation and chloracne have been reported in workers exposed to 0.1 mg/m^3 for several months. Levels of 10 mg/m^3 were unbearably irritating, presumably to mucous membranes and skin.[1] Chloracne does not appear to occur at concentrations below 0.1 mg/m^3.

Chloracne is usually persistent and affects the face, ears, neck, shoulders, arms, chest, and abdomen (especially around the umbilicus and on the scrotum). The most sensitive areas are below and to the outer side of the eye (malar crescent) and behind the ear. The skin is frequently dry, with intense pruritis, noninflammatory comedones and pale yellow cysts containing sebaceous matter and keratin. Some evidence of liver disease is often seen in association with PCB-induced chloracne.[2]

Some studies of occupationally exposed groups have revealed evidence of liver injury by serum enzyme studies or other liver function tests. Adverse effects and exposure–effect relationships have not been consistent within and between studies, raising the possibility that other factors (*e.g.*, alcohol intake, other exposures) could be responsible.[1] Some liver effects may have occurred with repeated exposures at concentrations below 0.1 mg/m^3, assuming that PCBs were responsible. Several deaths resulting from toxic hepatitis have been reported among workers exposed to mixtures of PCBs with chlorinated naphthalenes; such effects have not been observed with PCB exposures alone.[1]

A cross-sectional survey of 205 capacitor manufacturing workers with a geometric mean serum PCB level of 18.2 ppb (SD 2.88) found no statistically significant correlations between PCB levels and clinical chemistry results, including SGOT, GGTP, and LDH levels.[3] The primary dielectric used in the plant was Arochlor 1242. However, another cross-sectional survey of 120 railroad transformer workers with mean plasma PCB levels of 33.4 ppb did reveal statistically significant correlations of PCB level with serum triglyceride and SGOT (but not SGPT or GGPT) levels.[4] There was a significant correlation between self-reported direct dermal contact with PCBs and plasma PCB level.

In a survey of 80 more heavily exposed capacitor or transformer manufacturing workers in Italy with mean blood PCB levels of about 340 ppb, there was a correlation between blood PCB levels and abnormal liver findings (including hepatomegaly and increased GGTP, SGOT, and SGPT levels).[5] Even in this latter group, except for a few cases of chloracne, no other symptoms or findings referable to PCB exposure were present. The biological significance of these generally mild elevations in serum enzymes is also unclear.

Industrial hygiene studies support the notion that the dermal and dermal/oral, rather than the respiratory route of exposure, are the predominant contributors to body burden among workers occupationally exposed to PCBs.[6]

The toxic effects of PCBs in humans are further illustrated by a 1968 outbreak of poisoning in Japan that involved more than 1000 people who ingested PCB-contaminated rice bran oil for a period of several months.[7,8] The contamination of the oil (estimated 1500 to 2000 ppm) occurred when heat transfer pipes immersed in the oil during processing developed pin-sized holes. The clinical aspects of the poisoning included chloracne, brown pigmentation of the skin and nails, distinctive hair follicles, increased eye discharge, swelling of eyelids, transient visual disturbance, and systemic gastrointestinal symptoms with jaundice. In some patients, symptoms persisted 3 years after PCB exposure was discontinued. Infants born to poisoned mothers had decreased birthweights and showed skin discoloration.

Chemical analysis of the contaminated rice oil has revealed significant amounts of polychlorinated dibenzofurans (PCDFs) as well as PCBs.[9] High concentrations of PCDFs were found in blood and adipose tissue of these so-called *Yusho* victims. In contrast, in a group of workers occupationally exposed to PCBs, PCB levels were higher than in the *Yusho* victims, but PCDFs were not generally detected.[9,10] Animal experiments have reproduced some findings seen in *Yusho* victims with the administration of PCDFs but not PCBs. Thus, it appears that PCDFs were the main causative agents in the induction of *Yusho* disease.[9,10]

PCBs are poorly metabolized and tend to accumulate in animal tissues, including humans. The accumulation, particularly in tissues and organs rich in lipids, appears to be higher in the case of penta- and more highly chlorinated biphenyls. Studies have revealed PCBs in human fat tissue and blood plasma. PCBs in amounts greater than 2 ppm were reported in 198 of 637 (31%) samples of human fat tissue taken from the general population of 18 states and the District of Columbia.[7] PCB residues ranging up to 29 ppb have also been found in 43% of 616 plasma samples collected from volunteers in a southeastern U.S. county.[7] A recent study of nonoccupationally exposed adults in Southern California found mean plasma PCB levels of 5 ± 4 ppm (±SD), with a range of 1 to 37 ppb.[11] Other studies of the general U.S. population have revealed mean plasma or serum levels of PCBs between 2.1 and 24.4 ppb.[11]

A cohort study of 544 male and 1556 female workers employed in a capacitor manufacturing plant using PCBs in Italy between 1946 and 1978 found statistically significant excesses of total cancer deaths in males (14 obs. vs. 7.6 exp.) and females (12 obs. vs. 5.3 exp.), cancer of the gastrointestinal tract in males (6 obs. vs. 2.2 exp.), and hematologic neoplasms in females (4 obs. vs. 1.1 exp.).[12] Of the six gastrointestinal tract malignancies in males, the primary sites were stomach (two), pancreas (two), liver (one), and biliary tract (one). There was an excess of hematologic neoplasms in males (3 obs. vs. 1.1 exp.), but this excess was not statistically significant. The authors qualified their conclusions regarding excess malignancies because of the small number of deaths in the cohort, the occurrence of some tumors in workers with minimal exposure or short latency intervals, and the disparate sites and types of tumors.[12]

An update of a retrospective cohort mortality study of 2599 workers exposed to PCBs in two capacitor manufacturing plants revealed no excess of all cancers (SMR 78), stomach cancer (SMR 36), pancreatic cancer (SMR 54), or lymphatic and hematopoietic cancer (SMR 68). The only statistically significant excess was for cancer of the liver and biliary passages (5 obs. vs. 1.9 exp., SMR 263).[13] Both Aroclor 1254 (54% chlorine) and 1242 (42% chlorine) had been used at different times in both plants. Although the workers studied had positions involving greater exposure to PCBs than other workers in the plants, historical levels of exposure were unknown. Four of the five cases of liver and biliary tract cancer occurred in women in plant 2. Although all five workers were first employed in the 1940s and early 1950s, when exposures were presumed to be highest, analysis did not reveal that risk was associated with time since first employment or length of employment in "PCB-exposed" jobs. Attempts to confirm the site of the origin of the cancer as liver by review of records were unsuccessful in two of the cases; in one case, no records were available, and, in the other case, records indicated that the primary site was unknown. The author tempered his conclusions because of these limitations and the small number of cases identified.[13]

Single inhalation exposures of brief duration to trivial concentrations of PCBs in air, or to surface residues, are generally of no clinical significance.

Rats exposed to 5.4 mg/m^3 for 7 hours daily for 4 months showed increased liver weight and injury to liver cells; 1.5 mg/m^3 for 7 months also produced histopathologic evidence of liver damage, which was considered to be of a reversible character.[14] When the liquid was applied to the skin of rabbits, the minimal lethal dose was 1.5 g/kg.[15] The vapor and the liquid are moderately irritating to the eye; contact with skin leads to removal of natural fats and oils, with subsequent drying and cracking of the skin.[15]

After application of radiolabeled PCB, 54% chlorine, to the skin of guinea pigs, 56% of the applied dose was absorbed.[16]

Administration of a PCB mixture (mean chlorine content 54%) twice a week for 6 weeks by means of stomach tube to rats at relatively low dose levels led to histopathologic changes in the liver, increases in cholesterol and triglyceride levels, and serum enzyme increases. At the 2 mg/kg dose, centrilobular hepatic necrosis and elevated cholesterol levels were observed. Increases in bilirubin and triglyceride levels occurred only at 50 mg/kg, and increases in SGOT (AST) and SGPT (ALT) occurred only at doses above 50 mg/kg.[17]

All PCB mixtures adequately tested in mice and rats have shown carcinogenic activity.[1] For example, hepatomas developed in 9 of 22 BALB/cj male mice fed Arochlor 1254 at 300 ppm for 11 months. Of 27 rats fed Arochlor 1254 at 100 ppm in the diet for 24 months, 19 developed liver tumors, six of which were hepatomas, compared with one neoplastic nodule in 23 controls. Evidence from bioassays suggests that the more highly chlorinated PCBs (*e.g.*, Arochlor 1254) have more carcinogenic potential than the less highly chlorinated mixtures (*e.g.*, Arochlor 1242).[1]

IARC concluded in 1982 (prior to publication of the two aforementioned cohort studies) that there was inadequate evidence for carcinogenicity of PCBs to humans, but that there was sufficient evidence for carcinogenicity to animals, based on a number of studies in mice and rats in which oral administration resulted in benign and malignant liver neoplasms.[18]

Adverse reproductive effects have been observed in animals fed PCB in the diet.[1] Fetal resorptions were common, and dose-related incidences of terata were found in pups and piglets when bitches and sows were fed Arochlor 1254 at 1 mg/kg/day or more. PCBs have been observed in human cord blood and in tissues of newborn humans and animals.[1]

REFERENCES

1. National Institute for Occupational Safety and Health: Criteria for a Recommended Standard . . . Occupational Exposure to Polychlorinated Biphenyls. DHEW (NIOSH) Pub No 77–225. Washington, DC, US Government Printing Office, 1977
2. von Oettingen WF: The Halogenated Aliphatic, Olefinic, Cyclic, Aromatic, and Aliphatic-Aromatic Hydrocarbons Including the Halogenated Insecticides, Their Toxicity and Potential Dangers. US Public Health Service Pub No 414, pp 311–313. Washington, DC, US Government Printing Office, 1955
3. Acquavella JF et al: Assessment of clinical, metabolic, dietary, and occupational correlations with serum polychlorinated biphenyl levels among employees at an electrical capacitor manufacturing plant. J Occup Med 28:1177–1180, 1986
4. Chase KH et al: Clinical and metabolic abnormalities associated with occupational exposure to polychlorinated biphenyls (PCBs). J Occup Med 24:109–114, 1982
5. Maroni M et al: Occupational exposure to polychlorinated biphenyls in electrical workers. II. Health effects. Br J Ind Med 38:55–60, 1981
6. Lees PSJ, Corn M, Breysse P: Evidence for dermal absorption as the major route of body entry during exposure of transformer maintenance and repairmen to PCBs. Am Ind Hyg Assoc J 48:257–264, 1987
7. Lloyd JW, Moore RM Jr, Woolf BS, Stein HP: Polychlorinated biphenyls. J Occup Med 18:109–113, 1976
8. Kuratsune M, Yoshimura T, Matsuzaka J, Yamasuchi A: Yusho, a poisoning caused by rice oil contaminated with polychlorinated biphenyls. HSMHA Health Rep 36:1083–1091, 1971
9. Masuda Y, Yoshimura H: Chemical analysis and toxicity of polychlorinated biphenyls and dibenzofurans in relation to Yusho. J Toxicol Sci 7:161–175, 1982
10. Kunita N et al: Causal agents of Yusho. Am J Ind Med 5:45–58, 1984
11. Sahl JD et al: Polychlorinated biphenyl concentrations in the blood plasma of a selected sample of non-occupationally exposed Southern California working adults. Sci Total Environ 46:9–18, 1985
12. Bertazzi PA et al: Cancer mortality of capacitor manufacturing workers. Am J Ind Med 11:165–176, 1987
13. Brown DP: Mortality of workers exposed to polychlorinated biphenyls—an update. Cincinnati, National Institute for Occupational Safety and Health (NIOSH), US Department of Health and Human Services, 1986
14. Treon JF, Cleveland FP, Cappel JW, Atchley RW: The toxicity of the vapors of Arochlor

1242 and Arochlor 1254. Am Ind Hyg Assoc Q 17:204–213, 1956
15. Hygenic Guide Series: Chlorodiphenyls. Am Ind Hyg Assoc J 26:92–94, 1965
16. Werter RC et al: Polychlorinated biphenyls (PCBs): Dermal absorption, systemic elimination, and dermal wash efficiency. J Toxicol Environ Health 12:511–519, 1983
17. Baumann M et al: Effects of polychlorinated biphenyl at low dose levels in rats. Arch Environ Contam Toxicol 12:509–515, 1983
18. IARC Monographs on the Evaluation of the Carcinogenic Risk of Chemicals to Humans: Chemicals, Industrial Processes, and Industries Associated with Cancer in Humans, Suppl 4, pp 217–219. Lyon, International Agency for Research on Cancer, 1982

CHLOROFORM

CAS: 67-66-3

$CHCl_3$ 1987 TLV = 10 ppm

Synonym: Trichloromethane

Physical Form. Colorless liquid

Uses. Manufacture of fluorocarbons for refrigerants, aerosol propellants, plastics; purifying antibiotics; solvent; photographic processing; dry cleaning

Exposure. Inhalation

Toxicology. Chloroform is a central nervous system depressant and hepatotoxin; renal and cardiac damage may also occur. It is carcinogenic in experimental animals.

Chloroform was abandoned as an anesthetic agent because of the frequency of cardiac arrest during surgery and of delayed death owing to hepatic injury.[1] Concentrations used for the induction of anesthesia were in the range of 20,000 to 40,000 ppm, followed by lower maintenance levels.[2] Continued exposure to 20,000 ppm results in respiratory failure, cardiac arrhythmia, and death.[1] Effects of damage to the liver typically are not observed for 24 to 48 hours postexposure.[1] Symptoms include progressive weakness, prolonged vomiting, delirium, coma, and death. Increased serum bilirubin, ketosis, lowered blood prothrombin, and fibrinogen are reported. Death usually occurs on the fourth or fifth day, and autopsy shows massive hepatic necrosis.

In experimental human exposures, 14,000 to 16,000 ppm caused rapid loss of consciousness in humans; 4100 ppm or less caused serious disorientation, whereas single exposures of 1000 ppm caused dizziness, nausea, and aftereffects of fatigue and headache.[2] Prolonged exposure ranging from 77 ppm to 237 ppm caused lassitude; digestive disturbances; frequent, burning urination; and mental dullness; whereas 20 to 70 ppm produced milder symptoms.[3] Of 68 chemical workers exposed regularly to concentrations of 10 to 200 ppm for 1 to 4 years, nearly 25% had hepatomegaly.[1] However, another group exposed repeatedly to about 50 ppm experienced no signs or symptoms.[1]

High concentrations of vapor cause conjunctival irritation and blepharospasm.[4] Liquid chloroform splashed in the eye causes immediate burning pain and conjunctival irritation; the corneal epithelium may be injured, but regeneration is prompt, and the eye returns to normal in 1 to 3 days.[4] Application of chloroform to the skin causes burning pain, erythema, and vesiculation.[1]

Several substances alter the toxicity of chloroform in animals—most probably by modifying the metabolism to a reactive intermediate.[1] Factors that potentiate chloroform's toxic effects include ethanol, polybrominated biphenyls, steroids, and ketones. Disulfiram, its metabolites, and high carbohydrate diet appear to protect somewhat against chloroform toxicity.[1]

In animals, chloroform causes some fetal loss and delays in fetal development when administered during gestation at levels of 100 ppm or greater.[5,6] Teratogenic effects such as cleft palate were observed in the mouse at doses associated with maternal toxicity.[5]

Evidence for the carcinogenicity of chloroform in experimental animals following chronic oral administration includes statistically significant increases in renal epithelial tumors in male rats, hepatocellular carcinomas in mice, and renal tumors in male mice.[7-9] Chloroform has also been shown to promote growth and metastasis of murine tumors.[10] In these studies, the carcinogenicity of chloroform is organ specific to primarily the liver and kidneys; these organs are also the target of acute chloroform toxicity and covalent binding by reactive intermediates (phosgene, carbene, chlorine ion) of chloroform metabolism.[1]

Small increases in rectal, bladder, and colon cancer have been observed in several studies of human populations with chlorinated drinking water. Because other possible carcinogens were present along with chloroform, it is impossible to identify chloroform as the sole carcinogenic agent. Based on sufficient animal evidence and limited epidemiologic evidence, IARC regards chloroform as a probable human carcinogen.[11]

REFERENCES

1. US Environmental Protection Agency: Health Assessment Document for Chloroform. Final Report. Washington, DC, Office of Health and Environmental Assessment, September 1985
2. National Institute for Occupational Safety and Health: Criteria for a Recommended Standard . . . Occupational Exposure to Chloroform. DHEW (NIOSH) Pub No 75–114. Washington, DC, US Government Printing Office, 1974
3. Challen PJ, Hickish DE, Bedford J: Chronic chloroform intoxication. Br J Ind Med 15:243–249, 1958
4. Winslow SG, Gerstner HB: Health aspects of chloroform—a review. Drug Chem Toxicol 1:259–275, 1978
5. Murray FA et al: Toxicity of inhaled chloroform in pregnant mice and their offspring. Toxicol Appl Pharmacol 50:515–522, 1979
6. Schwetz BA et al: Embryo- and fetotoxicity of inhaled chloroform in rats. Toxicol Appl Pharmacol 25:442–451, 1974
7. National Cancer Institute (NCI): Report on Carcinogenesis Bioassay of Chloroform. PB–264018. Springfield, Virginia, National Technical Information Service, 1976
8. Jorgenson TA et al: Carcinogenicity of chloroform in drinking water to male Osborne Mendel rats and female B6C3F1 mice. Fund Appl Toxicol 5:760–769, 1985
9. Roe FJC et al: Safety evaluation of toothpaste containing chloroform. I. Long-term studies in mice. J Environ Pathol Toxicol 2:799–819, 1979
10. Capel ID et al: The effect of chloroform ingestion on the growth of some murine tumors. Eur J Cancer 15:1485–1490, 1979
11. IARC Monographs on the Evaluation of the Carcinogenic Risk of Chemicals to Man, Vol 20, Some Halogenated Hydrocarbons, pp 401–417. Lyon, International Agency for Research on Cancer, 1979

---

bis (CHLOROMETHYL) ETHER

CAS: 542-88-1

$ClCH_2OCH_2Cl$ 1987 TLV = 0.001 ppm; recognized human carcinogen

---

Synonyms: BCME; Chloromethyl ether; chloro(chloromethoxy)methane; dichloromethyl ether; symmetrical-dichloro-dimethyl ether; dimethyl–1–1′–dichloroether

Physical Form. Colorless liquid

Uses. Chemical intermediate

Exposure. Inhalation; skin absorption

Toxicology. Bis (chloromethyl) ether (BCME) exposure is associated with an increased incidence of lung cancer in humans; BCME is highly carcinogenic in mice and rats.

A retrospective study of 136 BCME workers employed at least 5 years revealed five cases of lung cancer, which represented a ninefold increase in lung cancer risks; 0.54 cases would have been expected to occur in the plant population.[1] The predominant histologic type of carcinoma was small cell–undifferentiated; exposure ranged from 7.5 to 14 years, and the mean induction period was 15 years.[1] In addition, abnormal sputum cytology was observed in 34% of 115 current workers with exposure to BCME for 5 or more years, as contrasted with 11% in a control group.

In another study, six cases of lung cancer occurred among 18 experimental technical department workers, a group known to experience very high BCME exposure; two other cases of lung cancer were reported among 50 production workers.[2] Oat-cell carcinomas occurred in five of eight cases.[2]

BCME is also found as an impurity (1% to 7%) in the related chloromethyl methyl ether (CMME); 14 cases of lung cancer, mainly of oat-cell type, were reported in a chemical plant where exposure to CMME occurred.[3] In the reported epidemiologic studies, insufficient evidence is available to differentiate the carcinogenic effects of the two compounds.[4]

It is noteworthy that in these studies, it was difficult to determine those workers actually exposed to BCME. Therefore, the entire production force had to be considered at risk, thus making the estimated incidence conservative.[1] In addition, many workers in the study had not experienced a sufficient latency period for lung cancer.[1] Most of the men who developed lung cancer were cigarette smokers, suggesting that smoking may interact with the primary carcinogen in a promotional or synergistic manner.[1]

Exposure to 1 ppm for 6 hours/day, 5 days/week for 82 days caused lung tumors in 26 of 47 animals, with an average of 5.2 tumors per tumor-bearing animal; 20 of 49 controls developed lung tumors, with 2.2 tumors per tumor-bearing animal.[5] In 19 rats exposed to 0.1 ppm BCME 6 hours/day for 101 exposures, five squamous cell carcinomas of the lung and five esthesioneuroepitheliomas arising from the olfactory epithelium were observed.[6] Cutaneous application of 2 mg BCME applied to mice three times per week for 325 days caused papillomas in 13 of 20 animals; 12 of these papillomas progressed to squamous cell carcinomas.[7] The IARC has concluded that there is sufficient evidence of carcinogenicity of BCME to both humans and animals.[8]

BCME is a mucous membrane and respiratory irritant in both humans and animals.[7,9] Acute exposure of rats and hamsters resulted in pulmonary edema, hemorrhage, and necrotizing bronchitis. In humans, concentrations of 3 ppm are reported to be distinctly irritating. A fatal case of accidental acute poisoning of a research chemist by BCME has been reported.[9] Increased frequency of chronic cough and low end–expiratory flowrates has been described to occur in a dose-related fashion with exposure to BCME and CMME.[10]

REFERENCES

1. Lemen RA et al: Cytologic observations and cancer incidence following exposure to BCME. Ann NY Acad Sci 271:71–80, 1976
2. Theiss AM, Hay W, Zeller H: Zur Toxikologie von Dichlorodimethylather-Verdacht, auf kanzerogene Wirking auch beim Menschen. (Toxicology of bis[chloromethyl]ether—suspicion of carcinogenicity in man). Zentralbl Arbeitsmed 23:97–102, 1973
3. Figueroa WG, Raszkowski R, Weiss W: Lung

cancer in chloromethyl methyl ether workers. N Engl J Med 288:1096–1097, 1973
4. IARC Monographs on the Evaluation of the Carcinogenic Risk of Chemicals to Man, Vol 4, Some Aromatic Amines, Hydrazine and Related Substances, N–Nitroso Compounds and Miscellaneous Alkylating Agents, pp 231–238. Lyon, International Agency for Research on Cancer, 1974
5. Leong BKJ, Macfarland HN, Reese WH Jr: Induction of lung adenomas by chronic inhalation of bis (chloromethyl) ether. Arch Environ Health 22:663–666, 1971
6. Laskin S et al: Tumors of the respiratory tract induced by inhalation of bis(chloromethyl)-ether. Arch Environ Health 23:135–136, 1971
7. Van Duuren BL et al: Alpha-haloethers: A new type of alkylating carcinogen. Arch Environ Health 16:472–476, 1968
8. IARC Monographs on the Evaluation of the Carcinogenic Risk of Chemicals to Humans: Chemicals, Industrial Processes and Industries Associated with Cancer in Humans, Suppl 4, pp 64–66. Lyon, International Agency for Research on Cancer, 1982
9. Environmental Protection Agency: Ambient Water Quality Criteria for Chloroalkyl Ethers, pp C–25, C–27. Springfield, Virginia, NTIS (USEPA), 1980
10. Weiss W: Chloromethyl ethers, cigarettes, cough and cancer. J Occup Med 18:194–199, 1976

CHLOROMETHYL METHYL ETHER

CAS: 107-30-2

$ClCH_2OCH_3$ — 1987 TLV = none established; suspected human carcinogen

Synonyms: CMME; dimethylchloroether; methyl chloromethyl ether

Physical Form. Colorless liquid

Uses. Chemical intermediate; preparation of ion-exchange resins

Exposure. Inhalation

Toxicology. Chloromethyl methyl ether (CMME) exposure has been associated with an increased incidence of human lung cancer.

Among 111 CMME workers observed during a 5-year period, there were four cases of lung cancer; this was 8 times the incidence of a control group of plant workers with similar smoking histories.[1] Evidence of a lung cancer risk was further supported by the retrospective identification of a total of 14 cases among chemical operators in a plant engaged in synthesis of CMME. Except for one case of doubtful exposure, duration of exposure was 3 to 14 years, and age at diagnosis ranged from 33 to 55 years. During the synthetic process, fumes were often visible. The employees considered it to be a good day if the entire building had to be evacuated only three or four times per 8-hour shift because of noxious fumes. Three of the men had never smoked, and one had smoked a pipe only; the other 10 workers had smoked one or more packs of cigarettes per day. Oat-cell carcinoma was histologically confirmed in 12 cases, whereas the doubtful exposure case was squamous-cell carcinoma; cell type in one case was not determined.[1]

In another study of 669 workers exposed from 1948 to 1972, 19 died of lung cancer, while only 5.6 cases were expected.[2] There were higher relative risks for workers exposed to intermediate to high levels of CMME for 1 or more years.[2]

It should be noted that commercial CMME contains 1% to 7% of highly carcinogenic bis (chloromethyl) ether (BCME); in the reported epidemiologic studies, insufficient evidence is available to differentiate the carcinogenic effects of the two compounds.[3] Further, when CMME is hydrolyzed, HCl and formaldehyde are produced, which may recombine to form BCME. Therefore, although findings may reflect the carcinogenicity of BCME, commercial grade CMME must also be considered to be a carcinogen, although perhaps of a lower potency than BCME.

In a study of 276 men exposed to CMME and followed through 1980 at a plant in the United Kingdom in operation since 1948, there were 10 deaths from lung cancer, with a relative risk of 10.97 compared with an unexposed group.[4] Occurrence of lung cancer appeared to be related to both estimated exposure level and duration of exposure. Among a subgroup of 51 workers who began work after the process was enclosed in 1972, no deaths from lung cancer had been observed through 1980. In another factory where 394 men had been exposed to CMME at lower estimated exposure levels, no excess of lung cancer was observed.[4]

CMME is a mucous membrane and respiratory irritant in humans and animals.[5,6] Human exposure to CMME has been reported to cause breathing difficulties, sore throat, fever, and chills.[5] Acute exposure of rats and hamsters resulted in pulmonary edema and hemorrhage and necrotizing bronchitis.[6] An increased frequency of chronic cough and low end–expiratory flow rates has been observed in a dose-related fashion with exposure to CMME and BCME.[7]

Technical-grade CMME (contaminated with BCME) on subcutaneous injection in mice has produced local sarcomas.[3] Dermal application of mice, followed by a phorbol ester promoter, resulted in an apparent excess of skin papillomas and carcinomas. Inhalation studies in mice showed an equivocally increased occurrence of lung tumors compared with unexposed controls.[3]

The IARC has concluded that there is sufficient evidence for carcinogenicity of technical-grade CMME to humans and animals.[8]

REFERENCES

1. Figueroa WG, Raszkowski R, Weiss W: Lung cancer in chloromethyl methyl ether workers. N Engl J Med 288:1096–1097, 1973
2. DeFonso LR, Kelton SC Jr: Lung cancer following exposure to chloromethyl methyl ether. Arch Environ Health 31:125–130, 1976
3. IARC Monographs on the Evaluation of the Carcinogenic Risk of Chemicals to Man, Vol 4, Some Aromatic Amines, Hydrazine and Related Substances, N–nitroso Compounds and Miscellaneous Alkylating Agents, pp 239–244. Lyon, International Agency for Research on Cancer, 1974
4. McCallum RI, Woolley V, Petrie A: Lung cancer associated with chloromethyl methyl ether manufacture: An investigation at two factories in the United Kingdom. Br J Ind Med 40:384–389, 1983
5. Van Duuren BL et al: Alpha-haloethers: A new type of alkylating carcinogen. Arch Environ Health 16:472–476, 1968
6. Ambient Water Quality Criteria for Chloroalkyl Ethers, pp C–25, C–27. Springfield, Virginia, National Technical Information Service, US Environmental Protection Agency, 1980
7. Weiss W: Chloromethyl ethers, cigarettes, cough and cancer. J Occup Med 18:194–199, 1976
8. IARC Monographs on the Evaluation of the Carcinogenic Risk of Chemicals to Humans. Suppl 4, pp 64–66. Lyon, International Agency for Research on Cancer, 1982

1–CHLORO–1–NITROPROPANE

CAS: 600-25-9

$CH_3CH_2CHClNO_2$ 1987 TLV = 2 ppm

Synonyms: None

Physical Form. Liquid

Uses. Fungicide

Exposure. Inhalation

Toxicology. 1–Chloro–1–nitropropane is an irritant of the eyes and mucous membranes. It is a pulmonary irritant in animals, and severe exposure is expected to cause the same effects in humans. Systemic effects in humans have not been reported.

It is five times more toxic to rabbits than the unchlorinated mononitropropane.[1] Rabbits exposed to 2600 ppm for 2 hours died, but 2200 ppm for 1 hour was nonlethal. Effects included irritation of eyes and mucous membranes; autopsy revealed pulmonary edema and cellular necrosis of the heart, liver, and kidneys.[2,3]

REFERENCES

1. Stokinger HE: Aliphatic nitro compounds, nitrates, nitrites. In Clayton GD, Clayton FE (eds): Patty's Industrial Hygiene and Toxicology, 3rd ed, rev, Vol 2C, Toxicology, pp 4143–4162. New York, Wiley–Interscience, 1982
2. Machle W et al: The physiological response of animals to certain mononitroparaffins. J Ind Hyg Toxicol 27:95–102, 1945
3. Browning E: Toxicity and Metabolism of Industrial Solvents, pp 292–293. Amsterdam, Elsevier, 1965

CHLOROPICRIN

CAS: 76-06-2

CCl_3NO_2 1987 TLV = 0.1 ppm

Synonyms: Trichloronitromethane; nitrochloroform

Physical Form. Colorless, slightly oily liquid

Uses. Fumigant for cereals and grains; a soil insecticide; war gas

Exposure. Inhalation

Toxicology. Chloropicrin is a severe irritant of the eyes, mucous membranes, skin, and lungs.

A lethal exposure for humans is stated to be 119 ppm for 30 minutes, death usually resulting from pulmonary edema. Particular injury occurs in the medium and small bronchi.[1,2] In addition to pulmonary irritation, human exposure results in lacrimation, cough, nausea, vomiting, and skin irritation; persons injured by inhalation of chloropicrin vapor are said to be more susceptible to subsequent exposures.[1]

A concentration of 15 ppm could not be tolerated longer than 1 minute, even by persons acclimated to chloropicrin; exposure to 4 ppm for a few seconds is temporarily disabling because of irritant effects.[2] Concentrations of 0.3 to 0.37 ppm have resulted in painful eye irritation in 3 to 30 seconds.[1]

A man accidentally exposed to residual spray of undetermined concentration had dry cough, and his nasal and pharyngeal mucosa were red and edematous.[3]

In mice, exposure to 9 ppm caused a 50% decrease in respiratory rate. Lesions included ulceration and necrosis of the respiratory epithelium and moderate damage to lung tissue.[4]

REFERENCES

1. Stokinger HE: Aliphatic nitro compounds, nitrates, nitrites. In Clayton GD, Clayton FE (eds): Patty's Industrial Hygiene and Toxicology, 3rd ed, Vol 2C, Toxicology, pp 4164–4166. New York, Wiley–Interscience, 1982
2. Chloropicrin. Documentation of TLVs and BEIs, 5th ed, p 134. Cincinnati, American Conference of Governmental Industrial Hygienists (ACGIH), 1986
3. TeSlaa G et al: Chloropicrin toxicity involving animal and human exposure. Vet Hum Toxicol 28:323–324, 1986
4. Buckley LA et al: Respiratory tract lesions induced by sensory irritants at the LD_{50} concentrations. Toxicol Appl Pharmacol 74:417–429, 1984

CHLOROPRENE

CAS: 126-99-8

$CH_2{=}CClCH{=}CH_2$ 1987 TLV = 10 ppm; skin

Synonyms: Chlorobutadiene; 2-chlorobuta-1,3-diene

Physical Form. Colorless liquid

Uses. In manufacture of synthetic rubber

Exposure. Inhalation; skin absorption

Toxicology. Chloroprene causes central nervous system abnormalities as well as skin and eye irritation. It may have reproductive, mutagenic, embryotoxic, and/or carcinogenic effects.

Exposure of workers to high concentrations for short periods of time led to temporary unconsciousness; one fatality occurred after a 3- to 4-minute exposure inside an unventilated polymerization vessel containing chloroprene vapor.[1] Experimental exposure of humans to 973 ppm led to nausea and giddiness in resting subjects in 15 minutes and in subjects performing light work in 5 to 10 minutes.[1] Extreme fatigue and unbearable chest pain occurred after approximately 1 month of exposure to levels ranging from 56 ppm to greater than 334 ppm. Irritability, personality changes, and reversible hair loss were also reported.

A significant rise in the number of chromosome aberrations was observed in a number of studies of workers exposed to chloroprene at and below 5 ppm.[2] Functional disturbances in spermatogenesis and morphologic abnormalities of sperm were observed among workers occupationally exposed to 0.28 to 1.94 ppm chloroprene.[3] A threefold excess of miscarriages by the wives of these workers was also reported.

Two Russian studies suggested an increased incidence of lung and skin cancers in chloroprene-exposed workers compared with a variety of control groups.[4] A U.S. study of cancer mortality among two cohorts of males engaged in the production and/or polymerization concluded that there was no significant excess of lung cancer deaths.[5] However, there was a disproportionately high incidence of lung cancer cases in maintenance workers who had potentially high exposure to chloroprene. Numerous limitations in these studies preclude evaluation of the carcinogenic risk to humans from chloroprene.[2] Because most reported effects have involved mixed exposures to multiple substances and to short-chain polymers of chloroprene, all the reported symptoms cannot be assigned to the monomer alone.[4]

Contact with skin may cause chemical burns. Conjunctivitis and focal necrosis of the cornea have been reported from eye exposure.[4]

In acute animal studies, the concentrations that killed at least 70% of animals with an 8-hour exposure were 170 ppm for mice, 700 ppm for cats, 2000 ppm for rabbits, and 4000 to 6000 ppm for rats.[6] Symptoms included inflammation of the mucous membranes of the eyes and nose, followed by central nervous system depression and death from respiratory failure. Repeated exposure of rats 6 hours/day, 5 days/week for 4 weeks caused skin and eye irritation and growth depression at 40 ppm; at 160 and 625 ppm, it resulted in loss of hair, morphologic liver damage, and increased mortality.[7]

Exposure of male rats at concentrations of 120 to 6227 ppm and of male mice at concentrations of 12 to 152 ppm for 8 hours resulted in sterility or impotence in 13 of 19 rats and in 8 of 14 mice vs. a mean of 0.5 in two control groups.[6] Degenerative changes in the testes were observed in some of the exposed animals. A significant increase in embryonic mortality was observed in female rats fertilized by males exposed to 1 ppm 4 hours/day for 48 days.[2,4]

Hydrocephalus and cerebral herniation occurred in all fetuses from rat dams given oral doses of 0.5 mg/kg during 14 days of pregnancy. Inhalation of 1.11 ppm for 2 days of the pregnancy also caused increases in these anomalies.[2,4] In another study, neither embryotoxic nor convincing teratologic effects were found after exposing rats to 1, 10, or 25 ppm.[8]

No carcinogenic effects were found in rats following oral, subcutaneous, or intratracheal administration, or in mice by skin application.[2] The IARC Working Group considered the studies inadequate to evaluate the carcinogenic risk in animals.[2]

REFERENCES

1. Nystrom AE: Health hazards in the chloroprene industry and their prevention. Acta Med Scand Suppl 132:5–125, 1948
2. IARC Monographs on the Evaluation of the Carcinogenic Risk of Chemicals to Humans. Some Monomers, Plastics and Synthetic Elastomers, and Acrolein, Vol 19, pp 131–156. Lyon, International Agency for Research on Cancer, 1979
3. Sanotskii IV: Aspects of the toxicology of chloroprene: Immediate and long-term effects. Environ Health Perspect 17:85–93, 1976
4. National Institute for Occupational Safety and Health: Criteria for a Recommended Standard . . . Occupational Exposure to Chloroprene. DHEW (NIOSH) Pub No 77–210, pp 1–176. Washington, DC, US Government Printing Office, 1977
5. Pell S: Mortality of workers exposed to chloroprene. J Occup Med 20:21–29, 1978
6. von Oettingen WF et al: 2–Chloro-butadiene (chloroprene): Its toxicity and pathology and the mechanism of its action. J Ind Hyg Toxicol 18:240–270, 1936
7. Clary JJ et al: Toxicity of B–Chloroprene (2–chlorobutadiene–1,3): Acute and subacute toxicity. Toxicol Appl Pharmacol 46:375–384, 1978
8. Culik R et al: B–Chloroprene (2–chlorobutadiene–1,3) embryotoxic and teratogenic studies in rats (abstr no 194). Toxicol Appl Pharmacol 37:172, 1976

CHLORPYRIFOS

CAS: 2921-88-2

$(C_2H_5O)_2P(S)OC_6H_2Cl_3$

1987 TLV = 0.2 mg/m^3; skin

Synonyms: O,O–Diethyl–O–(3,5,6–trichloro–2–pyridinyl) phosphorothioate; Dursban; Dowco 179; ENT 27311; Eradex; Lorsban; NA 2783; OMS–0971; Pyrinex

Physical Form. White crystalline solid

Uses. Insecticide

Exposure. Inhalation; skin absorption; ingestion

Toxicology. Chlorpyrifos is an anticholinesterase agent.

Signs and symptoms of overexposure are caused by the inactivation of the enzyme cholinesterase, which results in the accumulation of acetylcholine at synapses in the nervous system, skeletal and smooth muscles, and secretory glands. The sequence of the development of systemic effects varies with the route of entry. The onset of signs and symptoms is usually prompt but may be delayed up to 12 hours.[1–4]

Chlorpyrifos does not have enough vapor pressure to present a vapor hazard; however, if dispersed as a mist, particulate inhalation is possible.

After inhalation, respiratory and ocular effects are the first to appear, often within a few minutes of exposure. Respiratory effects include tightness in the chest, wheezing resulting from bronchoconstriction, and excessive bronchial secretion; laryngeal spasms and excessive salivation may add to the respiratory distress; cyanosis may also occur. Ocular effects include miosis, blurring of distant vision, tearing, rhinorrhea, and frontal headache.

After ingestion, gastrointestinal effects such as anorexia, nausea, vomiting, abdominal cramps, and diarrhea appear within 15 minutes to 2 hours. After skin absorption, localized sweating and muscular fasciculations in the immediate area usually occur within 15 minutes to 4 hours; skin absorption is somewhat greater at higher ambient temperatures and is increased by the presence of dermatitis.

With severe intoxication by all routes, an excess of acetylcholine at the neuromuscular junctions of skeletal muscles causes weakness aggravated by exertion, involuntary twitchings, fasciculations, and, eventually, paralysis. The most serious consequence is paralysis of the respiratory muscles. Effects on the central nervous system include giddiness, confusion, ataxia, slurred speech, Cheyne–Stokes respiration, convulsions, coma, and loss of reflexes. The blood pressure may fall to low levels, and cardiac irregularities, including complete heart block, may occur.

Complete symptomatic recovery usually occurs within 1 week; increased susceptibility to the effects of anticholinesterase agents persists for up to several weeks after exposure.[5] Daily exposure to concentrations that are insufficient to produce symptoms following a single exposure may result in the onset of symptoms. Continued daily exposure may be followed by increasingly severe effects.[6]

Human subjects who ingested chlorpyrifos once daily for 4 weeks showed depression of plasma cholinesterase but were symptomless at a dose of 0.1 mg/kg. When four repeated doses were applied to the skin of human volunteers for 12 hours each, doses of 25 mg/kg depressed plasma cholinesterase but caused no symptoms. Chlorpyrifos and its principal metabolite, 3,5,6–trichloro–2–pyridinyl are rapidly eliminated and therefore have a low potential to accumulate in humans on repeated exposures.[9]

There was no evidence of teratologic or reproductive effects in rats fed 1.0 mg/kg/day during a three-generation study; it was not teratogenic in mice at gavage doses of up to 25 mg/kg/day.[10]

The persistent strong odor is most likely due to the sulfur content of the pesticide.

Phosphorothioate anticholinesterases owe their activity to an oxidative reaction in which the P=S group is converted to P=O, which increases the potency considerably. This metabolite inactivates cholinesterase by phosphorylation of the active site of the enzyme to form the "diethylphosphoryl enzyme." Over the following 24 to 48 hours, there is a process, termed *aging*, of conversion to the "monoethylphosphoryl enzyme." Aging is of clinical interest in the treatment of poisoning, because cholinesterase reactivators such as pralidoxime (2–PAM, Protopam) chloride are ineffective after aging has occurred.

Diagnosis. Initial signs and symptoms include headache, blurred vision, pallor, weakness, sweating, abdominal pain, nausea, vomiting, and diarrhea.

Moderate to severe intoxication includes the following signs and symptoms: miosis, lacrimation, excessive salivation, muscle fasciculations, dyspnea, cyanosis, convulsions, shock, cardiac arrhythmias, and coma.

Differential Diagnosis: Diagnosis is primarily based on a history of exposure and clinical evidence of diffuse parasympathetic stimulation. Careful observation of the effects of atropine and pralidoxime may be valuable. Patients with organophosphate poisoning are resistant to the action of atropine at moderate dosages; failure of 1 to 2 mg of atropine administered parenterally to produce signs of atropinization (flushing, mydriasis, tachycardia, or dryness of mouth) indicates organophosphate poisoning. Intravenous injection of 1 g pralidoxime generally results in some recovery from signs and symptoms.

Special Tests: Two types of cholinesterase are clinically significant (1) true acetylcholinesterase, found principally in the nervous system and the red blood cell; and (2) pseudo- or butyrylcholinesterase, found in the plasma, liver, and the nervous system. Although the action of both types is inhibited by organophosphates, the level of depression of red blood cell cholinesterase is a better indicator of clinically significant reduction of cholinesterase activity in the nervous system.

Laboratory evidence of depression of red blood cell cholinesterase to a level substantially below pre-exposure levels (at least 50% and usually much lower) is verification of poisoning. There is an imperfect correlation between the degree of cholinesterase enzymes and the occurrence of symptoms. With a rapid drop in cholinesterase activity, generally reflecting an acute heavy exposure, there may be symptoms with only a 30% depression, whereas with slower drops to 70% depression, reflecting chronic low level exposure, there may be no symptoms.[7]

If no pre-exposure baseline has been performed but symptoms are not sufficient to justify treatment with atropine, repeated testing during the recovery period demonstrating progressively increasing plasma and red blood cell cholinesterase levels over several days and weeks, respectively, suggests the diagnosis of anticholinesterase poisoning.

There are many different methods of estimating cholinesterase content of blood, and associated with each method is a different set of normal values and a different set of reporting units. The laboratory report of a cholinesterase determination should state the units involved, along with the appropriate normal range. Based upon the Michel method, the normal range of red blood cell cholinesterase activity (delta *p*H/hour) is 0.39 to 1.02 for men, and 0.34 to 1.10 for women.[8] The normal range of the enzyme activity (delta *p*H/hour) of plasma is 0.44 to 1.63 for men, and 0.24 to 1.54 for women.

Treatment. Treatment of organophosphate poisoning ranges from simple removal from exposure in very mild cases to the provision of rigorous supportive and antidotal measures in severe cases.[2,3,9,10] In moderate to severe cases, because of pulmonary involvement, there may be need for artificial respiration using a positive-pressure method. Careful attention must be paid to removal of secretions and to maintenance of a patent airway. Anticonvulsants such as thiopental sodium may be necessary. Maintenance of respiration is critical, because death usually results from weakness of the muscles of respiration and accumulation of excessive secretions in the respiratory tract.[2]

As soon as cyanosis has been overcome, 2 to 4 mg of atropine should be given intravenously. (Atropine may induce ventricular fibrillation in the presence of cyanosis.) *This dose of atropine is approximately ten times the amount that is administered for other conditions in which atropine is considered therapeutic.* This dose should be repeated at 5- to 10-minute intervals until signs of atropinization appear (dry, flushed skin, tachycardia as high as 140 beats/minute, and pupillary dilatation). A mild degree of atropinization should be maintained for at least 48 hours.[2]

Pralidoxime (2–PAM, Protopam) chloride is a cholinesterase reactivator that complements the action of atropine. In moderate to severe cases, the dose for adults is 1 to 2 g injected intravenously at a rate not in excess of 500 mg/minute. After 1 hour, a second dose of 1 g is indicated if muscle weakness has not been relieved. Treatment with pralidoxime chloride will be most effective if given within 24 hours after poisoning.[2] Morphine, aminophylline, and phenothiazines are contraindicated because of documented experience of adverse reactions in cases of organophosphate poisoning.[10]

It is of great importance to decontaminate the patient. Contaminated clothing should be removed at once, and the skin should be washed with generous amounts of soap or detergent and a flood of water; this is best accomplished under a shower or by submersion in a pond or other body of water if the exposure occurred in the field. Careful attention should be paid to cleansing of the skin and hair.

The patient should be attended to and monitored continuously for at least 24 hours, because serious and sometimes fatal relapses have occurred as a result of continuing absorption of the toxin or dissipation of the effects of the antidote.

Regeneration of cholinesterase is primarily by synthesis of new enzyme and takes place at the rate of approximately 1% per day.[10] A patient who has recovered from the acute phase of poisoning remains hypersusceptible to anticholinesterases for up to several weeks.

Medical Control. Medical control includes preplacement and annual physical examination with determination of pre-exposure red blood cell and plasma cholinesterase activity. A person whose red blood cell cholinesterase falls to or below 40% of the pre-exposure baseline should be removed from further exposure until the activity returns to within 80% of the pre-exposure baseline.

REFERENCES

1. Koelle GB (ed): Cholinesterases and anticholinesterase agents. Handbuch der Experimentellen Pharmakologie, Vol 15, pp 989–1027. Berlin, Springer–Verlag, 1963
2. Koelle GB: Anticholinesterase agents. In Goodman LS, Gilman A (eds): The Pharmacological Basis of Therapeutics, 5th ed, pp 456–466. New York, Macmillan, 1975
3. Hayes WJ Jr: Clinical Handbook on Economic Poisons. Emergency Information for Treating Poisoning. US Public Health Service Pub No 476, pp 12–23, 35–37. Washington, DC, US Government Printing Office, 1963
4. Namba T, Nolte CT, Jackrel J, Grob D: Poisoning due to organophosphate insecticides. Am J Med 50:475, 1971
5. Milby TH: Prevention and management of organophosphate poisoning. JAMA 216:2131, 1971
6. Hayes WJ Jr: Toxicology of Pesticides, pp 379–428. Baltimore, Williams & Wilkins, 1975
7. Griffin TB et al: Abstr 32. Soc Toxicol, Atlanta, Georgia, March 1976
8. Chlorpyrifos. Documentation of TLVs and BEIs, 5th ed, p 138. Cincinnati, American Conference of Governmental Industrial Hygienists (ACGIH), 1986
9. Nolan RJ et al: Chlorpyrifos: Pharmacokinetics in human volunteers. Toxicol Appl Pharmacol 73:8–15, 1984
10. Deacon MM et al: Embryotoxicity and fetotoxicity of orally administered chlorpyrifos in mice. Toxicol Appl Pharmacol 54:31–40, 1980
11. Michel HO: Electrometric method for determination of red blood cell and plasma cholinesterase activity. J Lab Clin Med 34:1564, 1949

CHROMIUM
CAS: 7440-47-3 (metal)
Cr 1987 TLV
= 0.5 mg/m^3—metal
= 0.5 mg/m^3—Cr II compounds
= 0.5 mg/m^3—Cr III compounds
= 0.05 mg/m^3—water-soluble Cr VI compounds
= 0.05 mg/m^3— certain water-insoluble Cr VI compounds (recognized human carcinogen)
= 0.05 mg/m^3 — chromite ore processing (chromate), as Cr (recognized human carcinogen)

METAL AND INORGANIC COMPOUNDS, AS CR

Uses. In stainless and alloy steels, refractory products, tanning agents, pigments, electroplating, catalysts, and corrosion resistant products

Classification. Chromium can have a valence of 2, 3, or 6. Chromium compounds vary greatly in their toxic and carcinogenic effects. For this reason, the ACGIH divides chromium and its inorganic compounds into a number of groupings:

1. *Chromium metals and alloys*: Include chromium metal, stainless steels, and other chromium-containing alloys.
2. *Divalent chromium compounds* (Cr^{2+}) (chromous compounds): Include chromous chloride ($CrCl_2$) and chromous sulfate ($CrSO_4$).
3. *Trivalent chromium compounds* (Cr^{3+}) (chromic compounds): Include chromic oxide (Cr_2O_3), chromic sulfate ($Cr_2[SO_4]_3$), chromic chloride ($CrCl_3$), chromic potassium sulfate ($KCr[SO_4]_2$), and chromite ore ($FeO \cdot Cr_2O_3$).
4. *Hexavalent chromium compounds* (Cr^{6+}): Include chromium trioxide (CrO_3)—the anhydride of chromic acid chromates (*e.g.*, Na_2CrO_4), dichromates (*e.g.*, $Na_2Cr_2O_7$), and polychromates.

Certain hexavalent chromium compounds have been demonstrated to be carcinogenic on the basis of epidemiologic investigations on workers and experimental studies in animals. In general, these compounds tend to be of low solubility in water and thus are subdivided into two subgroups.

a. Water-soluble hexavalent chromium compounds: Include chromic acid and its anhydride, and the monochromates and dichromates of sodium, potassium, ammonium lithium, cesium, and rubidium.
b. Water-insoluble hexavalent chromium compounds: Include zinc chromate, calcium chromate, lead chromate, barium chromate, strontium chromate, and sintered chromium trioxide.[1]

CHROMIUM METAL AND DIVALENT AND TRIVALENT COMPOUNDS

Exposure. Inhalation

Toxicology. Chromium metal is relatively nontoxic. There is little evidence for significant toxicity from chromic or chromous salts, probably because of poor penetration of skin and mucous membranes. Dermatitis from some chromic salts has been reported.

Four workers engaged in the production of ferrochrome alloys developed a nodular type of pulmonary disease with impairment of pulmonary function; air concentrations of chromium averaged 0.26 mg/m^3, although other fumes and dusts were also present.[2] Chest roentgenograms are said to have revealed only "exaggerated pulmonic markings" in workers exposed to chromite dust.[1] The lungs of other workers exposed to chromite dust have been shown to be the seat of pneumoconiotic changes consisting of slight thickening of interstitial tissue and interalveolar septa, with histologic fibrosis and hyalinization.[3] A refractory plant using chromite ore to make chromite brick had no excess of lung cancer deaths among its workers over a 14-year period, and it was concluded that chromite alone probably is not carcinogenic.[4] Exposure to chromium metal does not give rise to pulmonary fibrosis.[1]

Chromite ore roast mixed with sheep fat implanted intrapleurally in rats produced sarcomas coexisting with squamous cell carcinomas of the lungs; the same material implanted in the thigh of rats produced fibrosarcomas.[5] However, the IARC has concluded that these studies were inadequate to fully evaluate the carcinogenicity of this compound.[6]

Unlike nickel, chromium metal does not produce allergic contact dermatitis.[7] Some patients exhibit positive patch tests to divalent chromium compounds, but these compounds are considerably less potent as sensitizers than hexavalent chromium compounds.

These compounds do not appear to cause other effects associated with the hexavalent chromium compounds, such as chrome ulcers, irritative dermatitis, or nasal septal perforation.[7] A case of chromium (chromic) sulfate–induced asthma in a plating worker, confirmed by specific challenge testing and the presence of IgE antibodies, has been reported.[8]

There is no evidence indicating carcinogenicity of trivalent chromium compounds in humans or in animals. Despite a number of animal and epidemiologic studies, the IARC has concluded that there is inadequate data to evaluate the carcinogenicity of these compounds.[6]

HEXAVALENT CHROMIUM

Exposure. Inhalation

Toxicology. The water-soluble hexavalent chromium compounds such as chromic acid mist and certain chromate dusts are severe irritants of the nasopharynx, larynx, lungs, and skin. Exposure to certain hexavalent chromium compounds, mainly water-insoluble, appears to be related to increased risk of lung cancer.

Hexavalent chromium compounds have been implicated as responsible for such effects as ulcerated nasal mucosa, perforated nasal septa, rhinitis, nosebleed, perforated eardrums, pulmonary edema, asthma, kidney damage, erosion and discoloration of the teeth, primary irritant dermatitis, sensitization dermatitis, and skin ulceration.[9]

Chromic Acid: Workers exposed to chromic acid or chromates in concentrations of 0.11 to 0.15 mg/m^3 developed ulcers of the nasal septum and irritation of the conjunctiva, pharynx, and larynx, as well as asthmatic bronchitis.[10] A worker exposed to unmeasured but massive amounts of chromic acid mist for 4 days developed severe frontal headache, wheezing, dyspnea, cough, and chest pain on inspiration; after 6 months, there was still chest pain on inspiration and cough.[10]

In an industrial plant where the airborne chromic acid concentrations measured from 0.18 to 1.4 mg/m^3, moderate irritation of the nasal septum and turbinates was observed after 2 weeks of exposure, ulceration of the septum was present after 4 weeks, and there was perforation of the septum after 8 weeks.[10] A worker exposed to an unmeasured concentration of chromic acid mist for 5 years developed jaundice and was found to be excreting significant amounts of chromium; liver function was mildly to moderately impaired in four other workers with high urinary chromium excretion.[10]

Erosion and discoloration of the teeth has been attributed to chromic acid exposure.[10] Papillomas of the oral cavity and larynx were found in 15 of 77 chrome platers exposed for an average of 6.6 years to chromic acid mist at air concentrations of chromium of 0.4 mg/m^3.[4] There are no reports of increased lung cancer from exposure to chromic acid alone.[10]

A concentrated solution of chromic acid in the eye causes severe corneal injury; chronic exposure to the mist causes conjunctivitis. Prolonged exposure to chromic acid mist causes dermatitis, which varies from a dry erythematous eruption to a weeping eczematous condition.

Chromates: An increased incidence of bronchogenic carcinoma occurs in workers exposed to certain chromate dusts. Nearly all the implications of carcinogenicity have arisen from studies of the worker population of the chromate–bichromate industry and from animal studies using the intermediates produced in that industry.[1] The relative risk of dying from respiratory cancer among chromate workers is more than 20 times the rate for a control population; the latent period is relatively short.[9]

Some less-soluble hexavalent chromium compounds (lead chromate and zinc chromate pigments; calcium chromate) are carcinogenic in rats, producing tumors at the sites of administration by several routes. Lead chromate also produces renal carcinomas following intramuscular administration in rats.[9] As evaluated by the IARC, studies of a number of other chromates, including the water-soluble chromates, were inadequate to evaluate the carcinogenicity of these compounds, but most of the animal data suggest that the water-soluble chromates are not carcinogenic.[6]

The IARC has concluded that there is sufficient evidence of respiratory carcinogenicity in men occupationally exposed during chromate production.[6] There is exposure to a mixture of hexavalent chromium compounds, both soluble and insoluble, in chromate production. Data from studies of the chromate pigment industry suggest an increased risk of lung cancer, comparable to that seen in chromate production. Results from studies in other occupations with chromium exposure are insufficient, according to the IARC, to establish an increased cancer risk.[6] Exposure to soluble hexavalent chromium compounds, including chromium trioxide, in chromium plating operations may be associated with an increased risk of lung cancer. The increased risk appears to occur in the "hard" chromium plating operations, where the greatest exposure occurs, rather than in the "bright" plating operations.[10]

Chrome ulcer, a penetrating lesion of the skin, occurs chiefly on the hands and forearms where there has been a break in the epidermis and is believed to be due to a direct necrotizing effect of the chromate ion. The ulcer is relatively painless, heals slowly, and produces a characteristic depressed scar. Sensitization dermatitis with varying degrees of eczema has been reported numerous times and is the single most common manifestation of chromium toxicity, affecting not only industrial workers but also the general population.[9]

REFERENCES

1. Chromium. Documentation of the TLVs and BEIs, 5th ed, p 139. Cincinnati, American Conference of Governmental Industrial Hygienists (ACGIH), 1986
2. Princi F, Miller LH, Davis A, Cholak J: Pulmonary disease of ferroalloy workers. J Occup Med 4:301–310, 1962
3. Mancuso TF, Hueper WC: Occupational cancer and other health hazards in a chromate plant: A medical appraisal. 1. Lung cancers in chromate workers. Ind Med Surg 20:358–363, 1951

4. Committee on Biologic Effects of Atmospheric Pollutants, Division of Medical Sciences, National Research Council: Chromium, pp 42–73, 125–145. Washington, DC, National Academy of Sciences, 1974
5. Hueper WC: Experimental studies in metal cancerigenesis. X. Cancerigenic effects of chromite ore roast deposited in muscle tissue and pleural cavity of rats. AMA Arch Ind Health 18:284–291, 1958
6. Chromium and chromium compounds. In IARC Monographs on the Evaluation of the Carcinogenic Risk of Chemicals to Humans: Some Metals and Metallic Compounds, Vol 23, pp 205–323. Lyons, International Agency for Research on Cancer, 1980
7. Burrows D: The dichromate problem. Int J Dermatol 21:215–220, 1984
8. Novey H et al: Asthma and IgE antibodies induced by chromium and nickel salts. J Allergy Clin Immunol 72:407–412, 1983
9. Enterline PE: Respiratory cancer among chromate workers. J Occup Med 16:523–526, 1974
10. Franchini I et al: Mortality experience among chromeplating workers. Scand J Work Environ Health 9:247–252, 1983

COAL DUST

1987 TLV = 2 mg/m^3, respirable dust fraction, for dust containing $< 5\%$ quartz; 0.1 mg/m^3, respirable dust fraction, for dust containing $> 5\%$ quartz

Synonyms: None

Physical Form. Solid

Uses. Coal is a fuel and is used in the production of coke, coal gas, and coal tar compounds

Exposure. Inhalation

Toxicology. The inhalation of coal dust causes coal workers' pneumoconiosis (CWP).

Simple CWP has no clinically distinguishing symptoms, because many miners have a slight cough and blackish sputum, which are of no help in establishing whether or not the disease is present.[1] Simple CWP is diagnosed according to the number of small opacities present in the chest film; the small opacities may be linear (irregular) or rounded (regular); however, the latter are more commonly seen in CWP and are most frequently located in the upper lung zones.[1] The primary lesion of simple CWP is the coal macule, a focal collection of coal dust particles with a little reticulin and collagen accumulation, measuring up to 5 mm in diameter. Focal emphysema may be associated.

Simple CWP often occurs concomitantly with chronic bronchitis and emphysema.[1] Although CWP is associated with several respiratory impairments, it is not associated with shortened life span; the importance of this benign condition is the fact that it is a precursor of progressive massive fibrosis (PMF).[1] However, simple CWP does not progress in the absence of further exposure.

Any opacity greater than 1 cm in a coal miner is classified as complicated pneumoconiosis or PMF unless there is evidence to suggest another disease such as tuberculosis.[1]

Complicated pneumoconiosis (PMF) is associated with a reduction in ventilatory capacity, low diffusing capacity, abnormalities of gas exchange, low arterial oxygen tension, pulmonary hypertension, and premature death; it may appear several years after exposure has ceased and may progress in the absence of further dust exposure.[1] Macroscopically, the lesions consist of a mass of black tissue that is often adherent to the chest wall. The lesions are of a rubbery consistency and are relatively well defined. Unlike conglomerate silicosis, which consists of matted aggregates of whorled silicotic nodules, the massive lesion of PMF is amorphous, irregular, and relatively homogeneous. In some instances, its center may contain a cavity filled with a black liquid. Cavitation is a consequence of ischemic necrosis or secondary infection by tuberculosis. In PMF, the vascular bed of the affected region is destroyed. Obstructive airway disease is common and is probably a consequence of the distortion and narrowing of the bronchi and bronchioles produced by the conglomerate mass.

The percentage of miners showing definite radiographic evidence of either simple or complicated pneumoconiosis has varied considerably in different geographic areas; factors responsible for this difference include respirable dust levels, physical and chemical composition of the coal, and the number of years of exposure.[2] A study of 9076 US miners from 1969 to 1971 showed an overall prevalence of CWP of 30%, with PMF occurring in 2.5% of the sample.[2] In Britain, the prevalence of all categories of CWP in working miners has fallen from 13.4% in 1959–1960 to 5.2% in 1978; for PMF, the rate in 1978 was 0.4%.[3]

REFERENCES

1. Morgan WKC: Coal workers' pneumoconiosis. In Morgan WKC, Seaton A (eds): Occupational Lung Diseases, 2nd ed, pp 377–488. Philadelphia, WB Saunders, 1984
2. Morgan WKC et al: The prevalence of coal workers' pneumoconiosis in US coal miners. Arch Environ Health 27:221–226, 1973
3. Parkes WR: Occupational Lung Disorders, 2nd ed, p 178. London, Butterworths, 1982

COAL TAR PITCH VOLATILES

CAS: 65996-93-2 — 1987 TLV = 0.2 mg/m³ as benzene-soluble fraction; recognized human carcinogen

Synonyms: CTPV; particulate polycyclic organic matter (PPOM); particulate polycyclic aromatic hydrocarbons (PPAH); polynuclear aromatics (PNAs)

Physical Form. As stated by the ACGIH,

> The pitch of coal tar is the black or dark brown amorphous residue that remains after the redistillation process. The volatiles contain a large quantity of lower molecular weight polycyclic hydrocarbons. As these hydrocarbons (naphthalene, fluorene, anthracene, acridine, phenanthrene) sublime into the air, there is an increase of benzo(a)pyrene (BaP or 3,4–benzpyrene) and other higher weight polycyclic hydrocarbons in the tar and in the fumes. Polycyclic hydrocarbons, known to be carcinogenic, are of this large molecular type.[1]

Sources. Emissions from coke ovens, from coking of coal tar pitch, and from Soderberg aluminum reduction electrolytic cells

Uses. Base for coatings and paints; for roofing and paving; and as a binder for carbon electrodes

Exposure. Inhalation

Toxicology. Epidemiologic evidence suggests that workers intimately exposed to the products of combustion or distillation of bituminous coal are at increased risk of cancer at many sites, including lungs, kidney, and skin.[2]

The chemical composition and particle size distribution of CTPV from different sources are significant variables in determining toxicity.[3,4]

In a study of 22,010 US male aluminum reduction workers with over 5 years' employment in the industry, there was a slight positive association with lung cancer (SMR* = 121), which was somewhat stronger in horizontal-Soderberg workers (SMR = 162).[5] There was a mild excess of leukemia (SMR = 170) and lymphoma (SMR = 125) in potroom workers, but neither was statistically significant.

In a more detailed analysis of the mortality experience of this cohort up to year-end 1977, the results of other studies relative to an excess of lung cancer were not confirmed, but there were indications of a higher than expected mortality in pancreatic cancer, lymphohematopoietic cancers, genitourinary cancer, nonmalignant respiratory disease, and benign and unspecified neoplasms.[6]

In a study in Canada of 5891 men in two aluminum reduction plants, the mortality from lung cancer was related to "tar-years" of exposure; the SMR for persons exposed for more than 21 years to the higher levels of tars was 2.3 times that of persons not exposed to tars.[7] A follow-up study of this cohort through 1977 showed excess deaths from respiratory disease; pneumonia and bronchitis; malignant neoplasms (all sites); malignant neoplasms of the stomach and esophagus, bladder, and lung; other malignant neoplasms; Hodgkin's disease; and other hypertensive disease. Mortality from malignant neoplasms of the bladder and lung was meaningfully related to number of tar-years and of years of exposure.[8]

Exposure to coke oven emissions is a cause of lung and kidney cancer. A major study of US coke oven workers showed that mortality from lung cancer for full topside workers is 9 times the expected rate, for partial topside workers it is almost 2.5 times the expected rate, and for side oven workers, it is 1.7 times the expected rate.[9] All these rates are based on 5 or more years of exposure in the job category. As the length of employment increases, so does the mortality experience. For example, for employees with 20 or more years of employment topside, the lung cancer rate is 20 times the expected rate.

In addition to the risk of lung cancer, the relative risk of mortality from kidney cancer for all coke oven workers is 7.5.[10]

Certain industrial populations exposed to coal tar products have a demonstrated risk of skin cancer. Substances containing polycyclic hydrocarbons or PNAs (polynuclear aromatics), which may produce skin cancer, also produce contact dermatitis (*e.g.*, coal tar pitch, cutting oils).[4] Although allergic dermatitis is readily induced by PNAs in guinea pigs, it is only rarely reported in humans from occupational contact with PNAs. Incidences in humans have resulted largely from the therapeutic use of coal tar preparations. Components of pitch and coal tar produce cutaneous photosensitization; skin eruptions are usually limited to areas exposed to ultraviolet light.[4,11,12] Most of the phototoxic agents will induce hypermelanosis of the skin; if chronic photodermatitis is severe and prolonged, leukoderma may occur.[11] Some oils containing PNAs have been associated with changes, commonly taking the form of acne, of follicular and sebaceous glands.[4]

* Standardized mortality ratio

REFERENCES

1. Coal Tar Pitch Volatiles. Documentation of TLVs and BEIs, 5th ed, p 143. Cincinnati, American Conference of Governmental Industrial Hygienists (ACGIH), 1986
2. National Institute for Occupational Safety and Health, US Department of Health, Education

and Welfare: Criteria for a Recommended Standard . . . Occupational Exposure to Coke Oven Emissions. (HSM) Pub No 73–11016, pp III–1 to III–14, V–1 to V–9. Washington, DC, US Government Printing Office, 1973
3. Hittle DC, Stukel JJ: Particle size distribution and chemical composition of coal tar fumes. Am Ind Hyg Assoc J 37:199–204, 1976
4. Scala RA: Toxicology of PPOM. J Occup Med 17:784–788, 1975
5. Cooper C, Gaffey W: A Mortality Study of Aluminum Workers. New York, The Aluminum Association, Inc, 1977
6. Rockette HE, Arena VC: Mortality studies of aluminum reduction plant workers: Pot room and carbon department. J Occup Med 25:549–557, 1983
7. Gibbs GW, Horowitz I: Lung cancer mortality in aluminum plant workers. J Occup Med 21:347–353, 1979
8. Gibbs GW: Mortality of aluminum reduction plant workers, 1950 through 1977. J Occup Med 27:761–770, 1985
9. Lloyd JW: Long-term mortality study of steelworkers. V. Respiratory cancer in coke plant workers. J Occup Med 13:53–68, 1971
10. Redmond CK, Ciocco A, Lloyd JW, Rush HW: Longterm mortality study of steelworkers. VI. Mortality from malignant neoplasms among coke oven workers. J Occup Med 14:621–629, 1972
11. Committee on Biologic Effects of Atmospheric Pollutants, Division of Medical Sciences, National Research Council: Particulate Polycyclic Organic Matter, pp 166–246, 307–354. Washington, DC, National Academy of Sciences, 1972
12. National Institute for Occupational Safety and Health: Criteria for a Recommended Standard . . . Occupational Exposure to Coal Tar Products. DHEW (NIOSH) Pub No 78–107. Washington, DC, US Government Printing Office, 1977

COBALT (Metal Fume and Dust)

CAS: 7440-48-4

Co 1987 TLV = 0.05 mg/m^3 as Co

Synonyms: None

Physical Form. Fume or dust

Uses. Alloys; carbides; high-speed steels; paints; electroplating

Exposure. Inhalation

Toxicology. Cobalt causes interstitial fibrosis, interstitial pneumonitis, and sensitization of the respiratory tract and skin.

Three types of lung disease have been reported in the cemented tungsten carbide industry: (1) an interstitial fibrotic process, (2) an interstitial pneumonitis that often disappears when exposure ceases, and (3) an obstructive airways syndrome. The latter may result from simple irritation, but, in addition, a distinct form of occupational asthma occurs.[1] Cobalt, which is used as a binder for the tungsten carbide crystals, has been implicated as the etiologic agent.[2,3]

Among 12 workers engaged in the manufacturing of, or grinding with, tungsten carbide tools and who developed interstitial lung disease, there were 8 fatalities; serial chest roentgenograms over a period of 3 to 12 years revealed gradually progressive densities of a linear and nodular nature, which gradually involved major portions of both lungs. Cough, production of scanty mucoid sputum, dyspnea on exertion, and reduced pulmonary function occurred early in the course of the disease.[3] Disease is seldom seen without at least 10 years of exposure, but shorter periods have been reported.[2]

The obstructive airways syndrome appears to be an allergic response and is characterized by wheezing, cough, and shortness of breath while at work.[1,3] There is no evidence that this type of disease progresses to interstitial fibrosis. In a report of nine cases, the syndrome did not develop until after 6 to 18 months of exposure.[4]

Cobalt and its compounds produce an allergic dermatitis of an erythematous papular type, which usually occurs in skin areas subjected to friction, such as the ankle, elbow flexures, and sides of the neck.[5]

Rhabdomyosarcomas developed in rats injected intramuscularly with the powder of either pure cobalt metal or cobalt oxide.[6] Guinea pigs develop acute pneumonitis from the intratracheal administration of cobalt metal.[7]

REFERENCES

1. Morton WKC, Seaton A: Occupational Lung Diseases, 2nd ed, pp 486–489. Philadelphia, WB Saunders, 1984
2. Miller CW, Davis MW, Goldman A, Wyatt JP: Pneumoconiosis in the tungsten-carbide tool industry. AMA Arch Ind Hyg Occup Med 8:453, 1953
3. Coates EO Jr, Watson JHL: Diffuse interstitial lung disease in tungsten-carbide workers. Ann Intern Med 75:709, 1971
4. Coates EO Jr: Hypersensitivity bronchitis in tungsten-carbide workers. Chest 64:390, 1973
5. Browning E: Toxicity of Industrial Metals, 2nd ed, pp 132–142. London, Butterworth, 1969
6. Heath JC: The histogenesis of malignant tumors induced by cobalt in the rat. Br J Cancer 14:478, 1960
7. Schepers GWH: The biological action of particulate cobalt metal. AMA Arch Ind Health 12:127–133, 1955

COPPER FUME

CAS: 7440-50-8

CuO 1987 TLV = 0.2 mg/m^3

Synonyms: None

Physical Form. Fume

Sources. Copper and brass manufacture; welding of copper-containing metals

Exposure. Inhalation

Toxicology. Copper fume causes irritation of the upper respiratory tract and metal fume fever, an influenza-like illness.

In humans, effects of copper fume include irritation of the upper respiratory tract, metallic or sweet taste, and, in some instances, discoloration of the skin and hair.[1] Exposure of workers to concentrations of 1 to 3 mg/m^3 for short periods of time resulted in altered taste response but no nausea; levels from 0.02 to 0.4 mg/m^3 produced no complaints.[1]

Typical metal fume fever, a 24- to 48-hour illness characterized by chills, fever, aching muscles, dryness in the mouth and throat, and headache, has been reported in several workers exposed to copper fume.[2,3] With metal fume fever, leukocytosis is usually present, which may amount to 12,000 to 16,000/mm^3; recovery is usually rapid, and there are no sequelae.[4] Most workers develop an immunity to these attacks, but it is quickly lost; attacks tend to be more severe on the first day of the workweek.[4]

Lung damage after chronic exposure to fumes in industry has not been described.[5] The higher incidence of respiratory cancer reported in copper smelters is due to the presence of arsenic in the ore.[5] Transient irritation of the eyes has followed exposure to a fine dust of oxidation products of copper produced in an electric arc.[6]

REFERENCES

1. Copper as Cu. Documentation of the TLVs and BEIs, 5th ed, p 146. Cincinnati, American Conference of Governmental Industrial Hygienists (ACGIH), 1986
2. Committee on Medical and Biologic Effects of Environmental Pollutants: Copper, pp 55–58. Washington, DC, National Academy of Sciences, 1977
3. Cohen SR: A review of the health hazards from copper exposure. J Occup Med 16:621–624, 1974
4. McCord CP: Metal fume fever as an immunological disease. Ind Med Surg 29:101–106, 1960
5. Triebig G, Schaller KH: Copper. In Alessio L et al (eds): Biological Indicators for the Assessment of Human Exposure to Industrial Chem-

icals, pp 57–66. Luxembourg, Office for Official Publications of the European Communities, 1984
6. Grant WM: Toxicology of the Eye, 3rd ed, pp 260–269. Springfield, Illinois, Charles C Thomas, 1986

COTTON DUST

1987 TLV = 0.2 mg/m^3

Synonyms: None

Physical Form. Fibers

Source. Cotton processing

Exposure. Inhalation

Toxicology. Raw cotton dust causes a respiratory syndrome termed *byssinosis*.

The initial symptoms are chest tightness, cough, wheezing, and dyspnea in varying degrees of severity on the first day of the workweek (Grade 1 byssinosis).[1–3] Symptoms usually disappear an hour or so after leaving work but may recur on the first day of each workweek. With continued exposure, the symptoms also appear on subsequent days of the week (Grade 2 byssinosis). There is usually a decrease in the FEV_1 and in vital capacity on the first day of the workweek after 3 to 4 hours of exposure; the changes in airway resistance and decreased flow rates have been attributed to narrowing of small airways owing to bronchoconstriction. Eventually, obstructive airway disease, which is irreversible, occurs (Grade 3 byssinosis).

Although a loss in lung function has been documented in cotton textile workers, no clear evidence of increased mortality has been reported.[4] A review of 2895 consecutive autopsies showed no significant differences in the prevalence of emphysema, interstitial fibrosis, or cor pulmonale between 283 employees of a cotton textile mill and the nontextile population.[5] In another postmortem study of 49 cotton workers, the incidence of emphysema was associated with cigarette smoking, with 16 of 36 smokers showing centrilobular emphysema vs. 1 of 13 nonsmokers.[6]

Another study of women with advanced byssinosis confirmed the association between emphysema and cigarette smoking rather than cotton dust exposure.[7]

A syndrome termed *mill fever*, which may or may not be related to the development of byssinosis, has been described in some persons unaccustomed to breathing cotton dust.[2] Shortly after exposure, there is development of malaise, cough, fever, chills, and upper respiratory symptoms; these may recur daily for days to months until acclimatization takes place and symptoms disappear. Tolerance may be lost temporarily after a period of absence from exposure or if exposure to a greater concentration of dust occurs. The exact prevalence of mill fever among new employees is unknown, but estimates range from 10% to 80%.[1]

Epidemiologic studies have indicated that prevalence of byssinosis among cotton workers can be correlated with the average concentration of lint-free dust of particle size under 15μ diameter and with the number of years of exposure.[2] There is little evidence of a threshold below which zero prevalence is found.[2] The slopes of the prevalence–dustiness curves obtained by different investigators vary considerably.[2]

Little is known about the identity of the byssinogenic agent or the underlying mechanisms; the active agent is thought to be water extractable, filterable through an 0.22 μ filter, nonvolatile at 40°C, nondialyzable, and present in the harvested cotton, rather than in the lint.[8]

REFERENCES

1. Harris TR, Merchant JA, Kilburn KH, Hamilton JD: Byssinosis and respiratory diseases of cotton mill workers. J Occup Med 14:199–206, 1972
2. National Institute for Occupational Safety and Health: Criteria for a Recommended Standard . . . Occupational Exposure to Cotton Dust. DHEW (NIOSH) Pub No 75–118, pp 12–60. Washington, DC, US Government Printing Office, 1974
3. Department of Labor: Standard for Exposure to Cotton Dust. Federal Register 41:56498–56525, 1976
4. Elwood PC et al: Respiratory disability in ex-cotton workers. Br J Ind Med 43:580–586, 1986
5. Moran TJ: Emphysema and other chronic lung disease in textile workers: An 18-year autopsy study. Arch Environ Health 38:267–276, 1983
6. Pratt PC et al: Epidemiology of pulmonary lesions in non-textile and cotton textile workers—a retrospective autopsy analysis. Arch Environ Health 35:133–138, 1980
7. Honeybourne D, Pickering CAC: Physiological evidence that emphysema is not a feature of byssinosis. Thorax 41:6–11, 1986
8. Brown DF, McCall ER, Piccolo B, Tripp VW: Survey of effects of variety and growing location of cotton on cardroom dust composition. Am Ind Hyg Assoc J 38:107–115, 1977

CRESOL (all isomers)

CAS: 1319-77-3

$CH_3C_6H_4OH$ 1987 TLV = 5 ppm; skin

Synonyms: Cresylic acid; tricresol; methylphenol; *o*-cresol; *m*-cresol; *p*-cresol

Physical Form. Para isomer is liquid; ortho and meta isomers are crystalline

Uses. Antiseptics; disinfectants; insecticides; resins; plasticizers

Exposure. Skin absorption; inhalation

Toxicology. All isomers of cresol cause skin and eye burns; exposure may also cause impairment of kidney and liver function, as well as central nervous system and cardiovascular disturbances.

Skin and eye contact are the major concerns of occupational exposure.[1] Signs and symptoms related to skin contact are a burning sensation, erythema, skin peeling, localized anesthesia, and, occasionally, ochronosis—a darkening of the skin.[1] Hypersensitivity has also been reported.[2]

Cresols are rapidly absorbed through the skin, producing systemic effects.[1] About 20 ml of a 90% cresol solution accidentally poured over an infant's head caused chemical burns, cyanosis, unconsciousness, and death within 4 hours.[3] Histopathologic examination revealed hepatic necrosis, cerebral edema, acute tubular necrosis of the kidneys, and hemorrhagic effusions from the peritoneum, pleura, and pericardium. The blood contained 12 mg cresol per 100 dl.

Inhalation of appreciable amounts of cresol vapor is unlikely under normal conditions because of the low vapor pressure; however, hazardous concentrations may be generated at elevated temperatures.[1] Seven workers exposed to cresol vapor at unspecified concentrations for 1.5 to 3 years had headaches, which were frequently accompanied by nausea and vomiting.[1] Four of the workers also had elevated blood pressure, signs of impaired kidney function, blood calcium imbalance, and marked tremors.

Eight of ten subjects exposed to 1.4 ppm *o*-cresol vapor experienced upper respiratory tract irritation.[1]

Several cases of ingestion have shown cresol to be corrosive to body tissues and to cause toxic effects on the vascular system, liver, kidneys, and pancreas.[1]

Rats survived 8 hours of inhaling air saturated with the vapor.[4] Irritation of the nose and eyes and some deaths were observed in mice exposed to saturated concentrations 1 hour/day for 10 days.[5] Animal experiments have produced varying results with regard to concentrations necessary to produce death.[1] In general, the *o*- and *p*-isomers are considered equal in toxicity, whereas the *m*-isomer is considered least toxic.[1]

Cresol isomers promoted dimethylbenzanthracene-induced papillomas in mice when applied as 20% solutions in benzene twice weekly for 11 weeks; no carcinomas were produced.[6]

In the eyes of rabbits, undiluted cresols caused permanent opacification and vascularization; a drop of a 33% solution applied to rabbit eyes and removed with irrigation within 60 seconds caused only moderate injury, which was reversible.[7]

The odor of cresol is recognized at concentrations as low as 5 ppm.[2]

REFERENCES

1. National Institute for Occupational Safety and Health: Criteria for a Recommended Standard . . . Occupational Exposure to Cresol. DHEW (NIOSH) Pub No 78–133, p 117. Washington, DC, US Government Printing Office, 1978
2. Deichmann WB, Keplinger ML: Phenols and phenolic compounds. In Clayton GD, Clayton FE (eds): Patty's Industrial Hygiene and Toxicology, 3rd ed, rev, Vol 2A, Toxicology, pp 2597–2601. New York, Wiley–Interscience, 1981
3. Green MA: A household remedy misused—fatal cresol poisoning following cutaneous absorption—(a case report). Med Sci Law 15:65–66, 1975
4. Smyth HF Jr: Improved communication—hygienic standards for daily inhalation. Am Ind Hyg Assoc Q 17:129–185, 1956
5. Campbell J: Petroleum cresylic acids—a study of their toxicity and the toxicity of cresylic disinfectants. Soap Sanit Chem 17:103–111, 1941
6. Boutwell RK, Bosch DK: The tumor-promoting action of phenol and related compounds for mouse skin. Cancer Res 19:413–424, 1959
7. Grant WM: Toxicology of the Eye, 3rd ed, pp 283–284. Springfield, Illinois, Charles C Thomas, 1986

CROTONALDEHYDE

CAS: 4170-30-3

$CH_3CH{=}CHCHO$ 1987 TLV = 2 ppm

Synonyms: β-Methyl acrolein; 2–butenal; crotonic aldehyde

Physical Form. Liquid

Uses. Manufacture of *n*-butyl alcohol; solvent in purification of oils

Exposure. Inhalation

Toxicology. Crotonaldehyde is an irritant of the eyes, mucous membranes, and skin.

Exposure of humans to 4 ppm for 10 minutes caused lacrimation and upper respiratory irritation; at 45 ppm, there was conjunctival irritation after a few seconds.[1,2]

In a series of eight cases of corneal injury from industrial exposure to crotonaldehyde, healing was complete in 48 hours; the severity of exposure was not specified.[3] A case of apparent sensitization of undetermined nature to crotonaldehyde has been reported in a laboratory worker.[4]

Rats did not survive exposure to 1650 ppm for 10 minutes; effects included respiratory distress, an excitatory stage, and terminal convulsions; autopsy revealed bronchiolar damage.[2] Pulmonary edema has also been observed in rats after fatal exposure to 1500 ppm for 30 minutes.[2]

REFERENCES

1. Pattle RE, Cullumbine H: Toxicity of some atmospheric pollutants. Br Med J 2:913–916, 1956
2. Rinehart WE: The effect of single exposures to crotonaldehyde vapor. Am Ind Hyg Assoc J 28:561–566, 1967
3. Grant WM: Toxicology of the Eye, 3rd ed, p 284–285. Springfield, Illinois, Charles C Thomas, 1986

4. Crotonaldehyde. Documentation of the TLVs and BEIs, 5th ed, p 149. Cincinnati, American Conference of Governmental Industrial Hygienists (ACGIH), 1986

CUMENE
CAS: 98-82-8
$(CH_3)_2CHC_6H_5$ 1987 TLV = 50 ppm; skin

Synonyms: Cumol; 2–phenylpropane; isopropylbenzene

Physical Form. Colorless liquid

Uses. Thinner for paints and lacquers; component of high-octane aviation fuel; production of styrene; organic synthesis

Exposure. Inhalation; skin absorption

Toxicology. Cumene is an eye and mucous membrane irritant; at high concentrations, it causes narcosis in animals; it is expected that severe exposure will produce the same effect in humans.

Concentrations lethal to humans are not expected to be encountered at room temperature because of the low volatility of cumene.[1] If inhalation of high vapor concentrations does occur, dizziness, incoordination, and unconsciousness could be expected.[2] In animals, cumene narcosis is characterized by slow induction and long duration, suggesting a cumulative action.[3] There are no reports of systemic effects in humans.

The LC_{50} for rats was 8000 ppm for a 4-hour exposure.[4] The LC_{50} was 2040 ppm for a 7-hour exposure to mice; the effect was central nervous system depression.[3]

Repeated inhalation by rabbits and rats of 2000 ppm caused ataxia and lethargy.[5] No significant changes were noted in rats exposed 8 hours/day to 500 ppm for 150 days.[2] The LD_{50} for penetration of rabbit skin was 12.3 ml/kg after 14 days.[4] It is generally agreed that cumene has no damaging effect on the hematopoietic system in spite of its chemical similarity to benzene.[5] Contact of the liquid with the skin causes erythema and irritation.[6] Eye contamination may produce conjunctival irritation.[1]

Cumene has a sharp, penetrating odor, but it may not be sufficient to give warning of hazardous concentrations.[6]

REFERENCES

1. AIHA Hygienic Guide Series: Cumene, Vol 1. Akron, Ohio, American Industrial Hygiene Association, 1978
2. Sandmeyer EE: Aromatic hydrocarbons. In Clayton GD, Clayton FE (eds): Patty's Industrial Hygiene and Toxicology, 3rd ed, Vol 2B, Toxicology, pp 3309–3310. New York, Wiley-Interscience, 1981
3. Werner HW, Dunn RC, von Oettingen WF: The acute effects of cumene vapors in mice. J Ind Hyg Toxicol 26:264–268, 1944
4. Smyth HF Jr, Carpenter CP, Weil CS: Range-finding toxicity data: List IV. AMA Arch Ind Hyg Occup Med 4:119–122, 1951
5. Snyder R (ed): Ethel Browning's Toxicity and Metabolism of Industrial Solvents, 2nd ed, Vol 1, p 102. New York, Elsevier, 1987
6. Gerarde HW: Toxicological studies on hydrocarbons. AMA Arch Ind Health 19:403–418, 1959

CYANIDES (Alkali)
NaCN: 143-33-9
KCN: 151-50-8
$Ca(CN)_2$: 592-01-8
1987 TLV = 5 mg/m^3, as CN; skin

Synonyms: Sodium cyanide; potassium cyanide; calcium cyanide ("black cyanide")

Physical Form. Powders, granules, or flakes

Uses. Extraction of gold and silver; electroplating; hardening of metals; coppering; zincing; bronzing; manufacture of mirrors; fumigant (calcium cyanide); photography

Exposure. Inhalation; skin absorption; ingestion

Toxicology. The alkali salts of cyanide can cause rapid death from metabolic asphyxiation.

Cyanide ion exerts an inhibitory action on certain metabolic enzyme systems, most notably cytochrome oxidase, the enzyme involved in the ultimate transfer of electrons to molecular O_2.[1] Since cytochrome oxidase is present in practically all cells that function under aerobic conditions, and since the cyanide ion diffuses easily to all parts of the body, cyanide quickly halts almost all cellular respiration. The venous blood of a patient dying of cyanide is bright red and resembles arterial blood, because the tissues have not been able to utilize the oxygen brought to them.[2] Cyanide intoxication produces lactic acidosis, the result of an increased rate of glycolysis and production of lactic acid.[3]

In the presence of even weak acids, HCN gas is liberated from cyanide salts. A concentration of 270 ppm HCN has long been quoted as being immediately fatal to humans. A more recent study, however, states that the estimated LC_{50} to humans for a 1-minute exposure is 3404 ppm.[1] Other investigators state that 270 ppm is fatal after 6 to 8 minutes, 181 ppm after 10 minutes, and 135 ppm after 30 minutes.[1]

If large amounts of cyanide have been absorbed, collapse is usually instantaneous, the patient falling unconscious, often with convulsions, and dying almost immediately. Symptoms of intoxication from less severe exposure include weakness, headache, confusion, and, occasionally, nausea and vomiting.[1,2] Respiratory rate and depth are usually increased initially and, at later stages, become slow and gasping. Coma and convulsions occur in some cases. If cyanosis is present, it usually indicates that respiration has either ceased or has been inadequate for a few minutes. In one case of nonfatal ingestion of 600 mg of KCN, the clinical course was marked by acute pulmonary edema and lactic acidosis.[3]

Hydrogen cyanide has recently been recognized, in significant concentrations, in some fires, as a combustion product of wool, silk, and many synthetic polymers; it may play a role in toxicity and deaths from smoke inhalation.[4]

Chronic cyanide poisoning from occupational exposure, at least in any serious or incapacitating form, is apparently rare, if it occurs at all.[1]

Most reported cases involved workers with a mixture of repeated acute or subacute exposures and chronic or prolonged low-level exposures, making it unclear whether symptoms simply resulted from multiple acute exposures with acute intoxication. Some symptoms persisted after cessation of such exposures, perhaps owing to the effect of anoxia from inhibition of cytochrome oxidase. Symptoms from alleged "chronic" exposure may include weakness, nausea, headache, and vertigo.[1] A study of 36 former workers in a silver-reclaiming facility, who were repeatedly exposed to cyanide, demonstrated some residual symptoms 7 or more months after cessation of exposure; frequent headache, eye irritation, easy fatigue, loss of appetite, and epistaxis occurred in at least 30% of these workers.[5]

Cyanide solutions or cyanide aerosols generated in humid atmospheres have been reported to cause irritation of the upper respiratory tract (primarily nasal irritation) and skin and to cause allergic contact dermatitis.[1] Skin contact with solutions of cyanide salts can cause itching, discoloration, or corrosion, which is most likely due to the alkalinity of the solutions. Skin irritation and mild systemic symptoms (*e.g.*, headache, dizziness) have been caused by solutions as dilute as 0.5% KCN.[1] Skin contact with aqueous cyanide solutions for long periods of time have caused caustic burns; these cases have generally been fatal.[1]

Cyanide is one of the few toxic substances for which a specific antidote exists, and it functions as follows.[6,7] First, amyl nitrite (by inhalation) and then sodium nitrite (intravenously) are administered to form methemoglobin, which binds firmly with free cyanide ions. This traps any circulating cyanide ions. The formation of 10% to 20% methemoglobin does not usually involve appreciable risk, yet it provides a large amount of cyanide-binding substance. Second, sodium thiosulfate is administered intravenously to increase the rate of conversion of cyanide to the less toxic thiocyanate. Although early literature suggests the use of methylene blue, it must not be administered, because it is a poor methemoglobin former and, moreover, promotes the conversion of methemoglobin back to hemoglobin.[7]

Note: The therapeutic regimen (amyl nitrite/sodium nitrite plus sodium thiosulfate) discussed previously and detailed in the Treatment section that follows is the "classical" one that has long been recommended. However, there is growing support for a new regimen of 100% oxygen together with intravenous hydroxocobalamin and/or sodium thiosulfate.[3,8] Hydroxocobalamin reverses cyanide toxicity by combining with cyanide to form cyanocobalamin (vitamin B_{12}); one advantage is that hydroxocobalamin is apparently of low toxicity. The only side effects have been occasional urticaria and a brown-red discoloration of the urine. Hydroxocobalamin is not currently available or approved for treatment of cyanide intoxication in the United States.[4] Nonspecific supportive therapy is an essential part of the treatment; in two cases, patients have recovered from massive cyanide poisoning with supportive therapy only.[3]

At high levels, cyanide acts so rapidly that its odor has no warning value.[2] At lower levels, the odor may provide some warning, although many people are unable to recognize the scent of "bitter almonds."[3]

Diagnosis. *Signs and Symptoms:* Asphyxia and death can occur from high-exposure levels; weakness, headache, confusion, nausea, and vomiting result from lesser exposures. Other signs and symptoms include increased rate and depth of respiration, slow and gasping respiration, pulmonary edema, eye, nose, and skin irritation, and lactic acidosis.

Differential Diagnosis: The diagnosis is usually self-evident from history, but even if it is not completely established and cyanide poisoning is suspected, the recommended therapy should be considered.

Special Tests: A blood level of cyanide in excess of 0.2 μg/ml suggests a toxic reaction, but, because of tight binding of cyanide to cytochrome oxidase, serious poisoning may occur with only modest blood levels, especially several hours after poisoning. In view of the likelihood of lactic acidosis, there should be measurement of blood *p*H, plasma bicarbonate, and blood lactic acid.[3]

Treatment. Preparedness and speed of action are prerequisites for successful treatment for any overexposure to cyanides.[1,2,6,7] All persons working with or around cyanides should be given specific and detailed instructions on the use of antidote kits containing amyl nitrite ampules for inhalation. Kits for use by medical personnel should contain the following:

2 boxes of amyl nitrite ampules, 0.3 ml/ampule
2 ampules of sterile sodium nitrite, 10 ml of 3% solution each
2 ampules of sterile sodium thiosulfate, 50 ml of 25% solution each
2 sterile 10-ml syringes with intravenous needles
1 sterile 50-ml syringe with intravenous needles
1 tourniquet
1 gastric tube (rubber)
1 nonsterile 100-ml syringe

A complete cyanide antidote package can be purchased from the Eli Lilly Co, stock number M76.

The exposed person should be removed as rapidly as possible to an uncontaminated area by someone with adequate respiratory protective equipment.

Resuscitate: If breathing has stopped, immediately institute resuscitation and continue until normal breathing is established.

Decontaminate: If liquid cyanide has contaminated the skin or clothing, the clothing should be removed and the skin flushed with copious amounts of water. If taken by mouth, gastric lavage should be performed, but this should be postponed until after specific therapy has been initiated.

Administer Amyl Nitrite: A pearl (ampule), if not provided with a fabric sleeve, should be wrapped lightly in a handkerchief or gauze pad, broken, and held about 1 inch from the patient's mouth and nostrils for 15 to 30 seconds of every minute while the sodium nitrite solution is being prepared. Repeat five times at 15-second intervals. Use a fresh pearl every 5 minutes.

Administer Sodium Nitrite and Sodium Thiosulfate: If the patient does not respond to the amyl nitrite administration or if severe exposure is suspected, administer 0.3 g sodium nitrite (10 ml of a 3% solution) intravenously for adults at the rate of 2.5 to 5 ml/minute followed by injection of 12.5 g sodium thiosulfate (50 ml of a 25% solution) over approximately a 10-minute period by way of the same needle and vein.

Observe Patient: The patient should be kept under observation for 24 to 48 hours. The blood levels of methemoglobin should be monitored and not allowed to exceed 40%. The desired methemoglobin level is approximately 25%. If signs of intoxication persist or reappear, the injection of nitrite and thiosulfate should be repeated in one half the aforementioned doses. Even if the patient appears well, a second injection may be given 2 hours after the first injection for prophylactic purposes.

Note: The use of oxygen should not be ignored, because the methemoglobinemia induced by this treatment reduces the ability of blood to carry oxygen to the brain.

REFERENCES

1. National Institute for Occupational Safety and Health: Criteria for a Recommended Standard . . . Occupational Exposure to Hydrogen Cyanide and Cyanide Salts (NaCN, KCN, and $CA(CN)_2$). DHEW (NIOSH) Pub No 77–108, pp 37–95, 106–114, 160–173, 178. Washington, DC, US Government Printing Office, 1976
2. Gosselin RE, Smith RP, Hodge HC: Clinical Toxicology of Commercial Products, Section III, 5th ed, pp 123–130. Baltimore, Williams & Wilkins, 1984
3. Graham DL, Laman D, Theodore J, Robin ED: Acute cyanide poisoning complicated by lactic acidosis and pulmonary edema. Arch Intern Med 137:1051–1055, 1977
4. Becker CE: The role of cyanide in fires. Vet Hum Toxicol 27:487–490, 1985
5. Blanc P et al: Cyanide intoxication among silver-reclaiming workers. JAMA 253:367–371, 1985

6. Chen KK, Rose CL: Nitrite and thiosulfate therapy in cyanide poisoning. JAMA 149:113–119, 1952
7. Wolfsie JH: Treatment of cyanide poisoning in industry. AMA Arch Ind Hyg Occup Med 4:417–425, 1951
8. Berlin C: Cyanide poisoning—a challenge. Arch Intern Med 137:993–994, 1977

CYCLOHEXANE

CAS: 110-82-7

C_6H_{12} 1987 TLV = 300 ppm

Synonyms: Hexahydrobenzene; benzene hexahydride; hexamethylene

Physical Form. Colorless liquid

Uses. Chemical intermediate; solvent for fats, oils, waxes, resins, and rubber

Exposure. Inhalation

Toxicology. Cyclohexane is irritating to the eyes and mucous membranes; at high concentrations it causes narcosis in animals, and it is expected that severe exposure will produce the same effect in humans.

A concentration of 300 ppm is detectable by odor and is somewhat irritating to the eyes and mucous membranes.[1] At higher concentrations, it may cause dizziness, nausea, and unconsciousness.[2]

Rabbits exposed to 786 ppm cyclohexane for 50 periods of 6 hours each showed minor microscopic changes in the liver and kidneys; lethargy, light narcosis, increased respiration, diarrhea, and some deaths were observed during a total of 60 hours of exposure to 7444 ppm; 1-hour exposure to 26,752 ppm caused rapid narcosis, tremor, and, rarely, opisthotonos and was lethal to all exposed rabbits.[3] In mice, exposure to 18,000 ppm produced tremors within 5 minutes, disturbed equilibrium by 15 minutes, and recumbency at 25 minutes.[2] Lethal concentrations administered by inhalation or orally to animals caused generalized vascular damage with severe degenerative changes in the heart, lung, liver, kidney, and brain.[2]

Unlike benzene, cyclohexane is not associated with hematologic changes.

Cyclohexane defats the skin on repeated contact.[2]

REFERENCES

1. Gerarde HW: The alicyclic hydrocarbons. In Fassett DW, Irish DD (eds): Industrial Hygiene and Toxicology, 2nd ed, Vol 2, Toxicology, pp 938–940. New York, Interscience, 1963
2. Sandmeyer EE: Alicyclic hydrocarbons. In Clayton GD, Clayton FE (eds): Patty's Industrial Hygiene and Toxicology, 3rd ed, Vol 2B, Toxicology, pp 3227–3228. New York, Wiley–Interscience, 1981
3. Treon JF, Crutchfield WE Jr, Kitzmiller KV: The physiological response of animals to cyclohexane, methylcyclohexane, and certain derivatives of these compounds. J Ind Hyg Toxicol 25:323–347, 1943

CYCLOHEXANOL

CAS: 108-93-0

$C_6H_{11}OH$ 1987 TLV = 50 ppm; skin

Synonyms: Hexahydrophenol; cyclohexyl alcohol

Physical Form. Colorless, viscous liquid

Uses. Solvent for oils, resins, ethyl cellulose; manufacture of soap, plastics

Exposure. Inhalation; skin absorption

Toxicology. Cyclohexanol causes irritation of the eyes, nose, and throat; at high concentrations, it causes narcosis in animals, and it is expected that severe exposure will produce the same effect in humans.

Human volunteers exposed to a vapor concentration of 100 ppm for 3 to 5 minutes experienced eye, nose, and throat irritation.[1] Headache and conjunctival irritation have resulted from prolonged exposure to "excessive" but undefined concentrations.[2]

Rabbits exposed 6 hours/day to 272 ppm over a 10-week period showed slight eye irritation; at 997 ppm, additional effects included salivation, lethargy, narcosis, mild convulsive movements, and some deaths.[3] Lethal doses of cyclohexanol produced slight necrosis of the myocardium and damage to the lungs, liver, and kidneys.[3] The application of 10 ml of cyclohexanol to the skin of a rabbit for 1 hour for 10 days induced narcosis, hypothermia, tremors, and athetoid movements; necrosis, exudative ulceration, and thickening of the skin occurred in the area of contact.[2]

Mice fed diets containing 1% cyclohexanol during gestation produced offspring with an increased mortality during the first 3 weeks of life.[2]

REFERENCES

1. Nelson KW: Sensory response to certain industrial solvent vapors. J Ind Hyg Toxicol 25:282–285, 1943
2. Rowe VK, McCollister SB: Alcohols. In Clayton GD, Clayton FE (eds): Patty's Industrial Hygiene and Toxicology, 3rd ed, Vol 2C, Toxicology, pp 4643–4649. New York, Wiley–Interscience, 1982
3. Treon JF, Crutchfield WE Jr, Kitzmiller KV: The physiological response of animals to cyclohexane, methylcyclohexane, and certain derivatives of these compounds. J Ind Hyg Toxicol 25:323–347, 1943

CYCLOHEXANONE

CAS: 108-94-1

$C_6H_{10}O$ — 1987 TLV = 25 ppm; skin

Synonyms: Pimelic ketone; hexanon; sextone

Physical Form. Colorless liquid

Uses. Metal degreaser; manufacture of nylon; solvent for lacquer, resins, and insecticides

Exposure. Inhalation; skin absorption

Toxicology. Cyclohexanone causes eye, nose, and throat irritation; at high concentrations, it produces lethargy and narcosis in animals, and it is expected that severe exposure will cause the same effect in humans.

In most human subjects, exposure to 25 ppm of the vapor for 5 minutes did not cause effects, while 50 ppm was irritating, especially to the throat; exposure to 75 ppm resulted in more noticeable eye, nose, and throat irritation.[1]

Rabbits exposed to 190 ppm for 6 hours/day for 50 days showed slight liver and kidney injury. At 309 ppm, there was slight conjunctival irritation, while at 1414 ppm, lethargy was observed. At 3082 ppm, effects were incoordination, salivation, labored breathing, narcosis, and some deaths.[2] Five of six rats survived exposure to 2000 ppm for 4 hours, but 4000 ppm caused coma and death to all six rats. Narcosis, hypothermia, and decreased respiration were observed in guinea pigs exposed to 4000 ppm for 6 hours.[4] Recovery from narcosis was slow, and 3 of 10 animals died within 4 days of exposure.

Eye contact with cyclohexanone liquid may cause corneal injury.[5] The liquid is a defatting agent, and prolonged or repeated skin contact may produce irritation or dermatitis.[5]

Cyclohexanone has an odor similar to peppermint, and harmful concentrations are not likely to be tolerated.

REFERENCES

1. Nelson KW et al: Sensory response to certain industrial solvent vapors. J Ind Hyg Toxicol 25:282, 1943
2. Treon JF, Crutchfield WE Jr, Kitzmiller KV: The physiological response of animals to cyclohexanone, methylcyclohexane and certain derivatives of these compounds. II. Inhalation. J Ind Hyg Toxicol 25:323, 1943
3. Smyth HF Jr et al: Range-finding toxicity data. List VII. Am Ind Hyg Assoc J 30:470–476, 1969
4. Specht H, Miller JW, Valaer PJ, Sayers RR: Acute Response of Guinea Pigs to the Inhalation of Ketone Vapors. National Institute of Public Health Bulletin No 176. Washington, DC, US Government Printing Office, 1940
5. Hygienic Guide Series: Cyclohexanone. Am Ind Hyg Assoc J 26:630, 1965

CYCLOHEXENE

CAS: 110-83-8

C_6H_{10} 1987 TLV = 300 ppm

Synonym: 1,2,3–Tetrahydrobenzene

Physical Form. Colorless liquid

Uses. Manufacture of adipic acid, maleic acid, hexahydrobenzoic acid, and aldehyde

Exposure. Inhalation

Toxicology. Cyclohexene is regarded as a mild respiratory irritant and central nervous system depressant by analogy to the observed effects of chemically similar substances.

No acute or chronic effects have been reported in humans.

Mice lost their righting reflex at approximately 9000 ppm, and approximately 15,000 ppm was lethal in single exposures.[1]

Dogs inhaling cyclohexene vapor (concentration not stated) exhibited symptoms characterized by muscular quivering and incoordination.[2] A 6-month inhalation study of various species repeatedly exposed to 75, 150, 300, and 600 ppm showed lower weight gain for rats at 600 ppm.[3] Increased alkaline phosphatase was found with exposures, but no other biochemical or hematologic abnormalities were observed.[3]

The liquid defats the skin on direct contact.

REFERENCES

1. Sandmeyer EE: Alicyclic hydrocarbons. In Clayton GD, Clayton FE (eds): Patty's Industrial Hygiene and Toxicology, 3rd ed, Vol 2B, Toxicology, pp 3233–3236. New York, Wiley–Interscience, 1981
2. Fairhall LT: Industrial Toxicology, pp 278–279. Baltimore, Williams & Wilkins, 1949
3. Laham S: Inhalation toxicity of cyclohexene (abstr no 152). Toxicol Appl Pharmacol 37:155–156, 1976

CYCLOHEXYLAMINE

CAS: 108-91-8

$C_6H_{11}NH_2$ 1987 TLV = 10 ppm

Synonyms: Aminocyclohexane; aminohexahydrobenzene; CHA; hexahydroaniline; hexahydrobenzenamine

Physical Form. Colorless to slightly yellow liquid with a strong, fishy odor

Uses. Production of rubber-processing chemical; corrosion inhibitor in boiler feed water; production of insecticides, plasticizers, and dry-cleaning soaps

Exposure. Inhalation; eye/skin contact

Toxicology. Cyclohexylamine is an irritant of the mucous membranes, eyes, and skin.

In three cases of acute human exposure from industrial accidents, symptoms included light-headedness, drowsiness, anxiety and apprehension, and nausea; slurred speech and vomiting also occurred in one case.[1] In human patch tests, a 23% solution caused severe irritation and possible sensitization. However, guinea pig sensitization tests did not confirm a potential for sensitization.[2]

In a multispecies study, rabbits, guinea pigs, and rats were exposed 7 hours/day, 5 days/week to levels of 1200, 800, and 150 ppm.[1] At 1200 ppm, all animals except for one rat died after a single exposure. At 800 ppm, fractional mortality occurred after repeated exposures. At 150 ppm, four of five rats and two guinea pigs survived 70 hours of exposure, but one rabbit died after only 7 hours. Effects were irritation of the respiratory tract and eyes with the development of corneal opacities.

When the undiluted liquid was applied to the skin of guinea pigs and kept in contact under an occluding cuff, the LD_{50} was between 1 and 5 ml/kg; edema, necrosis, and persistent eschars were observed.[3] One drop of a 50% aqueous solution in the eye of a rabbit caused complete destruction of the eye.

Cyclohexylamine has long been known to be pharmacologically active and has sympathomimetic effects on the heart and blood pressure.[4] However, it is not particularly potent.

Cyclohexylamine is a metabolite of the artificial sweetener sodium cyclamate, but the amount of conversion varies considerably from person to person.[5] Cyclohexylamine has been studied for carcinogenicity in two studies in mice, one of which was a multigeneration study, and in four studies in rats.[6] There were no differences in tumor incidence between treated and control animals. Several studies have shown no evidence of mutagenicity or teratogenicity.

Chromosome damage was induced in bone marrow cells of rats by intraperitoneal injection of 10 to 50 mg/kg per day for 5 days, and in peripheral blood cells of fetal lambs treated in utero with 50 to 250 mg/kg.[7,8]

REFERENCES

1. Watrous RM, Schulz HN: Cyclohexylamine, *p*-chloronitrobenzene, 2–aminopyridine: Toxic effects in industrial use. Ind Med Surg 19:317, 1950
2. Fassett DW: Laboratory of Industrial Medicine, Rochester, NY, Eastman Kodak Company (unpublished results)
3. Sutton WL: Aliphatic and alicyclic amines. In Fassett DW, Irish DD (eds): Industrial Hygiene and Toxicology, 2nd ed, Vol 2, Toxicology, p 2045. New York, Interscience, 1963
4. Eichelbaum M et al: Pharmacokinetics, cardiovascular and metabolic actions of cyclohexylamine in man. Arch Toxicol 31:243, 1974
5. Kojima S, Ichibagase H: Studies on synthetic sweetening agents: Vol III. Cyclohexylamine, a metabolite of sodium cyclamate. Chem Pharm Bull (Tokyo) 14:971, 1966
6. IARC Monographs on the Evaluation of the Carcinogenic Risk of Chemicals to Man, Vol 22, Some Non-nutritive Sweeting Agents, p 93. Lyon, International Agency for Research on Cancer, 1980
7. Legator MS et al: Cytogenetic studies in rats of cyclohexylamine, a metabolite of cyclamate. Science 165:1139, 1969
8. Turner JH, Hutchinson DL: Cyclohexylamine mutagenicity: An in vivo evaluation utilizing fetal lambs. Mutat Res 26:407, 1974

CYCLOPENTADIENE

CAS: 542-92-7

C_5H_6 1987 TLV = 75 ppm

Synonyms: 1,3–Cyclopentadiene; *p*-pentine; pentole; pyropentylene

Physical Form. Volatile liquid

Uses. In manufacture of resins; in organic synthesis

Exposure. Inhalation; minor skin absorption

Toxicology. Cyclopentadiene has an irritating, objectionable odor.

In human volunteers, the "sensory response was distinctly unfavorable" at both 250 and 500 ppm.[1] Rats exposed 7 hours/day to 500 ppm for 35 days (over a period of 53 days) developed mild injury to the liver and kidneys.[1] Dogs exposed 39 times to 400 ppm for 6 hours followed by 16 exposures at 800 ppm had no ill effects as determined by observation, clinical tests, or histologic examination.[1]

Concentrations significantly below 250 ppm are required for comfort, but systemic injury is not expected at this level.

REFERENCES

1. Cyclopentadiene. Documentation of the TLVs and BEIs, 5th ed, p 163. Cincinnati, American Conference of Governmental Industrial Hygienists (ACGIH), 1986

DDT
CAS: 50-29-3
$C_{14}H_9Cl_5$ 1987 TLV = 1 mg/m^3

Synonyms: 1,1,1–Trichloro–2,2–bis(*p*-chlorophenyl) ethane; dichlorodiphenyltrichloroethane

Physical Form. White crystalline solid

Uses. Insecticide; use banned in many temperate climate countries, but still widely used in the tropics

Exposure. Inhalation; skin absorption; ingestion

Toxicology. DDT affects the nervous system at high doses and causes paresthesias, tremor, and convulsions.

Ingestion of 10 mg/kg is sufficient to cause effects in some people, and convulsions have frequently occurred after ingestion of 16 mg/kg; 285 mg/kg has been taken without fatal results.[1] The onset of effects, usually occurring 2 or 3 hours after ingestion, is characterized by paresthesias of the tongue, lips, and face; the subject soon develops tremor, a sense of apprehension, dizziness, confusion, malaise, headache, and fatigue; in severe intoxication, convulsions may occur, and there may be paresis of the hands.[1] Ingestion of very large doses induces vomiting.[1] Recovery is well advanced or complete in 24 hours except in the most serious cases; three persons who each ingested an estimated 20 g of DDT showed a residual weakness of the hands after 5 weeks. Chronic poisoning in humans has not been described.

Large oral doses of DDT in rats caused focal and centrilobular necrosis of the liver.[1] However, in clinical evaluation and laboratory studies of 31 workers exposed to equivalent oral intakes of 3.6 to 18 mg daily for an average of 21 years, there was no evidence of hepatotoxicity; an observed increase in activity of hepatic microsomal enzymes was not accompanied by clinical evidence of detriment to general health.[2]

Continued absorption of DDT by humans results in storage of DDT and metabolites, including DDE [2,2–bis(*p*-chlorophenyl)–1,1–dichloroethylene] in fat.[3–5] In a study of 20 workers exposed to DDT for 11 to 19 years and with a calculated daily intake of 18 mg/person (calculated from DDA content in fat and DDA excretion in urine), the sum of isomers and metabolites of DDT in the fat was 38 to 647 ppm (compared with an average of 8 ppm for the general population); while DDE was the major excretory product in the general population, DDA [2,2-bis(*p*-chlorophenyl)acetic acid] was the major excretory product in DDT-exposed workers.[3]

Heavy exposure to DDT dust may result in eye and skin irritation.[1]

The hepatocarcinogenicity of DDT by the oral route has been demonstrated and confirmed in several strains of mice. Liver cell tumors have been produced in both sexes, and in CF mice the tumors were found to have metastasized to the lungs.[6]

REFERENCES

1. Hayes WJ Jr: Pesticides Studied in Man, pp 180–205. Baltimore, Williams & Wilkins, 1982
2. Laws ER Jr, Maddrey WC, Curley A, Burse VW: Long-term occupational exposure to DDT. Arch Environ Health 27:318–321, 1973
3. Laws ER Jr, Curley A, Biros FJ: Men with intensive occupational exposure to DDT—a clinical and chemical study. Arch Environ Health 15:766–775, 1967
4. Hayes WJ Jr, Dale WE, Pirkle CI: Evidence of safety of long-term, high, oral doses of DDT for man. Arch Environ Health 22:119–135, 1971
5. Wolfe HR, Armstrong JF: Exposure of formulating plant workers to DDT. Arch Environ Health 23:169–176, 1971
6. IARC Monographs on the Evaluation of the Carcinogenic Risk of Chemicals to Man, Vol 5, Some Organochlorine Pesticides, pp 90–124. Lyon, International Agency for Research on Cancer, 1974

DECABORANE

CAS: 17702-41-9

$B_{10}H_{14}$ — 1987 TLV = 0.05 ppm; skin

Synonyms: Decaboron tetradecahydride

Physical Form. White, crystalline solid

Uses. In rocket propellants; as catalyst in olefin polymerization

Exposure. Inhalation; skin absorption

Toxicology. Decaborane affects the nervous system and causes signs of both hyperexcitability and narcosis.

In humans, the onset of symptoms is frequently delayed for 24 to 48 hours after exposure; dizziness, headache, and nausea are common; other symptoms of mild intoxication include light-headedness, drowsiness, incoordination, and fatigue; more severe intoxication results in tremor, localized muscle spasm, and convulsive seizures.[1–3] Muscle spasm usually subsides after 24 hours, whereas light-headedness and fatigue may remain for up to 3 days.[3]

Effects on the eyes or skin of humans have not been reported; mice exposed to 26 ppm for 4 hours exhibited corneal opacity.[5] Exposure may impair the sense of smell.[2] The hazard from skin absorption is high.[4]

Exposure of rabbits to 56 ppm for 6 hours was fatal; effects included dyspnea, coarse movements of the head, weakness, rigid hindquarters, absence of eye reflexes, and convulsive seizures.[5] In dogs repeatedly given oral doses of 3 mg/kg, the effects on the central nervous system were not pronounced, but there was damage to the liver and kidneys.[2]

Intravenous administration of 4 to 10 mg/kg produced bradycardia and an initial transient hypertensive effect in the anesthetized dog.[6]

Toxicity is thought to occur from the decomposition of decaborane to a stable intermediate, which, in turn, inhibits intracellular pyridoxal phosphate–requiring enzymes.[7]

REFERENCES

1. Lowe HJ, Freeman G: Boron hydride (borane) intoxication in man. AMA Arch Ind Health 16:523–533, 1957
2. Chemical Safety Data Sheet SD–84, Boron Hydrides, pp 5–18. Washington, DC, Manufacturing Chemists' Association, Inc, 1961

3. Roush G Jr: The toxicology of the boranes. J Occup Med 1:46–52, 1959
4. Krackow EH: Toxicity and health hazards of boron hydrides. AMA Arch Ind Hyg Occup Med 8:335–339, 1953
5. Hughes RL, Smith IC, Lawless EW: Production of the Boranes and Related Research, pp 302–311, 329–331, 433–489. New York, Academic Press, 1967
6. Tadepalli AS, Buckley JP: Cardiac and peripheral vascular effects of decaborane. Toxicol Appl Pharmacol 29:210–222, 1974
7. Naeger LL, Leibman KC: Mechanisms of decaborane toxicity. Toxicol Appl Pharmacol 22:517–527, 1972

DEMETON

CAS: 8065-48-3

$(C_2H_5O)_2P(S)OC_2H_4SC_2H_5$

1987 TLV = 0.01 ppm; skin

Synonyms: Mixture of O,O–diethyl S–(and O)–2–[(ethylthio)ethyl]phosphorothioates

Physical Form. Pale-yellow to light-brown liquid

Uses. Acaricide; insecticide

Exposure. Inhalation; skin absorption; ingestion

Toxicology. Demeton is an anticholinesterase agent.

At least four fatal, several severe nonfatal, and a number of mild cases of demeton intoxication have been reported. Animal experiments and human exposures suggest that the toxicity and potency of demeton is similar to that of parathion.[1] Signs and symptoms of overexposure are caused by the inactivation of the enzyme cholinesterase, which results in the accumulation of acetylcholine at synapses in the nervous system, skeletal and smooth muscle, and secretory glands.[1–3] The sequence of the development of systemic effects varies with the route of entry. The onset of signs and symptoms is usually prompt but may be delayed up to 12 hours. After inhalation, respiratory and ocular effects are the first to appear, often within a few minutes of exposure. Respiratory effects include tightness in the chest and wheezing caused by bronchoconstriction and excessive bronchial secretion; laryngeal spasm and excessive salivation may add to the respiratory distress; cyanosis may also occur. Ocular effects include miosis, blurring of distant vision, tearing, rhinorrhea, and frontal headache.

After ingestion, gastrointestinal effects, such as anorexia, nausea, vomiting, abdominal cramps, and diarrhea, appear within 15 minutes to 2 hours. After skin absorption, localized sweating and muscular fasciculations in the immediate area occur usually within 15 minutes to 4 hours; skin absorption is somewhat greater at higher ambient temperatures and is increased by the presence of dermatitis.[1–3]

With severe intoxication by all routes, an excess of acetylcholine at the neuromuscular junctions of skeletal muscle causes weakness aggravated by exertion, involuntary twitchings, fasciculations, and, eventually, paralysis. The most serious consequence is paralysis of the respiratory muscles. Effects on the central nervous system include giddiness, confusion, ataxia, slurred speech, Cheyne–Stokes respiration, convulsions, coma, and loss of reflexes. The blood pressure may fall to low levels, and cardiac irregularities, including complete heart block, may occur. Complete symptomatic recovery usually occurs within 1 week; increased susceptibility to the effects of anticholinesterase agents persists for up to several weeks. Daily exposure to concentrations that are insufficient to produce symptoms following a single exposure may result in the onset of symptoms. Continued daily exposure may be followed by increasingly severe effects.

Demeton inactivates cholinesterase by phosphorylation of the active site of the enzyme to form the diethylphosphoryl enzyme. Over the following 24 to 45 hours, there is a process, termed *aging*, of conversion to the monoethylphosphoryl enzyme. Aging is of clinical interest in the treatment of poisoning because cholinesterase reactivators such as pralidoxime (2–PAM, Protopam) chloride are ineffective after aging has occurred.

Diagnosis. Initial signs and symptoms include headache, blurred vision, pallor, weakness, sweating, abdominal pain, nausea, vomiting, and diarrhea.

Signs and symptoms of *moderate-to-severe intoxication* include miosis, lacrimation, excessive salivation, muscle fasciculations, dyspnea, cyanosis, convulsions, shock, cardiac arrhythmias, and coma.

Differential Diagnosis: Diagnosis is based primarily on a history of exposure and clinical evidence of diffuse parasympathetic stimulation. Careful observation of the effects of atropine and pralidoxime may be valuable. Patients with organophosphate poisoning are resistant to the action of atropine at moderate dosages; failure of 1 to 2 mg of atropine administered parenterally to produce signs of atropinization (flushing, mydriasis, tachycardia, or dryness of mouth) indicates organophosphate poisoning. Intravenous injections of 1 g pralidoxime generally result in some recovery from signs and symptoms.

Special Tests: Two types of cholinesterase are clinically significant: (1) true acetylcholinesterase, found principally in the nervous system and the red blood cell; and (2) pseudo- or butyrylcholinesterase, found in the plasma, liver, and nervous system. Whereas the action of both types is inhibited by organophosphates, the level of depression of red blood cell cholinesterase is a better indicator of clinically significant reduction of cholinesterase activity in the nervous system.

Laboratory evidence of depression of red blood cell cholinesterase to a level substantially below pre-exposure levels (at least 50% and usually much lower) is verification of poisoning. There is an imperfect correlation between the degree of depression of cholinesterase enzymes and the occurrence of symptoms. With a rapid drop in cholinesterase activity, generally reflecting an acute heavy exposure, there may be symptoms with only a 30% depression, whereas with slower drops to 70% depression, reflecting chronic low-level exposure, there may be no symptoms.[4]

If no pre-exposure baseline has been performed but symptoms are not sufficient to justify treatment with atropine, repeated testing during the recovery period demonstrating progressively increasing plasma and red blood cell cholinesterase levels over several days and weeks, respectively, suggests the diagnosis of anticholinesterase poisoning.

There are many different methods for estimation of cholinesterase content of blood, and associated with each method is a different set of normal values and a different set of reporting units. The laboratory report of a cholinesterase determination should state the units involved along with the appropriate normal range. Based on the Michel method, the normal range of red blood cell cholinesterase activity (delta *p*H/hour) is 0.39 to 1.02 for men, and 0.34 to 1.10 for women.[5] The normal range of the enzyme activity (delta *p*H/hour) of plasma is 0.44 to 1.63 for men, and 0.24 to 1.54 for women.

Treatment. Treatment of organophosphate poisoning ranges from simple removal from exposure in very mild cases to the provision of very rigorous supportive and antidotal measures in severe cases.[1,3,6,7] In moderate-to-severe cases, because of pulmonary involvement, there may be need for artificial respiration using a positive-pressure method. Careful attention must be paid to removal of secretions and to maintenance of a patent airway. Anticonvulsants such as thiopental sodium may be necessary. Maintenance of respiration is critical, because death usually results from weakness of the muscles of respiration and accumulation of excessive secretions in the respiratory tract.[3]

As soon as cyanosis has been overcome, 2 to 4 mg of atropine should be given intravenously. (Atropine may induce ventricular fibrillation in the presence of cyanosis.) *This dose of atropine is approximately 10 times the amount that is administered for other conditions in which atropine is considered therapeutic.* This dose should be repeated at 5- to 10-minute intervals until signs of atropinization appear (dry, flushed skin, tachycardia as high as 140 beats/minute, and pupillary dilatation). A mild degree of atropinization should be maintained for at least 48 hours.[3]

Pralidoxime (2–PAM, Protopam) chloride is a cholinesterase reactivator that complements the action of atropine. It has its greatest effect in reversing the nicotinic action of anticholinesterase agents at skeletal neuromuscular junctions but virtually no effect on central nervous system manifestations. In moderate-to-severe cases, the dose for adults is 1 to 2 g injected intravenously at a rate not in excess of 500 mg/minute. After an hour, a second dose of 1 g is indicated if muscle weakness has not been relieved. Treatment with pralidoxime chloride will be most effective if given within 24 hours after poisoning.[3] Morphine, aminophylline, and phenothiazines are contraindicated because of documented experience of adverse reactions in cases of organophosphate poisoning.[6]

It is of great importance to decontaminate the patient. Contaminated clothing should be removed at once, and the skin should be washed with generous amounts of soap or detergent and a flood of water, which is best accomplished under a shower or by submersion in a pond or other body of water if the exposure occurred in the field. Careful attention should be paid to cleansing of the skin and hair.

The patient should be attended and monitored continuously for not less than 24 hours; serious and sometimes fatal relapses have occurred because of continuing absorption of the toxin or dissipation of the effects of the antidote.

Regeneration of cholinesterase occurs primarily by synthesis of new enzyme and takes place at the rate of approximately 1% per day.[6] A patient who has recovered from the acute phase of poisoning remains hypersusceptible to anticholinesterases for up to several weeks.

Medical Control. Medical control involves preplacement and annual physical examination with determination of pre-exposure red blood cell cholinesterase activity. A person whose red blood cell cholinesterase falls to or below 40% of the pre-exposure baseline should be removed from further exposure until the activity returns to within 80% of the pre-exposure baseline.

REFERENCES

1. Hayes WJ Jr: Organic phosphorus pesticides. In Pesticides Studied in Man, pp 284–435. Baltimore, Williams & Wilkins, 1982
2. Koelle GB (ed): Cholinesterases and anticholinesterase agents. In Handbuch der Experimen-

tellen Pharmakologie, Vol 15, pp 989–1027. Berlin, Springer-Verlag, 1963
3. Taylor P: Anticholinesterase agents. In Gilman AG et al (eds): Goodman and Gilman's The Pharmacological Basis of Therapeutics, 7th ed, pp 110–129. New York, Macmillan, 1985
4. Coye MJ, Lowe JA, Maddy KT: Biological monitoring of agricultural workers exposed to pesticides. I. Cholinesterase activity determinations. J Occup Med 28:619–627, 1986
5. Michel HO: Electrometric method for determination of red blood cell and plasma cholinesterase activity. J Lab Clin Med 34:1564–1568, 1949
6. Milby TH: Prevention and management of organophosphate poisoning. JAMA 216:2131–2133, 1971
7. Namba T, Nolte CT, Jackrel J, Grob D: Poisoning due to organophosphate insecticides. Am J Med 50:475–492, 1971

DIACETONE ALCOHOL

CAS: 123-42-2

$(CH_3)_2C(OH)CH_2COCH_3$

1987 TLV = 50 ppm

Synonyms: 4–Hydroxy–4–methyl–2–pentanone; diacetonyl alcohol; diacetone; dimethyl acetonyl carbinol

Physical Form. Liquid

Uses. Solvent for pigments, cellulose, resins, oils, fats, and hydrocarbons; hydraulic brake fluid; antifreeze

Exposure. Inhalation; minor skin absorption

Toxicology. Diacetone alcohol causes irritation of the eyes and respiratory tract; at high concentrations, it causes narcosis in animals, and it is expected that severe exposure will cause the same effect in humans.

Most human subjects exposed to 100 ppm for 15 minutes complained of eye, nose, and throat irritation; exposure to 400 ppm also caused chest discomfort.[1,2]

Animals exposed to 2100 ppm for 1 to 3 hours exhibited restlessness, mucous membrane irritation, and drowsiness.[3] Rats exposed to 1500 ppm for 8 hours survived.[4] Injection of 3 ml/kg or intragastric administration of 5 ml/kg diacetone alcohol in rabbits caused respiratory depression, narcosis, and death.[5] A temporary decrease in the number of erythrocytes in the blood of rats was observed for 1 to 4 days following intragastric administration of 2 ml/kg of diacetone alcohol; hepatic lesions characterized by vacuolization and granulation of the parenchymal cells were noted, but recovery was complete in 7 days.[6]

The liquid defats the skin and may produce dermatitis with prolonged or repeated contact; in the eyes, it causes moderate to marked irritation and transient corneal damage.[3]

REFERENCES

1. Silverman L, Schulte HF, First MW: Further studies on sensory response to certain industrial solvent vapors. J Ind Hyg Toxicol 28:262–266, 1946
2. Shell Chemical Corporation: Diacetone alcohol, SC 57–84. Ind Hyg Bull Toxicity Data Sheet, 1957
3. Krasavage WJ et al: Ketones. In Clayton GD, Clayton FE (eds): Patty's Industrial Hygiene and Toxicology, 3rd ed, Vol 2C, Toxicology, pp 4754–4756. New York, Wiley–Interscience, 1982
4. Smyth HF: Improved communication—hygienic standards for daily inhalation. Am Ind Hyg Assoc Q 17:129–185, 1956
5. Walton DC, Kehr EF, Lovenhart AS: A comparison of the pharmacological action of diacetone alcohol and acetone. J Pharmacol 33:175–183, 1928
6. Keith HM: Effect of diacetone alcohol on the liver of the rat. Arch Pathol 13:707–712, 1932

DIAZOMETHANE

CAS: 334-88-3

CH_2N_2 1987 TLV = 0.2 ppm

Synonyms: Azimethylene

Physical Form. Yellow gas

Uses. Powerful methylating agent for acidic compounds such as carboxylic acids, phenols and enols

Exposure. Inhalation

Toxicology. Diazomethane is a severe pulmonary irritant.

It is extremely dangerous, causing irritation of the eyes, chest pain, cough, fever, and severe asthmatic attacks. A chemist briefly exposed to an unknown concentration of diazomethane in a laboratory developed a violent cough and shortness of breath leading to severe pulmonary edema; symptoms completely subsided within 2 weeks.[1] In a fatal incident, another chemist exposed to an unknown concentration of diazomethane, as well as other irritant gases, experienced immediate respiratory distress leading to pneumonitis and death on the fourth day following exposure.[1]

A physician exposed to diazomethane from a laboratory spill noted only a faint odor but immediately experienced severe headache, cough, mild anterior chest pain, generalized aching of muscles, and a sensation of overwhelming tiredness.[2] Within 5 minutes he was stuporous, and, on early admission to a hospital, was markedly flushed and feverish; he recovered in approximately 48 hours.[2] Subsequent exposure to trace amounts of the gas produced wheezing, cough, and malaise, leading to the suspicion that this substance may also have a sensitizing effect upon the respiratory system.[2] Skin exposure has produced irritation and denudation.[3]

Exposure of cats to 175 ppm for 10 minutes resulted in pulmonary edema and hemorrhage, with death occurring in 3 days.[1] Limited animal studies indicate that diazomethane is carcinogenic in mice (increased incidence of lung tumors following skin application) and rats (exposure to the gas caused lung tumors). The carcinogenic risk to humans has not been determined.[4]

The warning properties of diazomethane are poor.[3]

REFERENCES

1. Reinhardt CF, Brittelli MR: Heterocyclic and miscellaneous nitrogen compounds. In Clayton GD, Clayton FE (eds): Patty's Industrial Hygiene and Toxicology, 3rd ed, rev, Vol 2A, Toxicology, pp 2784–2786. New York, Wiley–Interscience, 1981
2. Lewis CE: Diazomethane poisoning, report of a case suggesting sensitization reaction. J Occup Med 6:91–93, 1964
3. Sunderman WF: Diazomethane. In International Labour Office Encyclopaedia of Occupational Health and Safety, Vol I, A–K, pp 383–384. New York, McGraw-Hill, 1971
4. IARC Monographs On the Evaluation of the Carcinogenic Risk of Chemicals to Man, Vol 7, Some Anti-Thyroid and Related Substances, Nitrofurans and Industrial Chemicals, pp 223–230. Lyon, International Agency for Research on Cancer, 1974

DIBORANE

CAS: 19287-45-7

B_2H_6 1987 TLV = 0.1 ppm

Synonyms: Boroethane; boron hydride

Physical Form. Gas

Uses. High-energy fuel; reducing agent; initiator of polymerization of ethylene, vinyl, and styrene

Exposure. Inhalation

Toxicology. Diborane is a pulmonary irritant.

In humans, overexposure results in a sensation of tightness in the chest, leading to precordial pain, shortness of breath, nonproductive cough, and sometimes nausea.[1-4] Prolonged exposure to low concentrations causes headache, light-headedness, vertigo, chills, and, less frequently, fever. Fatigue or weakness occurs and may persist for several hours; tremor or muscular fasciculations occur infrequently and are usually localized and of short duration.[1-4] Diborane gas has not been found to have significant effects upon contact with skin or mucous membranes, although high concentrations may cause eye irritation.[5]

The LC_{50} for rats was 50 ppm for 4 hours; in other animal experiments, acute exposure caused pulmonary edema and hemorrhage and temporary damage to the liver and kidneys.[6] Repeated exposure of dogs at about 5 ppm for 6 hours/day resulted in death after 10 to 20 exposures; one of two animals survived repeated exposure at 1 to 2 ppm for 6 months.[7] Repeated respiratory insult was thought to be the underlying cause of death.

The threshold of odor detection is approximately 3.3 ppm; the repulsive odor is described as rotten eggs, sickly sweet, musty, or foul.[6]

REFERENCES

1. Lowe HJ, Freeman G: Boron hydride (boron) intoxication in man. AMA Arch Ind Health 16:523–533, 1957
2. Cordasco EM, Cooper RW, Murphy JV, Anderson C: Pulmonary aspects of some toxic experimental space fuels. Dis Chest 41:68–74, 1962
3. Roush G Jr: The toxicology of the boranes. J Occup Med 1:46–52, 1959
4. Rozendaal HM: Clinical observations on the toxicology of boron hydrides. AMA Arch Ind Hyg Occup Med 4:257–260, 1951
5. Chemical Safety Data Sheet, SD–84, Boron Hydrides, pp 5–7, 16–18. Washington, DC, MCA, Inc, 1961
6. Holzmann RT (ed): Production of the Boranes and Related Research, pp 289–294, 329–331, 433–489. New York, Academic Press, 1967
7. Comstock CC et al: Research Report No 258. Washington, DC, US Army Chemical Corps, Medical Laboratories, March 1954

1,2 DIBROMO-3-CHLOROPROPANE

CAS: 96-12-8

$CH_2BrCHBrCH_2Cl$

1977 Emergency Temporary Standard = 0.01 ppm

Synonyms: DBCP; dibromochloropropane

Physical Form. Yellow or amber liquid

Uses. Agricultural nematocide (use now banned in the United States)

Exposure. Inhalation; skin absorption

Toxicology. DBCP causes sterility in male workers owing to a selective effect on seminiferous tubules. It causes cancer in mice and rats, and it is an irritant of the eyes and mucous membranes, as well as a mild depressant of the central nervous system in animals.

DBCP has caused oligospermia and aspermia in male workers.[1] Initial documentation of these effects occurred in workers engaged in the production of DBCP at an agricultural chemical plant in Lathrop, California. Of 41 exposed workers, 3 were women and 11 were men with previous vasectomies. Of the 27 remaining men, 11 had abnormally low sperm counts of less than 1 million/ml; all had been exposed for at least 3 years. None with sperm counts above 40 million had been exposed for more than 3 months.[2]

Subsequent studies in this and three other DBCP plants showed a total of more than 100 cases of oligospermia or aspermia. Exposures in one plant were estimated at 100 to 600 ppb.[1]

A larger clinical-epidemiologic study of these men was undertaken to determine the exposure–effect relationships involved. Of 142 nonvasectomized men providing semen samples, 107 had been exposed to DBCP and 35 had not been exposed. There was a clear-cut difference in both the distribution of sperm counts and the median counts between the exposed men and the nonexposed men. Of the exposed men, 13.1% were azoospermic, 16.8% were severely oligospermic, and 15.8% were mildly oligospermic.[3] Sperm concentration and serum follicle-stimulating hormone (FSH) levels in 44 of these workers were reassessed 5 to 8 years after exposure was terminated in 1977. Two of the eight originally azoospermic workers produced sperm during the follow-up, although only one worker had normal sperm counts. No increase in sperm production could be detected in men who had low sperm counts in 1977, and elevated serum FSH levels did not drop in oligospermic or azoospermic men. These results suggest that permanent destruction of germinal epithelium occurs in most DBCP-sterile persons.[4]

The LC_{50} for rats was 368 ppm for 1 hour and 103 ppm for 8 hours.[5] Irritation of the eyes and respiratory tract was observed at levels of 60 ppm and higher. Moderate depression of the central nervous system was manifested as sluggishness and ataxia.

In rats of both sexes given 50 to 66 exposures to 12 ppm over 70 to 90 days, 40% to 50% of the animals died.[5] In most cases, death was attributed to lung infection. The most striking observation at autopsy was severe atrophy and degeneration of the testes. There was also degenerative changes of the seminiferous tubules, reduction in sperm count, and abnormal development of sperm cells. Other effects were mild damage to the liver and kidneys.

The liquid applied undiluted in the eye of a rabbit caused transient irritation.[5] An LD_{50} of 1.4 g/kg was obtained when the material was applied undiluted for 24 hours to the rabbit skin. Repeated application (20 times) to the skin of a rabbit caused slight crustiness. However, the dermis and subcutaneous tissue showed extensive necrosis.

In a study of carcinogenesis, DBCP was orally administered to rats and mice five times per week at maximally tolerated doses and at half those doses.[6–8] As early as 10 weeks after initiation of treatment, there was a high incidence of squamous cell carcinoma of the stomach in both species. In female rats, there were also mammary adenocarcinomas.

The odor of DBCP was detected at 1.7 ppm, the only level tested.[5]

REFERENCES

1. Department of Labor: Emergency temporary standard for occupational exposure to 1,2 dibromo–3–chloropropane (DBCP). Federal Register 42:45535, 1977
2. Whorton D et al: Infertility in male pesticide workers. Lancet 2:1259–1261, 1977
3. Whorton D et al: Testicular function in DBCP-exposed pesticide workers. J Occup Med 21:161–166, 1979
4. Eaton M et al: Seven-year follow-up of workers exposed to 1,2–dibromo–3–chloropropane. J Occup Med 28:1145–1150, 1986
5. Torkelson TR et al: Toxicologic investigations of 1,2–dibromo–3–chloropropane. Toxicol Appl Pharmacol 3:545–559, 1961
6. Olsen WA et al: Brief communication: Induction of stomach cancer in rats and mice by halogenated aliphatic fumigants. J Natl Cancer Inst 51:1993–1995, 1973
7. Ward JM, Habermann RT: Pathology of stomach cancer in rats and mice induced with the agricultural chemicals ethylene dibromide and dibromochloropropane. Bull Soc Pharmacol Environ Pathol 74 (series 2, issue 2):10–11, 1974
8. Powers MB et al: Carcinogenicity of ethylene dibromide (EDB) and 1,2–dibromo–3–chloropropane and oral administration in rats and mice (abstr). Toxicol Appl Pharmacol 33:171–172, 1975

DIBUTYL PHOSPHATE

CAS: 107-66-4

$(n\text{-}C_4H_9O)_2(OH)PO$ 1987 TLV = 1 ppm

Synonyms: None

Physical Form. Pale amber liquid

Uses. Organic catalyst; antifoaming agent

Exposure. Inhalation

Toxicology. Dibutyl phosphate is an irritant of the eyes and mucous membranes.

Data on effects in humans are sparse; workers exposed to unspecified concentrations of vapor complained of respiratory irritation and headache.[1] It is a moderately strong acid and could be expected to be irritating on contact.[1]

In rats, the oral LD_{50} is 3.2 g/kg.[2]

REFERENCES

1. Dibutyl phosphate. Documentation of the TLVs and BEIs, 5th ed, p 175. Cincinnati, American Conference of Governmental Industrial Hygienists (ACGIH), 1986
2. Christensen HE (ed): Registry of Toxic Effects of Chemical Substances, 1976 edition. DHEW (NIOSH) Pub No 76–191, p 412. Washington, DC, US Government Printing Office, 1976

DIBUTYL PHTHALATE

CAS: 84-74-2

$C_{16}H_{22}O_4$ 1987 TLV = 5 mg/m^3

Synonyms: Butyl phthalate; 1,2–benzenedicarboxylic acid dibutyl ester; phthalic acid dibutyl ester; DBP

Physical Form. Colorless or slightly colored oily liquid

Uses. Plasticizer in plastics; manometer fluid; insect repellent

Exposure. Inhalation; ingestion

Toxicology. Dibutyl phthalate (DBP) is of low-order acute toxicity.

A chemical worker who accidentally swallowed 10 g (about 140 mg/kg) DBP developed nausea, dizziness, headache, pain and irritation in the eyes, conjunctivitis, and toxic nephritis. He recovered completely after 2 weeks.[1]

There were no positive reactions to 5% DBP among 53 subjects given a 48-hour closed patch test.[2] Cosmetic formulations containing up to 9% DBP ranged from nonirritating to slightly irritating in various patch test procedures. Sensitization and photosensitization did not occur.[3]

Rats exposed to concentrations as low as 0.5 mg/m^3 of DBP mist for 6 hours/day for 6 months had smaller weight gains and greater brain and lung weights than controls.[4] At 50 mg/m^3, the effects were more pronounced. A 2-hour exposure of mice to an aerosol of DBP at 250 mg/m^3 resulted in eye and upper respiratory tract irritation. Higher levels caused labored breathing, ataxia, paresis, convulsions, and death from paralysis of the respiratory system.[5]

Undiluted DBP instilled in rabbit eyes caused no observable irritation up to 48 hours postinstillation.[6]

Reduced testes weights and histologic evidence of testicular injury was found in mice and guinea pigs but not in hamsters fed 2 g/kg/day DBP for 10 days.[7]

Oral or intraperitoneal administration of DBP in pregnant animals at high doses relative to the LD_{50} produced an increased number of resorptions, increased fetal deaths, neural tube defects, and skeletal abnormalities.[8–10] Carcinogenesis was not observed in 18-month or longer feeding studies in rats.[3]

REFERENCES

1. Krauskopf LG: Studies on the toxicity of phthalates via ingestion. Environ Health Perspect 3:61–72, 1973

2. Kaaber S et al: Skin sensitivity to denture base materials in the burning mouth syndrome. Contact Dermatitis 5:90–96, 1979
3. Final Report on the Safety Assessment of Dibutyl Phthalate, Dimethyl Phthalate, and Diethyl Phthalate. J Am Coll Toxicol 4:267–303, 1985
4. Kawano M: Toxicological studies on the phthalate esters. 1. Inhalation effects of dibutyl phthalate on rats. Japan J Hyg 35:684–692, 1980
5. Dibutyl phthalate. Documentation of the TLVs for Substances in Workroom Air, 4th ed, pp 124–125. Cincinnati, American Conference of Governmental Industrial Hygienists (ACGIH), 1984
6. Lawrence WH et al: Toxicological investigation of some acute, short-term, and chronic effects of administering di–2–ethylhexyl phthalate (DEHP) and other phthalate esters. Environ Res 9:1–11, 1975
7. Gangolli SD: Testicular effects of phthalate esters. Environ Health Perspect 45:77–84, 1982
8. Peters JW, Cook RM: Effect of phthalate esters on reproduction in rats. Environ Health Perspect 3:91–94, 1973
9. Singh AR, Lawrence WH, Autian J: Teratogenicity of phthalate esters in rats. J Pharm Sci 61:51–55, 1972
10. Shista K, Nishimura H: Teratogenicity of di(2–ethylhexyl) phthalate (DEHP) and di–n–butyl phthalate (DBP) in mice. Environ Health Perspect 45:65–70, 1982

o-DICHLOROBENZENE

CAS: 95-50-1

$C_6H_4Cl_2$ 1987 TLV = C 50 ppm

Synonyms: 1,2–Dichlorobenzene; dichlorobenzol; dichloricide

Physical Form. Clear liquid

Uses. Solvent; fumigant; insecticide; chemical intermediate

Exposure. Inhalation

Toxicology. *o*-Dichlorobenzene is a skin and eye irritant. At high doses, it causes central nervous system depression and liver and kidney damage in animals. Heavy exposure is expected to produce the same effects in humans.

In humans, eye irritation is not usually evident below 20 ppm but becomes noticeable at 25 to 30 ppm and painful to some at 60 to 100 ppm if exposures are for more than a few minutes duration.[1] Some acclimation may occur, but its extent is not great. Workers exposed to concentrations ranging from 1 to 44 ppm and averaging 15 ppm showed no indication of injury.[2] Accidental exposure of 26 subjects to unspecified levels 8 hours/day for 4 days caused eye, nose, and throat irritation.[3] Ten of the 26 subjects reported dizziness, severe headache, fatigue, and nausea. Chromosome studies showed significant alterations in the leukocytes of exposed workers; these changes appeared reversible 6 months later.

The liquid left on the skin may produce blistering, and, later, the area may become pigmented.[2] Sensitization dermatitis has been reported.[2]

Rats died from exposure to 977 ppm for 7 hours but survived when exposed for only 2 hours; animals survived exposure to 539 ppm for 3 hours but at necropsy showed marked centrolobular necrosis of the liver, as well as cloudy swelling of the tubular epithelium of the kidney.[2] Several species of animals exposed for periods of 6 or 7 months to 93 ppm for 7 hours daily showed no adverse effects.[2]

Repeated dermal application to rats was fatal.[4] The liquid instilled in the rabbit eye produced apparent distress and slight conjunctival irritation.[2]

There was no clear evidence of carcinogenicity in rats or mice receiving 60 or 120 mg/kg by gavage five times per week for 2 years.[5]

The odor of *o*-dichlorobenzene is perceptible to most people at 2 to 4 ppm.[1]

REFERENCES

1. Hygienic Guide Series: *o*-Dichlorobenzene. Am Ind Hyg Assoc J 25:320–323, 1964
2. Hollingsworth RL et al: Toxicity of *o*-dichlorobenzene. AMA Arch Ind Health 17:180–187, 1958
3. Zapata–Gayon C et al: Clastogenic chromosomal aberrations in 26 individuals accidentally exposed to ortho dichlorobenzene vapors in the National Medical Center in Mexico City. Arch Environ Health 37:231–235, 1982
4. US Environmental Protection Agency: Health Assessment Document for Chlorinated Benzenes, Final Report. Washington, DC, Office of Health and Environmental Assessment, January 1985
5. National Toxicology Program: Toxicology and Carcinogenesis Studies of 1,2–Dichlorobenzene (*o*-Dichlorobenzene) (CAS No 95-50-1) in F344/N Rats and B6C3F1 Mice (Gavage Studies). TRS-255. DHHS (NIH) Pub No 86–2511. Research Triangle Park, North Carolina, US Department of Health and Human Services, October 1985

---

p-DICHLOROBENZENE

CAS: 106-46-7

$C_6H_4Cl_2$ 1987 TLV = 75 ppm

Synonyms: Parazene; di-chloricide; paramoth; 1,4–dichlorobenzene; paracide; paradi; paradow; PDB; santochlor

Physical Form. Colorless or white crystals

Uses. Insecticide; disinfectant; chemical intermediate; used against clothes moths

Exposure. Inhalation

Toxicology. *p*-Dichlorobenzene vapor is an irritant of the eyes and upper respiratory tract and is toxic to the liver. Bioassays in mice and rats have indicated evidence of carcinogenicity.

In five cases of intoxication by inhalation from household or occupational exposure to *p*-dichlorobenzene used as a mothproofing agent, one person with only moderate exposure suffered severe headache, periorbital swelling, and profuse rhinitis, which subsided 24 hours after cessation of exposure.[1] The other four persons who had more prolonged and heavy exposure developed anorexia, nausea, vomiting, weight loss, and hepatic necrosis with jaundice; two persons died, and another developed cirrhosis. Although these five cases were temporarily associated with known exposure to *p*-dichlorobenzene in four different settings, it is unclear how thoroughly other potential causes for these findings were excluded.

In 58 workers exposed for an average of 4.8 years (range 8 months to 25 years) to *p*-dichlorobenzene at levels of 10 to 725 ppm, there was no evidence of hematologic effects in spite of the structural similarity to benzene, a potent bone marrow depressant. Painful irritation of the eyes and nose was noted at levels between 50 and 80 ppm, and pain was severe at 160 ppm.

Solid particles of *p*-dichlorobenzene in the human eye cause pain.[2] The solid material produces a burning sensation when held in contact with the skin, but the resulting irritation is slight; warm fumes or strong solutions of *p*-dichlorobenzene may irritate the intact skin slightly on prolonged or repeated contact.[2,3] A case of allergic purpura induced by *p*-dichlorobenzene has been reported.[4]

In a study of workers engaged in synthesizing or otherwise handling *p*-dichlorobenzene, it was concluded that urinary excretion of 2,5–dichlorophenol (a metabolite of *p*-dichlorobenzene) can serve as an index of exposure.[5]

In male rats given *p*-dichlorobenzene by gavage at 150 and 300 mg/kg for 2 years, there was a significant dose-related increased incidence of tubular cell adenocarcinomas of the kidney; no excess was observed in female rats.[6] There was evidence of nephropathy in both male and female rats. *p*-Dichlorobenzene increased the incidence of hepatocellular adenomas and carcinomas, as well as non-neoplastic liver lesions in male and female mice dosed at 600 mg/kg for 2 years. Conclusions from the National Toxicology Program study were that there was clear evidence of carcinogenicity for the male rats and the male and female mice.[6] In a long-term inhalation study in male and female rats and female mice, there was no evidence of carcinogenicity following exposure at 75 or 500 ppm for 5 hours/day, 5 days/week for 76 weeks (rats) or 57 weeks (mice).[7] Although there has been a report of five cases of blood dyscrasias, including leukemia, among persons exposed to *o*- or *p*-dichlorobenzene, the IARC has concluded that the human data are inadequate to evaluate the carcinogenicity of dichlorobenzenes.[8]

REFERENCES

1. Cotter LH: Paradichlorobenzene poisoning from insecticides. NY State J Med 53:1690–1692, 1953
2. Hollingsworth RL et al: Toxicity of paradichlorobenzene—determinations on experimental animals and human subjects. AMA Arch Ind Health 14:138–147, 1956
3. Hygienic Guide Series: *p*-Dichlorobenzene. Am Ind Hyg Assoc J 25:323–325, 1964
4. Nalbandian RM, Pearce JF: Allergic purpura induced by exposure to *p*-dichlorobenzene. JAMA 194:238–239, 1965
5. Pagnotto LD, Walkley JE: Urinary dichlorophenol as an index of para-dichlorobenzene exposure. Am Ind Hyg Assoc J 26:137–142, 1965
6. National Toxicology Program: Toxicology and Carcinogenesis Studies of 1,4-Dichlorobenzene (CAS No 106-46-7) in F344/N Rats and B6C3F1 Mice (Gavage Studies). DHHS (NIH) Pub No 87–2575. Research Triangle Park, North Carolina, US Department of Health and Human Services, 1987
7. Loeser F, Litchfield MH: Review of recent toxicology studies on *p*-Dichlorobenzene. Fd Chem Toxicol 21:825–832, 1983
8. IARC Monographs on the Evaluation of the Carcinogenic Risk of Chemicals to Humans: Some Industrial Chemicals and Dyestuffs, Vol 29, pp 213–238. Lyon, International Agency for Research on Cancer, 1982

3,3′-DICHLOROBENZIDINE

CAS: 91-94-1

$C_{12}H_{10}Cl_2N_2$

1987 TLV = none; suspected human carcinogen; skin

Synonyms: DCB, 4,4′-diamino-3,3′-dichlorobiphenyl; *o*-dichlorobenzidine

Physical Form. Colorless crystals

Uses. Chemical intermediate for the product of azo dyes

Exposure. Inhalation; skin absorption

Toxicology. 3,3′-Dichlorobenzidine (DCB) is carcinogenic in rats, mice, and hamsters.

There are no reports in which DCB has induced cancer in humans.[1] However, DCB exposure may have been a factor in some cases of bladder cancer attributed to benzidine, since these substances are often produced together, and DCB also bears a close structural similarity to benzidine.[2]

Of 111 rats given 20 mg DCB by injection or gastric intubation 6 days/week for 10 to 20 months, 17 had tumors of the zymbal gland (a specialized sebaceous gland adjacent to the external ear canal), 13 had mammary tumors, 8 had skin tumors, 5 had malignant lymphomas, 3 had urinary bladder tumors, 3 had salivary gland tumors, and 2 had intestinal tumors; no tumors were found in 130 control rats.[3]

Of 44 male rats fed 1000 ppm for 12 months, 9 developed granulocytic leukemia and 8 developed zymbal gland tumors; mammary gland tumors were found in rats of both sexes.[4]

In hamsters, 0.3% DCB in the diet produced transitional cell carcinomas of the bladder and some liver cell tumors.[5] Liver tumors were also found in mice exposed to DCB.[3]

Because of demonstrated potent carcinogenicity in animals, human exposure by any route should be avoided.[6]

REFERENCES

1. Gerarde HW, Gerarde DF: Industrial experience with 3,3′–dichlorobenzidine: An epidemiological study of a chemical manufacturing plant. J Occup Med 16:322–344, 1974
2. IARC Monographs on the Evaluation of the Carcinogenic Risk of Chemicals to Man, Vol 4, Some Aromatic Amines, Hydrazine and Related Substances, N-nitroso Compounds and Miscellaneous Alkylating Agents, pp 49–55. Lyon, International Agency for Research on Cancer, 1974
3. Pliss GB: Dichlorobenzidine as a blastomogenic agent. Vop Onkol 5:524–533, 1959
4. Stula EF, Sherman H, Zapp JA Jr, Clayton JW Jr: Experimental neoplasia in rats from oral administration of 3,3′–dichlorobenzidine, 4,4′–methylene–bis(2–chloroaniline), and 4,4′–methylene–bis(2–methylaniline). Toxicol Appl Pharmacol 31:159–176, 1975
5. Sellakumar AR, Montesano R, Saffiotti U: Aromatic amines carcinogenicity in hamsters. Proc Am Assoc Cancer Res 10:78, 1969
6. 3,3′–Dichlorobenzidine. Documentation of TLVs and BEIs, 5th ed, p 180. Cincinnati, American Conference of Governmental Industrial Hygienists (ACGIH), 1986

DICHLORODIFLUOROMETHANE

CAS: 75-71-8

CCl_2F_2 1987 TLV = 1,000 ppm

Synonyms: Freon 12; Refrigerant 12; Isotron; Halon; Genetron 12; Frigen 12

Physical Form. Colorless gas

Uses. Refrigerant gases; propellant for insect aerosol bombs

Exposure. Inhalation

Toxicology. Dichlorodifluoromethane causes central nervous system depression at very high concentrations.

Volunteers exposed to 200,000 ppm for a short period of time experienced significant eye irritation as well as central nervous system effects.[1] The effects disappeared within minutes after return to fresh air. Exposure at 110,000 ppm for 11 minutes caused a marked decrease in consciousness, amnesia, and cardiac arrhythmias; at 40,000 ppm for 80 minutes, there was generalized paresthesia, tinnitus, apprehension, and slurred speech.[1] Two volunteers exposed to 10,000 ppm for 2.5 hours showed slight psychomotor impairment.[2]

Chronic exposure of volunteers to 1000 ppm 8 hours/day for 17 days caused no subjective symptoms, no cardiac abnormalities, and no pulmonary function abnormalities.[3]

Sniffing aerosols of fluorochlorinated hydrocarbons has caused sudden death from cardiac arrest probably owing to cardiac arrhythmias from sensitization of the myocardium to epinephrine.[4]

In rats, when dichlorodifluoromethane was administered at various concentrations with 20% oxygen for 30 minutes, the following effects were observed: 200,000 ppm, no observable effects; 300,000 ppm, muscular twitching and tremor; 800,000 ppm, coma, corneal reflexes absent; 800,000 ppm for 4 and 6 hours was not lethal, and the animals suffered no permanent effects.[5]

Chronic exposure of rats 6 hours/day for 90 days at 10,000 ppm and of dogs at 5,000 ppm caused no adverse effects as determined by observation, clinical tests, or histologic examination.[6]

A variety of reproductive and carcinogenic studies have found no significant effects.[1]

REFERENCES

1. Haskell Laboratory, Dupont: Toxicity Review Freon 12, pp 1–26. November 1982
2. Azar A et al: Experimental human exposures to fluorocarbon 12 (dichlorodifluoromethane). Am Ind Hyg Assoc J 33:207–216, 1972
3. Stewart RD et al: Physiological response to aerosol propellants. Environ Health Perspect 26:275–285, 1978
4. Reinhardt CF et al: Cardiac arrhythmias and aerosol "sniffing." Arch Environ Health 22:265–279, 1971
5. Lester D, Greenberg LA: Acute and chronic toxicity of some halogenated derivatives of methane and ethane. Arch Ind Hyg Occup Med 2:335–344, 1950
6. Leuschner F et al: Report of subacute toxicological studies with several fluorocarbons in rats and dogs by inhalation. Drug Res 33:1475–1476, 1983

1,3-DICHLORO-5,5-DIMETHYLHYDANTOIN

CAS: 118-52-5

$C_5H_6Cl_2O_2N_2$ 1987 TLV = 0.2 mg/m^3

Synonyms: Dactin; Halane

Physical Form. White powder

Uses. Chlorinating agent; disinfectant; laundry bleach; in water treatment; intermediate for drugs; insecticides; polymerization catalyst

Exposure. Inhalation

Toxicology. 1,3-Dichloro-5,5-dimethylhydantoin powder in contact with water yields hypochlorous acid, which is an irritant of the eyes and mucous membranes.

There is a single report of workmen exposed to an average concentration of 1.97 mg/m^3 experiencing marked respiratory irritation.[1] The LD_{50} for rats when administered orally as a 10% aqueous suspension was 542 mg/kg; at necropsy, gastrointestinal hemorrhages were found.

REFERENCE

1. 1,3-Dichloro-5,5-Dimethylhydantoin. Documentation of the TLVs and BEIs, 5th ed, p 183. Cincinnati, American Conference of Governmental Industrial Hygienists (ACGIH), 1986

1,1-DICHLOROETHANE

CAS: 75-34-3

CH_3CHCl_2 1987 TLV = 200 ppm

Synonyms: Ethylidene dichloride

Physical Form. Colorless liquid

Uses. Cleansing agent; degreaser, solvent for plastics, oils, and fats; grain fumigant; chemical intermediate; formerly used as an anesthetic

Exposure. Inhalation; skin absorption

Toxicology. At high concentrations, 1,1-dichloroethane causes narcosis in animals, and it is expected that severe exposure will cause the same effect in humans.

There have been no reported cases of human overexposure by inhalation; prolonged or repeated skin contact can produce a slight burn.[1]

Rats exposed to 32,000 ppm for 30 minutes survived but died after 2.5 hours of exposure.[1] The most consistent findings in animals exposed to concentrations of above 8000 ppm for up to 7 hours were pathologic changes in the kidney and liver and, at much higher concentrations, near 64,000 ppm, damage to the lungs as well. Repeated daily exposure of several species to 1000 ppm resulted in no pathologic or hematologic changes.

The liquid applied to the intact or abraded skin of rabbits produced slight edema and very slight necrosis after the sixth of 10 daily applications. When instilled in the eyes of rabbits, there was immediate, moderate conjunctival irritation and swelling, which subsided within a week.

Although the liquid may be absorbed through the skin, it is apparently not absorbed in amounts sufficient to produce systemic injury.

Exposure of rats to 6000 ppm 7 hours/day on days 6 through 15 of gestation was associated with an increased incidence of delayed ossification of sternebrae.[2] Maternal toxicity was limited to decreased weight gain.

A significant increase in endometrial stromal polyps, a benign neoplasm, occurred in female mice administered 1,1-dichloroethane by gavage for 78 weeks.[3] High mortality in all rat groups obscured results. The National Cancer Institute determined that there was no conclusive evidence for carcinogenicity, but 1,1-dichloroethane should be treated with caution by analogy to other chloroethanes shown to be carcinogenic in laboratory animals.[3,4]

Odor cannot be relied upon to provide warning of overexposure.

REFERENCES

1. Hygienic Guide Series: 1,1-Dichloroethane (ethylidene chloride). Am Ind Hyg Assoc J 32:67–71, 1971
2. Schwetz BA et al: Embryo and fetotoxicity of inhaled carbon tetrachloride, 1,1-dichloroethane and methyl ethyl ketone in rats. Toxicol Appl Pharmacol 28:452–464, 1974
3. National Cancer Institute: Carcinogenesis Technical Report Series No 66. Bioasssay of 1,1-Dichloroethane for Possible Carcinogenicity. NCI-CG-TR-66. DHEW (NIH) Pub No 78-1316, p 82. Washington, DC, US Government Printing Office, 1978
4. National Institute for Occupational Safety and Health: Current Intelligence Bulletin 27 Chloroethanes: Review of Toxicity. DHEW (NIOSH) Pub No 78-181, p 22. Washington, DC, US Government Printing Office, 1978

1,2-DICHLOROETHYLENE

CAS: 540-59-0

$C_2H_2Cl_2$ 1987 TLV = 200 ppm

Synonyms: Acetylene dichloride

Physical Form. Liquid

Uses. Solvent for acetyl cellulose, resins, and waxes; refrigerant

Exposure. Inhalation

Toxicology. 1,2-Dichloroethylene causes irritation of the eyes and, at high concentrations, central nervous system depression.

It has been used as a general anesthetic in humans. Exposure to the *trans*-isomer at 2200 ppm causes burning of the eyes, vertigo, and nausea.[1] There has been only one report of an industrial poisoning, a fatality caused by very high vapor inhalation in a small enclosure.[1]

The *cis*-isomer at 16,000 ppm anesthetized rats in 8 minutes and killed them in 4 hours; the *trans*-isomer was twice as potent.[2] No effects were observed in several species with repeated exposure for up to 6 months at 1000 ppm.[3] Dogs exposed to high concentrations of vapor developed superficial corneal turbidity, which was reversible.[4]

1,2-Dichloroethylene may be recognized by a slightly acrid odor.

REFERENCES

1. von Oettingen WF: The Halogenated Aliphatic, Olefinic, Cyclic, Aromatic, and Aliphatic-Aromatic Hydrocarbons Including the Halogenated Insecticides, Their Toxicity and Potential Dangers. US Public Health Service Pub No 414, pp 198–202. Washington, DC, US Government Printing Office, 1955
2. Smyth HF Jr: Improved communication—hy-

gienic standards for daily inhalation. Am Ind Hyg Assoc J 17:154, 1956
3. 1,2–Dichloroethylene. Documentation of the TLVs and BEIs, 5th ed, p 185. Cincinnati, American Conference of Governmental Industrial Hygienists (ACGIH), 1986
4. Grant WM: Toxicology of the Eye, 3rd ed, pp 325–326. Springfield, Illinois, Charles C Thomas, 1986

DICHLOROETHYL ETHER

CAS: 111-44-4 — 1987 TLV = 5 ppm; skin

$C_4H_8Cl_2O$

Synonyms: bis(2–Chloroethyl) ether; chlorex; 1–chloro–2–(*beta*-chloroethoxy)ethane

Physical Form. Colorless liquid

Uses. Solvent for resins, wax, oils, turpentine; insecticide

Exposure. Inhalation; skin absorption

Toxicology. Dichloroethyl ether is a severe respiratory and eye irritant; high levels cause narcosis in animals, and severe exposure is expected to cause the same effects in humans. The inhalation hazard is limited by its relatively low volatility; skin absorption is more hazardous.

In experimental human exposure, 500 ppm caused intolerable irritation to the eyes and nose with cough, lacrimation, and nausea; at 100 ppm, there was some irritation, while at 35 ppm, there were no effects.[1]

In guinea pigs, concentrations of 500 to 1000 ppm were fatal after 5 to 8 hours of exposure; effects were immediate lacrimation and nasal irritation, followed by unsteadiness, and coma; autopsy findings included pulmonary edema, pulmonary hemorrhage, and, occasionally, complete consolidation.[1]

Repeated oral administration of 300 mg/kg daily to both sexes of two strains of mice for 80 weeks induced a significant elevated incidence of tumors, mostly hepatomas.[2]

Both the liquid and a 10% solution dropped in the eye of a rabbit caused moderate discomfort, conjunctival irritation, and corneal injury, which generally healed within 24 hours.[3] On the skin of rabbits, the pure liquid had no local effect, but a sufficient amount penetrated the skin to cause death within a day; the 24-hour LD_{50} for skin absorption was 90 mg/kg. The oral LD_{50} in the rabbit was 126 mg/kg.[3]

REFERENCES

1. Schrenk HH, Patty FA, Yant WP: Acute response of guinea pigs to vapors of some new commercial organic compounds. VII. Dichloroethyl ether. Public Health Rep 48:1389–1398, 1933
2. Innes JRM et al: Bioassay of pesticides and industrial chemicals for tumorigenicity in mice: A preliminary note. J Natl Cancer Inst 42:1101–1114, 1969
3. Kirwin C, Sandmeyer E: Ethers. In Clayton GD, Clayton FE (eds): Patty's Industrial Hygiene and Toxicology, 3rd ed, Vol 2, Toxicology, pp 2517–2519. New York, Wiley–Interscience, 1981

DICHLOROMONOFLUOROMETHANE

CAS: 75-43-4

$CHFCl_2$ — 1987 TLV = 10 ppm

Synonyms: Dichlorofluoromethane; fluorodichloromethane; Freon 21; Refrigerant 21; FC–21

Physical Form. Colorless gas

Uses. Refrigerant gas; propellant gas

Exposure. Inhalation

Toxicology. Dichloromonofluoromethane causes asphyxia in animals at high concentrations; repeated or prolonged exposure to lower concentrations results in liver damage.

Acute or chronic effects from human exposure have not been reported. In liquid form, this substance may cause frostbite.

Exposure of guinea pigs to 400,000 ppm with 18% oxygen was fatal, and death was preceded by dyspnea, tremor, and convulsive movements, but not narcosis.[1] Animals died at 102,000 ppm with congested lungs, kidneys, and liver but survived 52,000 ppm showing tremor, incoordination, and irregular breathing.[1]

In rats, 90-day exposures to 1000 and 5000 ppm caused bilateral hair loss, extensive liver damage, and excessive mortality.[2] The chronic toxicity of dichloromonofluoromethane appears to be quite different from difluorinated methanes and more similar to the hepatotoxin chloroform.[3] In mice, 100,000 ppm induced arrhythmias and sensitized the heart to epinephrine.

After exposure at 10,000 ppm on days 6 through 15 of gestation, 15 of 25 pregnant female rats had no viable fetuses or implantation sites on the uterine wall.[3]

REFERENCES

1. von Oettingen WF: The Halogenated Aliphatic, Olefinic, Cyclic, Aromatic, and Aliphatic-Aromatic Hydrocarbons Including the Halogenated Insecticides, Their Toxicity and Potential Dangers. US Public Health Service Pub No 414, pp 73–75. Washington, DC, US Government Printing Office, 1955
2. Trochimowicz HJ et al: Ninety-day inhalation toxicity studies on two fluorocarbons. Toxicol Appl Pharmacol 41:299 (abstr), 1977
3. Dichlorofluoromethane. Documentation of TLVs and BEIs, 5th ed, p 187. Cincinnati, American Conference of Governmental Industrial Hygienists (ACGIH), 1986

1,1-DICHLORO-1-NITROETHANE

CAS: 594-72-9

$CH_3CCl_2NO_2$ 1987 TLV = 2 ppm

Synonyms: Ethide

Physical Form. Colorless liquid

Uses. Fumigant insecticide

Exposure. Inhalation

Toxicology. 1,1-Dichloro-1-nitroethane is a pulmonary, skin, and eye irritant in animals; it is expected that severe exposure will cause the same effects in humans.

No effects in humans have been reported.

Exposure of rabbits to 2500 ppm for 40 minutes was fatal, but exposure to 170 ppm for 30 minutes was nonlethal; autopsy revealed pulmonary edema and hemorrhage, with damage to the heart, liver, and kidneys. At high concentrations, effects included lacrimation, increased nasal secretion, sneezing, cough, pulmonary rales, and weakness. Application of the liquid to the skin of rabbits caused irritation and edema.[1,2] This compound is considerably more irritating to skin and mucous membranes of animals than is 1-chloro-1-nitropropane and exhibits greater toxicity by inhalation.[3]

REFERENCES

1. Machle W et al: The physiological response of animals to certain chlorinated mononitroparaffins. J Ind Hyg Toxicol 27:95–102, 1945
2. Negherbon WO (ed): Handbook of Toxicology, Vol III, Insecticides, pp 212–213. Philadelphia, WB Saunders, 1959
3. Stokinger HE: Aliphatic nitro compounds, nitrates, nitrites. In Clayton GD, Clayton FE (eds): Patty's Industrial Hygiene and Toxicology, 3rd ed, rev, Vol 2C, Toxicology, pp 4162–4164. New York, Wiley–Interscience, 1982

2,4-DICHLOROPHENOXYACETIC ACID

CAS: 94-75-7

$C_8H_6Cl_2O_3$ 1987 TLV = 10 mg/m^3

Synonyms: 2,4-D; Hedonal

Physical Form. Crystalline solid

Uses. Weed control; component of Agent Orange

Exposure. Inhalation; ingestion; skin absorption

Toxicology. 2,4-Dichlorophenoxyacetic acid (2,4-D) causes signs of both hypo- and hyperexcitation of the central nervous system.

One fatal case of poisoning involved a suicidal person who ingested not less than 6500 mg and experienced violent convulsions; there were no significant findings at autopsy.[1] In another fatality from suicidal ingestion of a mixture of 2,4-D and two other related herbicides, progressive hypotension, coma, tachypnea, and abdominal distention preceded death. An autopsy revealed nonspecific findings. Concentrations of 2,4-D measured in blood and urine were 520 and 670 mg/liter, respectively.[2] A single dose of 3.6 g of 2,4-D administered intravenously to a patient for treatment of disseminated coccidioidomycosis caused stupor, hyporeflexia, fibrillary twitching of some muscles, and urinary incontinence; 24 hours after the dose, the patient still complained of profound muscular weakness, which subsided after an additional 24 hours.[3,4] Contact of the material with the skin may cause dermatitis.[3,4] Dermal absorption and ingestion of aerosol droplets trapped in the nose appear to be the primary routes of entry in spraying operations.

Peripheral neuropathy has been reported to occur occasionally following exposure to 2,4-D, but more frequently following exposure to another phenoxyherbicide, 2,4,5-T (2,4,5-trichlorophenoxyacetic acid) or its contaminants, including 2,3,7,8-TCDD (2,3,7,8-tetrachlorodibenzo-p-dioxin).[5] A study of workers employed in the manufacture of 2,4-D and 2,4,5-T found a statistically significant increased frequency of mild slowing of nerve conduction velocity in the sural sensory and median motor nerves; there were no associated symptoms.[6]

Several case-control studies of soft-tissue sarcoma and lymphoma have suggested an increased risk among workers exposed to phenoxyacetic acid herbicides, including 2,4-D. The IARC has deemed the evidence implicating 2,4-D as inadequate.[7] Large cohort studies of agricultural and forestry workers exposed to these herbicides have not subsequently confirmed any increased incidence of malignancy.[7,8] A case-control study of Vietnam-era veterans with soft-tissue sarcoma did not find an association with potential exposure to Agent Orange (a 1:1 mixture of 2,4-D and 2,4,5-T).[9]

A two- to threefold increased risk of birth defects among children of Vietnam veterans exposed to Agent Orange has been suggested by several epidemiologic studies, but these studies have been criticized on a number of grounds, including exposure assessment, outcome verification, and potential for recall bias.[10] Animal studies have not demonstrated clear-cut adverse effects of phenoxyherbicide exposure on reproductive outcomes.[10,11]

REFERENCES

1. Nielsen K, Kaempe B, Jensen-Holm J: Fatal poisoning in man by 2,4-diphenoxyacetic acid (2,4-D): Determination of the agent in forensic materials. Acta Pharmacol Toxicol 22:224–234, 1965

2. Fraser AD, Isner AF, Perry RA: Toxicologic studies in a fatal overdose of 2,4–D, Mecoprop, and Dicamba. J Forensic Sci 29:1237–1241, 1984
3. Seabury JH: Toxicity of 2,4–dichlorophenoxyacetic acid for man and dog. Arch Environ Health 7:202–209, 1963
4. Hayes WJ Jr: Clinical Handbook on Economic Poisons, Emergency Information for Treating Poisoning, US Public Health Service Pub No 476, pp 106–109. Washington, DC, US Government Printing Office, 1963
5. Kolmodin-Hedman, B, Holund S, Akerblom M: Studies on phenoxy acid herbicides. I. Field study: Occupational exposure to phenoxy acid herbicides (MCPA, Dichloroprop, Mecoprop, and 2,4–D) in agriculture. Arch Toxicol 54:257–265, 1983
6. Singer R et al: Nerve conduction velocity studies of workers employed in the manufacture of phenoxy herbicides. Environ Res 29:297–311, 1982
7. IARC Monographs on the Evaluation of the Carcinogenic Risk of Chemicals to Humans: Chemicals, Industrial Processes and Industries Associated with Cancer in Humans, Suppl 4, pp 101–103, 211–212. Lyon, International Agency for Research on Cancer, 1982
8. Wiklund K, Holme L: Soft tissue sarcoma risk in Swedish agricultural and forestry workers. J Natl Cancer Inst 76:229–234, 1986
9. Kang HK et al: Soft tissue sarcomas and military service in Vietnam: A case comparison group analysis of hospital patients. J Occup Med 28:1215–1218, 1986
10. Hatch MC, Stein ZA: Agent Orange and risks to reproduction: The limits of epidemiology. Teratogenesis, Carcinogenesis and Mutagenesis 6:185–202, 1986
11. Pesticide Fact Sheet: 2,4–D. Washington, DC, US Environmental Protection Agency, 1986

DICHLOROTETRAFLUOROETHANE

CAS: 76-14-2

$F_2ClC-CClF_2$ 1987 TLV = 1000 ppm

Synonyms: Refrigerant 114

Physical Form. Colorless gas

Uses. Refrigerant gas; propellant gas

Exposure. Inhalation

Toxicology. Dichlorotetrafluoroethane causes asphyxia at extremely high concentrations.

Although dichlorotetrafluoroethane has not been directly implicated, sniffing aerosols of other fluorochlorinated hydrocarbons has caused sudden death owing to cardiac arrest, probably a result of sensitization of the myocardium to epinephrine.[1] The liquid spilled on the skin may produce frostbite.

Exposure to 200,000 ppm for 16 hours was fatal to dogs, whereas single 8-hour exposures produced tremor and convulsions but no fatalities; repeated exposures at 140,000 to 160,000 ppm for 8 hours caused incoordination, tremor, and, occasionally, convulsions, but all dogs survived.[2] At 47,000 ppm for 2 hours, guinea pigs developed respiratory irritation.[2] At 25,000 ppm, 1 of 12 dogs developed serious arrhythmia following intravenous epinephrine.[1]

Chronic administration of 10,000 ppm to rats and 5,000 ppm to dogs 6 hours/day for 90 days caused no effects as determined by clinical, biochemical, and histologic examinations.[3]

A 40% solution applied to rabbit skin was without effect. Repeated spraying caused irritation of the mucous membrane of rabbit eyes.[2]

REFERENCES

1. Reinhardt CF et al: Cardiac arrhythmias and aerosol "sniffing." Arch Environ Health 22:265–279, 1971
2. Dichlorotetrafluoroethane. Documentation of the TLVs and BEIs, 5th ed, p 191. Cincinnati, American Conference of Governmental Industrial Hygienists (ACGIH), 1986
3. Leuschner F et al: Report on subacute toxicological studies with several fluorocarbons in rats and dogs by inhalation. Drug Res 33:1475–1476, 1983

DICHLORVOS

CAS: 62-73-7 1987 TLV = 0.1 ppm; skin
$(CH_3O)_2P(O)OCH{=}CCl_2$

Synonyms: 2,2–Dichlorovinyl dimethyl phosphate; DDVP

Physical Form. Oily liquid

Uses. Insecticide; commodity or space fumigant; "pest strips"

Exposure. Inhalation; skin absorption; ingestion

Toxicology. Dichlorvos (DDVP) is an anticholinesterase agent.

Signs and symptoms of overexposure are caused by the inactivation of the enzyme cholinesterase, which results in the accumulation of acetylcholine at synapses in the nervous system, skeletal and smooth muscle, and secretory glands.[1,2] The sequence of the development of systemic effects varies with the route of entry. The onset of signs and symptoms is usually prompt but may be delayed up to 12 hours. After inhalation, respiratory and ocular effects are the first to appear, often within a few minutes of exposure. Respiratory effects include tightness in the chest and wheezing owing to bronchoconstriction and excessive bronchial secretion; laryngeal spasm and excessive salivation may add to the respiratory distress; cyanosis may also occur.[3] Ocular effects include blurring of distant vision, tearing, rhinorrhea, and frontal headache. After ingestion, gastrointestinal effects such as anorexia, nausea, vomiting, abdominal cramps, and diarrhea appear within 15 minutes to 2 hours. After skin absorption, localized sweating and muscular fasciculations in the immediate area usually occur within 15 minutes to 4 hours; skin absorption is somewhat greater at higher ambient temperatures and is increased by the presence of dermatitis.[1,2]

With severe intoxication by all routes, an excess of acetylcholine at the neuromuscular junctions of skeletal muscle causes weakness aggravated by exertion, involuntary twitchings, fasciculations, and, eventually, paralysis. The most serious consequence is paralysis of the respiratory muscles. Effects on the central nervous system include giddiness, confusion, ataxia, slurred speech, Cheyne–Stokes respiration, convulsions, coma, and loss of reflexes. The blood pressure may fall to low levels, and cardiac irregularities, including complete heart block, may occur.

Complete symptomatic recovery usually occurs within a week; increased susceptibility to the effects of anticholinesterase agents persists for up to several weeks after exposure. Daily exposure to concentrations that are insufficient to produce symptoms following a single exposure may result in the onset of symptoms. Continued daily exposure may be followed by increasingly severe effects.

In a study of 13 workers exposed for 12 months to an average concentration of DDVP of 0.7 mg/m^3, the erythrocyte cholinesterase activity was reduced by approximately 35%, whereas the serum cholinesterase activity was reduced by 60%. The results of other tests and of thorough medical examination conducted at regular intervals were entirely normal.[4]

DDVP inactivates cholinesterase by phosphorylation of the active site of the enzyme to form the "dimethylphosphoryl enzyme." Over the following 24 to 48 hours, there is a process, termed *aging*, of conversion to the "monomethylphosphoryl enzyme." Aging is of clinical interest in the treatment of poisoning, because cholinesterase reactivators such as pralidoxime (2–PAM,

Protopam) chloride are ineffective after aging has occurred.

DDVP has been shown to cause a persistent irritant contact dermatitis in one worker with negative patch tests and appears to be capable of inducing an allergic contact dermatitis.[5,6]

DDVP is an alkylating agent, causing methylation of DNA in vitro. However, there is no evidence of mutagenicity in humans or other mammals, presumably because of its rapid degradation. Several animal bioassays for carcinogenicity have not demonstrated any statistically significant excesses of tumors. However, because of limitations in these studies, the IARC has concluded that the data are inadequate to evaluate the carcinogenicity of this agent.[7]

Diagnosis. Initial signs and symptoms include headache, blurred vision, pallor, weakness, sweating, abdominal pain, nausea, vomiting, and diarrhea.

Signs of moderate-to-severe intoxication include miosis, lacrimation, excessive salivation, muscle fasciculations, dyspnea, cyanosis, convulsions, shock, cardiac arrhythmias, and coma.

Differential Diagnosis: Diagnosis is primarily based on a history of exposure and clinical evidence of diffuse parasympathetic stimulation. Careful observation of the effects of atropine and pralidoxime may be valuable. Patients with organophosphate poisoning are resistant to the action of atropine at moderate dosages; failure of 1 to 2 mg of atropine administered parenterally to produce signs of atropinization (flushing, mydriasis, tachycardia, or dryness of mouth) indicates organophosphate poisoning. Intravenous injection of 1 g pralidoxime generally causes some recovery from signs and symptoms.

Special Tests: Two types of cholinesterase are clinically significant: (1) true acetylcholinesterase, found principally in the nervous system and red blood cells; and (2) pseudo- or butyrylcholinesterase, found in the plasma, liver, and nervous system. Whereas the action of both types is inhibited by organophosphates, the level of depression of red blood cell cholinesterase is a better indicator of clinically significant reduction of cholinesterase activity in the nervous system.

Laboratory evidence of depression of red blood cell cholinesterase to a level substantially below pre-exposure levels (at least 50% and usually much lower) is verification of poisoning. There is an imperfect correlation between the degree of depression of cholinesterase enzymes and the occurrence of symptoms. With a rapid drop in cholinesterase activity, generally reflecting an acute heavy exposure, there may be symptoms with only a 30% depression, whereas with slower drops to 70% depression, reflecting chronic low level exposure, there may be no symptoms.[8]

If no pre-exposure baseline has been performed but symptoms are not sufficient to justify treatment with atropine, repeated testing during the recovery period demonstrating progressively increasing plasma and red blood cell cholinesterase levels over several days and weeks, respectively, suggests the diagnosis of anticholinesterase poisoning.

There are many different methods for estimation of cholinesterase content of blood, and associated with each method is a different set of normal values and a different set of reporting units. The laboratory report of a cholinesterase determination should state the units involved along with the appropriate normal range. Based on the Michel method, the normal range of red blood cell cholinesterase activity (delta *p*H per hour) is 0.39 to 1.02 for men, and 0.34 to 1.10 for women.[9] The normal range of the enzyme activity (delta *p*H per hour) of plasma is 0.44 to 1.63 for men, and 0.24 to 1.54 for women.

Treatment. Treatment of organophosphate poisoning ranges from simple removal from exposure in very mild cases to the provision of very rigorous supportive and antidotal measures in severe cases.[2,3,10,11] In the moderate to severe case, because of pulmonary involvement, there may be need for artificial respiration using a positive-pressure method. Careful attention must be paid to removal of secretions and to maintenance of a patent airway. Anticonvulsants such as thiopental sodium may be necessary. Maintenance of respiration is critical, because death usually results from weakness of the muscles of respiration and accumulation of excessive secretions in the respiratory tract.[2]

As soon as cyanosis has been overcome, 2 to 4 mg of atropine should be given intravenously. (Atropine may induce ventricular fibrillation in the presence of cyanosis.) *This dose of atropine is approximately 10 times the amount that is administered for other conditions in which atropine is considered therapeutic.* This dose should be repeated at 5- to 10-minute intervals until signs of atropinization appear (dry, flushed skin, tachycardia as high as 140 beats/minute, and pupillary dilatation). A mild degree of atropinization should be maintained for at least 48 hours.[2]

Pralidoxime (2–PAM, Protopam) chloride is a cholinesterase reactivator that complements the action of atropine. It has its greatest effect in reversing the nicotinic action of anticholinesterase agents at skeletal neuromuscular junctions but virtually no effect on central nervous system manifestations. In the moderate to severe case, the dose for adults is 1 to 2 g injected intravenously at a rate not in excess of 500 mg/minute. After an hour, a second dose of 1 g is indicated if muscle weakness has not been relieved. Treatment with pralidoxime chloride will be most effective if given within 24 hours after poisoning.[2] Morphine, aminophylline, and phenothiazines are contraindicated because of documented experience of adverse reactions in cases of organophosphate poisoning.[10]

It is important to decontaminate the patient. Contaminated clothing should be removed at once, and the skin should be washed with generous amounts of soap or detergent and a flood of water, which is best accomplished under a shower or by submersion in a pond or other body of water if the exposure occurred in the field. Careful attention should be paid to cleansing of the skin and hair.

The patient should be attended and monitored continuously for not less than 24 hours, because serious and sometimes fatal relapses have occurred owing to continuing absorption of the toxin or dissipation of the effects of the antidote.

Regeneration of cholinesterase is primarily by synthesis of new enzyme and takes place at the rate of approximately 1%/day.[9] A patient who has recovered from the acute phase of poisoning remains hypersusceptible to anticholinesterases for up to several weeks.

Medical Control. This involves preplacement and annual physical examination with determination of pre-exposure red blood cell cholinesterase activity. A person whose red blood cell cholinesterase falls to or below 40% of the pre-exposure baseline should be removed from further exposure until the activity returns to within 80% of the pre-exposure baseline.

REFERENCES

1. Koelle GB (ed): Cholinesterases and anticholinesterase agents. Handbuch der Experimentellen Pharmakologie, Vol 15, pp 989–1027. Berlin, Springer–Verlag, 1963
2. Taylor P: Anticholinesterase agents. In Gilman AG et al (eds): Goodman and Gilman's Pharmacological Basis of Therapeutics, 7th ed, pp 110–129. New York, Macmillan, 1985
3. Hayes WJ Jr: Toxicology of Pesticides, pp 379–428. Baltimore, Williams & Wilkins, 1975
4. Menz M, Luetkemeir H, Sachsse K: Long-term exposure of factory workers to dichlorvos (DDVP) insecticide. Arch Environ Health 28:72–76, 1971
5. Mathias CGT: Persistent contact dermatitis from the insecticide dichlorvos. Contact Dermatitis 9:217–218, 1983
6. Matsushita T et al: Allergic contact dermatitis from organophosphorus insecticides. Ind Health 23:145–153, 1985
7. IARC Monographs on the Evaluation of the Carcinogenic Risk of Chemicals to Humans: Some Halogenated Hydrocarbons, Vol 20, pp 97–123. Lyon, International Agency for Research on Cancer, 1979
8. Coye MJ, Lowe JA, Maddy KT: Biological monitoring of agricultural workers exposed to pesticides. I. Cholinesterase activity determinations. J Occup Med 28:619–627, 1986
9. Michel HO: Electrometric method for determination of red blood cell and plasma cholinesterase activity. J Lab Clin Med 34:1564–1568, 1949
10. Milby TH: Prevention and management of organophosphate poisoning. JAMA 216:2131–2133, 1971
11. Namba T, Nolte CT, Jackrel J, Grob D: Poisoning due to organophosphate insecticides. Am J Med 50:475–492, 1971

DIELDRIN

CAS: 60-57-1 — 1987 TLV = 0.25 mg/m^3; skin

$C_{12}H_8Cl_6O$

Synonyms: Compound 497; Octalox; HEOD

Physical Form. White odorless crystals

Uses. Insecticide

Exposure. Inhalation; skin absorption; ingestion

Toxicology. Dieldrin is a convulsant; it causes liver cancer in mice.

A number of poisonings have occurred among workers involved in spraying or manufacture of dieldrin. Early symptoms of intoxication may include headache, dizziness, nausea, vomiting, malaise, sweating, and myoclonic jerks of the limbs; clonic and tonic convulsions and sometimes coma follow and may occur without the premonitory symptoms.[1–3] In some patients following convulsions, agitation, hyperactivity, and temporary personality changes, including weeping, mania, and inappropriate behavior, have occurred. There are very few reports of fatalities resulting from occupational exposures. Recovery is generally prompt over several weeks and complete, although a few patients have been described with persistent symptoms for several months, and recurrent convulsions have rarely occurred.[1] The half-life of dieldrin in humans is reportedly as long as 0.73 years. Dieldrin is well absorbed dermally, which may be the primary route of occupational exposure.[1]

Electroencephalogram changes, including bilateral spikes, spike and wave complexes, and slow theta waves, occur in sufficiently exposed workers. Electroencephalograms have been used successfully in monitoring workers and justifying removal from exposure, but this test has been supplanted by measurement of blood levels.[1] In a study of five aldrin/dieldrin workers who had suffered one or more convulsive seizures and/or myoclonic limb movements, the probable concentration of dieldrin in the blood during intoxication ranged from 16 to 62 μg/100 g of blood; in healthy workers, the concentrations of dieldrin ranged up to 22 μg/100 g of blood.[4]

The hepatocarcinogenicity of dieldrin in mice has been confirmed in several experiments, and, in some cases, the liver cell tumors metastasized.[5] No excess of tumors has been observed in a number of bioassays in rats and one bioassay in Syrian golden hamsters.[5,6]

Single doses of 30 mg/kg administered orally to pregnant golden hamsters during the period of fetal organogenesis cause a high incidence of fetal deaths, congenital anomalies, and growth retardation.[7] The lethal oral dose for animals ranges from 20 to 70 mg/kg; effects include injury to the liver and kidneys.[8]

REFERENCES

1. Hayes WJ Jr: Pesticides Studied in Man, pp 237–246. Baltimore, Williams & Wilkins, 1982
2. Committee on Toxicology: Occupational dieldrin poisoning. JAMA 172:2077–2080, 1960
3. Hoogendam I, Versteeg JPJ, de Vlieger M: Nine years' toxicity control in insecticide plants. Arch Environ Health 10:441–448, 1965
4. Brown VKH, Hunter CG, Richardson A: A blood test diagnostic of exposure to aldrin and dieldrin. Br J Ind Med 21:283–286, 1964
5. IARC Monographs on the Evaluation of the Carcinogenic Risk of Chemicals to Man, Vol 5, Some Organochlorine Pesticides, pp 125–156. Lyon, International Agency for Research on Cancer, 1974
6. Ashwood-Smith MJ: The genetic toxicology of aldrin and dieldrin. Mutat Res 86:137–154, 1981
7. Ottolenghi AD, Haseman JK, Suggs F: Teratogenic effects of aldrin, dieldrin, and endrin in hamsters and mice. Teratology 9:11–16, 1974
8. Hodge HC, Boyce AM, Deichmann WB, Kraybill HF: Toxicology and no-effect levels of aldrin and dieldrin. Toxicol Appl Pharmacol 10:613–675, 1967

DIETHANOLAMINE

CAS: 111-42-2

$H(NCH_2CH_2OH)_2$ 1987 TLV = 3 ppm

Synonyms: DEA; 2,2–Iminodiethanol

Physical Form. Clear, colorless, viscous liquid with ammonia odor

Uses. Manufacture of emulsifiers and dispersing agents; cosmetic formulations; cleaners and detergents; alkalizing agents

Exposure. Inhalation

Toxicology. DEA is an irritant to the eyes and has caused liver and kidney damage in animals.

The oral LD_{50} in rats has ranged from 0.71 ml/kg to 2.83 g/kg.[1,2] The effects of intraperitoneal administration to rats of doses of 100 or 500 mg/kg were assessed at 4 and 24 hours after dosing.[3] In the liver and kidneys, there was cytoplasmic vacuolization. The high doses caused renal tubular degeneration. In rats fed 0.17 g/kg for 90 days, effects included cloudy swelling and degeneration of kidney tubules and fatty degeneration of the liver.[2,4]

The liquid applied to the skin of rabbits under semiocclusion for 24 hours on 10 consecutive days caused only minor, if any, irritation.[5] Clinical skin testing of cosmetic products containing DEA showed mild skin irritation in concentrations above 5%.[6]

When 0.2 ml of the liquid was dropped in the rabbit eye and rinsed out after 15 seconds, there was moderate to severe conjunctival irritation and corneal injury.[5]

In the presence of N-nitrosating agents, DEA may give rise to N-nitrosodiethanolamine, a known animal carcinogen.[6]

REFERENCES

1. Smyth HF Jr, Weil CS, West JS, Carpenter CP: An exploration of joint toxicity. II. Equitoxic versus equivolume mixtures. Toxicol Appl Pharmacol 17:498, 1970
2. Mellon Institute. Submission of Data by FDA. Mellon Institute of Industrial Research, University of Pittsburgh, Special Report on the Acute and Subacute Toxicity of Mono-, Di-, and Triethanolamine, Carbide and Carbon Chem Div, UCC Industrial Fellowship No 274–13 (Report 13–67), 1950
3. Grice HC et al: Correlation between serum enzymes, isozyme patterns, and histologically detectable organ damage. Food Cosmet Toxicol 9:847, 1971
4. Smyth HF Jr et al: Range-finding toxicity data, List IV. AMA Arch Ind Hyg Occup Med 4:119, 1951
5. CTFA. Submission of Data by CTFA. (2–5–24). CIR Safety Data Test Summary, Primary Skin Irritation and Eye Irritation of Diethanolamine, 1979
6. Beyer KH Jr et al: Final report on the safety assessment of triethanolamine, diethanolamine and monoethanolamine. J Am Coll Toxicol 2:193–235, 1983

DIETHYLAMINE

CAS: 109-89-7

$(C_2H_5)_2NH$ 1987 TLV = 10 ppm

Synonyms: Diethamine; *N*-ethyl-ethanamine

Physical Form. Colorless liquid

Uses. In the rubber and petroleum industry; in flotation agents; in resins, dyes, pharmaceuticals

Exposure. Inhalation

Toxicology. Diethylamine is an irritant of eyes, mucous membranes, and skin.

Exposure to high vapor concentrations will cause severe cough and chest pain; heavy, repeated, or prolonged exposure could result in pulmonary edema.[1,2] Contact of the liquid with eyes causes corneal damage. In one reported case, the liquid splashed into the eye causing intense pain.[3] In spite of emergency irrigation and treatment, the cornea became swollen and cloudy; some permanent visual impairment resulted. Prolonged or repeated contact of the eyes with the vapor at concentrations slightly below the irritant level often results in corneal edema with consequent foggy vision and the appearance of haloes around lights.[1] Dermal contact with the liquid causes vesiculation and necrosis of the skin.[2]

In rats, exposure to the saturated vapor is lethal in 5 minutes, and the 4-hour inhalation LC_{50} is 4000 ppm.[4]

Rabbits repeatedly exposed to 50 ppm for 7 hours daily showed corneal damage and pulmonary irritation.[5]

REFERENCES

1. Chemical Safety Data Sheet SD–97, Diethylamine, pp 15–16. Washington, DC, MCA, Inc, 1971
2. Hygienic Guide Series: Diethylamine. Am Ind Hyg Assoc J 21:266–267, 1960
3. Grant WM: Toxicology of the Eye, 3rd ed, p 333. Springfield, Illinois, Charles C Thomas, 1986
4. Smyth HF Jr et al: Range-finding toxicity data, List IV. AMA Arch Ind Hyg Occup Med 4:109–122, 1951
5. Brieger H, Hodes WA: Toxic effects of exposure to vapors of aliphatic amines. Arch Ind Hyg Occup Med 3:287–291, 1951

2–DIETHYLAMINOETHANOL

CAS: 100-37-8

$(C_2H_5)_2NCH_2CH_2OH$ 1987 TLV = 10 ppm; skin

Synonyms: Diethylethanolamine; DEAE; n,n-diethylethanolamine

Physical Form. Colorless liquid

Uses. Chemical intermediate for the production of emulsifiers, detergents, solubilizers, cosmetics, drugs, and textile finishing agents

Exposure. Inhalation

Toxicology. Diethylaminoethanol is an irritant of the eyes, mucous membranes, and skin in animals, and it is expected that severe exposure will cause the same effects in humans.

An attempt by a laboratory worker to remove animals from an inhalation chamber containing approximately 100 ppm resulted in nausea and vomiting within 5 minutes after a fleeting exposure; no irritation of the eyes or throat was noted during this brief exposure.[1] Other persons in the same room also complained of a nauseating odor but showed no ill effects.

The liquid is a severe skin irritant; in the guinea pig, it is a skin sensitizer.[2] It is also a severe eye irritant and may produce permanent eye injury.[2] No systemic effects from industrial exposure have been reported.

Rats exposed to 500 ppm 6 hours daily for 5 days exhibited marked eye and nasal irritation, and a number of animals had corneal opacity by the end of the third day; the mortality rate was 20%, and, at autopsy, findings were acute purulent bronchiolitis and bronchopneumonia.[1]

REFERENCES

1. Cornish HH: Oral and inhalation toxicity of 2–diethylaminoethanol. Am Ind Hyg Assoc J 26:479–484, 1965
2. Miller FA, Scherberger RF, Tischer KS, Webber AM: Determination of microgram quantities of diethanolamine, 2–methylaminoethanol, and 2–diethylaminoethanol in air. Am Ind Hyg Assoc J 28:330–334, 1967

DIETHYLHEXYL ADIPATE

CAS: 103-23-1 — 1987 TLV = none established

$C_{22}H_{42}O_4$

Synonyms: DEHA; bis(2–ethylhexyl) adipate; octyl adipate

Physical Form. Colorless or very pale amber liquid

Uses. Plasticizer in polyvinyl chloride films, sheeting, extrusions, and plastisols; solvent and emollient in cosmetics

Exposure. Inhalation; skin contact

Toxicology. Diethylhexyl adipate (DEHA) is of low acute toxicity but is carcinogenic in mice by the oral route.

There are no reports of effects in humans from specific exposure to DEHA, but fumes generated at high temperatures may cause throat and eye irritation in meat wrappers applying polyvinyl chloride film cut with a hot wire.[1]

DEHA has a low acute toxicity as indicated by the relatively high oral LD_{50} of 9.1 g/kg in rats.[2] Skin absorption is expected to be low, since the dermal LD_{50} in rabbits is 15 g/kg.

In an NTP carcinogenicity study, rats and mice of both sexes were fed DEHA at 12,000 and 25,000 ppm in the diet for 103 weeks.[3] It was noncarcinogenic in rats but caused hepatocellular carcinomas in female mice in both dose groups and hepatocellular adenomas in males at the higher dose.

The liquid in contact with the skin of rabbits under occlusion for 24 hours produced mild skin irritation.[4] In the eye, examination after 24 hours revealed no irritation.[4]

REFERENCES

1. Smith TF et al: Evaluation of emissions from simulated commercial meat wrapping operation using PVC wrap. Am Ind Hyg Assoc J 44:176, 1983

2. Smyth HF Jr, Carpenter CP, Weil CS: Range-finding toxicity data: List IV. Arch Ind Hyg Occup Med 4:119, 1951
3. National Toxicology Program: NTP Technical Report on the Carcinogenesis Bioassay of Di(2–ethylhexl) Adipate in F344 Rats and B6C3F1 Mice. NTP–80–29, DHHS (NIH) Pub No 81–1768. Washington, DC, US Government Printing Office, 1982
4. Wickhen Products Inc: FYI–OTS–0684–0286, Suppl Seq H. Di(2–ethylhexyl) Adipate: Animal Toxicology Studies. Report from Kolmar Research Center, May–August 1967. Washington, DC, Office of Toxic Substances, US Environmental Protection Agency, 1984

DI(2–ETHYLHEXYL) PHTHALATE

CAS: 117-81-7

$C_{24}H_{38}O_4$ 1987 TLV = 5 mg/m^3

Synonyms: DEHP; bis(2–ethylhexyl) phthalate; diethylhexyl phthalate; di-sec-octyl phthalate

Physical Form. Clear to slightly colored oily liquid

Uses. Solvent; plasticizer; in vacuum pumps

Exposure. Inhalation

Toxicology. The acute toxicity of di(2–ethylhexyl) phthalate (DEHP) is low; chronic exposure has been associated with liver damage, testicular injury, and teratogenic and carcinogenic effects in experimental animals.

Two male volunteers developed mild gastric disturbance after swallowing 5 or 10 g.[1] Dermally applied, it was judged to be moderately irritating and, at most, only slightly sensitizing to human skin.

The oral LD_{50} for rats is 26 g/kg and, for rabbits, is 34 g/kg.[1] A single oral dose of 2 g/kg of DEHP to dogs caused no toxicity. The lethal effects appear to be cumulative since the chronic LD_{50} value for intraperitoneal administration to mice five times per week for 10 weeks was 1.36 g/kg bodyweight, in comparison to a single-dose value of 37.8 g/kg bodyweight.[2] It is poorly absorbed through the skin, but two of six rabbits died several days after dermal exposure to 19.7 g/kg bodyweight.[3]

Growth retardation and increased liver and kidney weights, but no significant histopathologic findings, were noted in rats fed 60 to 200 mg/kg/day for 104 days.[4] Altered liver morphology as evidenced by excessive pigmentation, congestion, and some fatty degeneration occurred in rats fed diets containing 0.4% and 0.13% for 2 years.[5]

In 2-year chronic studies, DEHP caused a significant increase in hepatocellular carcinomas in female rats fed diets containing 6000 or 12000 ppm and in mice of both sexes fed 3000 or 6000 ppm.[6]

Investigators have suggested that the production of hepatic tumors with DEHP in rodents may be associated with the ability of this substance to induce proliferation of hepatic peroxisomes, because no significant genotoxicity of DEHP has been found.[1,7] Differences between rodent and human metabolism and susceptibility to peroxisome proliferation suggest that humans may be less susceptible to the carcinogenic effects of DEHP. The IARC has determined that there is sufficient evidence for the carcinogenicity in mice and rats and that no adequate data were available to evaluate the risk to humans.[3]

Embryolethal and teratogenic effects have been reported in animals administered high doses; daily doses of 4000 or 1000 mg/kg of diet to mice caused complete resorption; 1000 or 2000 mg/kg caused embryolethality, exencephaly, and spina bifida; no adverse effects were observed at 500 mg/kg.[8] Mice given diets containing 0.1% and 0.3% DEHP for 7 days prior to and during a 98-day cohabitation period had dose-dependent decreases in male and female fertility and in the number of pups born alive.[9]

DEHP-induced testicular injury has also been reported in a number of studies.[1] Administration of 20,000 mg/kg in the diet of rats produced seminiferous tubular degeneration and testicular atrophy within 7 days; 12,500 mg/kg produced similar effects within 90 days; and 6,000 ppm was effective by the end of 2 years of exposure. DEHP is thought to cause testicular atrophy by depleting testicular zinc levels. Use of a diet containing high zinc levels has been found to prevent or ameliorate the gonadotoxic effects of DEHP.[10]

REFERENCES

1. Thomas JA, Thomas MJ: Biological effects of di–(2–ethylhexyl) phthalate and other phthalic acid esters. Crit Rev Toxicol 13:283–317, 1984
2. Lawrence WH et al: A toxicological investigation of some acute, short-term and chronic effects of administering di–2–ethylhexyl phthalate (DEHP) and other phthalate esters. Environ Res 9:1–11, 1975
3. IARC Monographs on the Evaluation of the Carcinogenic Risk of Chemicals to Humans. Some Industrial Chemicals and Dyestuffs, Vol 29, pp 269–294. Lyon, International Agency for Research on Cancer, May 1982
4. Schaffer CB et al: Acute and subacute toxicity of di–(2–ethylhexyl) phthalate with note upon its metabolism. J Ind Hyg Toxicol 27:130–135, 1945
5. Carpenter D et al: Chronic oral toxicity of di–(2–ethylhexyl) phthalate for rats, guinea pigs and dogs. AMA Arch Ind Hyg Occup Med 8:219–227, 1953
6. National Toxicology Program: Carcinogenesis Bioassay of Di–(2–Ethylhexyl) Phthalate (CAS No 117-81-7) in Fischer 344 Rats and B6C3F1 Mice (feed study). DHHS (NIH) Pub No 82–1773, NTP–80–37, pp 1–127. Washington, DC, US Government Printing Office, March 1982
7. Turnbull D, Rodricks JV: Assessment of possible carcinogenic risk to humans resulting from exposure to di–(2–ethylhexyl) phthalate (DEHP). J Am Coll Toxicol 4:111–145, 1985
8. Shiota K et al: Embryotoxic effects of di–(2–ethylhexyl) phthalate (DEHP) and di–n–butyl phthalate (DBP) in mice. Environ Res 22:245–253, 1980
9. Lamb JC IV et al: Reproductive effects of four phthalic acid esters in the mouse. Toxicol Appl Pharmacol 88:255–269, 1987
10. Agarwal D et al: Relationship between zinc and testicular atrophy induced by di–2–(ethylhexyl) phthalate (DEHP). Pharmacology 25:226, 1983

DIETHYL PHTHALATE

CAS: 84-66-2

$C_6H_4(COOC_2H_5)_2$ 1987 TLV = 5 mg/m^3

Synonyms: Anazol; 1,2–benzenedicarboxylic acid, diethyl ester

Physical Form. Colorless, oily liquid

Uses. Solvent for cellulose acetate; manufacturing varnishes

Exposure. Inhalation

Toxicology. Diethyl phthalate is of low toxicity.

The oral LD_{50} in rats ranges from 0.5 to 31 g/kg.[1] There were no adverse effects in rats fed 1.25 g/kg/day or in dogs fed 2.5 g/kg/day for 6 weeks or more.

Exposure to the heated vapor may produce some transient irritation of the nose and throat.[2] No cumulative effects are known pertaining to its occupational use.

REFERENCES

1. Shibko SI, Blumenthal H: Toxicology of phthalic acid esters used in food-packaging material. Environ Health Perspect 3:131, 1973
2. Sandmeyer EE, Kirwin CJ Jr: Esters. In Clayton GD, Clayton FE (eds): Patty's Industrial Hygiene and Toxicology, 3rd ed, Vol 2A, Toxicology, p 2344. New York, Wiley–Interscience, 1981

DIFLUORODIBROMOMETHANE

CAS: 75-61-6

CF_2Br_2 1987 TLV = 100 ppm

Synonyms: Halon 1202; Freon 12–B2

Physical Form. Colorless liquid or gas

Uses. Fire-extinguishing agent

Exposure. Inhalation

Toxicology. Difluorodibromomethane causes respiratory irritation and narcosis in animals, and severe exposure is expected to produce the same effects in humans.

No effects have been reported from industrial exposures.

Rats exposed to 4000 ppm for 15 minutes showed pulmonary edema, whereas 2300 ppm daily for 6 weeks resulted in the death of more than half the animals.[1] At 2300 ppm, dogs showed rapid and progressive signs of intoxication after a few days of exposure, with weakness and loss of balance followed by convulsions. At autopsy, these dogs had pulmonary congestion, centrolobular necrosis of the liver, and evidence of central nervous system damage. Other dogs tolerated daily exposures of 350 ppm for 7 months without signs of intoxication.

REFERENCE

1. Difluorodibromomethane. Documentation of the TLVs and BEIs, 5th ed, p 201. Cincinnati, American Conference of Governmental Industrial Hygienists (ACGIH), 1986

DIGLYCIDYL ETHER

CAS: 2238-07-5

$C_6H_{10}O_3$ 1987 TLV = 0.1 ppm

Synonyms: Bis(2,3–epoxy propyl)ether; DGE; Di(2,3–epoxy propyl)ether

Physical Form. Colorless liquid

Uses. Diluent for epoxy resins; stabilizer of chlorinated organic compounds

Exposure. Inhalation; skin absorption

Toxicology. Diglycidyl ether causes severe irritation of the eyes, respiratory tract, and skin; hematopoietic effects have been observed in animals.

Because of its toxicity, DGE generally is not used outside experimental laboratories.[1] No systemic effects have been reported in humans.

The LC_{50} for mice was 30 ppm for 4 hours, but 8-hour exposure at 200 ppm was not lethal to rats.[2] Rabbits exposed to 24 ppm for 24 hours had leukocytosis at autopsy. There were acute changes in lung and kidneys as well as atrophied testes.[3] At 12 ppm, there was thrombocytopenia, and, at 6 ppm, basophilia.[3] In rats, three or four exposures at 20 ppm for 4 hours produced intense cytoplasmic basophilia, grossly distorted lymphocytic nuclei with indistinct cellular membranes, and lowered leukocyte and marrow cell counts.[3] Repeated exposure of rats to 3 ppm caused increased mortality, decreased bodyweight, and leukopenia. Exposures to 0.3 ppm did not cause significant changes.[3] Cutaneous applications greater than 100 mg/kg also caused leukopenia, weight loss, and death.[3]

The oral LD_{50} was 0.17 g/kg in mice and 0.45 g/kg in rats; following intragastric administration, effects were incoordination, ataxia, depressed motor activity, and coma.[2]

Diglycidyl ether is extremely damaging to skin, producing ecchymoses and necrosis. In one long-term study, skin painting three times per week for 1 year caused hyperkeratosis, epithelial hyperplasia, and skin papillomas.[4]

Instilled in rabbit eyes, it is a severe irritant. Exposure to vapor at 3 ppm for 24 hours produced erythema and edema of the conjunctiva in rabbits, and, at 24 ppm, corneal opacity was produced.[3]

REFERENCES

1. National Institute for Occupational Safety and Health: Criteria for a Recommended Standard . . . Occupational Exposure to Glycidyl Ethers. DHEW (NIOSH) Pub No 78–166. Washington, DC, US Government Printing Office, 1978
2. Hine CH et al: The toxicology of glycidol and some glycidyl ethers. Arch Ind Health 14:250–264, 1956
3. Hine CH et al: Effects of diglycidyl ether on blood of animals. Arch Environ Health 2:37/31–50/44, 1961
4. McCammon CJ, Kotkin P, Falk HL: The cancerogenic potency of certain diepoxides. Proc Assoc Cancer Res 2:229–230, 1957

DIISOBUTYL KETONE

CAS: 108-83-8

$[(CH_3)_2CHCH_2]_2CO$ 1987 TLV = 25 ppm

Synonyms: Isovalerone; 2,6–dimethyl–4–heptanone

Physical Form. Liquid

Uses. Solvent; dispersant for resins; intermediate in the synthesis of pharmaceuticals and insecticides

Exposure. Inhalation

Toxicology. Diisobutyl ketone vapor is an irritant of the eyes and mucous membranes; at high concentrations, it causes narcosis in animals, and it is expected that severe exposure will produce the same effect in humans.[1]

Human subjects exposed to 100 ppm for 3 hours noted slight lacrimation and throat irritation, and slight headache and dizziness on returning to fresh air.[2] In another study, the majority of subjects experienced eye irritation above 25 ppm, and nose and throat irritation above 50 ppm within 15 minutes.[3]

The liquid is a defatting agent, and prolonged or repeated skin contact may cause dermatitis.

Although it may be more toxic and irritative than lower molecular weight ketones at equivalent concentration, it poses less of an inhalation hazard because of its relatively low volatility.[1]

Exposure of female rats to 2000 ppm for 8 hours caused narcosis, and 7 of 12 rats died; however, male rats survived the same treatment as did both sexes of one other strain of rats.[2] Damage to the lungs, liver, and kidneys was observed at autopsy.[2] Repeated exposures to rats over 30 days resulted in increased liver and kidney weights at 920 and 530 ppm, but no effects at 125 ppm.[2]

REFERENCES

1. National Institute for Occupational Safety and Health: Criteria for a Recommended Standard . . . Occupational Exposure to Ketones, (NIOSH) 78–174, pp 79–80, 134, 187. Washington, DC, US Government Printing Office, June 1978
2. Carpenter CP, Pozzani UC, Weil CS: Toxicity and hazard of diisobutyl ketone vapors. AMA Arch Ind Hyg Occup Med 8:377–381, 1953
3. Silverman L, Schulte HF, First MW: Further studies on sensory responses to certain industrial solvent vapors. Ind Hyg Toxicol 28:262–266, 1946

DIISOPROPYLAMINE

CAS: 108-18-9 1987 TLV = 5 ppm; skin

$(CH_3)_2CHNHCH(CH_3)_2$

Synonyms: N-(1–Methylethyl)–2–propanamine

Physical Form. Liquid

Uses. Chemical intermediate

Exposure. Inhalation; skin absorption

Toxicology. Diisopropylamine is an eye irritant in humans; it is a pulmonary irritant in animals, and severe exposure is expected to produce the same effect in humans.

Workers exposed to concentrations between 25 and 50 ppm complained of disturbances of vision described as "haziness."[1] In two instances, there were also complaints of nausea and headache. Prolonged skin contact with an irritant of this nature is likely to cause dermatitis.

Exposure of several species of animals to 2207 ppm for 3 hours was fatal; effects were lacrimation, corneal clouding, and severe irritation of the upper respiratory tract; at autopsy, findings included pulmonary edema and hemorrhage.

Diisopropylamine has a strong odor of ammonia.

REFERENCE

1. Treon JF, Sigmon H, Kitzmiller KV, Heyroth FF: The physiological response of animals to respiratory exposure to the vapors of diisopropylamine. J Ind Hyg Toxicol 31:142–145, 1949

DIMETHOXYETHYL PHTHALATE

CAS: 117-82-8

$C_6H_4[COOCH_2CH_2OCH_3]_2$

1987 TLV = none established

Synonyms: DMEP; 1,2–benzenedicarboxylic acid bis(2–methoxyethyl)ester; bis(methoxyethyl)phthalate; dimethyl cellosolve phthalate

Physical Form. Light-colored, clear liquid

Uses. Plasticizer; solvent

Exposure. Inhalation

Toxicology. DMEP causes teratogenic, reproductive, and fetotoxic effects in animals.

Acute lethality data indicate that DMEP exhibits slight to moderate toxicity. The oral LD_{50} in rats ranged from 3.2 to 6.4 g/kg.[1] Exposure of rats to 1595 ppm for 6 hours caused deaths of all animals, whereas 770 ppm for 6 hours was not lethal.[2] The dermal LD_{50} in guinea pigs was greater than 10 ml/kg, suggesting very little absorption.

Fetotoxic and teratogenic effects were observed in rats given DMEP via intraperitoneal injection of 0.374 ml/kg on days 5, 10, and 15 of gestation.[3] Resorptions occurred at an incidence of 27.6%, and teratogenic effects included skeletal and gross abnormalities. In another study, pregnant rats were given a single intraperitoneal injection of 0.6 ml/kg on day 10, 11, 12, 13, or 14 of gestation.[4] The chief effect was skeletal malformations. A single intraperitoneal injection of 2.38 ml/kg in mice resulted in a marked reduction in the incidence of pregnancies and litter size per pregnancy.[5] Oral administration of 1000 mg/kg by gavage to male rats for a total of 12 treatments over 16 days caused reduced testes weight and increases in abnormal sperm heads.[6]

DMEP did not cause dermal sensitization in guinea pigs.[2] In the eyes of rabbits, it was not irritating.

REFERENCES

1. Eastman Kodak Co: FYI-OTS-0884-0329 Suppl, Seq C. Material Data Sheet from Eastman Kodak Co to T O'Bryan. Washington, DC, Office of Toxic Substances, US Environmental Protection Agency, 1984
2. Sandmeyer EE, Kirwin CJ: Esters. In Clayton GD, Clayton FE (eds): Patty's Industrial Hygiene and Toxicology, 3rd ed, Vol 2A, Toxicology, pp 2346–2351. New York, Wiley-Interscience, 1981
3. Singh AR, Lawrence WH, Autian J: Teratogenicity of phthalate esters in rats. J Pharmacol Sci 61(1):51–55, 1972
4. Parkhie MR, Webb M, Norcross MA: Dimethoxyethyl phthalate. Embryopathy, teratogenicity, fetal metabolism and the role of zinc in the rat. Environ Health Perspect 45:89, 1982
5. Singh AR, Lawrence WH, Autian J: Mutagenic and anti-fertility sensitivities of mice to di-2-ethylhexyl phthalate (DEHP) and dimethoxyethyl phthalate (DMEP). Toxicol Appl Pharmacol 29:35–46, 1974
6. Eastman Kodak Co: FYI-OTS-0385-0329 Seq D. Basic Toxicity of Bis(Methoxyethyl) Phthalate from Eastman Kodak Co to Document Control Officer. Washington, DC, Office of Toxic Substances, US Environmental Protection Agency, 1985

DIMETHYLACETAMIDE

CAS: 127-19-5 — 1987 TLV = 10 ppm; skin

$CH_3CON(CH_3)_2$

Synonyms: Acetyldimethylamine; N,N-dimethyl acetamide; DMAC

Physical Form. Colorless liquid

Uses. Commercial solvent

Exposure. Inhalation; skin absorption

Toxicology. Dimethylacetamide (DMAC) causes liver damage.

Workers repeatedly exposed to 20 to 25 ppm developed jaundice; appreciable skin absorption was thought to have occurred.[1] Nine patients with neoplastic disease were given daily doses of 400 mg/kg by an unspecified route for 3 or more days as a therapeutic trial; they experienced depression, lethargy, confusion, and disorientation; on the last (fourth or fifth) day of therapy or within 24 hours thereafter, the patients had visual and auditory hallucinations, perceptual distortions, and, at times, delusions; after 24 hours these gradually subsided.[2]

Rats exposed at 288 ppm 6 hours/day for 2 weeks showed nasal irritation, transient increase in blood cholesterol, and liver hypertrophy; testicular atrophy was evident 2 weeks postexposure.[3]

Repeated dermal application of the liquid to dogs at a dosage level of 4.0 mg/kg for 6 weeks caused severe fatty infiltration of the liver.[3] Repeated exposure of rats to a concentration of 195 ppm for 6 months resulted in focal necrosis of the liver; exposure to 40 ppm for the same period of time caused no adverse effects.[4] The skin absorption approximate lethal dose for pregnant rats and rabbits was 7.5 and 5.0 g/kg, respectively.[5] Application of DMAC resulted in a marked incidence of embryomortality at doses that did not affect maternal bodyweight or show any signs of toxicity. Teratogenic effects (3 of 34 fetuses with encephalocele; 1 of 8 with diffuse subcutaneous edema) were found in rats only when DMAC was applied on gestation days 10 and 11 at a total dose of 2400 mg/kg.[5] Dimethylacetamide has a significant antitumor effect in animals.[2]

In practice, the dermal factor is considered to be so significant that no air concentration however low will provide protection if skin contact with the liquid is permitted.

REFERENCES

1. Dimethyl acetamide. Documentation of TLVs and BEIs, 5th ed, p 205. Cincinnati, American Conference of Governmental Industrial Hygienists (ACGIH), 1986
2. Weiss AJ et al: Dimethylacetamide: A hitherto unrecognized hallucinogenic agent. Science 136:151–152, 1962
3. Kelly DP et al: Subchronic inhalation toxicity of dimethylacetamide in rats. The Toxicologist 4:65 (abstr), 1984
4. Horn HJ: Toxicology of dimethylacetamide. Toxicol Appl Pharmacol 3:12–24, 1961
5. Stula EF, Krauss WC: Embryotoxicity in rats and rabbits from cutaneous application of amide-type solvents and substituted ureas. Toxicol Appl Pharmacol 41:35–55, 1977

DIMETHYLAMINE

CAS: 124-40-3

$(CH_3)_2NH$ 1987 TLV = 10 ppm

Synonyms: *N*-Methylmethanamine; DMA

Physical Form. Gas, liquifying at 7°C

Uses. Manufacture of pharmaceuticals; stabilizer in gasoline; in production of insecticides and fungicides; in manufacture of soaps and surfactants

Exposure. Inhalation

Toxicology. Dimethylamine is an irritant of the skin, eyes, mucous membranes, and respiratory tract.

Dermatitis and conjunctivitis are occasionally observed in chemical workers after prolonged exposure.[1] No systemic effects from industrial exposure have been reported.

The LC_{50} for rats exposed 6 hours to dimethylamine and observed for 48 hours postexposure was 4540 ppm.[2] Clinical observations were characterized by signs of eye irritation immediately after onset of exposure. This was followed by gasping, secretion of bloody mucus from the nose, salivation, and lacrimation within 1 hour of exposure. Corneal opacity was generally observed after 3 hours of exposure. Death was often preceded by convulsions. Rats exposed to nonlethal concentrations (600 to 2500 ppm for 6 hours) showed signs of eye irritation, moderate gasping, and slight bloody nasal discharge. At autopsy, findings included severe congestion, ulcerative rhinitis, and necrosis of the nasal turbinates. At concentrations above 2500 ppm, peripheral emphysema, bronchopneumonia, hepatic necrosis, and corneal ulceration were noted.

At lower concentrations, 175 to 500 ppm, less damage to the lower respiratory tract occurred, but inflammation, ulcerative rhinitis, and early squamous metaplasia were observed in the respiratory nasal mucosa.[3]

Animals repeatedly exposed to concentrations of approximately 100 to 200 ppm for 18 to 20 weeks showed marked irritation of the respiratory tract with pulmonary edema, as well as hepatic injury, including centrolobular necrosis; corneal injury was observed in guinea pigs and rabbits after 9 days of exposure.[4] Various species survived 5 ppm of continuous exposure for 90 days without signs of toxicity, but, at autopsy, some showed mild inflammatory changes in the lungs.[5]

Skin contact with the liquid causes necrosis, and a drop in the eye may result in severe corneal injury or permanent corneal opacity.

A "fish" odor is detectable at 0.5 ppm.

REFERENCES

1. Chemical Safety Data Sheet SD–57, Methylamines, pp 17–19. Washington, DC, MCA, Inc, 1955
2. Steinhagen WH et al: Acute inhalation toxicity and sensory irritation of dimethylamine. Am Ind Hyg Assoc J 3:411–417, 1982

3. McNulty MJ: Biochemical toxicology of inhaled dimethylamine. CIIT Activities. Chemical Industry Institute of Toxicology, pp 1–4. August 1983
4. Hollingsworth RL, Rowe VK: Chronic inhalation toxicity of dimethylamine for laboratory animals (unpublished). The Dow Chemical Company, Midland, Michigan, 1964
5. Coon RA et al: Animal inhalation studies on ammonia, ethylene glycol, formaldehyde, dimethylamine, and ethanol. Toxicol Appl Pharmacol 16:646–655, 1970

4–DIMETHYLAMINOAZOBENZENE

CAS: 60-11-7 1987 TLV = none established
$C_6H_5N_2C_6H_4N(CH_3)_2$

Synonyms: *p*-Dimethylaminoazobenzene; butter yellow; DAB

Physical Form. Yellow solid

Uses. Coloring polishes and wax products

Exposure. Inhalation

Toxicology. 4–Dimethylaminoazobenzene (DAB) is a potent carcinogen in animals.

In humans, contact dermatitis was observed in 90% of the factory workers handling DAB.[1] There have been no reports of an increased cancer incidence among exposed persons.[2]

Two of ten dogs survived ingestion of 20 mg/kg/day for 38 months of continuous treatment followed by 48 months of intermittent treatment; both developed bladder papillomas.[3] Oral administration of 1, 3, 10, 20, or 30 mg/day produced liver tumors in rats; the induction time was inversely proportional to the daily dose ranging between 34 days for the 30 mg/day dose and 700 days for the 1 mg/day dose.[4] In rats fed 5 mg DAB/day for 40 to 200 days and then kept for lifespan on a normal diet, there was a 20% to 81% incidence of liver carcinoma.[5]

Cutaneous application of 1 ml of a 2% solution of DAB in acetone two times per week for 90 weeks caused skin tumors in all six male rats treated; squamous cell, basal cell, and anaplastic carcinomas were observed; there were no tumors in controls given acetone alone.[6]

Because of the demonstrated carcinogenicity in animals, human exposure of DAB by any route should be avoided. However, in recent years, this compound has been used only in laboratories as a model of tumorigenic activity in animals.[7] It is of little occupational health importance.

REFERENCES

1. National Research Council: Food Colors, p 7. Washington, DC, National Academy of Sciences, 1971
2. IARC Monographs on the Evaluation of the Carcinogenic Risk of Chemicals to Man, Vol 8, Some Aromatic Azo Compounds, pp 125–146. Lyon, International Agency for Research on Cancer, 1975
3. Nelson SA, Woodward G: Tumors of the urinary bladder, gall bladder and liver in dogs fed *o*-aminoazotoluene or *p*-dimethylaminoazobenzine. J Natl Cancer Inst 13:1497–1509, 1953
4. Druckrey H, Kupfmuller K: Quantitative analyse der krebsentstehung. Z Naturforsch 3b:254–266, 1948
5. Druckrey H: Quantitative aspects in chemical carcinogenesis. In Trichaut R (ed): Potential Carcinogenic Hazards from Drugs. UICC Monograph Series, 7:60–78, 1967
6. Fare G: Rat skin carcinogenesis by topical applications of some azo dyes. Cancer Res 26:2405–2408, 1966
7. Stokinger HE: In Clayton GD, Clayton FE (eds): Patty's Industrial Hygiene and Toxicology, 3rd ed, rev, p 2893. New York, Wiley–Interscience, 1981

DIMETHYLANILINE

CAS: 121-69-7 — 1987 TLV = 5 ppm; skin

$C_6H_5N(CH_3)_2$

Synonyms: Dimethylphenylamine; aminodimethylbenzene

Physical Form. Yellow liquid

Uses. Analytical reagent and synthetic precursor of many dyes

Exposure. Inhalation; skin absorption

Toxicology. Dimethylaniline absorption causes anoxia due to the formation of methemoglobin.

Few reports of industrial experience are available from which to form an accurate appraisal of its health hazards; it is said to be less potent than aniline as a cause of methemoglobin, but more of a central nervous system depressant.[1] The effects of methemoglobinemia are cyanosis (especially of the lips, nose, and earlobes), weakness, dizziness, and severe headache.

In dogs, the repeated subcutaneous injection of 1.5 g caused vomiting, weakness, cyanosis, methemoglobinemia, and hyperglobulinemia.[2] Rats survived an 8-hour exposure to concentrated vapor.[3] The single-dose oral LD_{50} for rats was 1.41 ml/kg, and the single-dose dermal LD_{50} for rabbits was 1.77 ml/kg. The liquid was slightly irritating to the clipped skin of rabbits within 24 hours of a 0.01 ml application, and 0.005 ml caused severe burns when instilled in rabbit eyes.[3]

Diagnosis. Signs and symptoms include headache; signs of anoxia, including cyanosis of lips, nose, and earlobes; and anemia.

Differential Diagnosis: Other causes of cyanosis, including hypoxia owing to lung disease, hypoventilation, and decreased cardiac output, must be differentiated from methemoglobinemia due to chemical exposure. Lung disease may be suspected from results of pulmonary function tests and arterial blood gas analysis. The arterial Po_2 may be normal in methemoglobinemia but tends to be decreased in cyanosis owing to lung disease. Hypoventilation will cause elevation of arterial Pco_2, which is not seen in chemical exposure. Decreased cardiac output states will cause cyanosis only when accompanied by arterial hypotension. If blood withdrawn from the vein shows the characteristic chocolate-brown coloration, the diagnosis of an abnormal pigment is almost certain, especially if the color remains after shaking the blood in air.[4]

Special Tests: Special tests include examination of urine for blood; determination of methemoglobin concentration in the blood when chemical intoxication is suspected and at regular intervals until the methemoglobin has been fully reduced to normal hemoglobin.[5] Methemoglobin may be differentiated from sulfhemoglobin by the addition of a few drops of 10% potassium cyanide, which results in the rapid production of bright red cyanomethemoglobin but has no effect on the color of sulfhemoglobin.[4] Spectrophotometry is required for the precise identification of the pigment and its quantitation. Normal acid methemoglobin has a characteristic absorption spectrum with peaks at 502 and 632 nm, which disappear with the addition of cyanide, whereas sulfhemoglobin has a peak at 620 nm, which does not disappear with cyanide.[4]

Treatment. All the contaminant on the body must be removed. Immediately remove all clothing and wash the entire body from head to foot with soap and water. Pay special attention to the hair and scalp, finger- and toenails, nostrils, and ear canals. Administer oxygen to alleviate the headache and general sense of weakness; confine the patient to bed. Determine the methemoglobin concentration in the blood, and repeat every 3 to 6 hours for 18 to 24 hours. Repeat skin cleansing if the methemoglobin concentration appears to rise after 3 to 4 hours. In general, patients will return to normal within 24 hours provided that all sources of further absorption are completely eliminated. The methemoglobin will be reduced spontaneously to ferrous hemoglobin in 2 to 3 days.[4]

Such therapy is not effective in subjects with glucose–6–dehydrogenase deficiency.[4]

The only justifiable use of methylene blue would be in cases of coma or stupor, usually at methemoglobin levels over 60%. In those patients in whom therapy is necessary, methylene blue, 1 to 2 mg/kg of a 1% solution in saline, may be given intravenously over a 10-minute period. If cyanosis has not disappeared within an hour, a second dose of 2 mg/kg may be administered.[4,5] The total dose should not exceed 7 mg/kg, because methylene blue may cause toxic effects such as dyspnea, precordial pain, restlessness, apprehension, red cell hemolysis, and changes in the electrocardiogram (reduction in the height or even reversal of the T wave, frequently with lowering of the R wave).[4]

REFERENCES

1. Dimethylaniline. Documentation of the TLVs and BEIs, 5th ed, p 207. Cincinnati, American Conference of Governmental Industrial Hygienists (ACGIH), 1986
2. von Oettingen WF: The Aromatic Amino and Nitro Compounds, Their Toxicity and Potential Dangers. US Public Health Service Pub No 271, pp 15–16. Washington, DC, US Government Printing Office, 1941
3. Smyth HF Jr et al: Range-finding toxicity data: List VI. Am Ind Hyg Assoc J 23:95–107, 1962
4. Rieder RF: Methemoglobinemia and sulfmethemoglobinemia. In Wyngaarden JB, Smith LH (eds): Cecil Textbook of Medicine, 16th ed, p 896. Philadelphia, WB Saunders Co, 1982
5. Mangelsdorff AF: Treatment of methemoglobinemia. AMA Arch Ind Health 14:148–153, 1956

DIMETHYLFORMAMIDE

CAS: 68-12-2 — 1987 TLV = 10 ppm; skin

$(CH_3)_2NCHO$

Synonyms: DMF; DMFA

Physical Form. Colorless liquid

Uses. Solvent

Exposure. Inhalation; skin absorption

Toxicology. Dimethylformamide is toxic to the liver.

Subjective complaints of exposed workers have included nausea, vomiting, and anorexia.[1] Air concentration measurements may not define the total exposure experience, because DMF is readily absorbed through the skin as well as the lungs. A worker who was splashed with the liquid over 20% of the body surface initially suffered only dermal irritation and hyperemia; abdominal pain began 62 hours after the exposure and became progressively more severe, with vomiting; the blood pressure was elevated to 190/100; the effects gradually subsided and were entirely abated by day 7 after the exposure.[2] Some workers have noted flushing of the face after inhalation of the vapor, especially with coincident ingestion of alcohol.[2] Prolonged or repeated skin contact with the liquid may cause dermatitis as a result of a defatting action.[3]

In both mice and rats exposed 6 hours/day 5 days/week for 12 weeks, the no-effect dose was below 150 ppm, and the maximal tolerated dose was below 600 ppm. At doses of up to 1200 ppm, there were few signs of overt toxicity, and, at necropsy, the only treatment-related lesions occurred in the liver.[4]

No conclusions on the teratogenicity or carcinogenicity can be made at this time.[1] In one isolated report, DMF administered by intraperitoneal injection to rats as a solvent control caused malignant tumors in the gastrointestinal tract.[1] However, in a number of short-term assays, DMF was not mutagenic or genotoxic, suggesting little likelihood for carcinogenic activity.[5]

REFERENCES

1. Hazard Data Bank: Sheet No 77. Dimethyl formamide. The Safety Practitioner, pp 48–49. May 1986
2. Potter HP: Dimethylformamide-induced abdominal pain and liver injury. Arch Environ Health 27:340–341, 1973
3. Hygienic Guide Series: Dimethylformamide. Am Ind Hyg Assoc J 18:279–280, 1957
4. Craig DK et al: Subchronic inhalation toxicity of dimethylformamide in rats and mice. Drug Chem Toxicol 7:551–571, 1984
5. Antoine JL et al: Lack of mutagenic activity of dimethylformamide. Toxicology 26:207–212, 1983

1,1-DIMETHYLHYDRAZINE

CAS: 57-14-7 — 1987 TLV = 0.5 ppm; skin

$(CH_3)_2NNH_2$

Synonyms: *asym*-Dimethylhydrazine; unsymmetrical dimethylhydrazine; UDMH; dimazine

Physical Form. Colorless liquid; fumes in air and turns yellow

Uses. Base in rocket fuel formulations; intermediate in organic synthesis

Exposure. Inhalation; skin absorption

Toxicology. 1,1-Dimethylhydrazine is an irritant and convulsant, and is carcinogenic in mice.

Accidental human exposures have resulted in eye irritation, a choking sensation, chest pain, dyspnea, lethargy, nausea, and skin irritation.[1] Based upon the results of exposure of dogs, the effects expected in humans from exposure for 60 minutes are: 100 ppm, irritation of eyes and mucous membranes; 200 ppm, marked central nervous system stimulation and perhaps death; and 900 ppm, convulsions and death.[2] Impairment of liver function (elevated SGPT levels) has been reported in 47 of 1193 workers exposed to 1,1-dimethylhydrazine under variable conditions; in a few of these cases, fatty infiltration of the liver was also demonstrated by liver biopsy, although alcohol intake may have been a factor in some.[3]

Exposure of dogs to 111 ppm for 3 hours caused vomiting, convulsions, and death; at autopsy, pulmonary edema and hemorrhage were present but were believed to be a secondary manifestation of the convulsive seizures rather than a primary effect of the agent.[4] Dogs repeatedly exposed to 25 ppm developed vomiting, diarrhea, ataxia, convulsions, and hemolytic anemia.[4] Applied to the shaved skin of dogs, the liquid was mildly irritating and rapidly absorbed; the dermal LD_{50} was between 1.2 and 1.7 g/kg.[5] In the eye, it caused mild conjunctivitis.[6]

Intraperitoneal administration of 10, 30, or 60 mg/kg/day in rats on days 6 to 15 of pregnancy caused a dose-dependent reduction in maternal weight gain and slight embryotoxicity in the form of reduced 20-day fetal weights in the high-dose group.[7] Administration of 0.1% in the drinking water of 50 male and 50 female Swiss mice resulted in a high incidence of angiosarcomas in various organs (79%); tumors of the lungs (71%), kidneys (10%), and liver (7%) were also observed.[8]

In another study, mice given daily

gavage doses of 0.5 mg 5 days/week for 40 weeks showed inconclusive evidence of lung tumor induction.[6]

The carcinogenic risk to humans has not been determined, but 1,1–dimethylhydrazine is classified as a suspected human carcinogen based upon animal results. National Institute for Occupational Safety and Health (NIOSH) has also noted that the dose of nitrosodimethylamine, a contaminant of 1,1–dimethylhydrazine, must be considered in evaluating the tumorigenicity of 1,1–dimethylhydrazine.[6]

The ammonical or fishy odor has variously been reported as detectable between 6 and 14 ppm, and perhaps below 0.3 ppm.[6]

REFERENCES

1. Shook BS, Cowart OH: Health hazards associated with unsymmetrical dimethylhydrazine. Ind Med Surg 26:333–336, 1957
2. 1,1–Dimethylhydrazine—emergency exposure limits. Am Ind Hyg Assoc J 25:582–584, 1964
3. Petersen P, Bredahl E, Lauritsen O, Laursen T: Examination of the liver in personnel working with liquid rocket propellant. Br J Ind Med 27:141–146, 1970
4. Jacobson KH et al: The acute toxicity of the vapors of some methylated hydrazine derivatives. AMA Arch Ind Health 12:609–616, 1955
5. Smith EB, Clark DA: Absorption of unsymmetrical dimethylhydrazine (UDMH) through canine skin. Toxicol Appl Pharmacol 18:649–659, 1971
6. National Institute for Occupational Safety and Health: Criteria for a Recommended Standard . . . Occupational Exposure to Hydrazines. DHEW (NIOSH) Pub No 78–172, p 269. Washington, DC, US Government Printing Office, 1974
7. Keller WC et al: Teratogenic assessment of three methylated hydrazine derivatives in the rat. J Toxicol Environ Health 13:125–131, 1984
8. IARC Monographs on the Evaluation of the Carcinogenic Risk of Chemicals to Man, Vol 4, Some Aromatic Amines, Hydrazine and Related Substances, N-nitroso Compounds and Miscellaneous Alkylating Agents, pp 137–143. Lyon, International Agency for Research on Cancer, 1974

DIMETHYL HYDROGEN PHOSPHITE

CAS: 868-85-9 — 1987 TLV = none established

$(CH_3O)_2POH$

Synonyms: DMHP; dimethoxyphosphine oxide; dimethyl phosphite

Physical Form. Colorless liquid

Uses. Fire-proofing agent in textiles; intermediate in the production of pesticides and herbicides

Exposure. Inhalation

Toxicology. DMHP is an irritant of the eyes, mucous membranes, and skin; it causes neurologic impairment and reversible cataracts in animals; it is carcinogenic in rats and causes testicular atrophy in mice. No human cases of intoxication have been reported.

Rats exposed to airborne levels of 934 ppm, 6 hours/day for 3 days died.[1] Effects observed included irritation of the skin, eyes, and mucous membranes; neuromuscular impairment; and lung congestion. Rats exposed to 431 ppm, 6 hours/day for 5 days survived but exhibited the same irritant effects as seen at 934 ppm.

In a month-long study, rats were exposed to 12, 35, 119, or 198 ppm for 6 hours/day, 5 days/week.[2,3] In the high-dose group, 27 of 40 animals were dead by day 27; in the 119 ppm group, two animals died on days 14 and 23. There was neurologic impairment at 198 ppm and 119 ppm, which usually resolved by the following morning. Necrosis and purulent inflammation of the skin were thought to be the only lesions that may have caused death. Although there was treatment-related irritation of the eyes and nares, there was no treatment-related irritation of the trachea or lungs.

Lenticular opacities occurred at 35 ppm and above, which progressed to cataracts in the 119- and 198-ppm groups. In rats killed 2 weeks post-treatment, the process of cataract formation had stopped; at 4 weeks, the formation of normal lens fibers was evident.

In a carcinogenic study, male and female rats were given DMHP by gavage 5 days/week for 103 weeks.[4] At 200 mg/kg, there were increases in alveolar/bronchiolar carcinomas, squamous cell carcinomas of the lung, and carcinomas of the stomach in male rats. Limited data indicate that DMHP may have testicular effects.[4] Focal calcification and atrophy of the testes were observed in mice in the course of chronic and subchronic oral studies at 200 mg/kg for 103 weeks, and 375 and 750 mg/kg for 13 weeks, respectively.

REFERENCES

1. Mobil Research and Development Corporation. TSCA sec. 8(e) Submission 8EHQ–0381–0366 Follow-up. A Five Day Inhalation Toxicity Study of MCTR–174–79 in the Rat. Performed by Bio/dynamics Inc. Washington, DC, Office of Toxic Substances, US Environmental Protection Agency, 1981
2. Mobil Oil Corporation. TSCA sec. 8(e) Submission 8EHQ–0381–0366 Follow-up. A Four Week Inhalation Toxicity Study in the Rat. Prepared by Bio/dynamics, Inc. Washington, DC, Office of Toxic Substances, US Environmental Protection Agency, 1981
3. Mobil Oil Corporation. TSCA sec. 8(e) Submission 8EHQ–0381–0366 Follow-up. Histopathologic Observations on a Four Week Inhalation Toxicity Study of MCTR–242–79 in the Rat. Prepared by Toxicity Research Laboratories, Ltd. Washington, DC, Office of Toxic Substances, US Environmental Protection Agency, 1981
4. National Toxicology Program: NTP Technical Report on the Toxicology and Carcinogenesis Studies of Dimethyl Hydrogen Phosphite (CAS No 868–85–9) in F344 Rats and B7CSF Mice (Gavage Studies). NTP TR 287. Research Triangle Park, North Carolina, National Toxicology Program, 1984

DIMETHYL PHTHALATE

CAS: 131-11-3

$C_{10}H_{10}O_4$ 1987 TLV = 5 mg/m^3

Synonyms: 1,2–benzenedicarboxylic acid dimethyl ester; phthalic acid dimethyl ester; methyl phthalate

Physical Form. Oily liquid

Uses. Solvent; plasticizer; insect repellent for application to the skin

Exposure. Inhalation (of spray or mist); skin absorption

Toxicology. Dimethyl phthalate is of low-order acute toxicity.[1]

A solution, including 2% DMP in petrolatum, was nonirritating to humans following 48-hour patch tests.[2]

Rats fed 4.0% and 8.0% in the diet showed slight but significant changes in growth; chronic nephritis was seen at the higher dose, but mortality rates were the same as those in controls.[3] Applied to 10% of the body surface of rabbits for 90 days, 4.0 ml/kg caused some deaths with pulmonary edema and slight renal damage; no skin irritation was observed.[4] The undiluted liquid instilled into rabbit eyes produced no grossly observable irritation for up to 48 hours.[5] Intraperitoneal injection of pregnant rats with 10%, 33%, or 50% of the LD_{50} (3.4 ml/kg) on days 5, 10, and 15 of gestation resulted in litters with a higher number of skeletal abnormalities.[6]

REFERENCES

1. Final Report on the Safety Assessment of Dibutyl Phthalate, Dimethyl Phthalate, and Diethyl Phthalate. J Am Coll Toxicol 4:267–303, 1985
2. Schulsinger C, Mollgaard K: Polyvinyl chloride dermatitis not caused by phthalates. Contact Dermatitis 6:477–480, 1980

3. Lehman AJ: Insect repellents. Assoc Food Drug Office US Q Bull 19:87–99, 1955
4. Draize J et al: Toxicological investigations of compounds proposed for use as insect repellents. J Pharmacol Exp Ther 93:26–39, 1948
5. Lawrence WH et al: Toxicological investigation of some acute, short-term, and chronic effects of administering di-2-ethylhexyl phthalate (DEHP) and other phthalate esters. Environ Res 9:1–11, 1975
6. Singh AR, Lawrence WH, Autian J: Teratogenicity of phthalate esters in rats. J Pharmacol Sci 61:51–55, 1972

DIMETHYL SULFATE

CAS: 77-78-1 — 1987 TLV = 0.1 ppm; skin

$(CH_3)_2SO_4$

Synonyms: Sulfuric acid dimethyl ester; DMS

Physical Form. Colorless, oily liquid

Uses. Methylating agent in the manufacture of many organic chemicals

Exposure. Inhalation; skin absorption

Toxicology. Dimethyl sulfate is a severe irritant of the eyes, mucous membranes, and skin; it is carcinogenic in experimental animals.

Several human deaths have occurred from occupational exposure, and it has been estimated that inhalation of 100 ppm for 10 minutes would be fatal.[1,2] Often, exposure of humans produces no immediate effects other than occasional slight eye and nose irritation; after a latent period of up to 10 hours or more, there is onset of headache and giddiness with intense conjunctival irritation, photophobia, and angioneurotic edema, followed by inflammation of the pharyngolaryngeal mucosa, dysphonia, aphonia, dysphagia, productive cough, oppression in the chest, dyspnea, and cyanosis.[1,3] Vomiting and diarrhea may intervene.[1,3] Dysuria may occur for 3 to 4 days; there may be persistence of laryngeal edema for up to 2 weeks and of photophobia for several months.[1,3] Other effects are delirium, fever, icterus, albuminuria, and hematuria.[1,4]

Contact of the liquid with the eyes or skin will cause very severe burns because of its powerful vesicant action.[1] In an incident of moderate skin contact with the liquid, generalized intoxication occurred even though there was prompt treatment of the skin; vapor inhalation lasted for a few minutes, at the most.[4]

In mice and rats, inhalation at 0.1 to 4.0 ppm throughout pregnancy caused preimplantation losses and embryotoxic effects, including anomalies of the cardiovascular system.[2]

Dimethyl sulfate is carcinogenic to animals following its inhalation or subcutaneous injection, producing mainly local tumors and, after prenatal exposure, producing tumors of the nervous system.[5] Of 15 rats surviving exposure to 10 ppm 1 hour/day for 19 weeks, 3 rats developed squamous cell carcinoma of the nasal cavity, 1 rat developed a glioma of the cerebellum, and another developed a lymphosarcoma of the thorax with metastases in the lungs. Several early deaths from inflammation of the nasal cavity and pneumonia were also reported.[5] A statistically significant increase in lung adenomas was observed in a group of 90 mice exposed at 4 ppm for 4 hours/day, 5 days/week.[5] A single intravenous dose of 20 mg/kg given to eight pregnant rats on day 15 of gestation induced malignant tumors, including three tumors of the nervous sytem, in 7 of 59 offspring observed for more than 1 year.[6] The LC_{50} in rats was 75 ppm for 18 minutes; autopsy findings were marked pulmonary edema, pulmonary emphysema, peribronchitis, and focal necrosis of the liver.[4]

Despite anecdotal case reports of

cancer in exposed individuals, no significant increase in mortality or deaths from lung cancer were found in a group of workers exposed for various periods between 1932 and 1972.[2,5]

The IARC has determined that there is sufficient evidence of carcinogenicity to animals; it should be assumed to be a potential human carcinogen.[2,5]

Dimethyl sulfate does not have any characteristic odor or other property that might warn of exposure.[2]

REFERENCES

1. Browning E: Toxicity and Metabolism on Industrial Solvents, pp 713–721. Amsterdam, Elsevier Publishing Company, 1965
2. World Health Organization: Environmental Health Criteria 48, Dimethyl Sulfate, p 55. Geneva, 1985
3. Dimethyl sulfate. Documentation of the TLVs and BEIs, 5th ed, pp 212-231. Cincinnati, American Conference of Governmental Industrial Hygienists (ACGIH), 1986
4. Fassett DW: Esters. In Patty FA (ed): Industrial Hygiene and Toxicology, 2nd ed, Vol 2, Toxicology, pp 1927–1930. New York, Interscience, 1963
5. IARC Monographs on the Evaluation of the Carcinogenic Risks of Chemicals to Humans. Suppl 4, pp 119–120. Lyon, International Agency for Research on Cancer, 1982
6. IARC Monographs on the Evaluation of the Carcinogenic Risk of Chemicals to Man, Vol 4, Some Aromatic Amines, Hydrazine and Related Substances, N-Nitroso Compounds and Miscellaneous Alkylating Agents, pp 271–276. Lyon, International Agency for Research on Cancer, 1974

DINITROBENZENE (all isomers)

CAS: 528-29-0; 99-65-0; 100-25-4

$C_6H_4(NO_2)_2$ 1987 TLV = 0.15 ppm; skin

Synonyms: Dinitrobenzol

Physical Form. Colorless or yellowish needles or plates

Uses. Synthesis of dyestuffs, explosives, celluloid production

Exposure. Inhalation; skin absorption

Toxicology. All isomers of dinitrobenzene cause anoxia owing to the formation of methemoglobin; moderate exposure causes respiratory tract irritation, and chronic exposure results in anemia.

Exposed workers have complained of a burning sensation in the mouth, dry throat, and thirst; somnolence, staggering gait, and coma have been observed with more intense exposures.[1] Most signs and symptoms of overexposure are due to the loss of oxygen carrying capacity of the blood.

The onset of symptoms of methemoglobinemia is insidious and may be delayed for up to 4 hours; headache is often the first symptom and may become quite intense as the severity of methemoglobinemia progresses.[2] Cyanosis occurs when the methemoglobin concentration is 15% or more; blueness in the lips, the nose, and the earlobes is usually recognized by fellow workers.[2]

The subject usually feels well, has no complaints, and is insistent that nothing is wrong until the methemoglobin concentration approaches approximately 40%.[2] At methemoglobin concentrations of over 40%, there is usually weakness and dizziness; of up to 70% concentration, there may be ataxia, dyspnea on mild exertion, tachycardia, nausea, vomiting, and drowsiness.[2] Coma may ensue with methemoglobin levels about 70%, and the lethal level is estimated to be 85% to 90%.[3]

The ingestion of alcohol aggravates the toxic effects of dinitrobenzene. In general, higher ambient temperatures increase susceptibility to cyanosis from exposure to methemoglobin-forming agents. Chronic exposure of workers causes anemia; there are scattered reports of liver injury. Visual impairment has occurred in the form of reduced visual acuity and central scotomas, particularly for red and green colors; yellow discoloration of the conjunctiva and sclera is common.[1,4] Yellow-brown discoloration of the hair and exposed skin of workers has been reported.[1]

Special Tests: Determine the methemoglobin concentration in the blood when dinitrobenzene intoxication is suspected and at regular intervals until the methemoglobin has been fully reduced to normal hemoglobin. Methemoglobin may be differentiated from sulfhemoglobin by the addition of a few drops of 10% potassium cyanide, which results in the rapid production of bright red cyanomethemoglobin but has no effect on the color of sulfhemoglobin.[5] Spectrophotometry is required for the precise identification of the pigment and its quantitation. Normal acid methemoglobin has a characteristic absorption spectrum with peaks at 502 and 632 nm, which disappear with the addition of cyanide, whereas sulfhemoglobin has a peak at 620 nm, which does not disappear with cyanide.[5]

Treatment. All dinitrobenzene on the body must be removed. Immediately remove all clothing and wash the entire body from head to foot with soap and water. Pay special attention to the hair and scalp, finger- and toenails, nostrils, and ear canals. Administer oxygen to alleviate the headache and general sense of weakness; confine to bed. Determine the methemoglobin concentration in the blood, and repeat every 3 to 6 hours for 18 to 24 hours. Repeat skin cleansing if the methemoglobin concentration appears to rise after 3 to 4 hours. In general, patients will return to normal within 24 hours provided that all sources of further absorption are completely eliminated.[2]

The only justifiable use of methylene blue would be in cases of coma or stupor, usually at methemoglobin levels over 60%; in those patients in whom therapy is necessary, methylene blue, 1 to 2 mg/kg, may be given intravenously over a 5-minute period as a 1% solution; if cyanosis has not disappeared within an hour, a second dose of 2 mg/kg should be administered.[5,6] The total dose should not exceed 7 mg/kg, because methylene blue may cause toxic effects such as dyspnea, precordial pain, restlessness, apprehension, red cell hemolysis, and changes in the electrocardiogram (reduction in the height or even reversal of the T wave, frequently with lowering of the R wave).[5,6] In case of eye splashes, flush with water.

REFERENCES

1. von Oettingen WE: The Aromatic Amino and Nitro Compounds, Their Toxicity and Potential Dangers, US Public Health Service Bulletin No 271, pp 94–103. Washington, DC, US Government Printing Office, 1941
2. Hamblin DP: Aromatic nitro and amino compounds. In Patty FA (ed): Industrial Hygiene and Toxicology, 2nd ed, Vol 2, Toxicology, pp 2105–2131, 2138–2140. New York, Interscience, 1963
3. Chemical Safety Data Sheet SD–21, Nitrobenzene, pp 5–6, 12–14. Washington, DC, MCA, Inc, 1967
4. Grant WM: Toxicology of the Eye, 3rd ed, pp 355–356. Springfield, Illinois, Charles C Thomas, 1986
5. Rieder RF: Methemoglobinemia and sulfhemoglobinemia. In Wyngaarden JB, Smith LH (eds): Cecil Textbook of Medicine, 16th ed, p 896. Philadelphia, WB Saunders Co, 1982
6. Mangelsdorff AF: Treatment of methemoglobinemia. AMA Arch Ind Health 14:184, 1956

DINITRO-*o*-CRESOL
CAS: 534-52-1 1987 TLV = 0.2 mg/m^3; skin
$CH_3C_6H_2OH(NO_2)_2$

Synonyms: DNOC; 2–methyl–4,6–dinitrophenol; dinitrol

Physical Form. Yellow, crystalline solid

Uses. Herbicide; insecticide; ovicide; fungicide

Exposure. Inhalation; skin absorption; ingestion

Toxicology. Dinitro-*o*-cresol causes an increase in metabolic rate that results in hyperpyrexia. Severe exposure may cause coma and death. Exposure also causes a yellow pigmentation of the skin, hair, sclera, and conjunctivae.

In a report of eight fatalities among agricultural sprayers, symptoms of intoxication included fatigue, profuse sweating, excessive thirst, and weight loss, which were incorrectly attributed to heat strain.[1] There was rapid decline with hyperpnea, tachycardia, and fever; death occurred within 48 hours of exposure. A number of other fatalities owing to hyperthermia have been reported.[2]

The risk of serious intoxication increased during hot weather.[1] A nonfatal case of intoxication resulting from exposure to dust concentration of 4.7 mg/m^3 resulted in fever, tachycardia, hyperpnea, profuse sweating, cough, shortness of breath, and a marked increase in basal metabolic rate.[2] The clinical picture resembled that of thyroid crisis.

Lethal doses may be absorbed through the skin; local irritation is usually slight. Skin application of 50 g of a 25% dinitro-*o*-cresol ointment to a 4-year-old boy caused vomiting, headache, yellow stained skin and sclera, elevated pulse and respiratory rate, unconsciousness, and death within 3.5 hours.[2] Autopsy showed diffuse petechial hemorrhages in the intestinal mucosa and brain, as well as pulmonary edema.

In two cases, ingestion of 50 g and 140 g was lethal.[2]

In human volunteers given 75 mg/day orally for 5 days, the earliest symptom was an exaggerated sense of well-being at blood levels of dinitro-*o*-cresol of approximately 20 μg/g.[3] At a level near 40 μg/g of blood, symptoms included headache, lassitude, and malaise; yellow coloration of the sclera appeared on the fourth day of exposure and persisted for 5 days; urinary excretion of unchanged dinitro-*o*-cresol was so slow that blood levels of 1 to 1.5 μg/g were still detectable 40 days after the last dose was administered.[4]

The development of bilateral cataracts has been reported in chronic intoxication owing to the repeated ingestion of dinitro-*o*-cresol for ill-advised therapeutic purposes; cataracts have not been observed following industrial or agricultural exposure.[2]

Blood levels appear to be associated with the severity of intoxication.[2] Persons with concentrations of 40 μg/g of whole blood or greater will most likely develop toxic effects. Those with ranges between 20 and 40 μg/g may or may not show adverse effects, and most persons with blood levels below 20 μg/g are not affected.[2]

Investigators have concluded that dinitro-*o*-cresol affects metabolism by uncoupling the oxidative phosphorylation process, resulting in increased cellular respiration (increased oxygen consumption) and decreased formation of adenosine triphosphate (ATP), which contains high-energy phosphate bonds.[2] Therefore, energy generated in the body cannot be converted to its usual form (ATP) and is released as heat instead. Toxicity is cumulative.[2]

REFERENCES

1. Bidstrup PL, Payne DJH: Poisoning by dinitro-*ortho*-cresol, report of eight fatal cases occurring in Great Britain. Br Med J 2:16–19, 1951
2. National Institute for Occupational Safety and Health: Criteria for a Recommended Standard . . . Occupational Exposure to Dinitro-*ortho*-cresol. DHEW (NIOSH) Pub No 78–131, pp 1–147. Washington, DC, US Government Printing Office, 1978
3. Bidstrup PL, Bonnell JAL, Harvey DG: Prevention of acute dinitro-*ortho*-cresol (DNOC) poisoning. Lancet 262:794–795, 1952
4. Harvey DG, Bidstrup PL, Bonnell JAL: Poisoning by dinitro-*ortho*-cresol, some observations on the effects of dinitro-*ortho*-cresol administered by mouth to human volunteers. Br Med J 2:13–16, 1951

DINITROTOLUENE

CAS: 121-14-2 — 1987 TLV = 1.5 mg/m^3; skin

$C_6H_3CH_3(NO_2)_2$

Synonyms: DNT, dinitrotoluol

Physical Form. Yellow crystals

Uses. Organic synthesis; explosives; dyes

Exposure. Inhalation; skin absorption

Toxicology. Dinitrotoluene absorption causes anoxia owing to the formation of methemoglobin; jaundice and anemia from chronic exposure have been reported.

Signs and symptoms of overexposure are caused by the loss of oxygen-carrying capacity of the blood. The onset of symptoms of methemoglobinemia is often insidious and may be delayed up to 4 hours; headache is often the first symptom and may become quite intense as the severity of methemoglobinemia progresses; fatigue, nausea, vomiting, chest pain, and loss of weight have also been reported. Cyanosis occurs when the methemoglobin concentration is 15% or more; blueness in the lips, the nose, and the earlobes is usually observed by fellow workers. The subject usually feels well, has no complaints, and is insistent that nothing is wrong until the methemoglobin concentration approaches approximately 40%. At methemoglobin concentrations of over 40%, there is usually weakness and dizziness; of up to 70% concentration, there may be ataxia, dyspnea on mild exertion, tachycardia, nausea, vomiting, and drowsiness.[1]

Ingestion of alcohol aggravates the toxic effects of dinitrotoluene.[2] No chronic effects have been reported from single exposures. Because dinitrotoluene is a solid at room temperatures, splashes in the eyes will not occur unless the substance is hot; then associated thermal burns can be expected. In general, higher ambient temperatures increase susceptibility to cyanosis from exposure to methemoglobin-forming agents.[4]

Special Tests: Determine the methemoglobin concentration in the blood when dinitrotoluene intoxication is suspected and at regular intervals until the methemoglobin has been fully reduced to normal hemoglobin. Methemoglobin may be differentiated from sulfhemoglobin by the addition of a few drops of 10% potassium cyanide, which results in the rapid production of bright red cyanomethemoglobin but has no effect on the color of sulfhemoglobin.[5] Spectrophotometry is required for the precise identification of the pigment and its quantitation. Normal acid methemoglobin has a characteristic absorption spectrum with peaks at 502 and 632 nm, which disappear with the addition of cyanide, whereas sulfhemoglobin has a peak at 620 nm, which does not disappear with cyanide.[5]

Determine the level of dinitrotoluene in the urine: urinary excretion of dinitrotoluene in excess of 25 mg/liter of urine indicates significant absorption.[3]

Treatment. All dinitrotoluene on the body must be removed. Immediately remove all clothing, and wash the entire body from head to foot with soap and water. Pay special attention to the hair and scalp, finger- and toenails, nostrils, and ear canals. Administer oxygen to alleviate the headache and general sense of weakness; confine the patient to bed. Determine the methemoglobin concentration in the blood and repeat every 3 to 6 hours for 18 to 24 hours. Repeat skin cleansing if the methemoglobin concentration appears to rise after 3 to 4 hours. In general, patients will return to normal within 24 hours provided that all sources of further absorption are completely eliminated.[1]

The only justifiable use of methylene blue would be in cases of coma or stupor, usually at methemoglobin levels over 60%; in those patients in whom therapy is necessary, methylene blue, 1 to 2 mg/kg, may be given intravenously over a 5-minute period as a 1% solution; if cyanosis has not disappeared within an hour, a second dose of 2 mg/kg should be administered.[5,6] The total dose should not exceed 7 mg/kg, because methylene blue may cause toxic effects such as dyspnea, precordial pain, restlessness, apprehension, red cell hemolysis, and changes in the electrocardiogram (reduction in the height or even reversal of the T wave, frequently with lowering of the R wave).[5,6]

REFERENCES

1. Hamblin DP: Aromatic nitro and amino comounds. In Patty FA (ed): Industrial Hygiene and Toxicology, 2nd ed, Vol 2, Toxicology, pp 2105–2119, 2130–2131. New York, Interscience, 1963
2. von Oettingen WF: The Aromatic Amino and Nitro Compounds, Their Toxicity and Potential Dangers, US Public Health Service Bulletin No 271, pp 108–111. Washington, DC, US Government Printing Office, 1941
3. Chemical Safety Data Sheet SD–93, Dinitrotoluenes, pp 5–6, 14–15. Washington, DC, MCA, Inc, 1966
4. Linch AL: Biological monitoring for industrial exposure to cyanogenic aromatic nitro and amino compounds. Am Ind Hyg Assoc J 35:426–432, 1974
5. Rieder RF: Methemoglobinemia and sulfhemoglobinemia. In Wyngaarden JB, Smith LH (eds): Cecil Textbook of Medicine, 16th ed, p 896. Philadelphia, WB Saunders Co, 1982
6. Mangelsdorff AF: Treatment of methemoglobinemia. AMA Arch Ind Health 14:148–153, 1956

DIOXANE

CAS: 123-91-1 — 1987 TLV = 25 ppm; skin

$C_4H_8O_2$

Synonyms: 1,4–Diethylene dioxide; diethylene ether; 1,4–dioxacyclohexane; 1,4–dioxane; *p*-dioxane; dioxyethylene ether

Physical Form. Flammable liquid

Uses. Solvent; stabilizer in chlorinated solvents

Exposure. Inhalation; skin absorption

Toxicology. Dioxane is an irritant of the eyes and mucous membranes; on prolonged exposure, it is toxic to the liver and kidneys. It is carcinogenic in experimental animals.

Human volunteers exposed for 15 minutes to 300 ppm reported mild transient irritation of the eyes, nose, and throat.[1] Exposure to 1600 ppm for 10 minutes caused immediate burning of the eyes with lacrimation, and, at 5,500 ppm for 1 minute, slight vertigo was also noted.[2] Five deaths due to heavy exposure for 5 weeks have been reported.[3] Signs and symptoms of poisoning included epigastric pain, anorexia, and vomiting, followed by oliguria, anuria, coma, and death. At autopsy, there was liver necrosis, kidney damage, and edema of the lungs and brain. Another fatal case involved a 1-week exposure to levels ranging from 208 to 605 ppm and possibly higher with concurrent skin exposure.[4] Epigastric pain, increased blood pressure, convulsions, and unconsciousness preceded death. Studies of workers exposed at levels up to 24 ppm for periods of up to 50 years showed no increase in chronic disease, no excess total deaths, no excess cancer deaths, or common cause of death.[5]

Applied to human skin, dioxane causes dryness without other signs of irritation; hypersensitivity has been reported.[5]

In animal experiments, guinea pigs exposed to 30,000 ppm for 3 hours exhibited narcosis after 87 minutes and died within 2 days.[6] The LC_{50} for rats was 14,000 ppm for 4 hours.[7] Repeated exposure of several animal species to 1,000 ppm produced damage to kidneys and liver, and repeated inhalation of 800 ppm over 30 days resulted in fatal kidney injury in some of the exposed rabbits.[6,8] No adverse effects were observed in rats exposed to 111 ppm 8 hours/day, 5 days/week for 2 years.[9]

The liquid applied to rabbit and guinea pig skin was rapidly absorbed and produced signs of incoordination and narcosis. Repeated applications caused liver and kidney damage.[6] Instilled in a rabbit's eye, dioxane produced hyperemia and purulent conjunctivitis.[10]

High doses of dioxane by oral administration produced malignant tumors of the nasal cavity and liver in rats, and tumors of the liver and gallbladder in guinea pigs.[5,11] The carcinogenic risk has not been established for humans.[11]

The warning properties are inadequate to prevent overexposure. Although it has a low odor threshold (3 to 6 ppm), dioxane is not unpleasant, and persons acclimatize within a few minutes.[5]

REFERENCES

1. Silverman L, Schulte HF, First MW: Further studies on sensory response to certain industrial solvent vapors. J Ind Hyg Toxicol 28:262–266, 1946
2. Yant WP et al: Acute response of guinea pigs to vapors of some new commercial organic compounds—VI. Dioxan. Pub Health Rep 45:2023–2032, 1930
3. Barber H: Haemorrhagic nephritis and necrosis of the liver from dioxane poisoning. Guys Hosp Rep 84:267–280, 1934
4. Johnstone RT: Death due to dioxane? Arch Ind Health 20:445–447, 1959
5. National Institute for Occupational Safety and Health: Criteria for a Recommended Standard . . . Occupational Exposure to Dioxane. DHEW (NIOSH) Pub No 77–226. Washington, DC, US Government Printing Office, 1977
6. Fairley A et al: The toxicity to animals of 1:4 dioxane. J Hyg 34:486–501, 1934
7. Pozzani UC et al: The toxicologic basis of threshold limit values—5. Am Ind Hyg Assoc J 20:364–369, 1959
8. Smyth HF Jr: Improved communication—Hygienic standards for daily inhalation. Am Ind Hyg Assoc J 17:129–185, 1956
9. Torkelson TR et al: 1,4–Dioxane—II. Results of a 2-year inhalation study in rats. Toxicol Appl Pharmacol 30:287–298, 1974
10. von Oettingen WF, Jirouch EA: The pharmacology of ethylene glycol and some of its derivatives in relation to their chemical constitution and physical chemical properties. J Pharmacol Exp Ther 42:355–372, 1931
11. IARC Monographs on the Evaluation of Carcinogenic Risk of Chemicals to Man. Cadmium, Nickel, Some Epoxides, Miscellaneous Industrial Chemicals and General Considerations on Volatile Anaesthetics, Vol 11, pp 247–253. Lyon, International Agency for Research on Cancer, 1976

DIPROPYLENE GLYCOL METHYL ETHER

CAS: 34590-94-8

$CH_3OC_3H_6OC_3H_6OH$

1987 TLV = 100 ppm

Synonyms: Dipropylene glycol monomethyl ether; DPGME; Dowanol DPM

Physical Form. Colorless liquid

Uses. Solvent for nitrocellulose and synthetic resins

Exposure. Inhalation

Toxicology. Dipropylene glycol methyl ether at very high concentrations causes narcosis in animals, and it is expected that severe exposure will produce the same effect in humans.

Concentrations expected to be hazardous to humans are disagreeable and not tolerated; in addition, concentrations above 200 ppm (40% saturated atmosphere) are difficult to attain, suggesting that these levels would not normally be encountered in the work environment.[1] Vapor concentrations of 300 ppm caused eye and nasal irritation in humans.[2] No evidence of skin irritation or sensitization was observed when the undiluted liquid was applied to the skin of 250 subjects for prolonged periods of time or after repeated applications.[2]

A single 7-hour exposure of rats to 500 ppm resulted in mild narcosis with rapid recovery.[2] Repeated daily inhalation exposures to 300 to 400 ppm for more than 100 days produced minor histopathologic liver changes in rabbits, monkeys, and guinea pigs; rats initially experienced slight narcosis but developed tolerance to this effect after a few weeks.[2] Daily exposure of rats and rabbits to 200 ppm for 13 weeks caused no effects.[1] Topical administration of 10 mg/kg dipropylene glycol methyl ether five times per week for 13 weeks to shaved rabbit skin caused six deaths among seven animals.[2]

The LD_{50} for rats was 5.4 ml/kg; the low oral toxicity indicates that it is unlikely that toxic amounts of these materials would be swallowed in ordinary handling and use.[2]

Direct contact of the eyes with the liquid or with high vapor concentrations may cause transient irritation.[1]

REFERENCES

1. Landry TD, Yano BL: Dipropylene glycol monomethyl ether: A 13-week inhalation toxicity study in rats and rabbits. Fund Appl Toxicol 4:612–617, 1984
2. Rowe VK et al: Toxicology of mono-, di-, and tri-propylene glycol methyl ethers. AMA Arch Ind Hyg Occup Med 9:509–525, 1954

DIQUAT

CAS: 85-00-7

$C_{12}H_{12}Br_2N_2$ 1987 TLV = 0.5 mg/m^3

Synonyms: 1,1–ethylene–2,2–dipyridylium dibromide

Physical Form. Yellow crystals; available commercially as aqueous solutions (15% to 25% w/v) and as water-soluble granules (2.5%)

Uses. Contact herbicide

Exposure. Inhalation; skin/eye contact

Toxicology. Diquat causes cataracts in animals, but no cataracts have been reported in exposed workers.

The oral LD_{50} in rats ranged from 230 to 440 mg/kg.[1] Effects included dilated pupils, lethargy, and labored respiration. Lung changes characteristic of paraquat were not observed. When applied daily to the skin of rabbits at 40 mg/kg, four of six rabbits died after 8 to

20 applications. Prior to death, effects included weight loss, incoordination, and muscular weakness. Prolonged exposure to diquat is necessary to produce cataracts, and a clear dose–response relationship has been established in chronic feeding studies in animals. Lens opacities developed within 11 months in dogs fed 15 mg/kg/day and within 17 months at 5 mg/kg/day.[1] Dogs tolerated 1.7 mg/kg/day for 4 years without developing cataracts. When dropped in the eyes of rabbits, one drop of a 20% solution caused slight irritation and hyperemia of the lids and conjunctivae, but these effects resolved after 2 days.

Rats exposed to 1.9 mg/m^3 for 4 hours/day, 6 days/week for 5 months showed inflammatory changes in the peribronchial and perivascular connective tissues.[2] Long-term studies have shown no carcinogenic potential.[1,2] In a multigeneration study of reproductive effects, levels of 500 ppm or 125 ppm did not effect fertility, litter production, or litter size and did not cause congenital abnormalities.[3] Lens opacities were found in the parents, F1 and F2 generation receiving 500 ppm, but not at 125 ppm.

There is one report of a human fatality from ingestion.[4] Initial effects included gastrointestinal symptoms, ulceration of mucous membranes, acute renal failure, and liver damage. Pulmonary signs included interalveolar exudation, but there was no evidence of proliferative or fibroplastic changes characteristic of paraquat intoxication.

REFERENCES

1. Clark DG, Hurst EW: The toxicity of diquat. Br J Ind Med 27:51, 1970
2. Bainova A, Zlateva M, Vulcheva VI: Chronic inhalation toxicity of dipyridilium herbicides. Khig Zdraveopazvane 15:25, 1972
3. World Health Organization Pesticide Residues Series, No 2, p 243. Geneva, 1973
4. Schonborn H, Schuster HP, Kossling FK: Klinik und morphologie der akuten peroralen diquat intoxikation. Arch Toxikol 27:204, 1971

ENDRIN

CAS: 72-20-8 — 1987 TLV = 0.1 mg/m^3; skin

$C_{12}H_8Cl_6O$

Synonyms: Compound 269; Experimental Insecticide 269

Physical Form. White, crystalline solid

Uses. Insecticide, avicide, and rodenticide

Exposure. Inhalation; skin absorption; ingestion

Toxicology. Endrin is a convulsant.

In humans, the first effect of endrin intoxication is frequently a sudden epileptiform convulsion that may occur from 30 minutes to up to 10 hours after overexposure; it lasts for several minutes and is usually followed by a stuporous state for 15 minutes to 1 hour.[1,2] Severe poisoning results in repeated violent convulsions and, in some cases, status epilepticus.[3] The electroencephalogram (EEG) may show dysrhythmic changes that frequently precede convulsions; withdrawal from exposure usually results in a normal EEG within 1 to 6 months.[2] In most cases, recovery is rapid, but headache, dizziness, lethargy, weakness, and anorexia may persist for 2 to 4 weeks.[2] In less severe cases of endrin intoxication, complaints include headache, dizziness, leg weakness, abdominal discomfort, nausea, vomiting, insomnia, agitation, and, occasionally, slight mental confusion.[1,3]

Poisonings resulting in convulsions have occurred in manufacturing workers. Recovery following occupational exposures is usually complete within 24 hours. Unlike dieldrin, which is persistent, endrin is rapidly eliminated from the body and apparently does not accumulate, even in fatty tissue.[3,4] However, endrin is the most acutely toxic of the cyclodiene compounds, which also include chlordane, heptachlor, dieldrin, and aldrin.[4]

There are reports of numerous fatalities from ingestion of endrin.[2,4] In one nonfatal incident, ingestion of bread made with endrin-contaminated flour resulted in sudden convulsions in three people; in one person, the serum endrin level was 0.053 ppm 30 minutes after the convulsion, and 0.038 ppm after 20 hours; in the other two cases, no endrin was detected in the blood at 8.5 or 19 hours, respectively, after convulsions.

Single doses of 2.5 mg/kg of endrin administered orally to pregnant golden hamsters during the period of fetal organogenesis caused a high incidence of fetal death, congenital anomalies, and growth retardation.[5] Rats fed a diet of 50 or 100 ppm endrin for 2 years developed degenerative changes in the liver.[1] The IARC has concluded that animal bioassays in mice and rats have been inadequate to evaluate the carcinogenicity of endrin.[6]

REFERENCES

1. Jager KW: Aldrin, Dieldrin, Endrin and Telodrin—An Epidemiological and Toxicological Study of Long-Term Occupational Exposure, pp 78–87, 217–218, 225–234. Amsterdam, Elsevier, 1970
2. Coble Y et al: Acute endrin poisoning. JAMA 202:489–493, 1967
3. Hayes WJ Jr: Pesticides Studied in Man, pp 247–251. Baltimore, Williams & Wilkins, 1982
4. Centers for Disease Control: Acute convulsions associated with endrin poisoning—Pakistan. MMWR 33:687–688, 693, 1984
5. Ottolenghi AD, Haseman JK, Suggs F: Teratogenic effects of aldrin, dieldrin, and endrin in hamsters and mice. Teratology 9:11–16, 1974
6. IARC Monographs on the Evaluation of the Carcinogenic Risk of Chemicals to Man, Vol 5, Some Organochlorine Pesticides, pp 157–171. Lyon, International Agency for Research on Cancer, 1974

EPICHLOROHYDRIN

CAS: 106-89-8 — 1987 TLV = 2 ppm; skin

C_3H_5OCl

Synonyms: 1–Chloro–2,3–epoxypropane; 3–chloro–1,2–epoxypropane; (chloromethyl)ethylene oxide; chloromethyloxirane; 3–chloro–1,2–propylene oxide; alpha-epichlorohydrin

Physical Form. Liquid

Uses. In the manufacture of epoxy and phenoxy resins

Exposure. Inhalation; skin absorption

Toxicology. Epichlorohydrin is a severe irritant of the skin, eyes, and respiratory tract. Repeated or prolonged exposure can cause lung, liver, and kidney damage. Epichlorohydrin causes chromosomal aberrations in lymphocytes and is carcinogenic in experimental animals.

According to one industrial report, exposure at 20 ppm for 1 hour caused temporary burning of the eyes and nasal passages.[1]

At 40 ppm, irritation was more persistent, lasting 48 hours.[1] Pulmonary edema and renal lesions may result from exposure to concentrations greater than 100 ppm. In one worker acutely exposed to unspecified but probably very high concentrations, immediate effects included nausea, vomiting, headache, and dyspnea, with conjunctival and upper respiratory irritation. During the 2 years following the incident, bronchitis, liver damage, and hypertension were observed.[1]

Exposed workers had a marked increase in percentage of lymphocytes with chromatid breaks, chromosome breaks, severely damaged cells, and abnormal cells.[2]

Skin contact causes itching, erythema, and severe burns, which appear after a latent period ranging from several minutes to days, depending upon the intensity of exposure. One worker who failed to remove contaminated shoes for 6 hours developed severe skin damage, with painful enlarged lymph nodes in the groin.[1] Skin sensitization has been reported.[3]

Mice showed signs of irritation, gradual development of cyanosis, and muscular relaxation of the extremities and finally died from depression of the respiratory system following repeated 1-hour exposures to 2370 ppm.[4]

Rats repeatedly exposed to 120 ppm 6 hours/day experienced labored breathing, profuse nasal discharge, weight loss, leukocytosis, and increased urinary protein excretion. At autopsy, lung, liver, and kidney damage were found.[5]

Respiratory distress in rats was observed at 56 ppm during multiple exposures, whereas 17 ppm for 19 days produced no effects. Function of the liver and kidney was altered in rats receiving 5.2 or 1.8 ppm for 4 hours.[1]

Male rats given five oral doses of 20 mg/kg had a temporary fertility loss, whereas a single 100 mg/kg dose caused spermatocele formation and probable permanent sterility.[6] Fifty inhalation exposures to 50 ppm for 6 hours caused transient infertility in male rats; no changes were observed in reproductive parameters of female rats; rabbits remained fertile.[7] No detrimental effect on fertility has been found in occupationally exposed workers where 8-hour TWA exposures are estimated to be less than 1 ppm.[8] There was no evidence of teratogenicity in rat fetuses at doses that caused death in some of the treated dams.[9]

A number of studies indicate that epichlorohydrin induces tumors of localization dependent upon the mode of application. A high incidence (100% for females, 81% for males vs none in controls) of squamous cell carcinomas of the forestomach occurred in rats administered 10 mg/kg five times per week for up to 2 years by gastric intubation.[10] When administered in the drinking water, epichlorohydrin also caused squamous cell carcinomas of the forestomach in rats.[11] Exposure to 100 ppm, 6 hours/day for 30 days produced a high incidence of malignant tumors of the nasal cavity in rats.[12] An increase in local sarcomas occurred in mice given weekly subcutaneous injections.[13] Historic studies of occupationally exposed workers have not found an increase in cancer mortality.[14] The carcinogenic risk to humans cannot be fully assessed, however, because of insufficient latency periods, limited number of deaths, and indeterminate levels and duration of exposure.

The IARC has determined that there is sufficient evidence of carcinogenicity in animals and inadequate evidence in humans.[15]

REFERENCES

1. National Institute for Occupational Safety and Health: Criteria for a Recommended Standard . . . Occupational Exposure to Epichlorohydrin. DHEW (NIOSH) Pub No 76–206, p 152. Washington, DC, US Government Printing Office, 1976
2. Picciano D: Cytogenic investigation of occupational exposure to epichlorohydrin. Mutat Res 66:169–173, 1979
3. Beck MH, King CM: Allergic contact dermatitis to epichlorohydrin in a solvent cement. Contact Dermatitis 9:315, 1983
4. Freuder E, Leake CD: The toxicity of epichlo-

rohydrin. Univ Calif Berk Pub in Pharmacol 2:69–77, 1941
5. Gage JC: The toxicity of epichlorohydrin vapour. Br J Ind Med 16:11–14, 1959
6. Cooper ERA et al: Effects of *alpha*-chlorohydrin and related compounds on the reproductive organs and fertility of the male rat. J Reprod Fertil 38:379–386, 1974
7. John JA et al: Inhalation toxicity of epichlorohydrin: Effects on fertility in rats and rabbits. Toxicol Appl Pharmacol 68:415–423, 1983
8. Venable JR et al: A fertility study of male employees engaged in the manufacture of glycerine. J Occup Med 22:87–91, 1980
9. Marks TA et al: Teratogenic evaluation of epichlorohydrin in the mouse and rat and glycidol in the mouse. J Toxicol Environ Health 9:87–96, 1982
10. Wester PW et al: Carcinogenicity study with epichlorohydrin (CEP) by gavage in rats. Toxicology 36:325–339, 1985
11. Konishi T et al: Forestomach tumors induced by orally administered epichlorohydrin in male Wistar rats. Gann 71:922–923, 1980
12. Laskin S et al: Inhalation carcinogenicity of epichlorohydrin in noninbred Spague-Dawley rats. J Natl Cancer Inst 65:751–757, 1980
13. Van Duuren BL et al: Carcinogenic activity of alkylating agents. J Natl Cancer Inst 53:695–700, 1974
14. Tassignon JP et al: Mortality in European cohort occupationally exposed to epichlorohydrin (ECH). Int Arch Occup Environ Health 51:325–336, 1983
15. IARC Monographs on the Evaluation of the Carcinogenic Risk of Chemicals to Humans. Suppl 4, pp 122–124. Lyon, International Agency for Research on Cancer, 1982

EPN

CAS: 2104-64-5 1987 TLV = 0.5 mg/m^3;
$C_2H_5O(C_6H_5)P(S)OC_6H_4NO_2$ skin

Synonyms: o-Ethyl o-p-nitrophenyl phenylphosphonothioate; EPN–300

Physical Form. Liquid

Uses. Acaricide; insecticide

Exposure. Inhalation; skin absorption; ingestion

Toxicology. EPN is an anticholinesterase agent.

A few deaths have been reported following poisoning by EPN, most resulting from suicidal ingestion, but at least one death has been associated with EPN spraying. It is moderately toxic to animals but less potent than parathion.[1]

Signs and symptoms of overexposure are caused by the inactivation of the enzyme cholinesterase, which results in the accumulation of acetylcholine at synapses in the nervous system, skeletal and smooth muscle, and secretory glands.[1–3] The sequence of the development of systemic effects varies with the route of entry. The onset of signs and symptoms is usually prompt but may be delayed up to 12 hours. After inhalation, respiratory and ocular effects are the first to appear, often within a few minutes of exposure. Respiratory effects include tightness in the chest and wheezing due to bronchoconstriction and excessive bronchial secretion; laryngeal spasm and excessive salivation may add to the respiratory distress; cyanosis may also occur. Ocular effects include miosis, blurring of distant vision, tearing, rhinorrhea, and frontal headache.

After ingestion, gastrointestinal effects, such as anorexia, nausea, vomiting, abdominal cramps, and diarrhea, appear within 15 minutes to 2 hours. After skin absorption, localized sweating and muscular fasciculations in the immediate area usually occur within 15 minutes to 4 hours; skin absorption is somewhat greater at higher ambient temperatures and is increased by the presence of dermatitis.[2,3]

With severe intoxication by all routes, an excess of acetylcholine at the neuromuscular junctions of skeletal muscle causes weakness aggravated by exertion, involuntary twitchings, fasciculations, and eventually paralysis. The most serious consequence is paralysis of the respiratory muscles. Effects on the central nervous system include giddiness, confusion, ataxia, slurred speech, Cheyne–Stokes respiration, convulsions, coma, and loss of reflexes. The blood pressure may fall to low levels, and cardiac irregularities, including complete heart block, may occur.[1–3]

Complete symptomatic recovery usually occurs within a week; increased susceptibility to the effects of anticholinesterase agents persists for up to several weeks after exposure. Daily exposure to concentrations that are insufficient to produce symptoms following a single exposure may result in the onset of symptoms. Continued daily exposure may be followed by increasingly severe effects.

EPN is neurotoxic to atropine-protected hens, producing polyneuropathy progressing to paralysis and some deaths following ingestion of 5 to 10 mg/kg/day. There are no reports of neurotoxicity from EPN in humans.[1]

Diagnosis. Initial signs and symptoms include headache, blurred vision, pallor, weakness, sweating, abdominal pain, nausea, vomiting, and diarrhea.

Moderate-to-severe intoxication includes miosis, lacrimation, excessive salivation, muscle fasciculations, dyspnea, cyanosis, convulsions, shock, cardiac arrhythmias, and coma.

Differential Diagnosis: Diagnosis is primarily based upon a history of exposure and clinical evidence of diffuse parasympathetic stimulation. Careful observation of the effects of atropine and pralidoxime may be valuable. Patients with organophosphate poisoning are resistant to the action of atropine at moderate dosages; failure of 1 to 2 mg of atropine administered parenterally to produce signs of atropinization (flushing, mydriasis, tachycardia, or dryness of mouth) indicates anticholinesterase poisoning. Intravenous injections of 1 g pralidoxime generally causes some recovery from signs and symptoms.

Special Tests: Two types of cholinesterase are clinically significant: (1) true acetylcholinesterase, found principally in the nervous system and the red blood cell; and (2) pseudo- or butyrylcholinesterase, found in the plasma, liver, and the nervous system. Whereas the action of both types is inhibited by organophosphates, the level of depression of red blood cell cholinesterate is a better indicator of clinically significant reduction of cholinesterase activity in the nervous system.

Laboratory evidence of depression of red blood cell cholinesterase to a level substantially below pre-exposure levels (at least 50% and usually much lower) is verification of poisoning. There is an imperfect correlation between the degree of depression of cholinestersase enzymes and the occurrence of symptoms. With a rapid drop in cholinesterase activity, generally reflecting an acute heavy exposure, there may be symptoms with only a 30% depression, whereas with slower drops to 70% depression, reflecting chronic low-level exposure, there may be no symptoms.[4]

If no pre-exposure baseline has been performed but symptoms are not sufficient to justify treatment with atropine, repeated testing during the recovery period demonstrating progressively increasing plasma and red blood cell cholinesterase levels over several days and weeks, respectively, suggests the diagnosis of anticholinesterase poisoning.

There are many different methods for estimation of cholinesterase content of blood, and associated with each method is a different set of normal values and a different set of reporting units. The laboratory report of a cholinesterase determination should state the units involved along with the appropriate normal range. Based on the Michel method, the normal range of red blood cell cholinesterase activity (delta *p*H per hour) is 0.39 to 1.02 for men, and 0.34 to 1.10 for women.[5] The normal range of the enyzme activity (delta *p*H per hour) of plasma is 0.44 to 1.63 for men, and 0.24 to 1.54 for women.

Treatment. Treatment of organophosphate poisoning ranges from simple removal from exposure in very mild cases to the provision of very rigorous supportive and antidotal measures in severe cases.[1,3,6,7] In moderate-to-severe cases, because of pulmonary involvement, there may be need for artificial respiration using a positive-pressure method. Careful attention must be paid to removal of secretions and to maintenance of a patent airway. Anticonvulsants such as thiopental sodium may be necessary. Maintenance of respiration is critical, since death usually results from weakness of the muscles of respiration and accumulation of excessive secretions in the respiratory tract.

As soon as cyanosis has been overcome, 2 to 4 mg of atropine should be given intravenously. (Atropine may induce ventricular fibrillation in the presence of cyanosis.) *This dose of atropine is approximately 10 times the amount that is administered for other conditions in which atropine is considered therapeutic.* This dose should be repeated at 5- to 10-minute intervals until signs of atropinization appear (dry, flushed skin, tachycardia as high as 140 beats/minute and pupillary dilatation). A mild degree of atropinization should be maintained for at least 48 hours.[3]

Pralidoxime (2–PAM, Protopam) chloride is a cholinesterase reactivator that complements the action of atropine. It has its greatest effect in reversing the nicotinic action of anticholinesterase agents at skeletal neuromuscular junctions but virtually no effect on central nervous system manifestations. In moderate-to-severe cases, the dose for adults is 1 to 2 g injected intravenously at a rate not in excess of 500 mg/minute. After an hour, a second dose of 1 g is indicated if muscle weakness has not been relieved. Treatment with pralidoxime chloride will be most effective if given within 24 hours after poisoning.[3] Morphine, aminophylline, and phenothiazines are contraindicated because of documented experience of adverse reactions in cases of organophosphate poisoning.[6]

It is of great importance to decontaminate the patient. Contaminated clothing should be removed at once, and the skin should be washed with generous amounts of soap or detergent and a flood of water, which is best accomplished under a shower or by submersion in a pond or other body of water if the exposure occurred in the field. Careful attention should be paid to cleansing of the skin and hair.

The patient should be attended and monitored continuously for not less than 24 hours, because serious and sometimes fatal relapses have occurred owing to continuing absorption of the toxin or dissipation of the effects of the antidote.

Regeneration of cholinesterase occurs primarily by synthesis of new enzyme and takes place at the rate of approximately 1% per day.[6] A patient who has recovered from the acute phase of poisoning remains hypersusceptible to anticholinesterases for up to several weeks.

Medical Control. Medical control involves preplacement and annual physical examination with determination of pre-exposure red blood cell cholinesterase activity. A person whose red blood cell cholinesterase falls to or below 40% of the pre-exposure baseline should be removed from further exposure until the activity returns to within 80% of the pre-exposure baseline.

REFERENCES

1. Hayes WJ Jr: Organic phosphorus pesticides. In Pesticides Studied in Man, pp 284–435. Baltimore, Williams & Wilkins, 1982
2. Koelle GB (ed): Cholinesterases and anticholinesterase agents. Hanbuch der Experimentellen Pharmakologie, Vol 15, pp 989–1027. Berlin, Springer-Verlag, 1963
3. Taylor P: Anticholinesterase agents. In Gilman AG et al (eds): Goodman and Gilman's The Pharmacological Basis of Therapeutics, 7th ed, pp 110–129. New York, Macmillan Publishing Co, 1985
4. Coye MJ, Lowe JA, Maddy KT: Biological monitoring of agricultural workers exposed to pesticides. I. Cholinesterase activity determinations. J Occup Med 28:619–622, 1986
5. Michel HO: Electrometric method for determination of red blood cell and plasma cholinesterase activity. J Lab Clin Med 34:1564–1568, 1949
6. Milby TH: Prevention and management of organophosphate poisoning. JAMA 216:2131–2133, 1971
7. Namba T, Nolte CT, Jackrel J, Grob D: Poisoning due to organophosphate insecticides. Am J Med 50:475–492, 1971

EPOXY RESINS

1987 TLV = no value set

Synonyms: Epoxies; Epon resins

Physical Form. Uncured resins are long-chained prepolymers that are viscous liquids or solids; the cured resins are strong, solid polymers.

Uses. Molding compounds; surface coatings; adhesives; laminating or reinforcing plastics

Epoxy resins are polymers containing more than one epoxide group (a three-membered ring containing two carbon atoms and one oxygen atom).[1] The term *epoxy* is a prefix that indicates the presence of an epoxide group in a compound. An epoxy resin system is composed of two primary components: (1) the uncured resin, and (2) the curing agent (also referred to as the hardener, catalyst, accelerator, activator, or cross-linking agent). Uncured resins are oligomers of relatively low molecular weight that may be a liquid or a solid. Before epoxy resins can become useful products, they must be cured, with the addition of a curing agent. Curing involves the cross-linkage by polymerization of the reactive epoxy groups into a three-dimensional matrix.[1]

In addition to the two primary components, there are several other components: diluents/solvents, fillers, and pigments. Diluents, which may represent 10% to 15% of resin volume, are added primarily to reduce viscosity. There are two types of diluents: reactive and nonreactive. Reactive diluents, primarily the glycidyl ethers, contain epoxy groups, which will take part in the curing process. Nonreactive diluents include a variety of organic solvents. The toxicity of epoxy resin systems results from the toxicity of the various components, each of which must be considered.[1]

Some uncured resins (liquids) are primary skin irritants or sensitizers or both. Toxicity generally decreases with increase in molecular weight and epoxy number. The resins with the greatest potential for sensitization are those with molecular weights below 500.[2] None of the uncured resins possess significant volatility; thus, inhalation poses little risk.[1] The majority of epoxy resins are manufactured by the reaction between epichlorohydrin and bisphenol A, producing DGEBA (diglycidyl ether of bisphenol A) resins.[2] After the initial manufacture of uncured resin, epichlorohydrin is probably not present during the subsequent mixing and polymerization steps.[1]

Exposure. Inhalation; skin contact

Toxicology. Epichlorohydrin is a severe skin irritant, a severe irritant to mucous membranes and the respiratory tract, a possible skin sensitizer, and a possible respiratory carcinogen.[1]

Bisphenol A causes minimal eye and skin irritation and may produce skin sensitization on repeated contact.[1] Skin irritation is the major hazard with DGEBA resins. In some cases, exposure results in allergic contact dermatitis.

Other types of resins include epoxy novolac resins (in which epichlorohydrin is linked with phenol formaldehyde, *o*-cresol, or other phenolic compounds), brominated epoxy resins, and cycloaliphatic epoxy resins. The epoxy novolac resins cause skin and minor eye irritation. Combustion of brominated resins may result in release of bromine and hydrogen bromide vapors.[1,2]

Curing agents account for much of the potential hazard associated with use of epoxy resins.[1] There are several major types of curing agents: aliphatic amines, aromatic amines, cycloaliphatic amines, acid anhydrides, polyamides, and catalytic curing agents. The latter two types are true catalysts in that they do not participate in the curing process.

The aliphatic amines, including triethylene tetramine (TETA) and diethylene triamine (DETA) are highly alkaline (*p*H 13 to 14), caustic, and volatile and may cause severe burns.[3] They can cause skin irritation and sensitization and respiratory tract irritation. Eye irritation with conjunctivitis and corneal edema (resulting in "halos" around lights) may occur. Asthmatic symptoms suggesting respiratory tract sensitization have been described.[1] In ski manufacturing workers using epoxy resins, 3–(dimethylamino)propylamine has been shown to cause cross-shift declines in FEV_1 and flow rates, and work-related respiratory symptoms (e.g., cough, chest tightness).[4]

Aromatic amines are generally solids and less irritating than aliphatic amines. 4,4′–Methylene dianiline (MDA) has caused outbreaks of reversible toxic hepatitis, apparently after skin absorption. Severe symptoms, including elevated AST, alkaline phosphatase, and bilirubin, and liver enlargement, have been observed in some workers using MDA as a curing agent with epoxy resins.[5] M-Phenylenediamine is a strong irritant and allergic sensitizer; like MDA, it stains the skin and nails yellow.[2]

Acid anhydrides can cause severe eye and skin irritation and burns, depending upon the concentration and duration of contact.[1] Inhalation of high concentrations can cause significant respiratory tract irritation. Phthalic anhydride (PA), tetrachlorophthalic anhydride (TCPA), and trimellitic anhydride can induce asthma in epoxy resin workers; frequently, a dual (immediate and late) asthmatic response has been documented. Specific IgE antibodies on RAST testing have been demonstrated in patients with TCPA asthma.[6] One worker developed asthma on grinding epoxy resin cured with phthalic anhydride, presumably owing to release of some unreacted residual phthalic anhydride during grinding of a cured moulding.[7]

Polyamides, reaction products of aliphatic amines and fatty acids, are considerably less toxic than the aliphatic amines but are moderately irritating to the skin and extremely irritating to the eyes.[1,2]

Isophorone diamine, a cycloaliphatic

amine, has been reported to cause skin sensitization.[7]

Glycidyl ethers, reactive diluents in epoxy resin systems, are characterized by the presence of the 2,3–epoxypropyl group and an ether linkage to another organic group. Virtually all of these substances are liquids with low vapor pressures at room temperature. Dermal contact is the major route of exposure. Vapor pressures become more appreciable at higher temperatures, which may occur during the curing process. Some glycidyl ethers commonly used in epoxy resin systems are allyl glycidyl ether (AGE), *n*-butyl glycidyl ether (BGE), *o*-cresyl glycidyl ether (CGE), isopropyl glycidyl ether (IGE), phenyl glycidyl ether (PGE), resorcinol diglycidyl ether, and 1,4–butanediol diglycidyl ether.[1,8] In humans exposed to glycidyl ethers, adverse effects have generally been limited to irritation and sensitization.[8] PGE and BGE have produced severe skin irritation in humans, causing burns and blistering. AGE has produced skin and eye irritation in humans. Skin sensitivity to AGE, BGE, and PGE has been documented in some humans occupationally exposed to epoxy resins.[1,8] In animals, glycidyl ethers have produced CNS effects, including muscular incoordination, reduced motor activity, agitation and excitement, deep depression, narcosis, and coma. PGE has produced CNS depression with dermal administration; BGE and AGE have produced depression after inhalation exposure.[8] Experimental inhalation of glycidyl ethers has resulted in pulmonary irritation and inflammation, including pneumonitis and peribronchiolitis. For example, rats exposed to PGE at 15 ppm for 7 hours/day, 5 days/week for 10 weeks had peribronchial and perivascular inflammatory infiltrates. Exposure to some glycidyl ethers, usually by injection, has been demonstrated to produce testicular abnormalities, alteration of leukocyte counts, atrophy of lymphoid tissue, and bone marrow cytotoxicity.[8]

Solvents used as nonreactive diluents include acetone, cellosolve, methyl ethyl ketone, methyl isobutyl ketone, methylene chloride, 1,1,1–trichloroethane, toluene, and xylene. Skin and eye irritation and, in higher concentrations, CNS depression and respiratory irritation may result from exposure to these solvents as diluents for epoxy resin systems. These solvents may dehydrate and defat the skin, which may render the skin more vulnerable to the irritating and sensitizing components of epoxy resin formulations.[1,2]

Fillers used in epoxy resins are normally inert, finely divided powders. Common fillers include calcium carbonate, clay (bentonite), talc, silica, diatomaceous earth, and asbestos. Workers exposed to excessive amounts of some of these dusts may experience lung damage.[1]

The curing process renders the resin essentially inert and nontoxic. At room temperature, full curing may take several days; incompletely cured resins may cause skin irritation and sensitization.[1] Respiratory symptoms may result from inhalation of cured epoxy dusts during grinding, presumably from release of residual curing agent.[1,7]

Dermatitis from epoxy resin components usually develops first on the hands, particularly between the fingers, in the finger webs, on the dorsum of the hands, and on the wrists. It may vary in severity from erythema to a marked bullous eruption.[2] When sensitization occurs, the eruption is typically pruritic, with small vesicles on the fingers and hands resembling dyshidrotic eczema. The eruption may spread to other areas of the body that accidentally contact resin components such as the face and neck. In highly sensitized persons, vapors from the curing agent or reactive diluents may cause recurrence of itching and redness in the absence of direct skin contact.[2]

Prevention of epoxy dermatitis requires meticulous attention to avoiding skin contact during mixing and application, use of protective clothing such as PVC gloves, good housekeeping, regular hand washing before eating and breaks, and prohibition of eating and smoking in the work area. In some cases, sensitized workers may need to be completely removed from the work area and further exposure.[2]

There are no reports of carcinogenic, mutagenic, teratogenic, or reproductive effects to humans of uncured resins, curing agents, or glycidyl ethers, but there are some positive findings in animal studies.[1,8] Animal experiments using DGEBA resins have generally indicated no carcinogenic activity but are inconclusive.[1] Mutagenicity tests using various liquid and solid epoxy resins have yielded some positive and some negative results.[1] Of the aromatic amine curing agents, diaminodiphenyl sulfone is tumorigenic in animal experiments, whereas 4,4'-methylenedianiline is a suspect animal carcinogen.[1] Many of the glycidyl ethers produce a mutagenic response in the Ames assay and in some other short-term tests.[8] Testicular degeneration has been noted in several animal species after exposure to AGE, DGE, and PGE, generally after high doses. Di(2,3-epoxypropyl) ether (DGE), a glycidyl ether with two epoxy groups that is not used in epoxy resin systems, has produced benign skin papillomas in mice after repeated skin painting.[8] Glycidyl ethers are rapidly metabolized to less cytotoxic substances and rapidly conjugate with skin proteins on dermal contact. Their low volatility decreases the possibility of significant systemic absorption by inhalation. Together, these factors reduce the likelihood of conjugation with nuclear macromolecules in somatic or germ cells, which otherwise might result in carcinogenic or teratogenic effects.[8]

REFERENCES

1. Acres Consulting Service: Occupational Health Survey—Worker Exposure to Epoxy Resins and Associated Substances. Niagara Falls, Ontario, Occupational Health and Safety Division, Ontario Ministry of Labor, 1984
2. Adams RM: Occupational Skin Disease, pp 241–250. Philadelphia, JB Lippincott, 1978
3. Birmingham DJ: Clinical observations on the cutaneous effects associated with curing epoxy resins. AMA Arch Ind Health 19:365–367, 1959
4. Brubaker RE et al: Evaluation and control of a respiratory exposure to 3-(dimethylamino)-propylamine. J Occup Med 21:688–690, 1979
5. Bastion PG: Occupational hepatitis caused by methylenedianiline. Med J Australia 141:533–535, 1984
6. Howe W et al: Tetrachlorophthalic anhydride asthma: Evidence for specific IgE antibody. J Allergy Clin Immunol 71:5–11, 1983
7. Ward MJ, Davies D: Asthma due to grinding epoxy resin cured with phthallic anhydride. Clin Allergy 12:165–168, 1982
8. National Institute for Occupational Safety and Health: Criteria for a Recommended Standard . . . Occupational Exposure to Glycidyl Ethers, DHEW (NIOSH) Pub 78-166. Washington, DC, US Government Printing Office, 1978

ETHANOLAMINE

CAS: 141-43-5

$NH_2CH_2CH_2OH$ 1987 TLV = 3 ppm

Synonyms: 2-Aminoethanol; 2 hydroxyethylamine; ethylolamine; colamine

Physical Form. Liquid

Uses. To remove CO_2 and H_2S from natural gas and other gases; synthesis of surface active agents; softening of hides

Exposure. Inhalation

Toxicology. Ethanolamine is a pulmonary irritant in animals; it is expected that severe exposure will cause the same effect in humans.

No systemic effects from industrial exposure have been reported. The liquid applied to the human skin for 1.5 hours caused marked erythema.[1]

Dogs and cats exposed to 990 ppm for 4 days survived, but four of six guinea pigs died from exposure to 233 ppm for 1 hour; pathologic changes were chiefly those of pulmonary irritation, with some nonspecific changes in the liver and kidneys.[1] In animals exposed repeatedly to 66 to 100 ppm, there was mortality in some cases during 24 to 30 days of exposure, and all animals were lethargic.[2] No mortality or pathology resulted from 90-day continuous exposure of dogs to 26 ppm, of rats to 12 ppm, or of guinea pigs to 15 ppm.[2] The liquid produced moderate irritation of the skin of rabbits and severe irritation in the eyes of rabbits.[1]

The odor is described as ammonia-like or musty at 25 ppm but is detected by means of a sensation at 3 ppm.[2]

REFERENCES

1. Beard RR, Noe JT: Aliphatic and alicyclic amines. In Clayton GD, Clayton FE (eds): Patty's Industrial Hygiene and Toxicology, 3rd ed, rev, Vol 2B, Toxicology, p 3168. New York, Wiley–Interscience, 1981
2. Weeks MH et al: The effects of continuous exposure of animals to ethanolamine vapor. Am Ind Hyg Assoc J 21:374–381, 1960

2–ETHOXYETHANOL

CAS: 110-80-5 — 1987 TLV = 5 ppm; skin

$C_2H_5OCH_2CH_2OH$

Synonyms: Ethylene glycol monoethyl ether; Cellosolve

Physical Form. Colorless liquid

Uses. Solvent for nitrocellulose lacquers and alkyd resins; in dyeing textiles and leather; in cleaners and varnish removers

Exposure. Inhalation; skin absorption

Toxicology. 2–Ethoxyethanol is a mild irritant of the eyes and has caused lung and kidney injury in animals in very high concentrations. Animal experiments indicate adverse effects on the bone marrow and testis as well as embryo/fetotoxicity and congenital malformations.

Dogs repeatedly exposed to 840 ppm for 12 weeks developed a slight decrease in red cells and hemoglobin and an increase in immature white cells. In female rats exposed to 125 ppm for 4 hours, there was an increase in erythrocyte osmotic fragility, an effect that has been noted from other glycol ethers and in other species.[1] In mice, the LC_{50} for 7 hours was 1820 ppm; death was attributed to pulmonary edema and kidney injury.[2]

Teratology studies in rats and rabbits have demonstrated both embryo/fetotoxicity and congenital malformations following exposure by oral, inhalational, or dermal routes. Exposures of pregnant rabbits at 160 ppm resulted in significant increases in cardiovascular, renal, and ventral body wall defects, minor skeletal changes, and fetal resorptions, with minimal maternal toxicity. Similarly, exposure of pregnant rats at 200 ppm resulted in fetal growth suppression and an increase in cardiovascular defects and wavy ribs in the absence of significant maternal toxicity. Dermal exposure of pregnant rats led to increased fetal resorptions, cardiovascular malformations, and skeletal variation.[3] A no-effect level of 10 to 50 ppm for reproductive effects in animals has been observed.[4]

Testicular atrophy with decreased testis weight, tubular atrophy, and changes in the germinal epithelium has been observed following exposure in rats and mice.[3] Decreased fertility is reversible following cessation of exposure. A concentration of 100 ppm appears to be a no-effect level on the testis.[4]

The liquid instilled in the eyes of animals caused immediate discomfort, some conjunctival irritation, and a slight transitory irritation of the cornea, which was readily reversible.[5] Repeated and prolonged contact of the liquid with the skin of rabbits caused only a mild irritation, but toxic amounts were readily absorbed through the skin.[5]

REFERENCES

1. Carpenter CP et al: The toxicity of butyl cellosolve solvent. AMA Arch Ind Health 14:114–131, 1956
2. Browning E: Toxicity and Metabolism of Industrial Solvents, pp 601–605. Amsterdam, Elsevier Publishing Company, 1965
3. Hardin B: Reproductive toxicity of the glycol ethers. Toxicology 27:91–102, 1983
4. Reproductive toxicity of the glycol ethers. Reproductive Toxicology 4:15–18, 1985
5. Rowe VK: Derivatives of glycols. In Patty FA (ed): Industrial Hygiene and Toxicology, 2nd ed, Vol 2, Toxicology, pp 1547–1550. New York, Interscience, 1963

2–ETHOXYETHYL ACETATE

CAS: 111-15-9 1987 TLV = 5 ppm; skin

$C_2H_5OCH_2CH_2OOCCH_3$

Synonyms: Cellosolve acetate; ethylene glycol monoethyl ether acetate; ethyl glycol acetate

Physical Form. Colorless liquid

Uses. Solvent for nitrocellulose and some resins

Exposure. Inhalation; skin absorption

Toxicology. 2–Ethyoxyethyl acetate is irritating to the eyes, nose, and throat; at very high concentrations, it has caused central nervous system depression in animals, and it is expected that severe exposure will cause the same effect in humans.

No effects have been reported in humans, probably because the vapor becomes objectionable before concentrations necessary to cause adverse systemic effects are reached.[1] It is a defatting agent, and prolonged or repeated contact may lead to dermatitis.[1]

Mice, guinea pigs, and a rabbit survived 12 8-hour exposures to 450 ppm, but another rabbit and two cats died before the end of the exposure period; kidney damage was observed at autopsy.[2] Dogs survived 120 daily exposures to 600 ppm with slight eye and nose irritation but without apparent systemic injury as determined by histopathology and hematologic tests.[3] Exposure for 8 hours to 1500 ppm was fatal to two of six rats. Guinea pigs survived exposure to saturated vapor concentrations (4000 ppm), but two such exposures of cats for 4 to 6 hours caused narcosis, kidney damage, and death.[1] Testicular atrophy and leukopenia were observed following oral administration of 2–ethoxyethyl acetate, 400 mg/kg, to mice over a 5-week period.

REFERENCES

1. Rowe VK, Wolf MA: Derivatives of glycols. In Clayton GD, Clayton FE (eds): Patty's Industrial Hygiene and Toxicology, 3rd ed, Vol 2C, Toxicology, pp 4024–4026. New York, Wiley–Interscience, 1982
2. Lehmann KB, Flury F: Toxicology and Hygiene of Industrial Solvents, p 289. Baltimore, Williams & Wilkins, 1943
3. Smyth HF Jr: Improved communication—hygienic standards for daily inhalation. Am Ind Hyg Assoc J 17:129–184, 1956
4. Carpenter CP: Cellosolve. JAMA 135:880, 1947
5. Grant WM: Toxicology of the Eye, 3rd ed, p 1034. Springfield, Illinois, Charles C Thomas, 1986

ETHYL ACETATE

CAS: 141-78-6

$CH_3COOC_2H_5$ 1987 TLV = 400 ppm

Synonyms: Acetic ether; ethyl ester; ethyl ethanoate

Physical Form. Liquid

Uses. Lacquer solvent; artificial fruit essences

Exposure. Inhalation

Toxicology. Ethyl acetate causes respiratory tract irritation; at very high concentrations, it produces narcosis in animals, and it is expected that severe exposure will cause the same effect in humans.

Unacclimated human subjects exposed to 400 ppm for 3 to 5 minutes experienced nose and throat irritation.[1] However, no adverse symptoms were observed in workmen exposed at 375 to 1500 ppm for several months.[2] In rare instances, exposure may cause sensitization resulting in inflammation of the mucous membranes and in eczematous eruptions.[3]

Cats exposed to 9000 ppm for 8 hours suffered irritation and labored breathing; 20,000 ppm for 45 minutes caused deep narcosis, whereas 43,000 ppm for 14 to 16 minutes was fatal; at autopsy, findings included pulmonary edema with hemorrhage and hyperemia of the respiratory tract.[3] Repeated exposure of rabbits to 4450 ppm resulted in secondary anemia with leukocytosis, hyperemia, and damage to the liver.[3]

Ethyl acetate has a fruity odor detectable at 10 ppm.[4]

REFERENCES

1. Nelson KW et al: Sensory response to certain industrial solvent vapors. J Ind Hyg Toxicol 25:282–285, 1943
2. Patty FA: Potential exposures in industry. In Patty FA (ed): Industrial Hygiene and Toxicology, 2nd ed, Vol 2, Toxicology, p 2278. New York, Interscience, 1963
3. von Oettingen WF: The aliphatic acids and their esters: Toxicity and potential dangers. AMA Arch Ind Health 21:28–65, 1960
4. Hygienic Guide Series: Ethyl acetate. Am Ind Hyg Assoc J 26:201–203, 1964

ETHYL ACRYLATE

CAS: 140-88-5 1987 TLV = 5 ppm;

$CH_2{=}CHCOOC_2H_5$ skin

Synonyms: Ethyl 2–propenoate; 2–propenoic acid ethyl ester; acrylic acid ethyl ester

Physical Form. Colorless, flammable liquid with an acrid odor; commercial form contains 1000 ppm hydroquinone or 15 or 200 ppm hydroquinone monomethyl ether to prevent polymerization

Uses. In manufacture of emulsion polymers used in paints, textiles, paper, and polishes, leather; in solution polymers as surface coatings; as comonomers in the production of acrylic fibers

Exposure. Inhalation; skin absorption

Toxicology. Ethyl acrylate is an irritant of the eyes, nose, throat, and skin; it has a history of dermal sensitization.

The vapor is moderately irritating at 4 ppm, and it is believed that workers would not tolerate 25 ppm for any length of time.[1] Skin sensitization has occurred from industrial exposure; a 4% concentration in petrolatum produced sensitization reactions in 10 of 24 volunteers.[2]

In rats, 2000 ppm for 4 hours was fatal, with death attributed to severe pulmonary irritation; 1000 ppm for 4 hours was not fatal but caused irritation of the skin.[3] Repeated exposure to 500 ppm was fatal to rats, whereas 275 ppm was lethal to rabbits and guinea pigs.[4] Irritation of the eyes, nose, and mouth as well as lethargy, dyspnea, and convulsive movements preceded death. At autopsy, there was pulmonary edema and degenerative changes in liver, kidneys, and heart muscle.

Exposure of pregnant rats to 150 ppm 6 hours/day during days 6 through 15 of gestation caused some maternal toxicity and a slight, but not statistically significant, increase in malformed fetuses; at 50 ppm, there was neither maternal toxicity nor an adverse effect on the fetus.[5]

Chronic exposure of mice and rats to 25, 75, or 225 ppm caused concentration-dependent lesions within the nasal cavity.[6] There was no indication of an oncogenic response in any organ or tissue.[6]

Ethyl acrylate applied to the skin of mice three times per week for life caused dermatitis, dermal fibrosis, epidermal necrosis, and hyperkeratosis; neoplastic changes were not observed.[7]

Ethyl acrylate was carcinogenic in rats and mice when administered by gavage in corn oil, producing tumors of the forestomach.[8] The relevance, if any, of this study to the occupational setting is minimal.

One drop of the liquid instilled in the eye of the rabbit caused corneal necrosis within 24 hours.[3]

The odor is detectable below 1 ppm and should serve as a good warning property.[1,3]

REFERENCES

1. Hygienic Guide Series: Ethyl acrylate. Am Ind Hyg Assoc J 27:571–574, 1966
2. Opdyke DLJ: Monographs on fragrance raw materials, ethyl acrylate. Food Cosmet Toxicol Suppl 13:801–802, 1975
3. Pozzani UC, Weil CS, Carpenter CP: Subacute vapor toxicity and range-finding data for ethyl acrylate. J Ind Hyg Toxicol 31:311–316, 1949
4. Treon JR et al: The toxicity of methyl and ethyl acrylate. J Ind Hyg 31:317–326, 1949
5. Murray JS et al: Teratological evaluation of inhaled ethyl acrylate in rats. Toxicol Appl Pharmacol 60:106–111, 1981
6. Miller RR et al: Chronic toxicity and oncogenicity bioassay of inhaled ethyl acrylate in Fischer 344 rats and B6C3F1 mice. Drug Chem Toxicol 8:1–42, 1985
7. DePass LR et al: Dermal oncogenicity bioassays of acrylic acid, ethyl acrylate and butyl acrylate. J Toxicol Environ Health 14:115–120, 1984
8. National Toxicology Program: Carcinogenesis Studies of Ethyl Acrylate (CAS No 140-88-5) in F344/N Rats and B6C3F Mice (Gavage Studies). Technical Report Series 259, DHHS (NIH) Pub No 87–2515, pp 1–224. Research Triangle Park, North Carolina, National Institutes of Health, 1986

ETHYL ALCOHOL

CAS: 64-17-5

C_2H_5OH 1987 TLV = 1000 ppm

Synonyms: Ethanol; algrain; anhydrol; ethyl hydrate; ethyl hydroxide; grain alcohol

Physical Form. Clear, colorless, mobile, flammable liquid

Uses. Solvent

Exposure. Inhalation; ingestion

Toxicology. Ethyl alcohol is an irritant of the eyes and mucous membranes and causes central nervous system depression.

Exposure of humans at 5,000 to 10,000 ppm has caused transient irritation of the eyes and nose, and cough.[1,2] At 15,000 ppm, effects included continuous lacrimation and cough. A level of 20,000 ppm was judged as just tolerable; above this level, the atmosphere was described as intolerable and suffocating on even brief exposure.[1]

Chronic exposure to the vapor may result in irritation of mucous membranes, headache, and symptoms of central nervous system depression such as lack of concentration and somnolence.[3] However, in current industrial practice, the vapor is considered to be practically devoid of systemic hazard from inhalation.

Ethanol is not appreciably irritating to skin, even with repeated or prolonged exposure.[2]

Splashed in the eye, the liquid causes immediate burning and stinging sensation with reflex closure of the lids and tearing.[4]

Some fetotoxicity has been observed in the offspring of mice treated with large doses of ethanol during gestation.[2]

REFERENCES

1. Lester D, Greenberg LA: The inhalation of ethyl alcohol by man. Q J Stud Alcohol 12:167–168, 1951
2. Rowe VK, McCollister SB: Alcohols. In Patty's Industrial Hygiene and Toxicology, 3rd ed, Vol 2, Toxicology, pp 4541–4556. New York, Wiley–Interscience, 1982
3. Hygienic Guide Series: Ethyl alcohol (ethanol). Am Ind Hyg Assoc J 17:94–95, 1956
4. Grant WM: Toxicology of the Eye, 3rd ed, pp 53–59. Springfield, Illinois, Charles C Thomas, 1986

ETHYLAMINE

CAS: 75-04-7

$C_2H_5NH_2$ 1987 TLV = 10 ppm

Synonyms: Monoethylamine; aminoethane; ethanamine

Physical Form. Liquid (BP 16.6°C)

Uses. In resin chemistry; stabilizer for rubber latex; intermediate for dyestuffs, pharmaceuticals; in oil refining

Exposure. Inhalation

Toxicology. Ethylamine is an irritant of the eyes, mucous membranes, and skin.

Eye irritation and corneal edema in humans have been reported from industrial exposure.[1]

Exposure of rats to 8000 ppm for 4 hours was fatal to two of six animals wtihin 14 days.[2] Rabbits survived exposures to 50 ppm daily for 6 weeks but showed pulmonary irritation and some myocardial degeneration; corneal damage was observed after 2 weeks of exposure.[3] In the rabbit eye, one drop of a 70% solution of ethylamine caused immediate, severe irritation. A 70% solution dropped on the skin of guinea pigs caused prompt skin burns leading to necrosis; when held in contact with guinea pig skin for 24 hours, there was severe skin irritation with extensive necrosis and deep scarring.[1]

The odor is like that of ammonia.

REFERENCES

1. Ethylamine. Documentation of the TLVs for Substances in Workroom Air, 5th ed, p 243. Cincinnati, American Conference of Governmental Industrial Hygienists (ACGIH), 1986
2. Smyth HF Jr et al: Range-finding toxicity data: List V. AMA Arch Ind Hyg Occup Med 10:61–68, 1954
3. Brieger H, Hodes WA: Toxic effects of exposure to vapors of aliphatic amines. AMA Arch Ind Hyg Occup Med 3:287–291, 1951

ETHYL AMYL KETONE

CAS: 541-85-5

$C_2H_5C(O)CH_2(CH_3)CH{=}C_2H_5$

1987 TLV = 25 ppm

Synonyms: 5–Methyl–3–heptanone; ethyl *sec*-amyl ketone

Physical Form. Colorless liquid

Uses. Solvent for resins; organic intermediate

Exposure. Inhalation

Toxicology. Ethyl amyl ketone is an irritant of the eyes and mucous membranes; at very high concentrations, it produces central nervous system depression in animals, and it is expected that severe exposure will cause the same effect in humans.

Humans exposed to 25 ppm experienced irritation of the eyes and respiratory tract and detected a strong odor; at 100 ppm, irritation of mucous membranes, headache, and nausea were too severe to tolerate for more than a few minutes.[1] Eye contact with the liquid causes transient corneal injury. Prolonged or repeated cutaneous contact may lead to drying and cracking of the skin.

Three of six mice and no rats died after a 4-hour exposure to 3000 ppm, whereas exposure to 6000 ppm for 8 hours caused death in all exposed mice and in four of six rats; all animals developed signs of eye and respiratory tract irritation; varying degrees of ataxia, prostration, respiratory distress, and narcosis were observed.[1] Surviving animals recovered with no apparent adverse effects.

REFERENCE

1. Krasavage WJ et al: Ketones. In Clayton GD, Clayton FE (eds): Patty's Industrial Hygiene and Toxicology, 3rd ed, Vol 2C, Toxicology, pp 4767–4768. New York, Wiley–Interscience, 1982

ETHYL BENZENE

CAS: 100-41-4

$C_6H_5(C_2H_5)$ 1987 TLV = 100 ppm

Synonyms: Ethylbenzol; phenylethane

Physical Form. Colorless liquid

Uses. Solvent; antiknock agent; intermediate in the production of styrene

Exposure. Inhalation; skin absorption

Toxicology. Ethyl benzene is an irritant of the skin and mucous membranes; at high concentrations, it causes narcosis in animals, and it is expected that severe exposure will cause the same effect in humans.

Humans exposed briefly to 1000 ppm experienced eye irritation, but tolerance developed rapidly; 2000 ppm caused lacrimation, nasal irritation, and vertigo; 5000 ppm produced intolerable irritation of the eyes and nose.[1] The rate of absorption of ethyl benzene through the skin of the hand and the forearm in human subjects was 22 to 33 $mg/cm^2/hour$.[2]

When chronic exposures exceeded 100 ppm, complaints included fatigue, sleepiness, headache, and mild irritation of the eyes and respiratory tract.[3]

In guinea pigs, exposure to 10,000 ppm caused immediate, intense eye and nose irritation, ataxia, narcosis, and death in 2 to 3 hours; 5,000 ppm was lethal during or after 8 hours of exposure; 2,000 ppm produced ataxia in 8 hours, and 1,000 ppm resulted in eye irritation.[1] Inhalation of ethyl benzene at 600 ppm for 186 days by rats and guinea pigs resulted in slight changes in liver and kidney weights and slight testicular histopathology in rabbits and monkeys.[4] Exposure of rabbits to 230 ppm 4 hours/day for 7 months resulted in changes in blood cholinesterase activity, leukocytosis, reticulocytosis, and dystrophic changes in the liver and kidneys.[5] Ethyl benzene does not cause damage to the hematopoietic system in spite of its chemical similarity to benzene.[6]

Pregnant rats exposed to 100 or 1000 ppm 6 hours/day on days 1 to 19 of gestation had offspring with a significant increase in extra rib formation; at the higher dose, maternal toxicity was indicated by increased liver, kidney, and spleen weights.[7]

Two drops of the liquid in the eyes of a rabbit caused slight conjunctival irritation but no corneal injury.[4] The liquid in contact with the skin of a rabbit caused erythema, exfoliation, and vesiculation.[4]

REFERENCES

1. Yant WP, Schrenk HH, Waite CP, Patty FA: Acute response of guinea pigs to vapors of some new commercial organic compounds. Public Health Rep 45:1241–1250, 1930
2. Dutkiewicz T, Tyras H: A study of the skin absorption of ethylbenzene in man. Br J Ind Med 24:330–332, 1967
3. Bardodej Z, Bardodejova E: Biotransformation of ethylbenzene, styrene and alpha-methylstyrene in man. Am Ind Hyg Assoc J 31:206–209, 1970
4. Wolf MA et al: Toxicological studies of certain alkylated benzenes and benzene. AMA Arch Ind Health 14:387–398, 1956
5. Haley TJ: A review of the literature on ethylbenzene. Dang Prop Ind Mat Rep pp 2–4, July/August 1981
6. Gerarde HW: Toxicological studies on hydrocarbons. AMA Arch Ind Health 13:468, 1956
7. Hardin BD et al: Testing of selected workplace chemicals for teratogenic potential. Scand J Work Environ Health 7(suppl 4):66–75, 1981

ETHYL BROMIDE

CAS: 74-96-4

C_2H_5Br 1987 TLV = 200 ppm

Synonyms: Bromoethane; hydrobromic ether; bromic ether

Physical Form. Colorless liquid

Uses. Ethylating agent in synthesis of pharmaceuticals; refrigerant

Exposure. Inhalation

Toxicology. Ethyl bromide is a respiratory irritant and a hepato- and renal toxin; at high concentrations it causes narcosis.

The former use of ethyl bromide as a human anesthetic (at concentrations approaching 100,000 ppm) produced respiratory irritation and caused some fatalities, either immediately, owing to respiratory or cardiac arrest, or delayed, owing to effects on the liver, kidneys, or heart.[1] At autopsy, findings were pulmonary edema and marked fatty degeneration of the liver, kidneys, and heart.[1] Relatively little experience with this substance in industry has been reported, but exposure of volunteers to 6500 ppm for 5 minutes produced vertigo, slight headache, and mild eye irritation.[2]

Guinea pigs exposed to 50,000 ppm for 98 minutes died within an hour following exposure.[2] Exposure to 24,000 ppm for 30 minutes was fatal within 3 days; at autopsy, findings included pulmonary edema and centrolobular necrosis of the liver; exposure to 3,200 ppm for 9 hours produced lung irritation, and death occurred after 1 to 5 days.

The 1-hour LC_{50} for male rats was 27,000 ppm; for mice, 16,200 ppm.[3]

Applied to the skin of mice, it produced local necrosis.[4] Instilled in rabbit eyes, it was an irritant.

The ether-like odor of ethyl bromide is detectable only at concentrations well above 200 ppm and therefore will not give warning of hazardous concentrations.[4]

REFERENCES

1. von Oettingen WF: The Halogenated Aliphatic, Olefinic, Cyclic, Aromatic, and Aliphatic-Aromatic Hydrocarbons Including the Halogenated Insecticides, Their Toxicity and Potential Dangers. US Public Health Service Pub No 414, pp 134–138. Washington, DC, US Government Printing Office, 1955
2. Sayers RR, Yant WP, Thomas BGH, Berger LB: Physiological response attending exposure to vapors of methyl bromide, methyl chloride, ethyl bromide and ethyl chloride. Public Health Bull 185:1–56, 1929

3. Vernot EH et al: Acute toxicity and skin corrosion data for some organic and inorganic compounds and aqueous solutions. Toxicol Appl Pharmacol 42:417, 1977
4. Hygienic Guide Series: Ethyl bromide. Am Ind Hyg Assoc J 26:192–195, 1978

ETHYL BUTYL KETONE

CAS: 106-35-4

$C_2H_5COC_4H_9$ 1987 TLV = 50 ppm

Synonyms: 3–Heptanone; EBK

Physical Form. Colorless liquid

Uses. Solvent and intermediate for organic materials

Exposure. Inhalation

Toxicology. Ethyl butyl ketone (EBK) is mildly irritating to the skin and eyes of rats and causes narcosis at high concentrations. No effects have been reported in humans.

Rats survived a 4-hour exposure to 2000 ppm, but 4000 for 4 hours was fatal.[1] The oral LD_{50} in rats was 2.76 g/kg, and the LD_{50} for penetration of rabbit skin was greater than 20 ml/kg.[1]

Rats given 1% EBK in drinking water for 120 days showed no signs of neurotoxicity.[2] Exposure of rats at 700 ppm 72 hours/week for 24 weeks was also without neurotoxic effect.[3] Extremely large gavage doses, 2 g/kg/day, 5 days/week for 14 weeks, were required to produce signs of neurotoxicity; two of two rats had hindlimb weakness and tail drag.[4] Neuropathology showed central-peripheral-distal axonopathy characterized by giant axonal swelling and neurofilamentous hyperplasia.[4]

When dropped into rabbit eyes or applied to skin, the liquid has caused mild irritation.[1]

REFERENCES

1. Smyth HF Jr, Carpenter CP, Weil CS: Range-finding toxicity data, List III. J Ind Hyg Toxicol 31:60–62, 1949
2. Homan ER, Maronpot RR: Neurotoxic evaluation of some aliphatic ketones (abstr). Toxicol Appl Pharmacol 45:312, 1978
3. Katz GV et al: Comparative neurotoxicity and metabolism of ethyl n-butyl ketone and methyl n-butyl ketone in rats. Toxicol Appl Pharmacol 52:153–158, 1980
4. O'Donoghue JL et al: Further studies on ketone neurotoxicity and interactions. Toxicol Appl Pharmacol 72:201–209, 1984

ETHYL CHLORIDE

CAS: 75-00-3

CH_3CH_2Cl 1987 TLV = 1000 ppm

Synonyms: Chloroethane; monochloroethane; hydrochloric ether; muriatic ether

Physical Form. Colorless liquid

Uses. Chemical intermediate; refrigerant; manufacture of ethyl cellulose; formerly used as an inhalation anesthetic agent

Exposure. Inhalation; skin absorption

Toxicology. At high concentrations, ethyl chloride causes central nervous system depression.

Inhalation of 40,000 ppm by human subjects caused dizziness, eye irritation, and abdominal cramps, whereas inhalation of 25,000 ppm caused incoordination.[1] Exposure to 19,000 ppm resulted in mild analgesia after 12 minutes, and 13,000 ppm caused slight symptoms of inebriation.[2] Sudden and unforeseen fatalities from ethyl chloride anesthesia have been reported.[2]

Chronic effects from industrial exposure have not been reported, although skin absorption is said to occur. In liquid form, this substance may cause frostbite.

Guinea pigs exposed to 40,000 ppm appeared uncoordinated in 3 minutes, had eye irritation, and were unable to stand after 40 minutes; some animals died from exposure for 9 hours, but exposure for 4.5 hours was nonfatal; histopathologic changes in the lungs, liver, and kidneys were observed in sacrificed animals of the latter group.[1]

Two-week repeated exposure of rats and dogs to 4,000 or 10,000 ppm caused no treatment-related effects except for slight increases in liver-to-bodyweight ratios in male rats.[3]

REFERENCES

1. Sayers RR, Yant WP, Thomas BH, Burger LB: Physiological response attending exposure to vapors of methyl bromide, methyl chloride, ethyl bromide and ethyl chloride. Public Health Bull 185:1–56, 1929
2. von Oettingen WF: The Halogenated Aliphatic, Olefinic, Cyclic, Aromatic, Aliphatic-Aromatic Hydrocarbons Including the Halogenated Insecticides, Their Toxicity and Potential Dangers. US Public Service Pub No 414, pp 128–134. Washington, DC, US Government Printing Office, 1955
3. Landry TD et al: Ethyl chloride: A two-week inhalation toxicity study and effects on liver non-protein sulfhydryl concentrations. Fund Appl Toxicol 2:230–234, 1982

ETHYLENE

CAS: 74-85-1

$H_2C{=}CH_2$ 1987 TLV = none established; simple asphyxiant

Synonyms: Ethene; acetene; bicarburetted hydrogen; olefiant gas

Physical Form. Colorless gas

Uses. Manufacture of polyethylene, ethylene oxide, ethylene dichloride, and ethyl benzene

Exposure. Inhalation

Toxicology. Ethylene is of low toxicity and conceivably could be a simple asphyxiant at extraordinarily high doses.

Ethylene inhaled at 11.5 g/m^3 (10,000 ppm) for 4 hours is hepatotoxic in rats pretreated with the polychlorinated biphenyl, Arochlor 1254, given orally to induce liver enzymes at a dose of 300 μmol/kg daily for 3 days. It is not toxic without such treatment.[1,2]

Male mice exposed to 19.6 mg/m^3 (17 ppm) C14-labeled ethylene metabolized it to ethylene oxide.[2]

REFERENCES

1. Connolly RB, Jaeger RJ: Acute hepatotoxicity of ethylene and halogenated ethylenes after PCB pretreatment. Environ Health Perspect 21:131, 1977
2. IARC Monographs on the Evaluation of the Carcinogenic Risk of Chemicals to Man, Vol 11, Cadmium, Nickel, Some Epoxides, Miscellaneous Industrial Chemicals and General Considerations on Volatile Anesthetics, p 157. Lyon, International Agency for Research on Cancer, 1976

ETHYLENE CHLOROHYDRIN

CAS: 107-07-3 1987 TLV = C 1 ppm; skin

CH_2OHCH_2Cl

Synonyms: beta-Chloroethyl alcohol; glycol chlorohydrin; 2–chloroethanol

Physical Form. Colorless liquid

Uses. Production of ethylene glycol and ethylene oxide; solvent for cellulose acetate, cellulose ethers, and various resins

Exposure. Inhalation; skin absorption

Toxicology. Ethylene chlorohydrin is an irritant and is toxic to the liver, kidneys, and central nervous system; it is rapidly absorbed through the skin.

Several human fatalities have resulted from inhalation, dermal contact, or ingestion of ethylene chlorohydrin. Typically, neurotoxic symptoms were described and death was attributed to cardiac and respiratory collapse.[1] One fatality was caused by exposure to an estimated 300 ppm for 2.25 hours.[2] In another fatal case, autopsy revealed pulmonary edema and damage to the liver, kidneys, and brain.[2]

Exposure to the vapor has caused irritation of the eyes, nose, and throat; visual disturbances; vertigo; incoordination, paresthesias; and nausea and vomiting.[2,3] More severe exposure has also caused headache, severe thirst, delirium, low blood pressure, cyanosis, collapse, shock, and coma.[2,3] In some cases, there have been albumin, casts, and red blood cells in the urine.[2,3]

Ethylene chlorohydrin is highly irritating to mucous membranes but produces little reaction upon contact with rabbit skin.[1] Toxic amounts can be absorbed through the skin without causing dermal irritation; the dermal LD_{50} for rabbits is 68 mg/kg.[4] This value extrapolated to humans suggests that a volume slightly more than a teaspoon could be lethal with prolonged contact.[4]

The liquid instilled in rabbit eyes caused moderately severe injury, but human eyes have recovered from corneal burns within 48 hours.[5]

Inhalation exposures of 15 minutes a day at concentrations of approximately 1000 ppm were fatal to rats within a few days.[6]

Significant levels of fetotoxicity and maternal toxicity but no teratogenicity were found in rabbits administered 36 mg/kg/day intravenously.[7]

In 2-year dermal studies, there was no evidence of carcinogenicity in rats given 50 or 100 mg/kg/day or mice given 7.5 or 15 mg per animal per day.[1]

Skin contact is particularly hazardous, because the absence of signs of immediate irritation prevents any warning when the skin is wetted by the substance.[3]

REFERENCES

1. National Toxicology Program: Toxicology and Carcinogenesis Studies of 2-Chloroethanol (Ethylene Chlorohydrin) (CAS NO 107-07-3) in F344/N Rats and Swiss CD-1 Mice (Dermal Studies). DHHS (NTP) TR 275, p 194. Washington, DC, US Government Printing Office, November 1985
2. Bush AF, Abrams HK, Brown HV: Fatality and illness caused by ethylene chlorohydrin in an agricultural occupation. J Ind Hyg Toxicol 26:352–358, 1949
3. Hygienic Guide Series: Ethylene chlorohydrin. Am Ind Hyg Assoc J 22:513–515, 1961
4. Lawrence W et al: Toxicity of ethylene chlorohydrin. I. Acute toxicity studies. J Pharm Sci 60:568–571, 1971
5. Grant WM: Toxicology of the Eye, 2nd ed, pp 266–267. Springfield, Illinois, Charles C Thomas, 1974
6. Goldblatt M, Chiesman W: Toxic effects of ethylene chlorohydrin. Part I. Clinical. Br J Ind Med 1:207–223, 1944
7. Research Triangle Institute: Teratologic Evaluation of Ethylene Chlorohydrin (CAS No 107-07-3) in New Zealand White Rabbits. Final Report. Washington, DC, National Institute of Environmental Health Sciences, 1983

ETHYLENEDIAMINE

CAS: 107-15-3

$NH_2CH_2CH_2NH_2$ 1987 TLV = 10 ppm

Synonyms: 1,2–Diaminomethane

Physical Form. Colorless, hygroscopic, fuming liquid

Uses. Catalytic agent in epoxy reins; dyes; solvent stabilizer; neutralizer in rubber products

Exposure. Inhalation

Toxicology. Ethylenediamine is a potent skin sensitizer and is an irritant of the eyes and mucous membranes.

In human subjects, inhalation of 400 ppm for 5 to 10 seconds caused intolerable nasal irritation; 200 ppm caused tingling of the face and slight nasal irritation; 100 ppm was inoffensive.[1]

Most of the information regarding the skin sensitization potential of ethylenediamine has come from its use as a stabilizer in pharmaceutical preparations, especially in Mycolog cream, in which it has caused many cases of sensitization.[2,3] Results of skin patch tests conducted between 1972 and 1974 showed that 6% of the 3216 patients tested exhibited sensitivity to 1% ethylenediamine-HCL solution.[4] Although ethylenediamine is a potent sensitizer, industrial exposure rarely leads to sensitization and dermatitis because exposure is not prolonged or intimate, and normal skin is usually involved.[2] In clinical practice, the ethylenediamine in Mycolog cream is often applied to damaged skin, which is more readily sensitized than the relatively normal skin of most industrial workers.[2]

In a case of asthma resulting from ethylenediamine exposure in a 30-year-old male, initial symptoms of sneezing, nasal discharge, and productive cough began 2.5 years after employment and progressed during the following 5 months. An inhalation provocation test with ethylenediamine produced chest tightness, cough, wheezing, and a 26% reduction in FEV_1 4 hours postexposure. The reaction was reproducible on a different day and was specific; a similar reaction was not demonstrated with other chemicals to which the subject was exposed.[5]

Exposure of rats to 4000 ppm for 8 hours was uniformly fatal, whereas 2000 ppm was not lethal.[6] Rats exposed daily for 30 days to 484 ppm did not survive; injury to lungs, liver, and kidneys was observed; at 132 ppm, there was no mortality.[1]

No reproductive toxicity was found in rats exposed to 0.50 g/kg/day for two generations.[7] A reduction in body-weight gain and changes in liver and kidney weights were observed in the F_0 and F_1 parent rats. A microscopic liver lesion occurred in the F_1 rats, with a greater prevalence and severity in the female rats. Ethylenediamine was not genotoxic in a variety of in vivo and in vitro tests, nor was it carcinogenic in lifetime skin painting studies in mice.[8]

In the eye of a rabbit, the liquid caused extreme irritation and corneal damage; partial corneal opacity was produced by a 5% solution.[6] The undiluted liquid applied to the shaved skin of rabbits and left uncovered produced severe irritation and necrosis.[6]

REFERENCES

1. Pozzani UC, Carpenter CP: Response of rats to repeated inhalation of ethylenediamine vapors. AMA Arch Ind Hyg Occup Med 9:223–226, 1954
2. Fisher AA: Contact Dermatitis, 2nd ed, pp 40–41. Philadelphia, Lea & Febiger, 1973
3. Baer R, Ramsey DL, Biondi E: The most common contact allergens. Arch Dermatol 108:74–78, 1973
4. North American Contact Dermatitis Group: The frequency of contact sensitivity in North America 1972–74. Contact Dermatitis 1:277–280, 1975
5. Lam S, Chan–Yeung M: Ethylenediamine-induced asthma. Am Rev Respir Dis 121:151–155, 1980
6. Smyth HF Jr et al: Range-finding toxicity data: List IV. AMA Arch Ind Hyg Occup Med 4:119–122, 1951
7. Yang RSH et al: Two-generation reproduction study of ethylenediamine in Fischer 344 rats. Fund Appl Toxicol 4:539–546, 1984
8. Slesinski RS et al: Assessment of genotoxic potential of ethylenediamine: In vitro and in vivo studies. Mutat Res 124:299–314, 1983

ETHYLENE DIBROMIDE

CAS: 106-93-4

CH_2BrCH_2Br TLV = none stated; suspect of carcinogenic potential for humans; skin

Synonyms: 1,2 Dibromethane; EDB

Physical Form. Clear liquid

Uses. A fumigant; in gasoline as a lead scavenger; chemical intermediate in the industrial synthesis of other brominated compounds

Exposure. Inhalation; skin absorption

Toxicology. Ethylene dibromide (EDB) is a severe mucous membrane, eye, and skin irritant. It causes liver and kidney damage and is carcinogenic in experimental animals.

In an early report, accidental use of EDB as a human anesthetic produced general weakness, vomiting, diarrhea, chest pain, cough, shortness of breath, cardiac insufficiency, and uterine hemorrhaging.[1] Death occurred 44 hours after inhalation. Postmortem examination showed upper respiratory tract irritation; swelling of the pulmonary lymph glands; advanced states of parenchymatous degeneration of the heart, liver, and kidneys; and hemorrhages in the respiratory tract.

Recently, two workers collapsed while inside a tank that was later found to contain a 0.1% to 0.3% EDB solution.[2] Removed after being in the tank 20 to 45 minutes, one man was intermittently comatose; the other man was delirious and combative. Both men experienced vomiting, diarrhea, abdominal pain, and burning of the eyes and throat. Metabolic acidosis and acute renal and hepatic failure ensued. Death occurred 12 and 64 hours later, respectively, despite supportive measures.

Skin contact produces intensive burning pain preceding hyperemia that develops into blisters. Skin sensitization has been reported.[1]

Acute exposure of experimental animals resulted in adverse effects similar to those described for humans. Rats did not survive for longer than 6 minutes when exposed to the vapor at 3000 ppm; minimal lethal concentration for an 8-hour exposure was 200 ppm; these exposures caused hepatic necrosis, pulmonary edema, and cloudy swelling of renal tubules.[3] Depression of the central nervous system was observed in rats exposed at higher concentrations, and deaths occurred within 24 hours from respiratory or cardiac failure. At lower concentrations, deaths owing to pneumonia occurred as a result of injury to the lungs and was delayed for up to 12 days postexposure.

Four species of animals tolerated daily inhalation of 25 ppm for 6 months without adverse effects.[3]

Application of a 10% solution or the undiluted liquid to rabbit skin caused marked central nervous system depression and death within 24 hours.[3] A dermal LD_{50} of 400 mg/kg was estimated.[1]

An increased incidence of skin carcinomas and lung tumors has been found in mice receiving repeated skin applications.[4] Rats and mice chronically exposed to 10 or 40 ppm had an increased incidence of a variety of tumors.[5] Animal studies have shown increased toxic effects when EDB is administered with Disulfiram, a widely used drug in alcoholism control programs.[6] Testicular atrophy was found in 90% of the male rats receiving a combination dose.

Other reported reproductive effects include abnormal spermatozoa and decreased spermatozoic concentration in bulls fed EDB. Intraperitoneal injection of 10 mg/kg for 5 days to male rats caused a decrease in average litter size in females mated after 3 weeks of exposure and in no litters at 4 weeks. Continuous exposure to 32 ppm during gestation caused minor skeletal anomalies in rats and mice.[9]

Human epidemiologic studies to observe carcinogenic effects are inconclusive owing to small cohort size, incomplete exposure data, and insufficient latencies.[10] No adverse effects were found in sperm counts of 50 workers exposed to less than 5.0 ppm.[11] A safe level of exposure has not been determined.

REFERENCES

1. National Institute for Occupational Safety and Health: Criteria for a Recommended Standard . . . Occupational Exposure to Ethylene Dibromide. DHEW (NIOSH) Pub No 77–221. Washington, DC, US Government Printing Office, 1977
2. Letz GA et al: Two fatalities after acute occupational exposure to ethylene dibromide. JAMA 252:2428–2431, 1984
3. Rowe VK et al: Toxicity of ethylene dibromide determined on experimental animals. AMA Arch Ind Hyg Occup Med 6:158–173, 1952
4. Van Duuren BL et al: Carcinogenicity of halogenated olefinic and aliphatic hydrocarbons in mice. J Natl Cancer Inst 63:1433–1439, 1979
5. National Cancer Institute: Carcinogenesis Bioassay of 1,2–Dicromoethane (Inhalation Study), TR–210 (CAS No 106-93-4), Carcinogenesis Testing Program. DHHS (NIH) Pub No 81–1766. Washington, DC, US Government Printing Office, 1981
6. Wong LCK et al: Carcinogenicity and toxicity of 1,2–dibromoethane in the rat. Toxicol Appl Pharmacol 63:155–165, 1982
7. Amir D, Colcani R: Effect of dietary ethylene dibromide on bull semen. Nature 206:99–100, 1965
8. Edwards K et al: Studies with alkylating esters—II. A chemical interpretation through metabolic studies of the infertility effects of ethylene dimethanesulphonate and ethylene dibromide. Biochem Pharmacol 19:1783–1789, 1970
9. Short RD Jr et al: Toxicity Studies of Selected Chemicals, Task I—The Developmental Toxicity of Ethylene Dibromide Inhaled by Rats and Mice During Organogenesis. Report No EPA–560/6–76–018. Washington, DC, US Environmental Protection Agency, Office of Toxic Substances, 1976
10. Ott MG et al: Mortality experience of 161 employees exposed to ethylene dibromide in two production units. Br J Ind Med 37:163–168, 1980
11. Ter Haar G: An Investigation of Possible Sterility and Health Effects from Exposure to Ethylene Dibromide. In Banbury Report 5—Ethylene Dichloride: A Potential Health Risk? pp 167–188. Cold Spring Harbor, New York, Cold Spring Harbor Laboratory, 1980

ETHYLENE DICHLORIDE

CAS: 107-06-2

CH_2ClCH_2Cl 1987 TLV = 10 ppm

Synonyms: 1,2–Dichlorethane; dichloroethane; ethylene chloride

Physical Form. Colorless liquid

Uses. Manufacture of vinyl chloride; antiknock agent; fumigant; insecticide; degreaser compounds; rubber cements

Exposure. Inhalation; ingestion

Toxicology. Ethylene dichloride is a central nervous system depressant and causes injury to the liver and kidneys.

In a recent fatality, exposure to concentrated vapor in a tank for 30 minutes caused drowsiness, nausea, and respiratory distress; coma developed 20 hours after initial exposure.[1] Serum levels of lactate and ammonia were increased, followed by elevation of glutamic transaminases, lactic dehydrogenase, and creatine phosphokinase. Ornithine carbamyl transferase and glutamic oxaloacetic transaminase of mitochondrial origin were remarkably high. Multiple organ failure developed, and the patient died in cardiac arrhythmia on the fifth day. At autopsy, the lungs were severely congested and edematous; diffuse degenerative changes of the myocardium, extensive centrilobular necrosis of the liver, and acute tubular necrosis of the kidneys were noted.

Workers exposed to 10 to 200 ppm complained of lacrimation, dizziness, insomnia, vomiting, constipation, and anorexia; liver tenderness upon palpation, epigastric pain, and elevated urobilinogen were observed.[2] Impairment of the central nervous system and increased morbidity, especially diseases of the liver and bile ducts, were found in workers chronically exposed to ethylene dichloride at concentrations below 40 ppm and averaging 10 to 15 ppm.[2]

Ingestion of quantities estimated at between 8 and 200 ml have been reported to be lethal, with a toxic response similar to that of cases of inhalation.[3,4]

Eye contact with either the liquid or with high concentrations of vapor causes immediate discomfort with conjunctival hyperemia and slight corneal injury; corneal burns from splashes recover quickly with no scarring. Prolonged skin exposure, as from contact with soaked clothing, produces severe irritation, moderate edema, and necrosis; systemic effects may ensue, as the liquid is readily absorbed through the skin.[2]

Animal studies indicate that ethylene dichloride has little ability to adversely affect the reproductive or developmental processes except at maternally toxic levels.[4]

In rats, chronic administration by gavage of 95 or 47 mg/kg bodyweight per day for 78 weeks caused a significant increase in hemangiosarcomas of the circulatory system in rats.[5] Squamous cell carcinomas of the forestomach were significantly increased in male rats, and high-dose females had increased incidences of mammary gland adenocarcinomas and fibroadenomas. A variety of tumors have been similarly induced in mice.[5] Intraperitoneal and inhalation studies in animals have not shown a significant carcinogenic response.

The IARC has determined that there is sufficient evidence of carcinogenicity in mice and rats, and, in the absence of adequate human data, ethylene dichloride should be regarded as presenting a carcinogenic risk to humans.[6]

Most subjects could detect ethylene dichloride at a concentration of 6 ppm.[1]

REFERENCES

1. Nouchi T et al: Fatal intoxication by 1,2–dichloroethane—a case report. Int Arch Occup Environ Health 54:111–113, 1984
2. National Institute for Occupational Safety and Health: Criteria for a Recommended Standard . . . Occupational Exposure to Ethylene Dichloride (1,2–dichloroethane). DHEW (NIOSH) Pub No 76–139. Washington, DC, US Government Printing Office, 1976
3. Yodaiken RE, Babcock JR: 1,2–Dichloroethane poisoning. Arch Environ Health 26:281–284, 1973
4. Health Assessment Document for 1,2–Dichloroethane (Ethylene Dichloride). Final Report—EPA/600/8–84/006F. Washington, DC, US Environmental Protection Agency, Office of Toxic Substances, September 1985
5. National Cancer Institute: Bioassay of 1,2–Dichloroethane for Possible Carcinogenicity, TR-55. DHEW (NIH) Pub No 78–1361. Washington, DC, US Government Printing Office, 1978
6. IARC Monographs on the Evaluation of the Carcinogenic Risk of Chemicals to Humans, Vol 20, pp 422–448. Lyon, International Agency for Research on Cancer, 1979

ETHYLENE GLYCOL

CAS: 107-21-1

CH_2OHCH_2OH 1987 TLV = C 50 ppm

Synonyms: 1,2 Dihydroxyethane; 1,2–ethanediol; ethylene alcohol; ethylene dihydrate

Physical Form. Clear, colorless, thick liquid

Uses. Antifreeze; solvent

Exposure. Inhalation; ingestion; eye contact

Toxicology. Ethylene glycol aerosol causes irritation of the upper respiratory tract; ingestion can cause central nervous system depression, severe metabolic acidosis, liver and kidney damage, and pulmonary edema.

Inhalation is not usually a hazard because the low vapor pressure precludes excessive vapor exposure. Exposure to the vapor from the liquid heated to 100°C has been reported to cause nystagmus and coma of 5- to 10-minute duration.[1] Human volunteers exposed to an aerosol of 12 ppm for 20 to 22 hours/day for 4 weeks complained of throat irritation and headache.[2] At 56 ppm, there was more pronounced irritation of the upper respiratory tract, and, at 80 ppm of aerosol, the irritation and cough were intolerable.

The chief risk associated with ethylene glycol is ingestion of large quantities in a single dose. Several metabolites are responsible for the clinical syndrome, which can be divided into three stages.[3] During the first 12 hours, central nervous system manifestations predominate. If the intoxication is mild, the patient appears to be drunk but without the breath odor of alcohol. In more severe cases, there will be convulsions and coma. Other signs may include nystagmus, ophthalmoplegia, papilledema, depressed reflexes, and tetanic convulsions. The central nervous system manifestations are related to the aldehyde metabolites of ethylene glycol, which reach their maximal concentrations 6 to 12 hours after ingestion.

In the second stage, cardiopulmonary symptoms become prominent; these consist of mild hypertension, tachypnea, and tachycardia. Widespread capillary damage is assumed to be the primary lesion. If the patient survives the first two stages, renal complications may be expected at 24 to 72 hours postingestion. Albuminuria and hematuria are common findings, and oxalate crystals are excreted in the urine. Glycoaldehyde, glycolic acid, and glyoxylate are the putative agents for kidney damage.[3]

The most significant laboratory findings in ethylene glycol intoxication are severe metabolic acidosis from the accumulation of glycolate and the presence of high anion gap.[3] Low arterial *p*H and low bicarbonate levels are often observed. Nonspecific findings include leukocytosis and increased amounts of protein in the cerebrospinal fluid. Chelation by calcium oxalate may cause hypocalcemia, which, when severe enough, can lead to tetany and cardiac dysfunction.[3] The minimal lethal dose is on the order of 100 ml in adults, although it has been reported that much higher doses have been survived.

The effects of the liquid in the eyes of rabbits are immediate signs of moderate discomfort with mild conjunctivitis but no significant corneal damage.[5] In one human incident of a splash in the eye, there was reversible conjunctival inflammation.[6] The liquid produces no significant irritant action on the skin.

Ethylene glycol is of historic interest in that the passage of the 1935 Federal Food, Drug, and Cosmetic Act was precipitated by the deaths of many persons who had ingested elixir of sulfanilamide, which contained ethylene glycol. The 1938 law required that all drugs entering the marketplace after that date be shown to be safe for human use before they could be marketed.

Ethylene glycol was found to be nonmutagenic in the salmonella/microsome mutagenicity test.[7]

REFERENCES

1. Troisis FM: Chronic intoxication by ethylene glycol vapour. Br J Ind Med 7:65, 1950
2. Wills JH et al: Inhalation of aerosolized ethylene glycol by man. Clin Toxicol 7:463, 1974
3. Linnanvuo–Laitinen M, Huttunen K: Ethylene glycol intoxication. Clin Toxicol 24:167–174, 1986
4. Parry MF, Wollach R: Ethylene glycol poisoning. Am J Med 57:143, 1974
5. McDonald TO, Roberts MD, Borgman AR: Ocular toxicity of ethylene chlorohydrin and ethylene glycol in rabbits eyes. Toxicol Appl Pharmacol 21:143, 1972
6. Sykowsky P: Ethylene glycol toxicity. Am J Ophthalmol 34:1599, 1951
7. McCann J, Choi E, Yamasaka E, Ames BN: Detection of carcinogens as mutagens in salmonella/microsome test: Assay of 300 chemicals. Proc Natl Acad Sci 72:5135, 1975

ETHYLENE GLYCOL DINITRATE

CAS: 628-96-6 — 1987 TLV = 0.05 ppm; skin

$CH_2NO_3CH_2NO_3$

Synonyms: EGDN; nitroglycol

Physical Form. Liquid

Uses. Explosive; mixed with nitroglycerin (NG) in the manufacture of dynamite

Exposure. Inhalation; skin absorption. Data on toxic effects are reported chiefly from industrial exposures to EGDN–NG mixed vapors.

Toxicology. Ethylene glycol dinitrate (EGDN) causes vasodilatation and may produce low levels of methemoglobin and the formation of Heinz bodies in erythrocytes.

Intoxication results in a characteristic intense throbbing headache, presumably from cerebral vasodilatation, often associated with dizziness and nausea, and occasionally with vomiting and abdominal pain.[1,2] More severe exposure also causes hypotension, flushing, palpitation, low levels of methemoglobinemia, delirium, and depression of the central nervous system. Aggravation of these symptoms after alcohol ingestion has been observed. Upon repeated exposure, a tolerance to headache develops but is usually lost after a few days without exposure. At times, persistent tachycardia, diastolic hypertension, and reduced pulse pressure have been observed. Rarely, a worker may have an attack of angina pectoris a few days after cessation of repeated exposures, a manifestation of cardiac ischemia. Sudden death owing to unheralded cardiac arrest has also been reported under these circumstances.[3]

Volunteers exposed to the vapor of a mixture of EGDN and nitroglycerin (NG) at a combined concentration of 2 mg/m^3 experienced headache and fall in blood pressure within 3 minutes of exposure; a mean concentration of 0.7 mg/m^3 for 25 minutes also produced lowered blood pressure and slight headache.[4]

REFERENCES

1. Trainor DC, Jones RC: Headaches in explosive magazine workers. Arch Environ Health 12:231–234, 1966
2. Einert CE et al: Exposure to mixtures of nitroglycerin and ethylene glycol dinitrate. Am Ind Hyg Assoc J 24:435–447, 1963
3. Carmichael P, Lieben J: Sudden death in explosives workers. Arch Environ Health 7:424–439, 1963
4. Lund RP, Haggendal J, Johnsson G: Withdrawal symptoms in workers exposed to nitroglycerin. Br J Ind Med 25:136–138, 1968

ETHYLENE GLYCOL MONOBUTYL ETHER

CAS: 111-76-2

$C_4H_9OCH_2CH_2OH$ 1987 TLV = 25 ppm

Synonyms: Butyl Cellosolve; 2–butoxy ethanol; EGBE

Physical Form. Liquid

Uses. Solvent

Exposure. Inhalation; skin absorption

Toxicology. Ethylene glycol monobutyl ether (EGBE) is an irritant of the eyes and mucous membranes, and, in animals, it is a hemolytic agent.

Exposure of humans to high concentrations (300 to 600 ppm) of the vapor for several hours would be expected to cause respiratory and eye irritation, narcosis, and damage to the kidney and liver.[1]

Human subjects exposed to 195 ppm for 8 hours experienced discomfort of the eyes, nose, and throat, although there were no objective signs of injury and no increase in erythrocytic fragility.[2] Similar symptoms occurred at 113 ppm for 4 hours.[2] No clinical signs of adverse effects nor subjective complaints occurred among seven male volunteers exposed to 20 ppm for 2 hours.[3]

The 4-hour LC_{50} values are 486 ppm for male rats and 450 ppm for female rats; toxic effects included narcosis, respiratory difficulty, and kidney damage.[4] Acute or prompt deaths are likely to be due to the narcotic effects of the substance, whereas delayed deaths are usually attributable to congested lungs and severely damaged kidneys. In a 9-day study, rats exposed to 245 ppm 6 hours/day had significant depression of red blood cell count and hemoglobin, with increases in nucleated erythrocytes, reticulocytes, and lymphocytes.[4] Decreased bodyweight gains and increased liver weights were also found. Toxic effects showed substantial reversal 14 days postexposure. In a 90-day study, only mild hematologic alterations were observed in rats exposed to 77 ppm 30 hours/week. The mechanism for EGBE-induced red blood cell depression in rats is unknown, but acid metabolites may be involved.[3,4] There appears to be strikingly different hematologic effects among different species; differences in metabolism are probably responsible for this.[4] It has been suggested that the hematologic effects are of lesser consequence in humans than in rodents, since acute exposures of 200 ppm produced no alterations in erythrocyte fragility.[2]

EGBE appears to be less hazardous than other monoalkyl ethers of ethylene glycol with regard to reproductive effects.[4] Mice treated orally with 1000 mg/kg for 5 weeks had no change in absolute or relative testis weights.[5] Exposure of pregnant rats at 100 ppm or rabbits at 200 ppm during organogenesis resulted in maternal toxicity and embryotoxicity, including decreased number of viable implantations per litter.[6] Slight fetotoxicity in the form of poorly ossified or unossified skeletal elements was also observed in rats. Teratogenic effects were not observed in either species.[6]

Daily skin application to rabbits of 150 mg/kg as a 43.8% aqueous solution for 13 weeks caused no adverse effects.[7] The LD_{50} for rabbits was 0.45 ml/kg (0.40 g/kg) when confined to the skin for 24 hours.[1] It is not significantly irritating to the skin; instilled directly into the eye it produces pain, conjunctival irritation, and transient corneal injury.[1]

REFERENCES

1. Rowe VK, Wolf MA: Derivatives of glycols. In Clayton GD, Clayton FE (eds): Patty's Industrial Hygiene and Toxicology, 3rd ed, rev, Vol

2C, Toxicology, pp 3931–3939. New York Wiley–Interscience, 1982
2. Carpenter CP et al: The toxicity of butyl cellosolve solvent. AMA Arch Ind Health 14:114–131, 1956
3. Johanson G et al: Toxicokinetics of inhaled 2-butoxyethanol (ethylene glycol monobutyl ether) in man. Scand J Work Environ Health 12:594–602, 1986
4. Dodd DE et al: Ethylene glycol monobutyl ether: Acute 9-day, and 90-day vapor inhalation studies in Fischer 344 rats. Toxicol Appl Pharmacol 68:405–414, 1983
5. Nagano K et al: Testicular atrophy of mice induced by ethylene glycol mono alkyl ether. Japan J Ind Health 21:29–35, 1979
6. Glycol Ethers Program Panel Research Status Report, p 5. Washington, DC, Chemical Manufacturers Association, April 22, 1985

ETHYLENE OXIDE

CAS: 75-21-8

C_2H_4O 1987 TLV = 1 ppm

Synonyms: 1,2–Epoxyethane; oxirane; dimethylene oxide

Physical Form. Colorless gas liquefying below 12°C

Uses. Fumigant; sterilizing agent; agricultural fungicide; reagent in organic chemical synthesis

Exposure. Inhalation

Toxicology. Ethylene oxide is an irritant of the eyes, respiratory tract, and skin; at high concentrations, it causes central nervous system depression; it is carcinogenic in female mice.

In humans, early symptoms include irritation of the eyes, nose, and throat and a peculiar taste; effects that may be delayed are headache, nausea, vomiting, dyspnea, cyanosis, pulmonary edema, drowsiness, weakness, and incoordination.[1]

Contact of solutions of ethylene oxide with the skin of human volunteers caused characteristic burns; after a latent period of 1 to 5 hours, effects were edema and erythema and progression to vesiculation with a tendency to coalescence into blebs, and desquamation. Complete healing without treatment usually occurred within 21 days, with, in some cases, residual brown pigmentation. Application of the liquid to the skin caused frostbite; three of the eight volunteers became sensitized to ethylene oxide solutions.[2] The undiluted liquid or solutions may cause severe eye irritation or damage.

Exposure of several species of animals to concentrations calculated to be greater than 1000 ppm for 2 hours caused lacrimation and nasal discharge followed by gasping and labored breathing; corneal opacity was observed in guinea pigs. Delayed effects occurred after several days and included vomiting, diarrhea, dyspnea, pulmonary edema, paralysis of hindquarters, convulsions, and death; at autopsy, findings included degenerative changes of the lungs, liver, and kidneys.[3] The LC_{50} for mice exposed 4 hours was 835 ppm.[4]

A number of cases of subacute sensory motor polyneuropathy have been described among sterilizing workers exposed to ethylene oxide.[5,6] Findings included weakness with bilateral foot drop, sensory loss, loss of reflexes, and neuropathologic changes on electromyography (EMG) in the lower extremities. In some cases, sural nerve biopsy showed axonal degeneration. Removal from exposure resulted in resolution of symptoms in 1 to 7 months. The abnormalities have been consistent with a distal "dying-back" axonopathy with secondary demyelination similar to that seen with other peripheral neurotoxins, such as n-hexane.[5] An animal model of distal axonal degeneration with pathologic confirmation has been described in rats exposed to 500 ppm ethylene oxide three times per week for 13 weeks.[7]

Two studies of workers exposed to ethylene oxide revealed increased rates of leukemia. In one study, two cases of leukemia (0.14 expected) and three cases of stomach cancers (0.4 expected) were observed. In the other study, three cases of leukemia (0.2 expected) were found. Since these workers had exposures to other potential carcinogens, the findings cannot, with certainty, be linked to ethylene oxide. A third study did not find an increased risk of leukemia or other tumors. The IARC has concluded that the available data are inadequate to determine the carcinogenicity of ethylene oxide to humans.[8]

In a chronic inhalation bioassay in rats exposed for 6 hours/day, 5 days/week for 2 years to 100, 33, or 10 ppm ethylene oxide, there was a dose-related increased occurrence of mononuclear cell leukemia in both sexes at all concentrations. There was also an increased occurrence of primary brain tumors at 100 and 33 ppm in both sexes and peritoneal mesotheliomas arising from the testicular serosa at 100 and 33 ppm in male rats.[9]

Hospital staff exposed to ethylene oxide in sterilizing operations during pregnancy were found to have a higher frequency of spontaneous abortions (16.7%) compared with a control group (5.6%) according to a questionnaire study. Analysis of a hospital discharge register confirmed the findings. The association persisted after analysis for potential confounding factors, such as age and smoking status.[10] There is animal evidence of adverse reproductive effects, including decreased fertility and reduced sperm count and motility in males, and increased fetal losses and malformed fetuses in females.[11] There is evidence of increased occurrence or chromosomal aberrations and sister chromatid exchanges (SCE) among workers exposed to ethylene oxide.[12,13] Ethylene oxide is active in a number of short-term tests for mutagenicity and is an alkylating agent that will bind irreversibly with DNA.[13]

Ethylene oxide has a characteristic ether-like odor.

REFERENCES

1. Glaser ZR: Special Occupational Hazard Review with Central Recommendations for the Use of Ethylene Oxide as a Sterilant in Medical Facilities. DHEW (NIOSH) Pub No 77–200. Washington, DC, US Government Printing Office, 1977
2. Sexton RJ, Henson EV: Experimental ethylene oxide human skin injuries. AMA Arch Ind Hyg Occup Med 2:549–564, 1950
3. Hollingsworth RL et al: Toxicity of ethylene oxide determined on experimental animals. AMA Arch Ind Health 13:217, 1956
4. Jacobson KH, Hackley EB, Feinsilver L: The toxicity of inhaled ethylene oxide and propylene oxide vapors. AMA Arch Ind Health 13:237–244, 1956
5. Finelli P et al: Ethylene oxide-induced polyneuropathy: A clinical and electrophysiologic study. Arch Neurol 40:419–421, 1983
6. Kuzuhara S: Ethylene oxide polyneuropathy: Report of 2 cases with biopsy studies of nerve and muscle. Clin Neurol 22:707–713, 1982
7. Ohnishi A et al: Ethylene oxide induces central peripheral distal axonal degeneration of the lumbar primary neurones in rats. Br J Ind Med 42:373–379, 1985
8. IARC Monographs on the Evaluation of the Carcinogenic Risk of Chemicals to Humans, Suppl 4, pp 126–128. Lyon, International Agency for Research on Cancer, 1982
9. Snellings W, Weil C, Maronpot R: A two-year inhalation study of the carcinogenic potential of ethylene oxide in Fischer 344 rats. Toxicol Appl Pharmacol 75:105–117, 1984
10. Hemminki K et al: Spontaneous abortion in hospital staff engaged in sterilizing instruments with chemical agents. Br Med J 285:1461–1463, 1982
11. Landrigan PJ et al: Ethylene oxide: An overview of toxicologic and epidemiologic research. Am J Ind Med 6:103–115, 1984
12. Yager JW, Hines CJ, Spear RC: Exposure to ethylene oxide at work increases sister chromatid exchanges in human peripheral lymphocytes. Science 219:1221–1223, 1983
13. NIOSH Current Intelligence Bulletin #35, Eth-

ylene Oxide: Evidence of Carcinogenicity. Springfield, Virginia, National Institute for Occupational Safety and Health, US Department of Health and Human Services, 1981

ETHYLENE THIOUREA

CAS: 96-45-7

$C_3H_6N_2S$ 1987 TLV = none established

Synonyms: ETU; imidazolidinethione; 2-imidazoline-2-thiol; 2-mercapto-imidazoline

Physical Form. White crystalline solid

Uses. Accelerator in the curing of polychloroprene (Neoprene) and polyacrylate rubber; intermediate in the manufacture of antioxidants, insecticides, fungicides, dyes, pharmaceuticals, and synthetic resins

Exposure. Inhalation

Toxicology. Ethylene thiourea (ETU) is an antithyroid substance and animal carcinogen.

Clinical examination and thyroid function tests carried out over a period of 3 years on 13 exposed workers showed one subgroup, the mixers, to have significantly lower levels of total thyroxine than other workers; one person had an appreciably raised level of thyroid-stimulating hormone and was considered to be hypothyroid.[1] There was no evidence of any clinical effect in any of the workers. Background air concentrations at the plants ranged up to 240 $\mu g/m^3$, whereas levels up to 330 $\mu g/m^3$ were registered on one individual's personal sampler.

In two previous studies, only slight differences in total thyroxine and triidothyronine in exposed workers (concentrations unspecified) were found.

In groups of rats fed 125 or 625 ppm for up to 90 days, marked increases in serum thyroid-stimulating hormone were found.[2] The high-dose group also exhibited a decrease in iodide uptake by the thyroid, in serum triidothyronine, and in thyroxine. The majority of rats at both these exposure levels had enlarged red thyroids. Clinical signs of poisoning included excessive salivation, hair loss, and scaly skin texture by day 8 in the 625-ppm group. The no-effect level for dietary ETU in rats was considered to be 25 ppm.

In rats, ETU produced a high incidence of follicular carcinoma of the thyroid after oral administration in three studies.[3-5] Doses in one of these studies were 5, 25, 125, 250, or 500 ppm.[2] At the two highest dose levels, animals of both sexes had thyroid carcinomas, although male rats had a higher incidence. The lower dose levels produced thyroid follicular hyperplasia.

ETU is believed to induce thyroid tumors through the suppression of thyroxin synthesis leading to hyperplasia of the thyroid gland.[1]

In mice, repeated oral administration of the maximal tolerated dose of 215 mg/kg ETU produced liver tumors.[6] It was teratogenic in rats at 40 and 80 mg/kg doses that produced no maternal toxicity or fetal deaths.[7]

At present, there is no threshold limit value for ETU.[1] Suggested precautions include exclusion of women of childbearing potential from exposure, encapsulation of the powder into master batch rubber, personal respiratory protection at a level of 30 ppm and above, local exhaust ventilation, and general hygiene.[1]

REFERENCES

1. Smith DA: Ethylene thiourea: Thyroid function in two groups of exposed workers. Br J Ind Med 41:362–366, 1984
2. Freudenthal RI et al: Dietary subacute toxicity

of ethylene thiourea in the laboratory rat. J Environ Pathol Toxicol 1:147–161, 1977
3. Graham SL et al: Effects of one year administration of ethylene thiourea upon the thyroid of the rat. J Agric Feed Chem 21:324, 1973
4. Graham SL et al: Effects of prolonged ethylene thiourea ingestion on the thyroid of the rat. Food Cosmet Toxicol 13:493, 1975
5. Weisburger EK et al: Carcinogenicity tests of certain environmental and industrial chemicals. J Natl Cancer Inst 67:75, 1981
6. Innes JRM et al: Bioassay of pesticides and industrial chemicals for tumorigenicity in mice: A preliminary note. J Natl Cancer Inst 42:1101, 1969
7. Khera KS: Ethylene thiourea: Teratogenicity study in rats and rabbits. Teratology 7:243, 1973

ETHYLENIMINE

CAS: 151-56-4 — 1987 TLV = 0.5 ppm; skin

$(CH_2)_2NH$

Synonyms: Aziridine; ethyleneimine; azirane; azacyclopropane; dihydroazirine

Physical Form. Colorless liquid

Uses. Organic syntheses; production of polyethylenimines used in the paper industry and as flocculation aids in the clarification of effluents

Exposure. Inhalation; skin absorption

Toxicology. Ethylenimine is a severe irritant of the eyes, mucous membranes, and skin and causes pulmonary edema; in experimental animals, it is carcinogenic.

More than 100 cases of significant acute effects following exposure have been reported in the past 30 years; these cases include fatalities from inhalation and skin contact.[1] The effects of overexposure in humans are usually delayed for 1/2 to 3 hours and include nausea, vomiting, headache, dizziness, irritation of eyes and nose, laryngeal edema, bronchitis, dyspnea, pulmonary edema, and secondary bronchial pneumonia.[1,2] In experimental human studies, eye and nose irritation occurred at concentrations of 100 ppm and above.[3]

Severe corneal damage and death resulted from placing 0.005 ml of the liquid in the eyes of rabbits; severe eye burns in humans have resulted from direct contact. On the skin, the liquid is a potent irritant and vesicant, which may produce sensitization.[4]

The LC_{50} in mice was 2236 ppm for 10 minutes; there were signs of irritation of eyes and nose, delayed onset of pulmonary edema and renal tubular damage with proteinuria, hematuria, and elevated blood urea nitrogen.[5] In other exposed animals, a decrease in the white blood cell count and a depression of all blood elements have also been observed.[1]

Animal studies have confirmed the carcinogenic potential of ethylenimine. In one study, rats given subcutaneous injections twice weekly for 33 weeks developed sarcomas at the injection site.[6] Ethylenimine administered to mice by gavage for 3 weeks followed by dietary administration for 77 weeks caused a significant increase in hepatomas and pulmonary tumors.[6]

Although animal studies have found ethylenimine to be carcinogenic, an epidemiologic study of 144 workers with up to 40 years experience showed no evidence of carcinogenicity.[7] It should, however, be handled as a suspected human carcinogen.[2]

The odor and irritant thresholds do not provide sufficient warning of overexposure.[4]

REFERENCES

1. Reinhardt CF, Brittelli MR: Heterocyclic and miscellaneous nitrogen compounds. In Clayton

GD, Clayton FE (eds): Patty's Industrial Hygiene and Toxicology, 3rd ed, rev, Vol 2A, Toxicology, pp 2672–2676. New York, Wiley–Interscience, 1981
2. Theiss AM et al: Aziridines. In International Labor Office: Encyclopaedia of Occupational Health and Safety, Vol I, A–K, pp 228–230. New York, McGraw-Hill, 1983
3. Carpenter CP, Smyth HF Jr, Shaffer CB: The acute toxicity of ethylene imine to small animals. J Ind Hyg Toxicol 30:2–6, 1948
4. Hygienic Guide Series: Ethyleneimine. Am Ind Hyg Assoc J 26:86–88, 1965
5. Silver SD, McGrath FP: A comparison of acute toxicities of ethylene imine and ammonia to mice. J Ind Hyg Toxicol 30:7–9, 1948
6. IARC Monographs on the Evaluation of the Carcinogenic Risk of Chemicals to Man, Vol 9, Some Aziridines, N-, S-, and O-mustards and Selenium, pp 37–46. Lyon, International Agency for Research on Cancer, 1975
7. Ethyleneimine. Documentation of the TLVs and BEIs, 5th ed, p 258. Cincinnati, American Conference of Governmental Industrial Hygienists (ACGIH), 1986

ETHYL ETHER

CAS: 60-29-7

$C_2H_5OC_2H_5$ 1987 TLV = 400 ppm

Synonyms: Diethyl ether; ethoxyethane; ethyl oxide; ether; anesthesia ether; sulfuric ether

Physical Form. Colorless liquid

Uses. Solvent in the manufacture of dyes, plastics, and cellulose acetate rayon; anesthetic agent

Exposure. Inhalation

Toxicology. Ethyl ether causes eye and respiratory irritation, and, at high concentrations, it produces central nervous system depression and narcosis.

Concentrations of ethyl ether ranging from 100,000 to 150,000 are required for induction of human anesthesia; however, exposure at this concentration may also produce fatalities from respiratory arrest.[1,2] Maintenance of surgical anesthesia is achieved at 50,000 ppm, and the lowest anesthetic limit is 19,000 ppm.[1] Continued inhalation of 2000 ppm in human subjects may produce dizziness; however, concentrations up to 7000 ppm have been tolerated by some workers for variable periods of time without untoward effects.[3] Initial symptoms of acute overexposure include vomiting, respiratory tract irritation, headache, and either depression or excitation. In some persons, chronic exposure results in anorexia, exhaustion, headache, drowsiness, dizziness, excitation, and psychic disturbances.[2] Albuminuria has been reported.[1] Tolerance may be acquired through repeated exposures.[2]

Ethyl ether is a mild skin irritant; repeated exposure causes drying and cracking.[2] The vapor is irritating to the eyes, and the undiluted liquid in the eyes causes painful inflammation of a transitory nature.[3] Human subjects found 200 ppm irritating to the nose but not to the eyes or throat.[4]

There is a large margin of safety between the concentration that causes nasal irritation and the concentrations that cause anesthesia, permanent damage, and death.

REFERENCES

1. Sandmeyer EE, Kirwin CJ Jr: Ethers. In Clayton GD, Clayton FE (eds): Patty's Industrial Hygiene and Toxicology, 3rd ed, rev, Vol 2A, Toxicology, pp 2507–2511. New York, Wiley–Interscience, 1981
2. Hygienic Guide Series: Ethyl ether. Am Ind Hyg Assoc J 27:85–87, 1966
3. Chemical Safety Data Sheet SD–29, Ethyl Ether, pp 17–18. Washington, DC, MCA, Inc, 1965
4. Nelson KW et al: Sensory response to certain industrial solvent vapors. J Ind Hyg Toxicol 25:282–285, 1943

ETHYL FORMATE

CAS: 109-94-4

$HC(O)OC_2H_5$ 1987 TLV = 100 ppm

Synonyms: Formic ether; ethyl methanoate

Physical Form. Clear liquid

Uses. Solvent for cellulose nitrate, oils, and greases; synthetic flavors; fumigant

Exposure. Inhalation

Toxicology. Ethyl formate causes irritation of the eyes and nose; at very high concentrations, it causes narcosis in animals, and it is expected that severe exposure will produce the same effect in humans.

In humans, a concentration of 330 ppm caused slight irritation of the eyes and rapidly increasing nasal irritation.[1] No chronic systemic effects have been reported in humans.[2]

Rats survived 4 hours inhalation of 4000 ppm, but 8000 ppm was fatal to five of six animals.[3] Cats exposed to 5,000 ppm for 20 minutes showed eye irritation; 10,000 ppm for 80 minutes caused narcosis followed by death.[1] Pulmonary edema and death were observed in dogs exposed to 10,000 ppm for 4 hours.[4]

Applied to the skin of mice, it was not tumorigenic in 10 weeks.[4]

The liquid is only slightly irritating to the skin, but dropped into the eye, it causes moderate injury to the cornea.[3]

REFERENCES

1. Ethyl formate. Documentation of the TLVs for Substances in Workroom Air, 5th ed, p 260. Cincinnati, American Conference of Governmental Industrial Hygienists (ACGIH), 1986
2. Smyth HF Jr: Improved communication—hygienic standards for daily inhalation. Am Ind Hyg Assoc J 17:129–185, 1956
3. Smyth HF Jr et al: Range-finding toxicity data: List V. AMA Arch Ind Hyg Occup Med 10:61–68, 1954
4. Sandmeyer EE, Kirwin CJ: Esters. In Patty's Industrial Hygiene and Toxicology, 3rd ed, Vol 2A, Toxicology, pp 2263–2267. New York, Wiley–Interscience, 1981

2–ETHYLHEXYL ACRYLATE

CAS: 103-11-7

$H_2C{=}CHC(O)OCH_2CH(C_2H_5)(CH_2)_3CH_3$

1987 TLV = none established

Synonyms: EHA; 2–propenoic acid 2–ethylhexyl ester

Physical Form. Colorless liquid; commercial form contains hydroquinone (1000 ppm) or hydroquinone methyl ether (15 or 200 ppm) to prevent polymerization

Uses. Plasticizing comonomer with vinyl acetate to produce latex paints; plastics production; paper treatment; protective coatings

Exposure. Inhalation; skin contact

Toxicology. By analogy to effects caused by other acrylates, 2–ethylhexyl acrylate (EHA) is expected to be an irritant of the eyes, nose, and skin.

Dermal sensitization to EHA has been documented from exposure to its presence in adhesive tape.[1] This potential has been confirmed in the guinea pig.[2]

In a lifetime dermal oncogenesis study in mice, 20 mg EHA in acetone was applied three times weekly for their lifespan.[3] There were 40 mice in the group at the start of the study. Two animals developed squamous cell carcinomas, and four other animals had squamous cell papillomas. The first tumor was observed after 11 months of treatment. None of the acetone-treated controls developed tumors. There was an apparent increase in the frequency of chronic nephritis in the EHA-treated mice (68%) compared with the controls (15%). Treatment with EHA may have exacerbated the onset and development of this condition, which is normally seen in aged mice.

REFERENCES

1. Jordan WP: Cross-sensitization patterns in acrylate allergies. Contact Dermatitis 1:13, 1975
2. Waegemaekers TH, Van Der Walle HB: The sensitizing potential of 2–ethylhexyl acrylate in the guinea pig. Contact Dermatitis 9:372, 1983
3. DePass LR, Maronpot RR, Weil CS: Dermal oncogenicity bioassays of monofunctional and multifunctional acrylates and acrylate-based oligomers. J Toxicol Environ Health 16:55, 1985

ETHYL MERCAPTAN

CAS: 75-08-1

C_2H_5SH 1987 TLV = 0.5 ppm

Synonyms: Ethanethiol; ethyl hydrosulfide; ethyl sulfhydrate; ethyl thioalcohol; thioethanol; thioethyl alcohol

Physical Form. Flammable gas liquefying at 34.7°C

Uses. Stenching agent for liquefied petroleum gases; adhesive stabilizer; manufacture of plastics, insecticides, and antioxidants

Exposure. Inhalation

Toxicology. Ethyl mercaptan causes irritation of mucous membranes; at high concentrations, it causes narcosis in animals, and it is expected that severe exposure will cause the same effects in humans.

At concentrations below those necessary to produce toxic effects, ethyl mercaptan is extremely malodorous, and voluntary exposure to high concentrations is unlikely to occur. Observations on human exposure are limited to a single brief report of exposure of workers to 4 ppm for 3 hours daily over 5 to 10 days; the workers experienced headache, nausea, fatigue, and irritation of mucous membranes.[1]

In animals, ethyl mercaptan vapor causes mucous membrane irritation, narcosis, and, at near lethal levels, by analogy to other mercaptans, it may produce pulmonary edema. It appears to be several-fold less acutely toxic than hydrogen sulfide or methyl mercaptan.

In rats, the LD_{50} for 4 hours was 4420 ppm; effects included irritation of mucous membranes, increased respiration, incoordination, staggering gait, weakness, partial skeletal muscle paralysis, light to severe cyanosis, and mild to heavy sedation.[2]

Animals that survived single near-lethal doses by the intraperitoneal and oral routes frequently had liver and kidney damage at autopsy up to 20 days post-treatment.[2] The liquid dropped in the eyes of rabbits caused slight to moderate irritation.[2] Chronic inhalation exposures in rats, mice, and rabbits over 5 months showed no significant effects at 40 ppm.[1]

The odor threshold of ethyl mercaptan is about 0.25 ppb.[3]

REFERENCES

1. Ethyl mercaptan. Documentation of the TLVs and BEIs, 5th ed, p 202. Cincinnati, American Conference of Governmental Industrial Hygienists (ACGIH), 1986
2. Fairchild EJ, Stokinger HE: Toxicologic studies on organic sulfur compounds—1. Acute toxicity of some aliphatic and aromatic thiols (mercaptans). Am Ind Hyg Assoc J 19:171–189, 1958
3. Windholz M, Budavari S, Stroumtsos LY, Fertig MN (eds): The Merck Index, 9th ed, p 4990. New Jersey, Merck and Company, Inc, 1976

N–ETHYLMORPHOLINE

CAS: 100-74-3 1987 TLV = 5 ppm; skin
$C_6H_{13}NO$

Synonyms: 4–Ethylmorpholine

Physical Form. Colorless liquid

Uses. Catalyst in polyurethane foam production

Exposure. Inhalation; skin absorption

Toxicology. *N*-Ethylmorpholine is an irritant of the eyes and mucous membranes.

In an experimental study, humans exposed to 100 ppm for 2.5 minutes experienced irritation of eyes, nose, and throat; 50 ppm produced lesser irritation.[1] Workers exposed to low vapor concentrations for several hours reported temporarily fogged vision with rings around lights; corneal edema was observed.[2] This effect is thought to occur when air concentrations of substituted morpholines are 40 ppm or higher; the symptoms usually appear at the end of the workday and clear within 3 to 4 hours after cessation of exposure.[3]

The liquid instilled in the eye of a rabbit caused corneal haziness, sloughing, and irregularities of the surface, characteristic of severe desiccation.[3] On the skin of a rabbit, the undiluted liquid produced no reaction, surprisingly unlike unsubstituted morpholine, which is a severe skin irritant.[4]

REFERENCES

1. N–Ethylmorpholine. Documentation of the TLVs and BEIs, 5th ed, p 263. Cincinnati, American Conference of Governmental Industrial Hygienists (ACGIH), 1986
2. Dernehl CU: Health hazards associated with polyurethane foams. J Occup Med 8:59–62, 1966
3. Mellerio J, Weale RA: Miscellanea: Hazy vision in amine plant operatives. Br J Ind Med 23:153–154, 1966
4. Smyth HF Jr et al: Range-finding toxicity data: List V. AMA Arch Ind Hyg Occup Med 10:61–68, 1954

ETHYL SILICATE

CAS: 78-10-4
$Si(OC_2H_5)_4$ 1987 TLV = 10 ppm

Synonyms: Tetraethyl silicate; ethyl orthosilicate; silicic acid, tetramethyl ester

Physical Form. Flammable, colorless liquid

Uses. For arresting decay and disintegration of stone; for manufacture of weatherproof and acidproof mortars and cement

Exposure. Inhalation

Toxicology. Ethyl silicate is an irritant of the eyes and mucous membranes.

In humans, the eyes and nose are affected by brief exposures as follows: 3000 ppm, extremely irritating and intolerable; 1200 ppm, lacrimation and stinging; 700 ppm, mild stinging; and 250 ppm, slight tingling.[1] Repeated or prolonged skin contact with the liquid may cause dermatitis owing to its solvent effect.[2]

Exposure of guinea pigs to 2530 ppm for 4 hours was lethal to more than half the animals; death was usually delayed and a result of pulmonary edema; effects were irritation of the eyes and nose, lacrimation, tremor, dyspnea, and narcosis; some surviving animals developed a delayed but profound anemia.[1] Exposure to 1000 ppm for up to three 7-hour periods was fatal to 4 of 10 rats; autopsy findings included marked tubular degeneration and necrosis of the kidneys, mild liver damage, and slight pulmonary edema and hemorrhage.[2] In rats exposed to 125 ppm for

15 to 20 7-hours periods, slight to moderate kidney damage was observed, but there were no pathologic changes detected in the liver or lungs. Instillation of the liquid into the rabbit eye caused immediate marked irritation, which was reversible.[2]

REFERENCES

1. Smyth HF Jr, Seaton J: Acute response of guinea pigs and rats to inhalation of the vapors of tetraethyl orthosilicate (ethyl silicate). J Ind Hyg Toxicol 23:288–296, 1940
2. Rowe VK, Spencer HC, Bass SL: Toxicological studies on certain commercial silicones and hydrolyzable silane intermediates. J Ind Hyg Toxicol 30:332–352, 1948

FERBAM

CAS: 14484-64-1

$[(CH_3)_2NCS_2]_3Fe$ 1987 TLV = 10 mg/m^3

Synonyms: Ferric dimethyldithiocarbamate

Physical Form. Black solid

Uses. Fungicide

Exposure. Inhalation

Toxicology. Ferbam is an irritant of the eyes and respiratory tract; in animals, it causes central nervous system depression, and it is expected that severe exposure and absorption will cause the same effect in humans.

In humans, the dust is irritating to the eyes and the respiratory tract; it causes dermatitis in some individuals. Large oral doses cause gastrointestinal disturbances.[1]

In guinea pigs given ferbam by stomach tube, the lethal range was 450 to 2000 mg/kg; the animals became stuporous and died in coma.[2] Ten of 20 rats died from a diet containing 0.5% ferbam for 30 days; there was a slight and ill-defined tendency toward anemia. At autopsy, there was no evidence of a regularly appearing tissue injury; minor abnormalities of the lung, liver, kidney, and bone marrow were observed in a few animals.[2] Animal experiments revealing an increased acetaldehyde level postingestion of alcohol suggest that ferbam, like other dithiocarbamates, may be capable of causing an Antabuse-like reaction.[3]

Following oral administration of ferbam to mice and rats, no carcinogenic effects were seen, but the IARC has concluded that insufficient data are available to fully evaluate the carcinogenicity of this compound.[3] Administration of ferbam to pregnant rats in high but sublethal doses yielded evidence of embryo/fetotoxicity.[3]

REFERENCES

1. AMA Council on Pharmacy and Chemistry: Outlines of information on pesticides. Part 1. Agricultural fungicides. JAMA 157:237–241, 1955
2. Hodge HC et al: Acute and short-term oral toxicity tests of ferric dimethyldithiocarbamate (ferbam) and zinc dimethyldithiocarbamate (ziram). J Am Pharm Assoc 41:662–665, 1952
3. Ferbam in IARC Monographs on the Evaluation of Carcinogenic Risk of Chemicals to Man, Vol 12, Some Carbamates, Thiocarbamates and Carbazides, pp 121–129. Lyon, International Agency for Research on Cancer, 1982

FERROVANADIUM DUST

CAS: 12604-58-9

FeV 1987 TLV = 1 mg/m^3

Synonyms: None

Physical Form. Gray to black dust

Uses. Added to steel to produce fineness of grain, toughness, and resistance to high temperature and torsion

Exposure. Inhalation

Toxicology. Ferrovanadium dust is a mild irritant of the eyes and respiratory tract.

Workers exposed to unspecified concentrations developed slight irritation of the eyes and respiratory tract.[1] Systemic effects have not been reported from industrial exposure.

Animals exposed for 1 hour on alternate days for 2 months to very high concentrations (1000 to 2000 mg/m^3) developed chronic bronchitis and pneumonitis.[2] No active intoxication occurred in animals exposed at concentrations as high as 10,000 mg/m^3.[2]

REFERENCES

1. Roberts WC: The ferroalloy industry—hazards of the alloys and semimetallics: Part II. J Occup Med 7:71–77, 1965
2. Ferrovanadium. Documentation of TLVs and BEIs, 5th ed, p 269. Cincinnati, American Conference of Governmental Industrial Hygienists (ACGIH), 1986

FLUORIDES

F 1987 TLV = 2.5 mg/m^3

Synonyms: None

Physical Form. Dust

Sources. Grinding, drying, and calcining of F-containing minerals and acidulation of these minerals; metallurgic processes such as aluminum reduction and steel-making, involving fluoride fluxes or melts; kiln firing of brick and ceramic materials; melting of raw material in glass-making; fluoridation of water supplies

Exposure. Inhalation; ingestion

Toxicology. Fluoride causes irritation of the eyes and respiratory tract; absorption of excessive amounts of fluoride over a long period of time results in increased radiographic density of bone.[1]

Workers exposed to an airborne fluoride concentration of 5 mg/m^3 complained of eye and respiratory tract irritation and nausea.[2] The lethal oral dose of sodium fluoride for humans is approximately 5 g; effects from ingestion include diffuse abdominal pain, diarrhea, and vomiting; excessive salivation, thirst, and perspiration; painful spasms of the limbs; and sometimes albuminuria.[3,4]

Most absorbed fluoride is excreted rapidly in the urine. A portion is stored in bone, but a nearly equal amount is mobilized from bone and excreted.[2,5] Some storage of fluoride occurs from the ingestion of as little as 3 mg/day.[6] Repeated exposure to excessive concentrations of fluoride over a period of years results in increased radiographic density of bone and eventually may cause crippling fluorosis (osteosclerosis due to deposition of fluoride)—now a rare phenomenon.[1] The gross changes in the skeleton are quite distinctive and characteristic; as the amount of fluoride in the bone increases, exostoses may develop, especially on the long bones; the sacrotuberous and sacrosciatic ligaments begin to calcify, vertebrae occasionally fuse together, and typical stiffness of the spinal column develops.[1]

The daily absorption of 20 to 80 mg of fluoride may be expected to lead to crippling fluorosis in 10 to 20 years; this condition has not been reported in the United States from industrial exposure.[7] Evidence from several sources indicates that urinary fluoride concentrations not exceeding 5 mg/liter in preshift samples taken after 2 days off work are not associated with detectable osteosclerosis and that such changes are unlikely at urinary levels of 5 to 8 mg/liter.[7] Preshift urinary fluoride concentration is considered to be a measure of the worker's body (skeletal) burden of fluoride, whereas the postshift sample is taken to be representative of exposure conditions during that workshift.[1]

NIOSH recommends that urinary postshift fluoride analysis be made available at an interval not exceeding every 3 months to at least one fourth of all workers subject to occupational exposure to fluoride.[8] Participating workers should be rotated to provide all exposed workers the opportunity for urinalysis every year. Spot urine samples are collected at the conclusion of the work shift after 4 or more consecutive days of exposure.

Mottled appearance and altered form of teeth are produced only when excessive amounts of fluoride are ingested during the period of formation and calcification of teeth, which occurs during the first 8 years of life in humans; after calcification has been completed, fluoride does not have an adverse effect on the teeth.[4]

REFERENCES

1. Hodge HC, Smith FA: Occupational fluoride exposure. J Occup Med 19:12–39, 1977
2. Fluorides. Documentation of TLVs and BEIs, 5th ed, pp 272–274. Cincinnati, American Conference of Governmental Industrial Hygienists (ACGIH), 1986
3. Fluorides and Human Health, pp 225–271. Geneva, World Health Organization, 1970
4. Committee on Biologic Effects of Atmospheric Pollutants, Division of Medical Sciences. National Research Council: Fluorides, pp 163–221. Washington, DC, National Academy of Sciences, 1971
5. Dinman BD et al: Prevention of bony fluorosis in aluminum smelter workers. J Occup Med 18:7–25, 1976
6. Largent EJ: Rates of elimination of fluoride stored in the tissues of man. AMA Arch Ind Hyg Occup Med 6:37–42, 1952
7. Biological Monitoring Guides: Fluorides. Am Ind Hyg Assoc J 32:274–279, 1971
8. National Institute for Occupational Safety and Health: Criteria for a Recommended Standard . . . Occupational Exposure to Inorganic Fluorides. DHEW (NIOSH) Pub No 76–103, pp 19–100. Washington, DC, US Government Printing Office, 1975

FLUORINE

CAS: 7782-41-4

F_2 — 1987 TLV = 1 ppm

Synonyms: None

Physical Form. Yellow gas

Uses. Conversion of uranium tetrafluoride to uranium hexafluoride; oxidizer in rocket fuel systems; manufacture of various fluorides and fluorocarbons

Exposure. Inhalation

Toxicology. Fluorine is a severe irritant of the eyes, mucous membranes, skin, and lungs.

Fluorine reacts with water to produce ozone and hydrofluoric acid.[1] In humans, the inhalation of high concentrations causes laryngeal spasm and bronchospasm, followed by the delayed onset of pulmonary edema.[2,3] At sublethal levels, severe local irritation and laryngeal spasm will preclude voluntary exposure to high concentration unless the person is trapped or incapacitated.[2] Two human subjects found momentary exposure to 50 ppm intolerable; 25 ppm was tolerated briefly, but both subjects developed sore throat and chest pain that persisted for 6 hours.[4]

The LC_{50} in mice for 60 minutes was 150 ppm; effects were irritation of the eyes and nose and the delayed onset of labored breathing and lethargy; autopsy findings included marked pulmonary congestion and hemorrhage.[5] Mice exposed to sublethal concentrations had pulmonary irritation and delayed development of focal necrosis in the liver and kidneys.[5]

A blast of fluorine gas on the shaved skin of a rabbit caused a second-degree burn; lower concentrations cause severe burns of insidious onset resulting in ulceration, similar to those produced by hydrogen fluoride.[1,4]

REFERENCES

1. Largent EJ: Fluorine and compounds. In International Labor Office: Encyclopaedia of Occupational Health and Safety, Vol 1, A–K, pp 557–559. New York, McGraw-Hill, 1971
2. Ricca PM: Exposure criteria for fluorine rocket propellants. Arch Environ Health 12:339–407, 1966
3. Ricca PM: A survey of the acute toxicity of elemental fluorine. Am Ind Hyg Assoc J 31:22–29, 1970
4. Hygienic Guide Series: Fluorine. Am Ind Hyg Assoc J 26:624–627, 1965
5. Keplinger ML, Suissa LW: Toxicity of fluorine short-term inhalation. Am Ind Hyg Assoc J 29:10–18, 1968

FORMALDEHYDE

CAS: 50-00-0

HCHO 1987 TLV = 1 ppm

Synonyms: Methanal; formic aldehyde; oxomethane; oxymethylene; methylene oxide; methyl aldehyde

Physical Form. Gas (*Note:* formalin is a 37% to 50% solution by weight of formaldehyde gas)

Uses. Ionizing solvent; manufacture of formic esters, resins, plastics, leather, rubber, metals, and wood; disinfection in dialysis units

Exposure. Inhalation; skin absorption

Toxicology. Formaldehyde gas is an irritant of the eyes and respiratory tract; solutions cause both primary irritation and sensitization dermatitis.

Mild eye irritation with lacrimation and other transient symptoms of mucous membrane irritation have been observed in some persons at concentrations of 0.1 to 0.3 ppm. At 2 or 3 ppm, a very mild tingling sensation in the eyes, nose, and posterior pharynx may be felt by most people.[1] Some tolerance occurs so that repeated 8-hour exposures at this level are possible.[2] At 4 or 5 ppm, the discomfort increases rapidly and some mild lacrimation occurs in most people. This level can be tolerated fairly well for periods of perhaps 10 to 30 minutes by some, but not all people. After 30 minutes, discomfort becomes quite pronounced. Concentrations of 10 ppm can be withstood for only a few minutes; profuse lacrimation occurs in all subjects, even those acclimated to lower levels.

Between 10 and 20 ppm, it becomes difficult to take a normal breath voluntarily; burning of the nose and throat becomes more severe and extends to the trachea, producing cough. Upon cessation of exposure, the lacrimation subsides promptly, but the nasal and respiratory irritation may persist for about an hour. It is not known at which levels serious inflammation of the bronchi and lower respiratory tract would occur in humans; it is likely that 5- or 10-minute exposures to levels of 50 to 100 ppm might cause very serious injury. Acute irritation of the human respiratory tract from inhalation of high levels of formaldehyde have caused pulmonary edema, pneumonitis, and death.[1]

Solutions of 25% to 44% splashed in the eyes have caused severe injury and corneal damage.

Strong formaldehyde solutions on the skin and exposure to the gas have caused primary skin irritation.[1] Allergic contact dermatitis owing to formaldehyde solutions or formaldehyde-containing resins is a well-recognized risk.

Three different types of clinical lesions are seen. In the first type, a worker may develop a sudden eczematous reaction of the eyelids, face, neck, scrotum, and flexor surfaces of the arms, which may appear only a few days after commencing work. A second type of reaction that may not appear for a number of years is one that appears first in the digital areas, back of the hands, wrists, forearms, and parts of the body that are exposed to friction from clothing. The third type of reaction is a combination of the first two types.

A number of case reports of asthma, apparently induced by formaldehyde exposure, have been published. Some of these reports have been supported by the finding of diminished pulmonary function, for example, decreased forced expired volume in 1 second, following inhalation challenge with formaldehyde.[3,4] Late asthmatic reactions have occurred, suggesting true sensitization, further supported by negative histamine or methacholine challenge. The concentrations used in these studies ranged from 1 to 3.2 ppm, at which effects rather than sensitization could have been responsible. However, a study of seven volunteers with mild extrinsic asthma exposed for a 10-minute interval to 1 to 3 ppm of formaldehyde at rest and during exercise did not demonstrate an increase in airway resistance (bronchoconstriction).[6] A true irritative immunologically mediated allergic response has not been documented. Despite this controversy, it appears that formaldehyde-induced asthma, if it occurs, is rare.[3–5]

Formaldehyde has been shown to be carcinogenic in two strains of rats, resulting in squamous cell cancers of the nasal cavity following repeated inhalation of about 14 ppm. In one study, 51 of 117 male and 42 of 115 female Fischer 344 rats developed this tumor, but no nasal tumors were seen at 0 or 2 ppm. No other neoplasm was increased significantly. In a similar study of mice, this nasal tumor occurred in two male mice at 14.3 ppm. None of the excesses were statistically significant except for the high-exposure data in rats.[7] A number of dose-dependent histopathologic changes, including rhinitis and squamous metaplasia, were observed in rats and mice in these studies.

An excess of deaths from cancer of the brain has been noted in professional workers (embalmers, anatomists, and pathologists) who have been exposed to formaldehyde, but this excess has not been observed in industrial workers. A small excess of leukemias has also been seen in professional workers exposed to formaldehyde.[5] No excess of nasal cancers has been identified in studies to date.[5] A recent large historic cohort study of more than 26,000 industrial workers with exposure to formaldehyde failed to identify excesses of leukemia, brain cancer, or nasal cancer. Small excesses were observed for Hodgkin's disease and cancers of the lung, nasopharynx, and prostate, but none of these excesses were statistically significant or related to indices of cumulative exposure. The authors concluded that there was little evidence to suggest an association between formaldehyde levels experienced (44% had highest estimated TWA exposure less than 0.5 ppm) and cancer mortality in this group.

The odor is perceptible to previously unexposed persons at or below 1 ppm.[1]

REFERENCES

1. National Institute for Occupational Safety and Health: Criteria for a Recommended Standard . . . Occupational Exposure to Formaldehyde. DHEW (NIOSH) Pub No 77–126, pp 21–81. Washington, DC, US Government Printing Office, 1976
2. Nordman H, Keskmen H, Tuppurainen M: For-

maldehyde asthma—rare or overlooked? J Allergy Clin Immunol 75(1):91–99, 1985
3. Brige P et al: Occupational asthma due to formaldehyde. Thorax 40:255–260, 1985
4. Sheppard D, Eschenbacher W, Epstein J: Lack of bronchomotor response to up to 3 ppm formaldehyde in subjects with asthma. Environ Res 35:133–139, 1984
5. Report on the Consensus Workshop on Formaldehyde. Environ Health Perspect 58:323–381, 1984
6. Kerns W et al: Carcinogenicity of formaldehyde in rats and mice after long-term inhalation exposure. Cancer Res 43:4382–4392, 1983
7. Blair A et al: Mortality among industrial workers exposed to formaldehyde. JNCI 76(6):1071–1084, 1986

FORMIC ACID

CAS: 64-18-6
HCOOH 1987 TLV = 5 ppm

Synonyms: Methanoic acid; formylic acid, hydrogen carboxylic acid

Physical Form. Colorless liquid

Uses. Decalcifier, reducer in dyeing wool; dehairing and plumping hides, tanning; electroplating; coagulating rubber latex, aid in regenerating old rubber

Exposure. Inhalation

Toxicology. Formic acid vapor is a severe irritant of the eyes, mucous membranes, and skin.

Exposure causes eye irritation with lacrimation, nasal discharge, throat irritation, and cough.[1] A worker splashed in the face with hot formic acid developed marked dyspnea with dysphagia and died within 6 hours.[2] Workers exposed to a mixture of formic and acetic acids at an average concentration of 15 ppm of each complained of nausea.[3]

The liquid on the skin causes burns with vesiculation; keloid formation at the site of the burn often results.[2] Although ingestion of the liquid is unlikely in ordinary industrial use, the highly corrosive nature of the substance may be expected to produce serious burns of the mouth and esophagus.[3]

Because formic acid is an inhibitor of cellular respiration, persons with cardiovascular disease may be considered a special risk group.[4]

REFERENCES

1. Henson EV: Toxicology of the fatty acids. J Occup Med 1:339–345, 1959
2. von Oettingen WF: The aliphatic acids and their esters—toxicity and potential dangers. AMA Arch Ind Health 20:517–522, 530–531, 1959
3. Guest D et al: Aliphatic carboxylic acids. In Clayton GD, Clayton FE (eds): Patty's Industrial Hygiene and Toxicology, 3rd ed, rev, Vol 2C, Toxicology, pp 4903–4909. New York, Wiley–Interscience, 1982
4. Liesivuori J, Kettunen A: Farmers' exposure to formic acid vapour in silage making. Ann Occup Hyg 27:327–329, 1983

FURFURAL

CAS: 98-01-1 1987 TLV = 2 ppm; skin
$C_4H_4O_2$

Synonyms: 2–Furaldehyde; pyromucic aldehyde

Physical Form. Colorless, oily liquid

Uses. Solvent refining of lubricating oils, resins, and other organic materials; as insecticide, fungicide, germicide; reagent in analytical chemistry

Exposure. Inhalation; skin absorption

Toxicology. Furfural is an irritant of eyes, mucous membranes, and skin.

Although the vapor is a potent irritant, the liquid has a relatively low volatility so that inhalation by workers of significant quantities is unlikely.[1] Exposure of workers to levels of 1.9 to 14 ppm caused complaints of eye and throat irritation and headache.[2] The liquid or vapor is irritating to the skin and may cause dermatitis, allergic sensitization, and photosensitization.[3]

Exposure of cats to 2800 ppm for 30 minutes resulted in fatal pulmonary edema.[1] Inhalation of 260 ppm for 6 hours was fatal to rats but produced no deaths in mice or rabbits.[3] Slight liver changes were seen in dogs exposed daily to 130 ppm for 4 weeks.[3] Symptoms following oral administration of 50 to 100 mg/kg in rats were weakness, ataxia, coma, and death.[1] Drops of a 10% aqueous solution on the eyes of animals caused immediate discomfort; the lids and conjunctivae became red and swollen, but these effects disappeared after 24 hours.[4]

REFERENCES

1. Brabec MJ: Aldehydes and acetals. In Clayton GD, Clayton FE (eds): Patty's Industrial Hygiene and Toxicology, 3rd ed, Vol 2A, Toxicology, pp 2665–2666. New York, Wiley–Interscience, 1981
2. Furfural. Documentation of the TLVs and BEIs, 5th ed, p 280. Cincinnati, American Conference of Governmental Industrial Hygienists (ACGIH), 1986
3. Hygienic Guide Series: Furfural. Am Ind Hyg Assoc J 1, 1978
4. Grant WM: Toxicology of the Eye, 3rd ed, p 449. Springfield, Illinois, Charles C Thomas, 1986

FURFURYL ALCOHOL

CAS: 98-00-0 1987 TLV = 10 ppm; skin

$C_6H_6O_2$

Synonyms: 2–Furyl carbinol; 2–furanmethanol; furfural alcohol

Physical Form. Colorless liquid that turns dark in air

Uses. Solvent for cellulose ethers, esters, resins, and dyes; liquid propellant; binder in foundry cores; manufacture of resins, including furfuryl alcohol resin (furan resin) and furfuryl alcohol–formaldehyde resins

Exposure. Inhalation; skin absorption

Toxicology. At high concentrations, furfuryl alcohol causes narcosis and mucous membrane irritation in animals, and it is expected that severe exposure will cause the same effects in humans.

Workers exposed to 8.6 and 10.8 ppm in a foundry core-making operation experienced no discomfort, but two persons exposed to 15.8 ppm (and 0.33 ppm formaldehyde) when hot sand was used experienced lacrimation and a desire to leave the area. In another foundry core-making operation, no ill effects were seen after exposures of about 6 to 16 ppm.[1]

Large doses injected subcutaneously in dogs caused depressed respiration, lowered body temperature, salivation, diarrhea, diuresis, and signs of narcosis.[2]

In rats exposed to 700 ppm, effects included excitement followed by eye irritation and drowsiness.[3] In rats, the LC_{50} for 4 hours was 233 ppm.[4] Repeated daily exposure of rats to an average of 19 ppm caused moderate respiratory irritation.[3] Intravenous injection into rabbits and cats caused depression of the central nervous system; death occurred at doses of 800 to 1400 mg/kg.[5] Eye contact in rabbits resulted in reversible inflammation and corneal injury with opacity.[1] Animal experiments indicated that the liquid is well absorbed through the skin with a dose-related mortality; mild skin irritation may also result from contact.[1] Prolonged inhalation exposures of rats to 25, 50, and 100 ppm resulted in decreased weight gain and, at the two highest doses, biochemical changes in the brain suggestive of mitochondrial damage, glial cell degeneration, and early demyelinization.[6]

The odor is detectable at 8 ppm.[4] Mixing with acids results in polymerization, a highly exothermic reaction that may result in explosions.[1]

REFERENCES

1. National Institute for Occupational Safety and Health: Criteria for a Recommended Standard . . . Occupational Exposure to Furfuryl Alcohol, DHEW (NIOSH) Pub No 79–133. Washington, DC, US Government Printing Office, 1979
2. Erdmann E: Uber das Kaffeeol und die Physiologische Wirkung des darin Enthaltenen Furfuralkols. Arch Exp Pathol Pharmacol 48:233–261, 1902
3. Comstock CC, Oberst FW: Inhalation Toxicity of Aniline, Furfuryl Alcohol and Their Mixtures in Rats and Mice. Chemical Corps Medical Laboratories Research Report No 139, October 1952
4. Jacobson KH et al: The toxicology of an aniline-furfuryl alcohol-hydrazine vapor mixture. Am Ind Hyg Assoc J 19:91–100, 1958
5. Fine EA, Wills JH: Pharmacologic studies of furfuryl alcohol. Arch Ind Hyg Occup Med 1:625–632, 1950
6. Savelainen H, Pfäffli P: Neurotoxicity of furfuryl alcohol vapor in prolonged inhalation exposure. Environ Res 31:420–427, 1983

GLUTARALDEHYDE

CAS: 111-30-8

$OCH(CH_2)_3CHO$ 1987 TLV = C 0.2 ppm

Synonyms: Cidex (2% alkaline glutaraldehyde aqueous solution); 1,5–pentanedial; 1,5–pentanedione; glutaric dialdehyde; glutaral

Physical Form. Colorless crystalline solid, soluble in water and organic solvents

Uses. Broad-spectrum antimicrobial cold sterilant/disinfectant for hospital equipment; tanning agent for leather; tissue fixative; cross-linking agent for proteins; preservative in cosmetics; therapeutic agent for warts, hyperhidrosis, and dermal mycotic infections; x-ray processing solutions and film emulsion

Exposure. Inhalation

Toxicology. Glutaraldehyde is an irritant of the skin and mucous membranes and can cause an allergic contact dermatitis; it may be capable of inducing asthma.

Glutaraldehyde has caused an allergic contact dermatitis in hospital workers using it as a cold sterilant or in handling of recently processed x-ray film. It appears to be a strong sensitizer.[1–3] In general, reactions present as a vesicular dermatitis of the hands and forearms. Rubber gloves do not appear to afford complete protection. It is a mild skin irritant.

Glutaraldehyde can also produce eye and skin irritation when solutions are aerosolized.[4] It has a low vapor pressure at room temperature, which reduces the potential for inhalation exposure.

Glutaraldehyde can apparently cause asthma and rhinitis. Four nurses who were sterilizing endoscopes with glutaraldehyde developed symptoms of asthma and rhinitis temporally related to exposures to glutaraldehyde. Three of the four nurses, however, had a prior history of mild seasonal asthma.[4] On specific provocation testing, one patient had an increase in nasal airways resistance, with a dual immediate and late response pattern. Another patient had a delayed 22% decline in FEV_1 80 minutes after the final exposure to glutaraldehyde. The occurrence of late reactions suggested that the underlying mechanism involved sensitization rather than an irritant effect.[4]

Animal experiments demonstrate that solutions containing 25% or more glutaraldehyde cause a significant degree of skin irritation and eye injury; dilute solutions (5% or less) have low acute toxicity.[5] Results of short-term tests have been variable; glutaraldehyde was weakly positive in the Ames test in Salmonella in one study and negative in several other studies. Glutaraldehyde tested negative in several genotoxicity tests in mammalian cells, including sister chromatid exchange and unscheduled DNA synthesis tests. The authors concluded that its in vitro reactivity with proteins, along with its nonreactivity with nucleic acids, made genotoxic effects unlikely.[5]

REFERENCES

1. Goncalo S et al: Occupational contact dermatitis to glutaraldehyde. Contact Dermatitis 10:183–184, 1984
2. Hansen KS: Glutaraldehyde occupational dermatitis. Contact Dermatitis 9:81–82, 1983
3. Fisher AA: Reactions to glutaraldehyde with particular reference to radiologists and x-ray technicians. Cutis 28:113–122, 1981
4. Corrado OJ et al: Asthma and rhinitis after exposure to glutaraldehyde in endoscopy units. Hum Toxicol 5:325–327, 1986
5. Slesinski RS et al: Mutagenicity evaluation of glutaraldehyde in a battery of in vitro bacterial and mammalian test systems. Fd Chem Toxicol 21:621–629, 1983

GLYCIDOL

CAS: 556-52-5

$C_3H_6O_2$ 1987 TLV = 25 ppm

Synonyms: 2,3–Epoxy–1–propanol

Physical Form. Colorless liquid

Uses. Preparation of glycerol, glycidyl ethers, esters, and amines; in pharmaceuticals; in sanitary chemicals

Exposure. Inhalation

Toxicology. Glycidol is an irritant of the eyes, upper respiratory tract, and skin; at high concentrations, it causes narcosis in animals, and it is expected that severe exposure will cause the same effect in humans.

The practical risk to humans from vapor exposure appears to be relatively slight, as ample warning in the form of eye, nose, and throat irritation occurs at low concentrations; no chronic effects have been reported in humans.[1]

The LC_{50} in mice was 450 ppm (4 hours); in rats, it was 580 ppm (8 hours); labored breathing, lacrimation, salivation, and nasal discharge were seen; and pneumonitis was observed at autopsy.[1] Rats repeatedly exposed to 400 ppm showed only slight eye irritation and mild respiratory distress, with no evidence of systemic toxicity.

The oral LD_{50} was 0.45 g/kg for mice and 0.85 g/kg for rats; the liquid administered by gastric tube resulted in nervous system depression characterized by incoordination, ataxia, coma, and death.[1] Animals surviving exposure showed reversible excitation and tremor; lacrimation and labored breathing were also observed.[1]

Application of the liquid to animal skin caused moderate irritation. Chronic topical administration of a 5% solution to mice did not cause skin tumors or any visible skin reaction.[2] In the eyes, it produced severe eye irritation; despite the severity of primary injury, no blindness or permanent defects in the cornea, lens, or iris resulted from the applications.

REFERENCES

1. Hine CH et al: The toxicology of glycidol and some glycidyl ethers. AMA Arch Ind Health 14:250–264, 1956

2. Van Duuren BL et al: Carcinogenicity of epoxides, lactones and peroxy compounds. VI. Structure and carcinogenic activity. J Natl Cancer Inst 39:1217–1228, 1967

GRAPHITE (natural)
CAS: 7782-42-5
C (with traces of Fe, SiO_2, etc)
1987 TLV = 2.5 mg/m^3 respirable dust

Synonyms: Plumbago; black lead; mineral carbon

Physical Form. Usually soft, black scales; crystals rare

Uses. For pencils, refractory crucibles, pigment, lubricant, polishing compounds, electroplating

Exposure. Inhalation

Toxicology. Natural graphite dust causes graphite pneumoconiosis.

The earliest roentgenologic changes may be the disappearance of normal vascular markings with the later appearance of pinpoint and nodular densities in all lung fields.[1,2] Massive lesions, when present, are caused by large cysts filled with black fluid. The pleura is often involved; hydrothorax, pneumothorax, and pleural thickening may occur.

At autopsy, the lungs are grey-black to black; histologically there are widely scattered particles, spicules, and plates of graphite, often within intra-alveolar phagocytes amidst diffuse interstitial fibrosis and occasionally pneumonitis. There are also interwoven bands of collagen, similar to those found in silicosis, which are frequently the most prominent feature of the fibrotic lesions occupying the lung and the bronchial lymph nodes. Symptoms include expectoration of black sputum, dyspnea, and cough.

Of 344 workers in a graphite mine in Ceylon, 78 had radiographic abnormalities, including small rounded and irregular opacities, large opacities, and enlargment of hilar shadows. Some affected workers had cough, dyspnea, or digital clubbing.[3]

It has generally been believed that the capacity of inhaled natural graphite dust to cause a disease is largely the result of its crystalline silica component.

REFERENCES

1. Pendergrass EP et al: Observations on workers in the graphite industry—Part One. Med Radiogr Photogr 43:70–99, 1967
2. Pendergrass EP et al: Observations on workers in the graphite industry—Part Two. Med Radiogr Photogr 44:2–17, 1968
3. Ranasinka KW, Uragoda CG: Graphite pneumoconiosis. Br J Ind Med 29:178–183, 1972

GRAPHITE (synthetic)
C (containing < 1% free quartz)
1987 TLV = 10 mg/m^3, total dust

Synonyms: None

Physical Form. A crystalline form of carbon made from high temperature treatment of coal or petroleum products; same properties as natural graphite; it is chemically inert

Uses. Similar to those of natural graphite in refractories and electrical products

Exposure. Inhalation

Toxicology. Pure synthetic graphite acts as an inert or nuisance dust.

In contrast to the several reports of pneumoconiosis in workers exposed to natural graphite (qv), there is only the rare anecdotal report of significant pulmonary findings owing to exposure to synthetic graphite. One man who had spent 17 years turning and grinding synthetic graphite bars developed simple pneumoconiosis with cough, dyspnea, reduced pulmonary function, and x-ray changes. At autopsy, there was emphysema with scattered fine black (microscopic to 5 mm) nodules with some strands of fibrous tissue. Many of the nodules consisted of almost acellular collagen. There were traces of iron in the nodules and the hilar nodes. Ashed material from the lung showed little or no birefringent particles, indicating the absence of siliceous material.[1] The lung contained 8.8% to 9.5% carbon by dry weight.

Synthetic graphite injected peritoneally in mice produces a reaction characteristic of an inert material. On the basis of experimental evidence and the rarity of reports of adverse effects of exposure in humans, it is concluded that pure synthetic graphite acts only as an inert dust.[2]

REFERENCES

1. Lister WB, Wimborne D: Carbon pneumoconiosis in a synthetic graphite worker. Br J Ind Med 29:108–110, 1972
2. Synthetic graphite. Documentation of TLVs and BEIs, 5th ed, p 291. Cincinnati, American Conference of Governmental Industrial Hygienists (ACGIH), 1986

HAFNIUM (and compounds)

CAS: 7440-58-6

Hf 1987 TLV = 0.5 mg/m^3

Synonyms: Hafnyl chloride ($HfOCl_2$); hafnium chloride ($HfCl_4$)

Physical Form. Hard, shiny, ductile stainless steel–colored metal; dull grey powder

Uses/Sources. In mining and purification of the metal; used in control rods in nuclear reactors; manufacture light bulb filaments; found in all zirconium-containing minerals

Exposure. Inhalation

Toxicology. Hafnium dust is very low in toxicity. No health hazards have been recognized from the industrial handling of hafnium powder other than those arising from fire or explosion.[1]

Hafnium salts are mild irritants of the eye and the skin and have produced liver damage in animals.[2] In mice, the LD_{50} of hafnyl chloride by intraperitoneal injection was 112 mg/kg.[2] In cats, intravenous administration of hafnyl chloride at 10 mg/kg was fatal.[2] Rats fed a diet containing 1% for 12 weeks showed slight changes in the liver, consisting of perinuclear vacuolization of the parenchymal cells and coarse granularity of the cytoplasm.[1] The application of 1 mg of hafnium chloride to the eyes of rabbits produced transient irritation. Topical application of hafnium chloride crystals to unabraded rabbit skin produced transient edema and erythema; application to abraded skin caused ulceration.[2]

REFERENCES

1. Chemical Safety Data Sheet SD–92, Zirconium and Hafnium Powder, pp 5–6, 10. Washington, DC, MCA, Inc, 1966
2. Haley TJ, Raymond K, Komesu N, Upham HC: The toxicologic and pharmacologic effects of hafnium salts. Toxicol Appl Pharmacol 4:238–246, 1962

HEPTACHLOR

CAS: 76-44-8 — 1987 TLV = 0.5 mg/m^3; skin

$C_{10}H_5Cl_7$

Synonyms: Drinox; E–3314, ENT 15,152: Velsicol 104

Physical Form. White, crystalline solid

Uses. Insecticide

Exposure. Inhalation; skin absorption; ingestion

Toxicology. Heptachlor is a convulsant in animals; it is expected that severe exposure would cause the same effect in humans.

There are no reported cases of human poisoning by heptachlor.[1]

In rats, the oral LD_{50} was 90 mg/kg; within 30 to 60 minutes following administration, effects were tremor and convulsions; liver necrosis was noted.[2] Multiple applications of a solution of 20 mg/kg to the skin of rats were toxic, indicating a marked cumulative action.[2] Reversible histologic changes in the rat liver have occurred following dosages of 0.35 mg/kg for 50 weeks.[1] Rats given heptachlor in the diet at 6 mg/kg bodyweight developed cataracts after 4.5 to 9.5 months of feeding.[3] This observation has not been replicated in a number of subsequent animal studies.[1] In animals, heptachlor is more potent than chlordane, to which it is closely related chemically.[4]

The IARC has concluded that there is sufficient evidence that heptachlor (containing 20% chlordane) is carcinogenic in mice, producing liver carcinomas; there is suggestive evidence for carcinogenicity in female rats.[5] A study of two cohorts of workers exposed to chlordane and heptachlor at two different production facilities failed to demonstrate any overall excess of cancer. There was one death from liver cancer, with 0.59 expected. There was a slight excess of lung cancer (12 observed, 9 expected), but this was not statistically significant.[6]

REFERENCES

1. Hayes WJ Jr: Pesticides Studied in Man, pp 233–234. Baltimore, Williams & Wilkins, 1982
2. von Oettingen WF: The Halogenated Aliphatic, Olefinic, Cyclic, Aromatic, and Aliphatic-Aromatic Hydrocarbons Including the Halogenated Insecticides, Their Toxicity and Potential Dangers, US Public Health Service Pub No 414, pp 326–327. Washington, DC, US Government Printing Office, 1955
3. Mestitzova M: On reproduction studies and the occurrence of cataracts in rats after long-term feeding of the insecticide heptachlor. Experientia 23:42–43, 1967
4. Council on Pharmacy and Chemistry: The present status of chlordane. JAMA 158:1364–1367, 1955
5. IARC Monographs on the Evaluation of the Carcinogenic Risk of Chemicals to Humans, Vol 20, Some Halogenated Hydrocarbons, pp 129–154. Lyon, International Agency for Research on Cancer, 1979
6. Wong HH, MacMahon B: Mortality of workers employed in the manufacture of chlordane and heptachlor. J Occup Med 21:745–748, 1979

n-HEPTANE

CAS: 142-82-5

$CH_3(CH_2)_5CH_3$ — 1987 TLV = 400 ppm

Synonyms: Dipropyl methane; heptyl hydride; heptane

Physical Form. Volatile, flammable liquid

Uses. As standard in testing knock of gasoline engines; solvent

Exposure. Inhalation

Toxicology. *n*-Heptane causes central nervous system depression.

Human subjects exposed to 1000

ppm for 6 minutes, or to 2000 ppm for 4 minutes, reported slight vertigo.[1] At 5000 ppm for 4 minutes, effects included marked vertigo, inability to walk a straight line, hilarity, and incoordination but no complaints of eye and upper respiratory tract or mucous membrane irritation.[1] In some subjects, a 15-minute exposure at 5000 ppm produced a state of stupor lasting for 30 minutes after exposure. These subjects also reported loss of appetite, slight nausea, and a taste resembling gasoline for several hours after exposure.[1]

Dermal application resulted in immediate irritation characterized by erythema and hyperemia. The subjects complained of painful burning sensation, and, after 5 hours, blisters formed on the exposed areas.[2]

Mice exposed to *n*-heptane at 32,000 ppm for 5 minutes developed irregular respiratory patterns.[3] At 48,000 ppm, three of four mice had respiratory arrest within 4 minutes.[3]

REFERENCES

1. Patty FA, Yant WP: Report of Investigations—Odor Intensity and Symptoms Produced by Commercial Propane, Butane Pentane, Hexane and Heptane Vapor, No 2979. US Department of Commerce, Bureau of Mines, 1929
2. National Institute for Occupational Safety and Health: Criteria for a Recommended Standard . . . Occupational Exposure to Alkanes, (C5–C8), DHEW (NIOSH) Pub No 77–151. Washington, DC, US Government Printing Office, 1977
3. Swann HE Jr, Kwon BK, Hogan GK, Snellings WM: Acute inhalation toxicology of volatile hydrocarbons. Am Ind Hyg Assoc J 35:511–518, 1974

HEXACHLOROETHANE

CAS: 67-72-1

CCl_3CCl_3 1987 TLV = 10 ppm

Synonyms: Carbon hexachloride; perchloroethane

Physical Form. Colorless crystals

Uses. Chemical intermediate in the manufacture of pyrotechnics, insecticides, and other chlorinated materials

Exposure. Inhalation; skin absorption

Toxicology. Hexachloroethane is an eye irritant and causes kidney and central nervous system effects in animals. At high doses, it is carcinogenic to mice.

Exposure of workmen to fumes from hot hexachloroethane resulted in blepharospasm, photophobia, lacrimation, and reddening of the conjunctiva but no corneal injury or permanent damage.[1] No chronic effects have been reported from industrial exposure, although significant skin absorption is said to occur.[2]

Rats exposed to 5900 ppm for 8 hours showed ataxia, tremor, and convulsions, and two of six rats died.[1] At 260 ppm for 8 hours, there were no toxic signs, but repeated exposure to this concentration 6 hours/day, 5 days/week caused tremor, red exudate around the eyes, and some deaths after 4 weeks. Dogs exposed at 260 ppm developed tremors, ataxia, hypersalivation, and facial muscular fasiculations and held their eyelids closed during the exposure; three of four dogs survived 6 weeks of repeated exposures. No treatment-related effects were found in a number of species repeatedly exposed at 48 ppm.[1]

Rats fed 62 mg/kg/day for 16 weeks exhibited no overt toxicity.[2] The dermal LD_{50} for male rabbits was greater than

32 g/kg.[7] Applied to rabbit skin for 24 hours, the dry material caused no skin irritation, whereas a water paste caused slight redness.[1] In the eyes of five of six rabbits, 1 g of the crystal overnight caused moderate corneal opacity, iritis, severe swelling, and discharge.[1]

Gavage administration of 590 and 1179 mg/kg/day to mice for 78 weeks caused a significant increase in the incidence of hepatocellular carcinomas, whereas no increase in these tumors was observed in rats given 212 or 423 mg/kg/day. A nonsignificant increase in renal tumors was seen in rats, and tubular nephropathy occurred in both species.[3] There is limited evidence that hexachloroethane is carcinogenic in experimental animals, according to the IARC.[4]

Hexachloroethane has a camphor-like odor, readily sublimes, and, when heated to decomposition, it emits phosgene.[1]

REFERENCES

1. Weeks MH et al: The toxicity of hexachloroethane in laboratory animals. Am Ind Hyg Assoc J 40:187–199, 1979
2. Gorzinski SJ et al: Subchronic oral toxicity, tissue distribution and clearance of hexachloroethane in the rat. Drug Chem Toxicol 8:155–169, 1985
3. National Cancer Institute: Bioassay of Hexachloroethane for Possible Carcinogenicity, TR-68. DHEW Pub No (NIH) 78–1318. Washington, DC, US Government Printing Office, 1978
4. IARC Monographs on the Evaluation of the Carcinogenic Risks of Chemicals to Humans, Some Halogenated Hydrocarbons, Vol 20, pp 467–473. Lyon, International Agency for Research on Cancer, 1979

HEXACHLORONAPHTHALENE

CAS: 1335-87-1 1987 TLV = 0.2 mg/m^3;
$C_{10}H_2Cl_6$ skin

Synonyms: Halowax 1014

Physical Form. Waxy, yellow-white solid

Uses. In synthetic wax; in electric wire insulation; in lubricants

Exposure. Inhalation; skin absorption

Toxicology. Hexachloronaphthalene is toxic to the liver and causes chloracne.

Human fatalities from acute yellow atrophy of the liver have occurred with repeated exposure to penta- and hexachloronaphthalene.[1,2] Air measurements showed concentrations averaging 1 to 2 mg/m^3. Other workers experienced jaundice, nausea, indigestion, and weight loss.

The most common problem, a severe acne form of dermatitis termed *chloracne,* typically occurs from long-term contact with the fume or dust or shorter contact with the hot vapor.[3] The reaction is usually slow to appear and may take months to return to normal.

Repeated exposure of rats to an average concentration of 8.9 mg/m^3 of a mixture of penta- and hexachloronaphthalene produced jaundice and was fatal; the liver showed a marked fatty degeneration and centrolobular necrosis.[3] At 1.16 mg/m^3, minor liver injury still occurred.[3]

REFERENCES

1. Hygienic Guide Series: Chloronaphthalenes. Am Ind Hyg Assoc J 27:89–92, 1966
2. Elkins HB: The Chemistry of Industrial Toxicology, 2nd ed, pp 151–152. New York, John Wiley & Sons, 1959
3. Deichmann WB: Halogenated cyclic hydrocarbons. In Clayton GD, Clayton FE (eds): Patty's Industrial Hygiene and Toxicology, 3rd ed, Vol

2B, Toxicology, pp 3669–3675. New York, Wiley–Interscience, 1981

n-HEXANE

CAS: 110-54-3

$CH_3(CH_2)_4CH_3$ 1987 TLV = 50 ppm

Synonym: Hexane

Physical Form. Colorless, very volatile liquid solvent and thinner

Uses. Solvent, in production of tires, glues, tape, and bandages

Exposure. Inhalation

Toxicology. *n*-Hexane is an upper respiratory irritant and causes central nervous system depression; chronic exposure causes peripheral neuropathy.

In human subjects, 2000 ppm for 10 minutes produced no effects, but 5000 ppm resulted in dizziness and confusion.[1] Other investigators reported the presence of slight nausea, headache, and irritation of the eyes and throat at 1500 ppm.[2] In industrial practice, mild symptoms of narcosis such as dizziness have been observed when concentrations of solvents containing various isomers of hexane exceeded 1000 ppm but not below 500 ppm.[3]

Dermal exposure to hexane caused immediate irritation characterized by erythema and hyperemia.[4] Subjects complained of painful burning sensations with itching, and, after 5 hours, blisters formed on the exposed areas.[4]

Polyneuropathy has been reported following chronic occupational exposure to vapors containing *n*-hexane at concentrations typically in the range of 400 to 600 ppm, with some ceiling exposures up to 2500.[4–6] One person developed polyneuropathy after 1 year of exposure at 54 to 200 ppm.[5] Initial symptoms may include sensation disturbances, muscle weakness, and distal symmetric pain in the legs after 2 to 6 months of exposure.[5] Clinical changes include muscle atrophy, hypotonic decreased muscle strength, footdrop, and paresthesias in the arms and legs. Decreased motor nerve conduction and electromyographic indications of neurogenic damage are found.[5] Peripheral nerve biopsies show significant swelling of the nerve with thinning of the myelin sheath.[5] Functional disturbances commonly progress for 2 to 3 months after cessation of exposure. Recovery may be expected within a year, but, in some cases, clinical polyneuropathy has remained after 2 years.[5] Changes in color vision, in retinal pigmentation, and in perifoveal capillaries were found in workers exposed to 420 to 1280 ppm for more than 5 years.[5]

Experimental animals continuously exposed to pure *n*-hexane developed the same clinical, electrophysiologic, and histopathologic changes found in humans exposed to mixed vapors containing *n*-hexane. Continuous inhalation by rats of 400 ppm caused axonapathy.[7] Interestingly, intermittent exposure of rats to 10,000 ppm 6 hours/day, 5 days/week for 13 weeks caused only slight paranodal axonal swelling.[8] It is postulated that 2,5–hexanedione, a metabolite of *n*-hexane and purported neurotoxic agent, must build to an effective concentration. With continuous exposure, there is no recovery during each day or week.[8]

In regard to reproductive effects, the only difference found in rats exposed to 1000 ppm during gestation was in their offspring, which weighed less than expected at ages 1 week through 6 weeks.[9]

The neurotoxic properties of *n*-hexane are potentiated by exposure to methyl ethyl ketone (*qv*). Because other compounds may also have this effect, human exposure to mixed solvents containing any neurotoxic hexacarbon compound should be minimized.[6]

REFERENCES

1. Patty FA, Yant WP: Report of Investigations—Odor Intensity and Symptoms Produced by Commercial Propane, Butane, Pentane, Hexane, and Heptane Vapor, No 2979. US Department of Commerce, Bureau of Mines, 1929
2. Drinker P et al: The threshold toxicity of gasoline vapor. J Ind Hyg Toxicol 25:225–232, 1943
3. Elkins HB: Chemistry of Industrial Toxicology, p 101. New York, John Wiley & Sons, 1959
4. National Institute for Occupational Safety and Health: Criteria for a Recommended Standard . . . Occupational Exposure to Alkanes, (C5–C8) DHEW (NIOSH) Pub No 77–151. Washington, DC, US Government Printing Office, 1977
5. Jorgensen NK, Cohr KH: *n*-Hexane and Its Toxicological Effects. Scand J Work Environ Health 7:157–168, 1981
6. Spencer PS et al: The enlarging view of hexacarbon neurotoxicity. RC Crit Rev Toxicol 7:279–356, 1980
7. Schaumburg NH, Spencer PS: Degeneration in central and peripheral nervous systems produced by pure *n*-hexane: An experimental study. Brain 99:183–192, 1976
8. Cavender FL et al: A 13-week vapor inhalation study of *n*-hexane in rats with emphasis on neurotoxic effects. Fund Appl Toxicol 4:191–201, 1984
9. Bus JS et al: Perinatal toxicity and metabolism of *n*-Hexane in Fischer-344 rats after inhalation exposure during gestation. Toxicol Appl Pharmacol 511:295–302, 1979

---

sec-HEXYL ACETATE

CAS: 108-84-9

$C_8H_{15}O_2$ — 1987 TLV = 50 ppm

Synonyms: Methyl amylacetate; 4-methyl pentyl 2-acetate; 1,3-dimethylbutyl acetate; methyl isoamyl acetate

Physical Form. Clear liquid

Uses. Lacquer industry; fragrances

Exposure. Inhalation

Toxicology. *sec*-Hexyl acetate causes irritation of the eyes and upper respiratory tract; at concentrations approaching saturation, it causes narcosis in animals, and it is expected that similar exposure will cause the same effect in humans.

Human volunteers exposed to 100 ppm for 15 minutes experienced eye irritation and objected to the odor and taste; nose and throat irritation occurred at levels greater than 100 ppm.[1] No chronic or systemic effects in humans have been reported.

Four of six rats survived exposure to 4000 ppm for 4 hours, but 8000 ppm was lethal to all animals.[2,3]

The liquid was poorly absorbed through rabbit skin but did cause moderate irritation.[2,4] Little corneal injury resulted from instillation in the eye.[4]

REFERENCES

1. Silverman L, Schulte HF, First MW: Further studies on sensory response to certain industrial solvent vapors. J Ind Hyg Toxicol 28:262–266, 1946
2. Smyth HF Jr, Carpenter CP, Weil CS, Pozzani UC: Range-finding toxicity data. Arch Ind Hyg Occup Med 10:61–68, 1954
3. *sec*-Hexyl acetate. Documentation of the TLVs and BEIs, 5th ed, p 308. Cincinnati, American Conference of Governmental Industrial Hygienists (ACGIH), 1986
4. Carpenter CP et al: Range-finding toxicity data: List VIII. Toxicol Appl Pharmacol 28:313–319, 1974

HYDRAZINE

CAS: 302-01-2 — 1987 TLV = 0.1 ppm; skin

NH_2NH_2

Synonyms: Hydrazine anhydrous

Physical Form. Colorless oily liquid, fuming in air

Uses. Reducing agent; organic hydrazine derivatives; rocket fuel

Exposure. Inhalation; skin absorption

Toxicology. Hydrazine is a severe skin and mucous membrane irritant, a convulsant, a hepatotoxin, and a moderate hemolytic agent; in animals, it is a carcinogen.

In humans, the vapor is immediately irritating to the nose and throat and causes dizziness and nausea; itching, burning, and swelling of the eyes develop over a period of several hours.[1] Severe exposure of the eyes to the vapor causes temporary blindness lasting for about 24 hours.[2] The liquid in the eyes or on the skin causes severe burns.[1] Hydrazine and its salts will also produce skin irritation and allergic reactions in humans.[1]

In humans, hydrazine is absorbed through the skin as well as by inhalation. In one case attributed to hydrazine hydrate exposure, systemic effects included weakness, vomiting, excited behavior, and tremor; the chief histologic findings were severe tracheitis and bronchitis, fatty degeneration of the liver, and nephritis.[3]

The LC_{50} values for rats and mice were 570 ppm and 252 ppm, respectively.[4] The exposed rodents were restless and had breathing difficulties and convulsions.[4] Exposure of mice, rats, dogs, and monkeys to 1.0 and 5.0 ppm 6 hours/day, 5 days/week or at levels of 0.2 and 1.0 ppm continuously caused a variety of effects. Increased mortality occurred in mice and was attributed to liver damage; rats showed a dose-related growth depression; dogs also had increased mortality and developed depressed erythrocyte counts, hematocrit values, and hemoglobin concentrations at higher doses; there were no effects in monkeys.[1] Lipid deposition in the kidneys of monkeys has been reported following intraperitoneal administration of hydrazine.[1]

Hydrazine or hydrazine salts are carcinogenic in mice after oral administration (pulmonary adenocarcinoma; hepatocarcinoma) or intraperitoneal injection (pulmonary carcinoma), and in rats after oral administration (pulmonary adenocarcinoma).[5] Hydrazine induced a significantly greater incidence of nasal tumors, primarily benign, in rats and hamsters after 1-year intermittent inhalation exposure at levels up to 5.0 ppm.[6]

A group of 427 men with varying degrees of occupational exposure to hydrazine showed no increase in overall mortality (49 deaths vs 61.47 expected) or deaths from lung cancer (5 vs 6.65 expected).[7]

The IARC has determined that there is sufficient evidence of carcinogenicity in animals and inadequate evidence in humans.[5]

REFERENCES

1. National Institute for Occupational Safety and Health: Criteria for a Recommended Standard . . . Occupational Exposure to Hydrazines, DHEW (NIOSH) 78–172, pp 1–269. Washington, DC, US Government Printing Office, June 1978
2. Comstock CC, Lawson LH, Greene EA, Oberst FW: Inhalation toxicity of hydrazine vapor. AMA Arch Ind Hyg Occup Med 10:476–490, 1954
3. Sotanieme E et al: Hydrazine toxicity in the human—report of a fatal case. Ann Clin Res 3:30–33, 1971
4. Jacobson KH et al: The acute toxicity of the vapors of some methylated hydrazine derivatives. Arch Ind Health 12:609–616, 1955
5. IARC Monographs on the Evaluation of the Carcinogenic Risk of Chemicals to Man, Suppl 4, Chemicals, Industrial Processes and Industries Associated with Cancer in Humans, pp 136–138. Lyon, International Agency for Research on Cancer, 1982
6. Vernot EH et al: Long-term inhalation toxicity of hydrazine. Fund Appl Toxicol 5:1050–1064, 1985
7. Wald N et al: Occupational exposure to hydra-

zine and subsequent risk of cancer. Br J Ind Med 41:31–34, 1984

HYDROGEN BROMIDE

CAS: 10035-10-6

HBr 1987 TLV = C 3 ppm

Synonyms: Hydrobromic acid; anhydrous hydrobromic acid

Physical Form. Colorless, nonflammable gas

Uses. Manufacture of organic and inorganic bromides; reducing agent, catalyst in oxidations; alkylation of aromatic compounds

Exposure. Inhalation

Toxicology. Hydrogen bromide gas is an irritant of the eyes, mucous membranes, and skin.

There are no systemic effects reported from industrial exposure. Experimental exposure of humans to 5 ppm for several minutes caused nose and throat irritation in most persons, and a few people were affected at concentrations of 3 to 4 ppm.[1] Contact of solutions with the eyes, skin, or mucous membranes may cause burns.[2]

The 1-hour inhalation LC_{50} is 2860 ppm for rats and 815 ppm for mice.[3]

REFERENCES

1. Hydrogen bromide. Documentation of the TLVs and BEIs, 5th ed, p 312. Cincinnati, American Conference of Governmental Industrial Hygienists (ACGIH), 1986
2. Alexandrov DD: Bromine and compounds. In International Labor Office: Encyclopaedia of Occupational Health and Safety, 3rd ed, rev, Vol I, A–K, p 327. Geneva, 1983
3. Registry of Toxic Effects of Chemical Substances (RTECS). DHHS (NIOSH) Pub No 86–103, p 1024. Washington, DC, US Government Printing Office, Nov 1985

HYDROGEN CHLORIDE

CAS: 7647-01-0

HCl 1987 TLV = C 5 ppm

Synonyms: HCl; hydrogen chloride, anhydrous; hydrochloric acid, aqueous; muriatic acid

Physical Form. Gas (aqueous solution is hydrochloric acid)

Uses. Production of chlorinated organic chemicals; production of dyes and dye intermediates; steel-pickling; oil well acidizing operations to dissolve subsurface dolomite or limestone

Exposure. Inhalation

Toxicology. Hydrogen chloride is a strong irritant of the eyes, mucous membranes, and skin.

The major effects of acute exposure are usually limited to the upper respiratory tract and are sufficiently severe to encourage prompt withdrawal from a contaminated atmosphere.[1] Exposure to the gas immediately causes cough, burning of the throat, and a choking sensation. Effects are usually limited to inflammation and occasionally ulceration of the nose, throat, and larynx.[2] Acute exposures causing significant trauma are usually limited to people who are prevented from escaping; in such cases, laryngeal spasm or pulmonary edema may occur.

In workers, exposure to 50 to 100 ppm for 1 hour was barely tolerable; short exposure to 35 ppm caused irritation of the throat, and 10 ppm was considered the maximal concentration allowable for prolonged exposure.[3] In one study, workers chronically exposed to hydrogen chloride did not exhibit the pulmonary function changes observed in naive subjects exposed to similar concentrations; this observation suggests acclimatization of the workers to hydrogen chloride.[4]

Exposure of the skin to a high concentration of the gas or to a concentrated solution of the liquid (hydrochloric acid) will cause burns; repeated or prolonged exposure to dilute solutions may cause dermatitis.[2] Erosion of exposed teeth may occur from repeated or prolonged exposure. Although ingestion is unlikely, hydrochloric acid causes severe burns of the mouth, esophagus, and stomach, with consequent pain, nausea, and vomiting.[5]

Chronic exposure to 10 ppm, 6 hours/day for life did not cause any neoplastic lesions or serious irritating effects in the nasal epithelium of rats.[6]

Warning properties are good; most people can detect 5 ppm.[1]

REFERENCES

1. Committee on Medical and Biologic Effects of Environmental Pollutants: Chlorine and Hydrogen Chloride, pp 138–144. Washington, DC, National Academy of Sciences, 1976
2. Chemical Safety Data Sheet SD–39, Hydrochloric Acid, pp 5–6, 24–26. Washington, DC, MCA, Inc, 1970
3. Henderson Y, Haggard HW: Noxious Gases, p 126. New York, Reinhold, 1943
4. Toyama T, Kondo T, Nakamura K: Environments in acid aerosol producing workplaces and maximum flow rate of workers. Jap J Ind Health 4:15–22, 1962
5. Poteshman NL: Corrosive gastritis due to hydrochloric acid ingestion. Am J Roentgenol Radium Ther Nucl Med 99:182–185, 1967
6. Sellakumar AR et al: Carcinogenicity of formaldehyde and hydrogen chloride in rats. Toxicol Appl Pharmacol 81:401–406, 1985

HYDROGEN CYANIDE

CAS: 74-90-8 1987 TLV = C 10 ppm;
HCN skin

Synonyms: Hydrocyanic acid; aero liquid HCN; prussic acid; formonitrile

Physical Form. Colorless gas liquefying at 26°C (may be found in the workplace both as a liquid and a gas)

Uses. Rodenticide and insecticide; fumigant; chemical intermediate for the manufacture of synthetic fibers, plastics, and nitrites

Exposure. Inhalation; skin absorption; ingestion

Toxicology. Hydrogen cyanide can cause rapid death due to metabolic asphyxiation.

Cyanide ion exerts an inhibitory action on certain metabolic enzyme systems, most notably cytochrome oxidase, the enzyme involved in the ultimate transfer of electrons to molecular oxygen.[1] Since cytochrome oxidase is present in practically all cells that function under aerobic conditions, and since the cyanide ion diffuses easily to all parts of the body, cyanide quickly halts practically all cellular respiration. The venous blood of a patient dying of cyanide is bright red and resembles arterial blood because the tissues have not been able to utilize the oxygen brought to them.[2] Cyanide intoxication produces lactic acidosis, probably the result of increased rate of glycolysis and production of lactic acid.[3]

A concentration of 270 ppm hydrogen cyanide has long been quoted as being immediately fatal to humans. A more recent study, however, states that the estimated LC_{50} to humans for a 1-minute exposure is 3404 ppm.[1] Others state that 270 ppm is fatal after 6 to 8 minutes, 181 ppm after 10 minutes, and 135 ppm after 30 minutes.[1]

If large amounts of cyanide have been absorbed, collapse is usually instantaneous, the patient falling unconscious, often with convulsions, and dying almost immediately.[1,2] Symptoms of intoxication from less severe exposure include weakness, headache, confusion, vertigo, fatigue, anxiety, dyspnea, and occasionally nausea and vomiting.[1,2] Respiratory rate and depth is usually increased initially and at later stages becomes slow and gasping. Coma and convulsions occur in some cases. If cyanosis is present, it usually indicates that respiration has either ceased or has been very inadequate for a few minutes.

Hydrogen cyanide has recently been recognized in significant concentrations in some fires, as a combustion product of wool, silk, and many synthetic polymers; it may play a role in toxicity and deaths from smoke inhalation.[4]

Chronic cyanide poisoning from occupational exposure, at least in any serious or incapacitating form, is apparently rare, if it occurs at all.

Most reported cases involved workers with a mixture of repeated acute or subacute exposures and chronic or prolonged low-level exposures, making it unclear whether symptoms resulted simply from multiple acute exposures with acute intoxication. Some symptoms persisted after cessation of such exposures, perhaps owing to the effect of anoxia from inhibition of cytochrome oxidase. Symptoms from chronic exposure are similar to those reported after acute exposures, such as weakness, nausea, headache, and vertigo.[1] A study of 36 former workers in a silver-reclaiming facility who were chronically exposed to cyanide demonstrated some residual symptoms 7 or more months after cessation of exposure; frequent headache, eye irritation, easy fatigue, loss of appetite, and epistaxis occurred in at least 30% of these workers.[5]

Liquid hydrogen cyanide, hydrogen cyanide in aqueous solution (hydrocyanic acid), and concentrated vapor are absorbed rapidly through the intact skin and may cause poisoning with little or no irritant effect on the skin itself.[1] The liquid in the eye may cause some local irritation; the attendant absorption may be hazardous.[6]

Cyanide is one of the few toxic substances for which a specific antidote exists, and it functions as follows.[7,8] First, amyl nitrite (by inhalation) and then sodium nitrite (intravenously) are administered to form methemoglobin, which binds firmly with free cyanide ions. This traps any circulating cyanide ions. The formation of 10% to 20% methemoglobin does not usually involve appreciable risk yet provides a large amount of cyanide-binding substance. Second, sodium thiosulfate is administered intravenously to increase the rate of conversion of cyanide to the less toxic thiocyanate. Although early literature suggests the use of methylene blue, it must not be administered because it is a poor methemoglobin-former, and it promotes the conversion of methemoglobin back to hemoglobin.[7]

Note: The therapeutic regimen (amyl nitrite/sodium nitrite plus sodium thiosulfate) discussed previously and detailed in the *Treatment* section that follows is the "classical" one that has long been recommended. However, there is growing support for a new regimen of 100% oxygen together with intravenous hydroxocobalamin and/or sodium thiosulfate.[3,9] Hydroxocobalamin reverses cyanide toxicity by combining with cyanide to form cyanocobalamin (vitamin B_{12}); one advantage is that hydroxocobalamin is apparently of low toxicity. The only side effects have been occasional urticaria and a brown-red discoloration of the urine. Hydroxocobalamin is not currently available or approved for treatment of cyanide intoxication in the United States.[4] Nonspecific supportive therapy is an essential part of the treatment; in two cases, patients have recovered from massive cyanide poisoning with supportive therapy only.[3]

At high levels, cyanide acts so rapidly that its odor has no value as forewarning.[1] At lower levels, the odor may provide some forewarning, although many persons are unable to recognize the bitter almond scent.[3]

Diagnosis. *Signs and Symptoms:* Asphyxia and death can occur from high exposure levels; weakness, headache, confusion, nausea, and vomiting result from lesser exposures; increased rate and depth of respiration, slow and gasping respiration, and pulmonary edema are also seen.

Differential Diagnosis: The diagnosis is usually self-evident from history, but even if it is not completely established and cyanide poisoning is suspected, the recommended therapy should be administered.[7]

Special Tests: A blood level of cyanide in excess of 0.2 μg/ml suggests a toxic reaction, but because of tight binding of cyanide to cytochrome oxidase, serious poisoning may occur with only modest blood levels, especially several hours after poisoning.[9] In view of the likelihood of lactic acidosis, there should be measurement of blood *p*H, plasma bicarbonate, and blood lactic acid.[3]

Treatment. Preparedness and speed of action are prerequisites for successful treatment for any overexposure to cyanides.[1,2,7,8] All persons working with or around cyanides should be given specific and detailed instructions on the use of antidote kits containing amyl nitrite ampules for inhalation. Kits for use by medical personnel should contain the following:

2 boxes of amyl nitrite ampules, 0.3 ml/ampule
2 ampules of sterile sodium nitrite, 10 mg of 3% solution each
2 ampules of sterile sodium thiosulfate, 50 ml of 25% solution each
2 sterile 10-ml syringes with intravenous needles
1 sterile 50-ml syringe with intravenous needles
1 tourniquet
1 gastric tube (rubber)
1 nonsterile 100-ml syringe

A complete cyanide antidote package can be purchased from the Eli Lilly Co, stock number M76.

The exposed person should be removed as rapidly as possible to an uncontaminated area by someone with adequate respiratory protective equipment.

Resuscitate: If breathing has stopped, immediately institute resuscitation and continue until normal breathing is established.

Decontaminate: If liquid cyanide has contaminated the skin or clothing, the clothing should be removed and the skin flushed with copious amounts of water. If taken by mouth, gastric lavage should be performed but should be postponed until after specific therapy has been initiated.

Administer Amyl Nitrite: A pearl (ampule), if not provided with a fabric sleeve, should be wrapped lightly in a handkerchief or gauze pad, broken, and held about 1 inch from the patient's mouth and nostrils for 15 to 30 seconds of every minute while the sodium nitrite solution is being prepared. Use a fresh pearl every 5 minutes.

Administer Sodium Nitrite and Sodium Thiosulfate: If the patient does not respond to the amyl nitrite administration or if severe exposure is suspected, administer 0.3 g sodium nitrite (10 ml of a 3% solution) intravenously at the rate of 2.5 to 5 ml/minute followed by injection of 12.5 g sodium thiosulfate (50 ml of a 25% solution) over about 10 minutes by the same needle and vein.

Observe Patient: The blood levels of methemoglobin should be monitored and not allowed to exceed 40%.[9] The patient should be kept under observation for 24 to 48 hours. If signs of intoxication persist or reappear, the injection of nitrite and thiosulfate should be repeated in one half the aforementioned doses. Even if the patient appears well, a second injection may be given 2 hours after the first injection for prophylactic purposes. *Note:* The use of oxygen should not be ignored, because the methemoglobinemia induced by this treatment reduces the ability of blood to carry oxygen to the brain.

REFERENCES

1. National Institute for Occupational Safety and Health: Criteria for a Recommended Standard . . . Occupational Exposure to Hydrogen Cyanide Salts (NaCN, KCN, and $Ca[CN]_2$). DHEW (NIOSH) Pub No 77–108, pp 37–95, 106–114, 170–173, 178. Washington, DC, US Government Printing Office, 1976
2. Gosselin RE, Smith RP, Hodge HC: Clinical Toxicology of Commercial Products, Section III, 5th ed, pp 123–130. Baltimore, Williams & Wilkins, 1984
3. Graham DL, Laman D, Theodore J, Robin ED: Acute cyanide poisoning complicated by lactic acidosis and pulmonary edema. Arch Intern Med 137:1051–1055, 1977
4. Becker CE: The role of cyanide in fires. Vet Hum Toxicol 27:487–490, 1985
5. Blanc P et al: Cyanide intoxication among silver-reclaiming workers. JAMA 253:367–371, 1985
6. Hygienic Guide Series: Hydrogen Cyanide. Am Ind Hyg Assoc J 31:116–119, 1970
7. Chen KK, Rose CL: Nitrite and thiosulfate therapy in cyanide poisoning. JAMA 149:113–119, 1952
8. Wolfsie JH: Treatment of cyanide poisoning in industry. AMA Arch Ind Hyg Occup Med 4:417–425, 1951
9. Berlin C: Cyanide poisoning—a challenge. Arch Intern Med 137:993–994, 1977

HYDROGEN FLUORIDE

CAS: 7664-39-3

HF 1987 TLV = C 3 ppm

Synonyms: None

Physical Form. Gas, liquefying at 19.5°C; aqueous solution is hydrofluoric acid

Uses. Catalyst for production of high-octane gasoline; aqueous solution for frosting, etching, and polishing glass, for removing sand from metal casings, and for etching silicon wafers in semiconductor manufacture

Exposure. Inhalation; skin contact

Toxicology. Hydrogen fluoride (HF) as a gas is a severe respiratory irritant and, in solution, causes severe and painful burns of the skin and eyes.

From accidental, occupational, and volunteer exposures, it is estimated that the lowest lethal concentration for a 5-minute human exposure to hydrogen fluoride is in the range of 50 to 250 ppm.[1] The LC_{50} for 5, 15, and 60 minutes are considered to be 500 to 800 ppm, 450 to 1000 ppm, and 30 to 600 ppm, respectively.[1]

Inhalation of HF produces transient choking and coughing. After an asymptomatic period of several hours to 1 to 2 days, fever, cough, dyspnea, cyanosis, and pulmonary edema may develop. Death from pulmonary edema occurred within 2 hours in three of six workers splashed with 70% solution despite prompt showering with water. The HF concentration in the breathing zone was estimated to be above 10,000 ppm.[2] A chemist exposed to HF splashes on the face and upper extremities developed pulmonary edema 3 hours after exposure and died 10 hours later.[3] Significant systemic absorption by dermal or inhalation exposure may result in hypocalcemia and hypomagnesemia; cardiac arrhythmias may result as a consequence.[4,5] Persistent respiratory symptoms, including hoarseness, coughing fits, and nose bleeds, yet with normal pulmonary function, have been observed in one subject who survived a massive exposure. Acute renal failure of uncertain cause has also been documented after an ultimately fatal inhalation exposure.[6]

In human subjects, exposure to 120 ppm for 1 minute caused conjunctival and respiratory irritation with stinging of skin.[7]

At 30 ppm for several minutes, there was mild irritation of the eyes, nose, and respiratory tract; 2.6 to 4.8 ppm for periods up to 50 days caused slight irritation of nose, eyes, and skin but no signs or symptoms of pulmonary irritation.[7,8]

Repeated exposure to excessive concentrations of fluoride over a period of years results in increased radiographic density of bone and eventually may cause crippling fluorosis (osteosclerosis owing to deposition of fluoride in bone).[7] The early signs of increased bone density from fluoride deposition are most apparent in the lumbar spine and pelvis and can be detected by x-ray.

Biological monitoring of urinary fluoride concentration provides an indication of total fluoride intake. Data indicate that a postshift urinary fluoride level of less than 8 mg/liter, averaged over an extended period of time, will not lead to osteosclerosis, although a minimal or questionable increase in bone density might develop after many years of occupational exposure.[7]

HF solutions (hydrofluoric acid) in contact with skin result in marked tissue destruction; undissociated HF readily penetrates skin and deep tissue where the corrosive fluoride ion can cause necrosis of soft tissues and decalcification of bone; the destruction produced is excruciatingly painful.[5,9–11] Tendinitis and tenosynovitis may result. Fluoride ion also attacks enzymes (e.g., of glycolysis) and cell membranes. The process of tissue destruction and neutralization of the hydrofluoric acid is prolonged for days, unlike other acids that are rapidly neutralized.[9–11] Because of the insidious manner of penetration, a relatively mild or minor exposure can cause a serious burn. When there is skin contact with solutions of less than 20%, the burn manifests itself by pain and erythema, with a latent period of up to 24 hours; with 20% to 50% solutions, the burn becomes apparent 1 to 8 hours following exposure; solutions above 50% cause immediate pain, and tissue destruction is rapidly apparent.[9] Delayed recognition of contact with dilute solutions with consequently delayed irrigation often results in more severe burns.[5] Depending upon the severity of the burn, it may demonstrate erythema alone, central blanching with peripheral erythema, swelling, vesiculation, and serous crusting; with more serious burns, ulceration, blue-gray discoloration, and necrosis may be noted.[4,5]

Severe eye injuries from splashes may occur. In one case of eye burns from a fine spray of hydrofluoric acid in the face, considerable loss of epithelium occurred despite immediate and copious flushing with water and irrigation for 3 hours with a 0.5% solution of benzethonium chloride; within 19 days, there was recovery of normal vision.[12]

Diagnosis. *Signs and Symptoms:* Severe eye, nose, and throat irritation; delayed fever, cyanosis, and pulmonary edema; and severe and painful skin and eye burns from splashes of solutions may be present. Prolonged or repeated exposure to low concentrations of the gas may cause nasal congestion and bronchitis.

Differential Diagnosis: In mild exposure resulting in mucous membrane irritation, the symptoms may mimic a viral upper respiratory tract infection. The latter may be characterized by fever, myalgias, and lymphocytosis.

As the tracheobronchial tree and pulmonary parenchyma become involved, the symptoms and signs must be differentiated from cardiogenic pulmonary edema, severe viral or bacterial pneumonia, and adult respiratory distress syndrome.

Special Tests: Diagnostic studies should include electrocardiogram, sputum gram stain and culture, serum calcium and magnesium, differential white blood count, and arterial blood gas analysis. The determination of fluoride concentration in the urine may be used to gauge the degree of absorption of HF following suspected overexposure. Dietary intake of water high in fluorides may increase the urinary fluoride level.[6]

Treatment. The high risk of either immediate or delayed onset of pulmonary edema following inhalation of the gas requires that oxygen be administered immediately after a severe exposure and continued as long as necessary; close observation should be continued for 24 to 48 hours. A chest x-ray and arterial blood gases should be obtained. Some authors recommend the administration of a 2% to 3% calcium gluconate solution as a nebulized mist for significant inhalations, although the value of this therapy is, as yet, unproven.[5] Along with oxygen administration, intubation and mechanical ventilation with positive end expiratory pressure may be required in cases of noncardiogenic pulmonary edema. Treatment is similar to that of noncardiogenic pulmonary edema from other causes; the value of a short course of systemic steroids has not been established.[5]

Persons who have had contact with hydrofluoric acid should be subjected immediately to a drenching shower of water for at least 10 minutes. Contaminated clothing should be removed as rapidly as possible, even while the person is under the shower. It is essential that the exposed area be washed with copious quantities of water for a sufficient period of time to remove all hydrofluoric acid from the skin or eyes. Flushing is effective in removed surface HF, but not that which has penetrated to deeper tissues. The affected area is immediately soaked with a soft, bulky dressing that has been immersed in iced solutions of the quaternary ammonium compounds, 0.13% benzalkonium chloride (Zephiran) or 0.2% benzethonium chloride (Hyamine). An iced 25% solution of magnesium sulfate (epsom salts) or a calcium gluconate gel (if available) is also acceptable for topical therapy. The magnesium sulfate solution may be prepared by adding ½ to 1 cup of epsom salts to 1 quart of water. If the offending agent contained less than 20% hydrofluoric acid and if pain is relieved, treatment need not go beyond iced topical soaks for 1 to 4 hours.[4,5] Meticulous and frequent follow-up is recommended.

If the burns appear to be deep or if there is exquisite pain, particularly if the concentration of acid was greater than 20%, the painful areas should be cautiously injected with 10% calcium gluconate. Similarly, if topical therapy fails to eliminate pain within 30 to 60 minutes, injection should be considered. The calcium gluconate should be injected in small quantities in order not to distend the tissues and by a 25- or 30-gauge needle. By utilizing the patient's pain as a monitor, the smallest effective amount of calcium gluconate may be determined. Thus, local anesthesia should be avoided unless it is absolutely necessary. Without anesthesia, the patient can accurately localize the areas requiring treatment. The calcium gluconate is infiltrated directly into the affected dermis and subcutaneous tissue using a technique similar to the infiltration of a local anesthetic agent. Approximately 0.5 ml of calcium gluconate/cm^2 of burned surface area is a rough guide to the usual effective dose.[5] The infiltration is carried 0.5 cm away from the margin of the obviously injured tissue into the surrounding apparently uninjured area. After the calcium gluconate injection, the burnt area of patients with severe burns may be carefully débrided. All bullae should be débrided, cleansed, and probably infiltrated with calcium gluconate. The physician should not hesitate to remove the fingernail if there is any question of serious subungual exposure. If débridement is performed, the patient should probably be hospitalized. The hand should be dressed in a soft, bulky dressing, elevated, and observed carefully for the next 48 hours. If the pain recurs, additional calcium gluconate injections should be given. Regional intra-arterial infusion of calcium gluconate is an investigational therapy for severe burns, which has the advantage of not resulting in local tissue distention from subcutaneous infiltration.

REFERENCES

1. Halton DM et al: Toxicity Levels to Humans During Acute Exposure to Hydrogen Fluoride, p 40. Ottawa, Canada, Atomic Energy Control Board, November 28, 1984
2. Mayer L, Geulich J: Hydrogen fluoride (HF) inhalation and burns. Arch Environ Health 7:445–447, 1963
3. Kleinfeld M: Acute pulmonary edema of chemical origin. Arch Environ Health 10:942–946, 1965
4. White JW: Hydrofluoric acid burns. Cutis 34:241–244, 1984
5. Edelman P: Hydrofluoric acid burns. State of the Art Reviews: Occupational Medicine—The Microelectronics Industry. 1:89–103, 1986
6. Braun J et al: Intoxication following the inhalation of hydrogen fluoride. Arch Toxicol 56:50–54, 1984
7. National Institute for Occupational Safety and Health: Criteria for a Recommended Standard . . . Occupational Exposure to Hydrogen Fluoride. DHEW (NIOSH) Pub No 76–143, pp 106–115. Washington, DC, US Government Printing Office, 1976
8. Largent EJ: Fluorosis—The Health Aspects of Fluorine Compounds, pp 34–39, 43–48. Columbus, Ohio State University Press, 1961
9. Dibbell DG et al: Hydrofluoric acid burns of the hand. J Bone Joint Surg 52A:931–936, 1970
10. Reinhardt CF, Hume WG, Linch AL, Wetherhold JM: Hydrofluoric acid burn treatment. Am Ind Hyg Assoc J 27:166–171, 1966
11. Wetherhold JM, Shepherd FP: Treatment of hydrofluoric acid burns. J Occup Med 7:193–195, 1965
12. Grant WM: Toxicology of the Eye, 3rd ed, pp 490–492. Springfield, Illinois, Charles C Thomas, 1986

HYDROGEN PEROXIDE (90%)

CAS: 7722-84-1

90% H_2O_2 1987 TLV = 1 ppm

Synonyms: None

Physical Form. Liquid

Uses. Synthesis of compounds; bleaching agent, especially for textiles and paper; disinfectant; rocket fuel

Exposure. Inhalation

Toxicology. Hydrogen peroxide is an irritant of the eyes, mucous membranes, and skin.

In humans, inhalation of high concentrations of vapor or mist may cause extreme irritation and inflammation of the nose and throat.[1,2] Severe systemic poisoning may also cause headache, dizziness, vomiting, diarrhea, tremors, numbness, convulsions, pulmonary edema, unconsciousness, and shock.[3] Exposure for a short period of time to mist or diffused spray may cause stinging of the eyes and lacrimation.[1,2] Splashes of the liquid in the eyes may cause severe damage, including ulceration of the cornea; there may be a delayed appearance of damage to the eyes, and corneal ulceration has, on rare occasions, appeared even a week or more after exposure.[1]

Skin contact with the liquid for a short period of time will cause a temporary whitening or bleaching of the skin; if splashes on the skin are not removed, erythema and the formation of vesicles may occur.[1] Although ingestion is unlikely to occur in industrial use, it may cause irritation of the upper gastrointestinal tract; decomposition of the hydrogen peroxide will result in the rapid liberation of oxygen, which may distend the esophagus or stomach and cause severe damage.

Repeated exposure of dogs to 7 ppm for 6 months caused sneezing, lacrimation, and bleaching of hair; at autopsy, there was local atelectasis.[4]

A number of investigators have shown that hydrogen peroxide in vitro leads to genetic damage through the formation of free radicals.[5] It is not known whether such damage presents a danger to the mammalian organism or if various enzymes protect against damage.

An additional hazard is the possibility of explosion when higher strength hydrogen peroxide is mixed with organic compounds, and violent decomposition if contaminated by metallic ions or salts.[3] Since hydrogen peroxide is such a strong oxidizer, it can set fire to combustible materials when spilled on them.[3]

REFERENCES

1. Chemical Safety Data Sheet SD–53, Hydrogen Peroxide, pp 5, 30–31. Washington, DC, MCA, Inc, 1969
2. Hygienic Guide Series: Hydrogen Peroxide (90%). Am Ind Hyg Assoc J 18:275–276, 1957
3. Woodbury CM: Hydrogen peroxide. In Encyclopaedia of Occupational Health and Safety, 3rd ed, rev, Vol 1, A–K, pp 1088–1090. Geneva, International Labour Office, 1983
4. Oberst FW, Comstock CC, Hackley EB: Inhalation toxicity of ninety per cent hydrogen peroxide vapor—acute, subacute, and chronic exposures of laboratory animals. AMA Arch Ind Hyg Occup Med 10:319–327, 1954
5. Speit G et al: Characterization of sister chromatid exchange induction by hydrogen peroxide. Environ Mutat 4:135–142, 1982

HYDROGEN SELENIDE

CAS: 7783-07-5

H_2Se 1987 TLV = 0.05 ppm

Synonyms: Selenium hydride

Physical Form. Colorless gas

Sources. Produced by reaction of acids or water with metal selenides

Exposure. Inhalation

Toxicology. Hydrogen selenide gas is an irritant of the eyes and mucous membranes; it also causes pulmonary irritation and liver damage in animals.

In humans, a concentration of 1.5 ppm is said to produce intolerable irritation of the eyes and nose.[1] Five workers exposed to hydrogen selenide (and possibly other selenium compounds as well) at concentrations of less than 0.2 ppm for 1 month developed nausea, vomiting, diarrhea, metallic taste, garlic odor of the breath, dizziness, lassitude, and fatigability; following cessation of exposure, there was a gradual regression of symptoms during the succeeding months.[2] Urinary selenium levels of the workers ranged from 0 to 13.1 μg selenium/100 ml urine; there was no correlation between symptoms and urinary levels of selenium.[2]

Guinea pigs exposed to 10 ppm for 2 hours exhibited immediate irritation of the eyes and nose; a high percentage of the animals died, apparently from pneumonitis.[3] In guinea pigs, the LC_{50} for 8 hours was 1 mg/m^3 (0.3 ppm); pulmonary irritation and liver damage were observed.[3]

REFERENCES

1. Grant WM: Toxicology of the Eye, 2nd ed, p 560. Springfield, Illinois, Charles C Thomas, 1974
2. Buchan RF: Industrial selenosis. J Occup Med 3:439–456, 1947
3. Hygienic Guide Series: Hydrogen selenide. Am Ind Hyg Assoc J 20:514–515, 1959

HYDROGEN SULFIDE

CAS: 7783-06-4

H_2S 1987 TLV = 10 ppm

Synonyms: Sulfureted hydrogen; hydrosulfuric acid

Physical Form. Gas

Sources. By-product of many industrial processes; around oil wells and in areas where petroleum products are processed, stored, or used; decay of organic matter

Exposure. Inhalation

Toxicology. Hydrogen sulfide is an irritant of the eyes and respiratory tract at low concentrations; at higher levels, it causes respiratory paralysis with consequent asphyxia, and it is rapidly fatal.

Inhalation of 1000 ppm can cause coma after a single breath and can be rapidly fatal owing to respiratory paralysis.[1–5] At lower levels, neurologic effects may include nervousness, headache, fatigue, weakness of extremities, spasms, vertigo, convulsions, agitation, and delirium.

Direct damage to the cardiac muscle has been suggested from electrocardiographic changes following overexposure.

Pulmonary edema is common after exposure to 250 ppm for prolonged periods of time. Symptoms of gastrointestinal disturbances, including nausea, abdominal cramps, vomiting, and severe diarrhea, have been reported and frequently occur in subacute intoxication.

Reports of adverse effects of hydrogen sulfide on humans owing to chronic intoxication are less well established. It has been postulated that exposures below 50 ppm over long periods of time may cause certain neuroasthenic symptoms such as fatigue, headache, dizziness, and irritability. Other investigators suggest that the signs and symptoms referred to as chronic poisoning are actually the result of recurring acute exposures or the sequelae of acute poisoning.

Carcinogenic, teratogenic, and reproductive effects have not been studied.

Exposure to levels above 50 ppm for 1 hour can produce acute conjunctivitis with pain, lacrimation, and photophobia; in severe form, this can progress to keratoconjunctivitis and vesiculation of the corneal epithelium. Prolonged exposure to 50 ppm also causes rhinitis, pharyngitis, bronchitis, and pneumonitis.

The odor is offensive and characterized as "rotten eggs"; it is unreliable as a warning signal because olfactory fatigue occurs at 150 ppm.

REFERENCES

1. National Institute for Occupational Safety and Health: Criteria for a Recommended Standard . . . Occupational Exposure to Hydrogen Sulfide. DHEW (NIOSH), pp 22–64. Washington, DC, US Government Printing Office, 1977
2. Stine RJ, Slosberg B, Beacham BE: Hydrogen sulfide intoxication—a case report and discussion of treatment. Ann Intern Med 85:756, 1976
3. Milby TH: Hydrogen sulfide intoxication—review of the literature and report of unusual accident resulting in two cases of nonfatal poisoning. J Occup Med 4:431, 1962
4. Adelson L, Sunshine I: Fatal hydrogen sulfide intoxication. Arch Pathol 81:375, 1966
5. Beauchamp RO Jr et al: A critical review of the literature on hydrogen sulfide toxicity. CRC Crit Rev Toxicol 13:25–97, 1984

HYDROQUINONE

CAS: 123-31-9

$C_6H_4(OH)_2$ 1987 TLV = 2 mg/m^3

Synonyms: 1,4–Benzenediol; *p*-hydroxybenzene; hydroquinol; quinol; Tecquinol

Physical Form. Crystals

Uses. Photographic reducer and developer; antioxidant

Exposure. Inhalation

Toxicology. Hydroquinone primarily affects the eyes.

Acute exposure to quinone vapor and hydroquinone dust causes conjunctival irritation, whereas chronic exposure produces changes characterized as: (1) brownish discoloration of the conjunctiva and cornea confined to the interpalpebral tissue; (2) small opacities of the cornea; and (3) structural changes in the cornea that result in loss of visual acuity.[1,2] The pigmentation changes are reversible, but the more slowly developing structural changes in the cornea may progress. Although pigmentation may appear with less than 5 years of exposure, this is uncommon and usually not associated with serious injury to the eye.[2]

The lethal oral dose has not been accurately evaluated, because in most cases of ingestion, chemicals in addition to hydroquinone have been present.[3] In one nonfatal case of hydroquinone ingestion of approximately 12 g, tinnitus, dyspnea, cyanosis, and extreme sleepiness were observed.[3] Although acute, high-dose oral ingestion produces noticeable central nervous system effects in humans, no effects have been observed in workers exposed to lower concentrations in actual industrial situations.[3] No signs of toxicity were found in subjects who ingested 300 to 500 mg hydroquinone daily for 3 to 5 months.[4]

Repeated skin contact with hydroquinone creams (generally 5% or more hydroquinone) produced skin irritation, allergic sensitization, dermatitis, and depigmentation.[3]

Pellets of cholesterol containing 2 mg hydroquinone implanted in mice bladders caused an excessive number of bladder carcinomas.[5] In other studies, rats fed up to 1% hydroquinone in their diets for 2 years did not develop tumors, nor did hydroquinone initiate significant numbers of tumors in mice skin painting studies.[3,4] The IARC has determined that the available data do not allow an evaluation of carcinogenicity.[5]

REFERENCES

1. Anderson B, Oglesby F: Corneal changes from quinone-hydroquinone exposure. AMA Arch Ophthalmol 59:495–501, 1958
2. Sterner JH, Oglesby FL, Anderson B: Quinone

vapors and their harmful effects. I. Corneal and conjunctival injury. J Ind Hyg Toxicol 29:60–73, 1947
3. National Institute for Occupational Safety and Health: Criteria for a Recommended Standard . . . Occupational Exposure to Hydroquinone. DHEW (NIOSH) Pub No 78–155, p 182. Washington, DC, US Government Printing Office, 1978
4. Carlson AJ, Brewer NR: Toxicity studies on hydroquinone. Proc Soc Exp Biol Med 84:684–688, 1953
5. IARC Monographs on the Evaluation of the Carcinogenic Risk of Chemicals to Man, Vol 15, Some Fumigants, the Herbicides, 2,4–D and 2,4,5–T Chlorinated Dibenzodioxins and Miscellaneous Industrial Chemicals, pp 155–175. Lyon, International Agency for Research on Cancer, 1977

HYDROXYLAMINE AND SALTS

CAS:		
	Hydroxylamine	7803-49-8
	Hydroxylamine HCl	5470-11-1
	Hydroxylamine sulfate	10039-54-0
NH_2OH	1987 TLV = none established	

Synonyms: Oxammonium; hydroxyl ammonium

Physical Form. Colorless flakes or crystals

Uses. Reducing agents; production of synthetic rubbers; photographic developer solutions; hydroxylamine sulfate used in production of synthetic fibers

Exposure. Inhalation; skin/eye contact

Toxicology. Hydroxylamine and its salts are irritants of eyes, mucous membranes, and skin; high levels can cause methemoglobinemia.

Workers exposed to hydroxylamine sulfate for 1 day at unspecified air levels showed blood methemoglobinemia concentrations of 25%.[1] Dusts and mists of the sulfate are irritants of the mucous membranes and eyes. Although details are lacking, repeated exposure to the sulfate is reported to have caused respiratory sensitization with asthma-like symptoms.

Hydroxylamine hydrochloride is highly irritating to the skin, eyes, and mucous membranes and has caused contact dermatitis in workers exposed for 2 to 60 days.[2] Hydroxylamine itself is only moderately irritating to the skin. Hydroxylamine sulfate on the skin of rabbits was irritating at levels as low as a 10-mg dose.[3] It is considered to be a potential skin sensitizer.

Carcinogenicity has not been demonstrated. Several studies have shown a decreased incidence of spontaneous mammary tumors in mice exposed to the sulfate and hydrochloride.[3–6] There was some indication of an increase in the incidence of spontaneous mammary tumors when the sulfate was administered to older animals whose mammary glands were already well developed.

Embryotoxic effects have occurred in rabbits exposed to hydroxylamine hydrochloride by intracoelomic injection.[7]

REFERENCES

1. Hydroxylamine Sulfate; Product Safety Data Sheet. Morristown, New Jersey, Allied Corporation, 1983
2. Folesky Von H, Nickel H, Rothe A, Zschunke E: Allergisches Ekzem durch Salze des Hydroxylamins (Oxammonium). Z Gesamte Hyg Grenzgeb 17:353–356, 1971
3. Yamamoto RS, Weisburger EK, Korzis J: Chronic administration of hydroxylamine and derivatives in mice. Proc Soc Exp Biol Med 124:1217–1220, 1967
4. Harman D: Prolongation of the normal lifespan and inhibition of spontaneous cancer by antioxidants. J Gerontol 16:247–257, 1961
5. Evarts RP, Brown CA: Morphology of mammary gland, ovaries, and pituitary gland of hydroxylamine-fed C3H/HeN mice. Lab Invest 37:53–63, 1977
6. Evarts RP, Brown CA, Atta GJ: The effects of hydroxylamine on the morphology of the rat mammary gland and on the induction of mammary tumors by 7,12–dimethylbenz(a)anthracene. Exp Mol Pathol 30:337–348, 1979

7. DeSesso JM: Demonstration of the embryotoxic effects of hydroxylamine in the New Zealand white rabbit. Anat Rec 196:45A–46A, 1980

IODINE

CAS: 7553-56-2

I_2 1987 TLV = C 0.1 ppm

Synonyms: None

Physical Form. Crystalline solid, blue-black scales or plates

Uses. Synthesis of organic chemicals; photographic film

Exposure. Inhalation

Toxicology. Iodine is an irritant of the eyes, mucous membranes, and skin; it is a pulmonary irritant in animals, and it is expected that severe exposure will cause the same effect in humans.

Exposed workers (concentration and time unspecified) experienced a burning sensation in the eyes, lacrimation, blepharitis, rhinitis, stomatitis, and chronic pharyngitis; after brief accidental exposure in a laboratory, technicians reported headache and a feeling of tightness in the chest.[1,2]

Iodine absorbed by the lungs is changed to iodide and eliminated mainly in the urine; iodine is an essential element in the nutrition and is required by the thyroid.[3] Ingestion of as little as 2 to 3 g may be fatal; chronic absorption of iodine causes "iodism," a syndrome characterized by insomnia, conjunctivitis, rhinitis, bronchitis, tremor, tachycardia, parotitis, diarrhea, and weight loss.[4,5]

In an experimental investigation, four human subjects tolerated 0.57 ppm iodine vapor for 5 minutes without eye irritation, but all experienced eye irritation in 2 minutes at 1.63 ppm.[3] In patients exposed to air saturated with iodine vapor for 3 to 4 minutes for therapeutic purposes, there was brown staining of the corneal epithelium and subsequent spontaneous loss of the layer of tissue; recovery occurred within 2 to 3 days.[6] Iodine in crystalline form or in strong solutions is a severe skin irritant; it is not easily removed from the skin, and the lesions resemble thermal burns with brown staining.[5] Hypersensitivity to iodine characterized by a skin rash has been reported.[4,5] Iodine is absorbed through the skin in small amounts from a tincture or from vapor applied to the skin.[3]

Intratracheal administration of the vapor to dogs at 36 mg iodine/kg body-weight was fatal after about 3 hours; the animals developed cough, difficulty in breathing, and rales; autopsy findings included pulmonary edema, subpleural hemorrhage, and an increased iodine content of the thyroid and urine.[1]

REFERENCES

1. Luckhardt AB, Koch FC, Schroeder WF, Weiland AH: The physiological action of the fumes of iodine. J Pharmacol Exp Ther 15:1–21, 1920
2. Heyroth F: Halogens. In Patty FA (ed): Industrial Hygiene and Toxicology, 2nd ed, Vol 2, Toxicology, pp 854–856. New York, Wiley–Interscience, 1963
3. Hygienic Guide Series: Iodine. Am Ind Hyg Assoc J 26:423–426, 1965
4. Seymour WB Jr: Poisoning from cutaneous application of iodine, a rare aspect of its toxicologic properties. Arch Intern Med 59:952–966, 1937
5. Peterson JE: Iodine. In International Labour Office: Encyclopaedia of Occupational Health and Safety, 3rd ed, rev, Vol I, A–K, pp 1153–1154. New York, McGraw-Hill, 1983
6. Grant WM: Toxicology of the Eye, 3rd ed, pp 519–520. Springfield, Illinois, Charles C Thomas, 1986

IRON OXIDE FUME
CAS: 1309-37-1
Fe_2O_3 1987 TLV = 5 mg/m^3

Synonyms: Ferric oxide fume

Physical Form. Fume

Source. Result of welding and silver finishing

Exposure. Inhalation

Toxicology. Inhalation of iron oxide fume or dust causes a benign pneumoconiosis (siderosis).

Iron oxide alone does not cause fibrosis in the lungs of animals, and it is probable that this is also true in humans.[1] Exposures of 6 to 10 years are usually required before changes recognizable by x-ray occur; the retained dust produces x-ray shadows that may be indistinguishable from fibrotic pneumoconiosis.[2,3] Of 25 welders exposed chiefly to iron oxide for an average of 18.7 (range 3 to 32) years, 8 had reticulonodular shadows on chest x-ray consistent with siderosis, but there was no reduction in pulmonary function; exposure levels ranged from 0.65 to 47 mg/m^3.[4]

In another study, the x-rays of 16 welders with an average exposure of 17.1 (range 7 to 30) years also suggested siderosis; their spirograms were normal. However, the static and functional compliance of the lungs was reduced; some of the welders were smokers.[5] The welders with the lowest compliance complained of dyspnea.

Welders are typically exposed to a complicated mixture of dust and fume of metallic oxides, as well as irritant gases, and are subject to mixed-dust pneumoconiosis with possible loss of pulmonary function; this should not be confused with benign pneumoconiosis caused by iron oxide.[1] Although an increased incidence of lung cancer has been observed among hematite miners exposed to iron oxide, presumably owing to concomitant radon gas exposure, there is no evidence that iron oxide alone is carcinogenic to humans or animals.[6]

REFERENCES

1. Jones JG, Warner CG: Chronic exposure to iron oxide, chromium oxide, and nickel oxide fumes of metal dressers in a steelworks. Br J Ind Med 29:169–177, 1972
2. Sentz FC Jr, Rakow AB: Exposure to iron oxide fume at arcair and power-burning operations. Am Ind Hyg Assoc J 30:143–146, 1969
3. Harding HE, McLaughlin AIG, Doig AT: Clinical, radiographic, and pathological studies of the lungs of electric arc and oxyacetylene welders. Lancet 2:394–398, 1958
4. Kleinfeld M, Messite J, Kooyman O, Shapiro J: Welders' siderosis. Arch Environ Health 19:70–73, 1969
5. Stanescu DC et al: Aspects of pulmonary mechanics in arc welders' siderosis. Br J Ind Med 24:143–147, 1967
6. Stokinger HE: A review of world literature finds iron oxides noncarcinogenic. Am Ind Hyg Assoc J 45(2):127–133, 1984

ISOAMYL ACETATE
CAS: 123-92-2
$CH_3C(O)OCH_2CH(CH_3)C_2H_5$
1987 TLV = 100 ppm

Synonyms: Amyl acetate; banana oil; pear oil; amylacetic ester; 3–methyl butyl acetate; 3–methyl–1–butanol acetate

Physical Form. Colorless liquid

Uses. Solvent; flavor in water and syrups

Exposure. Inhalation

Toxicology. Isoamyl acetate is an irritant of the eyes and mucous membranes; at high concentrations, it causes narcosis in animals, and it is expected that severe exposure will cause the same effect in humans.

Several grades of technical amyl acetate are known; isoamyl acetate is the major component of some grades, whereas other isomers predominate in other grades.[1]

Men exposed to 950 ppm isoamyl acetate for 30 minutes experienced irritation of the nose and throat, headache, and weakness.[1]

Cats exposed to 1900 ppm for six 8-hour exposures showed irritation of the eyes, salivation, weakness, and loss of weight; lung irritation was noted at necropsy. A 24-hour exposure to 7200 ppm caused light narcosis and delayed death due to pneumonia.[2,3] Dogs exposed to 5000 ppm for 1 hour had nasal irritation and drowsiness.[2] Isoamyl acetate may cause skin irritation.

Amyl acetate has a banana or pear-like odor detectable at 7 ppm.[1]

REFERENCES

1. Hygienic Guide Series: Amyl acetate. Am Ind Hyg Assoc J 26:199–202, 1965
2. Sandmeyer EE, Kirwin CJ: Esters. In Clayton GD, Clayton FE (eds): Patty's Industrial Hygiene and Toxicology, 3rd ed, rev, Vol 2A, Toxicology, p 2274. New York, Wiley–Interscience, 1981
3. Isoamyl acetate. Documentation of the TLVs and BEIs, 5th ed, p 329. Cincinnati, American Conference of Governmental Industrial Hygienists (ACGIH), 1986

ISOAMYL ALCOHOL

CAS: 123-51-3

$(C_2H_5)_2CHOH$ 1987 TLV = 100 ppm

Synonyms: 3–Methybutanol–1; isobutyl carbinol; isopentyl alcohol

Physical Form. Colorless liquid

Uses. Solvent; chemical synthesis; manufacture of smokeless powders, artificial silk, and lacquers

Exposure. Inhalation

Toxicology. Isoamyl alcohol is an irritant of the eyes and mucous membranes; at high concentrations, it causes narcosis in animals, and it is expected that severe exposure will produce the same effect in humans.

Human volunteers exposed to 100 ppm for 3 to 5 minutes experienced throat irritation, and, at 150 ppm, there was also eye and nose irritation.[1] No chronic systemic effects have been reported in humans.

Rats survived 8-hour exposure to 2000 ppm. Oral administration of 0.7 g/kg produced stupor and loss of voluntary movement in half the treated rabbits; the LD_{50} was 3.4 g/kg.[3]

Instilled in rabbit eyes, isoamyl alcohol caused severe burns with moderately severe corneal necrosis.[4] Topical application produced minimal skin irritation.[4]

A total of 10 malignant tumors were found in 24 rats injected subcutaneously with 0.04 ml/kg isoamyl alcohol for 95 weeks; control animals had no malignancies.

Isoamyl alcohol has a disagreeable pungent odor.

REFERENCES

1. Nelson KW et al: Sensory response to certain industrial solvent vapors. J Ind Hyg Toxicol 25:282–285, 1943
2. Smyth HF Jr: Improved communication—hygienic standards for daily inhalation. Am Ind Hyg Assoc J 17:129–185, 1956
3. Munch JC: Aliphatic alcohols and alkyl esters: Narcotic and lethal potencies to tadpoles and to rabbits. J Ind Med 41:31–33, 1972
4. Rowe VK, McCollister SB: Alcohols. In Clayton GD, Clayton FE (eds): Patty's Industrial Hygiene and Toxicology, 3rd ed, rev, Vol 2C, Toxicology, pp 4594–4599. New York, Wiley–Interscience, 1982

ISOBUTYL ACETATE

CAS: 110-19-0

$CH_3C(O)OCH_2CH(CH_3)_2$

1987 TLV = 150 ppm

Synonyms: Acetic acid, isobutyl ester; 2–methylpropyl acetate

Physical Form. Colorless liquid

Uses. Solvent; flavoring

Exposure. Inhalation

Toxicology. At high concentrations, isobutyl acetate causes narcosis in animals, and it is expected that severe exposure will cause the same effect in humans. It is considered to be a respiratory tract and eye irritant by analogy with *n*-butyl acetate.

Rats survived exposure to 4000 ppm, but 8000 ppm for 4 hours was fatal to four of six rats.[1] Exposure of rats to 21,000 ppm for 150 minutes was fatal to all animals exposed; no symptoms were observed at 3,000 ppm for 6 hours.[2]

Isobutyl acetate has a fruity odor.

REFERENCES

1. Smyth HF Jr et al: Range-finding toxicity data: List VI. Am Ind Hyg Assoc J 23:95–107, 1962
2. Sandmeyer EE, Kirwin CJ: Esters. In Clayton GD, Clayton FE (eds): Patty's Industrial Hygiene and Toxicology, 3rd ed, Vol 2A, Toxicology, p 2273. New York, Wiley–Interscience, 1981

ISOBUTYL ALCOHOL

CAS: 78-83-1

$(CH_3)_2CHCH_2OH$ 1987 TLV = 50 ppm

Synonyms: 2–Methylpropanol–1; 2–methyl–1–propanol; isopropylcarbinol

Physical Form. Colorless liquid

Uses. Lacquers, paint removers, cleaners, and hydraulic fluids; manufacture of isobutyl esters

Exposure. Inhalation

Toxicology. At high concentrations, isobutyl alcohol causes narcosis in animals, and it is expected that severe exposure in humans would produce the same effect.

The liquid on the skin of a human subject was a mild irritant and caused slight erythema and hyperemia.[1] No evidence of eye irritation was noted in humans with repeated 8-hour exposures to 100 ppm.[1] No chronic systemic effects have been reported in humans.

Intermittent exposure of mice to 6400 ppm for 136 hours produced narcosis; exposure to 10,600 ppm for 300 minutes or 15,950 ppm for 250 minutes was fatal.[1]

Rats survived a 2-hour exposure to 16,000 ppm, but two of six rats died following a 4-hour exposure to 8000 ppm.[2]

One drop of isobutyl alcohol in a rabbit eye caused moderate-to-severe irritation without permanent corneal injury.[1]

A variety of malignant tumors developed in rats dosed twice weekly for life by oral intubation or subcutaneous injection.[1] Control animals had no malignancies. The carcinogenic risk to humans has not been determined.

REFERENCES

1. Rowe VK, McCollister SB: Alcohols. In Clayton GD, Clayton FE (eds): Patty's Industrial Hygiene and Toxicology, 3rd ed, Vol 2C, Toxicology, pp 4578–4582. New York, Wiley–Interscience, 1982
2. Smyth HF Jr, Carpenter CP, Weil CS, Pozzani UC: Range-finding toxicity data: List V. Arch Ind Hyg Occup Med 10:61–68, 1954

ISOPHORONE

CAS: 78-59-1
$C_9H_{14}O$ 1987 TLV = C 5 ppm

Synonyms: Isoacetophorone; Isoforon; trimethyl cyclohexenone

Physical Form. Water-white liquid

Uses. Solvent for lacquers, resins, and plastics

Exposure. Inhalation, dermal

Toxicology. Isophorone is an irritant of the eyes and mucous membranes.

In an early experimental study of human exposure to isophorone, irritation of the eyes, nose, and throat were noted, as well as symptoms of narcosis. This work has been criticized for improper analytical techniques and the use of impure isophorone.[1] In another study, human subjects exposed briefly to 25 ppm experienced irritation of the eyes, nose, and throat.[2] Workers exposed to 5 to 8 ppm for 1 month complained of fatigue and malaise, which disappeared when air levels were reduced to 1 to 4 ppm.[1] Repeated or prolonged skin contact with the liquid may cause dermatitis because of its defatting action.[1] Although it may be more toxic and irritative than lower molecular weight ketones at equivalent concentrations, it poses less of an inhalation hazard because of its relatively low volatility.[1]

Repeated exposures of animals at concentrations of 50 ppm or more resulted in evidence of damage to kidney and lung and, to a lesser extent, liver damage. No effects, however, were seen at 25 ppm. More recent feeding studies with pure compound in rats, mice, and beagle dogs have not demonstrated specific toxicity.[3]

A 2-year gavage study at 250 and 500 mg/kg demonstrated a dose-related statistically significant excess of tubular cell adenomas and adenocarcinomas of the kidney in male rats, a number of preputial gland tumors in dosed male rats, and a probable increased incidence of hepatocellular neoplasms in high-dose male mice.[3]

REFERENCES

1. National Institute for Occupational Safety and Health: Criteria for a Recommended Standard . . . Occupational Exposure to Ketones. DHEW (NIOSH) Pub No 78–173, pp 44, 86–87, 126–134, 176, 189–190, 242. Washington, DC, US Government Printing Office, June 1978
2. Silverman L, Schulte HF, First MW: Further studies on sensory response to certain industrial solvent vapors. J Ind Hyg Toxicol 28:262–266, 1946
3. Bucher J, Huff J, Kluwe W (NTP): Toxicology and carcinogenesis studies of isophorone in F344 rats and B6C3F1 mice. Toxicology 39:208–219, 1986

ISOPROPYL ACETATE

CAS: 108-21-4
$CH_3C(O)OCH(CH_3)_2$
1987 TLV = 250 ppm

Synonyms: 2–Propyl acetate; acetic acid, isopropyl ester

Physical Form. Colorless liquid

Uses. Solvent

Exposure. Inhalation

Toxicology. Isopropyl acetate is an irritant of the eyes; at extremely high concentrations, it causes narcosis in animals, and it is expected that severe exposure will produce the same effect in humans.

Human subjects exposed to 200 ppm for 15 minutes experienced some degree of eye irritation; there was little objection to the odor.[1] No systemic effects have been reported in humans.

Exposure to 32,000 ppm was fatal to five of six rats after 4 hours; 16,000 ppm for 4 hours was fatal to one of six rats.[2] The oral LD_{50} for rats was 6.75 g/kg.[2]

REFERENCES

1. Silverman L, Schulte HF, First MW: Further studies on sensory response to certain industrial solvent vapors. J Ind Hyg Toxicol 28:262–266, 1946
2. Smyth HF Jr, Carpenter CP, Weil CS, Pozzani UC: Range-finding toxicity data: List V. Arch Ind Hyg Occup Med 10:61–68, 1954

ISOPROPYL ALCOHOL

CAS: 67-63-0

$CH_3CHOHCH_3$ 1987 TLV = 400 ppm

Synonyms: Isopropanol; propanol–2; dimethyl carbinol

Physical Form. Colorless liquid

Uses. Manufacture of acetone; solvent; in skin lotions, cosmetics, and pharmaceuticals

Exposure. Inhalation; ingestion

Toxicology. Isopropyl alcohol is an irritant of the eyes and mucous membranes; at high doses, it causes central nervous system depression.

Human subjects exposed to 400 ppm for 3 to 5 minutes experienced mild irritation of the eyes, nose, and throat; at 800 ppm, the irritation was not severe, but the majority of subjects considered the atmosphere uncomfortable.[1] Inhalation of high levels of isopropyl alcohol has occurred in poorly ventilated areas during sponge baths intended to reduce fever. In several cases, symptoms included respiratory distress, severe stupor, and coma. Recovery was complete in 36 hours.[2]

Occupational poisoning by isopropyl alcohol has not been reported. Toxicity in humans is based largely on accidental ingestion. An oral dose of 25 ml in 100 ml of water produced hypotension, facial flushing, bradycardia, and dizziness. Other symptoms following ingestion have included vomiting, depression, headache, coma, and shock.[2] Renal insufficiency, including anuria followed by oliguria, nitrogen retention, and edema, may be a complication of isopropyl alcohol poisoning. Estimates of fatal doses are between 160 and 240 ml. Death following ingestion often occurs in 24 to 36 hours from respiratory paralysis.[2]

The dermatotoxic potential of isopropyl alcohol is low, although a few cases of hypersensitivity characterized by delayed, eczematous reactions have been observed.[2]

Rats exposed to 12,000 ppm for 4 hours survived, but exposure for 8 hours was lethal to half the animals.[4] Mice exposed to 3250 ppm for 460 minutes developed ataxia, prostration, and, finally, narcosis. In the eye of a rabbit, 70% isopropyl alcohol caused conjunctivitis, iritis, and corneal opacity.[4]

Pretreatment of animals with isopropyl alcohol has been shown to enhance the acute toxicity of carbon tetrachloride.[2] The metabolite acetone may be responsible for this effect. Extra caution is in order when isopropyl alcohol is used concurrently with carbon tetrachloride in an industrial setting.

Early epidemiologic studies suggested an association between the manufacture of isopropyl alcohol and paranasal sinus cancer.[3] The increased cancer incidence, however, appears to be associated with the strong-acid manufacturing process rather than the isopropyl alcohol itself.

The odor threshold is 40 to 200 ppm.[4]

REFERENCES

1. Nelson KW et al: Sensory response to certain industrial solvent vapors. J Ind Hyg Toxicol 25:282–285, 1943
2. Zakhari S et al: Isopropanol and ketones in the environment, pp 3–54. Cleveland, CRC Press, Inc, 1977
3. Weil CS, Smyth HF Jr, Nale TW: Quest for a suspected industrial carcinogen. J Ind Hyg Occup Med 5:535–547, 1952
4. Isopropyl alcohol. Documentation of TLVs and BEIs, 5th ed, p 337. Cincinnati, American Conference of Governmental Industrial Hygienists (ACGIH), 1986

ISOPROPYLAMINE

CAS: 75-31-0

$(CH_3)_2CHNH_2$ 1987 TLV = 5 ppm

Synonyms: 2–Aminopropane

Physical Form. Liquid

Uses. Chemical synthesis of dyes, pharmaceuticals

Exposure. Inhalation

Toxicology. Isopropylamine is an irritant of the eyes, mucous membranes, and skin.

Human subjects experienced irritation of the nose and throat after brief exposure to 10 to 20 ppm.[1] Workers complained of transient visual disturbances (haloes around lights) after exposure to the vapor for 8 hours, probably owing to mild corneal edema, which usually cleared within 3 to 4 hours.[2] The liquid is capable of causing severe eye burns, which may cause permanent visual impairment.[2] Isopropylamine in both liquid and vapor forms is irritating to the skin and may cause skin burns; repeated lesser exposures may result in dermatitis.[2]

All rats exposed to 8000 ppm for 4 hours died within 14 days, but six of six rats survived a 4-hour exposure at 4000 ppm.[3]

The odor is like ammonia and becomes definite at 5 to 10 ppm.[1]

REFERENCES

1. Beard RR, Noe JT: Aliphatic and alicyclic amines. In Clayton GD, Clayton FE (eds): Patty's Industrial Hygiene and Toxicology, 3rd ed, rev, Vol 2B, Toxicology, pp 3154–3155. New York, Wiley–Interscience, 1981
2. Chemical Safety Data Sheet SD–72, Isopropylamine, pp 13–15. Washington, DC, MCA, Inc, 1959
3. Smyth HJ Jr et al: Range-finding toxicity data: List IV. AMA Arch Ind Hyg Occup Med 4:119–122, 1951

ISOPROPYL ETHER

CAS: 108-20-3

$(CH_3)_2CHOCH(CH_3)_2$

1987 TLV = 250 ppm

Synonyms: Diisopropyl ether; 2–isopropoxypropane

Physical Form. Colorless liquid

Uses. Solvent; chemical intermediate

Exposure. Inhalation

Toxicology. Isopropyl ether is a mild irritant of the eyes and mucous membranes; at high concentrations, it causes narcosis in animals, and it is expected that severe exposure will produce the same effect in humans.

Human subjects exposed to 800 ppm for 5 minutes reported irritation of the eyes and nose, and the most sensitive subjects reported respiratory discomfort.[1] Thirty-five per cent of the volunteers exposed to 300 ppm for 15 minutes objected to the odor rather than the irritation.[2]

Animals (monkey, rabbit, and guinea pig) survived a 1-hour exposure to 30,000 ppm with signs of anesthesia; 60,000 ppm for 1 hour was lethal.[1] The lethal concentration for rats was 16,000 ppm for a 4-hour exposure.[1] In rabbits, repeated skin application of the liquid for 10 days caused dermatitis.[1] The liquid dropped in the eye of a rabbit caused minor injury.[1]

REFERENCES

1. Kirwin C, Sandmeyer E: Ethers. In Clayton GD, Clayton FE (eds): Patty's Industrial Hygiene and Toxicology, 3rd ed, Vol 2, Toxicology, pp 2511–2512. New York, Interscience, 1981
2. Silverman L, Schulte HF, First MW: Further studies on sensory response to certain industrial solvent vapors. J Ind Hyg Toxicol 28:262–266, 1946

ISOPROPYL GLYCIDYL ETHER

CAS: 4016-14-2

$C_6H_{12}O_2$ 1987 TLV = 50 ppm

Synonyms: IGE

Physical Form. Colorless liquid

Uses. Reactive diluent for epoxy resins; stabilizer for organic compounds; chemical intermediate for synthesis of ethers and esters

Exposure. Inhalation

Toxicology. Isopropyl glycidyl ether causes both primary irritation and sensitization dermatitis; in animals, it causes irritation of the eyes and mucous membranes, and it is expected that severe exposure will cause the same effects in humans.

Systemic effects have not been demonstrated in workers exposed to isopropyl glycidyl ether.[1]

A technician who handled both isopropyl glycidyl ether and phenyl glycidyl ether developed localized dermatitis on the back of the hands; patch testing showed sensitization to both substances.[2] Dermatitis has occurred in workers with repeated skin contact.[3]

In mice, the LC_{50} was 1500 ppm for 4 hours.[2] Rats repeatedly exposed to levels of 400 ppm exhibited slight eye and respiratory irritation. Large oral doses produced central nervous system depression, but this effect was not seen from inhalation exposure.

Moderate irritation resulted from instillation of the liquid in the eyes of rabbits and from application to the skin of rabbits.[2]

REFERENCES

1. National Institute for Occupational Safety and Health: Criteria for a Recommended Standard . . . Occupational Exposure to Glycidyl Ethers. DHEW (NIOSH) Pub No 78–166, p 197. Washington, DC, US Government Printing Office, 1978
2. Hine CH et al: The toxicology of glycidol and some glycidyl ethers. AMA Arch Ind Health 14:250–264, 1956
3. Hine CH, Rowe VK: Epoxy compounds. In Patty FA (ed): Industrial Hygiene and Toxicology, 2nd ed, Vol 2, Toxicology, pp 1637–1638. New York, Wiley–Interscience, 1963

KETENE

CAS: 463-51-4

$CH_2{=}CO$ 1987 TLV = 0.5 ppm

Synonyms: Ethenone; carbomethane; keten

Physical Form. Gas

Uses. Organic chemical synetheses; conversion of higher acids into their anhydrides; for acetylation in the manufacture of cellulose acetate and aspirin

Exposure. Inhalation

Toxicology. Ketene is a severe pulmonary irritant in animals and is expected to produce the same effect in humans; some human cases have been reported.

For mice, monkeys, cats, and rabbits, the least concentrations that caused death after a 10-minute exposure were 50, 200, 750, and 1000 ppm, respectively.[1] Few signs appeared during the exposure period, but, after a latent period of variable duration, there was dyspnea, cyanosis, and signs of severe pulmonary damage; death was often preceded by convulsions. Significant pathologic changes were confined to the lungs and consisted of generalized alveolar edema and congestion. Several species tolerated exposure to 1 ppm for 6 hours/day for 6 months without apparent chronic injury.[1] Exposure of mice to concentrations in excess of 5 ppm for 10 minutes protected mice 3 to 14 days later against otherwise lethal exposures to pulmonary edema–producing agents.[2] A high degree of tolerance to the acute effects of ketene itself has also been reported.[3] By analogy to effects on the skin caused by other severe irritants, repeated or prolonged exposure is expected to cause dermatitis.

REFERENCES

1. Treon JF et al: Physiologic response of animals exposed to air-borne ketene. J Ind Hyg Toxicol 31:209–218, 1949
2. Stokinger HE: Toxicologic interaction of mixtures of air pollutants. Int J Air Poll 2:313–326, 1960
3. Mendenhall RM, Stokinger HE: Tolerance and cross-tolerance development to atmospheric pollutants ketene and ozone. J Appl Physiol 14:923–926, 1959

LEAD (inorganic compounds)

CAS: 7439-92-1

Pb 1987 TLV = 0.15 mg/m³ (as Pb)

Synonyms: Metallic lead; lead oxide; lead salts, inorganic

Physical Form. Solid

Uses. Storage batteries; paint; ink; ceramics; ammunition

Exposure. Inhalation; ingestion

Toxicology. Prolonged absorption of lead or its inorganic compounds results in severe gastrointestinal disturbances and anemia; with more serious intoxication, there is neuromuscular dysfunction; the most severe lead exposure may result in encephalopathy.

The onset of symptoms of lead poisoning or plumbism is often abrupt; presenting complaints often include weakness, weight loss, lassitude, insomnia, and hypotension.[1–4] Associated with these complaints is a disturbance of the gastrointestinal tract, which includes constipation, anorexia, and abdominal discomfort, or actual colic, which may be excruciating.[1–3] Physical signs are usually facial pallor, malnutrition, abdominal tenderness, and pallor of the eye grounds.[1–4] The anemia often associated with lead poisoning is of the hypochromic, normocytic type, with reduction in mean corpuscular hemoglobin; stippling of erythrocytes and reticulocytosis is evident.[1–3] On gingival tissues, a line or band of punctate blue or blue-black pigmentation (lead line) may appear, but only in the presence of poor dental hygiene; this is not pathognomonic of lead poisoning.[3]

Occasionally, the alimentary symptoms are relatively slight and are overshadowed by neuromuscular dysfunction accompanied by signs of motor weakness, which may progress to paralysis of the extensor muscles of the wrist ("wrist drop") and less often of the ankles ("footdrop").[2,3] Encephalopathy, the most serious result of lead poisoning, frequently occurs in children owing to the ingestion of inorganic lead compounds, but rarely in adults except from exposure to organic lead.[1-4]

Subtle, often subclinical, neurologic effects have been demonstrated in workers with relatively low blood lead levels, below 40 to 60 μg/100 ml blood. Performance of lead workers on various neuropsychological tests was mildly reduced relative to a control group at mean levels of 49 μg/100 ml blood[5,6] and, in a prospective follow-up study, at levels between 30 and 40 μg/100 ml blood.[7] In some of these studies, the lead-exposed workers reported significantly more complaints of nonspecific subjective symptoms, such as anxiety, depressed mood, poor concentration, and forgetfulness.[5] Mild neurophysiologic changes, including reductions in motor and sensory nerve conduction velocities (sometimes still within the normal range), have been documented in lead-exposed workers compared with control groups, with blood lead levels less than 40 μg/100 ml blood.[8] A prospective follow-up study of workers with blood lead levels of 30 to 50 μg/100 ml blood demonstrated mild slowing of conduction velocities.[9]

Nephropathy has been associated with chronic lead poisoning.[2,3,10,11] A recent study of two large cohorts of heavily exposed lead workers followed through 1980 demonstrated a nearly threefold excess of deaths attributed to chronic nephritis or "other hypertensive disease," primarily kidney disease.[12] Most of the excess deaths occurred before 1970 among men who began work before 1946, suggesting that current lower levels of exposure may reduce the risk.

Following absorption, inorganic lead is distributed in the soft tissues, with the highest concentrations being in the kidneys and the liver.[4] In the blood, nearly all circulating inorganic lead is associated with the erythrocytes.[4] Over a period of time, the lead is redistributed, being deposited mostly in bone and also in teeth and hair.[3,4] Lead absorption is cumulative; elimination of lead from the body is slow, requiring considerably longer than the period of storage of toxic amounts.[1,4] Asymptomatic lead workers, when subjected to a sudden increase in exposure to and absorption of lead, often respond with an episode of typical lead poisoning.[1] Removal of the worker from exposure to abnormal quantities of lead often leads to a seemingly sudden and apparent complete recovery; this has occurred even when the worker has a considerable quantity of residual lead in the body.[1]

Epidemiologic studies have not shown a relation between lead exposure and the incidence of cancer.[13,15] A study of 437 lead-exposed workers in a copper smelter failed to demonstrate any significant excess of neoplasms.[6] A study of two large cohorts of lead workers (3519 battery plant workers and 2300 lead production workers) followed through 1980 demonstrated statistically significant elevation in the standardized mortality ratio (SMR) for gastric (SMR = 168) and lung cancer (SMR = 125) in the battery plant workers only. Citing the absence of prior evidence from other studies for these associations and their inability to assess and correct for possible confounding factors (such as diet, alcohol, and smoking), the authors considered these findings to be quite tentative. There were no excess deaths from malignancies of the kidney or other sites in either cohort.[12]

There are several reports that certain lead compounds, including lead acetate and lead phosphate, administered to animals in high doses are carcinogenic, primarily producing renal tumors.[3,17–19] The IARC has concluded that the evidence for carcinogenicity of lead to humans is inadequate, although there is sufficient evidence of carcinogenicity of some lead salts to animals. *Note:* Those salts demonstrating carcinogenicity in animals are soluble, whereas humans are primarily exposed to insoluble metallic lead and lead oxide.[19]

Reproductive effects from lead exposure have been documented in animals and humans of both sexes. Lead penetrates the placental barrier and has caused congenital abnormalities in animals.[3,20] Excessive exposure to lead during pregnancy has resulted in neurologic disorders in infants, although the possibility of postnatal exposure of the infants to lead could not be excluded.[21,22] In battery workmen with a mean occupational exposure to lead of 8.5 (1 to 23) years and with blood lead concentrations of 53 to 75 μg/100 ml of blood, there was an increased frequency of abnormalities of sperm, including hypospermia, as compared with a control group.[23]

Diagnosis. Signs and symptoms include weakness, lassitude, and insomnia; facial pallor and pallor of the eye grounds; anorexia, weight loss, and malnutrition; constipation, abdominal discomfort and tenderness, and colic; anemia; lead line on gingival tissues; signs of motor weakness, including paralysis of the extensor muscles of wrist and less often of the ankles; encephalopathy; and nephropathy. A detailed neurologic examination with electromyography and nerve conduction velocity may be useful when peripheral nerve damage is suspected.

Special Tests: The concentration of lead in the blood is the best single indication of exposure to inorganic lead; blood lead is less variable than urinary lead.[2] Blood lead determination is an exacting laboratory procedure requiring constant attention to quality control. The upper limits of lead levels traditionally classified as normal are 40 μg/100 ml blood and 80 μg/liter of urine, although the 50th and 95th percentiles of blood lead values in most nonoccupationally exposed populations are less than 20 and 35 μg/100 ml blood, respectively. Classical findings of lead intoxication probably do not occur below blood lead levels of 80 μg/100 ml.[24–26]

Other indicators of lead exposure relate to the inhibition by lead of the synthesis of heme. The inhibition of delta-aminolevulinic acid dehydrase (ALA-D), an enzyme involved in porphyrin synthesis, leads to an increase in levels of delta-aminolevulinic acid (ALA) in blood and urine.[3] The blood and urine levels of coproporphyrin III and free erythrocyte protoporphyrins (FEP) are also usually elevated.[3] FEP combines with zinc in the blood to form zinc protoporphyrin (ZPP), which is the moiety assayed.[26]

The ZPP test is now widely used in biological monitoring for lead absorption, in conjunction with blood lead levels. One advantage of the ZPP test is its ability to "average" the effects of lead absorption over a time period of several months, reflecting the 120-day average life span of the red blood cell. Although there is a delay in the increase in ZPP after initial lead exposure, the ZPP will remain elevated longer than the blood lead level following cessation of exposure. The blood lead level is a better indicator for acute lead intoxication, whereas the ZPP is a better indicator of chronic intoxication.[4] The ZPP is not completely specific for lead effects in that iron deficiency will also result in an increase in the ZPP level. The normal ZPP level is generally below 40 to 50 μg/100 ml blood in nonoccupationally exposed populations.[26]

ALA-D, ALA, coproporphyrin III assays, and blood examinations for hemoglobin, reticulocytes, and stippled red cells are useful in the assessment of worker health, but no one of these measurements alone is an accepted specific index of lead absorption.[27]

The Ca EDTA mobilization test for lead is used for estimating both current and previous absorption of increased amounts of lead.[3,27,28] In this test, a single dose of 1 g Ca EDTA in 250 ml of 5% dextrose is infused intravenously over a period of 1 hour. Urine is then collected quantitatively for 24 hours (4 days in subjects with renal insufficiency). The upper limit of normal in healthy adult subjects is 500 to 600 μg of lead excreted in the urine.

Treatment. In acute lead poisoning or acute exacerbations of chronic lead poisoning with severe neurologic or gastrointestinal symptoms, administer Ca EDTA (edetate calcium disodium, Versenate) intravenously 2 g/day (mild cases) or 3 to 4 g/day (cautiously in severe cases). The concentration of Ca EDTA in 5% D/W or NS should not exceed 0.5%. The drug is usually given in a dose of 1 g in 250 ml of 5% dextrose by slow infusion over an hour twice daily for 3 to 5 days. After several days, the course of therapy may be repeated if necessary. Less severe manifestations or asymptomatic elevations in blood lead level are treated by removal from further exposure.

Precautions: Ca EDTA has caused proteinuria, microscopic hematuria, large epithelial cells in the urinary sediment, renal failure from proximal tubule damage, hypercalcemia, and fever. It should not be used during periods of anuria. Safe administration of Ca EDTA requires the following determinations on the first, third, and fifth day of each course of therapy: serum electrolytes; urea nitrogen, creatinine, calcium, phosphorus, and alkaline phosphatase measurements in blood; and routine urinalysis. The patient should also be monitored for irregularities of cardiac rhythm.

Note: The oral administration of Ca EDTA to workers exposed to lead as prophylactic therapy has been considered and practiced by some. This practice is condemned on two counts: (1) Only 2% to 5% of the administered dose of Ca EDTA is absorbed from the gastrointestinal tract, (2) Ca EDTA given orally increases the absorption of lead from the bowel and may precipitate or aggravate symptoms of lead poisoning.[29]

REFERENCES

1. Kehoe RA: Occupational lead poisoning. Clinical types. J Occup Med 14:298–300, 1972
2. National Institute for Occupational Safety and Health: Criteria for a Recommended Standard . . . Occupational Exposure to Inorganic Lead. DHEW (HSM) Pub No 73–22020. Washington, DC, US Government Printing Office, 1972
3. Committee on Biologic Effects of Atmospheric Pollutants, Division of Medical Sciences, National Research Council: Lead-Airborne Lead in Perspective. Washington, DC, National Academy of Sciences, 1972
4. Klaassen CD: Heavy metals and heavy-metal antagonists. In Goodman LS, Gilman AG (eds): Goodman and Gilman's The Pharmacological Basis of Therapeutics, 6th ed, pp 1616–1622. New York, Macmillan Publishing Co, 1980
5. Jeyaratnam J et al: Neuropsychological studies on lead workers in Singapore. Br J Ind Med 43:626–629, 1986

6. Williamson AM, Teo RKC: Neurobehavioral effects of occupational exposure to lead. Br J Ind Med 43:374–380, 1986
7. Mantere P, Hänninen H, Hernberg S, Luukkonen R: A prospective follow-up study on psychological effects in workers exposed to low levels of lead. Scand J Work Environ Health 10:43–50, 1984
8. Zi-giang, Chen et al: Peripheral nerve conduction velocity in workers occupationally exposed to lead. Scand J Work Environ Health 11(suppl 4):26–28, 1985
9. Seppäläinem AM et al: Early neurotoxic effects of occupational lead exposure: A prospective study. Neurotoxicology 4:181–192, 1983
10. Haley TJ: Air Quality Monograph No 69–7, A Review of the Toxicology of Lead, American Petroleum Institute, pp 15–20, 31–53. New York, 1969
11. Vitale LF, Joselow MM, Wedeen RP, Pawlow M: Blood lead—an inadequate measure of occupational exposure. J Occup Med 17:155–156, 1975
12. Cooper WC, Wong O, Kheifets L: Mortality among employees of lead battery plants and lead-producing plants. 1947–1980. Scand J Work Environ Health 11:331–345, 1985
13. Cooper WC, Gaffey WR: Mortality of lead workers. J Occup Med 17:100–107, 1975
14. Lane RE: Health control in inorganic lead industries. Arch Environ Health 8:243–255, 1964
15. Department of Labor: Occupational exposure to lead. Federal Register 40:45934–45948, 1975
16. Gerhardsson L et al: Mortality and lead exposure: A retrospective cohort study of Swedish smelter workers. Br J Ind Med 43:707–712, 1986
17. Mao P, Molnar JJ: The fine structure and histochemistry of lead-induced renal tumors in rats. Am J Pathol 50:571–581, 1967
18. Boyland E, Dukes CE, Grover PL, Mitchley BCV: The induction of renal tumors by feeding lead acetate to rats. Br J Cancer 16:283–288, 1962
19. IARC Monographs on the Evaluation of the Carcinogenic Risk of Chemicals to Humans: Chemicals, Industrial Processes and Industries Associated with Cancer in Humans, Suppl 4, pp 149–150. Lyon, International Agency for Research on Cancer, 1982
20. Ferm VH, Carpenter SJ: Developmental malformations resulting from the administration of lead salts. Exp Mol Pathol 7:208–213, 1967
21. Palmisano PA, Sneed RC, Cassady G: Untaxed whiskey and fetal lead exposure. J Pediatr 75:868–872, 1969
22. Angle CR, McIntire MS: Lead poisoning during pregnancy. Am J Dis Child 108:436–439, 1964
23. Lancranjan I et al: Reproductive ability of workmen occupationally exposed to lead. Arch Environ Health 30:396–401, 1975
24. Lane RE et al: Diagnosis of inorganic lead poisoning: A statement. Br Med J 4:501, 1968
25. Kehoe RA: Occupational lead poisoning. 2–Chemical signs of the absorption of lead. J Occup Med 14:390–396, 1972
26. Lauwerys R: Industrial Chemical Exposure: Guidelines for Biological Monitoring, pp 27–34. Davis, California, Biomedical Publications, 1982
27. Selander S: Treatment of lead poisoning. A comparison between the effects of sodium calciumedate and penicillamine administered orally and intravenously. Br J Ind Med 24:272–283, 1967
28. Emmerson BT: Chronic lead nephropathy: The diagnostic use of calcium EDTA and the association with gout. Aust Ann Med 12:310–324, 1963
29. Lilis R, Fischbein A: Chelation therapy in workers exposed to lead. JAMA 235:2823–2824, 1976

LEAD ARSENATE

CAS: 10102-48-4

$Pb_3(AsO_4)_2$ 1987 TLV = 0.15 mg/m^3

Synonyms: Arsenate of lead

Physical Form. White, heavy powder

Uses. Insecticide; control of tapeworms in cattle, goats, and sheep

Exposure. Inhalation; ingestion

Toxicology. Lead arsenate may cause lead and/or arsenic intoxication; arsenic symptoms are likely to predominate in acute intoxication, whereas prolonged inhalation of lead arsenate may induce the symptoms of lead intoxication.[1]

Some of the effects of acute arsenic intoxication are nausea, vomiting, diarrhea, and irritation; inflammation and ulceration of the mucous membranes and skin; and kidney damage.[2] Among the effects of chronic arsenic poisoning are increased pigmentation and keratinization of the skin, dermatitis, and epidermoid carcinoma. Other effects seen after ingestion but not common from industrial exposure are muscular paralysis, visual disturbances, and liver and kidney damage.[2]

Effects of lead intoxication include damage to the central and peripheral nervous systems, to the kidneys, and to the blood-forming mechanism, which may lead to anemia.[3] Symptoms include colic, loss of appetite, and constipation; excessive tiredness and weakness; and nervous irritability.[3] In peripheral neuropathy, the distinguishing clinical feature of lead intoxication is a predominance of motor impairment, with minimal or no sensory abnormalities.[3] There is a tendency for the extensor muscles of the hands and feet to be affected.[3] Lead intoxication has also resulted in kidney damage, with few, if any, symptoms appearing until permanent damage has occurred.[3]

In a follow-up mortality study in 1973 of a cohort of 1231 persons (primarily orchardists) who had participated in a 1938 mortality study, it was concluded that excess mortality did not occur consistently from exposure to lead arsenate spray.[4,5] In contrast, two other independent studies reported a significant excess of lung cancer among other cohorts of this same population.[2] In a study of workers engaged in the formulation and packaging of lead arsenate and calcium arsenate, there was an excess of lung cancer, which was dose related.[6] In vineyard workers chronically exposed to lead, calcium, and copper arsenate dust in Germany and France, there are numerous reports of skin cancer, including basal cell and squamous cell carcinomas and Bowen's disease, as well as lung cancer.[7] The IARC has concluded that there is sufficient evidence that inorganic arsenic compounds, including lead arsenate, are skin and lung carcinogens in humans.[7]

REFERENCES

1. Lead arsenate. Documentation of TLVs and BEIs, 5th ed, p 346. Cincinnati, American Conference of Governmental Industrial Hygienists (ACGIH), 1986
2. Department of Labor: Standard for exposure to inorganic arsenic. Federal Register 40:3392–3404, 1975
3. Department of Labor: Occupational exposure to lead. Federal Register 40:45934–45948, 1975
4. Nelson WC et al: Mortality among orchard workers exposed to lead arsenate spray: A cohort study. J Chron Dis 26:105–118, 1973
5. Neal PA et al: A Study of the Effect of Lead Arsenate Exposure on Orchardists and Consumers of Sprayed Fruit. US Public Health Service Bull No 267, pp 47–165, 171–181. Washington, DC, US Government Printing Office, 1941
6. Ott MG, Holder BB, Gordon HL: Respiratory cancer and occupational exposure to arsenicals. Arch Environ Health 29:250–255, 1974
7. IARC Monographs on the Evaluation of the Carcinogenic Risk of Chemicals to Humans—Some Metals and Metallic Compounds, Vol 23, pp 39–41. Lyon, International Agency for Research on Cancer, 1980

LINDANE

CAS: 58-89-9

$C_6H_6Cl_6$ 1987 TLV = 0.5 mg/m^3; skin

Synonyms: 1,2,3,4,5,6–Hexachlorocyclohexane, gamma isomer; gamma HCH; gamma benzene hexachloride; gamma BHC; Kwell

Physical Form. crystalline solid

Uses. Insecticide

Exposure. Inhalation; skin absorption; ingestion

Toxicology. Lindane is a convulsant; in mice, it is a liver carcinogen.

Exposure to the vapor causes irritation of the eyes, nose, and throat, severe headache, and nausea.[1] Lindane levels in the blood do not appear to increase with increased duration of exposure but primarily reflect recent lindane absorption.[2] Production workers exposed to air levels of 31 to 1800 μg/m^3 had blood levels of 1.9 to 8.3 ppb.[1]

Lindane has been suspected as a cause of aplastic or hypoplastic anemia in a number of cases reported from various countries.[3] Although a recent report tabulated 46 case reports of bone marrow injury temporally associated with environmental exposure to lindane, the authors questioned the association on several grounds.[3] In 17 cases, there was exposure to other toxic agents, including benzene and chloramphenicol in two cases. In eight cases, investigation of the bone marrow did not reveal aplasia or hypoplasia. In some cases, documentation of exposure was limited. Moreover, no cases have been reported following the therapeutic use of lindane (Kwell) as a scabicide in children or adults, despite the fact that lindane is well absorbed dermally. Eleven of the 46 cases were associated with home use of vaporizers, 18 with sprays, and 7 with powder or dust.[3] Cross-sectional studies of workers chronically exposed to lindane during manufacture have failed to reveal any hematologic conditions or significant differences in hemoglobin or total leukocyte count relative to a control population.[4] Although some statistically significant differences were found in some hematologic parameters, such as increases in polymorphonuclear leukocyte counts and reticulocyte counts, compared with the control group, the results were still largely within the reference range and of questionable biological significance. No significant differences were observed for transaminases (AST, ALT) or other liver function studies.[4]

Accidental ingestion has caused fatalities; effects were repeated, violent, clonic convulsions, sometimes superimposed on a continuous tonic spasm. Respiratory difficulty and cyanosis secondary to the convulsions were common.[5] Following nonfatal accidental ingestions, symptoms have included malaise, dizziness, nausea, and vomiting. Agitation, collapse, convulsions, loss of consciousness, muscle tremor, fever, and cyanosis have commonly been observed. Most patients who survive recover completely over 1 to 3 days; protracted illness is rare.[2]

Minor liver lesions have been reported in rats at dosages as low as 2.6 to 5.0 mg/kg/day and in mice at dosages of 6 to 10 mg/kg/day. After repeated high doses, degenerative changes have been reported in the kidney, pancreas, and testes of rodent species.[2] Feeding of 1500 ppm in the diet to rats for 90 days, a maximally tolerated dose, resulted in testicular atrophy with spermatogenic arrest and apparent inhibition of androgen synthesis by Leydig cells.[6] Lindane has exhibited embryofetotoxicity in female rats following daily oral doses of 9.5 mg/kg bodyweight for 4 months; teratogenic effects have not been observed following doses as high as 20 mg/kg bodyweight.[7] Based on several oral bioassays in which excess liver tumors were noted, the IARC has concluded that there is sufficient evidence that lindane is carcinogenic in mice. Studies in rats have been inadequate for reaching a conclusion.[7]

REFERENCES

1. Hygienic Guide Series: Hexachlorocyclohexane, gamma isomer-lindane. Am Ind Hyg Assoc J 33:36–59, 1972
2. Hayes WJ Jr: Pesticides Studied in Man, pp 211–228. Baltimore, Williams & Wilkins, 1982
3. Morgan DP, Stockdale EM, Roberts RJ, Walter AW: Anemia associated with exposure to lindane. Arch Environ Health 35:307–310, 1980
4. Brassow HL, Baumann K, Hehnert G: Occupational exposure to hexachlorocyclohexane. II. Health conditions of chronically exposed workers. Int Arch Occup Environ Health 48:81–87, 1981
5. Hayes WJ Jr: Clinical Handbook on Economic Poisons, Emergency Information for Treating Poisoning, US Public Health Service Pub No 476, pp 50–55. Washington, DC, US Government Printing Office, 1963
6. Shivanandappa T, Krishnakumari MK: Hexachlorocyclohexane-induced testicular dysfunction in rats. Acta Pharmacol Toxicol 52:12–17, 1983
7. IARC Monographs on the Evaluation of the Carcinogenic Risk of Chemicals to Humans: Some Halogenated Hydrocarbons, Vol 20, pp 195–223. Lyon, International Agency for Research on Cancer, 1979

LIQUIFIED PETROLEUM GAS

CAS: 68476-85-7

Mixture of C_3H_6, C_3H_8, C_4H_8, and C_4H_{10} 1987 TLV = 1000 ppm

Synonyms: LPG; bottle gas; liquified hydrocarbon gas

Physical Form. Gas or liquid

Uses. Fuel; in production of chemicals

Exposure. Inhalation

Toxicology. Liquified petroleum gas is practically nontoxic below the explosive limits but may cause asphyxia by oxygen displacement at extremely high concentrations.[1]

No chronic systemic effects have been reported from occupational exposure. The vapor is not irritating to the eyes, nose, or throat.[2] Direct contact with the liquid may cause burns or frostbite to the eyes and skin.[3] Olefinic impurities may lend a narcotic effect. At extremely high concentrations, the limiting toxicologic factor is available oxygen. Minimal oxygen content should be 18% by volume under normal atmospheric pressure. Generally, flammability and explosive hazards outweigh the biological effects.[1]

REFERENCES

1. Deichmann WB, Gerarde HW: Toxicology of Drugs and Chemicals. New York, Academic Press, 1969
2. Weiss G: Hazardous Chemical Data Book. Noyes Data Corporation, p 568. Park Ridge, New Jersey, 1980
3. Sandmeyer EE: Aliphatic hydrocarbons. In Clayton GD, Clayton FE (eds): Patty's Industrial Hygiene and Toxicology, 3rd ed, rev, Vol 2B, Toxicology, pp 3175–3220. New York, Wiley–Interscience, 1981

LITHIUM HYDRIDE

CAS: 7580-67-8

LiH 1987 TLV = 0.025 mg/m^3

Synonyms: None

Physical Form. White crystals that darken on exposure to light

Uses. Reducing agent; condensing agent with ketones and acid esters; desiccant; in hydrogen generators

Exposure. Inhalation; ingestion

Toxicology. Lithium hydride is a severe irritant of the eyes, mucous membranes, and skin.

The toxicity of lithium hydride differs markedly from that of the soluble salts because of its vigorous chemical reactivity with water, producing acute irritation and corrosion of biological tissues.[1]

The explosion of a cylinder of lithium hydride led to eye contact and swallowing of a small amount of the dust by a technician.[2] The resulting burns caused scarring of both corneas and strictures of the larynx, trachea, bronchi, and esophagus; death occurred 10 months later.

Exposure of humans to concentrations in the range of 0.025 to 0.1 mg/m^3 caused some nasal irritation to which tolerance was acquired with continuous exposure; at 0.5 to 1.0 mg/m^3, severe nasal irritation, cough, and some eye irritation was noted; in the range of 1.0 to 5.0 mg/m^3, all effects were severe, and skin irritation was felt.[3]

Exposure of animals to concentrations of 5 to 55 mg/m^3 of lithium hydride caused sneezing and cough with secondary pulmonary emphysema; levels of 10 mg/m^3 corroded the body fur and skin of the legs, and there was occasionally severe inflammation of the eyes and nasal septum.[1] The lesions of the nose and legs were attributed to the alkalinity of lithium hydroxide, the hydrolysis product of lithium hydride. Since powdered lithium hydride may ignite spontaneously in humid air or on contact with most mucous surfaces, resulting tissue effects may have features of both thermal and alkali burns.[4]

REFERENCES

1. Spiegl CJ et al: Acute inhalation toxicity of lithium hydride. AMA Arch Ind Health 14:468–470, 1956
2. Cracovaner AJ: Stenosis after explosion of lithium hydride. Arch Otolaryngol 80:87–92, 1964
3. Stokinger HE: The metals. In Clayton GD, Clayton FE (eds): Patty's Industrial Hygiene and Toxicology, 3rd ed, rev, Vol 2A, pp 1728–1740. New York, Wiley–Interscience, 1981
4. Gosselin RE et al: Clinical Toxicology of Commercial Products, 5th ed. Baltimore, Williams & Wilkins, 1984

MAGNESIUM OXIDE FUME

CAS: 1309-48-4

MgO 1987 TLV = 10 mg/m^3

Synonyms: None

Physical Form. Fume

Sources. From manufacture of refractory crucibles, fire bricks, magnesia cements, boiler scale compounds

Exposure. Inhalation

Toxicology. Magnesium oxide fume is a mild irritant of the eyes and nose.

Examination of 95 workers exposed to an unspecifed concentration of magnesium oxide dust revealed slight irritation of the eyes and nose; the magnesium level in the serum of 60% of those examined was above the normal upper limit of 3.5 mg/dl.[1]

Experimental subjects exposed to fresh magnesium oxide fume developed metal fume fever, an illness similar to influenza; effects were fever, cough, oppression in the chest, and a leukocytosis.[2] There are no reports of metal fume fever resulting from industrial exposure to magnesium oxide fume.[2,3]

REFERENCES

1. Stokinger HE: The metals. In Clayton GD, Clayton FE (eds): Patty's Industrial Hygiene and Toxicology, 3rd ed, Vol 2, Toxicology, pp 1740–1748. New York, Wiley–Interscience, 1981
2. Drinker KR, Thomson RM, Finn JL: Metal fume fever. The effects of inhaled magnesium oxide fume. J Ind Hyg 9:187–192, 1927
3. Hygienic Guide Series: Magnesium. Am Ind Hyg Assoc J 21:97–98, 1960

MALATHION

CAS: 121-75-5

$C_{10}H_{19}O_6PS_2$ 1987 TLV = 10 mg/m^3; skin

Synonyms: Diethyl mercaptosuccinate, S-ester with 0,0-dimethyl phosphorodithioate; Malathon; carbophos; 4049; CYTHION

Physical Form. Colorless to light amber liquid

Uses. Insecticide

Exposure. Inhalation; skin absorption; ingestion

Toxicology. Malathion is an anticholinesterase agent, but it is of a relatively low order of toxicity in comparison with other organophosphates.

Signs and symptoms of intoxication by anticholinesterase agents are caused by the inactivation of the enzyme cholinesterase, which results in the accumulation of acetylcholine at synapses in the nervous system, skeletal and smooth muscle, and secretory glands.[1-4] After inhalation of extremely high concentrations of malathion, ocular and respiratory effects may appear simultaneously. Ocular effects include miosis, blurring of distant vision, tearing, rhinorrhea, and frontal headache. Respiratory effects include tightness in the chest, wheezing, laryngeal spasms, and excessive salivation. Peripheral effects include excessive sweating, muscular fasciculations, and weakness. Effects on the central nervous system include giddiness, confusion, ataxia, slurred speech, and convulsions. After ingestion, anorexia, nausea, vomiting, abdominal cramps, and diarrhea also appear.

Malathion itself has only a slight direct inhibitory action on cholinesterase, but one of its metabolites, malaoxon, is an active inhibitor.[4] Both malathion and malaoxon are rapidly detoxified by esterases in the liver and other organs.[4] This rapid metabolism is the apparent reason for the lower toxicity of malathion compared with other organophosphates. Malaoxon inactivates cholinesterase by phosphorylation of the active site of the enzyme to form the "dimethylphosphoryl enzyme." Over the following 24 to 48 hours, there is a process, termed *aging*, of conversion to the "monomethylphosphoryl enzyme." Aging is of clinical interest in the treatment of poisoning, because cholinesterase reactivators such as pralidoxime (2–PAM, Protopam) chloride are ineffective after aging has occurred.

The relative safety of malathion to humans has been repeatedly demonstrated. In a group of workers with an average exposure of 3.3 mg/m^3 for 5 hours (maximum of 56 mg/m^3), the cholinesterase levels in the blood were not significantly lowered, and no one exhibited signs of cholinesterase inhibition.[5] In a human experiment in which four men were exposed 1 hour daily for 42 days to 84.8 mg/m^3, there was moderate irritation of nose and conjunctiva, but there were no cholinergic signs or symptoms.[6]

Almost all reports of fatalities from malathion have involved ingestion.[4] The acute oral lethal dose is estimated to be somewhat below 1.0 g/kg.[4] Nonlethal intoxication has occurred in agricultural workers but has usually been the result of gross exposures with concomitant skin absorption.[4]

Malathion has caused skin sensitization, and dermatitis may occur under conditions of heavy field use.[7]

In rats, malathion was not teratogenic when administered by gastric intubation on days 6 to 15 of gestation at doses as high as 300 mg/kg.[8]

National Cancer Institute studies showed that administration of 4700 or 8150 mg/kg for 80 weeks, or 2000 or 4000 mg/kg for 103 weeks, in the diets of rats was not carcinogenic.[9–11] Subsequent data re-evaluation by NTP confirmed these conclusions.[12] Mice fed diets containing 8,000 or 16,000 mg/kg for 80 weeks also had no significant increase in tumor incidence.[9] The IARC has determined that no evidence is available to suggest that malathion is likely to present a carcinogenic risk to humans.[13]

Diagnosis. *Signs and Symptoms:* Initial signs and symptoms include headache, blurred vision, pallor, weakness, sweating, abdominal pain, nausea, vomiting, and diarrhea.

Moderate-to-severe intoxication involves miosis, lacrimation, excessive salivation, muscle fasciculations, dyspnea, cyanosis, convulsions, shock, cardiac arrhythmias, and coma.

Differential Diagnosis: Diagnosis is based primarily on a history of exposure and clinical evidence of diffuse parasympathetic stimulation.[4] Careful observation of the effects of atropine and pralidoxime may be valuable. Patients with organophosphate poisoning are resistant to the action of atropine at moderate dosages; failure of 1 to 2 mg of atropine administered parenterally to produce signs of atropinization (flushing, mydriasis, tachycardia, or dryness of mouth) indicates anticholinesterase poisoning. Intravenous injection of 1 g pralidoxime generally causes some recovery from signs and symptoms.

Special Tests: Two types of cholineterase are clinically significant: (1) true acetylcholinesterase, found principally in the nervous system and the red blood cell; and (2) pseudo- or butyrylcholinesterase, found in the plasma, liver, and nervous system. Whereas the action of both types is inhibited by organophosphates, the level of depression of red blood cell cholinesterase is a better indicator of clinically significant reduction of cholinesterase activity in the nervous system.

Laboratory evidence of depression of red blood cell cholinesterase to a level substantially below pre-exposure levels (at least 50% and usually much lower) is verification of poisoning. There is an imperfect correlation between the degree of depression of cholinesterase enzymes and the occurrence of symptoms. With a rapid drop in cholinesterase activity, generally reflecting an acute heavy exposure, there may be symptoms with only a 30% depression, whereas with slower drops to 70% depression, reflecting chronic low-level exposure, there may be no symptoms.[14]

If no pre-exposure baseline has been performed but symptoms are not sufficient to justify treatment with atropine, repeated testing during the recovery period demonstrating progressively increasing plasma and red blood cell cholinesterase levels over several days and weeks, respectively, suggests the diagnosis of anticholinesterase poisoning.

There are many different methods for estimation of cholinesterase content of blood, and associated with each method is a different set of normal values and a different set of reporting units. The laboratory report of a cholinesterase determination should state the units involved along with the appropriate normal range. Based on the Michel method, the normal range of red blood cell cholinesterase activity (delta *p*H per hour) is 0.39 to 1.02 for men, and 0.34 to 1.10 for women.[15] The normal range of the enzyme activity (delta *p*H per hour) of plasma is 0.44 to 1.63 for men, and 0.24 to 1.54 for women.

Treatment. Treatment of organophosphate poisoning ranges from simple removal from exposure in very mild cases to the provision of rigorous supportive and antidotal measures in severe cases.[2,3,16,17] In moderate-to-severe cases, because of pulmonary involvement, there may be need for artificial respiration using a positive-pressure method. Careful attention must be paid to removal of secretions and to maintenance of a patent airway. Anticonvulsants such as diazepam or thiopental sodium may be necessary. Maintenance of respiration is critical, because death usually results from weakness of the muscles of respiration and accumulation of excessive secretions in the respiratory tract.[2]

As soon as cyanosis has been overcome, 2 to 4 mg of atropine should be given intravenously. (Atropine may induce ventricular fibrillation in the presence of cyanosis.) *This dose of atropine is approximately 10 times the amount that is administered for other conditions in which atropine is considered therapeutic.* This dose should be repeated at 5- to 10-minute intervals until signs of atropinization appear (dry, flushed skin, tachycardia as high as 140 beats/minute, and pupillary dilatation). If muscarinic symptoms reappear, the dose of atropine should be repeated. A mild degree of atropinization should be maintained for at least 48 hours.[2]

Pralidoxime (2–PAM, Protopam) chloride is a cholinesterase reactivator that complements the action of atropine. It has its greatest effect in reversing the nicotinic action of anticholinesterase agents at skeletal neuromuscular junctions but virtually no effect on central nervous system manifestations. In moderate-to-severe cases, the dose for adults is 1 to 2 g injected intravenously at a rate not in excess of 500 mg/minute. After an hour, a second dose of 1 g is indicated if muscle weakness has not been relieved. Treatment with pralidoxime chloride will be most effective if given within 24 hours after poisoning.[2] Morphine, aminophylline, and phenothiazines are contraindicated because of documented experience of adverse reactions in cases of organophosphate poisoning.[16]

It is of great importance to decontaminate the patient. Contaminated clothing should be removed at once, and the skin should be washed with generous amounts of soap or detergent and a flood of water, which is best accomplished under a shower or by submersion in a pond or other body of water if the exposure occurred in the field. Careful attention should be paid to cleansing of the skin and hair.

The patient should be attended and monitored continuously for not less than 24 hours, because serious and sometimes fatal relapses have occurred owing to continuing absorption of the toxin or dissipation of the effects of the antidote.

Regeneration of cholinesterase is primarily by synthesis of new enzyme and takes place at the rate of approximately 1% per day.[16] A patient who has recovered from the acute phase of poisoning remains hypersusceptible to anticholinesterases for up to several weeks.

Medical Control. This involves preplacement and annual physical examination with determination of pre-exposure red blood cell and plasma cholinesterase activity. A person whose red blood cell cholinesterase falls to or below 40% of the pre-exposure baseline should be removed from further exposure until the activity returns to within 80% of the pre-exposure baseline.

REFERENCES

1. Grob D: Anticholinesterase intoxication in man and its treatment. In Koelle GB (ed): Handbuch der Experimentellen Pharmakologie, Vol 15, Cholinesterases and Anticholi-

nesterase Agents, pp 989–1027. Berlin, Springer-Verlag, 1963
2. Taylor P: Anticholinesterase agents. In Gilman AG et al (eds): Goodman and Gilman's The Pharmacological Basis of Therapeutics, 7th ed, pp 110–129. New York, Macmillan Publishing Co, 1985
3. Hayes WJ Jr: Clinical Handbook on Economic Poisons. Emergency Information for Treating Poisoning, US Public Health Service Pub No 476, pp 12–23. Washington, DC, US Government Printing Office, 1963
4. National Institute for Occupational Safety and Health: Criteria for a Recommended Standard . . . Occupational Exposure to Malathion. DHEW (NIOSH) Pub No 76-205. Washington, DC, US Government Printing Office, 1976
5. Culver D, Caplan P, Batchelor GS: Studies of human exposure during aerosol application of malathion and chlorthion. AMA Arch Ind Health 13:37–50, 1956
6. Golz HH: Controlled human exposures to malathion aerosols. AMA Arch Ind Health 19:516–523, 1959
7. Milby TH, Epstein WL: Allergic contact sensitivity to malathion. Arch Environ Health 9:434–437, 1964
8. Khera KC et al: Teratogenicity studies on linuron, malathion, and methoxychlor in rats. Toxicol Appl Pharmacol 45:435–444, 1978
9. National Cancer Institute: Bioassay of Malathion for Possible Carcinogenicity, TR-24. DHEW (NIH) Pub No 78-824. Washington, DC, US Government Printing Office, 1978
10. National Cancer Institute: Bioassay of Malathion for Possible Carcinogenicity, TR-192. DHEW (NIH) Pub No 78-1748. Washington, DC, US Government Printing Office, 1979
11. National Cancer Institute: Bioassay of Malathion for Possible Carcinogenicity, TR-135. DHEW (NIH) Pub No 79-1390. Washington, DC, US Government Printing Office, 1979
12. Huff JE et al: Malathion and malaoxon: Histopathology reexamination of the National Cancer Institute's carcinogenesis studies. Environ Res 37:154–173, 1985
13. IARC Monographs on the Evaluation of the Carcinogenic Risk of Chemicals to Humans. Miscellaneous Pesticides, Vol 30, pp 103–129. Lyon, International Agency for Research on Cancer, 1983
14. Coye MJ, Lowe JA, Maddy KT: Biological monitoring of agricultural workers exposed to pesticides. I. Cholinesterase activity determinations. Occup Med 27:619–627, 1986
15. Michel HO: Electrometric method for determination of red blood cell and plasma cholinesterase activity. J Lab Clin Med 34:1564–1568, 1949
16. Milby TH: Prevention and management of organophosphate poisoning. JAMA 216:2131–2133, 1971
17. Namba T, Greenfield M, Grob D: Malathion poisoning. Arch Environ Health 21:533–541, 1970

MALEIC ANHYDRIDE

CAS: 108-31-6

$C_4H_2O_3$ 1987 TLV = 0.25 ppm

Synonyms: 2,5–Furandione; cis-butenedioic anhydride; toxilic anhydride

Physical Form. White, crystalline solid

Uses. Chemical intermediate; alkyd resin ester and polyester resins

Exposure. Inhalation

Toxicology. Maleic anhydride is a severe irritant of the eyes; it is an irritant and sensitizer of both the skin and respiratory tract and may produce asthma on repeated exposures.

Workers exposed to vapors from heated maleic anhydride developed an intense burning sensation in the eyes and throat, with cough and vomiting; exposure to high fume concentrations caused photophobia, double vision, and a visual phenomenon of seeing rings around lights.[1,2] Exposure of humans to a concentration of 1.5 to 2 ppm resulted in nasal irritation within 1 minute and eye irritation after 15 to 20 minutes.[3] Among workers repeatedly exposed to 1.25 to 2.5 ppm, effects included ulceration of nasal mucous membranes, chronic bronchitis, and, in some cases, asthma.[3]

The dust on dry skin may result in a delayed burning sensation, but, on moist skin, the sensation is almost immediate, producing erythema, which may progress to vesiculation.[3] Prolonged or repeated exposure may also cause dermatitis.[3]

In a study in which rats were injected subcutaneously with 1 mg maleic anhydride in oil twice weekly for 61 weeks, two of three surviving animals developed fibrosarcomas, which appeared 80 weeks after the start of the experiment.[4]

REFERENCES

1. Grant WM: Toxicology of the Eye, 3rd ed, pp 574–575. Springfield, Illinois, Charles C Thomas, 1986
2. Chemical Safety Data Sheet SD–88, Maleic Anhydride, pp 5–6, 11–13. Washington, DC, MCA, Inc, 1962
3. Hygienic Guide Series: Maleic anhydride. Am Ind Hyg Assoc J 31:391–394, 1970
4. Dickens F, Jones HEH: Further studies on the carcinogenic and growth-inhibitory activity of lactones and related substances. Br J Cancer 17:100–108, 1963

MANGANESE (and compounds)

CAS: 7439-96-5

Mn

1987 TLV: Dust and compounds = C 5 ppm; fume = 1 ppm

Synonym: Manganese dioxide (pyrolusite ore)

Physical Form. Solid

Uses/Sources. Manufacture of alloys, dry-cell batteries; glass; inks; ceramics; paints, welding rods; rubber and wood preservatives; mining and processing of manganese ores

Exposure. Inhalation

Toxicology. Manganese affects the central nervous system, and intoxication occurs mostly in chronic form (manganism); inhalation of high concentrations of nascent manganese oxide causes an influenza-like illness (metal fume fever).

The neurologic disorder known as chronic manganese poisoning occurs after variable periods of heavy exposure ranging from 6 months to 3 years.[1,2] The disease begins insidiously with headache, asthenia, irritability, and, occasionally, psychotic behavior.[1] The latter, manganese psychosis, occurs most frequently in miners rather than in industrial workers and consists of transitory psychological disturbances such as hallucinations, compulsive behavior, and emotional instability.[3] Severe somnolence followed by insomnia is often found early in the disease.[7] As manganese exposure continues, symptoms include generalized muscle weakness, speech impairment, incoordination, and impotence. Tremor, paresthesia, and muscle cramps have been noted.[1,3,4] In the advanced stage, the subject exhibits excessive salivation, inappropriate emotional reactions, and Parkinson-like symptoms, such as mask-like facies, severe muscle rigidity, and gait disorders.[1] Manganism is reversible if it is limited to psychological disturbances and if the subject is removed from exposure. Established neurologic signs and symptoms tend to persist or even progress in the absence of additional exposure.[5] Evidence is accumulating from animal studies that the central dopaminergic system is disturbed in manganese neurointoxication.[1]

Exposure levels associated with advanced manganism have typically been very high; 150 cases were found in three mines where levels reached 450 mg/m^3.[2] More recent studies report cases showing neurologic symptoms and a few signs at lower concentrations. Of 36 workers exposed to MnO_2 ore dust ranging from 6.8 to 42.2 mg/m^3, 8 workers exhibited symptoms of manganism.[6] Neurologic screening of 117 workers with exposures greater than 5 mg/m^3 revealed 7 cases with definite signs and symptoms.[7] Comparison of 369 workers exposed to 0.3 to 20 mg/m^3 suggested that slight neurologic disturbances may occur at exposures less than 5 mg/m^3 but seem to be more prevalent at higher exposures.[8] Factors contributing to inconsistencies in the dose–response relationship include broad exposure ranges, different chemical forms and particle size, and a lack of good biological indicators of exposure.[1]

An association between manganese exposure and pulmonary effects, including pneumonia, chronic bronchitis, and airway disability, has been observed. Extrapolation from animal studies suggests that it is unlikely that manganese could be the sole etiologic agent responsible for serious pathologic changes in the lungs. Instead, it is possible that susceptibility to infection is increased.[1]

Acute poisoning by manganese is rare but may occur following ingestion of large amounts of manganese compounds or from inhalation.

Freshly formed manganese oxide fumes at high concentrations may cause metal fume fever. This influenza-like illness is characterized by chills, fever, sweating, nausea, and cough. The syndrome begins 4 to 12 hours after exposure and lasts for 24 hours without causing permanent damage.[9]

Repeated subcutaneous or intraperitoneal injection of manganese dichloride caused increased incidences of lymphosarcomas in mice.[10] There is no information relating manganese exposure to cancer occurrence in humans.[1]

REFERENCES

1. Health Assessment Document for Manganese. Final Report—PB84-229954, pp 1–353. Washington, DC, US Environmental Protection Agency, 1984
2. Rodier J: Manganese poisoning in Moroccan miners. Br J Ind Med 12:21–35, 1955
3. Cook DG, Fahn S, Brait KA: Chronic manganese intoxication. Arch Neurol 30:59–64, 1974
4. Hine CH, Pasi A: Manganese intoxication. West J Med 123:101–107, 1975
5. Barbeau A et al: Role of manganese in dystonia. Adv Neurol 14:339–352, 1976
6. Emara AM et al: Chronic manganese poisoning in the dry battery industry. Br J Ind Med 28:78–82, 1971
7. Tanaka S, Lieben J: Manganese poisoning and exposure in Pennsylvania. Arch Environ Health 19:674–684, 1969
8. Saric M, Lucic-Palaic S: Possible synergism of exposure to airborne manganese and smoking habit in occurrence of respiratory symptoms. In Walton WH (ed): Inhaled Particles, IV, pp 773–779. New York, Pergamon Press, 1977
9. Piscator M: Health hazards from inhalation of metal fumes. Environ Res 11:268–270, 1976
10. DiPaolo JA: The potentiation of lymphosarcomas in mice by manganese chloride. Fed Proc 23:393(abstr), 1964

MERCURY

CAS: 7439-97-6

Hg

1987 TLV = 0.05 mg/m^3, as Hg—mercury vapor; 0.10 mg/m^3, as Hg—aryl and inorganic compounds; skin

Synonyms: Quicksilver; mercury vapor; mercury liquid; mercury salts

Physical Form. Silver-white, heavy, liquid metal

Uses. Electric apparatus; measurement and control systems such as thermometers and sphygmomanometers; agricultural and industrial poisons; catalyst; antifouling paint; dental practice

Exposure. Inhalation; skin absorption; ingestion

Toxicology. Acute exposure to high concentrations of mercury vapor causes severe respiratory damage, whereas chronic exposure to lower levels is primarily associated with central nervous system damage.

Inhalation of mercury vapor may produce a metal fume–fever-like syndrome, including chills, nausea, general malaise, tightness in the chest, and respiratory symptoms.[1] High concentrations cause corrosive bronchitis and interstitial pneumonitis.[2] In the most severe cases, the patient will succumb because of respiratory insufficiency.[2] In one episode involving four workers, it was estimated that mercurial pneumonitis resulted from exposure for several hours to concentrations ranging between 1 and 3 mg/m^3.[3]

With chronic exposure to mercury vapor, early signs are nonspecific and include weakness, fatigue, anorexia, loss of weight, and disturbances of gastrointestinal function.[2] This syndrome has been termed *asthenic-vegetative syndrome*, or *micromercurialism*. At higher exposure levels, a characteristic mercurial tremor appears, beginning with intentional tremor of fingers, eyelids, and lips, and may progress to generalized trembling of the entire body and violent chronic spasms of the extremities.[2,4] Parallel to the development of tremor, mercurial erethism develops. This is characterized by behavioral and personality changes, increased excitability, loss of memory, insomnia, and depression. In severe cases, delirium and hallucination may occur. Another characteristic feature of mercury intoxication is severe salivation and gingivitis. Chronic changes in the cornea and lens have also been described.[5]

It has been estimated that the probability of manifesting typical mercurialism with tremor and behavioral changes will increase with exposures to concentrations of 0.1 mg/m^3 or higher.[2] There is no evidence of effects at concentrations below 0.01 mg/m^3.

Investigations of exposed populations have shown only a weak association between blood or urinary concentrations and the occurrence of clinical signs. On a group basis, however, high levels in blood and urine may be associated with prolonged exposure to high concentrations and thus a greater likelihood of signs of poisoning. Mercury vapor exposure levels that give rise to group mean values of mercury excretion in urine between 50 and 100 μg/liter may give rise to detectable health effects.[2]

Ingestion of mercuric salts causes corrosive ulceration, bleeding, and necrosis of the gastrointestinal tract, usually accompanied by shock and circulatory collapse.[2,4] If the patient survives the gastrointestinal damage, renal failure occurs within 24 hours owing to necrosis of the proximal tubular epithelium, followed by oliguria, anuria, and uremia.

Chronic low-dose exposure to mercuric salts, or probably even elemental mercury vapor, may also induce an immunologic glomerular disease.[2,4]

Applied locally, mercury may cause sensitization dermatitis.[1,2]

An increased frequency of aneuploidy in lymphocytes has been reported in mercury vapor–exposed workers.[2] Mercury vapor transverses the placenta, but fetal effects are not known.[2]

Intraperitoneal injection of metallic mercury in rats has produced sarcomas.[1] The sarcomas develop without exception at those sites in direct contact with the metal, suggesting a foreign reaction rather than chemical carcinogenesis.

Treatment. For systemic mercury intoxication, therapy with BAL is indicated.[6] Inject BAL 10% in oil IM 5 mg/kg immediately and repeat in 2 hours. Then administer 2.5 mg/kg every 6 hours for three doses, followed by two doses on the 2nd day and one dose on the 3rd and subsequent 10 days or until recovery. Side effects, often encountered at this dosage of BAL, include lacrimation, salivation, rhinorrhea, sense of constriction in throat and chest, flushing of the face, nausea, vomiting, fall in blood pressure and pulmonary edema.

The use of penicillamine may also be considered.

REFERENCES

1. National Institute for Occupational Safety and Health: Criteria for a Recommended Standard . . . Occupational Exposure to Inorganic Mercury. DHEW (NIOSH) Pub No 73–11024. Washington, DC, US Government Printing Office, 1973
2. Berlin M: Mercury. In Friberg L et al (eds): Handbook on the Toxicology of Metals, 2nd ed, Vol II, Specific Metals, pp 387–445. Amsterdam, Elsevier, 1986
3. Milne J, Christophers A, De Silva P: Acute mercurial pneumonitis. Br J Ind Med 27:334–338, 1970
4. Goyer RA: Toxic effects of metals. In Klaassen CD et al (eds): Casarett and Doull's Toxicology. The Basic Science of Poisons, 3rd ed, pp 605–609. New York, Macmillan Publishing Co, 1986
5. Rosenman KD et al: Sensitive indicators of inorganic mercury toxicity. Arch Environ Health 41:208–215, 1986
6. Arena JM: Poisoning, 3rd ed, pp 25–26, 128–129. Springfield, Illinois Charles C. Thomas, 1974

MERCURY (alkyl compounds)

CAS: Varies with compound

R-Hg-X — 1987 TLV = 0.01 mg/m^3, as Hg; skin

Compounds: Methyl mercury; ethyl mercury chloride; dimethyl mercury

Physical Form. Colorless liquids

Uses. Fungicides in seed dressings, folial sprays; preservative solutions for wood, paper pulp, textiles, and leather

Exposure. Inhalation; skin absorption; ingestion

Toxicology. Organo (alkyl) mercury compounds cause dysfunction of the central nervous system, and kidneys and are irritants of the eyes, mucous membranes, and skin; methyl mercury causes developmental effects in humans.

Methyl and ethyl mercury compounds have similar toxicologic properties, and there is no sharp demarcation between acute and chronic poisoning.[1] Once a toxic dose has been absorbed and retained for a period of time, functional disturbances and damage occur. The latency period for a single toxic dose may vary from one to several weeks; longer latency periods on the order of years have been reported for chronic exposures.[1,2]

Symptoms of poisoning include numbness and tingling of the lips, hands, and feet (paresthesia); ataxia; dysarthria; concentric constriction of the visual fields; impairment of hearing; and emotional disturbances.[3]

With severe intoxication, clonic seizures may occur, and the symptoms are usually irreversible.[1,3] Severe intoxication also results in incontinence; periods of spasticity and jerking movements of the limbs, head, or shoulders; and bouts of groaning, moaning, shouting, or crying. Less frequent symptoms include dizziness, hypersalivation, lacrimation, nausea, vomiting, and diarrhea or constipation.[4] The pathologic changes in the central nervous system are characterized by general neuron degeneration in the cerebral cortex, especially the visual areas of the occipital cortex and gliosis.[1]

An epidemic of intoxication from ingestion of fish contaminated with methyl mercury occurred in the Minamata district in Japan, and, as a result, methyl mercury intoxication is often referred to as Minamata disease.[4] Infants born to mothers with exposure to large amounts of methyl mercury had microencephaly, mental retardation, and cerebral palsy with convulsions. In an incidence in Iraq, ingestion of wheat products contaminated with methyl mercury fungicide by pregnant women caused similar symptoms of neurologic damage and mental retardation.[2] The fetus is particularly sensitive to the effects of methyl mercury, which interferes with organ development. Toxic concentrations inhibit the normal migration of nerve cells from the central parts of the neurotube toward the peripheral parts of the brain cortex and thus inhibit the normal development of the fetal brain.[1] Differences between fetal and adult hematocrits may result in differing mercury concentrations in the two; studies suggest that the difference in sensitivity between the fetus and adult organism is close to a factor of 2.[1] It has been suggested that women of childbearing age should have no occupational exposure to alkyl mercury.[2]

In humans, the biological half-life for methyl mercury is about 70 days; since elimination is slow, irregular, and individualized, there is considerable risk of an accumulation of mercury to toxic levels.[3] A precise relationship between atmospheric levels of alkyl mercury and concentrations of mercury in blood or urine has not been shown.[3] Clinical observations indicate that concentrations of 50 to 100 μg mercury/100 ml of whole blood may be associated with symptoms of intoxication; concentrations of about 10 to 20 μg mercury/100 ml are not associated with symptoms.[3] In a study of 20 workers engaged in the manufacture of organic mercurials and exposed for 6 years to mercury concentrations in air between 0.01 and 0.1 mg/m^3, there was no evidence of physical impairment or clinical laboratory abnormalities.[5]

The alkyl mercury halides are irritating to the eyes, mucous membranes, and skin and may cause severe dermatitis and burns; skin sensitization has occasionally occurred.[6,7]

Epidemiologic studies of methyl mercury–exposed populations have not shown any evidence of a carcinogenic effect.[8]

Methyl mercury vapor is detectable by smell at concentrations well below that which on intermittent exposure could prove hazardous.[8]

Diagnosis. The following signs and symptoms may occur: paresthesias of the lips, hands, and feet; ataxia; dysarthria; impairment of vision and hearing; emotional disturbances; spasticity, jerking movements of the limbs, head, or shoulders; dizziness; hypersalivation, lacrimation; nausea, vomiting, diarrhea, and constipation; kidney damage; dermatitis; and skin burns.

Differential Diagnosis: Effects seen with alkyl mercury may mimic organic lead syndromes, Parkinsonism, hereditary ataxias, cerebrocerebellar degeneration, neurosyphilis, and metastatic lesions of the central nervous system.

Special Tests: An electromyograph may be useful in determining the extent of nerve dysfunction. The mercury content of the blood may indicate the extent of absorption.

Treatment. Treatment is aimed at removing mercury compound from the body.[1] In severe cases, hemodialysis and infusion of chelating agents for mercury, such as cysteine, N-acetylcysteine, or N-acetylpenicillamine is required. Excretion of methyl mercury can be increased by catheterization and drainage of the choledochal duct or surgical establishment of gallbladder drainage, thereby breaking the enterohepatic circulation of mercury.[1]

Medical Control. Medical control involves preplacement and annual physical examination, with emphasis on the central nervous system, kidneys, and skin; and urinalysis.

REFERENCES

1. Berlin M: Mercury. In Friberg L et al: Handbook on the Toxicology of Metals, 2nd ed, Vol II, Specific Metals, pp 418–445. Amsterdam, Elsevier, 1986
2. Inskip MJ, Piotrowski JK: Review of the health effects of methylmercury. J Appl Toxicol 5:113–133, 1985
3. Report of an International Committee: Maximum allowable concentrations of mercury compounds. Arch Environ Health 19:891–905, 1969
4. Rustam H, Von Burg R, Amin-Zaki L, El Hassani S: Evidence for a neuromuscular disorder in methylmercury poisoning—clinical and electrophysiological findings in moderate to severe cases. Arch Environ Health 30:190–195, 1975
5. Dinman BD, Evans EE, Linch AL: Organic mercury—environmental exposure, excretion, and prevention of intoxication in its manufacture. AMA Arch Ind Health 18:248–260, 1958
6. Dales LG: The neurotoxicity of alkyl mercury compounds. Am J Med 53:219–232, 1972
7. American National Standards Institute, Inc: American National Standard Acceptable Concentrations of Organo (Alkyl) Mercury, ANSI Z37.30–1969. New York, American National Standards Institute, Inc, 1970
8. Junghans RP: A review of the toxicity of methylmercury compounds with application to occupational exposures associated with laboratory uses. Environ Res 31:1–31, 1983

MESITYL OXIDE

CAS: 141-79-7

$(CH_3)_2C\text{-}CHCOCH_3$ 1987 TLV = 15 ppm

Synonyms: Methyl isobutenyl ketone; isopropylideneacetone; 4–methyl–3–pentene–2–one

Physical Form. Oily, colorless liquid

Uses. Solvent; chemical intermediate

Exposure. Inhalation; skin absorption

Toxicology. Mesityl oxide is an irritant of the eyes and mucous membranes; at high concentrations, it causes narcosis in animals, and it is expected that severe exposure will produce the same effect in humans.

Human subjects exposed to 25 ppm for 15 minutes experienced eye irritation; at 50 ppm, there was also nasal irritation and a persistent unpleasant taste that remained with many subjects 3 to 6 hours after the exposure.[1] Liquid mesityl oxide produces dermatitis with sustained skin contact.[2]

Rats and guinea pigs exposed 8 hours/day to 500 ppm for 10 days had nose and eye irritation and developed slight kidney injury; slight liver and lung injury were observed in a few animals; 13 of 20 animals died from 30 exposures of 8 hours each at 500 ppm, while all animals tested at 250 ppm survived.[4] Guinea pigs exposed to 2000 ppm for up to 422 minutes died during or following exposure.[4] Signs of eye and respiratory tract irritation with gradual loss of corneal and auditory reflexes preceded coma and death.[4]

The strong peppermint or honey-like odor is detectable at 12 ppm; severe overexposure is unlikely because of local irritation and odor; however, olfactory fatigue may occur.[3,5]

The irritation and systemic effects resulting from mesityl oxide exposure appear to be more serious than those produced by the lower ketones.[6]

REFERENCES

1. Silverman L, Schulte HF, First MW: Further studies on sensory response to certain industrial solvent vapors. J Ind Hyg Toxicol 28:262–266, 1946
2. Shell Chemical Corporation: Safety Data Sheet SC: 57–105. Mesityl oxide, pp 1–3. Ind Hyg Bull, 1957
3. Smyth HF Jr, Seaton J, Fischer L: Response of guinea pigs and rats to repeated inhalation of vapors of mesityl oxide and isophorone. J Ind Hyg Toxicol 24:46–50, 1942
4. Specht H, Miller JW, Valaer PJ, Sayers RR: Acute response of guinea pigs to the inhalation of ketone vapors. Natl Inst Health Bull No 176, 1940
5. Shell Chemical Corporation: Toxicity Data Sheet SC: 57–106. Mesityl oxide. Ind Hyg Bull, 1957
6. National Institute for Occupational Safety and Health: Criteria for a Recommended Standard . . . Occupational Exposure to Ketones. DHEW (NIOSH) Pub No 78–173. Washington, DC, US Government Printing Office, 1978

METHACRYLIC ACID

CAS: 79-41-4

$H_2C{=}C(CH_3)COOH$ 1987 TLV = 20 ppm

Synonyms: 2–Methyl–2–propenoic acid; 2–methylenepropionic acid; alphamethacrylic acid

Physical Form. Colorless liquid

Uses. Manufacture of methacrylic resins and plastics

Exposure. Inhalation; skin contact

Toxicology. Methacrylic acid is an irritant of the eyes, nose, throat, and skin.

Rats exposed to 1300 ppm for 5 hours/day for 5 days showed nose and eye irritation but no adverse findings in blood and urine tests.[1] Exposure of rats to 300 ppm for 6 hours/day for 20 days resulted in no clinical signs, although histopathologic findings showed slight renal congestion.

The liquid applied to the depilated guinea pig abdomen for 24 hours under an occlusive wrap produced severe irritation.[2] The liquid is irritating to the eyes.

REFERENCES

1. Gage JC: The subacute inhalation toxicity of 109 industrial chemicals. Br J Ind Med 27:1, 1970
2. Unpublished Data: Health, Safety and Human Factors Laboratory, Eastman Kodak Co, Rochester, New York, 1980

METHOXYCHLOR

CAS: 72-43-5

$C_{16}H_{15}Cl_3O_2$ 1987 TLV = 10 mg/m^3

Synonyms: 1,1,1–Trichloro–2,2–bis(paramethoxyphenyl)ethane; dimethoxy—DDT

Physical Form. Crystalline solid

Uses. Insecticide

Exposure. Inhalation

Toxicology. Methoxychlor is a convulsant of low toxicity.

No adverse effects on health or clinical laboratory data were found in groups of volunteers given 2 mg/kg/day for 8 weeks.[1]

The oral LD_{50} in rats is 6 g/kg; the fatal oral dose in humans is estimated to be 450 g.[2] Dogs fed a daily diet containing 4 g/kg bodyweight developed signs of chlorinated hydrocarbon intoxication, including fasciculations, tremor, hyperesthesia, tonic seizures, and tetanic convulsions after 5 to 8 weeks. Most of the dogs died within 3 weeks after onset of effects.[3] Rabbits given oral daily doses of 200 mg/kg died after 4 to 15 doses; autopsy findings included mild liver damage and nephrosis.[4] In mice given 5 mg orally over 3 days and in rats given 20 mg, there was a uterotrophic effect manifested as a marked increase in weight of the uterus.[5]

Fetotoxic effects have been shown in animals. Administration of 1000 mg/kg in the diet to pregnant rats caused vaginal defects in their offspring.[6] Reduced fertility in both sexes was also noted when the offspring reached maturity.[6] Intragastric administration of 200 mg/kg/day to mice on days 6 to 15 of pregnancy caused a decrease in the number and weight of the fetuses and delayed bone ossification.[7]

Female mice fed up to 2000 mg/kg and males given 3500 mg/kg in the diet for 78 weeks showed no statistically significant increase in the yield of benign and malignant tumors that could be attributed to methoxychlor.[8] Chronic feeding studies in rats at 850 and 1400 mg/kg for males and females, respectively, also showed no significant carcinogenic responses, although high tumor rates in controls may have masked detection.[8] Based upon NCI results and several earlier animal studies, the IARC has determined that no evidence has been provided that methoxychlor is carcinogenic in experimental animals.[9] Evaluating these same studies, another reviewer has stated that methoxychlor is carcinogenic, especially to the livers of experimental animals.[10] It may be most appropriate to state that current studies are inadequate to assess the carcinogenicity of methoxychlor.

REFERENCES

1. Stein AA et al: Safety evaluation of methoxychlor in human volunteers. Toxicol Appl Pharmacol 7:499 (abstr), 1965
2. Methoxychlor. Documentation of the TLVs and BEIs, 5th ed, p 364. Cincinnati, American Conference of Governmental Industrial Hygienists (ACGIH), 1986
3. Tegeris AS, Earl FL, Smalley HE Jr, Curtis JM: Methoxychlor toxicity. Arch Environ Health 13:776–787, 1966
4. Negherbon WO: Handbook of Toxicology, Vol 3, pp 467–469. Philadelphia, WB Saunders Co, 1957
5. Tullner WW: Uterotrophic action of the insecticide methoxychlor. Science 133:647–648, 1961
6. Harris SJ et al: Effect of several dietary levels of technical methoxychlor on reproduction in rats. J Agric Food Chem 22:969–973, 1974
7. Khera KS et al: Teratogenicity studies on linuron, malathion and methoxychlor in rats. Toxicol Appl Pharmacol 45:435–444, 1978
8. National Cancer Institute: Bioassay of Methoxychlor for Possible Carcinogenicity, TR-35. DHEW (NIH) Pub No 78–835, p 91. Washington, DC, US Government Printing Office, 1978
9. IARC Monographs on the Evaluation of the Carcinogenic Risks of Chemicals to Humans. Vol 20, Some Halogenated Hydrocarbons, pp 259–281. Lyon, International Agency for Research on Cancer, 1979
10. Reuber MW: Carcinogenicity and toxicity of methoxychlor. Environ Health Perspect 36:205–219, 1980

2-METHOXYETHANOL

CAS: 109-86-4

$CH_3OCH_2CH_2OH$ 1987 TLV = 5 ppm; skin

Synonyms: Ethylene glycol monomethyl ether; methyl cellosolve; Dowanol EM

Physical Form. Liquid

Uses. Solvent; jet fuel anti-icing additive

Exposure. Inhalation; skin absorption

Toxicology. 2-Methoxyethanol (2-ME)

affects the central nervous system (CNS) and depresses the hematopoietic system; in animals, it causes adverse reproductive effects, including teratogenesis, testicular atrophy, and infertility.

Cases of toxic encephalopathy and macrocytic anemia have been reported from industrial exposures that may have been as low as 60 ppm.[1] Symptoms included headache, drowsiness, lethargy, and weakness. Manifestations of CNS instability included ataxia, dysarthria, tremor, and somnolence. These effects were usually reversible. In acute exposures, the CNS effects were the more pronounced, whereas prolonged exposure to lower concentrations primarily produced evidence of depression of erythrocyte formation. When exposure was reduced to 20 ppm, no further cases occurred.

Two workers exposed primarily through skin contact showed signs of encephalopathy; one had bone marrow depression, while the other had pancytopenia.[2]

The LC_{50} for 7-hour exposure of rats was 1480 ppm; death was due to lung and kidney injury.[3] Rabbits exposed to 800 ppm and 1600 ppm for 4 to 10 days showed irritation of the upper respiratory tract and lungs, severe glomerulonephritis, hematuria, and albuminuria.[3] Instilled in rabbit eyes, 2–ME caused immediate pain, conjunctival irritation, and slight corneal cloudiness, which cleared in 24 hours.[3]

Adverse reproductive effects have been reported in a number of species.[4] Testicular atrophy was observed in rats and mice exposed at 1000 ppm for 9 days and in rabbits exposed for 13 weeks at 300 ppm.[4,5] Slight-to-severe microscopic testicular changes occurred at 30 to 100 ppm in rabbits. At 500 ppm for 5 days, there was temporary infertility in male rats and abnormal spermhead morphology in mice.[4]

Exposure of pregnant rabbits to 50 ppm 6 hours/day on gestational days 6 through 18 induced significant increases in the incidence of malformation, especially of the skeletal and cardiovascular systems and in the number of resorptions.[6] At this exposure level, decreases in maternal bodyweight gain as well as decreased fetal weight occurred.[6] Only slight fetotoxicity was observed in mice and rats similarly exposed. In another study, fetal cardiovascular and skeletal defects occurred in rats exposed at the 50 ppm level on days 7 to 15 of gestation.[4] By gavage, 250 mg/kg on days 7 to 14 caused increased embryonic deaths and gross fetal defects in mice.[7] Further studies on developmental phase-specific effects in mice showed exencephaly to be related to exposure between gestation days 7 to 10, whereas paw anomalies were maximal after administration on gestational day 11.[8] In rabbits, the most sensitive species tested to date, the minimally toxic fetal dose is 10 ppm, and the no observed effect level is 3 ppm.[4]

In one epidemiologic study, a small group of occupationally exposed workers showed no clinically significant differences in fertility or hematologic indices.[9]

NIOSH recommends that 2–ME be regarded as having the potential to cause adverse reproductive effects in workers, including teratogenesis in the offspring of exposed pregnant females.[4]

REFERENCES

1. Zavon MR: Methyl cellosolve intoxication. Am Ind Hyg Assoc J 24:36–41, 1963
2. Ohi G, Wegman DH: Transcutaneous ethylene glycol monomethyl ether poisoning in the work setting. J Occup Med 20:675–676, 1978
3. Rowe VK, Wolf MA: Derivatives of glycols. In Clayton GD, Clayton FE (eds): Patty's Industrial Hygiene and Toxicology, 3rd ed, rev, Vol 2C, Toxicology, pp 3911–3919. New York, Wiley–Interscience, 1982

4. National Institute for Occupational Safety and Health: NIOSH Current Intelligence Bulletin 39, Glycol Ethers. DHHS (NIOSH) Pub No 83–112, p 22. Washington, DC, US Government Printing Office, May 2, 1983
5. Miller RR et al: Comparative short-term inhalation toxicity of ethylene glycol monomethyl ether and propylene glycol monomethyl ether in rats and mice. Toxicol Appl Pharmacol 61:368–377, 1981
6. Hanley TR Jr et al: Comparison of the teratogenic potential of inhaled ethylene glycol monomethyl ether in rats, mice and rabbits. Toxicol Appl Pharmacol 75:409–422, 1984
7. Nagano K et al: Embryotoxic effects of ethylene glycol monomethyl ether in mice. Toxicology 20:335–343, 1981
8. Horton VL et al: Developmental phase-specific and dose-related teratogenic effects of ethylene glycol monomethyl ether in CD–1 mice. Toxicol Appl Pharmacol 80:108–118, 1985
9. Cook RR et al: A cross-sectional study of ethylene glycol monomethyl ether process employees. Arch Environ Health 37:346–351, 1982

2–METHOXYETHYL ACETATE

CAS: 110-49-6

$CH_3COOCH_2CH_2OCH_3$

1987 TLV = 5 ppm; skin

Synonyms: Ethylene glycol monomethyl ether acetate; methyl cellosolve acetate; methyl glycol acetate

Physical Form. Colorless liquid

Uses. Lacquer industry; textile printing; manufacture of photographic film, coatings, and adhesives

Exposure. Inhalation; skin absorption

Toxicology. 2–Methoxyethyl acetate affects the central nervous system, the hematopoietic system, and reproductive system in animals. Similar effects are to be expected in humans.

Mice and rabbits tolerated 1-hour exposure to 4500 ppm with only irritation of mucous membranes; guinea pigs survived the 1-hour exposure but succumbed days later.[1] Repeated exposure to 500 ppm for 8 hours/day caused narcosis and death in cats, and 1000 ppm for 8 hours/day was lethal to rabbits; all animals showed kidney injury.[1] Anemia was observed in cats repeatedly exposed to 200 ppm for 4 to 6 hours.

Testicular atrophy and leukopenia have been reported in mice following oral gavage of 63 to 200 mg/kg 5 days/week for 5 weeks.[2]

2–methoxyethyl acetate is hydrolyzed in vivo to form 2–methoxyethanol.[3] Consequently, the acetate is expected to show similar profiles of developmental and reproductive toxicity as 2–methoxyethanol (qv).

The liquid is mildly irritating to the eyes of rabbits but not to the skin; prolonged contact can result in significant absorption.[1]

REFERENCES

1. Rowe VK, Wolf MA: Derivatives of glycols. In Clayton GD, Clayton FE (eds): Patty's Industrial Hygiene and Toxicology, 3rd ed, Vol 2C, Toxicology, pp 4022–4024. New York, Wiley–Interscience, 1982
2. Nagano K et al: Experimental studies on toxicity of ethylene glycol alkyl ethers in Japan. Environ Health Perspect 57:75–84, 1984
3. Hardin BD: Reproductive toxicity of the glycol ethers. Toxicology 27:91–102, 1983

METHYL ACETATE

CAS: 79-20-9

$CH_3C(O)OCH_3$ 1987 TLV = 200 ppm

Synonyms: Acetic acid, methyl ester

Physical Form. Colorless, highly volatile liquid

Uses. Solvent for lacquers, oils, and resins

Exposure. Inhalation

Toxicology. Methyl acetate is irritating to the eyes and mucous membranes; at high concentrations, it causes narcosis in animals, and it is expected that severe exposure will produce the same effect in humans.

Human exposure to 10,000 ppm for a short period of time resulted in eye, nose, and throat irritation, which persisted after cessation of exposure.[1] In a man exposed to unmeasured concentrations, effects were general central nervous system depression, headaches, and dizziness, followed by blindness of both eyes caused by atrophy of the optic nerve.[2] The toxic action on the optic nerve is possibly related to the presence of methanol after hydrolysis of methyl acetate.[3]

Cats exposed to 5,000 ppm showed eye irritation and salivation; at 18,500 ppm, there was dyspnea, convulsions, and narcosis; 54,000 ppm was lethal within minutes.[1] Repeated exposure at 6,600 ppm resulted in weight loss and weakness.[1]

Prolonged contact with the liquid may cause dryness, cracking, and irritation of the skin.

REFERENCES

1. Sandmeyer EE, Kirwin CJ: Esters. In Clayton GD, Clayton FE (eds): Patty's Industrial Hygiene, 3rd ed, p 2272. New York, Wiley–Interscience, 1981
2. Lund A: Toxic amblyopia after inhalation of methyl acetate. J Ind Hyg Toxicol 28:35 (abstr), 1946. From Laeger 106:408–422, 1944
3. Hygienic Guide Series: Methyl acetate. Am Ind Hyg Assoc J 25:317–319, 1964

METHYL ACETYLENE

CAS: 74-99-7

CH_3CCH 1987 TLV = 1000 ppm

Synonyms: Allylene; propyne; propine

Physical Form. Colorless gas

Uses. Propellant, welding

Exposure. Inhalation

Toxicity. At high concentrations, methyl acetylene causes narcosis in animals, and it is expected that severe exposure will produce the same effect in humans.

Rats exposed to 42,000 ppm became hyperactive within the first 7 minutes, and, at the end of 7 minutes, they appeared lethargic and ataxic. After 95 minutes, the animals were completely anesthetized. There was no mortality when the exposure was terminated at the end of 5 hours, and most of the animals recovered completely within 40 minutes.[1] Edema and alveolar hemorrhage were present in animals killed at termination of the single exposure, whereas bronchiolitis and pneumonitis were observed in rats sacrificed 9 days postexposure.

Two dogs and 20 rats were exposed to 28,700 ppm, 6 hours/day, 5 days/week for 6 months; after 7 minutes of exposure, ataxia was noted in the rats, and, after 13 minutes, ataxia and mydriasis were observed in the dogs. Within 15 minutes, the dogs also exhibited staggering, marked salivation, and muscular fasciculations. There was a 40% mortality rate among exposed rats vs a 10% mortality rate in the control animals.

Methyl acetylene has a "sweet" odor similar to acetylene.[1]

REFERENCE

1. Horn HJ, Weir RJ Jr, Reese WH: Inhalation toxicology of methylacetylene. AMA Arch Ind Health 15:20–25, 1957

METHYL ACRYLATE

CAS: 96-33-3

$CH_2{=}CHC(O)OCH_3$ 1987 TLV = 10 ppm; skin

Synonyms: 2–Propenoic acid methyl ester; acrylic acid methyl ester; methyl propenoate

Physical Form. Liquid

Uses. As a monomer, polymer, and copolymer, especially in dental, medical, and pharmaceutical sciences

Exposure. Inhalation; skin absorption

Toxicology. Methyl acrylate is a lacrimating agent and an irritant of the mucous membranes.

The lowest dose reported to have any irritant effect in humans is 75 ppm.[1] The liquid is readily absorbed by mucous membranes and through the skin.[1]

A single oral dose of 280 mg/kg to rabbits caused dyspnea, cyanosis, convulsions, and death.[2]

Exposure of rabbits 7 hours/day for 11 days to 237 ppm caused conjunctival irritation, lacrimation, nasal irritation, and labored respiration.[2] A level of 31 ppm caused no significant effect. The liquid applied on rabbit skin caused marked irritation, and contact sensitivity has been induced in guinea pigs using a variety of protocols.[2,3]

REFERENCES

1. Sandmeyer EE, Kirwin CJ: Esters. In Clayton GD, Clayton FE (eds): Patty's Industrial Hygiene and Toxicology, 3rd ed, Vol 2B, Toxicology, pp 2293–2296. New York, Wiley–Interscience, 1981
2. Treon JF, Sigmon H, Wright H, Kitzmiller KV: The toxicity of methyl and ethyl acrylate. J Ind Hyg Toxicol 31:317–326, 1949
3. Parker D, Turk JL: Contact sensitivity to acrylate compounds in guinea pigs. Contact Dermatitis 9:55–60, 1983

METHYLACRYLONITRILE

CAS: 126-98-7

$CH_2{=}C(CH_3)CN$ 1987 TLV = 1 ppm; skin

Synonyms: 2–Methyl–2–propenenitrile; 2–cyanopropene–1; isopropene cyanide; isopropenylnitrile; methacrylonitrile

Physical Form. Colorless liquid

Uses. Production of plastic elastomers and coatings

Exposure. Inhalation, skin absorption

Toxicology. Methylacrylonitrile is a metabolic asphyxiant with an action similar to cyanide.

The approximate LC_{50} for mice exposed to airborne concentrations for 1 hour was 630 ppm; for a 4-hour exposure, it was 400 ppm.[1] Exposure to 75 ppm for 8 hours caused no deaths, but respiratory difficulties and convulsions were observed.[1] In a recent study with rats exposed at concentrations between 3180 and 5700 ppm, the clinical symptoms, rapid unconsciousness with convulsions and lethality, suggested that the acute toxicity is predominantly caused by metabolically formed cyanide.[2] Cyanide reacts readily with cytochrome oxidase in mitochondria and inhibits cellular respiration. Cyanide antidotes were also effective against methlyacrylonitrile toxicity.[2]

The liquid was rapidly absorbed through the skin of a rabbit and caused death after 3 hours at a dose of 2.0 ml/kg.[1] The acute dermal LD_{50} in rabbits was 0.32 ml/kg.[3] Skin irritation at the site of application was negligible. One drop in the eye of a rabbit caused transient irritation.[1]

The oral LD_{50} in mice was 20 to 25 mg/kg; in rats, it was 25 to 50 mg/kg.[4] When beagle dogs were exposed 7 hours/day, 5 days/week to 13.5 ppm over a period of 90 days, two of three animals exhibited convulsions and loss of motor control in the hind limbs about halfway through the exposure period.[3] No effects occurred at 3.2 ppm.[3]

Treatment. Treatment must be rapid to be effective.[5] An effective mechanism is to administer nitrite, which oxidizes hemoglobin to methemoglobin, which in turn competes with cytochrome oxidase for the cyanide ion. Amyl nitrite may be administered by inhalation while sodium nitrite is prepared for intravenous administration (10 ml of a 3% solution). Alternatively, 4–dimethylaminophenol, which also oxidizes hemoglobin to methemoglobin, can be used in a dose of 3 mg/kg intravenously or intramuscularly to accelerate detoxification; following nitrite therapy, thiosulfate is administered intravenously (50 ml of a 25% aqueous solution), and the thiocyanate formed is readily excreted in the urine.

REFERENCES

1. McOmie WA: Comparative toxicities of methacrylonitrile and acrylonitrile. J Ind Hyg Toxicol 31:113, 1949
2. Peter H, Bolt HM: Effect of antidotes of the acute toxicity of methacrylonitrile. Int Arch Occup Environ Health 55:175–177, 1985
3. Pozzani UC, Kinhead ER, King JJ: The mammalian toxicity of methacrylonitrile. Am Ind Hyg Assoc J 29:202, 1968
4. Hartung R: Cyanides and nitriles. In Clayton GD, Clayton FE (eds): Patty's Industrial Hygiene and Toxicology, 3rd ed, rev, Vol 2C, Toxicology, pp 4867–4868. New York, Wiley–Interscience, 1972
5. Klaasen CD: Nonmetallic environmental toxicants: Air pollutants, solvents and vapors and pesticides. In Gilman AG et al (eds): Goodman and Gilman's The Pharmacological Basis of Therapeutics, 7th ed, pp 1642–1643. New York, Macmillan Publishing Co, 1985

METHYLAL

CAS: 109-87-5

$CH_2(OCH_3)_2$ 1987 TLV = 1000 ppm

Synonyms: Dimethoxymethane; formal; methylene dimethyl ether

Physical Form. Colorless liquid

Uses. Solvent; fuel

Exposure. Inhalation

Toxicology. Methylal is an irritant of the eyes and mucous membranes, and, at high concentrations, it causes central nervous system depression.

In humans, methylal has been used as an anesthetic in a number of surgical operations; however, anesthesia was produced more slowly than with ether, and the effect of methylal was more transitory.[1]

In guinea pigs exposed to a concentration near 154,000 ppm, effects included vomiting, lacrimation, sneezing, cough, and nasal discharge; coma occurred in 20 minutes and death occurred in 2.5 hours.[1]

The LC_{50} for a 7-hour exposure of mice was 18,354 ppm.[1] Chronic exposure 7 hours/day to 11,300 ppm for 1 week caused mild eye and nose irritation, incoordination, and light narcosis after 4 hours; 6 of the 50 mice died.[1] Animals exposed to toxic concentrations often developed marked fatty changes in the liver, kidney, and heart and inflammatory changes in the lungs.[1] Rats were unaffected by eight 6-hour exposures to 4000 ppm.[2]

Methylal can cause superficial irritation of the eyes.[1] Frequent or prolonged skin contact with the liquid may cause dermatitis owing to a defatting action.

The liquid has a chloroform-like odor and pungent taste.

REFERENCES

1. Weaver FL Jr, Hough AR, Highman B, Fairhall LT: The toxicity of methylal. Br J Ind Med 8:279–283, 1951
2. Gage JC: The subacute inhalation toxicity of 109 industrial chemicals. Br J Ind Med 27:1–18, 1970

METHYL ALCOHOL

CAS: 67-56-1

CH_3OH 1987 TLV = 200 ppm; skin

Synonyms: Methanol; wood spirit; carbinol; wood alcohol; wood naphtha; methylol; Columbian spirit; colonial spirit

Physical Form. Colorless liquid

Uses. Production of formaldehyde; paints, varnishes, cements, inks, and dyes

Exposure. Inhalation; skin absorption

Toxicology. Methyl alcohol causes optic neuropathy, metabolic acidosis, and respiratory depression.

Although methyl alcohol poisoning has occurred primarily from the ingestion of adulterated alcoholic beverages, symptoms can also occur from inhalation or absorption through the skin.[1,2] Impairment of vision and death from absorption by the latter routes have been reported in the older literature.[2] Typically, within 18 to 48 hours after ingestion, patients develop nausea, abdominal pain, headache, and abnormally slow deep breathing. This is accompanied by visual symptoms ranging from blurred or double vision and changes in color perception to constricted visual fields and complete blindness.[1,3] The most severely poisoned patients become comatose and may die; those who recover from coma may be found blind.[3] One of the most striking features of methyl alcohol poisoning is acidosis; the degree of acidosis has been found to closely parallel the severity of poisoning.[1] Accumulated evidence suggests that chronic exposure to 1200 to 8300 ppm can lead to impaired vision.[1] Exposure to vapor concentrations ranging from 365 to 3080 ppm resulted in blurred vision, headache, dizziness, and nausea.[4]

The liquid in the eyes has caused superficial lesions of the cornea that were of a nonserious nature.[1] Prolonged or repeated skin contact will cause dermatitis, erythema, and scaling.[1]

The presence of an asymptomatic latent period following ingestion suggests that methyl alcohol needs to be metabolized before toxicity is fully manifest.[1] This concept also explains the discrepancy between plasma concentrations of methyl alcohol and clinical signs of toxicity.[5] Furthermore, methyl alcohol poisoning is ameliorated by ethanol, a substance with greater affinity than methyl alcohol for alcohol dehydrogenase, which is responsible for the initial step in metabolism.[5] The metabolite formate appears to be the mediator of ocular injury and acidosis.[6] The individual variations in activity of the alcohol dehydrogenase systems, which are responsible for the oxidative metabolism of methyl (and ethyl) alcohol, may account for the wide variation in the individual responses observed with methyl alcohol poisoning.[1]

Diagnosis. Signs and symptoms include headache; blurred vision, constricted visual fields; shortness of breath; dizziness and vertigo.

Differential Diagnosis: The combination of visual disturbances and metabolic acidosis together with a history of exposure and the presence of formic acid in the urine is confirmation of methyl alcohol intoxication.

Special Tests: Assay of formic acid in the urine and measurement of blood *p*H and plasma bicarbonate should be done.

Treatment. Treatment includes (1) emesis or gastric lavage, rehydration, correction of acidosis, and folate to enhance formate oxidation; (2) intravenous ethanol when plasma methanol concentrations are higher than 20 mg/dl, when ingested doses are greater than 30 ml, or when there is evidence of acidosis or visual abnormalities; and (3) hemodialysis when plasma methanol concentrations are greater than 50 mg/dl or when metabolic acidosis is unresponsive to bicarbonate given intravenously.[5] Serum formate concentration may be a more direct indicator of toxicity; levels exceeding 20 mg/dl may be expected to produce acidosis or ocular injury.[6]

REFERENCES

1. National Institute for Occupational Safety and Health: Criteria for a Recommended Standard . . . Occupational Exposure to Methyl Alcohol. DHEW (NIOSH) Pub No 76–148, pp 68–75. Washington, DC, US Government Printing Office, 1976
2. Henson EV: The toxicology of some aliphatic alcohols. Part II. J Occup Med 1:497–502, 1960
3. Grant WM: Toxicology of the Eye, 3rd ed, pp 591–596. Springfield, Illinois, Charles C Thomas, 1986
4. Frederick LJ et al: Investigation and control of occupational hazards associated with the use of spirit duplicators. Am Ind Hyg Assoc J 45:51–55, 1984
5. Ekins BR et al: Standardized treatment of severe methanol poisoning with ethanol and hemodialysis. West J Med 142:337–340, 1985
6. Osterloh JD et al: Serum formate concentrations in methanol intoxication as a criterion for hemodialysis. Ann Intern Med 104:200–203, 1986

METHYLAMINE

CAS: 74-89-5

CH_3NH_2 1987 TLV = 10 ppm

Synonyms: Monomethylamine; aminomethane

Physical Form. Gas

Uses. Tanning; organic syntheses

Exposure. Inhalation

Toxicology. Methylamine is a severe irritant of the eyes, mucous membranes, and skin.

One case of bronchitis in a chemical worker has been reported; concentrations measured in the workroom ranged from 2 to 60 ppm; the duration of the exposure was not known.[1] Brief exposure to 20 to 100 ppm is said to produce transient irritation of the eyes, nose, and throat.[2] No symptoms of irritation are produced from longer exposures at less than 10 ppm.[2] On the basis of the irritant properties of methylamine, it is possible that severe exposure may cause pulmonary edema.

In rats, when administered orally as the base in a 40% aqueous solution, the LD_{50} was 0.1 to 0.2 g/kg.[2] In the eyes of animals, 1 drop of 5% aqueous solution caused conjunctival hemorrhage, superficial corneal opacities, and edema; a 40% solution in the eyes of rabbits caused corneal damage.[3] On the skin of a rabbit, a 40% solution caused necrosis.[2]

The ammonia-like odor is detectable at less than 10 ppm.[2]

REFERENCES

1. Methylamine. Documentation of the TLVs and BEIs, 5th ed, p 373. Cincinnati, American Conference of Governmental Industrial Hygienists (ACGIH), 1986
2. Beard RR, Noe JT: Aliphatic and alicyclic amines. In Clayton GD, Clayton FE (eds): Patty's Industrial Hygiene and Toxicology, 3rd ed, rev, Vol 2B, Toxicology, pp 3135–3173. New York, Wiley–Interscience, 1981
3. Grant WM: Toxicology of the Eye, 3rd ed, pp 606–607. Springfield, Illinois, Charles C Thomas, 1986

METHYL AMYL KETONE

CAS: 110-43-0

$CH_3COC_5H_{11}$ 1987 TLV = 50 ppm

Synonym: 2–Heptanone

Physical Form. Liquid

Uses. Solvent

Exposure. Inhalation

Toxicology. Methyl amyl ketone is irritating to the eyes and mucous membranes; at high concentrations, it causes narcosis in animals, and it is expected that severe exposure will produce the same effect in humans.

There have been no reports of effects in humans, and the concentration at which irritation may be produced is not known.[1]

In guinea pigs, exposure to 4800 ppm caused narcosis and death in 4 to 8 hours; 2000 ppm was strongly narcotic, and 1500 ppm was irritating to the mucous membranes.[2] Rats and monkeys exposed to 1025 ppm methyl *n*-amyl ketone for 6 hours/day, 5 days/week for 9 months showed no evidence of neuropathy or clinical signs of illness. Microscopic examination revealed no tissue damage.[3]

The liquid has a marked fruity odor and a pear-like flavor.[1]

REFERENCES

1. National Institute for Occupational Safety and Health: Criteria for a Recommended Standard . . . Occupational Exposure to Ketones. DHEW (NIOSH) Pub No 78–173. Washington, DC, US Government Printing Office, 1978
2. Specht H, Miller JW, Valaer RJ, Sayers RR: Acute response of guinea pigs to the inhalation of ketone vapors. Natl Inst Health Bull No 176, 1940
3. Johnson BL et al: Neurobehavioral effects of methyl *n*-butyl ketone and methyl *n*-amyl ketone in rats and monkeys: A summary of NIOSH investigations. J Environ Pathol Toxicol 2:113–133, 1979

METHYL BROMIDE

CAS: 74-83-9 1987 TLV = 5 ppm; skin

CH_3Br

Synonyms: Bromomethane; monobromomethane; isobrome

Physical Form. Colorless gas liquefying at 3°C

Uses. Fumigant of soil and stored foods; methylating agent; previously used as a refrigerant and fire-extinguishing agent

Exposure. Inhalation; skin absorption

Toxicology. Methyl bromide is a neurotoxin and causes convulsions; very high concentrations cause pulmonary edema; chronic exposure causes peripheral neuropathy.

There are numerous reports of human intoxication from accidental exposure associated with its use in fire extinguishers and as a fumigant.[1] Estimates of concentrations that have caused human fatalities range from 8000 ppm for a few hours to 60,000 ppm for a brief exposure. The onset of toxic symptoms is usually delayed and the fatal period may be from 30 minutes to several hours. Early symptoms include headache, visual disturbances, nausea, vomiting, and malaise.[2] In some instances, there is eye irritation, vertigo, and intention tremor of the hands; the tremor may progress to twitchings and finally to convulsions of the Jacksonian type, being first restricted to one extremity, but gradually spreading to the entire body.[1,3] Severe exposure may lead to pulmonary edema.[4] Tubular damage in the kidneys has been observed in fatal cases.[2] Some of those who have recovered from severe intoxication have had persistent central nervous system effects, including vertigo, depression, hallucinations, anxiety, and inability to concentrate.[2]

Eight of 14 workers repeatedly exposed to the vapor (concentration unmeasured) for 3 months developed peripheral neuropathy; all recovered within 6 months.[5]

It is unlikely that bromide ion resulting from metabolic conversion of methyl bromide plays a significant role in the toxicity of methyl bromide.[6] Blood bromide levels following methyl bromide poisoning are much lower than those associated with intoxication by inorganic bromide salts. Concentrations of 100 mg/liter have been associated with death following methyl bromide exposure, whereas blood bromide levels of 1000 mg/liter or greater have been observed following therapeutic administration of inorganic bromides in the absence of signs of intoxication.[6] A recent report of six methyl bromide poisonings showed serum bromide concentrations at the time of hospital admission to be a poor predictor of survival.[7] One fatal case had an antemortem bromide level of 108 mg/liter, while a survivor measured 321 mg/liter. The usefulness of this assay in occupational settings to detect hazardous exposures has not been determined.

Contact with the eye by the gas or liquid results in transient irritation and conjunctivitis.[8]

Minor skin exposure to the liquid produced erythema and edema.[9] Prolonged or repeated contact resulted in deeper burns with delayed vesiculation.[9] It is doubtful that significant cutaneous absorption occurs. Although victims of skin exposure may show symptoms of neurotoxicity, inhalation is considered the likely cause.[9]

Repeated exposure at 70 ppm of female rats before and during pregnancy did not cause maternal or embryo toxicity, but severe neurotoxicity and mortality was produced in the rabbit.[10]

In a 90-day study, 50 mg/kg administered by gavage 5 days/week caused squamous cell carcinomas of the forestomach in 13 of 20 rats; a dose-related incidence of hyperplasia was observed at the 2- and 10-mg/kg level.[11]

Methyl bromide has poor warning properties.

REFERENCES

1. von Oettingen WF: The Halogenated Aliphatic, Olefinic, Cyclic, Aromatic, and Aliphatic-Aromatic Hydrocarbons Including the Halogenated Insecticides, Their Toxicity and Potential Dangers. US Public Health Service Pub No 414, pp 15–30. Washington, DC, US Government Printing Office, 1955
2. Hine CH: Methyl bromide poisoning. J Occup Med 11:1–10, 1969
3. Greenberg JO: The neurological effects of methyl bromide poisoning. Ind Med 40:27–29, 1971
4. Rathus EM, Landy PJ: Methyl bromide poisoning. Br J Ind Med 18:53–57, 1961
5. Kantarjian AD, Shaheen AS: Methyl bromide poisoning with nervous system manifestations resembling polyneuropathy. Neurology 13:1054–1058, 1963
6. Hayes WJ Jr: Pesticides Studied in Man, pp 140–142. Baltimore, Williams & Wilkins, 1982

7. Marraccini JV et al: Death and injury caused by methyl bromide, an insecticide fumigant. J Forensic Sci 28:601–607, 1983
8. Grant WM: Toxicology of the Eye, 3rd ed, pp 607–610. Springfield, Illinois, Charles C Thomas, 1986
9. Jarowenko DG, Mancusi–Ungaro HR: The care of burns from methyl bromide (case report). J Burn Care Rehabil 6:119–123, 1985
10. Sikov MR et al: Teratologic assessment of butylene oxide, styrene oxide and methyl bromide. DHHS (NIOSH) Pub No 81–124. Washington, DC, US Government Printing Office, July 1981
11. Danse LHJC et al: Methylbromide: Carcinogenic effects in the rat forestomach. Toxicol Appl Pharmacol 72:262–271, 1984

METHYL BUTYL KETONE

CAS: 591-78-6

$CH_3C(O)C_4H_9$ 1987 TLV = 5 ppm

Synonyms: 2–Hexanone; *n*-butyl methyl ketone; MBK; MNBK; propylacetone

Physical Form. Colorless liquid

Uses. Solvent

Exposure. Inhalation; skin absorption

Toxicology. Methyl butyl ketone at high concentrations may produce ocular and respiratory irritation followed by central nervous system depression and narcosis. Chronic inhalation causes peripheral neuropathy.

Human volunteers exposed to a vapor concentration of 1000 ppm for several minutes developed moderate eye and nasal irritation.[1] Workers exposed to the mixed vapor of methyl butyl ketone (mean concentrations of 147 to 516 ppm, peaks to 763 ppm) for 6 to 12 months with extensive skin exposure developed peripheral neuropathy.[2–5] The neurologic pattern was one of a distal motor and sensory disorder, with minimal loss of tendon reflexes.[4] In those with prominent motor involvement, initial symptoms included slowly developing weakness of the hands, with difficulty in pincer movement on the grasping of heavy objects, or weakness of the ankle extensors, resulting in a slapping gait. In other cases, the initial symptoms were intermittent tingling and paresthesias in the hands or feet. In some cases, the condition progressed slowly for several months after cessation of exposure; in moderate-to-severe cases, improvement occurred over a period of up to 8 months.[4]

In guinea pigs, exposure to 10,000 to 20,000 ppm was potentially lethal in 30 to 60 minutes; concentrations greater than 20,000 ppm killed the animals within a few minutes; at 6000 ppm, there were signs of narcosis after 30 minutes, deep anesthesia after 1 hour, and deaths after approximately 6.5 hours.[1] A maximum of 3000 ppm for 1 hour did not cause serious disturbances.

Animals continually exposed to concentrations between 100 and 600 ppm developed signs of peripheral neuropathy after 4 to 8 weeks; in cats, the conduction velocity of the ulnar nerve was less than half of normal after exposure for 7 to 9 weeks.[3] In these animals, histologic examination revealed focal denudation of myelin from nerve fibers with or without axonal swelling.[3] In rats and monkeys, adverse effects on neurophysiologic indicators of nervous system integrity were found with 9 month exposure to 100 ppm 6 hours/day, 5 days/week.[6] Methyl butyl ketone neuropathies, however, occurred only after 4 months exposure at 1000 ppm.[6] Four months of intermittent respiratory exposure of rats to 1300 ppm caused severe symmetric weakness in the hind limbs.[7]

Damage resulting from hexacarbons such as MBK has also been found in the optic tract and hypothalamus of the cat. These findings are significant owing to the possibility that such central nervous system damage is permanent, whereas the peripheral nervous system shows regeneration.[8]

Testicular atrophy of the germinal epithelium was seen in male rats administered 660 mg/kg by gavage for 90 days.[9] A reduction in total circulating white blood cells has also been reported following MBK exposure.[1] 2,5–Hexanedione was found to be a major metabolite of methyl butyl ketone in several animal species; peripheral neuropathy occurred in rats after daily subcutaneous injection of 2,5–hexanedione at a dose of 340 mg/kg, 5 days/week for 19 weeks.[10,11] Non-neurotoxic aliphatic monoketones such as methyl ethyl ketone enhance the neurotoxicity of MBK. In one rat study, the longer the carbon chain length of the non-neurotoxic monoketone, the greater the potentiating effect on MBK. It is expected that exposure to a subneurotoxic dose of MBK plus high doses of some aliphatic monoketones would also produce neurotoxicity.

MBK can cause mild eye irritation and minor transient corneal injury. Repeated skin contact may be irritating because of the ability of MBK to defat the skin resulting in dermatitis.[1]

REFERENCES

1. Krasave WJ et al: Ketones. In Clayton GD, Clayton FE (eds): Patty's Industrial Hygiene and Toxicology, 3rd ed, Vol 2C, Toxicology, pp 4741–4747. New York, Wiley–Interscience, 1982
2. Billmaier R et al: Peripheral neuropathy in a coated fabrics plant. J Occup Med 16:665–671, 1974
3. Mendell JR et al: Toxic polyneuropathy produced by methyl *n*-butyl ketone. Science 185:787–789, 1974
4. Allen N et al: Toxic polyneuropathy due to methyl *n*-butyl ketone. Arch Neurol 32:209–218, 1975
5. Gilchrist MA et al: Toxic peripheral polyneuropathy. Morbidity Mortality Weekly Report 23:9–10, 1974
6. Johnson BL et al: Neurobehavioral effects of methyl *n*-butyl ketone and methyl *n*-amyl ketone in rats and monkeys: A summary of NIOSH investigations. J Environ Pathol Toxicol 2:113–133, 1979
7. Spencer PS et al: Nervous system degeneration produced by the industrial solvent methyl *n*-butyl ketone. Arch Neurol 32:219–222, 1975
8. Schaumberg HH, Spencer PS: Environmental hydrocarbons produce degeneration in cat hypothalamus and optic tract. Science 199:199–200, 1978
9. Krasavage WJ et al: The relative neurotoxicity of methyl-*n*-butyl ketone and *n*-hexane and their metabolites. Toxicol Appl Pharmacol 52:433–441, 1980
10. Scala RA: Hydrocarbon neuropathy. Ann Occup Hyg 19:293–299, 1976
11. Raleigh RL, Spencer PS, Schaumberg HH: Methyl-*n*-butyl ketone. J Occup Med 17:286, 1975

METHYL CHLORIDE

CAS: 74-87-3

CH_3Cl 1987 TLV = 50 ppm

Synonyms: Chloromethane; monochloromethane; Artic

Physical Form. Colorless gas

Uses. Manufacture of silicones; antiknock fuel additive; production of butyl rubber; fungicide, pesticide

Exposure. Inhalation

Toxicology. Methyl chloride causes kidney and liver damage and is a central nervous system depressant. In experimental animals, it is a reproductive toxin, a teratogen, and a carcinogen.

Human fatalities have occurred from a single severe exposure or prolonged exposures to lower concentrations.[1] Acute poisoning in humans is characterized by a latent period of several hours, followed by dizziness, drowsiness, staggering gait, and slurred speech; nausea, vomiting, and diarrhea; double vision; and weakness, paralysis, convulsions, cyanosis, and coma.[1–3] Renal or hepatic damage and anemia also occur.[1–3] Recovery from an acute exposure usually occurs within 5 to 6 hours but may take as long as 30 days or more in massive exposures.[1] In one study, however, 10 survivors of methyl chloride poisoning experienced mild neurologic or psychiatric sequelae 13 years after the incident.[4] Recurrence of symptoms after apparent recovery without further exposure has been observed in the immediate postexposure period.

Six workers chronically exposed to 200 to 400 ppm for 2 to 3 weeks developed symptoms of intoxication, including confusion, blurring of vision, slurred speech, and staggering gait; symptoms disappeared over a period of 1 to 3 months after removal from exposure.[1]

Mice exposed continuously to 100 ppm or intermittently to 400 ppm for 11 days had histopathologic evidence of brain lesions characterized by degeneration and atrophy of the granular layer of the cerebellum.[5] Daily exposure of mice to 1000 ppm for 2 years induced a functional limb muscle impairment and atrophy of the spleen.[6] At 2400 ppm, administered daily, there were renal and hematopoietic effects, and the mice were moribund by day 9.[5] For rats exposed to 3500 ppm 6 hours/day for up to 12 days, clinical signs included severe diarrhea, incoordination of the forelimbs, and, in a few animals, hind limb paralysis and convulsions.[7]

Daily exposure of male rats to 1500 ppm for 10 weeks caused severe testicular degeneration; no males sired litters during a subsequent 2-week breeding period.[8]

An increase in fetal heart defects was observed in mice following 12 days of repeated exposure in utero to 500 ppm.[9] Repeated exposure of male mice to 1000 ppm for 2 years caused a significant increase in renal tumors.[10]

NIOSH recommends that methyl chloride be considered a potential occupational teratogen and carcinogen.[10]

The IARC states that there is inadequate evidence for the carcinogenicity of methyl chloride to experimental animals and humans.[11]

REFERENCES

1. Scharnweber HC, Spears GN, Cowles SR: Chronic methyl chloride intoxication in six industrial workers. J Occup Med 16:112–113, 1974
2. Spevak L, Nadj V, Felle D: Methyl chloride poisoning in four members of a family. Br J Ind Med 33:272–274, 1976
3. Hansen H, Weaver NK, Venable FS: Methyl chloride intoxication. AMA Arch Ind Hyg Occup Med 8:328–334, 1953
4. Gudmundsson G: Letter to the editor. Methyl chloride poisoning 13 years later. Arch Environ Health 32:236–237, 1977
5. Landry TD et al: Neurotoxicity of methyl chloride in continuously versus intermittently exposed female C57BL/6 mice. Fund Appl Toxicol 5:87–98, 1985
6. Pavokv KL et al: Major findings in a twenty-four month inhalation toxicity study of methyl chloride in mice and rats. Toxicologist 2:161, 1982
7. Morgan KT et al: Histopathology of acute toxic response in rats and mice exposed to methyl chloride by inhalation. Fund Appl Toxicol 2:293–299, 1982
8. Hamm TE Jr et al: Reproduction in Fischer 344 rats exposed to methyl chloride by inhalation for two generations. Fund Appl Toxicol 5:568–577, 1985
9. Wolkowski–Tyl R et al: Evaluation of heart malformations in $B_6C_3F_1$ mouse fetuses induced by in utero exposure to methyl chloride. Teratology 27:197–206, 1983
10. National Institute for Occupational Safety and Health: Current Intelligence Bulletin 43. Monohalomethanes. Methyl Chloride CH_3CL Methyl Bromide CH_3Br Methyl Iodide CH_3I. DHHS(NIOSH) Pub No 84–117, p 22. Wash-

ington, DC, US Government Printing Office, September 27, 1984

11. IARC Monographs on the Evaluation of the Carcinogenic Risk of Chemicals to Man. Some Halogenated Hydrocarbons and Pesticide Exposures, Vol 41, p 176. Lyon, International Agency for Research on Cancer, 1986

METHYLCYCLOHEXANE

CAS: 108-87-2

C_7H_{14} 1987 TLV = 400 ppm

Synonyms: Cyclohexylmethane; hexahydrotoluene

Physical Form. Colorless liquid

Uses. Solvent; organic synthesis

Exposure. Inhalation

Toxicology. At high concentrations, methylcyclohexane causes narcosis in animals, and it is expected that severe exposure will produce the same effect in humans.

No effects have been reported in humans.

Rabbits did not survive exposure for 70 minutes to 15,227 ppm; conjunctival congestion, dyspnea, rapid narcosis, and severe convulsions preceded death.[1] Exposure to 10,000 ppm 6 hours/day for a total of 10 days resulted in convulsions, narcosis, and death.[1] There were no signs of intoxication in rabbits exposed to 2880 ppm for a total of 90 hours, but slight cellular injury was observed in the liver and kidneys.[1] The liquid on the skin of a rabbit caused local irritation, thickening, and ulceration.[2]

REFERENCES

1. Treon JF, Crutchfield WE Jr, Kitzmiller KV: The physiological response of animals to cyclohexane, methylcyclohexane, and certain derivatives of these compounds. II. Inhalation. J Ind Hyg Toxicol 25:323–347, 1943
2. Treon JF, Crutchfield WE Jr, Kitzmiller KV: The physiological response of rabbits to cyclohexane, methylcyclohexane, and certain derivatives of these compounds. I. Oral administration and cutaneous application. J Ind Hyg Toxicol 25:199–214, 1943

METHYLCYCLOHEXANOL

CAS: 25639-42-3

$CH_3C_6H_{10}OH$ 1987 TLV = 50 ppm

Synonyms: Hexahydrocresol; hexahydromethylphenol; methylhexalin

Physical Form. Colorless, viscous liquid

Uses. Solvent for lacquers; blending agent in textile soaps; antioxidant in lubricants

Exposure. Inhalation; skin absorption

Toxicology. In animals, methycyclohexanol is a mild irritant of the eyes and mucous membranes, and, at high concentrations, it causes signs of narcosis. It is expected that severe exposure will produce the same effects in humans.

Headache and irritation of the ocular and upper respiratory membranes may result from prolonged exposure to excessive concentrations of the vapor.[1]

Rabbits exposed 6 hours/day to 503 ppm for 10 weeks had conjunctival irritation and slight lethargy.[2] There were no signs of intoxication at 232 ppm for a total exposure of 300 hours.[2]

The minimal lethal dose for rabbits by oral administration was 2 g/kg; rapid narcosis and convulsive movements resulted.[3] Sublethal doses caused narcosis with spasmodic head jerking; salivation and lacrimation were also observed; hepatocellular degeneration was apparent at autopsy.[3]

Repeated cutaneous applications to rabbits of large doses of methylcyclohexanol caused skin irritation and thickening, weakness, tremor, narcosis, and death.[3]

Methylcyclohexanol can be detected by its odor at 500 ppm, a concentration capable of causing upper respiratory irritation.[1]

REFERENCES

1. Rowe VK, McCollister SB: Alcohols. In Clayton GD, Clayton FE (eds): Patty's Industrial Hygiene and Toxicology, 3rd ed, rev, Vol 2C, pp 4649–4652. New York, Wiley–Interscience, 1982
2. Treon JF, Crutchfield WE Jr, Kitzmiller KV: The physiological response of rabbits to cyclohexane, methylcyclohexane, and certain derivatives of these compounds. II. Inhalation. J Ind Hyg Toxicol 25:323–347, 1943
3. Treon JF, Crutchfield WE Jr, Kitzmiller KV: The physiological response of rabbits to cyclohexane, methylcyclohexane, and certain derivatives of these compounds. I. Oral administration and cutaneous application. J Ind Hyg Toxicol 25:199–214, 1943

---

o-METHYLCYCLOHEXANONE

CAS: 583-60-8

$CH_3O_5H_9CO$ 1987 TLV = 50 ppm; skin

Synonym: 2–Methylcyclohexanone

Physical Form. Clear to pale yellow liquid

Uses. Solvent; rust remover

Exposure. Inhalation; skin absorption

Toxicology. In animals, *o*-methylcyclohexanone is an irritant of the eyes and mucous membranes, and, at high concentrations, it causes narcosis; it is expected that severe exposure would produce the same effects in humans.

Several species of animals exposed to 3500 ppm suffered marked irritation of the mucous membranes, became incoordinated after 15 minutes of exposure, and became prostrate after 30 minutes.[1] Conjunctival irritation, lacrimation, salivation, and lethargy were observed in rabbits exposed to 1822 ppm for 6 hours/day for 3 weeks.[2] Exposure of mice to 450 ppm for an unspecified time period resulted in severe irritation of the eyes and respiratory tract.[3]

Repeated cutaneous application of large doses of the liquid to rabbits caused irritation of the skin, tremor, narcosis, and death; the minimal lethal dose was between 4.9 and 7.2 g/kg.[3]

There are no reports of chronic or systemic effects in humans, probably because of its irritant property and warning acetone-like odor at levels below those causing serious effects. Furthermore, lethal concentrations of vapors are not expected at temperatures commonly encountered in the workplace.[1]

REFERENCES

1. Krasavage WJ et al: Ketones. In Clayton GD, Clayton FE (eds): Patty's Industrial Hygiene and Toxicology, 3rd ed, rev, Vol 2C, Toxicology, pp 4782–4784. New York, Wiley–Interscience, 1982
2. Treon JF, Crutchfield WE Jr, Kitzmiller KV: The physiological response of rabbits to cyclohexane, methylcyclohexane, and certain derivatives of these compounds. II. Inhalation. J Ind Hyg Toxicol 25:323–347, 1943
3. Treon JF, Crutchfield WE Jr, Kitzmiller KV: The physiological response of rabbits to cyclohexane, methylcyclohexane, and certain derivatives of these compounds. I. Oral administration and cutaneous application. J Ind Hyg Toxicol 25:199–214, 1943

4,4′–METHYLENE BIS (2–CHLOROANILINE)

CAS: 101-14-4

$C_{13}H_{12}Cl_2N_2$

1987 TLV = 0.02 ppm ($0.22\ mg/m^3$); suspected carcinogen; skin

Synonyms: Methylenebis (*ortho*-chloroaniline); DACPM; MOCA

Physical Form. Colorless crystals

Uses. Curing agent for polyurethanes and epoxy resins

Exposure. Inhalation; skin absorption

Toxicology. 4,4′-Methylene bis (2–chloroaniline), or MOCA, is carcinogenic in rats and mice.

Rats fed 1000 ppm MOCA in a standard diet for 2 years developed lung tumors; there were 25 adenomatoses and 48 adenocarcinomas in 88 rats.[1] Accompanying liver changes included hepatocytomegaly, necrosis, bile duct proliferation, and fibrosis.[1] In 88 control animals, there were 2 lung adenomatoses. MOCA in a low-protein diet caused lung tumors in rats of both sexes, liver tumors in males, and malignant mammary tumors in females.[1]

Repeated subcutaneous injection of MOCA to 34 rats (total dose, 25 g/kg for 620 days) resulted in nine liver cell carcinomas and seven lung carcinomas; 13 of 50 control animals developed tumors, but no malignant tumors of the liver or lungs were observed.[2]

MOCA was fed to male and female mice for 18 months at a dose of either 1 g or 2 g/kg bodyweight; in female mice, but not in males, a statistically significant incidence of hepatoma was observed.[3] In addition, a higher incidence of hemangiosarcomas and hemangiomas was observed in treated animals than in controls.[3]

There was no evidence that MOCA was tumorigenic in a study of 31 active workers exposed from 6 months to 16 years.[4] Quantitative analysis of the workers' urine confirmed exposure to the chemical. In addition, the records of 178 employees who had at one time worked with MOCA but who had thereafter had no further exposure for at least 10 years were reviewed. The general health of exposed workers with respect to illness, absenteeism, and medical history was similar to that of the total plant population. Two deaths in this group owing to malignancy had been diagnosed prior to any work with or exposure to MOCA.[4] For the plant population in general, there were 115 cancer deaths/100,000 population over a 15-year period compared with the national death rate for cancer of 139/100,000 population.

In another study, exposure to MOCA was believed to be the cause of urinary frequency and mild hematuria in two of six exposed workers; however, a variety of other materials, including toluene diisocyanate, polyester resins, polyether resins, and isocyanate-containing resins, were also present.[5]

Based on animal experiments, MOCA is suspect of carcinogenic potential in humans; worker exposure by all routes should be carefully controlled.[6] However, it is considered by the IARC to show sufficient evidence of carcinogenicity in experimental animals only.[7]

REFERENCES

1. Stula EF, Sherman H, Zapp JA Jr, Wesley–Clayton J Jr: Experimental neoplasia in rats from oral administration of 3,3′–dichlorobenzidine, 4,4′methylenebis (2–chloroaniline), and 4,4′–methylene-bis (2–methylaniline). Toxicol Appl Pharmacol 31:159–176, 1975
2. Steinhoff D, Grundmann E: Zur Cancerogen

Wirkung von 3.3'-Dichlor-4,4'-diaminodiphenylmethan bei Ratten. Naturwissenschaften 58:578, 1971
3. Russfield AB et al: The carcinogenic effect of 4,4'-methylen-bis (2-chloroaniline) in mice and rats. Toxicol Appl Pharmacol 31:47–54, 1975
4. Linch AL et al: Methylene-bis-Ortho-Chloroaniline (MOCA): Evaluation of hazards and exposure control. Am Ind Hyg Assoc J 32:802–819, 1971
5. Mastromatteo E: Recent occupational health experiences in Ontario. J Occup Med 7:502–511, 1965
6. 4,4'-Methylene bis (2-chloroaniline). Documentation of TLVs and BEIs, 5th ed, p 392. Cincinnati, American Conference of Governmental Industrial Hygienists (ACGIH), 1986
7. IARC Monographs on the Evaluation of the Carcinogenic Risk of Chemicals to Man. Some Aromatic Amines, Hydrazine and Related Substances, N-Nitroso Compounds and Miscellaneous Alkylating Agents, Vol 4, p 65. Lyon, International Agency for Research on Cancer, 1974

METHYLENE BISPHENYL ISOCYANATE

CAS: 101-68-8

$CH_2(C_6H_4NCO)_2$ 1987 TLV = C 0.02 ppm

Synonyms: Methylene diisocyanate; MDI; diphenylmethane diisocyanate

Physical Form. Liquid; aerosol

Uses. Production of polyurethane foams and plastics

Exposure. Inhalation

Toxicology. Methylene bisphenyl isocyanate (MDI) is an irritant of the eyes and mucous membranes and a sensitizer of the respiratory tract.

If the breathing zone concentration reaches 0.5 ppm, the possibility of respiratory response is imminent.[1] Depending upon the length of exposure and level of concentration above 0.5 ppm, respiratory symptoms may develop with a latent period of 4 to 8 hours. Symptoms include increased secretion, cough, pain on respiration, and, if severe enough, some restriction of air movement owing to a combination of secretions, edema, and pain. Upon removal from exposure, the symptoms may persist for 3 to 7 days.[1]

A second type of response to isocyanates is allergic sensitization of the respiratory tract. It usually develops after some months of exposure.[1–4] The onset of symptoms may be insidious, becoming progressively more pronounced with continued exposure. Initial symptoms are often nocturnal dyspnea and/or nocturnal cough with progression to asthmatic bronchitis.[3] Asthma characterized by bronchial hyperreactivity, cough, wheeze, chest tightness, and dyspnea was observed in 12 of 78 foundry workers exposed to MDI concentrations greater than 0.02 ppm.[5] Inhalation provocation tests on six of nine of the asthmatics resulted in specific asthmatic reaction to MDI.[5] Persons who are sensitized must not be exposed to any concentration of MDI and must be removed from any work involving potential exposure to MDI.

MDI is not a significant eye or skin irritant, but it may produce skin sensitization.[3]

REFERENCES

1. Rye WA: Human response to isocyanate exposure. J Occup Med 15:306–307, 1973
2. Woolrich PF, Rye WA: Urethanes. J Occup Med 11:184–190, 1969
3. National Institute for Occupational Safety and Health: Criteria for a Recommended Standard . . . Occupational Exposure to Di-isocyanates. DHEW (NIOSH) Pub No 78–215. Washington, DC, US Government Printing Office, 1978
4. Tanser AR, Bourke MP, Blandford AG: Isocyanate asthma: Respiratory symptoms caused by diphenyl-methane-diisocyanate. Thorax 28:596–600, 1973
5. Johnson A et al: Respiratory abnormalities among workers in an iron and steel foundry. Br J Ind Med 42:94–100, 1985

METHYLENE CHLORIDE
CAS: 75-09-2
CH_2Cl_2 1987 TLV = 100 ppm

Synonyms: Dichloromethane; methylene dichloride; methylene bichloride

Physical Form. Colorless liquid

Uses. Multipurpose solvent; paint remover; manufacture of photographic film; aerosol propellants

Exposure. Inhalation; skin absorption

Toxicology. Methylene chloride is a mild central nervous system depressant and an eye, skin, and respiratory tract irritant; it is carcinogenic in experimental animals.

Concentrations in excess of 50,000 ppm are thought to be immediately life threatening.[1] Four workers exposed to unmeasured but high levels of methylene chloride for 1 to 3 hours had eye and respiratory tract irritation and reduced hemoglobin and red blood cell counts; all became comatose and one died.[2]

A chemist repeatedly exposed to concentrations ranging from 500 to 3600 ppm developed signs of toxic encephalopathy.[3] A healthy young worker engaged in degreasing metal parts had a brief exposure to an undetermined but very high concentration of vapor; he complained of excessive fatigue, weakness, sleepiness, light-headedness, chills, and nausea; pulmonary edema developed after several hours, but all signs and symptoms had cleared within 18 hours of terminating the exposure.[4]

In human experiments, inhalation of 500 to 1000 ppm for 1 or 2 hours resulted in light-headedness; there was sustained elevation of carboxyhemoglobin level in each of 11 nonsmoking subjects.[5] Carbon monoxide is metabolized from methylene chloride, and the elevation of COHb may persist longer than from CO exposure alone owing to continued metabolism after exposure has ceased.[3] In one case report, it was suggested that people with cardiovascular disease may be at greater risk because of the induced hypoxia.[6]

Volunteers exposed at 300 to 800 ppm for at least 40 minutes had altered responses to various sensory and psychomotor tests.[6] No effects were seen in volunteers exposed to 250 ppm for up to 7.5 hours.[6] Although an excess in self-reported neurologic symptoms was found in workers repeatedly exposed at 75 to 100 ppm, no significant deleterious effects were observed on clinical examination, which included measurement of motor conduction velocity, electrocardiogram, and psychological tests.[7] Limited epidemiologic studies have not found any specific cause for excess deaths in workers exposed to methylene chloride.[6] There is no clear evidence of liver or kidney damage in humans in spite of many reports of fatty degeneration in the liver and tubular degeneration in the kidneys of exposed animals.[8]

Contact with the liquid is irritating to the skin, and prolonged contact may cause severe burns.[9] In a thumb immersion experiment, an intense burning sensation was noted within 2 minutes, and mild erythema and exfoliation were observed after 30 minutes of immersion; the erythema and paresthesia subsided within an hour postexposure.[10] Marked irritative conjunctivitis and lacrimation were noted at concentrations sufficient to produce unconsciousness.[2] Splashed in the eye, it is painfully irritating but is not likely to cause serious injury.[1]

Methylene chloride was fetotoxic to rats and mice exposed at 1225 ppm on days 6 to 15 of gestation.[11] Mice exposed to 4000 ppm 6 hours/day, 5 days/week for 2 years had a significant increase in lung and liver adenomas and carcinomas.[12] In rats similarly exposed, there was a significant increase in the number of benign mammary gland neoplasms. NIOSH recommends that methylene chloride be considered a potential human carcinogen.[13]

Methylene chloride has a pleasant odor detectable at about 200 ppm.[9]

REFERENCES

1. Hygienic Guide Series: Dichloromethane. Am Ind Hyg Assoc J 26:633–636, 1965
2. Moskowitz S, Shapiro H: Fatal exposure to methylene chloride vapor. AMA Arch Ind Hyg Occup Med 6:116–123, 1952
3. Methylene chloride. Documentation of the TLVs and BEIs, 5th ed, pp 390–392. Cincinnati, American Conference of Governmental Industrial Hygienists (ACGIH), 1986
4. Hughes JP: Hazardous exposure to some so-called safe solvents. JAMA 156:234–237, 1954
5. Stewart RD et al: Experimental human exposure to methylene chloride. Arch Environ Health 25:342–348, 1972
6. Illing HPA, Shillaker RO: Toxicity Review 12. Dichloromethane (Methylene Chloride). Health and Safety Executive, p 87. London, Her Majesty's Stationery Office, 1985
7. Cherry N et al: Some observations on workers exposed to methylene chloride. Br J Ind Med 38:351–355, 1981
8. Environmental Health Criteria 32: Methylene Chloride, p 55. Geneva, World Health Organization, 1984
9. National Institute for Occupational Safety and Health: Criteria for a Recommended Standard . . . Occupational Exposure to Methylene Chloride. DHEW (NIOSH) Pub No 76–138. Washington, DC, US Government Printing Office, 1976
10. Stewart RD, Dodd HC: Absorption of carbon tetrachloride, trichloroethylene, tetrachloroethylene, methylene chloride, and 1,1,1–trichloroethane through the human skin. Am Ind Hyg Assoc J 25:439–446, 1964
11. Schwetz BA et al: The effect of maternally inhaled trichloroethylene, perchloroethylene, methyl chloroform and methylene chloride on embryonal and fetal development in mice and rats. Toxicol Appl Pharmacol 32:84–96, 1975
12. National Toxicology Program: Toxicology and Carcinogenesis Studies of Dichloromethane (Methylene Chloride) (CAS No 75-09-2) In F344/N Rats and $B_6C_3F_1$ Mice (Inhalation Studies). TR-306, DHHS (NIH) Pub No 86–2562. Washington, DC, US Government Printing Office, 1986
13. National Institute for Occupational Safety and Health: NIOSH: Current Intelligence Bulletin 46, Methylene Chloride. DHHS (NIOSH) Pub No 86–114. Washington, DC, US Government Printing Office, April 18, 1986

METHYL ETHYL KETONE

CAS: 78-93-3

$CH_3C(O)C_2H_5$ 1987 TLV = 200 ppm

Synonyms: 2–Butanone; MEK

Physical Form. Colorless liquid

Uses. Solvent

Exposure. Inhalation

Toxicology. Methyl ethyl ketone (MEK) is an irritant of the eyes, mucous membranes, and skin; at high concentrations, it causes narcosis in animals, and it is expected that severe exposure in humans will produce the same effect.[1]

In humans, short-term exposure to 300 ppm was "objectionable," causing headache and throat irritation; 200 ppm caused mild irritation of the eyes; 100 ppm caused slight nose and throat irritation.[2] Several workers exposed to both the liquid and vapor at 300 to 600 ppm for an unspecified time period complained of numbness of the fingers and arms; one worker complained of numbness in the legs and a tendency for them to "give way under him."[3] Many workers in this plant developed dermatitis from contact with the liquid, and two workers developed dermatitis of the face from vapor exposure alone.[3]

Three cases of polyneuropathy occurred in shoe factory workers exposed to combined methyl ethyl ketone and acetone vapors, as well as methyl ethyl ketone and toluene vapors at concentrations below 200 ppm.[4] Skin absorption also occurred.[4] Although not highly neurotoxic itself, MEK may potentiate substances known to cause neuropathy.[4]

In guinea pigs exposed to 10,000 ppm, signs of eye and nose irritation developed rapidly, and narcosis occurred after 5 hours.[5]

Exposure of rats to 6000 ppm 8 hours/day, 7 days/week did not result in any obvious motor impairment; all animals died from bronchopneumonia during the seventh week.[6] Animal studies have shown MEK to enhance the development of, or increase the severity of, effects, owing to methyl *n*-butyl ketone, ethyl butyl ketone, *n*-hexane, and 2,5–hexanedione.[7–10]

Rats exposed to 3000 ppm during days 6 through 15 of gestation produced litters with an increased incidence of a minor skeletal variation and delay in ossification of fetal bones.[11]

Methyl ethyl ketone can be recognized at 25 ppm by its odor, which is similar to acetone but more irritating; the warning properties should prevent inadvertent exposure to toxic levels.[5]

REFERENCES

1. National Institute for Occupational Safety and Health: Criteria for a Recommended Standard . . . Occupational Exposure to Ketones. DHEW (NIOSH) Pub No 78–173. Washington, DC, US Government Printing Office, 1978
2. Nelson KW et al: Sensory response to certain industrial solvent vapors. J Ind Hyg Toxicol 25:282–285, 1943
3. Smith AR, Mayers MR: Poisoning and fire hazards of butanone and acetone. Ind Hyg Bull 23:175–176, 1944
4. Dyro FM: Methyl ethyl ketone polyneuropathy in shoe factory workers. Clin Toxicol 13:371–376, 1978
5. Krasavage WJ et al: Ketones. In Clayton GD, Clayton FE (eds): Patty's Industrial Hygiene and Toxicology, 3rd ed, Vol 2C, Toxicology, pp 4728–4733. New York, John Wiley & Sons, 1982
6. Altenkirch H et al: Experimental studies on hydrocarbon neuropathies induced by methyl-ethyl-ketone (MEK). J Neurol 219:159–170, 1978
7. Saida K et al: Peripheral nerve changes induced by methyl *n*-butyl ketone and potentiation by methyl ethyl ketone. J Neuropathol Exp Neurol 35:207–225, 1976
8. O'Donoghue JL et al: Further studies on ketone neurotoxicity and interactions. Toxicol Appl Pharmacol 72:201–209, 1984
9. Altenkirch H et al: Potentiation of hexacarbon—neurotoxicity by methyl ethyl ketone. Neurobehav Toxicol Teratol 4:623–627, 1982
10. Ralston WH et al: Potentiation of 2,5–hexanedione neurotoxicity by methyl ethyl ketone. Toxicol Appl Pharmacol 81:319–327, 1985
11. Deacon MM et al: Embryo and fetotoxicity of inhaled methyl ethyl ketone in rats. Toxicol Appl Pharmacol 59:620–622, 1981

METHYL FORMATE

CAS: 107-31-3

$HC(O)OCH_3$ 1987 TLV = 100 ppm

Synonym: Methyl methanoate

Physical Form. Colorless liquid

Uses. Solvent; chemical intermediate; insecticide, fumigant; refrigerant

Exposure. Inhalation; skin absorption

Toxicology. Methyl formate is an irritant of the eyes and respiratory tract; at high concentrations, it causes narcosis in animals, and it is expected that severe exposure will produce the same effect in humans.

Workers exposed to the vapor of a solvent containing 30% methyl formate, in addition to ethyl formate and methyl and ethyl acetate, complained of irritation of mucous membranes, oppression in the chest, dyspnea, symptoms of central nervous system depression, and temporary visual disturbances; air concentrations were not determined.[1] No effects were noted from experimental human exposures to 1500 ppm for 1 minute.[2]

Exposure of guinea pigs to 10,000 ppm for 3 hours was fatal; effects included eye and nose irritation, incoordination, and narcosis; autopsy revealed pulmonary edema.[2]

Methyl formate has a distinct and pleasant odor, but an odor threshold has not been reported.[2]

REFERENCES

1. von Oettingen WF: The aliphatic acids and their esters—toxicity and potential dangers. AMA Arch Ind Health 20:517–531, 1959
2. Schrenk HH, Yant WP, Chornyak J, Patty FA: Acute response of guinea pigs to vapors of some new commercial organic compounds. XIII. Methyl formate. Public Health Rep 51:1329–1337, 1936

METHYL HYDRAZINE

CAS: 60-34-4

CH_3NHNH_2 1987 TLV = C 0.2 ppm; skin

Synonyms: Monomethylhydrazine; MMH

Physical Form. Liquid

Uses. Rocket fuel

Exposure. Inhalation; skin absorption

Toxicology. Methyl hydrazine causes blood dyscrasias and is a convulsant; it is carcinogenic in experimental animals.

Volunteers exposed to 90 ppm for 10 minutes had slight redness in the eyes and experienced a tickling sensation of the nose.[1] The only clinical abnormality found during the 60-day follow-up period was the presence of Heinz bodies in 3% to 5% of the erythrocytes by the seventh day.[1]

As a reducing agent, methyl hydrazine causes characteristic oxidative damage to human erythrocytes in vitro. Effects include formation of Heinz bodies and production of methemoglobin.[1]

Exposure of dogs to 21 ppm for 4 hours resulted in convulsions and some deaths; postmortem examination revealed no lesions attributable primarily to methyl hydrazine, although secondary manifestation, probably owing to convulsions, included pulmonary hemorrhage and edema; convulsions but not death occurred at 15 ppm.[2] In dogs that survived exposure, there was evidence of moderately severe intravascular hemolysis.[2] The hemolytic effect was most pronounced 4 to 8 days after exposure, and blood values returned to normal within 3 weeks.[2] In another study, additional signs, including eye irritation, tremor, ataxia, diarrhea, and cyanosis were noted in dogs.[3] Dogs exposed at 5 ppm 6 hours/day for 6 months had at least a twofold increase in methemoglobin and reductions in erythyrocytes, hemoglobin concentrations, and hematocrit values; the effect was reversible and was not observed at the 1 ppm level.[4]

Applied to the shaved skin of dogs, the liquid was rapidly absorbed, producing toxic signs; at the site of application, the skin became red and edematous.[5]

Administered intraperitoneally to rats on days 6 through 15 of pregnancy, 10 mg/kg/day caused slight maternal toxicity in the form of reduced weight gain but was not selectively embryotoxic or teratogenic.[6]

Results from various long-term animal cancer studies are equivocal. Mice administered 0.001% methyl hydrazine sulfate in drinking water for life developed an increase in lung tumors, whereas 0.01% methyl hydrazine enhanced the development of lung tumors by shortening latent periods; control incidences were not clearly defined in this study.[7] In two other mice studies that may not have allowed for a sufficient latency period, no evidence of carcinogenicity was found.[8,9] Results from hamster studies have also been varied.[1]

Although animal studies are inconclusive, the ACGIH has designated methyl hydrazine as a suspected human carcinogen.[10]

The odor threshold is 1 to 3 ppm; it is described as ammoniacal or fishy.[1]

REFERENCES

1. National Institute for Occupational Safety and Health: Criteria for a Recommended Standard . . . Occupational Exposure to Hydrazines. DHEW (NIOSH) Pub No 78–172, p 269. Washington, DC, US Government Printing Office, 1978
2. Jacobson KH et al: The acute toxicity of the vapors of some methylated hydrazine derivatives. AMA Arch Ind Health 12:609–616, 1955
3. Haun CC, MacEwen JD, Vernot EH, Eagan GF: Acute inhalation toxicity of monomethylhydrazine vapor. Am Ind Hyg Assoc J 31:667–677, 1970
4. MacEwen JD, Haun CC: Chronic Exposure Studies with Monomethylhydrazine, p 15. NTIS AD 751 440. Springfield, Virginia, National Technical Information Service, US Department of Congress, 1971
5. Smith EB, Clark DA: The absorption of monomethylhydrazine through canine skin. Proc Soc Exp Biol Med 131:226–232, 1969
6. Keller WC et al: Teratogenic assessment of three methylate hydrazine derivatives in the rat. J Toxicol Environ Health 13:125–131, 1984
7. Toth B: Hydrazine, methylhydrazine and methylhydrazine sulfate carcinogenesis in Swiss mice—failure of ammonium hydroxide to interfere in the development of tumors. Int J Cancer 9:109–118, 1972
8. Kelly MG et al: Comparative carcinogenicity of *n*-isopropyl-a-(2-methylhydrazino)-*p*-toluamide HCL (procarbazine hydrochloride), its degradation products, other hydrazines, and isonicotinic acid hydrazide. J Natl Cancer Inst 42:337–344, 1969
9. Roe FJC et al: Carcinogenicity of hydrazine and 1,1-dimethylhydrazine for mouse lung. Nature 216:375–376, 1967
10. Methyl hydrazine. Documentation of the TLVs and BEIs, 5th ed, p 398. Cincinnati, American Conference of Governmental Industrial Hygienists (ACGIH), 1986

METHYL IODIDE

CAS: 74-88-4

CH_3I — 1987 TLV = 2 ppm; skin

Synonyms: Iodomethane; monoiodomethane

Physical Form. Colorless liquid

Uses. Chemical intermediate; in microscopy because of its high refractive index

Exposure. Inhalation; skin absorption

Toxicology. Methyl iodide is a neurotoxin and a convulsant and has caused pulmonary edema. It is carcinogenic in experimental animals.

The latency period between exposure and onset of symptoms ranges from hours to days.[1] Initial symptoms include lethargy, somnolence, slurred speech, ataxia, dysmetria, and visual disturbances. Neurologic dysfunction may progress to convulsions, coma, and death. If recovery occurs, neurologic findings recede over several weeks and are followed by psychiatric disturbances such as paranoia, delusions, and hallucination.

A chemical worker accidentally exposed to an unknown concentration of the vapor developed giddiness, diarrhea, sleepiness, and irritability, with recovery in a week; when reexposed 3 months later, he experienced drowsiness, vomiting, pallor, incoordination, slurred speech, muscular twitching, oliguria, coma, and death.[2] At autopsy, there was bronchopneumonia and pulmonary hemorrhages, with accumulation of combined iodine in the brain.

Experimental application of the liquid to human skin produced a stinging sensation and slight reddening in 10 minutes; after 6 hours of contact, there was spreading erythema followed by formation of vesicles.[3] Absorption through the skin is said to occur.[4] Splashed in the eye, the liquid causes conjunctivitis.[4]

In rats, reported LC_{50} values are 1750 ppm, 900 ppm, and 232 ppm for 0.5-, 1-, and 4-hour exposures, respectively.[3–5]

Local sarcomas occurred in rats following subcutaneous injection with 10 mg/kg bodyweight weekly for 1 year or with a single 50-mg/kg bodyweight dose.[6] Tumors occurred between 500 and 700 days after the first injection, and, in most cases, pulmonary metastases were observed.[6] Repeated intraperitoneal injection of 44 mg/kg bodyweight in mice reduced survival and caused an increased incidence of lung tumors.[1]

NIOSH has determined that there is sufficient evidence of carcinogenicity in animals to indicate a potential for human carcinogenicity.[4]

REFERENCES

1. Appel GB, Galen R, O'Brien J, Schoenfeldt R: Methyl iodide intoxication—a case report. Ann Intern Med 82:534–536, 1975
2. von Oettingen WF: The Halogenated Aliphatic, Olefinic, Cyclic, Aromatic, and Halogenated Insecticides, Their Toxicity and Potential Dangers, US Public Health Service Pub No 414, pp 30–32. Washington, DC, US Government Printing Office, 1955
3. Buckell M: The toxicity of methyl iodide. I. Preliminary survey. Br J Ind Med 7:122–124, 1950
4. National Institute for Occupational Safety and Health: NIOSH Current Intelligence Bulletin 43, Monohalomethanes. DHHS (NIOSH) Pub No 84–117. Washington, DC, US Government Printing Office, 1984
5. Deichmann WB, Gerarde HW: Toxicology of drugs and chemicals, p 756. New York, Academic Press, 1969
6. Preussman R: Direct alkylating agents as carcinogens. Food Cosmet Toxicol 6:576–577, 1968
7. Poirier LA et al: Bioassay of alkyl halides and nucleotide base analogs by pulmonary tumor response in strain A mice. Cancer Res 35:1411–1415, 1975

METHYL ISOBUTYL CARBINOL

CAS: 108-11-2

$CH_3CHOHCH_2CH(CH_3)_2$

1987 TLV = 25 ppm; skin

Synonyms: Methyl amyl alcohol; 4–methyl–2–pentanol

Physical Form. Colorless liquid

Uses. Solvent; organic syntheses; brake fluids

Exposure. Inhalation; skin absorption

Toxicology. Methyl isobutyl carbinol is an eye irritant; at high concentrations, it causes narcosis in animals, and it is expected that severe exposure in humans would produce the same effect.

Human subjects exposed to 50 ppm for 15 minutes had eye irritation.[1] No acute, chronic, or systemic effects have been reported in humans.

Five of six rats died following exposure to 2000 ppm for 8 hours; there were no deaths after exposure for 2 hours to the saturated vapor.[2] The single-dose oral toxicity for rats was 2.6 g/kg; the dermal LD_{50} in rabbits was 3.6 ml/kg.[2]

REFERENCES

1. Silverman L, Schulte HF, First MW: Further studies on sensory response to certain industrial solvent vapors. J Ind Hyg Toxicol 28:262–266, 1946
2. Smyth HF Jr, Carpenter CP, Weil CS: Range-finding toxicity data: List IV. AMA Arch Ind Hyg Occup Med 4:119–122, 1951

METHYL ISOBUTYL KETONE

CAS: 108-10-1

$C_6H_{12}O$ 1987 TLV = 50 ppm

Synonyms: Hexone; MIBK; 4–methyl–2–pentanone

Physical Form. Colorless liquid

Uses. Solvent for gums, resins, and nitrocellulose

Exposure. Inhalation

Toxicity. Methyl isobutyl ketone is an irritant of the eyes, mucous membranes, and skin; high concentrations cause narcosis in animals, and it is expected that severe exposure will cause the same effect in humans.

Exposures to 80 to 500 ppm produced weakness, loss of appetite, headache, eye irritation, sore throat, and nausea.[1] At 200 ppm, the eyes of most persons were irritated, and 100 ppm was the highest concentration most volunteers estimated to be acceptable for an 8-hour exposure.[2]

Exposure of rats to 4000 ppm for 4 hours caused death; 2000 ppm for 4 hours was not fatal.[3] A 2-week exposure of rats to 200 ppm produced toxic nephrosis of the proximal tubules and increased liver weights.[4] A 90-day continuous exposure at 100 ppm produced no significant changes.[4] The liquid splashed in the eyes may cause pain and irritation. Repeated or prolonged skin contact may cause defatting of the skin with primary irritation and desquamation.[5]

Methyl isobutyl ketone has a characteristic camphor-like odor detectable at 100 ppm.[1]

REFERENCES

1. National Institute for Occupational Safety and Health: Criteria for a Recommended Standard . . . Occupational Exposure to Ketones. DHEW (NIOSH) Pub No 78–173. Washington, DC, US Government Printing Office, 1978
2. Silverman L, Schulte HF, First MW: Further studies on sensory response to certain industrial solvent vapors. J Ind Hyg Toxicol 28:262–266, 1946
3. Smyth HF Jr, Carpenter CP, Weil CS: Range-finding toxicity data: List IV. AMA Arch Ind Hyg Occup Med 4:119–122, 1951
4. MacEwen JD et al: Effect of 90-day continuous exposure to methylisobutylketone on dogs, monkeys and rats, p 23. NTIS AD 730 291. Springfield, Virginia, National Technical Information Service, US Department of Commerce, 1971
5. Hygienic Guide Series: Methyl isobutyl ketone. Am Ind Hyg Assoc J 27:209–211, 1966

METHYL ISOCYANATE

CAS: 624-83-9

CH_3CNO 1987 TLV = 0.02 ppm; skin

Synonyms: Isocyanic acid, methyl ester; MIC

Physical Form. Liquid; aerosol

Uses. Production of polyurethane foams and plastics; chemical intermediate

Exposure. Inhalation; skin absorption

Toxicology. Methyl isocyanate is an irritant of the eyes, mucous membranes, and skin; it can cause pulmonary irritation and sensitization.

Exposure of humans to high concentrations causes cough, dyspnea, increased secretions, and chest pain.[1] Isocyanates cause pulmonary sensitization in susceptible individuals; if this occurs, further exposure should be avoided, because extremely low levels of exposure may trigger an asthmatic episode; cross-sensitization to unrelated materials probably does not occur.

Experimental exposure of four human subjects for 1 to 5 minutes caused the following effects: 0.04 ppm, no effects; 2 ppm, lacrimation and irritation of the nose and throat; 4 ppm, symptoms of irritation more marked; 21 ppm, unbearable irritation of eyes, nose, and throat.[2]

In rats exposed for 4 hours, the LC_{50} was 5 ppm; effects were injury to the lungs and subsequent pulmonary edema.[2] A cotton plug saturated with the liquid was applied to the ear of a rabbit for 30 minutes and caused erythema, edema, necrosis, and perforation; a few drops on the ear of a rabbit causes destruction of tissue.[2] The liquid in contact with the eye may cause permanent damage.[1]

REFERENCES

1. Rye WA: Human responses to isocyanate exposure. J Occup Med 15:306–307, 1973
2. Methyl isocyanate. Documentation of TLVs and BEIs, 5th ed, pp 403–404. Cincinnati, American Conference of Governmental Industrial Hygienists (ACGIH), 1986

METHYL MERCAPTAN

CAS: 74-93-1

CH_3SH 1987 TLV = 0.5 ppm

Synonyms: Methanethiol; mercaptomethane; thiomethyl alcohol; methyl sulfhydrate

Physical Form. Flammable gas liquefying at 6°C; odor of rotten cabbage

Uses. Intermediate in manufacturing of jet fuels, pesticides, fungicides, plastics; synthesis of methionine; emission from paper pulp mills; odoriferous additive to natural gas

Exposure. Inhalation

Toxicology. Methyl mercaptan affects the nervous system and causes both convulsions and narcosis; high concentrations cause paralysis of the respiratory center; lower levels are stated to produce respiratory irritation and pulmonary edema.[1]

In a fatal human exposure, a worker engaged in emptying metal gas cylinders of methyl mercaptan was found comatose at the worksite; he developed expiratory wheezes, elevated blood pressure, tachycardia, and marked rigidity of extremities.[2] Methemoglobinemia and severe hemolytic anemia developed with hematuria and proteinuria but were brief in duration; deep coma persisted until death due to pulmonary embolus 28 days after exposure. It was determined that the person was deficient in erythrocyte glucose–6–phosphate dehydrogenase, which was the likely cause of the hemolysis and formation of methemoglobin.

In a nonfatal incident, a worker in a refinery inhaled methyl mercaptan and was comatose for 9 hours. Although not dyspneic, the person was cyanotic and experienced convulsions; recovery occurred by the fourth day. Ten days later, the worker was treated successfully for a lung abcess.[2]

Although details are lacking, one report states that effects in animals exposed to methyl mercaptan were restlessness and muscular weakness, progressing to paralysis, convulsions, respiratory depression, and cyanosis.[2] Exposure of rats to various concentrations for 4 hours allowed a determination of an LC_{50} of 675 ppm, thus making it slightly less acutely toxic than hydrogen sulfide (LC_{50} 444 ppm). A subchronic toxicity study at 2-, 17-, and 57-ppm interrupted exposures for 3 months in young male rats showed a dose-related decreased weight gain (about 15% at 57 ppm) in exposed animals but no clear pathologic or biochemical test alterations. There were some minor microscopic hepatic alterations in the exposed animals that were of questionable significance.[3]

REFERENCES

1. Methyl mercaptan (methanethiol). Documentation of the TLVs and BEIs, 5th ed, p 405. Cincinnati, American Conference of Governmental Industrial Hygienists (ACGIH), 1986
2. Shults WT, Fountain EN, Lynch EC: Methanethiol poisoning. JAMA 211:2153–2154, 1970
3. Tansy M et al: Acute and subchronic toxicity studies of rats exposed to vapors of methyl mercaptan and other reduced-sulfur compounds. J Toxicol Environ Health 8:71–88, 1981

METHYL METHACRYLATE

CAS: 80-62-6

$H_2C{=}C(CH_3)COOCH_3$

1987 TLV = 100 ppm

Synonyms: Methacrylic acid, methyl ester; methyl 2–methylpropenoic acid; methyl α-methylacrylate; methyl methylacrylate

Physical Form. Colorless liquid; commercial form contains a small amount of hydroquinone or hydroquinone monomethyl ether to inhibit spontaneous polymerization

Uses. Production of polymethyl methacrylate polymers for use in acrylic sheet and acrylic molding, extrusion powder, acrylic surface coatings

Exposure. Inhalation

Toxicology. Methyl methacrylate is an irritant of eyes, skin, and mucous membranes.

Human exposure to concentrations of 170 to 250 ppm has caused irritation (presumably of eyes and nose).[1] A level of 2300 ppm was unbearable.

In a study of 91 exposed and 43 nonexposed workers at five plants producing polymethyl methacrylate sheets, exposures ranged from 4 to 49 ppm, and there were no detectable clinical signs or symptoms.[2] In another survey of 152 workers exposed to concentrations ranging from 0.5 to 50 ppm, 78% reported a high incidence of headache, 30% reported pain in the extremities, 10% reported irritability, 20% reported loss of memory, and 21% reported excessive fatigue and sleep disturbances.[3]

Handlers of methyl methacrylate cement have developed paresthesia of the fingers.[4] Dental technicians who used bare hands to mold methyl methacrylate putty had significantly slower distal sensory conduction velocities from the digits, implicating mild axonal degeneration in the area of contact with methyl methacrylate.[5] The toxic effect on the nervous tissue may be due to diffusion into the nerve cells causing lysis of the membrane lipids and destruction of the myelin sheath. Humans have developed strong skin reactions when rechallenged with the liquid.[6]

Acute inhalation exposure of dogs to 11,000 ppm led to central nervous system depression, a drop in blood pressure, liver and kidney damage, and death due to respiratory arrest.[6] Mice exposed to 1520 ppm for 2 hours twice daily for 10 days showed no significant histologic changes in heart, liver, kidney, or lungs.[7] In male rats exposed to methyl methacrylate vapor at 116 ppm, 7 hours/day, 5 days/week for 5 months, the tracheal mucosa was denuded of cilia, and the number of microvilli on the epithelium was reduced.[8]

Exposure of pregnant rats to 27,500 ppm on days 6 through 15 of gestation for 54 minutes resulted in maternal toxicity and significant increases in fetal deaths, hematomas, and skeletal anomalies.[9]

In a 2-year inhalation study, there was no evidence of carcinogenicity of methyl methacrylate for male rats exposed at 500 or 1000 ppm, for female rats exposed at 250 or 500 ppm, or for male and female mice exposed at 500 or 1000 ppm.[10] There was inflammation of the nasal cavity and degeneration of the olfactory sensory epithelium in rats and mice; epithelial hyperplasia of the nasal cavity was also observed in exposed mice.

The toxic effects are due to the monomer; the polymer appears inert. The severity of effects is believed to be inversely proportional to the degree of polymerization.

REFERENCES

1. Methyl hydrazine. Documentation of TLVs and BEIs, 5th ed, p 406. Cincinnati, American Conference of Governmental Industrial Hygienists (ACGIH), 1986
2. Cromer J, Kronoveter K: A study of methyl methacrylate exposures and employee health. US Department of Health, Education, and Welfare, National Institute of Occupational Safety and Health, Cincinnati, Ohio. DHEW (NIOSH) Pub No 77–119. Washington, DC, US Government Printing Office, 1976
3. Blagodatin VM et al: Establishing the maximum permissible concentration of the methyl ester of methacrylic acid in the air of a work area. Gig Tr Prot Zabol 6:5–8, 1976
4. Kassis V et al: Contact dermatitis to methyl methacrylate. Contact Dermatitis 11:26–28, 1984
5. Seppalainen A, Rajaniemi R: Local neurotoxicity of methyl methacrylate among dental technicians. Am J Ind Med 5:471–477, 1984
6. Speakman CR et al: Monomeric methyl methacrylate. Studies on toxicity. J Ind Med 14:292, 1945
7. McLaughlin et al: Pulmonary toxicity of methyl methacrylate vapors: An environmental study. Arch Environ Health 34:336–338, 1979
8. Tansy M et al: Chronic biological effects of methyl methacrylate vapor. III. Histopathology, blood chemistry, and hepatic and ciliary function in the rat. Environ Res 21:117–125, 1980
9. Nicholas CA et al: Embryotoxicity and fetotoxicity from maternal inhalation of methyl methacrylate monomer in rats. Toxicol Appl Pharmacol 50:451–458, 1979
10. National Toxicology Program: NTP Technical Report on the Toxicology and Carcinogenesis Studies of Methyl Methacrylate (CAS No 80-62-6) in F344/N Rats and B6C3F Mice (Inhalation Studies). DHHS (NTP) TR-314, pp 1–202. Research Triangle Park, North Carolina, US Department of Health and Human Services, October 1986

METHYL PARATHION

CAS: 298-00-0

$(CH_3O)_2P(S)OC_6H_4NO_2$

1987 TLV = 0.2 mg/m^3; skin

Synonyms: O,O-Dimethyl O-*p*-nitrophenyl phosphorothioate; Metron; Nitrox; parathion-methyl; Metacide; metaphos; Wofatox; BAY E–601–BLADAN M; Folidol M

Physical Form. White solid (pure); tan-to-brown solid (technical)

Uses. Insecticide

Exposure. Inhalation; skin absorption; ingestion

Toxicology. Methyl parathion is an anticholinesterase agent.

Signs and symptoms of overexposure are caused by the inactivation of the enzyme cholinesterase, which results in the accumulation of acetylcholine at synapses in the nervous system, skeletal and smooth muscle, and secretory glands. The sequence of the development of systemic effects varies with the route of entry. The onset of signs and symptoms is usually prompt but may be delayed up to 12 hours.[1–5] After inhalation, respiratory and ocular effects are the first to appear, often within a few minutes of exposure. Respiratory effects include tightness in the chest and wheezing owing to bronchoconstriction and excessive bronchial secretion. Laryngeal spasms and excessive salivation may add to the respiratory distress; cyanosis may also occur. Ocular effects include miosis, blurring of distant vision, tearing, rhinorrhea, and frontal headache.

After ingestion, gastrointestinal effects, such as anorexia, nausea, vomiting, abdominal cramps, and diarrhea appear within 15 minutes to 2 hours. After skin absorption, localized sweating and muscular fasciculations in the immediate area usually occur within 15 minutes to 4 hours; skin absorption is somewhat greater at higher ambient temperatures and is increased by the presence of dermatitis.[1–3]

With severe intoxication by all routes, an excess of acetylcholine at the neuromuscular junctions of skeletal muscle causes weakness aggravated by exertion, involuntary twitchings, fasciculations, and, eventually, paralysis. The most serious consequence is paralysis of the respiratory muscles. Effects on the central nervous system include giddiness, confusion, ataxia, slurred speech, Cheyne–Stokes respiration, convulsions, coma, and loss of reflexes. The blood pressure may fall to low levels, and cardiac irregularities, including complete heart block, may occur.[2]

Complete symptomatic recovery usually occurs within a week; increased susceptibility to the effects of anticholinesterase agents persists for up to several weeks after exposure.[4] Daily exposure to concentrations that are insufficient to produce symptoms following a single exposure may result in the onset of symptoms. Continued daily exposure may be followed by increasingly severe effects.

Deaths from occupational exposure have been reported, usually following massive accidental exposures.[1]

Data from human poisonings by methyl parathion are not sufficiently detailed to identify the range between the doses producing first symptoms and those producing severe or fatal intoxication.[4] The minimal lethal oral dose for adults appears to be less than 1.84 g. Most animal data and limited human data indicate that methyl parathion is somewhat less acutely toxic than parathion.[4]

Methyl parathion itself is not a strong cholinesterase inhibitor, but one of its metabolites, methyl paraoxon, is an active inhibitor. Methyl paraoxon inactivates cholinesterase by phosphorylation of the active site of the enzyme to form the "dimethylphosphoryl enzyme." Over the following 24 to 48 hours, there is a process, termed *aging*, of conversion to the "monomethylphosphoryl enzyme." Aging is of clinical interest in the treatment of poisoning because cholinesterase reactivators such as pralidoxime (2–PAM, Protopam) chloride are ineffective after aging has occurred. Measurement of metabolites of methyl parathion, *p*-nitrophenol, and dimethylphosphate in the urine has been used to monitor exposure to workers.[6]

A 2-year bioassay of methyl parathion in mice and rats did not demonstrate any increased incidence of tumors in dosed animals.[7] The IARC has concluded that there is no evidence that methyl parathion is carcinogenic to experimental animals.[6] Methyl parathion, administered intraperitoneally at maternally lethal doses, was teratogenic to mice, producing cleft palate and rib abnormalities. High dose administration to rats, sometimes producing maternal toxicity, resulted in evidence of embryofetotoxicity with increased resorptions and growth retardation.[6] There is no evidence that methyl parathion can induce delayed peripheral neuropathy in humans or in experimental animals.[6]

Diagnosis. *Signs and Symptoms:* Initial signs and symptoms include headache, blurred vision, pallor, weakness, sweating, abdominal pain, nausea, vomiting, and diarrhea.

Moderate-to-severe intoxication includes miosis, lacrimation, excessive salivation, muscle fasciculations, dyspnea, cyanosis, convulsions, shock, cardiac arrhythmias, and coma.

Differential Diagnosis: Diagnosis is based primarily on a history of exposure and clinical evidence of diffuse parasympathetic stimulation. Careful observation of the effects of atropine and pralidoxime may be valuable. Patients with organophosphate poisoning are resistant to the action of atropine at moderate dosages; failure of 1 to 2 mg of atropine administered parenterally to produce signs of atropinization (flushing, mydriasis, tachycardia, or dryness of mouth) indicates organophosphate poisoning. Intravenous injections of 1 g pralidoxime generally cause some recovery from signs and symptoms.

Special Tests: Two types of cholinesterase are clinically significant: (1) true acetylcholinesterase, found principally in the nervous system and the red blood cell; and (2) pseudo- or butyrylcholinesterase, found in the plasma, liver, and nervous system. Whereas the action of both types is inhibited by organophosphates, the level of depression of red blood cell cholinesterase is a better indicator of clinically significant reduction of cholinesterase activity in the nervous system.

Laboratory evidence of depression of red blood cell cholinesterase to a level substantially below pre-exposure levels (at least 50% and usually much lower) is verification of poisoning. There is an imperfect correlation between the degree of depression of cholinesterase enzymes and the occurrence of symptoms. With a rapid drop in cholinesterase activity, generally reflecting an acute heavy exposure, there may be symptoms with only a 30% depression, whereas with slower drops to 70% percent depression, reflecting chronic low-level exposure, there may be no symptoms.[8]

If no pre-exposure baseline has been performed but symptoms are not sufficient to justify treatment with atropine, repeated testing during the recovery period, demonstrating progressively increasing plasma and red blood cell cholinesterase levels over several days and weeks, respectively, suggests the diagnosis of anticholinesterase poisoning.

There are many different methods for estimation of cholinesterase content of blood, and associated with each method is a different set of normal values and a different set of reporting units. The laboratory report of a cholinesterase determination should state the units involved along with the appropriate normal range. Based upon the Michel method, the normal range of red blood cell cholinesterase activity (delta *p*H per hour) is 0.39 to 1.02 for men, and 0.34 to 1.10 for women.[9] The normal range of the enzyme activity (delta *p*H per hour) of plasma is 0.44 to 1.63 for men, and 0.24 to 1.54 for women.

Treatment. Treatment of organophosphate poisoning ranges from simple removal from exposure in very mild cases to the provision of very rigorous supportive and antidotal measures in severe cases.[1,2,5,10] In moderate-to-severe cases, because of pulmonary involvement, there may be need for artificial respiration using a positive-pressure method. Careful attention must be paid to removal of secretions and to maintenance of a patent airway. Anticonvulsants such as thiopental sodium may be necessary. Maintenance of respiration is critical, because death usually results from weakness of the muscles of respiration and accumulation of excessive secretions in the respiratory tract.

As soon as cyanosis has been overcome, 2 to 4 mg of atropine should be given intravenously. (Atropine may induce ventricular fibrillation in the presence of cyanosis.) *This dose of atropine is approximately 10 times the amount that is administered for other conditions in which atropine is considered therapeutic.* This dose should be repeated at 5- to 10-minute intervals until signs of atropinization appear (dry, flushed skin, tachycardia as high as 140 beats/minute, and pupillary dilatation). A mild degree of atropinization should be maintained for at least 48 hours.[2]

Pralidoxime (2–PAM, Protopam) chloride is a cholinesterase reactivator that complements the action of atropine. It has its greatest effect in reversing the nicotinic action of anticholinesterase agents at skeletal neuromuscular junctions but has virtually no effect on central nervous system manifestations. In moderate-to-severe cases, the dose for adults is 1 to 2 g injected intravenously at a rate not in excess of 500 mg/minute. After an hour, a second dose of 1 g is indicated if muscle weakness has not been relieved. Treatment with pralidoxime chloride will be most effective if given within 24 hours after poisoning.[2] Morphine, aminophylline, and phenothiazines are contraindicated because of documented experience of adverse reactions in cases of organophosphate poisoning.[10]

It is of great importance to decontaminate the patient. Contaminated clothing should be removed at once, and the skin should be washed with generous amounts of soap or detergent and a flood of water, which is best accomplished under a shower or by submersion in a pond or other body of water if the exposure occurred in the field. Careful attention should be paid to cleansing of the skin and hair.

The patient should be attended to and monitored continuously for not less than 24 hours, because serious and sometimes fatal relapses have occurred from continuing absorption of the toxin or dissipation of the effects of the antidote.

Regeneration of cholinesterase is primarily by synthesis of new enzyme and takes place at the rate of approximately 1% per day.[9] A patient who has recovered from the acute phase of poisoning remains hypersusceptible to anticholinesterases for up to several weeks.

Medical Control. Medical control involves preplacement and annual physical examination with determination of pre-exposure red blood cell cholinesterase activity. A person whose red blood cell cholinesterase falls to or below 40% of the pre-exposure baseline should be removed from further exposure until the activity returns to within 80% of the pre-exposure baseline.

REFERENCES

1. Hayes WJ Jr: Pesticides Studied in Man, pp 284–435. Baltimore, Williams & Wilkins, 1982
2. Taylor P: Anticholinesterase agents. In Gilman AG et al (eds): Goodman and Gilman's The Pharmacological Basis of Therapeutics, 7th ed, pp 116–129. New York, Macmillan Publishing Co, 1985
3. Koelle GB (ed): Cholinesterases and anticholinesterase agents. Handbuch der Experimentellen Pharmakologie, Vol 15, pp 989–1027. Berlin, Springer-Verlag, 1963
4. National Institute for Occupational Safety and Health: Criteria for a Recommended Standard . . . Occupational Exposure to Methyl Parathion. DHEW (NIOSH) Pub No 77–106, pp

31–68. Washington, DC, US Government Printing Office, 1976
5. Namba T, Nolte CT, Jackrel J, Grob D: Poisoning due to organophosphate insecticides. Am J Med 50:475–492, 1971
6. IARC Monographs on the Evaluation of the Carcinogenic Risk of Chemicals to Humans: Miscellaneous Pesticides, pp 131–152. Lyon, International Agency for Research on Cancer, 1983
7. National Cancer Institute: Bioassay of Methyl Parathion for Possible Carcinogenicity, TR-157. DHEW (NIH) Pub No 79–1713. Washington, DC, US Government Printing Office, 1979
8. Coye MJ, Lowe JA, Maddy KT: Biological monitoring of agricultural workers exposed to pesticides. I. Cholinesterase activity determination. J Occup Med 28:619–627, 1986
9. Michel HP: Electrometric method for determination of red blood cell and plasma cholinesterase activity. J Lab Clin Med 34:1564–1568, 1949
10. Milby TH: Prevention and management of organophosphate poisoning. JAMA 216:2131–2133, 1971

METHYL PROPYL KETONE

CAS: 107-87-9

$CH_3COC_3H_7$ 1987 TLV = 200 ppm

Synonyms: 2–Pentanone; ethyl acetone

Physical Form. Colorless liquid

Uses. Solvent

Exposure. Inhalation

Toxicology. Methyl propyl ketone is an irritant of the eyes and mucous membranes; at high concentrations, it causes narcosis in animals, and it is expected that severe exposure will produce the same effect in humans.

Brief exposures of humans to 2000 to 4000 ppm were very irritating; 1500 ppm had a strong odor and caused irritation of the eyes and nose.[1] There have been no reports of chronic or systemic effects in humans. In guinea pigs, exposure to 50,000 ppm for 50 minutes or 13,000 ppm for 300 minutes was fatal.[2] Animals survived 810 minutes at 5000 ppm, but narcosis occurred in 460 to 710 minutes.[2] Applied to the skin of rabbits, the undiluted liquid was only slightly irritating within 24 hours.[3]

REFERENCES

1. Henson EV: Toxicology of some aliphatic ketones. J Occup Med 1:607–613, 1959
2. Yant WP et al: Acute response of guinea pigs to vapors of some new commercial organic compounds. Public Health Rep 51:392–399, 1936
3. National Institute for Occupational Safety and Health: Criteria for a Recommended Standard . . . Occupational Exposure to Ketones. DHEW (NIOSH) Pub No 78–173, p 244. Washington, DC, US Government Printing Office, 1978

N-METHYL–2–PYRROLIDONE

CAS: 872-50-4

C_5H_9NO 1987 TLV = none established

Synonyms: NMP; M-Pyrol; methylpyrrolidone

Physical Form. Almost colorless liquid with a mild amine-like odor

Uses. Chemical reaction medium; solvent for high-temperature resins, petrochemicals, and industrial gases; water-immiscible formulating agent for coating, stripping, or cleaning compounds

Exposure. Inhalation

Toxicology. N-Methyl–2–pyrrolidone (NMP), a heterocyclic amide solvent, may produce mild skin irritation and severe eye irritation on contact; it is not known to produce allergic contact dermatitis.

NMP produced no skin irritation on patch testing for 24 hours in 50 volunteers.[1] Some mild transient skin irritation has occurred after repeated and prolonged contact. There is no evidence for contact sensitization.[1] The very low vapor pressure at room temperature reduces the likelihood of significant inhalation exposures.

In rats, the oral LD_{50} was approximately 4.2 ml/kg. In rabbits, the dermal LD_{50} was between 4 and 8 gm/kg.[1] Repeated skin application in lower doses, 0.4 and 0.8 ml/kg/day, resulted in mild skin irritation in rabbits. NMP is a severe eye irritant in rabbits, producing conjunctivitis and corneal opacity after instillation, but did not appear to produce permanent eye damage. Rats exposed to vapor from NMP heated for 6 hours or saturated room temperature air for 6 hours/day for 10 days showed no evidence of toxic effects.[1]

Subacute 90-day feeding studies in rats, mice, and beagle dogs demonstrated no apparent clinically significant toxic effects in the treated animals; several minor statistically significant changes in laboratory parameters, such as GGTP and platelet counts, were noted at high doses in some treated groups but not consistently in all three studies.[1]

Dermal application studies in female rats showed no evidence of teratogenic effects, although lower weight gains in the maternal animals and skeletal variations in the offspring were observed at the highest dose (750 mg/kg/day); the latter effect was thought to be due to maternal toxicity.[1] No mutagenic activity was observed in the Ames test using several strains of *Salmonella typhimurium*.[1]

REFERENCE

1. *N*-Methylpyrrolidone—Summary of Toxicity Information. Wayne, New Jersey, GAF Corporation, 1983

α-METHYL STYRENE

CAS: 98-83-9

$C_6H_5C(CH_3){=}CH_2$ 1987 TLV = 50 ppm

Synonyms: 1–Methyl–1–phenyl ethylene; isopropenylbenzene; β-phenylpropylene

Physical Form. Colorless liquid

Uses. In modified polyester and alkyd resin formulations; plasticizers

Exposure. Inhalation

Toxicology. α-Methyl styrene is an irritant of the eyes and mucous membranes; severe exposure may result in central nervous system depression.

Humans briefly exposed to 600 ppm experienced strong eye and nasal irritation; at 200 ppm, the odor was objectionable, whereas at 100 ppm, the odor was strong but tolerated without excessive discomfort.[1]

Guinea pigs and rats exposed to 3000 ppm 7 hours/day for 3 to 4 days died; at 800 ppm for 27 days, there were slight changes in liver and kidney weight and some reduction in growth.[1] Exposure to 200 ppm 7 hours/day for 139 days caused no adverse effects in several species.[1]

The liquid dropped in the eyes of rabbits caused slight conjunctival irritation; applied to rabbit skin, it produced erythema.[1]

The odor of α-methyl styrene is detectable at 50 ppm; the odor and irritant properties provide good warning of toxic levels.[1]

REFERENCE

1. Wolf MA et al: Toxicological studies of certain alkylated benzenes and benzene. Arch Ind Health 14:387–398, 1956

MEVINPHOS
CAS: 7786-34-7
$(CH_3O)_2PO_2(CH_3)C{=}CHC(O)OCH_3$
1987 TLV = 0.01 ppm; skin

Synonyms: Methyl 3-hydroxy-α-crotonate, dimethyl phosphate

Physical Form. Light yellow to orange liquid

Uses. Insecticide

Exposure. Inhalation; skin absorption; ingestion

Toxicology. Mevinphos is an anticholinesterase agent.

Signs and symptoms of overexposure are caused by the inactivation of the enzyme cholinesterase, which results in the accumulation of acetylcholine at synapses in the nervous system, skeletal and smooth muscle, and secretory glands. The sequence of the development of systemic effects varies with the route of entry. The onset of signs and symptoms is usually prompt but may be delayed up to 12 hours.[1-3] After inhalation, respiratory and ocular effects are the first to appear, often within a few minutes after exposure. Respiratory effects include tightness in the chest and wheezing due to bronchoconstriction and excessive bronchial secretion. Laryngeal spasms and excessive salivation may add to the respiratory distress; cyanosis may also occur. Ocular effects include miosis, blurring of distant vision, tearing, rhinorrhea, and frontal headache.

After ingestion, gastrointestinal effects such as anorexia, nausea, vomiting, abdominal cramps, and diarrhea appear within 15 minutes to 2 hours. After skin absorption, localized sweating and muscular fasciculations in the immediate area usually occur within 15 minutes to 4 hours; skin absorption is somewhat greater at higher ambient temperatures and is increased by the presence of dermatitis.[1-3]

With severe intoxication by all routes, an excess of acetylcholine at the neuromuscular junctions of skeletal muscle causes weakness aggravated by exertion, involuntary twitchings, fasciculations, and eventually paralysis. The most serious consequence is paralysis of the respiratory muscles. Effects on the central nervous system include giddiness, confusion, ataxia, slurred speech, Cheyne–Stokes respiration, convulsions, coma, and loss of reflexes. The blood pressure may fall to low levels, and cardiac irregularities, including complete heart block, may occur. Complete symptomatic recovery usually occurs within 1 week; increased susceptibility to the effects of anticholinesterase agents persists for up to several weeks following exposure. Daily exposure to concentrations that are insufficient to produce symptoms following a single exposure may result in the onset of symptoms. Continued daily exposure may be followed by increasingly severe effects.

A group of 31 farm workers who inadvertently entered a field only 2 hours after it was sprayed with mevinphos developed a variety of initial symptoms, including eye irritation, headache, visual disturbances, dizziness, nausea, vomiting, chest pain, shortness of breath, pruritis, eyelid and arm fasciculations, excessive sweating, and diarrhea.[4] Headache, dizziness, visual disturbances, and nausea persisted for 5 to 8 weeks or more in a significant number of field workers following cessation of exposure. Despite symptoms suggesting moderate organophosphate intoxication, mean plasma and red blood cell cholinesterase depression was only 16%

and 6% respectively, when compared against a presumed baseline obtained in these workers long after the exposure.[4] Another study of 16 cauliflower workers poisoned by residues of mevinphos and phosphamidon (a less potent organophosphate) demonstrated persistent headaches, blurred vision, and weakness in a number of workers 5 to 9 weeks or more following the exposure.[5]

In two cases of moderate intoxication from mevinphos, urinary excretion of dimethylphosphate (a metabolite of mevinphos) was almost complete 50 hours after exposure.[6] Although a number of other organophosphorus pesticides also yield dimethyl phosphate, the presence of significant amounts of this metabolite in the urine may be useful in estimating the absorption of mevinphos.

Mevinphos inactivates cholinesterase by phosphorylation of the active site of the enzyme to form the "dimethylphosphoryl enzyme." Over the following 24 to 48 hours, there is a process, termed *aging*, of conversion to the "monomethylphosphoryl enzyme." Aging is of clinical interest in the treatment of poisoning because cholinesterase reactivators such as pralidoxime (2–PAM, Protopam) chloride are ineffective after aging has occurred.

Diagnosis. *Signs and Symptoms:* Initial symptoms include headache, blurred vision, pallor, weakness, sweating, abdominal pain, nausea, vomiting, and diarrhea.

Moderate-to-severe intoxication includes miosis, lacrimation, excessive salivation, muscle fasciculations, dyspnea, cyanosis, convulsions, shock, cardiac arrhythmias, and coma.

Differential Diagnosis: Diagnosis is based primarily on a history of exposure and clinical evidence of diffuse parasympathetic stimulation. Careful observation of the effect of atropine and pralidoxime may be valuable. Patients with organophosphate poisoning are resistant to the action of atropine at moderate dosages; failure of 1 to 2 mg of atropine administered parenterally to produce signs of atropinization (flushing, mydriasis, tachycardia, or dryness of mouth) indicates anticholinesterase poisoning. Intravenous injection of 1 g pralidoxime generally causes some recovery from signs and symptoms.

Special Tests: Two types of cholinesterase are clinically significant: (1) true acetylcholinesterase, found principally in the nervous system and the red blood cell; and (2) pseudo- or butyrylcholinesterase, found in the plasma, liver, and the nervous system. Whereas the action of both types is inhibited by organophosphates, the level of depression of red blood cell cholinesterase is a better indicator of clinically significant reduction of cholinesterase activity in the nervous system.

Laboratory evidence of depression of red blood cell cholinesterase to a level substantially below pre-exposure levels (at least 50% and usually much lower) is verification of poisoning. There is an imperfect correlation between the degree of depression of cholinesterase enzymes and the occurrence of symptoms. With a rapid drop in cholinesterase activity, generally reflecting an acute heavy exposure, there may be symptoms with only a 30% depression, whereas with slower drops to 70% depression, reflecting chronic low-level exposure, there may be no symptoms.[7]

If no pre-exposure baseline has been performed but symptoms are not sufficient to justify treatment with atropine, repeated testing during the recovery period demonstrating progressively increasing plasma and red blood cell cholinesterase levels over several days and weeks, respectively, suggests the diagnosis of anticholinesterase poisoning.

There are many different methods for estimation of cholinesterase content of blood, and associated with each method is a different set of normal values and a different set of reporting units. The laboratory report of a cholinesterase determination should state the units involved, along with the appropriate normal range. Based on the Michel method, the normal range of red blood cell cholinesterase activity (delta *p*H per hour) is 0.39 to 1.02 for men, and 0.34 to 1.10 for women.[8] The normal range of the enzyme activity (delta *p*H per hour) of plasma is 0.44 to 1.63 for men, and 0.24 to 1.54 for women.

Treatment. Treatment of organophosphate poisoning ranges from simple removal from exposure in very mild cases to the provision of rigorous supportive and antidotal measures in severe cases.[2,3,9,10] In moderate-to-severe cases, because of pulmonary involvement, there may be need for artificial respiration using a positive-pressure method. Careful attention must be paid to removal of secretions and to maintenance of a patent airway. Anticonvulsants such as diazepam or thiopental sodium may be necessary. Maintenance of respiration is critical, because death usually results from weakness of the muscles of respiration and accumulation of excessive secretions in the respiratory tract.[2]

As soon as cyanosis has been overcome, 2 to 4 mg of atropine should be given intravenously. (Atropine may induce ventricular fibrillation in the presence of cyanosis.) *This dose of atropine is approximately 10 times the amount that is administered for other conditions in which atropine is considered therapeutic.* This dose should be repeated at 5- to 10-minute intervals until signs of atropinization appear (dry, flushed skin, tachycardia as high as 140 beats/minute, and pupillary dilatation). If muscarinic symptoms reappear, the dose of atropine should be repeated. A mild degree of atropinization should be maintained for at least 48 hours.[2]

Pralidoxime (2–PAM, Protopam) chloride is a cholinesterase reactivator that complements the action of atropine. It has its greatest effect in reversing the nicotinic action of anticholinesterase agents at skeletal neuromuscular junctions but virtually no effect on central nervous system manifestations. In moderate-to-severe cases, the dose for adults is 1 to 2 g injected intravenously at a rate not in excess of 500 mg/minute. If muscle weakness has not been relieved or if it recurs within 20 minutes, a second dose of 1 g is indicated. Treatment with pralidoxime chloride will be most effective if given within 24 hours after poisoning.[2] Morphine, aminophylline, and phenothiazines are contraindicated because of documented experience of adverse reactions in cases of organophosphate poisoning.[10]

It is of great importance to decontaminate the patient. Contaminated clothing should be removed at once, and the skin should be washed with generous amounts of soap or detergent and a flood of water, which is best accomplished under a shower or by submersion in a pond or other body of water if the exposure occurred in the field. Careful attention should be paid to cleansing of the skin and hair.

The patient should be attended to and monitored continuously for not less than 24 hours, because serious and sometimes fatal relapses have occurred as a result of continuing absorption of the toxin or dissipation of the effects of the antidote.

Regeneration of cholinesterase is primarily by synthesis of new enzyme and takes place at the rate of approximately 1% per day.[10] A patient who has recovered from the acute phase of poisoning remains hypersusceptible to anticholinesterases for up to several weeks.

Medical Control. Medical control involves preplacement and annual physical examination with determination of pre-exposure red blood cell and plasma cholinesterase activity. A person whose red blood cell cholinesterase falls to or below 40% of the pre-exposure baseline should be removed from further exposure until the activity returns to within 80% of the pre-exposure baseline.[9]

REFERENCES

1. Koelle GB (ed): Cholinesterases and Anticholinesterase Agents. Handbuch der Experimentellen Pharmakologie, Vol 15, pp 989–1027. Berlin, Springer-Verlag, 1963
2. Taylor P: Anticholinesterase agents. In Gilman AG et al (eds): Goodman and Gilman's The Pharmacological Basis of Therapeutics, 7th ed, pp 110–129. New York, Macmillan Publishing Co, 1985
3. Hayes WJ Jr: Organic phosphorus pesticides. Pesticides Studied in Man, pp 284–435. Baltimore, Williams & Wilkins, 1982
4. Coye MJ et al: Clinical confirmation of organophosphate poisoning of agricultural workers. Am J Ind Med 10:399–409, 1986
5. Midtling JE, Barnett PG, Coye MJ: Clinical management of field worker organophosphate poisoning. West J Med 142:514–518, 1985
6. Holmes JHG, Starr HG Jr, Hanisch RC, von Kaulla KN: Short-term toxicity of mevinphos in man. Arch Environ Health 29:84–89, 1974
7. Coye MJ, Lowe JA, Maddy KT: Biological monitoring of agricultural workers exposed to pesticides. I. Cholinesterase activity determinations. J Occup Med 28:619–627, 1986
8. Michel HO: Electrometric method for determination of red blood cell and plasma cholinesterase activity. J Lab Clin Med 34:1564–1568, 1949
9. Namba T, Nolte CT, Jackrel J, Grob D: Poisoning due to organophosphate insecticides. Am J Med 50:475–492, 1971
10. Milby TH: Prevention and management of organophosphate poisoning. JAMA 216:2131–2133, 1971

MICA

CAS: 12001-26-2

$K_2Al_4(Al_2Si_6O_{20})(OH)_4$—Muscovite

Containing <1% quartz

1987 TLV = 3 mg/m^3 respirable dust

Synonyms: Mica is a nonfibrous silicate occurring in plate form and includes nine different species; muscovite is a hydrated aluminum potassium silicate also called white mica; phlogopite is an aluminum potassium magnesium silicate also called amber mica; other forms include biotite, lepidolite, zinnwaldite, and roscoelite.

Physical Form. Light grey- to dark-colored flakes or particles

Uses. Insulation in electrical equipment; manufacture of roofing shingles and wallpaper; in oil refining; in rubber manufacture

Exposure. Inhalation

Toxicology. Mica dust causes pneumoconiosis.

In a study of 57 workers exposed to mica dust, five of six workers exposed to concentrations in excess of 25 mppcf for more than 10 years had pneumoconiosis.[1] The most characteristic finding by chest x-ray was fine granulation of uneven density; there was a tendency, in some cases, to a coalescence of shadows. The symptoms most frequently reported were chronic cough and dyspnea; complaints of weakness and weight loss were less frequent.[1] Only one of six workers exposed at concentrations in excess of 25 mppcf for more than 10 years failed to show evidence of pneumoconiosis.

A group of mica miners were said to show a higher incidence of pneumoconiosis than were miners of other minerals, but some quartz was present in the dust to which they were all exposed.[2]

REFERENCES

1. Dreessen WC et al: Pneumoconiosis among mica and pegmatite workers. Public Health Bull 250:1–74, 1940
2. Vestal TF, Winstead JA, Joliet PV: Pneumoconiosis among mica and pegmatite workers. Ind Med 12:11–14, 1943

MOLYBDENUM (soluble compounds)

CAS: 7439-98-7

Mo 1987 TLV = 5 mg/m^3 soluble compounds

Synonyms: Soluble compounds include ammonium molybdate and sodium molybdate.

Physical Form. Silver white metallic element of a grey-blackish powder

Uses. Manufacture of special purpose steel; catalyst; additive to pigments; additive to fertilizers

Exposure. Inhalation

Toxicology. Molybdenum and its compounds are of low toxicity.

Human molybdenum intoxications are rare.[1]

In livestock, chronic molybdenum poisoning known as "teart disease" is caused by a diet high in molybdenum and low in copper.[2] Symptoms include anemia, gastrointestinal disturbances, bone disorders, and growth retardation.

The mechanism of molybdenum toxicity is not fully elucidated.[2] It is assumed that the primary factor is the formation of a copper–tetrathiomolybdate complex that reduces the biological utility of copper.[2]

REFERENCES

1. Lener J, Bibr B: Effects of molybdenum on the organisms (a review). J Hyg Epidemiol Microbiol Immunol 28:405–418, 1984
2. Friberg L, Lener J: Molybdenum. In Friberg L et al (eds): Handbook on the Toxicology of Metals, Vol II. Specific Metals, pp 445–461. Amsterdam, Elsevier, 1986

MOLYBDENUM (insoluble compounds)

CAS: 7439-98-7

Mo 1987 TLV = 10 mg/m^3

Synonyms: Calcium molybdate; molybdenum trioxide; molybdenum halides; molybdenum disulfide

Physical Form. Powder

Uses. Lubricant; hydrogenation catalyst; reagent; decorating ceramics; detecting and determining some inorganics; manufacture of pigments; corrosion inhibitor

Toxicology. Insoluble compounds of molybdenum have a low toxicity; however, molybdenum trioxide is an irritant of the eyes and mucous membranes.[1]

There have been no reports of effects from industrial exposure except for a report from the USSR of an increased incidence of nonspecific symptoms, including weakness, fatigue, anorexia, headaches, joint, and muscle pains, among mining and metallurgy workers exposed to 60 to 600 mg/m^3 molybdenum.[2] Signs of gout and elevated uric acid concentrations have been observed among inhabitants of areas of Armenia where the soil is rich in molybdenum. This effect apparently results from the induction of the enzyme, xanthine oxidase, for which molybdenum is a cofactor.

Guinea pigs exposed to molybdenum trioxide dust at a concentration of 200 mg molybdenum/m^3 for 1 hour daily for 5 days developed nasal irritation, diarrhea, weight loss, and incoordination.[3] Animals receiving daily oral doses of up to 500 mg molybdenum/day in the form of molybdenum trioxide or calcium molybdate showed anorexia, listlessness, and weight loss.

The metabolism of molybdenum is closely associated with that of copper; molybdenum toxicity in animals can be alleviated by the administration of copper.[4] High intake of molybdenum in rats resulted in a substantial reduction in activity of sulfide oxidase in the liver.[5] The reduced activity of this enzyme leads to accumulation of sulfide in the tissues and subsequent formation of highly undissociated copper sulfide, thus removing copper from metabolic activity. This is a possible explanation for the induction of copper deficiency by molybdate.

REFERENCES

1. Stokinger HE: The metals. In Clayton GD, Clayton FE (eds): Patty's Industrial Hygiene and Toxicology, 3rd ed, Vol 2, Toxicology, pp 1807–1819. New York, Wiley–Interscience, 1981
2. Lener J, Bibr B: Effects of molybdenum on the organism (a review). J Hyg Epidemiol Microbiol Immunol 29:405–419, 1984
3. Browning E: Toxicity of Industrial Materials, 2nd ed, pp 243–248. London, Butterworths, 1969
4. Molybdenum. Documentation of the TLVs and BEIs, 3rd ed, 3rd printing, pp 173–174. Cincinnati, American Conference of Governmental Industrial Hygienists (ACGIH), 1976
5. Halverson AW, Phifer JH, Monty KJ: A mechanism for the copper-molybdenum interrelationship. J Nutr 71:79–100, 1960

MONOMETHYLANILINE

CAS: 100-61-8

$C_6H_5NHCH_3$ 1987 TLV = 0.5 ppm; skin

Synonym: N-Methylaniline

Physical Form. Colorless or slightly yellow liquid that becomes brown on exposure to air

Uses. Chemical syntheses

Exposure. Inhalation; skin absorption

Toxicology. Monomethylaniline causes anoxia in animals due to the formation of methemoglobin.

Although there are no reports of human intoxication from exposure to monomethylaniline, the effects of methemoglobinemia include cyanosis (especially in the lips, nose, and earlobes), weakness, dizziness, and severe headache.

Animal fatalities occurred from daily exposure to 7.6 ppm; signs of intoxication included prostration, labored breathing, and cyanosis. Methemoglobinemia developed promptly in rabbits and cats; the rabbits also exhibited mild anemia and bone marrow hyperplasia.[1] Animals that died had pulmonary involvement ranging from edema to interstitial pneumonia, as well as occasional centrolobular hepatic necrosis, and moderate kidney damage.[1] The liquid readily caused poisoning in animals by absorption through the skin.[2]

Monomethylaniline (1.95 g/kg of food) given together with sodium nitrite (1.0 g/liter of drinking water) to Swiss mice resulted in a 17% incidence of lung adenomas and a 14% incidence of malignant lymphomas; there were no carcinogenic effects in animals treated with monomethylaniline alone, suggesting that in vivo nitrosation is necessary for forming carcinogenic nitrosamines.[3]

Diagnosis. *Signs and Symptoms:* Signs and symptoms include headache, signs of anoxia, including cyanosis of lips, nose, and earlobes; and anemia.

Differential Diagnosis: Other causes of cyanosis must be differentiated from methemoglobinemia due to chemical exposure. These include hypoxia from lung disease, hypoventilation, and decreased cardiac output. Lung disease may be suspected from results of pulmonary function tests and arterial blood gas analysis. The arterial Po_2 may be normal in methemoglobinemia but tends to be decreased in cyanosis due to lung disease. Hypoventilation will cause elevation of arterial Pco_2, which is not seen in chemical exposure. Decreased cardiac output states will cause cyanosis only when accompanied by arterial hypotension. If blood withrawn from the vein shows the characteristic chocolate-brown coloration, the diagnosis of an abnormal pigment is almost certain, especially if the color remains after shaking the blood in air.[4]

Special Tests: Special tests include examination of urine for blood; determination of methemoglobin concentration in the blood when chemical intoxication is suspected and at regular intervals until the methemoglobin has been fully reduced to normal hemoglobin.[5] Methemoglobin may be differentiated from sulfhemoglobin by the addition of a few drops of 10% potassium cyanide, which results in the rapid production of bright red cyanomethemoglobin but has no effect on the color of sulfhemoglobin.[4] Spectrophotometry is required for the precise identification of the pigment and its quantitation. Normal acid methemoglobin has a characteristic absorption spectrum with peaks at 502 and 632 nm, which disappear with the addition of cyanide, whereas sulfhemoglobin has a peak at 620 nm, which does not disappear with cyanide.[4]

Treatment. All the contaminant on the body must be removed. Immediately remove all clothing, and wash the entire body from head to foot with soap and water. Pay special attention to the hair and scalp, finger and toenails, nostrils, and ear canals. Administer oxygen to alleviate the headache and general sense of weakness; confine the patient to bed. Determine the methemoglobin concentration in the blood, and repeat every 3 to 6 hours for 18 to 24 hours. Repeat skin cleansing if the methemoglobin concentration appears to rise after 3 to 4 hours. In general, patients will return to normal within 24 hours provided that all sources of further absorption are completely eliminated. The methemoglobin will be reduced spontaneously to ferrous hemoglobin in 2 to 3 days.[4]

Such therapy is not effective in subjects with glucose–6–dehydrogenase deficiency.[4]

The only justifiable use of methylene blue would be in cases of coma or stupor, usually at methemoglobin levels over 60%. In those patients in whom therapy is necessary, methylene blue may be given intravenously 1 to 2 mg/kg of a 1% solution in saline over a 10-minute period. If cyanosis has not disappeared within an hour, a second dose of 2 mg/kg may be administered.[4,5] The total dose should not exceed 7 mg/kg, because methylene blue may cause toxic effects such as dyspnea, precordial pain, restlessness, apprehension, red cell hemolysis, and changes in the electrocardiogram (reduction in the height or even reversal of the T wave, frequently with lowering of the R wave).[4]

REFERENCES

1. Treon JF et al: The toxic properties of xylidine and monomethylaniline. AMA Arch Ind Hyg Occup Med 1:506–524, 1950
2. N-Methyl aniline. Documentation of the TLVs and BEIs, 5th ed, p 375. Cincinnati, American Conference of Governmental Industrial Hygienists (ACGIH), 1986
3. Greenblatt M, Mirvish S, So BT: Nitrosamine studies: Induction of lung adenomas by concurrent administration of sodium nitrite and secondary amines in Swiss mice. J Natl Cancer Inst 46:1029–1034, 1971
4. Rieder RF: Methemoglobinemia and sulfmethemoglobinemia. In Wyngaarden JB, Smith LH (eds): Cecil Textbook of Medicines, 16th ed, p 896. Philadelphia, WB Saunders, 1982
5. Mangelsdorff AF: Treatment of methemoglobinemia. AMA Arch Ind Health 14:148–153, 1956

MORPHOLINE

CAS: 110-91-8

C_4H_9NO — 1987 TLV = 20 ppm; skin

Synonyms: Diethyleneimide oxide; diethylene imidoxide

Physical Form. Mobile hygroscopic liquid

Uses. Solvent for resins, waxes, casein, dyes; morpholine compounds used as corrosion inhibitors, insecticides, antiseptics

Exposure. Inhalation; skin absorption

Toxicology. Morpholine vapor is an irritant of eyes, nose and throat.

In industry, some instances of skin and respiratory tract irritation have been observed, but no chronic effects have been reported.[1] A human exposure to 12,000 ppm for 1.5 minutes in a laboratory produced nose irritation and cough; mouth pipetting of the liquid caused a severe sore throat and reddened mucous membranes.[2] Workers exposed for several hours to low vapor concentrations complained of foggy vision with rings around lights, the results of corneal edema, which cleared within 3 to 4 hours after cessation of exposure.[3]

Repeated daily exposure of rats to 18,000 ppm for 8 hours was lethal to some animals; those dying had damage to lungs, liver, and kidneys.[2] Rats and guinea pigs survived an 8-hour exposure at 12,000 ppm. Sublethal signs from inhalation include lacrimation, rhinitis, and inactivity.[1] Oral doses of undiluted unneutralized morpholine caused irritation of the intestinal tract with hemorrhage. Applied to the skin of rabbits, it caused skin burns and systemic injury; the LD_{50} was 0.5 ml/kg. The liquid dropped in the eye of a rabbit caused moderate injury with ulceration of the conjunctiva and corneal clouding.[3]

Rats given 10 g morpholine/kg in the diet plus 0.2% sodium nitrite in the drinking water had a significantly increased incidence of liver tumors compared with controls.[4] The carcinogenic response is attributed to the *in vitro* production of N-nitrosomorpholine. In another study, morpholine alone produced a low number of tumors, and it was suggested that an unknown nitrate source reacted with the morpholine to form the carcinogenic N-nitrosomorpholine.[5]

REFERENCES

1. Reinhardt CF, Brittelli MR: Heterocyclic and miscellaneous nitrogen compounds. In Clayton GD, Clayton FE (eds): Patty's Industrial Hygiene and Toxicology, 3rd ed, rev, Vol 2A, Toxicology, pp 2693–2696. New York, Wiley–Interscience, 1981
2. Morpholine. Documentation of the TLVs and BEIs, 5th ed, p 417. Cincinnati, American Conference of Governmental Industrial Hygienists (ACGIH), 1986

3. Grant WM: Toxicology of the Eye, 3rd ed, p 642. Springfield, Illinois, Charles C Thomas, 1986
4. Mirvish SS et al: Liver and forestomach tumors and other forestomach lesions in rats treated with morpholine and sodium nitrite, with and without sodium ascorbate. J Natl Cancer Inst 71:81–84, 1983
5. Shank RC, Newberne PM: Dose–response study of the carcinogenicity of dietary sodium nitrite and morpholine in rats and hamsters. Fd Cosmet Toxicol 14:1–8, 1976

NALED

CAS: 300-76-5

$(CH_3O)_2P(O)PCHBrCBrCl_2$

1987 TLV = 3 mg/m^3; skin

Synonyms: DIBROM: 1,2–dibromo–2,2–dichloroethyl dimethyl phosphate

Physical Form. Light straw-colored liquid with slightly pungent odor

Uses. Acaricide; insecticide

Exposure. Inhalation; skin absorption; ingestion

Toxicology. Naled is an anticholinesterase agent.

Signs and symptoms of overexposure are caused by the inactivation of the enzyme cholinesterase, which results in the accumulation of acetylcholine in the nervous system, skeletal and smooth muscle, and secretory glands.[1–3] The sequence of the development of systemic effects varies with the route of entry. The onset of signs and symptoms is usually prompt but may be delayed up to 12 hours. After inhalation of the vapor, respiratory and ocular effects are the first to appear, often within a few minutes of exposure. Respiratory effects include tightness in the chest and wheezing due to bronchoconstriction and excessive bronchial secretion; laryngeal spasm and excessive salivation may add to the respiratory distress; cyanosis may also occur. Ocular effects include miosis, blurring of distant vision (due to spasm of accommodation), tearing, rhinorrhea, and frontal headache.

After ingestion of the liquid, gastrointestinal effects such as anorexia, nausea, vomiting, abdominal cramps, and diarrhea appear within 15 minutes to 2 hours. After skin absorption of the liquid, localized sweating and muscular fasciculations in the immediate area usually occur within 15 minutes to 4 hours; skin absorption is somewhat greater at higher ambient temperatures and is enhanced by the presence of dermatitis.

With severe intoxication, an excess of acetylcholine at the neuromuscular junctions of skeletal muscle causes weakness aggravated by exertion, involuntary twitchings, fasciculations, and, eventually, paralysis. The most serious consequence is paralysis of the respiratory muscles. Effects on the central nervous system include giddiness, confusion, ataxia, slurred speech, Cheyne–Stokes respiration, convulsions, coma, and loss of reflexes. The blood pressure may fall to low levels, and cardiac irregularities, including complete heart block, may occur. Complete symptomatic recovery usually occurs within a week; increased susceptibility to the effects of anticholinesterase agents persists for up to several weeks after exposure. Daily exposure to concentrations that are insufficient to produce symptoms following a single exposure may result in the onset of symptoms. Continued daily exposure may be followed by increasingly severe effects.

Dermatitis occurred on the arms, face, neck, and abdomen of 9 of 12 persons working in a field of flowers that had been freshly sprayed with a solution of Dibrom; three of four workers who were patch tested were positive to a 60% solution of Dibrom in xylene and were negative to xylene alone.[4] The liquid in the eye may be expected to cause injury.

Naled inactivates cholinesterase by phosphorylation of the active site of the enzyme to form the "dimethylphosphoryl enzyme." Over the following 24 to 48 hours, there is a process, termed *aging*, of conversion to the "monomethylphosphoryl enzyme." Aging is of clinical interest in the treatment of poisoning because cholinesterase reactivators such as pralidoxime (2–PAM, Protopam) chloride are ineffective after aging has occurred.

Diagnosis. *Signs and Symptoms:* Initial symptoms include headache, blurred vision, pallor, weakness, sweating, abdominal pain, nausea, vomiting, and diarrhea.

Moderate-to-severe intoxication includes miosis, lacrimation, excessive salivation, muscle fasciculations, dyspnea, cyanosis, convulsions, shock, cardiac arrhythmias, and coma.

Differential Diagnosis: Diagnosis is based primarily on a history of exposure and clinical evidence of diffuse parasympathetic stimulation. Careful observation of the effects of atropine and pralidoxime may be valuable. Patients with organophosphate poisoning are resistant to the action of atropine at moderate dosages; failure of 1 to 2 mg of atropine administered parenterally to produce signs of atropinization (flushing, mydriasis, tachycardia, or dryness of mouth) indicates organophosphate poisoning. Intravenous injection of 1 g pralidoxime generally causes some recovery from signs and symptoms.

Special Tests: Two types of cholinesterase are clinically significant: (1) true acetylcholinesterase, found principally in the nervous system and the red blood cell; and (2) pseudo- or butyrylcholinesterase, found in the plasma, liver, and nervous system. Whereas the action of both types is inhibited by organophosphates, the level of depression of red blood cell cholinesterase is a better indicator of clinically significant reduction of cholinesterase activity in the nervous system.

Laboratory evidence of depression of red blood cell cholinesterase to a level substantially below pre-exposure levels (at least 50% and usually much lower) is verification of organophosphate poisoning. There is an imperfect correlation between the degree of depression of cholinesterase enzymes and the occurrence of symptoms. With a rapid drop in cholinesterase activity, generally reflecting an acute heavy exposure, there may be symptoms with only a 30% depression, whereas with slower drops to 70% depression, reflecting chronic low level exposure, there may be no symptoms.[5]

If no pre-exposure baseline has been performed but symptoms are not sufficient to justify treatment with atropine, repeated testing during the recovery period demonstrating progressively increasing plasma and red blood cell cholinesterase levels over several days and weeks, respectively, suggests the diagnosis of anticholinesterase poisoning.

There are many different methods for estimation of cholinesterase content of blood, and, associated with each method is a different set of normal values and a different set of reporting units. The laboratory report of a cholinesterase determination should state the units involved, along with the appropriate normal range. Based on the Michel method, the normal range of red blood cell cholinesterase activity (delta *p*H per hour) is 0.39 to 1.02 for men, and 0.34 to 10.10 for women.[6] The normal range of the enzyme activity (delta *p*H per hour) of plasma is 0.44 to 1.63 for men, and 0.24 to 1.54 for women.

Treatment. Treatment of organophosphate poisoning ranges from simple removal from exposure in very mild cases to the provision of very rigorous supportive and antidotal measures in severe cases.[2,3,5,8] In moderate-to-severe cases, because of pulmonary involvement, there may be need for artificial respiration using a positive-pressure method. Careful attention must be paid to removal of secretions and to maintenance of a patent airway. Anticonvulsants such as thiopental sodium may be necessary. Maintenance of respiration is critical, because death usually results from weakness of the muscles of respiration and accumulation of excessive secretions in the respiratory tract.

As soon as cyanosis has been overcome, 2 to 4 mg of atropine should be given intravenously. (Atropine may induce ventricular fibrillation in the presence of cyanosis.) *This dose of atropine is approximately 10 times the amount that is administered for other conditions in which atropine is considered therapeutic.* This dose should be repeated at 5- to 10-minute intervals until signs of atropinization appear (dry, flushed skin, tachycardia as high as 140 beats/minute, and pupillary dilatation). A mild degree of atropinization should be maintained for at least 48 hours.[2]

Pralidoxime (2–PAM, Protopam) chloride is a cholinesterase reactivator that complements the action of atropine. It has its greatest effect in reversing the nicotinic action of anticholinesterase agents at skeletal neuromuscular junctions but virtually no effect on central nervous system manifestations. In moderate-to-severe cases, the dose for adults is 1 to 2 g injected intravenously at a rate not in excess of 500 mg/minute. After an hour, a second dose of 1 g is indicated if muscle weakness has not been relieved. Treatment with pralidoxime chloride will be most effective if given within 24 hours after poisoning.[2] Morphine, aminophylline, and phenothiazines are contraindicated because of documented experience of adverse reactions in cases of organophosphate poisoning.[2]

It is of great importance to decontaminate the patient. Contaminated clothing should be removed at once, and the skin should be washed with generous amounts of soap or detergent and a flood of water, which is best accomplished under a shower or by submersion in a pond or other body of water if the exposure occurred in the field. Careful attention should be paid to cleansing of the skin and hair.

The patient should be attended to and monitored continuously for not less than 24 hours, because serious and sometimes fatal relapses have occurred as a result of continuing absorption of the toxin or dissipation of the effects of the antidote.

Regeneration of cholinesterase is primarily by synthesis of new enzyme and takes place at the rate of approximately 1% per day.[8] A patient who has recovered from the acute phase of poisoning remains hypersusceptible to anticholinesterases for up to several weeks.

Medical Control. Medical control involves preplacement and annual physical examination with determination of pre-exposure red blood cell cholinesterase activity. A person whose red blood cell cholinesterase falls to or below 40% of the pre-exposure baseline should be removed from further exposure until the activity returns to within 80% of the pre-exposure baseline.

REFERENCES

1. Koelle GB (ed): Cholinesterases and anticholinesterase agents. Handbuch der Experimentellen Pharmakologie, Vol 15, pp 989–1027. Berlin, Springer-Verlag, 1963
2. Taylor P: Anticholinesterase agents. In Gilman AG et al (eds): Goodman and Gilman's The Pharmacological Basis of Therapeutics, 7th ed, pp 110–129. New York, Macmillan, 1985
3. Hayes WJ Jr: Pesticides Studied in Man, pp 312–313. Baltimore, Williams & Wilkins, 1982
4. Edmundson WF, Davies JE: Occupational dermatitis from naled—a clinical report. Arch Environ Health 15:89–91, 1967
5. Coye MJ, Lowe JA, Maddy KT: Biological monitoring of agricultural workers exposed to pesticides. I. Cholinesterase activity determinations. J Occup Med 38:619–627, 1986
6. Michel HO: Electrometric method for determination of red blood cell and plasma cholinesterase activity. J Lab Clin Med 34:1564–1568, 1949
7. Namba T, Nolte CT, Jackrel J, Grob D: Poisoning due to organophosphate insecticides. Am J Med 50:475–492, 1971
8. Milby TH: Prevention and management of organophosphate poisoning. JAMA 216:2131–2133, 1971

NAPHTHA (coal tar)

1987 TLV = none established

Synonyms: Naphtha solvent, high flash naphtha, refined naphtha, and heavy naphtha describe various fractions and grades.

Physical Form. Light yellow liquid with boiling ranges between 110°C and 190°C

Uses. Solvent

Exposure. Inhalation

Toxicology. Coal tar naphtha is a central nervous system depressant.

Coal tar naphtha is primarily a mixture of toluene, xylene, cumene, benzene, and other aromatic hydrocarbons; it is distinguished from petroleum naphtha, which is comprised mainly of aliphatic hydrocarbons.[1]

There are no well-documented reports of industrial injury resulting from the inhalation of coal tar naphtha.[1] However, severe exposure is expected to cause light-headedness, drowsiness, and possibly irritation of the eye, nose, and throat. Skin contact with the liquid may result in drying and cracking due to defatting action. Coal tar naphtha, a mixture of hydrocarbons, has been deleted from the ACGIH listing of TLVs in favor of reference to its chemical components.

REFERENCE

1. Browning E: Toxicity and Metabolism of Industrial Solvents, pp 141–144. New York, Elsevier, 1965

NAPHTHALENE

CAS: 91-20-3

$C_{10}H_8$ 1987 TLV = 10 ppm

Synonyms: Naphthalin; tar camphor; white tar

Physical Form. White crystalline solid with a characteristic "moth ball" odor

Uses. Moth repellant; chemical and dye manufacturing

Exposure. Inhalation; ingestion

Toxicology. Naphthalene is a hemolytic agent and an irritant of the eyes; it may cause cataracts.

Severe intoxication from ingestion results in characteristic manifestations of marked intravascular hemolysis and its consequences, including potentially fatal hyperkalemia.[1,2] Initial symptoms include eye irritation, headache, confusion, excitement, malaise, profuse sweating, nausea, vomiting, abdominal pain, and irritation of the bladder; there may be progression to jaundice, hematuria, hemoglobinuria, renal tubular blockade, and acute renal shutdown.[1,2] Hematologic features include red cell fragmentation, icterus, severe anemia with nucleated red cells, leukocytosis, and dramatic decreases in hemoglobin, hematocrit, and red cell count; sometimes there is formation of Heinz bodies and methemoglobin.[3] Naphthalene itself is nonhemolytic; several metabolites, including α-naphthol, are hemolytic.[3] Persons with a hereditary deficiency of the enzyme, glucose–6–phosphate dehydrogenase, in red blood cells (and consequently decreased concentrations of reduced glutathione) are particularly susceptible to the hemolytic properties of naphthalene.[3]

The vapor causes eye irritation at 15 ppm; eye contact with the solid may result in conjunctivitis, superficial injury to the cornea, chorioretinitis, scotoma, and diminished visual acuity. Cataracts and ocular irritation have been produced experimentally in animals and have been described in humans.[4] Of 21 workers exposed to high concentrations of fume or vapor for 5 years, eight had peripheral lens opacities. In other studies, no abnormalities of the eyes have been detected in workers exposed to naphthalene for several years.[4]

Reportedly, headache, nausea, and confusion may occur after inhalation of vapor. Occupational poisoning from vapor exposure is rare.[3] Naphthalene on the skin may cause hypersensitivity dermatitis; chronic dermatitis is rare.[1]

REFERENCES

1. Hygienic Guide Series: Naphthalene. Am Ind Hyg Assoc J 28:493–496, 1967
2. Gidron E, Leurer J: Naphthalene poisoning. Lancet 1:228–230, 1956
3. Gosselin RE, Smith RP, Hodge HC: Clinical Toxicology of Commercial Products, Section III, 5th ed, pp 307–311. Baltimore, Williams & Wilkins, 1984
4. Grant WM: Toxicology of the Eye, 3rd ed, pp 650–653. Springfield, Illinois, Charles C Thomas, 1986

β-NAPHTHYLAMINE

CAS: 91-59-8

$C_{10}H_9N$ 1987 TLV = none established; recognized carcinogen

Synonyms: 2–Aminonaphthalene; BNA; 2–naphthylamine

Physical Form. Colorless crystals that darken upon oxidation

Uses. Previously used in the manufacture of dyes and antioxidants; rarely used for industrial research purposes

Exposure. Inhalation

Toxicology. β-Naphthylamine (BNA) is a known human bladder carcinogen and therefore has no TLV.

The probability of a worker who is exposed to BNA of developing a bladder tumor is 61 times greater than that of the general population.[1]

A cohort study of a factory population revealed seven cases of bladder or kidney cancer in 735 person-years of exposure, an attack rate of 952 per 100,000 among BNA workers.[2] Of 48 BNA workers employed in a coal tar dye plant, 12 developed bladder tumors.[3] The time elapsed from first exposure to first abnormal signs or symptoms (dysuria, frequency, hematuria) ranged from 1 to 35 years, with a mean of 18 years.[3] The time elapsed from first exposure to diagnosis of bladder malignancy ranged from 2 to 42 years, with a mean of 23 years.[3]

Bladder tumors were induced in 24 of 34 dogs that were fed 6.25 to 50 mg/kg/day for 6 to 26 months; carcinomas were present in 9 of 11 dogs that received 100 to 200 g BNA, whereas 6 of 22 carcinomas occurred in dogs receiving total doses less than 100 g.[4] All dogs treated with the carcinogen had multiple tumors.[4]

In monkeys, intragastric administration of 37 to 2400 mg/kg/week for up to 250 weeks caused nine transitional cell carcinomas of the bladder and three papillary adenomas.[5]

Because of demonstrated high carcinogenicity in humans and animals, exposure by any route should be avoided.

REFERENCES

1. β-Naphthylamine. Documentation of TLVs and BEIs, 5th ed, p 421. Cincinnati, American Conference of Governmental Industrial Hygienists (ACGIH), 1986.
2. Mancuso TF, El-Attar AA: Cohort study of workers exposed to β-naphthylamine and benzidine. J Occup Med 9:277–285, 1967
3. Goldwater LJ, Rossa AJ, Kleinfeld M: Bladder tumors in a coal tar dye plant. Arch Environ Health 11:814–817, 1965
4. Conzelman GM Jr, Moulton JE: Dose–response relationships of the bladder tumorigen 2-naphthylamine: a study in beagle dogs. J Natl Cancer Inst 49:193–205, 1972
5. Conzelman GM Jr, et al: Induction of transitional cell carcinomas of the urinary bladder in monkeys fed 2-naphthylamine. J Natl Cancer Inst 42:825–836, 1969

NICKEL (and compounds)

CAS: 7440-02-0
Ni 1987 TLV = 0.1 mg/m^3—soluble inorganic compounds; as Ni
= 1.0 mg/m^3—metal
= 1.0 mg/m^3—insoluble compounds, as Ni

Compounds: Nickel chloride; nickel nitrate; nickel sulfate; nickel oxide; nickel subsulfide

Physical Form. Silver white metal; salts are crystals

Uses. Corrosion resistant alloys, electroplating, production of catalysts, nickel–cadmium batteries

Exposure. Inhalation

Toxicology. Metallic nickel and certain nickel compounds cause sensitization dermatitis. Nickel refining has been associated with an increased risk of nasal and lung cancer.

"Nickel itch" is a dermatitis resulting from sensitization to nickel; the first symptom is usually pruritis, which occurs up to 7 days before skin eruption appears.[2] The primary skin eruption is erythematous, or follicular; it may be followed by superficial discrete ulcers, which discharge and become crusted, or by eczema. The eruptions may spread to areas related to the activity of the primary site such as the elbow flexure, eyelids, or sides of the neck and face.[2] In the chronic stages, pigmented or depigmented plaques may be formed. Nickel sensitivity, once acquired, is apparently not lost; recovery from the dermatitis usually occurs within 7 days of cessation of exposure but may take several weeks.[1]

A worker who had developed cutaneous sensitization also developed apparent asthma from inhalation of nickel sulfate. Immunologic studies showed circulating antibodies to the salt, and controlled exposure to a solution of nickel sulfate resulted in decreased pulmonary function and progressive dyspnea; the possibility of hypersensitivity pneumonitis could not be excluded.[3]

Pneumoconiosis has been reported among workers exposed to nickel dust, but exposure to known fibrogenic substances could not be excluded.[4] Nasal irritation, damage to the nasal mucosa, perforation of the nasal septum, and loss of smell have only occasionally been reported in workers exposed to nickel aerosols and other contaminants.[5]

The severe acute systemic effects found with nickel carbonyl exposure are not associated with inorganic nickel.[4]

Epidemiologic studies have shown an increased incidence of cancers among nickel refinery workers.[5–7]

A recent mortality update of a cohort of 967 Clydach, Wales, refinery workers employed for at least 5 years and followed to 1971 showed significant risks in both lung and nasal cancers among those hired before 1930.[8] The SMR for lung cancer was 623 (O/E = 137/21.98), and, for nasal cancer, it was 28718 (O/E = 567/0.195). No case of nasal cancer occurred among those entering employment after 1930, and lung cancer rates dropped steeply after this date. The reduction was attributed to industrial hygiene improvements and process changes made in the 1920s.

An excess of sinus cancers occurred in a cohort of 1852 West Virginia nickel alloy workers employed prior to 1948 when calcining of nickel sulfide matte was done at the plant.[9]

In one of the largest studies, an excess of lung and nasal cancers was found in a cohort of 54,724 Canadian workers.[5–7] The respiratory cancer risk was confined to the sintering, calcining, and leaching occupational group. There was no excess among miners, concentrators, smeltors, or other groups.

Other cancers, including prostatic and laryngeal, have been significantly elevated in certain studies but are less convincingly associated with nickel refinery work.[7]

The IARC has determined that there is sufficient evidence for carcinogenicity to humans for nickel refining.[10] They further suggest that metallic nickel seems less likely to be carcinogenic than nickel subsulfide or nickel oxides.[10]

Others suggest that although nickel refinery workers have had an increased mortality from lung and nasal cancer, no specific agent(s) can be singled out as being responsible for the observed excesses.[7] Animal studies suggest, in general, an inverse relationship between solubility and carcinogenic potential; nickel metal, nickel oxide, and nickel subsulfide may exert variable degrees of carcinogenic potential in vivo and in vitro, whereas most nickel salts are noncarcinogenic.[6] The differences in activity may result from the ability of different compounds to enter the cell and be converted to nickel ion, the purported carcinogenic species.[6]

In experimental animals, a range of reproductive effects can be induced by nickel; in male rats, exposure to nickel salts results in degenerative changes in the testes and epididymis and in effects on spermatogenesis.[6] Exposure of pregnant animals has been associated with delayed embryonic development, increased resorptions, and an increase in structural malformations.[11] It has been noted, however, that doses used are high and may not relate at all to human exposure.[11] There are no reports indicating that exposure to nickel has caused malformations in humans.[4] Reproductive effects do not seem likely to occur as a result of occupational exposures if other toxic effects are prevented.[11]

REFERENCES

1. Browning E: Toxicity of Industrial Metals, 2nd ed, pp 249–260. London, Butterworths, 1969
2. Fisher AA: Contact Dermatitis, 2nd ed, pp 96–102. Philadelphia, Lea & Febiger, 1973
3. McConnel LH et al: Asthma caused by nickel sensitivity. Ann Intern Med 78:888–890, 1973
4. Norseth T: Nickel. In Friberg L et al (eds): Handbook on the Toxicology of Metals, 2nd ed, Vol II, Specific Metals, pp 462–481. Amsterdam, Elsevier, 1986
5. Mastromatteo E: Nickel. Am Ind Hyg Assoc J 10:589–601, 1986
6. US Environmental Protection Agency: Health Assessment Document for Nickel and Nickel Compounds. Final Report. Washington, DC, Office of Health and Environmental Assessment, September 1986
7. Wong O et al: Critical Evaluation of Epidemiologic Studies of Nickel-Exposed Workers—Final Report, pp 1–99. Berkeley, California, Environmental Health Associates, Inc, 1983
8. Doll R et al: Cancers of the lung and nasal sinuses in nickel workers: A reassessment of the period of risk. Br J Ind Med 34:102–105, 1977
9. Enterline PE, Marsh GM: Mortality among workers in a nickel refinery and alloy manufacturing plant in West Virginia. J Natl Cancer Inst 68:925–933, 1982
10. IARC Monographs on the Evaluation of the Carcinogenic Risk of Chemicals to Humans, Suppl 4, pp 167–170. Lyon, International Agency for Research on Cancer, 1982
11. Health Effects Document on Nickel, pp 1–204. Department of Environmental Medicine, Odense University, Odense, Denmark. Submitted to Ontario Ministry of Labour, 1986

NICKEL CARBONYL

CAS: 13463-39-3

$Ni(CO)_4$ 1987 TLV = 0.05 ppm

Synonym: Nickel tetracarbonyl

Physical Form. Colorless liquid

Uses. Purification intermediate in refining nickel; catalyst in the petroleum, plastic, and rubber industries

Exposure. Inhalation

Toxicology. Nickel carbonyl is a severe pulmonary irritant.

Initial symptoms usually include frontal headache, vertigo, nausea, vomiting, and sometimes substernal and epigastric pain; generally these early effects disappear when the subject is removed to fresh air.[1,2] It is estimated that exposure to 30 ppm for 30 minutes may be lethal to humans.[2]

There may be an asymptomatic interval between recovery from initial symptoms and the onset of delayed symptoms, which tend to develop 12 to 36 hours following exposure. Constrictive pain in the chest is characteristic of the delayed onset of pulmonary effects, followed by cough, hyperpnea, and cyanosis, leading to profound weakness; gastrointestinal symptoms may also occur. The temperature seldom rises above 101°F, and leukocytosis above 12,000/cmm is infrequent. Physical signs are compatible with pneumonitis or bronchopneumonia. Except for the pronounced weakness and hyperpnea, the physical findings and symptoms resemble those of a viral or influenzal pneumonia.[2,3]

Terminally, delirium and convulsions frequently occur; death has occurred from 3 to 13 days after exposure to nickel carbonyl. In subjects who recover from nickel carbonyl intoxication, convalescence is usually protracted (2 to 3 months) and is characterized by excessive fatigue on slight exertion.

A close correlation exists between the clinical severity of acute nickel carbonyl intoxication and the urinary concentration of nickel during the first 3 days after exposure; hospitalization should be considered in all cases in which the urinary nickel content exceeds 0.5 mg/liter of urine.[2]

Controversy as to whether nickel carbonyl causes cancer arose from observation of increased incidence of cancer of the paranasal sinuses and lungs of workers in nickel refineries. Suspicion of carcinogenicity focused primarily on nickel carbonyl vapor, although there were concurrent exposures to respirable particles of nickel, nickel subsulfide, and nickel oxide.[1] Subsequent studies have shown an increased risk of lung and sinus cancer in nickel refineries where nickel carbonyl was not used in the process.[4] Furthermore, the incidence of respiratory cancer decreased greatly by 1930 despite continued exposure of workers to the same levels of nickel carbonyl through 1957.

Administration of nickel carbonyl to rats by repeated intravenous injection was associated with an increased incidence of various malignant tumors.[5] Inhalation exposure of rats was associated with a few pulmonary malignancies not reaching statistical significance.

The IARC has determined that nickel in some form(s) is carcinogenic to humans.[5]

Treatment. If the concentration of nickel in the first 8-hour collection of urine is above 50 μg/100 ml, the exposure is classified as severe. Hospitalization is indicated. If the patient's condition is critical, sodium diethyldithiocarbamate (Dithiocarb) may be administered parenterally in an initial dosage of 25 mg/kg.[6] If clinically indicated, the total amount given during the first 24 hours may be increased to 100 mg/kg.

In less severe cases, Dithiocarb may be given orally. An initial dose of 2 g (ten 0.2-g capsules) should be given (one capsule every 2 minutes with sodium bicarbonate to prevent nausea). Subsequent doses are 1 g at 4 hours, 0.6 g at 8 hours, 0.4 g at 16 hours; on subsequent days, administer 0.4 g every 8 hours until the patient is free of symptoms and the concentration of nickel in the urine has decreased to the normal range (less than 5 μg/100 ml). If the concentration of nickel in the first 8-hour collection of urine is 10 to 50 μg/100 ml, the patient should be given Dithiocarb orally on the dosage schedule previously mentioned. Such patients should be under careful observation for at least a week, because delayed symptoms may develop.

REFERENCES

1. Committee on Medical and Biologic Effects of Environmental Pollutants, Division of Medical Sciences, National Research Council: Nickel, pp 113–128, 164–171, 231–268. Washington, DC, National Academy of Sciences, 1975
2. Hygienic Guide Series: Nickel carbonyl. Am Ind Hyg Assoc J 29:304–307, 1968
3. Jones CC: Nickel carbonyl poisoning. Arch Environ Health 26:245–248, 1973
4. Nickel carbonyl. Documentation of the TLVs and BEIs, 5th ed, p 424. Cincinnati, American Conference of Governmental Industrial Hygienists (ACGIH), 1986
5. IARC Monographs on the Evaluation of the Carcinogenic Risk of Chemicals to Man, Vol 11, p 104. Lyon, International Agency for Research on Cancer, 1976.
6. Sunderman FW: The treatment of acute nickel carbonyl poisoning with sodium diethyldithiocarbamate. Ann Clin Res 3:182–185, 1971

NICOTINE

CAS: 54-11-5

$C_{10}H_{14}N_2$ 1987 TLV = 0.5 mg/m^3; skin

Synonyms: 1-Methyl-2-(3-pyridyl) pyrrolidine; Black Leaf

Physical Form. Colorless to pale yellow, oily liquid; turns brown on exposure to air or light

Uses. Insecticide

Exposure. Inhalation; skin absorption; ingestion

Toxicology. Nicotine is a potent and rapid-acting poison; it is rapidly absorbed from all routes of entry, including the skin.

It acts on the central nervous system, autonomic ganglia, adrenal medulla, and neuromuscular junctions; initial stimulation is followed by a depressant phase of action.[1,2] The resulting physiologic effects are often complex and unpredictable. Small doses of nicotine cause nausea, vomiting, diarrhea, headache, dizziness, and neurologic stimulation resulting in tachycardia, hypertension, hyperpnea, tachypnea, sweating, and salivation.[1,2] With severe intoxication, there are convulsions and cardiac arrhythmias. In fatal cases, death nearly always occurs within 1 hour and has occurred within a few minutes.[3]

Nicotine, absorbed dermally, is probably the cause of "green-tobacco sickness," a self-limited illness consisting of pallor, vomiting, and prostration seen in men handling tobacco leaves in the field.[3] Currently, nicotine is not frequently used as an insecticide.[3]

Nicotine is teratogenic in mice; skeletal system malformations occurred in the offspring of pregnant mice injected subcutaneously with nicotine between days 9 and 11 of pregnancy.[4]

REFERENCES

1. Friedman PA: Poisoning and its management. In Petersdorf RG et al (eds): Harrison's Principles of Internal Medicine, 10th ed, p 1270. New York, McGraw-Hill, 1983
2. Taylor P: Ganglionic stimulating and blocking agents. In Gilman AG et al (eds): Goodman and Gilman's The Pharmacological Basis of Therapeutics, 7th ed, pp 217–218. New York, Macmillan, 1985
3. Gosselin RE, Smith RP, Hodge HC: Clinical Toxicology of Commercial Products, 5th ed, Section III, pp 311–314. Baltimore, Williams & Wilkins, 1984
4. Nishimura H, Nakai K: Developmental anomalies in offspring of pregnant mice treated with nicotine. Science 127:877–878, 1958

NITRIC ACID

CAS: 7697-37-2

HNO_3 1987 TLV = 2 ppm

Synonym: Aquafortis

Physical Form. Liquid

Uses. Production of fertilizers in the form of ammonium nitrate; manufacture of explosives

Exposure. Inhalation

Toxicology. Nitric acid causes corrosion of the skin and other tissues from topical contact and acute pulmonary edema or chronic obstructive pulmonary disease from inhalation.

When nitric acid is exposed to air or comes in contact with organic matter, it decomposes to yield a mixture of oxides of nitrogen, including nitric oxide and nitrogen dioxide, the latter being more hazardous than nitric acid.[1] Exposure to high concentrations of nitric acid vapor and nitrogen oxides causes pneumonitis and pulmonary edema, which may be fatal; onset of symptoms such as dryness of the throat and nose, cough, chest pain, and dyspnea may or may not be delayed.[2]

In contact with the eyes, the liquid produces severe burns, which may result in permanent damage and visual impairment.[2] On the skin, the liquid or concentrated vapor produces immediate, severe, and penetrating burns; concentrated solutions cause deep ulcers and stain the skin a bright yellow or yellowish-brown color.[1,2] Dilute solutions of nitric acid produce mild irritation of the skin and tend to harden the epithelium without destroying it.

The vapor and mist may erode exposed teeth. However, in cases of dental erosion attributed to nitric acid, there was concomitant exposure to sulfuric acid, a potent cause of dental erosion. Ingestion of the liquid will cause immediate pain and burns of the gastrointestinal tract.

REFERENCES

1. National Institute for Occupational Safety and Health: Criteria for a Recommended Standard . . . Occupational Exposure to Nitric Acid. DHEW (NIOSH) Pub No 76–141, pp 35–36. Washington, DC, US Government Printing Office, 1976
2. Hygienic Guide Series: Nitric acid. Am Ind Hyg Assoc J 25:426–428, 1964

NITRIC OXIDE

CAS: 10102-43-9

NO 1987 TLV = 25 ppm

Synonyms: Nitrogen monoxide; mononitrogen monoxide

Physical Form. Colorless gas

Uses. Manufacture of nitric acid; bleaching of rayon; as a stabilizer

Exposure. Inhalation

Toxicology. Nitric oxide causes cyanosis in animals, apparently from the formation of methemoglobin.

Exposure of mice to 5000 ppm for 6 to 8 minutes was lethal, as was 2500 ppm for 12 minutes; cyanosis occurred after a few minutes, the red eyegrounds became gray-blue, and then breathlessness appeared with paralysis and convulsions; spectroscopy of the blood showed methemoglobin.[1,2]

No effects in humans have been reported.

Nitric oxide is converted spontaneously in air to nitrogen dioxide; hence, some of the latter gas is invariably present whenever nitric oxide is found in the air.[1] At concentrations below 50 ppm, however, this reaction is slow, and substantial concentrations of nitric oxide may occur with negligible quantities of nitrogen dioxide.[1] It is likely that the effects of concomitant exposure to nitrogen dioxide will become manifest before the methemoglobin effects from nitric oxide can occur. Nitrogen dioxide may cause irritation of the eyes, nose, and throat and delayed pulmonary edema.[1]

REFERENCES

1. National Institute for Occupational Safety and Health: Criteria for a Recommended Standard . . . Occupational Exposure to Oxides of Nitrogen (Nitrogen Dioxide and Nitric Oxide). DHEW (NIOSH) Pub No 76–149, pp 46–50, 75–76. Washington, DC, US Government Printing Office, 1976
2. Pflesser G: [Nitric Oxide and Nitrous (Gas) Poisoning.] (Ger) Proceedings of the German Pharmacological Society, 12th Session. Munich, October 20–23, 1935. Naunyn-Schmiedebergs Arch Exp Pathol Pharmakol 181:145–146, 1936

p-NITROANILINE

CAS: 100-01-6

$NH_2C_6H_4NO_2$ 1987 TLV = 3 mg/m^3; skin

Synonyms: PNA; 1-amino-4-nitrobenzene

Physical Form. Yellow crystals

Uses. Synthesis of dyestuffs and other intermediates

Exposure. Inhalation; skin absorption

Toxicology. *p*-Nitroaniline absorption, whether from inhalation of the vapor or absorption of the solid through skin, causes anoxia owing to the formation of methemoglobin; jaundice and anemia have been reported from chronic exposure.

Signs and symptoms of overexposure are due to the loss of oxygen-carrying capacity of the blood. The onset of symptoms of methemoglobinemia is often insidious and may be delayed for up to 4 hours; headache is commonly the first symptom and may become quite intense as the severity of methemoglobinemia progresses.[1] Cyanosis develops early in the course of intoxication; blueness in the lips, nose, and earlobes is usually recognized by fellow workers.[1] Cyanosis occurs when the methemoglobin concentration is 15% or more.[1] The person usually feels well, has no complaints, and is insistent that nothing is wrong until the methemoglobin concentration approaches approximately 40%.[1] At methemoglobin concentrations of over 40%, there is usually weakness and dizziness; methemoglobin levels above 50% are rarely observed with *p*-nitroaniline exposure; however, with concentrations up to 70%, there may be ataxia, dyspnea on mild exertion, tachycardia, nausea, vomiting and drowsiness. Methemoglobin of about 75% usually results in collapse, coma, and even death.[1,2] Ingestion of alcohol aggravates the toxic effects of *p*-nitroaniline.[2]

In general, higher ambient temperatures increase susceptibility to cyanosis from exposure to methemoglobin-forming agents.[3] There are no chronic effects from single exposures, but prolonged or excessive exposures may cause liver damage.[1,2] *p*-Nitroaniline is mildly irritating to the eyes and may cause some corneal damage.[2]

Diagnosis. Signs and symptoms include headache; signs of anoxia, including cyanosis of lips, nose, and earlobes; eye irritation; anemia; and jaundice.

Differential Diagnosis: Other causes of cyanosis must be differentiated from *p*-nitroaniline exposure. These include hypoxia from lung disease, hypoventilation, and decreased cardiac output. Lung disease may be suspected from results of pulmonary function tests and arterial blood gas analysis. The arterial PO_2 may be normal in methemoglobinemia but will be decreased in cyanosis resulting from lung disease. Hypoventilation will cause elevation of arterial PCO_2, which is not seen in *p*-nitroaniline exposure. Decreased cardiac output states will cause cyanosis only when accompanied by arterial hypotension. If blood withdrawn from the vein shows the characteristic chocolate-brown coloration, the diagnosis of an abnormal pigment is almost certain, especially if the color remains after shaking the blood in air.[4]

Special Tests: Determine the methemoglobin concentration in the blood when *p*-nitroaniline intoxication is suspected and at regular intervals until the methemoglobin has been fully reduced to normal hemoglobin.[5] Methemoglobin may be differentiated from sulfhemoglobin by the addition of a few drops of 10% potassium cyanide, which results in the rapid production of bright red cyanomethemoglobin but has no effect on the color of sulfhemoglobin. Spectrophotometry is required for the precise identification of the pigment and its quantitation. Normal acid methemoglobin has a characteristic absorption spectrum with peaks at 502 and 732 nm, which disappear with the addition of cyanide, whereas sulfhemoglobin has a peak at 620 nm, which does not disappear with cyanide.[4]

Treatment. All *p*-nitroaniline on the body must be removed. Immediately remove all clothing, and wash the entire body from head to foot with soap and water. Pay special attention to the hair and scalp, finger and toenails, nostrils, and ear canals. Administer oxygen to alleviate the headache and general sense of weakness; confine the patient to bed. Determine the methemoglobin concentration in the blood, and repeat every 3 to 6 hours for 18 to 24 hours. Repeat skin cleansing if the methemoglobin concentration appears to rise after 3 to 4 hours. Provided that all sources of further absorption are completely eliminated, mildly affected patients require no other treatment, and the methemoglobin will be reduced spontaneously to ferrous hemoglobin in 2 to 3 days.[4]

The only justifiable use of methylene blue would be in cases of coma or stupor, usually at methemoglobin levels over 60%; in those patients in whom therapy is necessary, methylene blue may be given intravenously, 1 to 2 mg/kg, over a 5-minute period as a 1% solution; if cyanosis has not disappeared within an hour, a second dose of 2 mg/kg should be administered.[4,5] The total dose should not exceed 7 mg/kg, because methylene blue may cause toxic effects such as dyspnea, precordial pain, restlessness, apprehension, red cell hemolysis, and changes in the electrocardiogram (reduction in the height or even reversal of the T wave, frequently with lowering of the R wave).[5]

Exchange transfusion, hemodialysis, and hyperbaric oxygen treatment have all been proposed in severe cases of intoxication, but effectiveness has not been established.[6]

Medical Control: Medical control involves preplacement and annual physical examinations with emphasis on the blood, liver, and cardiovascular system; complete blood count; and liver function tests.

REFERENCES

1. Beard RR, Noe JT: Aromatic nitro and amino compounds. In Clayton GD, Clayton FE (eds): Patty's Industrial Hygiene and Toxicology, 3rd ed, Vol 2A, Toxicology, pp 2413–2489. New York, Wiley–Interscience, 1981
2. Chemical Safety Data Sheet SD–94, para-NITROANILINE, pp 5–6, 11–13. Washington, DC, MCA, Inc, 1966
3. Linch AL: Biological monitoring for industrial exposure to cyanogenic aromatic nitro and amino compounds. Am Ind Hyg Assoc J 35:426–432, 1974
4. Reider RF: Methemoglobinemia and sulfhemoglobinemia. In Wyngaarden JB, Smith LH (eds): Cecil Textbook of Medicine, 16th ed, pp 894–896. Philadelphia, WB Saunders, 1982
5. Mangelsdorff AF: Treatment of methemoglobinemia. AMA Arch Ind Health 14:148–153, 1956
6. Kearney TE et al: Chemically induced methemoglobinemia from aniline poisoning. West J Med 140:282–286, 1984

NITROBENZENE
CAS: 98-95-3
$C_6H_5NO_2$ 1987 TLV = 1 ppm; skin

Synonyms: Nitrobenzol; oil of mirbane

Physical Form. Almost water-white, oily liquid, turning yellow with exposure to air

Uses. Chemical intermediate in organic syntheses

Exposure. Inhalation; skin absorption

Toxicology. Nitrobenzene causes anoxia due to the formation of methemoglobin; chronic exposure produces a reversible anemia.

Exposure of workers to 40 ppm for 6 months resulted in some cases of intoxication and anemia; concentrations ranging from 3 to 6 ppm caused headache and vertigo in 2 of 39 workers; increased methemoglobin and sulfhemoglobin levels and Heinz bodies were observed in the blood.[1]

Signs and symptoms of overexposure are due to the loss of oxygen-carrying capacity of the blood. The onset of symptoms of methemoglobinemia is often insidious and may be delayed up to 4 hours; headache is commonly the first symptom and may become quite intense as the severity of methemoglobinemia progresses.[2] Cyanosis develops early in the course of intoxication. It is characterized by blueness of the lips, nose, and earlobes; it is usually recognized first by fellow workers, and occurs when the methemoglobin level is 15% or more. The person usually feels well, has no complaints, and will insist that nothing is wrong until the methemoglobin concentration approaches 40%. At methemoglobin concentrations ranging from 40% to 70%, there is headache, weakness, dizziness, ataxia, dyspnea on mild exertion, tachycardia, nausea, vomiting, and drowsiness.[2,3] Coma may ensue with methemoglobin levels above 70%, and the lethal level is estimated to be 85% to 90%.[3]

Ingestion of alcohol aggravates the toxic effects of nitrobenzene.[3] In general, higher ambient temperatures increase susceptibility to cyanosis from exposure to methemoglobin-forming agents.[4] *p*-Nitrophenol and *p*-aminophenol are metabolites of nitrobenzene; their presence in the urine is an indication of exposure.[5] Nitrobenzene is mildly irritating to the eyes; it may produce dermatitis due to primary irritation or sensitization.[3] Hepatotoxicity, manifested by alterations in liver function, including hyperbilirubinemia and decreased prothrombin activity, is associated with exposure in both animals and humans.[6] Degenerative testicular lesions occurred in rats exposed to single oral doses of 50 to 450 mg/kg.[6]

Diagnosis. Signs and symptoms include headache, signs of anoxia, including cyanosis of lips, nose, and earlobes, and anemia.

Differential Diagnosis: Other causes of cyanosis must be differentiated from methemoglobinemia resulting from chemical exposure. These include hypoxia from lung disease, hypoventilation, and decreased cardiac output. Lung disease may be suspected from results of pulmonary function tests and arterial blood gas analysis. The arterial PO_2 may be normal in methemoglobinemia but tends to be decreased in cyanosis owing to lung disease. Hypoventilation will cause elevation of arterial PCO_2, which is not seen in chemical exposure. Decreased cardiac output states will cause cyanosis only when accompanied by arterial hypotension. If blood withdrawn from the vein shows the characteristic chocolate-brown coloration, the diagnosis of an abnormal pigment is almost certain, especially if the color remains after shaking the blood in air.[7]

Special Tests: Determine the methemoglobin concentration in the blood when chemical intoxication is suspected and at regular intervals until the methemoglobin has been fully reduced to normal hemoglobin.[8] Methemoglobin may be differentiated from sulfhemoglobin by the addition of a few drops of 10% potassium cyanide, which results in the rapid production of bright red cyanomethemoglobin but has no effect on the color of sulfhemoglobin.[7] Spectrophotometry is required for the precise identification of the pigment and its quantitation. Normal acid methemoglobin has a characteristic absorption spectrum with peaks at 502 and 632 nm, which disappear with the addition of cyanide, whereas sulfhemoglobin has a peak at 620 nm, which does not disappear with cyanide.[7]

Treatment. All the contaminant on the body must be removed. Immediately remove all clothing, and wash the entire body from head to foot with soap and water. Pay special attention to the hair and scalp, finger and toenails, nostrils, and ear canals. Administer oxygen to alleviate the headache and general sense of weakness; confine the patient to bed. Determine the methemoglobin concentration in the blood, and repeat every 3 to 6 hours for 18 to 24 hours. Repeat skin cleansing if the methemoglobin concentration appears to rise after 3 to 4 hours. In general, patients will return to normal within 24 hours provided that all sources of further absorption are completely eliminated. The methemoglobin will be reduced spontaneously to ferrous hemoglobin in 2 to 3 days.[7]

Such therapy is not effective in subjects with glucose–6–dehydrogenase deficiency.[7]

The only justifiable use of methylene blue would be in cases of coma or stupor, usually at methemoglobin levels over 60%. In those patients in whom therapy is necessary, methylene blue, 1 to 2 mg/kg of a 1% solution in saline, may be given intravenously over a 10-minute period. If cyanosis has not disappeared within an hour, a second dose of 2 mg/kg may be administered.[7,8] The total dose should not exceed 7 mg/kg, because methylene blue may cause toxic effects such as dyspnea, precordial pain, restlessness, apprehension, red cell hemolysis, and changes in the electrocardiogram (reduction in the height or even reversal of the T wave, frequently with lowering of the R wave).[8]

REFERENCES

1. Pacseri I, Magos L, Batskor IA: Threshold and toxic limits of some amino and nitro compounds. AMA Arch Ind Health 18:1–8, 1958
2. Hamblin DO: Aromatic nitro and amino compounds. In Patty FA (ed): Industrial Hygiene and Toxicology, 2nd ed, Vol 2, Toxicology, pp 2105–2147. New York, Wiley–Interscience, 1963
3. Chemical Safety Data Sheet SD–21, Nitrobenzene, pp 5–6, 12–14. Washington, DC, MCA, Inc, 1967
4. Linch AL: Biological monitoring for industrial exposure to cyanogenic aromatic nitro and amino compounds. Am Ind Hyg Assoc J 35:426–432, 1974
5. Ikeda M, Kita A: Excretion of *p*-nitrophenol and *p*-aminophenol in the urine of a patient exposed to nitrobenzene. Br J Ind Med 21:210–213, 1964
6. Beauchamp RO et al: A critical review of the literature on nitrobenzene toxicity. CRC Crit Rev Toxicol 11:33–84, 1983
7. Reider RF: Methemoglobinemia and sulfmethemoglobinemia. In Wyngaarden JB, Smith LH (eds): Cecil Textbook of Medicine, 16th ed, pp 894–896. Philadelphia, WB Saunders Co, 1982
8. Mangelsdorff AF: Treatment of methemoglobinemia. AMA Arch Ind Health 14:148–153, 1956

o-NITROCHLOROBENZENE

CAS: 88-73-3

$ClC_6H_4NO_2$ 1987 TLV = none established

Synonyms: 2–Chloronitrobenzene; 1–chloro–2–nitrobenzene; 2–CNB

Physical Form. Yellow solid

Uses. Chemical intermediate in manufacture of dyes, picric acid, lumber preservatives, and diaminophenol hydrochloride (a photographic developer)

Exposure. Inhalation; skin absorption

Toxicology. *o*-Nitrochlorobenzene (2–CNB) absorption causes anoxia owing to formation of methemoglobin, which cannot transport oxygen.

Numerous cases of cyanosis in workers exposed to 2–CNB and related compounds occurred in the period 1935 to 1965.[1]

Signs and symptoms of overexposure are caused by the loss of oxygen-carrying capacity of the blood. The onset of symptoms of methemoglobinemia is often insidious and may be delayed up to 4 hours; headache is commonly the first symptom and may become quite intense as the severity of methemoglobinemia progresses.[1] Cyanosis develops early in the course of intoxication; it is characterized by blueness of the lips, nose, and earlobes, usually recognized first by fellow workers, and it occurs when the methemoglobin concentration approaches 40%. At concentrations ranging from 40% to 70%, there is headache, weakness, dizziness, ataxia, dyspnea on mild exertion, tachycardia, nausea, vomiting, and drowsiness.[1] Coma may ensue with methemoglobin levels about 70%, and the lethal level is estimated to be 85% to 90%.[2]

In general, higher ambient temperatures increase susceptibility to cyanosis from exposure to methemoglobin-forming agents.[1]

The acute LD_{50} values for 2–CNB in rats and mice by oral administration are approximately 110 to 400 mg/kg.[2]

In carcinogenicity bioassays, it was found that 2–CNB produced an increase in the incidence of multiple tumors in male rats at the low dose (1000 mg/kg diet) but not at the high dose (2000 mg/kg diet).[3] 2–CNB produced an increase in hepatocellular carcinomas in female mice at high- (6000 mg/kg diet) and low-dose (3000 mg/kg diet) levels, and in male mice at low-dose levels but not at high-dose levels. Because of the inconsistency of the dose–response effects, the high doses used, and the long latent periods before tumor development, 2–CNB was not regarded as a very potent carcinogen under the conditions of the test.

REFERENCES

1. Linch AL: Biological monitoring for industrial exposure to cyanogenic aromatic nitro and amino compounds. Am Ind Hyg Assoc J 35:426, 1974
2. Vernot EH, MacEwen ID, Haun CC, Kinkead ER: Acute toxicity and skin corrosion data for some organic and inorganic compounds and aqueous solution. Toxicol Appl Pharmacol 42:417, 1977
3. Weisburger EK et al: Testing of 21 environmental aromatic amines or derivatives for long-term toxicity or carcinogenicity. J Environ Pathol Toxicol 2:325, 1978

p-NITROCHLOROBENZENE

CAS: 100-00-5

$NO_2C_6H_4Cl$ 1987 TLV = 0.5 ppm mg/m^3; skin

Synonyms: *p*-Chloronitrobenzene; PCNB

Physical Form. Liquid or yellowish crystals

Uses. Manufacture of dyes, rubber, and agricultural chemicals

Exposure. Inhalation; skin absorption

Toxicology. Absorption of *p*-nitrochlorobenzene causes anoxia owing to the formation of methemoglobin.

Signs and symptoms of overexposure are due to the loss of oxygen-carrying capacity of the blood. The onset of symptoms of methemoglobinemia is often insidious and may be delayed for up to 4 hours; headache is commonly the first symptom and may become quite intense as the severity of methemoglobinemia progresses.[1] Cyanosis develops early in the course of intoxication; blueness occurs first in the lips, nose, and earlobes and is usually recognized by fellow workers.[1] Cyanosis occurs when the methemoglobin concentration is 15% or more. The subject usually feels well, has no complaints, and is insistent that nothing is wrong until the methemoglobin concentration approaches approximately 40%. At methemoglobin concentrations over 40%, there is weakness and dizziness; closer to 70% concentration, there may be ataxia, dyspnea on mild exertion, tachycardia, nausea, vomiting, and drowsiness.[1] The ingestion of alcohol aggravates the toxic effects of *p*-nitrochlorobenzene.[2] In general, higher ambient temperatures increase susceptibility to cyanosis from exposure to methemoglobin-forming agents.

Four workers exposed to an unmeasured concentration of the vapor for a period of 2 to 4 days developed methemoglobinemia; in these cases, there was an initial collapse, a slate-grey appearance, dyspnea, and a mild anemia 1 week after exposure.[2]

Cats exposed to 2.8 ppm for 198 hours survived, but 3.3 ppm for 24 hours was lethal; all animals lost weight, had severe methemoglobinemia, and showed liver and kidney damage at autopsy.[3]

Application of the solid dissolved in olive oil to the skin of rabbits caused methemoglobinemia, the formation of Heinz bodies in erythrocytes, anemia, hematuria, and hemoglobinuria.[4]

p-Nitrochlorobenzene has a pleasant, aromatic odor.

Diagnosis. *Signs and Symptoms:* Symptoms include headache; signs of anoxia, including cyanosis of lips, nose, and earlobes; and anemia.

Differential Diagnosis: Other causes of cyanosis must be differentiated from methemoglobinemia due to *p*-nitrochlorobenzene exposure. These include hypoxia resulting from lung disease, hypoventilation, and decreased cardiac output. Lung disease may be suspected from results of pulmonary function tests and arterial blood gas analysis. The arterial Po_2 may be normal in methemoglobinemia but tends to be decreased in cyanosis due to lung disease. Hypoventilation will cause elevation of arterial Pco_2, which is not seen in *p*-nitrochlorobenzene exposure. Decreased cardiac output states will cause cyanosis only when accompanied by arterial hypotension. If blood withdrawn from the vein shows the characteristic chocolate-brown coloration, the diagnosis of an abnormal pigment is almost certain, especially if the color remains after shaking the blood in the air.[5]

Special Tests: Determine the methemoglobin concentration in the blood when *p*-nitrochlorobenzene intoxication is suspected and at regular intervals until the methemoglobin has been fully reduced to normal hemoglobin.[6] Methemoglobin may be differentiated from sulfhemoglobin by the addition of a few drops of 10% potassium cyanide, which results in the rapid production of bright red cyanomethemoglobin but has no effect on the color of sulfhemoglobin.[5] Spectrophotometry is required for the precise identification of the pigment and its quantitation. Normal acid methemoglobin has a characteristic absorption spectrum with peaks at 502 and 632 nm, which disappear with the addition of cyanide, whereas sulfhemoglobin has a peak at 620 nm, which does not disappear with cyanide.[5]

Treatment. All the contaminant on the body must be removed. Immediately remove all clothing, and wash the entire body from head to foot with soap and water. Pay special attention to the hair and scalp, finger and toenails, nostrils, and ear canals. Administer oxygen to alleviate the headache and general sense of weakness; confine the patient to bed. Determine the methemoglobin concentration in the blood, and repeat every 3 to 6 hours for 18 to 24 hours. Repeat skin cleansing if the methemoglobin concentration appears to rise after 3 to 4 hours. In general, patients will be improved within 24 hours provided that all sources of further absorption are completely eliminated. The methemoglobin will be reduced spontaneously to ferrous hemoglobin in 2 to 3 days.[5]

Such therapy is not effective in subjects with glucose–6–phosphate dehydrogenase deficiency.[5]

The only justifiable use of methylene blue is in cases of coma or stupor, usually at methemoglobin levels over 60%. In those patients in whom therapy is necessary, methylene blue, 1 to 2 mg/kg of a 1% solution in saline, may be given intravenously over a 10-minute period. If cyanosis has not disappeared within an hour, a second dose of 2 mg/kg may be administered.[5,6] The total dose should not exceed 7 mg/kg, because methylene blue may cause toxic effects such as dyspnea, precordial pain, restlessness, apprehension, red cell hemolysis, and changes in the electrocardiogram (reduction in the height or even reversal of the T wave, frequently with lowering of the R wave).[6]

REFERENCES

1. Hamblin DO: Aromatic nitro and amino compounds. In Patty FA (ed): Industrial Hygiene and Toxicology, 2nd ed, Vol 2, Toxicology, pp 2105–2119, 2130–2131. New York, Wiley–Interscience, 1963
2. Renshaw A, Ashcroft GV: Four cases of poisoning by mononitrochlorobenzene and one by acetanilide occurring in a chemical works: With an explanation of the toxic symptoms produced. J Ind Hyg 8:67–73, 1926
3. Watrous RM, Schulz HN: Cyclohexylamine, *p*-chloronitrobenzene, 2–aminopyridine: Toxic effects in industrial use. Ind Med Surg 19:317–320, 1950
4. *p*-Nitrochlorobenzene. Documentation of the TLVs and BEIs, 5th ed, p 432. Cincinnati, American Conference of Governmental Industrial Hygienists (ACGIH), 1986
5. Rieder RF: Methemoglobinemia and sulfmethemoglobinemia. In Wyngaarden JB, Smith LH (eds): Cecil Textbook of Medicine, 16th ed, p 896. Philadelphia, WB Saunders, 1982
6. Mangelsdorff AF: Treatment of methemoglobinemia. AMA Arch Ind Health 14:148–153, 1956

4–NITRODIPHENYL

CAS: 92-93-3

$C_{12}H_9NO_2$ 1987 TLV = none established; recognized carcinogen

Synonyms: 4–Nitrobiphenyl; *p*-nitrobiphenyl; PNB

Physical Form. White crystals

Uses. Plasticizer, fungicide, wood preservative

Exposure. Inhalation; skin absorption

Toxicology. 4–Nitrodiphenyl, or *p*-nitrobiphenyl (PNB), is a urinary bladder carcinogen in dogs.

There are no reports of carcinogenicity of PNB in humans.[1] However, PNB was used as an intermediate in the preparation of 4–aminobiphenyl, a recognized human bladder carcinogen, and bladder tumors found in men exposed to 4–aminobiphenyl may have been partially due to PNB.[2]

Three of four dogs fed 0.3 g of PNB (in capsule) three times per week for up to 33 months developed bladder tumors.[2] The total dose administered ranged from 7 to 10 g/kg in the affected dogs; the animal that did not develop bladder tumors was the largest and therefore had received less of the compound per kg of bodyweight (5.5 g/kg).[2] The tumors produced by PNB were identical histologically to those produced by 4–aminobiphenyl.[2]

The case for the carcinogenicity of PNB is supported by (1) the induction of urinary bladder cancer in dogs after administration of PNB, (2) the evidence that PNB is metabolized in vivo, to 4–aminobiphenyl (a potent carcinogen), and (3) the possibility that the cases of human urinary bladder cancer attributed to 4–aminobiphenyl may also have been induced by exposure to PNB.[3]

Human exposure by any route, respiratory, oral, or skin, should be avoided.[3]

REFERENCES

1. IARC Monographs on the Evaluation of the Carcinogenic Risk of Chemicals to Man, Vol 4, Some Aromatic Amines, Hydrazine and Related Substances, N-Nitroso Compounds and Miscellaneous Alkylating Agents, pp 113–117. Lyon, International Agency for Research on Cancer, 1974
2. Deichmann WB et al: Para-nitrobiphenyl—a new bladder carcinogen in the dog. Ind Med Surg 27:634–637, 1958
3. 4–Nitrodiphenyl. Documentation of the TLVs and BEIs, 5th ed, p 433. Cincinnati, American Conference of Governmental Industrial Hygienists (ACGIH), 1986

NITROETHANE

CAS: 79-24-3

$C_2H_5NO_2$ 1987 TLV = 100 ppm

Synonyms: None

Physical Form. Colorless, oily liquid

Uses. Propellant; solvent; organic synthesis

Exposure. Inhalation

Toxicology. In animals, nitroethane is a respiratory irritant, and, at high concentrations, it causes narcosis and liver damage; it is expected that severe exposure will produce the same effect in humans.

No systemic effects have been reported in humans. The liquid is a mild skin irritant owing to solvent action.

Rabbits died from exposure to 5000 ppm for 3 hours but survived 3 hours at 2500 ppm.[1,2] Exposure to the higher concentrations caused irritation of mucous membranes, lacrimation, dyspnea, pulmonary rales, and, in a few animals, pulmonary edema; convulsions were rare and of brief duration.[1] Autopsy of animals exposed to lethal concentrations showed mild to severe liver damage and nonspecific changes in the kidneys.

The odor of nitroethane is detectable at 163 ppm; the odor and irritant properties do not provide sufficient warning of toxic concentrations.[1,2]

REFERENCES

1. Machle W, Scott EW, Treon J: The physiological response of animals to some simple mononitroparaffins and to certain derivatives of these compounds. J Ind Hyg Toxicol 22:315–332, 1940
2. AIHA Hygienic Guide Series: Nitroethane. Akron, OH, American Industrial Hygiene Association, 1978

NITROGEN DIOXIDE

CAS: 10102-44-0

NO_2 1987 TLV = 3 ppm

Synonyms: None

Physical Form. Red gas liquefying at 21°C

Uses. Intermediate in nitric and sulfuric acid production; nitration of organic compounds and explosives

Exposure. Inhalation

Toxicology. Nitrogen dioxide is a respiratory irritant; it causes pulmonary edema and, rarely, with extremely severe exposure, bronchiolitis obliterans.

Brief exposure of humans to concentrations of about 250 ppm causes cough, production of mucoid or frothy sputum, and increasing dyspnea.[1,2] Within 1 to 2 hours, the person may develop pulmonary edema with tachypnea, cyanosis, fine crackles and wheezes through the lungs, and tachycardia. Alternatively, there may be only increasing dyspnea and cough over several hours, with symptoms then gradually subsiding over a 2- to 3-week period. The condition may then enter a second stage of abruptly increasing severity; fever and chills precede a relapse, with increasing dyspnea, cyanosis, and recurring pulmonary edema. Death may occur either in the initial or second stage of the disease; a severe second stage may follow a relatively mild initial stage. The person who survives the second stage usually recovers over 2 to 3 weeks; however, some people do not return to normal but experience varying degrees of impaired pulmonary function.

The radiographic features in the acute initial stage vary from normal to those of typical pulmonary edema; most reports mention a pattern of nodular shadows on the chest film at the outset.[1,2] The roentgenogram may then clear, only to show miliary mottling as the second stage commences, progressing to the development of a confluent pattern. Results of pulmonary function tests in the acute stage show reduction in lung volume and diffusing capacity; similar findings are recorded in the second stage.

Pathologic examination of the acute lesion shows extensive mucosal edema and inflammatory cell exudation. The delayed lesion shows the histologic appearance of bronchiolitis obliterans; small bronchi and bronchioles contain an inflammatory exudate that tends to undergo fibrinous organization, eventually obliterating the lumen.

Humans exposed to nitrogen dioxide for 60 minutes can expect the following effects: 100 ppm, pulmonary edema and death; 50 ppm, pulmonary edema with possible subacute or chronic lesions in the lungs; and 25 ppm, respiratory irritation and chest pain.[3] A concentration of 50 ppm is moderately irritating to the eyes and nose; 25 ppm is irritating to some people.[1] Exposure of healthy and asthmatic volunteers at 4 ppm for 75 minutes caused a small but significant decrease in systolic blood pressure; there were no significant effects on airway resistance, heart rate, skin conductance, or self-reported emotional state.[4]

Most reported cases of severe illness due to nitrogen dioxide have been from accidental exposures to explosion or combustion of nitroexplosives, nitric acid, the intermittent process of arc or gas welding (especially in a confined space), or entry into an agricultural silo that was not vented.[1,5]

Animal experimentation has indicated that in addition to irritation and pathologic changes, nitrogen dioxide exposure may decrease host resistance to infection.[6] An increased mortality in mice infected with pneumonia-causing organisms was found subsequent to exposure at 0.5 ppm for 7 days and longer.[7]

Nitrogen dioxide does not appear to be teratogenic, mutagenic, or directly carcinogenic.[6]

REFERENCES

1. National Institute for Occupational Safety and Health: Criteria for a Recommended Standard . . . Occupational Exposure to Oxides of Nitrogen (Nitrogen Dioxide and Nitric Oxide). DHEW (NIOSH) Pub No 76–149, pp 76–85. Washington, DC, US Government Printing Office, 1976
2. Morgan WKC, Seaton A: Occupational Lung Diseases, pp 330–335, 344–345. Philadelphia, WB Saunders, 1975
3. Emergency Exposure Limits: Nitrogen dioxide. Am Ind Hyg Assoc J 25:580–582, 1964
4. Linn WS et al: Effects of exposure to 4 ppm nitrogen dioxide in healthy and asthmatic volunteers. Arch Environ Health 40:234–239, 1985
5. Scott EG, Hunt WB Jr: Silo filler's disease. Chest 63:701–706, 1973
6. State of California Air Resources Board: Short-Term Ambient Air Quality Standard for Nitrogen Dioxide. PO Box 2815, Sacramento, California, September 1985
7. Gardner DE et al: Influence of exposure mode on the toxicity of NO_2. Environ Health Perspect 30:23–29, 1979

NITROGEN TRIFLUORIDE

CAS: 7783-54-2

NF_3 1987 TLV = 10 ppm

Synonym: Nitrogen fluoride

Physical Form. Colorless gas

Uses. Oxidizing agent in fuel combustion

Exposure. Inhalation

Toxicology. Nitrogen trifluoride causes anoxia in animals as a result of the formation of methemoglobin.

Although there are no reports of human intoxication from nitrogen trifluoride, the effects of methemoglobinemia include cyanosis (especially in the lips, nose, and earlobes), weakness, dizziness, and severe headache.[1]

Rats died from exposure to 10,000 ppm for 60 to 70 minutes; the methemoglobin concentrations at the time of death were equivalent to 60% to 70% of available hemoglobin.[2] Animals exposed to nearly lethal concentrations suffered severe respiratory distress and cyanosis from methemoglobinemia; severely affected animals experienced incoordination, collapse, and convulsions. Rats repeatedly exposed to 100 ppm for 4.5 months appeared normal, but autopsy findings indicated injury to the liver and kidneys.[3] Dogs surviving exposure to 9600 ppm for 60 minutes exhibited Heinz body anemia, decreased hematocrit levels, decreased hemoglobin levels, reduced red blood cell count, and clinical signs consistent with anoxia owing to methemoglobin formation; some eye irritation was observed during exposure.[4]

Nitrogen trifluoride provides no odor-warning properties at potentially dangerous levels.

Diagnosis. Signs and symptoms include headache; signs of anoxia, including cyanosis of lips, nose, and earlobes; anemia; and hematuria.

Differential Diagnosis: Other causes of cyanosis must be differentiated from methemoglobinemia resulting from chemical exposure. These causes include hypoxia due to lung disease, hypoventilation, and decreased cardiac output. Lung disease may be suspected from results of pulmonary function tests and arterial blood gas analysis. The arterial Po_2 may be normal in methemoglobinemia but tends to be decreased in cyanosis resulting from lung disease. Hypoventilation will cause elevation of arterial Pco_2, which is not seen in chemical exposure. Decreased cardiac output states will cause cyanosis only when accompanied by arterial hypotension. If blood withdrawn from the vein shows the characteristic chocolate-brown coloration, the diagnosis of an abnormal pigment is almost certain, especially if the color remains after shaking the blood in air.[4]

Special Tests: Examine the urine for blood. Also determine the methemoglobin concentration in the blood when chemical intoxication is suspected and at regular intervals until the methemoglobin has been fully reduced to normal hemoglobin.[5] Methemoglobin may be differentiated from sulfhemoglobin by the addition of a few drops of 10% potassium cyanide, which results in the rapid production of bright red cyanomethemoglobin but has no effect on the color of sulfhemoglobin.[4] Spectrophotometry is required for the precise identification of the pigment and its quantitation. Normal acid methemoglobin has a characteristic absorption spectrum with peaks at 502 and 632 nm, which disappear with the addition of cyanide, whereas sulfhemoglobin has a peak at 620 nm, which does not disappear with cyanide.[4]

Treatment. All the contaminant on the body must be removed. Immediately remove all clothing, and wash the entire body from head to foot with soap and water. Pay special attention to the hair and scalp, finger and toenails, nostrils, and ear canals. Administer oxygen to alleviate the headache and general sense of weakness; confine the patient to bed. Determine the methemoglobin concentration in the blood, and repeat every 3 to 6 hours for 18 to 24 hours. Repeat skin cleansing if the methemoglobin concentration appears to rise after 3 to 4 hours. In general, patients will return to normal within 24 hours provided that all sources of further absorption are completely eliminated. The methemoglobin will be reduced spontaneously to ferrous hemoglobin in 2 to 3 days.[4]

Such therapy is not effective in subjects with glucose–6–dehydrogenase deficiency.[4]

The only justifiable use of methylene blue would be in cases of coma or stupor, usually at methemoglobin levels over 60%. In those patients in whom therapy is necessary, methylene blue, 1 to 2 mg/kg of a 1% solution in saline, may be given intravenously over a 10-minute period. If cyanosis has not disappeared within an hour, a second dose of 2 mg/kg may be administered.[4,5] The total dose should not exceed 7 mg/kg, because methylene blue may cause toxic effects such as dyspnea, precordial pain, restlessness, apprehension, red cell hemolysis, and changes in the electrocardiogram (reduction in the height or even reversal of the T wave, frequently with lowering of the R wave).[5,6]

REFERENCES

1. Hamblin DO: Aromatic nitro and amino compounds. In Patty FA (ed): Industrial Hygiene

and Toxicology, 2nd ed, Vol 2, Toxicology, pp 2105–2119. New York, Wiley–Interscience, 1963
2. Dost FN, Reed DJ, Wang CH: Toxicology of nitrogen trifluoride. Toxicol Appl Pharmacol 17:585–595, 1970
3. Torkelson TR, Oyen F, Sadek SE, Rowe VK: Preliminary toxicologic studies on nitrogen trifluoride. Toxicol Appl Pharmacol 4:770–781, 1962
4. Vernot EH, Haun CC, MacEwen JD, Egan GF: Acute inhalation toxicology and proposed emergency exposure limits of nitrogen trifluoride. Toxicol Appl Pharmacol 26:1–13, 1973
5. Reider RF: Methemoglobinemia and sulfmethemoglobinemia. In Wyngaarden JB, Smith LH (eds): Cecil Textbook of Medicine, 16th ed, p 896. Philadelphia, WB Saunders, 1982
6. Mangelsdorff AF: Treatment of methemoglobinemia. AMA Arch Ind Health 14:148–153, 1956

NITROGLYCERIN

CAS: 55-63-0

$CH_2NO_3CHNO_3CH_2NO_3$

1987 TLV = 0.05 ppm; skin

Synonyms: Glycerol trinitrate; nitroglycerol; trinitroglycerol; NG; trinitrin

Physical Form. Liquid

Uses. Manufacture of dynamite (mixed with EGDN, qv)

Exposure. Inhalation; skin absorption; data on toxic effects are reported chiefly from industrial exposures to EGDN–NG mixed vapors

Toxicology. Nitroglycerin (NG) causes vasodilatation and may produce low levels of methemoglobin and the formation of Heinz bodies in erythrocytes.

Intoxication results in a characteristic intense throbbing headache, presumably due to cerebral vasodilatation, often associated with dizziness and nausea, and occasionally with vomiting and abdominal pain.[1,2] More severe exposure also caused hypotension, flushing, palpitation, low levels of methemoglobinemia, delirium, and depression of the central nervous system. Aggravation of these symptoms after alcohol ingestion has been observed. Upon repeated exposure, a tolerance to headache develops but is usually lost after a few days without exposure. At times, persistent tachycardia, diastolic hypertension, and reduced pulse pressure have been observed. Rarely, a worker may have an attack of angina pectoris a few days after cessation of repeated exposures, a manifestation of cardiac ischemia. Sudden death due to unheralded cardiac arrest has also been reported under these circumstances.[2]

Volunteers exposed to the vapor of a mixture of NG and EGDN at a combined concentration of 2 mg/m^3 in an explosives magazine suffered headache and fall in blood pressure within 3 minutes of exposure; a mean concentration of 0.7 mg/m^3 for 25 minutes also produced lowered blood pressure and slight headache. Workers intermittently exposed (two or three times per week for an unspecified time period) to NG vapor at concentrations of 0.03 to 0.11 ppm complained of headache; complaints ceased when concentrations were reduced to below 0.01 ppm.[4]

Hemolytic crises may occur in workers with glucose–6–phosphate dehydrogenase deficiency within a few days after initial exposure to TNT.[5]

An excess frequency of ischemic heart disease and stroke among dynamite workers was observed in Sweden[6]; an excess of deaths from acute myocardial infarction in a similar group was reported in Scotland.[7]

REFERENCES

1. Trainor DC, Jones RC: Headaches in explosive magazine workers. Arch Environ Health 12:231–234, 1966
2. Carmichael P, Lieben J: Sudden death in ex-

plosives workers. Arch Environ Health 7:424–439, 1963
3. Lund RP, Haggendal J, Johnsson G: Withdrawal symptoms in workers exposed to nitroglycerin. Br J Ind Med 25:136–138, 1968
4. Fredrick WG et al: Annual Report, Bureau of Industrial Hygiene, Detroit Department of Health, Detroit, Michigan, 1961
5. Djerassi LS, Vitany L: Hemolytic episode in G6 PD Deficient workers exposed to TNT. Br J Ind Med 32:54–58, 1975
6. Hogstedt C, Axelson O: Nitroglycerin-nitroglycol exposure and mortality in cardio-cerebrovascular diseases among dynamite workers. J Occup Med 19:675, 1977
7. Craig R et al: Sixteen-year follow-up of workers in an explosives factory. J Soc Occup Med 35:107–110, 1985

NITROMETHANE

CAS: 75-52-5

CH_3NO_2 1987 TLV = 100 ppm

Synonym: Nitrocarbol

Physical Form. Oily liquid

Uses. Solvent; chemical synthesis; rocket fuel

Exposure. Inhalation

Toxicology. Nitromethane, in animals, affects the central nervous system (convulsions and narcosis), is a mild pulmonary irritant, and causes liver damage.

No systemic effects have been reported in humans; the liquid is a mild irritant because of solvent action.[1]

Rabbits died from exposure to 10,000 ppm for 6 hours; effects were weakness, ataxia, and muscular incoordination followed by convulsions.[2,3] The same concentration for 3 hours was not fatal. Autopsy of animals exposed to lethal concentrations showed focal necrosis in the liver and moderate kidney damage. Lower concentrations produced slight irritation of the respiratory tract, followed by mild narcosis, weakness, and salivation, but no evidence of eye irritation. However, a single monkey exposed to 1000 ppm for eight 6-hour exposures died.

REFERENCES

1. Hygienic Guide Series: Nitromethane. Am Ind Hyg Assoc J 2:518–520, 1961
2. Stokinger HE: Aliphatic nitro compounds, nitrates, nitrites. In Clayton GD, Clayton FE (eds): Patty's Industrial Hygiene and Toxicology, 3rd ed, rev, Vol 2C, Toxicology, pp 4153–4155. New York, Wiley–Interscience, 1982
3. Machle W, Scott EW, Treon J: Physiological response of animals to some simple mononitroparaffins and to certain derivatives of these compounds. J Ind Hyg Toxicol 22:315–332, 1940

1-NITROPROPANE

CAS: 108-03-2

$CH_3CH_2CH_2NO_2$ 1987 TLV = 25 ppm

Synonym: 1–NP

Physical Form. Liquid

Uses. Solvent for organic materials; propellant

Exposure. Inhalation

Toxicology. 1–Nitropropane vapor is an irritant of the eyes; in animals, it also causes liver damage and mild respiratory tract irritation.[1] There are no reports of systemic effects from industrial exposures.

Rabbits died from exposure to 5000 ppm for 3 hours, but 10,000 ppm for 1 hour was not lethal.[2] Effects included conjunctival irritation, lacrimation, slow respiration with some rales, incoordination, ataxia, and weakness.[2] Autopsy of animals exposed to lethal concentrations revealed severe fatty infiltration of the liver and moderate kidney damage.[2]

Rats exposed 7 hours/day, 5 days/week at 100 ppm for up to 21 months showed no gross organ or tissue effects attributable to 1-nitropropane, and no microscopic liver changes were noted.[3]

Although there is no evidence that 1-nitropropane is carcinogenic, its other toxic properties are similar to its isomer, 2–nitropropane, which is considered an animal carcinogen.

REFERENCES

1. Silverman L, Schulte HF, First MW: Further studies on sensory response to certain industrial solvent vapors. J Ind Hyg Toxicol 28:262, 1946
2. Machle W, Scott EW, Treon JF: The physiological response of animals to some simple mononitroparaffins and to certain derivatives of these compounds. J Ind Hyg Toxicol 22:315, 1940
3. 1–Nitropropane. Documentation of the TLVs and BEIs, 5th ed, p 440. Cincinnati, American Conference of Governmental Industrial Hygienists (ACGIH), 1986

2–NITROPROPANE

CAS: 79-46-9

$CH_3CHNO_2CH_3$ 1987 TLV = 10 ppm

Synonyms: Isonitropropane; nitroisopropane; 2–NP; NiPar S–20; NiPar S–30

Physical Form. Liquid

Uses. In solvent systems to improve drying time and to alter various other characteristics such as solvent release and pigment dispersion

Exposure. Inhalation

Toxicology. 2–Nitropropane is a pulmonary irritant; in animals, it has also caused liver damage, and it is expected that severe exposure will cause the same effect in humans. Inhalation of vapor produces hepatocellular carcinomas in rats; it is a suspected human carcinogen.

Workers exposed to hot vapor containing an unspecified concentration of xylene and 20 to 45 ppm of 2–nitropropane experienced occipital headache, anorexia, nausea, vomiting, and, in some cases, diarrhea.[1] Substitution of methyl ethyl ketone for 2–nitropropane eliminated the problem. Workers exposed to 30 to 300 ppm of 2–nitropropane complained of irritation of the respiratory tract.[2]

Chronic health effects in humans from exposure to 2–nitropropane have not been adequately determined, although a retrospective mortality study of 1481 employees and former employees of a 2–nitropropane production facility with up to 27 years of exposure found no increase in cancer of the liver or other organs and no unusual disease mortality pattern.[3]

Rabbits died from exposure to a concentration near 2400 ppm for 4.5 hours, but 1400 ppm was not lethal.[4] High concentrations caused lethargy, weakness, difficult breathing, cyanosis, prostration, and occasional convulsions; low levels of methemoglobin and the formation of Heinz bodies in erythrocytes were observed. Autopsy of animals exposed to lethal concentrations revealed pulmonary edema and hemorrhage, and liver damage.[4] The 6-hour LC_{50} in the male rat is 400 ppm.[5]

Rats exposed to 207 ppm daily for 6 months developed hepatic neoplasms; hepatocellular hyperplasia and necrosis occurred after 3 months exposure at this concentration.[5] In another series of inhalation experiments on rats, 200 ppm produced hepatocellular carcinomas in both sexes; 100 ppm resulted in liver tumors in males after 12 months exposure, and in females after 18 months. At 25 ppm for up to 22 months exposure, no tumors or other hepatic lesions were produced.[6]

The IARC has determined that there is sufficient evidence for carcinogenicity in rats.[7] There is suspected carcinogenic potential for humans.

REFERENCES

1. Skinner JB: The toxicity of 2–nitropropane. Ind Med 16:441–443, 1947
2. AIHA Hygienic Guide Series: Nitropropane. Akron, Ohio, American Industrial Hygiene Association, 1978
3. Bolender FL: 2–NP Mortality Epidemiology Study of the Sterlington, LA Employees: An Update Report to the International Minerals & Chemical Corp., Northbrook, Illinois, 1983
4. Treon JF, Dutra FR: Physiological response of experimental animals to the vapor of 2–nitropropane. AMA Arch Ind Hyg Occup Med 5:52–61, 1952
5. Lewis TR, Ulrich CE, Busey WM: Subchronic inhalation toxicity of nitromethane and 2–nitropropane. J Environ Pathol Toxicol 2:233–249, 1979
6. Griffin TB, Stein AA, Coulston F: Inhalation exposure of rats to vapors of 1–nitropropane at 100 ppm. Ecotoxicol Environ Safety 6:268–282, 1982
7. IARC Monograph on the Evaluation of the Carcinogenic Risk of Chemicals to Humans, 2–Nitropropane, Vol 29, pp 331–343. Lyon, International Agency for Research on Cancer, 1982

N-NITROSODIMETHYLAMINE

CAS: 62-75-9

$(CH_3)_2NN{=}O$

1987 TLV = none established; suspected carcinogen; skin

Synonyms: Dimethylnitrosamine; DMNA; DMN

Physical Form. Yellow liquid

Uses. Industrial solvent; in synthesis of rocket fuel

Exposure. Inhalation; skin absorption

Toxicology. *N*-Nitrosodimethylamine (DMN) is a highly toxic compound and is carcinogenic in animals.

Two men accidentally exposed to DMN developed toxic hepatitis.[1] There are no reports of chronic effects from human DMN exposure.[2]

The LC_{50} for rats exposed to DMN vapor for 4 hours (and observed for 14 days) was 78 ppm; for similarly exposed mice, the LC_{50} was 57 ppm.[3] Dogs exposed for 4 hours to 16 to 144 ppm developed vomiting, polydipsia, and anorexia; most exposed dogs died, but one survivor showed residual liver damage 7 months after exposure.[3]

Swiss mice fed a diet containing 0.005% DMN for 1 week developed tumors of the kidney and lung.[4] Hamsters fed a diet containing 0.0025% for 11 weeks developed liver tumors.[5] A consistent observation following oral administration of DMN in rats has been that long-term treatment with doses compatible with a favorable survival rate leads to liver tumors, whereas short-term treatment with high doses produces renal tumors.[2]

Hamsters receiving weekly subcutaneous injections of DMN for life developed tumors; 3 of 10 females receiving weekly injections of 4.3 mg/kg developed liver tumors; at 21.5 mg/kg/week, there were eight liver tumors and five kidney tumors; in 10 male animals receiving 2.8 mg/kg/week, there were five liver tumors and one kidney tumor.[6]

Intraperitoneal injection of 6 mg/kg once weekly for 10 weeks in mice resulted in a statistically significant increase of vascular tumors, mainly in the retroperitoneum in females. There was a low incidence of hepatic vascular tumors in both sexes.[7]

REFERENCES

1. Freund HA: Clinical manifestation and studies in parenchymatous hepatitis. Ann Intern Med 10:1144–1155, 1937
2. IARC Monographs on the Evaluation of the Carcinogenic Risk of Chemicals to Man. Vol 1,

pp 95–106. Lyon, International Agency for Research on Cancer, 1971
3. Jacobson KH, Wheelwright HJ Jr, Clem JH, Shannon RN: Studies on the toxicology of N-nitrosodimethylamine Vapor. AMA Arch Ind Health 12:617–622, 1955
4. Terracini B, Palestro G, Gigliardi RM, Montesano R: Carcinogenicity of dimethylnitrosamine in Swiss mice. Br J Cancer 20:871–876, 1966
5. Tomatis L, Magee PN, Shubik P: Induction of liver tumors in the Syrian golden hamster by feeding dimethylnitrosamine. J Natl Cancer Inst 33:341–345, 1964
6. Mohr U, Haas H, Hilfrich J: The carcinogenic effects of dimethylnitrosamine and nitrosomethylurea in European hamsters (*Cricetus cricetus* L.). Br J Cancer 29:359–364, 1974
7. Cardesa A, Pour P, Althoff J, Mohr U: Vascular tumors in female Swiss mice after intraperitoneal injection of dimethylnitrosamine. J Natl Cancer Inst 51:201–205, 1973

N-NITROSODIPHENYLAMINE

CAS: 86-30-6

$(C_6H_5)_2NN{=}O$ 1987 TLV = none established

Synonyms: NDPhA; diphenyl nitrosamine

Physical Form. Yellow to brown or orange powder or flakes

Uses. Vulcanization retarder in the rubber industry

Exposure. Inhalation

Toxicology. *N*-Nitrosodiphenylamine (NDPhA) is an animal carcinogen and causes bladder tumors in male and female rats.

Early carcinogenic studies in rats and mice in which NDPhA was administered orally or by intraperitoneal injection showed no evidence of carcinogenicity.[1–5] However, a more recent study demonstrated carcinogenesis in rats.[6,7] NDPhA was administered in the diet to rats and mice at the maximal tolerated dose for each species and at one half that amount. A significant incidence of bladder tumors occurred in male (40%) and female (90%) rats at 240 mg/kg and 320 mg/kg, respectively. Few bladder tumors were seen in the mice.

The IARC has determined that there is limited evidence for carcinogenicity in experimental animals and that no evaluation of the carcinogenicity to humans can be made.[8]

REFERENCES

1. Argus MF, Hoch–Ligeti C: Comparative study of the carcinogenic activity of nitrosamines. J Natl Cancer Inst 27:695–709, 1961
2. Boyland E, Carter RL, Gorrod JW, Roe FJC: Carcinogenic properties of certain rubber additives. Eur J Cancer 4:233–239, 1968
3. National Cancer Institute. Evaluation of Carcinogenic, Teratogenic, and Mutagenic Activities of Selected Pesticides and Industrial Chemicals, Vol I. Carcinogenic Study, 1968
4. Innes JRM et al: Bioassay of pesticides and industrial chemicals for tumorigenicity in mice: A preliminary note. J Natl Cancer Inst 42(6):1101–1106, 1969
5. Druckrey H, Preussmann R, Ivankovic S, Schmahl D: Organotrope carcinogene Wirkungen bei 65 Verschiedenen N-nitroso-verbindungen an BD-Ratten. Z Krebsforsch 69:103–201, 1967
6. National Cancer Institute: Bioassay of N-Nitrosodiphenylamine for Possible Carcinogenicity. DHEW (NIH) Pub No 79–1720. Washington, DC, US Government Printing Office, 1979
7. Cardy RH, Lijinsky W, Hilderbrandt PW: Neoplastic and non-plastic urinary bladder lesions induced in Fischer 344 rats and B6C3F, hybrid mice by N-nitrosodiphenylamine. Ectotoxicol Environ Safety 3:29–35, 1979
8. IARC Monographs on the Evaluation of the Carcinogenic Risk of Chemicals to Humans, Vol 27, pp 213–225. Lyon, International Agency for Research on Cancer, 1982

NITROTOLUENE
CAS: 88-72-1; 99-08-1; 99-99-0
$CH_3C_6H_4NO_2$ 1987 TLV = 2 ppm; skin

Synonyms: Methylnitrobenzene; nitrotoluol; nitrophenylmethane

Physical Form. Ortho and meta isomers are yellowish liquid; para isomer is a yellow solid.

Uses. All isomers are used in the synthesis of dyestuffs and explosives and as chemical intermediates.

Exposure. Inhalation; skin absorption

Toxicology. Nitrotoluene has a low potency for producing methemoglobin and subsequent anoxia. Chronic exposure to other aromatic nitro compounds has caused anemia, and it is expected that nitrotoluene may cause the same effect.

Signs and symptoms of overexposure are due to the loss of oxygen-carrying capacity of the blood. The onset of symptoms of methemoglobinemia is often insidious and may be delayed up to 4 hours; headache is commonly the first symptom and may become quite intense as the severity of methemoglobinemia progresses.[1] Cyanosis develops when the methemoglobin concentration is 15% or more; blueness develops first in the lips, nose, and earlobes and is usually recognized by fellow workers. Until the methemoglobin concentration approaches approximately 40%, the individual feels well, has no complaints, and typically may insist that nothing is wrong. At methemoglobin concentrations over 40%, there is usually weakness and dizziness; at closer to 70% concentration, there may be ataxia, dyspnea on mild exertion, tachycardia, nausea, vomiting, and drowsiness.[1]

In general, higher ambient temperatures increase susceptibility to cyanosis from exposure to methemoglobin-forming agents.[2,3]

Administered by oral gavage to rats for 6 months, all three isomers produced splenic lesions.[4] The meta and para isomers produced testicular atrophy, whereas ortho-nitrotoluene caused renal lesions.[4] Only the ortho isomer induces DNA excision repair in the *in vivo–in vitro* hepatocyte unscheduled DNA synthesis assay.[5] Furthermore, ortho-nitrotoluene binds to hepatic DNA to a much greater extent than does meta- or para-nitrotoluene, and investigators suggest that it may act similarly to the rodent hepatocarcinogen 2,6–dinitrotoluene.[6]

Diagnosis. *Signs and Symptoms:* Signs and symptoms include headache; signs of anoxia, including cyanosis of lips, nose, and earlobes; and anemia.

Differential Diagnosis: Other causes of cyanosis must be differentiated from methemoglobinemia due to severe nitrotoluene exposure. These include hypoxia due to lung disease, hypoventilation, or decreased cardiac output. Lung disease may be suspected from results of pulmonary function tests and arterial blood gas analysis. The arterial Po_2 may be normal in methemoglobinemia but tends to be decreased in cyanosis due to lung disease. Hypoventilation will cause elevation of arterial Pco_2, which is not seen in nitrotoluene exposure. Decreased cardiac output states will cause cyanosis only when accompanied by arterial hypotension. If blood withdrawn from the vein shows the characteristic chocolate-brown coloration, the diagnosis of an abnormal pigment is almost certain, especially if the color remains after shaking the blood in the air.[7]

Special Tests: Determine the methemoglobin concentration in the blood when nitrotoluene intoxication is suspected and at regular intervals until the methemoglobin has been fully reduced to normal hemoglobin.[8] Methemoglobin may be differentiated from sulfhemoglobin by the addition of a few drops of 10% potassium cyanide, which results in the rapid production of bright red cyanomethemoglobin but has no effect on the color of sulfhemoglobin.[7] Spectrophotometry is required for the precise identification of the pigment and its quantitation. Normal acid methemoglobin has a characteristic absorption spectrum with peaks at 502 and 632 nm, which disappear with the addition of cyanide, whereas sulfhemoglobin has a peak at 620 nm, which does not disappear with cyanide.[7] The presence of *o*-aminophenol and *p*-aminophenol in the urine is an indication of exposure.[3]

Treatment. All nitrotoluene on the body must be removed. Immediately remove all clothing, and wash the entire body from head to foot with soap and water. Pay special attention to the hair and scalp, finger and toenails, nostrils, and ear canals. Administer oxygen to alleviate the headache and general sense of weakness; confine the patient to bed. Determine the methemoglobin concentration in the blood, and repeat every 3 to 6 hours for 18 to 24 hours. Repeat skin cleansing if the methemoglobin concentration appears to rise after 3 to 4 hours. In general, patients will return to normal within 24 hours provided that all sources of further absorption are completely eliminated.[1]

The only justifiable use of methylene blue would be in cases of coma or stupor, usually at methemoglobin levels over 60%; in those patients in whom therapy is considered necessary, methylene blue may be given intravenously 1 to 2 mg/kg over a 5-minute period as a 1% solution; if cyanosis has not disappeared within 1 hour, a second dose of 2 mg/kg should be administered.[7,8] The total dose should not exceed 7 mg/kg, because methylene blue may cause toxic effects such as dyspnea, precordial pain, restlessness, apprehension, red cell hemolysis, and changes in the electrocardiogram (reduction in the height or even reversal of the T wave, frequently with lowering of the R wave.)[7,8]

REFERENCES

1. Hamblin DO: Aromatic nitro and amino compounds. In Patty FA (ed): Industrial Hygiene and Toxicology, 2nd ed, pp 2105–2119, 2148–2149. New York, Wiley–Interscience, 1963
2. Linch AL: Biological monitoring for industrial exposure to cyanogenic aromatic nitro and amino compounds. Am Ind Hyg Assoc J 35:426–432, 1974
3. Linch AL: Nitro-compounds, aromatic. In International Labor Office: Encyclopaedia of Occupational Health and Safety, Vol II, L–Z, pp 942–944. New York, McGraw-Hill, 1972
4. Ciss M et al: Toxicological study of nitrotoluenes: Long-term toxicity. Dakar Med 25:293, 1980
5. Doolittle DJ et al: The influence of intestinal bacteria, sex of the animal, and position of the nitro group on the hepatic genotoxicity of nitrotoluene isomers in vivo. Cancer Res 43:2836, 1983
6. Rickert DE et al: Hepatic macromolecular covalent binding of mononitrotoluenes in Fischer-344 rats. Chem Biol Interact 52:131–139, 1984
7. Rieder RF: Methemoglobinemia and sulfhemoglobinemia. In Wyngaarden JB, Smith LH (eds): Cecil Textbook of Medicine, 16th ed, p 896. Philadelphia, WB Saunders, 1982
8. Mangelsdorff AF: Treatment of methemoglobinemia. AMA Arch Ind Health 14:148–153, 1956

NONYLPHENOL

CAS: 25154-52-3 (mixed isomers)
136-83-4 (2–nonylphenol)
104-40-5 (4–nonylphenol)

$C_9H_{19}(C_6H_4)OH$ 1987 TLV = none established

Synonyms: 2–nonylphenol; *o*-nonylphenol; 4–nonylphenol; *p*-nonylphenol

Physical Form. Clear, straw-colored liquid; technical grade is a mixture of isomers, predominantly para-substituted

Uses. Principal use as an intermediate in the production of nonionic ethoxylated surfactants; as an intermediate in the manufacture of phosphite antioxidants used for the plastics and rubber industries

Exposure. Inhalation

Toxicology. Nonylphenol is a severe irritant of the eyes and skin.

Reports of the oral LD_{50} in rats for the mixed isomers have ranged from 580 to 1537 mg/kg; the dermal LD_{50} in rabbits was between 2000 and 3160 mg/kg.[1–4] Nonylphenol is considered to be a corrosive agent that may cause burns and blistering of the skin.[3] When the liquid was applied to the shaved skin of a rabbit and left in place for 4 hours, there was skin necrosis 48 hours after the application.[5] No skin sensitization occurred in tests with guinea pigs.[6] When tested on black guinea pigs and black mice, irritation was observed, but nonylphenol did not induce depigmentation.[7]

The liquid in the eye of the rabbit as a 1% solution caused severe corneal damage.[1,2]

Leukoderma was reported in two women engaged in degreasing metal parts with synthetic detergents containing polyoxyethylene (3 to 16), nonyl-, or octylphenylether. Analysis revealed contamination with free alkylphenol, possibly octylphenol, or nonylphenol. Although a relationship between the cases of leukoderma and octyl- and nonylphenol exposure was suggested, it could not be confirmed.[8]

REFERENCES

1. Smyth HF Jr et al: Range-finding toxicity data: List VI. Am Ind Hyg Assoc J 23:95–107, 1962
2. Smyth HF Jr et al: Range-finding toxicity data: List VII. Am Ind Hyg Assoc J 30:470–476, 1969
3. Texaco Chemical Co. FYI–OTS–06845–0402 FLWP, Seq 1. Material Safety Data Sheet. Washington, DC, US Environmental Protection Agency, Office of Toxic Substances, 1985
4. Monsanto Industrial Chemicals Co. FYI–OTS–0685–0402 FLWP Seq G. Material Safety Data Sheet. Washington, DC, US Environmental Protection Agency, Office of Toxic Substances, 1985
5. Texaco Chemical Co. FYI–OTS–0685–0402 FLWP Seq I. DOT Corrosivity Study in Rabbits. Washington, DC, US Environmental Protection Agency, Office of Toxic Substances, 1985
6. Texaco Chemical Co. FYI–OTS–0685–0402 FLWP Seq I. Dermal Sensitization Study. Washington, DC, US Environmental Protection Agency, Office of Toxic Substances, 1985
7. Gellin GA, Maibach HI, Misiaszek MH, Ring M: Detection of environmental depigmenting substances. Contact Dermatitis 5:201–213, 1979
8. Ikeda M, Ohtsuji H, Miyahara S: Two cases of leukoderma, presumably due to nonyl- or octylphenol in synthetic detergents. Ind Health 8:192–196, 1970

NUISANCE PARTICULATES

Containing <1% quartz)
1987 TLV = 10 mg/m^3—total dust

Physical Form. Total dust as here described includes air-suspended particles of greater than respirable diameter.

Source. Ubiquitous

Exposure. Inhalation

Toxicology. As stated by the ACGIH,

In contrast to fibrogenic dusts which cause scar tissue to be formed in lungs when inhaled in excessive amounts, so-called nuisance dusts have a long history of little adverse effect on lungs and do not produce significant organic disease or toxic effect when exposures are kept under reasonable control. The nuisance dusts have also been called biologically inert dusts, but the latter term is inappropriate to the extent that there is no dust which does not evoke some cellular response in the lung when inhaled in sufficient amount. However, the lung-tissue reaction caused by inhalation of nuisance dusts has the following characteristics: the architecture of the air spaces remains intact; collagen (scar tissue) is not formed to a significant extent; the tissue reaction is potentially reversible.

Excessive concentrations of nuisance dusts in the workroom air may seriously reduce visibility, may cause unpleasant deposits in the eyes, ears and nasal passages (Portland Cement dust), or cause injury to the skin or mucous membranes by chemical or mechanical action per se or by the rigorous skin cleansing procedures necessary for their removal.[1]

REFERENCE

1. Nuisance particulates. Documentation of the TLVs and BEIs, 5th ed, p 445. Cincinnati, American Conference of Governmental Industrial Hygienists (ACGIH), 1986

Toxicology. In toxic doses, the higher chlorinated naphthalenes are likely to cause severe injury to the liver.[1]

Exposure of workers by inhalation or skin absorption to lower chlorinated naphthalenes (penta- and hexachloro) causes a severe acne-form dermatitis chloracne.[2] Surprisingly, on human volunteers, octachloronaphthalene was entirely nonacneigenic.[2] There is no information on systemic effects in humans. In animals, systemic toxicity from chorinated naphthalenes appears to be limited to liver injury characterized as acute yellow atrophy.[1] Ingestion experiments on cattle suggest that the octachlor is less readily absorbed than the hexachlor derivative, although the greater degree of chlorination, the more toxic the compound may be.[3]

REFERENCES

1. Deichmann WB: Halogenated cyclic hydrocarbons. In Clayton GD, Clayton FE (eds): Patty's Industrial Hygiene and Toxicology, 3rd ed, Vol 2B, Toxicology, pp 3669–3675. New York, Wiley–Interscience, 1981
2. Shelley WB, Kligman AM: The experimental production of acne by penta- and hexachloronaphthalenes. Arch Dermatitis 75:689–695, 1957
3. Octachloronaphthalene. Documentation of the TLVs and BEIs, 5th ed, p 447. Cincinnati, American Conference of Governmental Industrial Hygienists (ACGIH), 1986

OCTACHLORONAPHTHALENE

CAS: 2234-13-1

$C_{10}Cl_8$ 1987 TLV = 0.1 mg/m^3; skin

Synonym: Halowax 1051

Physical Form. Waxy solid

Uses. In electric cable insulation; additive to lubricants

Exposure. Inhalation; skin absorption

OCTANE

CAS: 111-65-9

$CH_3(CH_2)_6CH_3$ TLV = 300 ppm

Synonyms: None

Physical Form. Clear liquid

Uses. n-Octane is used as a solvent and in organic synthesis. Iso-octane is a standard for anti-knock properties of gasoline. Octanes are present in gasolines and naphthas.

Exposure. Inhalation

Toxicology. In animals, octane is a mucous membrane irritant, and, at high concentrations, it causes narcosis; it is expected that severe exposure in humans will produce the same effects.

The narcotic potency of octane is approximately that of heptane but does not appear to exhibit the central nervous system effect as do the two lower homologues.[1]

There was no narcosis in mice exposed to iso-octane at 8000 ppm for 5 minutes; at 16,000 ppm, there was sensory irritation throughout the 5-minute exposure and one of four mice died; at 32,000 ppm, effects included irritation and irregular respiration, and all four mice died within 4 minutes of exposure.[2]

REFERENCES

1. Sandmeyer EE: Aliphatic hydrocarbons. In Clayton GD, Clayton FE (eds): Patty's Industrial Hygiene and Toxicology, 3rd ed, rev, Vol 2B, Toxicology, p 3190. New York, Wiley–Interscience, 1981
2. Swann HE Jr, Kwon BK, Hogan GK, Snellings WM: Acute inhalation toxicology of volatile hydrocarbons. Am Ind Hyg Assoc J 35:511, 1974

OIL MIST (mineral)

CAS: 8012-95-1 1987 TLV = 5 mg/m^3

Synonyms: Petrolatum liquid; mineral oil; paraffin oil

Physical Form. Colorless, oily, odorless, and tasteless liquid

Uses. Mineral oil is a lubricant and is used as a solvent for inks in the printing industry.

Exposure. Inhalation

Toxicology. Mineral oil mist is of low toxicity.

A single case of lipoid pneumonitis suspected to result from repeated exposure to very high concentrations of oil mist was reported in 1950; this case occurred in a cash register serviceman whose heavy exposure occurred over 17 years of employment.[1]

A review of exposures to mineral oil mist averaging below 15 mg/m^3 (but higher in some jobs) in several industries disclosed a striking lack of reported cases of illness related to these exposures.[2] A study of oil mist exposures in machine shops, at mean concentrations of 3.7 mg/m^3 and maximal concentrations of 110 mg/m^3, showed no increase in respiratory symptoms or decrement in respiratory performance attributable to oil mist inhalation among men employed for many years.[3] Similar results were found in a 5-year study of 460 printer pressmen exposed to a respirable concentration of less than 5 mg/m^3.[4,5]

Early epidemiologic studies linked cancers of the skin and scrotum with exposure to mineral oils.[6] These effects have been attributed to contaminants such as polycyclic aromatic hydrocarbons (PAH) and/or additives with carcinogenic properties present in the oil. Solvent refining, and to some extent hydroprocessing, selectively extract PAH and reduce carcinogenicity.[7] More recent studies, which have also reported excess numbers of scrotal cancer, have failed to characterize the composition of the mineral oil and the exposure levels.[8] The IARC has determined that there is no evidence that the fully solvent refined oils are carcinogenic to experimental animals in feeding or skin painting studies.[9] The IARC's determination that there is sufficient evidence for carcinogenicity in humans is based on epidemiologic studies of uncharacterized mineral oils containing additives and impurities.[9] Most mineral oils in use today present no hazard because of refining techniques; however, because individual oils may vary in composition, an assessment must be made on each product.[10]

REFERENCES

1. Proudfit JP, Van Ordstrand HS, Miller CW: Chronic lipid pneumonia following occupational exposure. AMA Arch Ind Hyg Occup Med 1:105–111, 1950
2. Hendricks NV et al: A review of exposures to oil mist. Arch Environ Health 4:139–145, 1962
3. Ely TS, Pedley SF, Hearne FT, Stille WT: A study of mortality, symptoms, and respiratory function in humans occupationally exposed to oil mist. J Occup Med 12:253–261, 1970
4. Lippman M, Goldstein DH: Oil mist studies, environmental evaluation and control. Arch Environ Health 21:591–599, 1970
5. Goldstein DH, Benoit JN, Tyroler HA: An epidemiologic study of an oil mist exposure. Arch Environ Health 21:600–603, 1970
6. IARC Monographs on the Evaluation of the Carcinogenic Risk of Chemicals to Humans, Suppl 4, pp 227–228. Lyon, International Agency for Research on Cancer, October 1982
7. Bingham E et al: Carcinogenic potential of petroleum hydrocarbons. A critical review of the literature. J Environ Pathol Toxicol 3:483–563, 1980
8. Jarvolm B et al: Cancer morbidity among men exposed to oil mist in the metal industry. J Occup Med 23:333–337, 1981
9. IARC Monographs on the Evaluation of the Carcinogenic Risk of Chemicals to Humans, Vol 33, pp 87–168. Lyon, International Agency for Research on Cancer, April 1984
10. Kane ML et al: Toxicological characteristics of refinery streams used to manufacture lubricating oils. J Ind Med 5:183–200, 1984

OSMIUM TETROXIDE

CAS: 20816-12-0

OsO_4 1987 TLV = 0.0002 ppm (0.002 mg/m^3)

Synonym: Osmic acid

Physical Form. Colorless crystals or yellow crystalline mass with acrid chlorine-like odor

Uses. Oxidizing agent

Exposure. Inhalation

Toxicology. Osmium tetroxide is an irritant of the eyes, mucous membranes, and skin.

A laboratory investigator briefly exposed to a high concentration of vapor experienced a sensation of chest constriction and difficulty in breathing.[1] Irritation of the eyes is usually the first symptom of exposure to low concentrations of the vapor; lacrimation, a gritty feeling in the eyes, and the appearance of rings around lights are frequently reported. In most cases, recovery occurs within a few days.[2] Workers exposed to fume concentrations of up to 0.6 mg/m^3 developed lacrimation, visual disturbances, and, in some cases, frontal headache, conjunctivitis, and cough.[1]

Rabbits exposed for 30 minutes to vapor at estimated concentrations of 130 mg/m^3 developed irritation of mucous membranes and labored breathing; at autopsy, there was bronchopneumonia, as well as slight kidney damage.[1] A 4-hour exposure to 400 mg/m^3 was lethal to rats.[3] Application of a drop of 1% solution of osmium tetroxide to a rabbit eye caused severe corneal damage, permanent opacity, and superficial vascularization.[4] Osmium compounds have a caustic action on the skin that results in eczema and dermatitis.[2]

REFERENCES

1. McLaughlin AIG, Milton R, Perry KMA: Toxic manifestations of osmium tetroxide. Br J Ind Med 3:183–186, 1946
2. Hygienic Guide Series: Osmium and its compounds. Am Ind Hyg Assoc J 29:621–623, 1968
3. Registry of Toxic Effects of Chemical Substances. DHHS (NIOSH) Pub No 86-103, p 1315. Washington, DC, US Department of Health and Human Services, 1985
4. Grant WM: Toxicology of the Eye, 3rd ed, p 682. Springfield, Illinois, Charles C Thomas, 1986

OXALIC ACID

CAS: 144-62-7

HOOCCOOH 1987 TLV = 1 mg/m^3

Synonym: Ethanedioic acid

Physical Form. Crystalline solid

Uses. Chemical synthesis; bleaches; metal polish; rust remover

Exposure. Inhalation; ingestion

Toxicology. Oxalic acid is an irritant of the eyes, mucous membranes, and skin; severe intoxication results in convulsions.

There is little reported information on industrial exposure, although chronic inflammation of the upper respiratory tract has been described in a worker exposed to hot vapor arising from oxalic acid.[1] Ingestion of as little as 5 g has caused fatalities; there is rapid onset of shock, collapse, and convulsions. The convulsions are thought to be the result of hypocalcemia owing to the calcium-complexing action of oxalic acid, which depresses the level of ionized calcium in body fluids. Marked renal damage from deposition of calcium oxalate may occur.[1]

Gross contact of the hands with solutions of oxalic acid (5.3% and 11.5% in two reported cases) used as cleaning solutions caused tingling, burning, soreness, and cyanosis of the fingers.[2]

Splashes of solutions in the eyes have produced epithelial damage from which recovery has been prompt.[3]

The single oral LD_{50} for a 5% by weight oxalic acid solution was 9.5 ml/kg for male rats and 7.5 ml/kg for female rats.[4] Applied to rabbit skin, a single exposure of 20 g/kg of the solution was not lethal.

REFERENCES

1. Fassett DW: Oxalic acid and derivatives. In International Labor Office: Encyclopaedia of Occupational Health and Safety, Vol II, L–Z, p 984. New York, McGraw-Hill, 1972
2. Klauder JV, Shelanski L, Gabriel K: Industrial uses of compounds of fluorine and oxalic acid. AMA Arch Ind Health 12:412–419, 1955
3. Grant WM: Toxicology of the Eye, 3rd ed, pp 685–686. Springfield, Illinois, Charles C Thomas, 1986
4. Vernot EH et al: Acute toxicity and skin corrosion data for some organic and inorganic compounds and aqueous solutions. Toxicol Appl Pharmacol 42:417–423, 1977

OXYGEN DIFLUORIDE

CAS: 7783-41-7

OF_2 1987 TLV = C 0.05 ppm

Synonyms: Fluorine monoxide; oxygen fluoride; fluorine oxide

Physical Form. Colorless gas

Uses. Oxidant in missile propellant systems

Exposure. Inhalation

Toxicology. Oxygen difluoride is a severe pulmonary irritant in animals; exposure is expected to cause the same effect in humans.

In humans, inhalation of the gas at

fractions of a ppm produced intractable headache.[1] Although there are no reports of effects on the eyes or skin of humans, it would be expected that the gas under pressure impinging upon the eyes or skin would produce serious burns.[2]

In monkeys and dogs, the LC_{50} was 26 ppm for 1 hour; signs of toxicity included lacrimation, dyspnea, muscular weakness, and vomiting; at autopsy, massive pulmonary edema and hemorrhage were observed.[3] In mice, exposure to a low concentration (1 ppm for 60 minutes) produced tolerance to subsequent exposures 8 days later at levels that would otherwise have been fatal (4.25 ppm for 60 minutes).[3]

REFERENCES

1. Oxygen difluoride. Documentation of the TLVs and BEIs, 5th ed, p 452. Cincinnati, American Conference of Governmental Industrial Hygienists (ACGIH), 1986
2. Hygienic Guide Series: Oxygen difluoride. Am Ind Hyg Assoc J 28:194–196, 1967
3. Davis UV: Acute Toxicity of Oxygen Difluoride, Proceedings of the First Annual Conference on Environmental Toxicology, AMRL–TR–70–102, pp 329–340. Aerospace Medical Research Laboratory, Wright–Patterson Air Force Base, Ohio, 1970

OZONE

CAS: 10028-15-6

O_3 1987 TLV = 0.1 ppm

Synonym: Triatomic oxygen

Physical Form. Blue gas

Sources. Inert-gas–shielded arc welding; around ozoning devices used for air and water purification; around high voltage electric equipment

Exposure. Inhalation

Toxicology. Oxone is an irritant of the mucous membranes and the lungs

The primary target of ozone exposure is the respiratory tract.[1] Symptoms range from nose and throat irritation to cough, dyspnea, and chest pain. By analogy to animal studies, severe exposure may cause pulmonary edema and hemorrhage. Except for one report, the threshold for effects in humans appears to be between 0.2 and 0.4 ppm.[1] Exposure to 0.5 ppm 3 hours/day, 6 days/week for 12 weeks caused a significant reduction in 1-second forced expiratory volume without subject symptoms.[2] Bronchial irritation, slight dry cough, and substernal soreness were reported in subjects exposed to 0.6 to 0.8 ppm ozone for 2 hours. Marked changes in lung function lasting up to 24 hours were also found.[3] A single 2-hour exposure to 1.5 to 2 ppm caused impaired lung function, chest pain, loss of coordinating ability, and difficulty in articulation. Cough and fatigue persisted for 2 weeks.[4] No deaths from exposure to ozone have been reported.[5]

Extrapulmonary toxic effects mentioned as potentially attributable to ozone exposure include hematologic changes, chromosomal effects in circulating lymphocytes, alterations of hepatic metabolism, reproductive effects, central nervous system effects, changes in visual acuity, and altered susceptibility to infectious agents.[1,6] Some extrapulmonary effects may be secondary to respiratory system damage.

The toxic effects of ozone can be attributed to its strong oxidative capacity. Cell membranes are thought to be a likely site for oxidation and the formation of free radicals. If damage is severe, the cell dies; necrosis is commonly reported in the lungs of heavily exposed animals.[1] In animal studies, a characteristic ozone lesion occurs at the junction of the conducting airways and the gas exchange region of the lung following acute ozone exposure. This anatomic site is probably also affected in humans.[1]

One of the principal modifiers of the magnitude of response to ozone is minute ventilation (Vt), which increases proportionally with increase in exercise work load.[7]

Surprisingly, patients with mild to moderate respiratory disease do not appear to be more sensitive than normal persons to threshold ozone concentrations.[8]

The effects of ozone appear to be cumulative for initial exposures followed by adaptation. Five of six subjects exposed to 0.5 ppm ozone 2 hours/day for 4 days showed cumulative effects of symptoms and lung function tests for the first 3 days followed by a return to near control values on day 4.[9] In animals, exposure to 0.3 to 3 ppm for up to 1 hour permits the animals to withstand multilethal doses for months afterward.[10] It is not known how variations in the length, frequency, or magnitude of exposure modify the time course for tolerance.

REFERENCES

1. Menzel DB: Ozone: An overview of its toxicity in man and animals. J Toxicol Environ Health 13:183–204, 1984
2. Bennett G: Ozone contamination of high altitude aircraft cabins. Aerospace Med 33:969–973, 1962
3. Young WA et al: Effect of low concentrations of ozone on pulmonary function. J Appl Physiol 19:765–768, 1964
4. Griswold SS et al: Report of a case exposure to high ozone concentrations for two hours. Arch Ind Health 15:108–118, 1957
5. Nasr ANM: Ozone poisoning in man: Clinical manifestations and differential diagnosis—a review. Clin Toxicol 4:461–466, 1971
6. Lagerwerff JM: Prolonged ozone inhalation and its effects on visual parameters. Aerospace Med 34:479–486, 1963
7. Kagawa J: Exposure-effect relationship of selected pulmonary function measurements in subjects exposed to ozone. Int Arch Occup Environ Health 53:345–358, 1984
8. Kulle TJ et al: Pulmonary function adaptation to ozone in subjects with chronic bronchitis. Environ Res 34:55–63, 1984
9. Hackney JD et al: Adaptation to short-term respiratory effects of ozone in men exposed repeatedly., J Appl Physiol 43:82–85, 1977
10. Stokinger HE, Scheel LD: Ozone toxicity: Immunochemical and tolerance-producing aspects. Arch Environ Health 4:327–334, 1962

PARAQUAT

CAS: 4685-14-7

$C_{12}H_{14}N_2$ 1987 TLV (respirable) = 0.1 mg/m^3

Synonyms: 1,1′-Dimethyl-4,4′-bipyridinium dichloride or dimethosulfate; gramoxone; methylviologen

Physical Form. Yellow solid

Uses. Herbicide

Exposure. Inhalation; skin absorption; ingestion

Toxicology. Paraquat is an irritant of the eyes, mucous membranes, and skin; by contrast, ingestion causes fibroblastic proliferation in the lungs.

In a study of 30 workers engaged in spraying paraquat over a 12-week period, approximately 50% of them had minor irritation of the eyes or nose; one worker had an episode of epistaxis.[1] Of 296 spray operators with skin exposure described as "gross and prolonged," 55 had damaged fingernails. The most common lesion was transverse white bands of discoloration, but other lesions involved loss of nail surface, transverse ridging and gross deformity of the nail plate, and, in some cases, loss of the nail.[2]

Paraquat is commonly combined in commercial herbicides with diquat, a related compound; in several instances, the commercial preparations splashed in the eyes have caused serious injury.[3,4] Effects have been loss of corneal and conjunctival epithelium, mild iritis, and residual corneal scarring.[4] In contrast, in the eye of a rabbit, one drop of a 50% aqueous solution of pure paraquat caused slow development of mild conjunctival inflammation, and pure diquat proved even less irritating.[5] Presumably, the surfactants present in the commercial preparations are responsible for the severe eye injuries to humans.[4]

In a survey of 36 paraquat formulation workers, acute skin rashes and burns from a delayed caustic effect, eye injuries with conjunctivitis from splash injuries, nail damage, and minor epistaxis were common clinical complaints.[6] Despite a mean exposure period of 5 years, there was no evidence of chronic effects on skin, mucous membranes, or general health. Comparison of a group of 27 Malaysian plantation spraymen with a mean of 5.3 years of heavy paraquat use to unexposed groups did not demonstrate any significant differences in pulmonary, renal, liver, and hematologic functions.[7] No abnormalities were attributable to paraquat exposure.[7]

In rats exposed to aerosols of paraquat, the LC_{50} for 6 hours was 1 mg/m^3; death was delayed and resulted from pulmonary hemorrhage and edema.[8] In practice, the large particle size of agricultural sprays probably mitigates against this occurring in exposed workers.[9]

The results from ingestion by humans or injection in animals are in marked contrast to the irritant effects usually encountered in industrial exposure. There are numerous reports of fatal accidental and suicidal ingestion by humans.[9–11] In two cases, one person ingested approximately 114 ml of a 20% solution; the other person was believed to have taken only a mouthful of the liquid, most of which was rejected immediately. The former person died after 7 days; the latter person died after 15 days.[10] Initial symptoms included burning in the mouth and throat, nausea, vomiting, and abdominal pain with diarrhea. After 2 to 3 days, signs of liver and kidney toxicity developed, including jaundice, oliguria, and albuminuria; electrocardiogram changes were suggestive of toxic myocarditis with conduction defects. Shortly before death, respiratory distress occurred; at autopsy, findings in the lung included hemorrhage, edema, and massive solid areas that were airless owing to fibroblastic proliferation in the alveolar walls and elsewhere.[10] Early deaths from massive poisonings usually result from a combination of acute pulmonary edema, acute oliguric renal failure, and hepatic failure. Deaths from less massive poisonings typically result from pulmonary fibrosis that develops 1 to 3 weeks after ingestion.[9]

Intraperitoneal injection or oral administration to rats at doses that caused delayed death resulted in the same proliferative lesion in the lung; findings were alveolar, perivascular, and peribronchial edema, with cellular proliferation into the alveolar walls, resulting in large solid areas of the lung with no air-containing cavities.[5]

There is no evidence that inhalation exposures in occupational settings cause the rapid progressive pulmonary fibrosis and injury to the heart, liver, and kidneys that occur from ingestion. Rarely, dermal exposure to paraquat has resulted in systemic poisonings and deaths with renal and pulmonary damage.[9] Such episodes occurred with prolonged skin contact during spraying and exposure to concentrated solutions or exposure to areas of pre-existing dermatitis; all episodes could have been prevented by the use of recommended work practices.[9]

Workers involved in the manufacture of paraquat were found to have a high prevalence of hyperpigmented macules and hyperkeratosis, both of which may be premalignant skin lesions. Analysis of the data suggested that exposure to bipyridine precursors along with sunlight, rather than paraquat itself, was responsible.[12]

A mouse bioassay involving dietary exposure to 25, 50, and 75 mg/kg/day for 80 weeks yielded no evidence of carcinogenicity, despite the occurrence of some deaths from respiratory disease.[9] A 2-year bioassay in rats exposed to paraquat in drinking water at 1.3 and 2.6 mg/kg similarly resulted in lung pathology but no increased tumor incidence.[9]

REFERENCES

1. Swan AAB: Exposure of spray operators to paraquat. Br J Ind Med 26:322–329, 1969
2. Hearn CED, Keir W: Nail damage in spray operators exposed to paraquat. Br J Ind Med 28:399–403, 1971
3. Cant JS, Lewis DRH: Ocular damage due to paraquat and diquat. Br Med J 2:224, 1968
4. Grant WM: Toxicology of the Eye, 3rd ed, pp 699–700. Springfield, Illinois, Charles C Thomas, 1986
5. Clark DG, McElligott TF, Hurst EW: The toxicity of paraquat. Br J Ind Med 23:126–132, 1966
6. Howard JK: A clinical survey of paraquat formulation workers. Br J Ind Med 36:220–223, 1979
7. Howard JK, Sabapathy NN, Whitehead PA: A study of the health of Malaysian plantation workers with particular reference to paraquat spraymen. Br J Ind Med 38:110–116, 1981
8. Gage JC: Toxicity of paraquat and diquat aerosols generated by a size-selective cyclone: Effect of particle size distribution. Br J Ind Med 25:304–314, 1968
9. International Programme on Chemical Safety, Environmental Health Criteria 39-Paraquat and Diquat, pp 13–128. Geneva, World Health Organization, 1984
10. Bullivant CM: Accidental poisoning by paraquat: Report of two cases in man. Br Med J 1:1272–1273, 1966
11. Toner PG, Vetters JM, Spilg WGS, Harland WA: Fine structure of the lung lesion in a case of paraquat poisoning. J Pathol 102:182–185, 1970
12. Wang JD et al: Occupational risk and the development of premalignant skin lesions among paraquat manufacturers. Br J Ind Med 44:196–200, 1987

PARATHION

CAS: 56-38-2

$(C_2H_5O)_2P(S)OC_6H_4NO_2$

1987 TLV = 0.1 mg/m^3; skin

Synonyms: O,O-Diethyl O-p-nitrophenyl phosphorothioate; Akron; Niran; Amer Cyan 3422; BAY E-605; BLADAN; FOLIDOL E605

Physical Form. Brown or yellowish liquid

Uses. Acaracide; insecticide

Exposure. Inhalation; skin absorption; ingestion

Toxicology. Parathion is an anticholinesterase agent.

Hundreds of deaths associated with parathion exposure have been reported. These deaths have resulted from accidental, suicidal, and homicidal poisonings. It has been the cause of most cropworker poisonings in the United States.[1] Fatal human poisonings have resulted from ingestion, skin exposure, and inhalation (with varying degrees of skin exposure.)

Signs and symptoms of overexposure are caused by the inactivation of the enzyme cholinesterase, which results in the accumulation of acetylcholine at synapses in the nervous system, skeletal and smooth muscle, and secretory glands. The sequence of the development of systemic effects varies with the route of entry. The onset of signs and symptoms is usually prompt but may be delayed up to 12 hours.[1–4] After inhalation, respiratory and ocular effects are the first to appear, often within a few minutes after exposure. Respiratory effects include tightness in the chest and wheezing due to bronchoconstriction and excessive bronchial secretion; laryngeal spasms and excessive salivation may add to the respiratory distress, and cyanosis may also occur. Ocular effects include miosis, blurring of distant vision, tearing, rhinorrhea, and frontal headache.

After ingestion, gastroinestinal effects, such as anorexia, nausea, vomiting, abdominal cramps, and diarrhea, appear within 15 minutes to 2 hours. After skin absorption, localized sweating and muscular fasciculations in the immediate area usually occur within 15 minutes to 4 hours; skin absorption is somewhat greater at higher ambient temperatures and is increased by the presence of dermatitis.[3]

With severe intoxication by all routes, an excess of acetylcholine at the neuromuscular junctions of skeletal muscle causes weakness aggravated by exertion, involuntary twitchings, fasciculations, and, eventually, paralysis. The most serious consequence is paralysis of the respiratory muscles. Effects on the central nervous system include giddiness, confusion, ataxia, slurred speech, Cheyne–Stokes respiration, convulsions, coma, and loss of reflexes. The blood pressure may fall to low levels, and cardiac irregularities, including complete heart block, may occur.[2]

Complete symptomatic recovery usually occurs within 1 week; increased susceptibility to the effects of anticholinesterase agents persists for up to several weeks after exposure. Daily exposure to concentrations that are insufficient to produce symptoms following a single exposure may result in the onset of symptoms. Continued daily exposure may be followed by increasingly severe effects.

The minimal lethal oral dose of parathion for humans has been estimated to range from less than 10 mg up to 120 mg.[6] In a study of 115 workers exposed to parathion under varying conditions, the majority of them excreted significant amounts of *p*-nitrophenol (a metabolite of parathion) in the urine, whereas only those with heavier exposures had a measurable decrease in blood cholinesterase.[7] Measurement of urinary *p*-nitrophenol can be useful in assessing parathion absorption in occupational or other settings.[1]

With dermal exposure in the occupational setting, onset of symptoms may be delayed for several hours up to as long as 12 hours. This delay in onset, which is unusual for other organophosphate compounds, may occur even with poisonings that prove to be serious.[1]

Parathion itself is not a strong cholinesterase inhibitor, but one of its metabolites, paraoxon, is an active inhibitor. Paraoxon inactivates cholinesterase by phosphorylation of the active site of the enzyme to form the "diethylphosphoryl enzyme." Over the following 24 to 48 hours, there is a process, termed *aging*, of conversion to the "monoethylphosphoryl enzyme." Aging is of clinical interest in the treatment of poisoning, because cholinesterase reactivators such as pralidoxime (2–PAM, Protopam) chloride are ineffective after aging has occurred.

In the field, parathion is converted to varying degrees of paraoxon, which may persist on foliage and in soil. Exposure to paraoxon from weathered parathion residues by the dermal route upon re-entry by fieldworkers has resulted in anticholinesterase poisonings.[8]

In an animal bioassay, a dose-related increase in the incidence of adrenal cortical adenomas (with a few carcinomas at this site as well) has been observed in one strain of rats of both sexes. The significance of these lesions in aged rats is unclear. Other bioassays in mice and rats had sufficient limitations, such that the IARC deemed them inadequate for evaluation and concluded that there are insufficient data to evaluate the carcinogenicity of parathion for animals and no data for humans.[9] There are no documented cases of peripheral neuropathy from parathion exposure.[9]

Diagnosis. *Signs and Symptoms:* Initial symptoms include headache, blurred vision, pallor, weakness, sweating, abdominal pain, nausea, vomiting, and diarrhea.

Moderate-to-severe intoxication includes miosis, lacrimation, excessive salivation, muscle fasciculations, dyspnea, cyanosis, convulsions, shock, cardiac arrhythmias, and coma.

Differential Diagnosis: Diagnosis is primarily based on a history of exposure and clinical evidence of diffuse parasympathetic stimulation. Careful observation of the effects of atropine and pralidoxime may be valuable. Patients with organophosphate poisoning are resistant to the action of atropine at moderate dosages; failure of 1 to 2 mg of atropine administered parenterally to produce signs of atropinization (flushing, mydriasis, tachycardia, or dryness of mouth) indicates anticholinesterase poisoning. Intravenous injection of 1 g pralidoxime generally causes some recovery from signs and symptoms.

Special Tests: Two types of cholinesterase are clinically significant: (1) true acetylcholinesterase, found principally in the nervous system and the red blood cell; and (2) pseudo- or butyrylcholinesterase, found in the plasma, liver, and nervous system. Whereas the action of both types is inhibited by organophosphates, the level of depression of red blood cell cholinesterase is a better indicator of clinically significant reduction of cholinesterase activity in the nervous system.

Laboratory evidence of depression of red blood cell cholinesterase to a level substantially below pre-exposure levels (at least 50% and usually much lower) is verification of poisoning. There is an imperfect correlation between the degree of depression of cholinesterase enzymes and the occurrence of symptoms. With a rapid drop in cholinesterase activity, generally reflecting an acute heavy exposure, there may be symptoms with only a 30% depression, whereas with slower drops to 70% depression, reflecting chronic low-level exposure, there may be no symptoms.[7]

If no pre-exposure baseline has been performed but symptoms are not sufficient to justify treatment with atropine, repeated testing during the recovery period demonstrating progressively increasing plasma and red blood cell cholinesterase levels over several days and weeks, respectively, suggests the diagnosis of anticholinesterase poisoning.

There are many different methods for estimation of cholinesterase content of blood, and associated with each method is a different set of normal values and a different set of reporting units. The laboratory report of a cholinesterase determination should state the units involved along with the appropriate normal range. Based on the Michel method, the normal range of red blood cell cholinesterase activity (delta *p*H per hour) is 0.39 to 1.02 for men, and 0.34 to 1.10 for women.[11] The normal range of the enzyme activity (delta *p*H per hour) of plasma is 0.44 to 1.63 for men, and 0.24 to 1.54 for women.

Treatment. Treatment of organophosphate poisoning ranges from simple removal from exposure in very mild cases to the provision of rigorous supportive and antidotal measures in severe cases.[2,3,9,10] In moderate-to-severe cases, because of pulmonary involvement, there may be need for artificial respiration using a positive-pressure method. Careful attention must be paid to removal of secretions and to maintenance of a patent airway. Anticonvulsants such as diazepam or thiopental sodium may be necessary. Maintenance of respiration is critical, because death usually results from weakness of the muscles of respiration and accumulation of excessive secretions in the respiratory tract.[2]

As soon as cyanosis has been overcome, 2 to 4 mg of atropine should be given intravenously. (Atropine may induce ventricular fibrillation in the presence of cyanosis.) *This dose of atropine is approximately 10 times the amount that is administered for other conditions in which atropine is considered therapeutic.* This dose should be repeated at 5- to 10-minute intervals until signs of atropinization appear (dry, flushed skin, tachycardia as high as 140 beats/minute, and pupillary dilatation). If muscarinic symptoms reappear, the dose of atropine should be repeated. A mild degree of atropinization should be maintained for at least 48 hours.[2]

Pralidoxime (2–PAM, Protopam) chloride is a cholinesterase reactivator that complements the action of atropine. It has its greatest effect in reversing the nicotinic action of anticholinesterase agents at skeletal neuromuscular junctions but virtually no effect on central nervous system manifestations. In moderate-to-severe cases, the dose for adults is 1 to 2 g injected intravenously at a rate not in excess of 500 mg/minute. If muscle weakness has not been relieved or if it recurs within 20 minutes, a second dose of 1 g is indicated. Treatment with pralidoxime chloride will be most effective if given within 24 hours after poisoning.[2] Morphine, aminophylline, and phenothiazines are contraindicated because of documented experience of adverse reactions in cases of organophosphate poisoning.[10]

It is of great importance to decontaminate the patient. Contaminated clothing should be removed at once, and the skin should be washed with generous amounts of soap or detergent and a flood of water, which is best accomplished under a shower or by submersion in a pond or other body of water if the exposure occurred in the field. Careful attention should be paid to cleansing of the skin and hair.

The patient should be attended and monitored continuously for not less than 24 hours, because serious and sometimes fatal relapses have occurred as a result of continuing absorption of the toxin or dissipation of the effects of the antidote.

Regeneration of cholinesterase is primarily by synthesis of new enzyme and takes place at the rate of approximately 1% per day.[5,10] A patient who has re covered from the acute phase of poisoning remains hypersusceptible to anticholinesterases for up to several weeks.

Medical Control. Medical control involves preplacement and annual physical examination with determination of pre-exposure red blood cell and plasma cholinesterase activity. A person whose red blood cell cholinesterase falls to or below 40% of the pre-exposure baseline should be removed from further exposure until the activity returns to within 80% of the pre-exposure baseline.[12]

REFERENCES

1. Hayes WJ Jr: Organic phosphorus pesticides. In Pesticides Studied in Man, pp 284–435, Baltimore, Williams & Wilkins, 1982
2. Taylor P: Anticholinesterase agents. In Gilman AG et al (eds): Goodman and Gilman's The Pharmacological Basis of Therapeutics, 7th ed, pp 110–129. New York, Macmillan, 1985
3. Koelle GB (ed): Cholinesterases and anticholinesterase agents. Handbuch der Experimentellen Pharmakologie, Vol 15, pp 989–1027. Berlin, Springer-Verlag, 1963
4. Namba T, Nolte CT, Jackrel J, Grob D: Poisoning due to organophosphate insecticides. Am J Med 50:475–492, 1971
5. Milby TH: Prevention and management of organophosphate poisoning. JAMA 216:2131–2133, 1971
6. Hygienic Guide Series: Parathion. Am Ind Hyg Assoc J 30:308–312, 1969
7. Arterberry JD, Durham WF, Elliott JW, Wolfe HR: Exposure to parathion—measurement by blood cholinesterase level and urinary *p*-nitrophenol excretion. Arch Environ Health 3:476–485, 1961
8. Spear RC et al: Worker poisonings due to paraoxon residues. J Occup Med 19:411–414, 1977
9. IARC Monographs on the Evaluation of the Carcinogenic Risk of Chemicals to Humans: Miscellaneous Pesticides, Vol 30, pp 153–181. Lyon, International Agency for Research on Cancer, 1983
10. Coye MJ, Lowe JA, Maddy KT: Biological monitoring of agricultural workers exposed to pesticides. I. Cholinesterase activity determinations. J Occup Med 28:619–627, 1986
11. Michel HO: Electrometric method for determination of red blood cell and plasma cholinesterase activity. J Lab Clin Med 34:1564–1568, 1949
12. National Institute for Occupational Safety and Health: Criteria for a Recommended Standard . . . Occupational Exposure to Parathion, DHEW (NIOSH) Pub No 76-190, pp 13–36. Washington, DC, US Government Printing Office, 1976

PENTABORANE

CAS: 19624-22-7

B_5H_9 1987 TLV = 0.005 ppm

Synonyms: Dihydropentaborane (9); pentaboron undecahydride

Physical Form. Volatile liquid

Uses. Reducing agent in propellant fuels

Exposure. Inhalation; ingestion

Toxicology. Pentaborane is extremely toxic; it affects the nervous system and causes signs of narcosis and hyperexcitation.

In humans, the onset of symptoms may be delayed for up to 24 hours.[1] Minor intoxication causes lethargy, confusion, fatigue, inability to concentrate, headache, and feelings of constriction of the chest.[1] With moderate intoxication, effects are more obvious and include thick, slurred speech; confused, sleepy appearance; transient nystagmus and drooping of the eyelids; and euphoria.[1] With severe intoxication, there are signs of muscular incoordination; tremor, tonic spasms of the muscles of the face, neck, abdomen, and extremities; and convulsions and opisthotonos.[1–4] In a fatal case involving extremely heavy accidental exposure with direct skin contact, there was rapid onset of seizures and opisthotonic spasms accompanied by severe metabolic acidosis without respiratory compensation.[5] The patient expired on day 8, and autopsy revealed severe bilateral necrotizing pneumonia, widespread fatty change of the liver with centrilobular degeneration, widespread degeneration of the brain, and absence of mature spermatozoa in the testes. Another worker who was exposed while in an adjacent building survived but sustained severe neurologic damage. After 6 months, he demonstrated marked muscular weakness, incoordination, and spasticity. He could see only shapes and colors. CT scan showed marked cortical atrophy and ventricular dilation. Institutionalization was required. Of 14 persons with mild exposure to pentaborane from the same accident, 8 were judged to have mild cognitive deficits as determined by various neuropsychological tests 2 months postexposure.[6]

The concentrations of vapor and duration of exposures that cause mild, moderate, or severe intoxication are not documented. It has been estimated, on the basis of animal studies, that exposure for 60 minutes will cause slight signs of toxicity at 8 ppm, convulsions at 15 ppm, and death at 30 ppm. It was the clinical impression of one investigator that, in humans, a transient "wafting" odor did not produce symptoms (median odor threshold is 1 ppm), but a "strong whiff, producing a penetrating feeling in the nose, usually produced symptoms."[1,7] Olfactory fatigue also occurs, so that dangerous levels of pentaborane may not be readily detected.[1]

Severe irritation and corneal opacity of the eyes of test animals occurred from exposure to the vapor; the liquid on the skin of animals caused acute inflammation.[8] Because the liquid may ignite spontaneously, fire and consequent burn damage may be a greater hazard than toxicity on contact with the liquid.[4]

REFERENCES

1. Mindrum G: Pentaborane intoxication. Arch Intern Med 114:367–374, 1964
2. Lowe HJ, Freeman G: Boron hydride (Borane) intoxication in man. AMA Arch Ind Health 16:523–533, 1957
3. Rozendaal HM: Clinical observations on the toxicology of boron hydrides. AMA Arch Ind Hyg Occup Med 4:257–260, 1951
4. Hygienic Guide Series: Pentaborane-9. Am Ind Hyg Assoc J 27:307–310, 1966
5. Yarbrough BE et al: Severe central nervous system damage and profound acidosis in persons exposed to pentaborane. Clin Toxicol 23:519–536, 1985–1986
6. Hart RP et al: Neuropsychological function following mild exposure to pentaborane. Am J Ind Med 6:37–44, 1984
7. Emergency Exposure Limits: Pentaborane-9. Am Ind Hyg Assoc J 27:193–195, 1966
8. Hughes RL, Smith IC, Lawless EW: III. Pentaborane (9) and derivatives. In Holzmann RT (ed): Production of the Boranes and Related Research, pp 294–302, 329–331, 433–489. New York, Academic Press, 1967

PENTACHLORONAPHTHALENE

CAS: 1321-64-8

$C_{10}H_3Cl_5$ 1987 TLV = 0.5 mg/m^3

Synonym: Halowax 1013

Physical Form. Waxy white solid

Uses. In electric wire insulation; in lubricants

Exposure. Inhalation; skin absorption

Toxicology. Pentachloronaphthalene is toxic to the liver and skin.

The most striking human response to prolonged skin contact with the solid, or to shorter-term inhalation of hot vapor, is chloracne.[1–3] This is an acneform skin eruption characterized by papules, large comedones and pustules, chiefly affecting the face, neck, arms, and legs. Pruritic erythematous and vasculoerythematous reactions have also been reported. The reaction is usually slow to appear and may take months to return to normal. Skin lesions are often accompanied by symptoms of systemic effects, including headache, vertigo, and anorexia. Liver damage characterized by toxic jaundice, which may progress to fatal hepatic necrosis, results from the repeated inhalation of higher concentrations of the hot fumes of the molten substance.[4]

Rats exposed to the vapor of a mixture of hexa- and pentachloronaphthalene at average concentrations of 1.16 mg/m^3 for 16 hours daily for up to 4.5 months showed definite liver injury, whereas 8.8 mg/m^3 produced some mortality and severe liver injury.[3] Animal experiments have confirmed the greater toxicity of the more highly chlorinated members of the chloronaphthalene series up to and including hexachloronaphthalene.[3] Skin absorption has been demonstrated in animals and is suspected in humans.

REFERENCES

1. Kleinfeld M, Messite J, Swencicki R: Clinical effects of chlorinated naphthalene exposure. J Occup Med 14:377–379, 1972
2. Greenburg L, Mayers MR, Smith AR: The systemic effects resulting from exposure to certain chlorinated hydrocarbons. J Ind Hyg Toxicol 21:29–38, 1939
3. Deichmann WB: Halogenated cyclic hydrocarbons. In Clayton GD, Clayton FE (eds): Patty's Industrial Hygiene and Toxicology, 3rd ed, Vol 2B, Toxicology, pp 3669–3675. New York, Interscience, 1981
4. Cotter LH: Pentachlorinated naphthalenes in industry. JAMA 125:273–274, 1944

PENTACHLOROPHENOL

CAS: 87-86-5

C_6Cl_5OH 1987 TLV = 0.5 mg/m^3; skin

Synonyms: Penta; PCP; penchlorol; Santophen 20; Dowicide 7

Physical Form. Solid

Uses. Insecticide for termite control; preharvest defoliant; general herbicide; preservation of wood products, dextrins, and starches

Exposure. Inhalation; skin absorption

Toxicology. Pentachlorophenol causes irritation of the eyes and upper respiratory tract; absorption results in an increase in metabolic rate and hyperpyrexia. Prolonged skin exposure causes an acneform dermatitis. It is fetotoxic in experimental animals.

Human exposure to dust or mist concentrations greater than 1 mg/m^3 causes pain in the nose and throat, violent sneezing, and cough; 0.3 mg/m^3 may cause some nose irritation; persons acclimated to pentachlorophenol can tolerate concentrations up to 2.4 mg/m^3.[1] Pentachlorophenol readily penetrates the skin; systemic intoxication is cumulative and has been fatal.[2]

Symptoms of poisoning can include rapid onset of profuse diaphoresis, hyperpyrexia, tachycardia, tachypnea, generalized weakness, nausea, vomiting, abdominal pain, anorexia, headache, intense thirst, pain in the extremities, progressive coma, and death within hours of the onset of symptoms.[3]

Postmortem examination in a fatal case has shown extreme rigor in the muscles of the thighs and legs, edema and intra-alveolar hemorrhage in the lungs, cerebral edema, and liver and kidney damage.[4] The risks of serious intoxication are increased during hot weather.[5]

The acute toxicity results from the uncoupling of oxidative phosphorylation causing stimulation of cell metabolism and accompanying heat dissipation.[6] Acute poisoning can affect both renal and hepatic function, with elevations of the alkaline phosphatase, serum creatinine, and blood urea nitrogen.[3] Metabolic acidosis with an increased anion gap has also been observed.

Chronic exposure is associated with an increased prevalence of conjunctivitis, chronic sinusitis, bronchitis, polyneuritis, and dermatitis. Chloracne has been reported but is probably due to the dioxin contaminants in commercial grade pentachlorophenol.[3] On the skin, solutions of pentachlorophenol as dilute as 1% may cause irritation if contact is repeated or prolonged.

Animal studies have shown that the immune system is sensitive to exposure.[7] Mice fed diets containing 50 or 500 ppm technical grade pentachlorophenol showed greatly reduced immunocompetence in the form of increased susceptibility to the growth of transplanted tumors.

Given orally to pregnant rats at doses ranging from 5 to 50 mg/kg bodyweight per day, pentachlorophenol produced dose-related signs of fetotoxicity, including resorptions, subcutaneous edema, dilated ureters, and anomalies of the skull, ribs, and vertebrae.[8]

Animal studies have not shown a statistically significant carcinogenic response, and the IARC has determined that there is insufficient evidence for complete evaluation of carcinogenic risk.[9]

REFERENCES

1. Pentachlorophenol. Documentation of the TLVs and BEIs, 5th ed, p 461. Cincinnati, American Conference of Governmental Industrial Hygienists (ACGIH), 1986
2. Hayes WJ Jr: Clinical Handbook on Economic Poisons, Emergency Information for Treating Poisoning. US Public Health Service Pub No 476, pp 97–100. Washington, DC, US Government Printing Office, 1963
3. Wood S et al: Pentachlorophenol poisoning. J Occup Med 25:527–530, 1983
4. Bergner H, Constantinidis P, Martin JH: Industrial pentachlorophenol poisoning in Winnipeg. Can Med Assoc J 92:448–451, 1965
5. Williams PL: Pentachlorophenol, an assessment of the occupational hazard. Am Ind Hyg Assoc J 43:799–810, 1982
6. Gray RE et al: Pentachlorophenol intoxication: Report of a fatal case, with comments on the clinical course and pathologic anatomy. Arch Environ Health 40:161–164, 1985
7. Kerkvliet NI et al: Immunotoxicity of pentachlorophenol (PCP): Increased susceptibility to tumor growth in adult mice fed technical PCP-contaminated diets. Toxicol Appl Pharmacol 62:55–64, 1982
8. Schwetz BA et al: The effect of purified and commercial grade pentachlorophenol on rat embryonal and fetal development. Toxicol Appl Pharmacol 28:151–161, 1974
9. IARC Monographs on the Evaluation of the Carcinogenic Risk of Chemicals to Humans, Some Halogenated Hydrocarbons, Vol 20, pp 303–325. Lyon, International Agency for Research on Cancer, 1979

PENTANE

CAS: 109-66-0

$CH_3(CH_2)_3CH_3$ 1987 TLV = 600 ppm

Synonym: Amyl hydride

Physical Form. Colorless liquid

Uses. Fuel, solvent; chemical synthesis

Exposure. Inhalation

Toxicology. In animals, pentane is a mucous membrane irritant, and, at extremely high concentrations, it causes narcosis; it is expected that severe exposure will produce the same effects in humans.

Human subjects exposed to 5000 ppm for 10 minutes did not experience mucous membrane irritation or other symptoms.[1]

Topical application of pentane to volunteers caused painful burning sensations accompanied by itching; after 5 hours, blisters formed on the exposed areas.[2]

A 5-minute exposure at 128,000 ppm produced deep anesthesia in mice; respiratory arrest occurred in one of the four animals during exposure.[3] Mice exposed to 32,000 or 64,000 ppm for 5 minutes showed signs of respiratory irritation and became lightly anesthetized during the recovery period.[3] No effects were observed for 5-minute exposures at 16,000 ppm or below.

The concentration of an alkane required for acute toxic effects decreases as the carbon number increases, and it is possible that this trend also applies to the effects of long-term exposure. Therefore, although there is no documentation of neurotoxic effects of pentane, higher exposure levels for longer periods of time may be necessary to demonstrate toxicity. Furthermore, most cases of neurotoxicity attributable to alkanes have involved mixed exposures; identification of a single causative chemical is difficult, and there may be an additive toxic effect from mixed exposures.[2]

The odor of pentane is readily detectable at 5000 ppm.[1]

REFERENCES

1. Patty FA, Yant WP: Report of Investigations—Odor Intensity and Symptoms Produced by Commercial Propane, Butane, Pentane, Hexane, and Heptane Vapor, Pub No 2979. US Department of Commerce, Bureau of Mines, 1929
2. National Institute for Occupational Safety and Health: Criteria for a Recommended Standard . . . Occupational Exposure to Alkanes (C5–C8), DHEW (NIOSH) Pub No 77–151. Washington, DC, US Government Printing Office, 1977
3. Swann HE Jr, Kwon BK, Hogan GK, Snellings WM: Acute inhalation toxicology of volatile hydrocarbons. Am Ind Hyg Assoc J 35:511–518, 1974

PERCHLOROETHYLENE

CAS: 127-18-4

$Cl_2C{=}CCl_2$ 1987 TLV = 50 ppm

Synonyms: Tetrachloroethylene; ethylene tetrachloride; Nema; Tetracap; Tetropil; Perclene; Ankilostin; Didakene

Physical Form. Liquid

Uses. Solvent and cleaning agent

Exposure. Inhalation

Toxicology. Perchloroethylene causes central nervous system depression and liver damage. Chronic exposure has caused peripheral neuropathy, and it is carcinogenic in experimental animals.

Occupational exposure has caused signs and symptoms of central nervous system depression, including dizziness, light-headedness, "inebriation," and difficulty in walking.[1]

Four human subjects exposed to 5000 ppm left a chamber after 6 minutes to avoid being overcome; they experienced vertigo, nausea, and mental confusion during the 10 minutes following cessation of exposure.[2] In an industrial exposure to an average concentration of 275 ppm for 3 hours, followed by 1100 ppm for 30 minutes, a worker lost consciousness; there was apparent clinical recovery 1 hour after exposure; the monitored concentration of perchloroethylene in the patient's expired air diminished slowly over a 2-week period.[3] During the second and third postexposure weeks, the results of liver function tests became abnormal. Additional instances of liver injury following industrial exposure have been reported.[4]

Other effects on humans from inhalation of various concentrations of perchloroethylene are as follows: 2000 ppm, mild central nervous system depression within 5 minutes; 600 ppm, sensation of numbness around the mouth, dizziness, and some incoordination after 10 minutes.[5] In human experiments, 7-hour exposures at 100 ppm resulted in mild irritation of the eyes, nose, and throat; flushing of the face and neck; headache; somnolence; and slurring speech.[6] Prolonged exposure has caused impaired memory, numbness of extremities, and peripheral neuropathy, including impaired vision.[1]

Of 40 dry-cleaning workers, 16 showed signs of central nervous system depression, and, in 21 cases, the autonomic nervous system was also affected.[7] Liver enzyme activities were not altered; exposures were found to range up to 300 ppm.[7] Twenty dry-cleaning workers exposed for an average of 7.5 years to concentrations between 1 and 40 ppm had altered electrodiagnostic and neurologic rating scores.[7] Abnormal EEG recordings were found in 4 of 16 factory employees exposed to concentrations ranging from 60 ppm to 450 ppm for periods of 2 years to more than 20 years.[7]

The liquid on the skin for 40 minutes resulted in a progressively severe burning sensation beginning within 5 to 10 minutes and marked erythema, which subsided after 1 to 2 hours.[2]

Rats did not survive when exposed to 12,000 ppm for longer than 12 to 18 minutes. When exposed repeatedly to 470 ppm, they showed liver and kidney injury.[2] Cardiac arrhythmias owing to sensitization of the myocardium to epinephrine have been observed with certain other chlorinated hydrocarbons, but exposure of dogs to perchloroethylene concentrations of 5,000 and 10,000 ppm did not produce this phenomenon.[8]

Rats exposed to 300 ppm 7 hours/day on days 6 to 15 of pregnancy showed reduced bodyweight and a slightly increased number of resorptions. Among litters of mice similarly exposed, the incidences of delayed ossification of skull bones, subcutaneous edema, and split sternebrae were significantly increased compared with those in controls.[9]

Large gavage doses, approximately 500 and 1000 mg/kg per day for 78 weeks, caused a statistically significant increase in the incidence of hepatocellular carcinomas in mice.[10] Inhalation exposure by rats to 200 or 400 ppm for 2 years caused an increased incidence of mononuclear cell leukemia; a dose-related trend for a rare renal tubular neoplasm was observed in males.[11] An increased incidence of hepatocellular adenomas and carcinomas was produced in mice with repeated exposure at 100 and 200 ppm.[11]

Evidence from epidemiologic studies among dry-cleaning and laundry workers was determined by the IARC to be inadequate to assess the carcinogenic risk to humans.[12]

REFERENCES

1. National Institute for Occupational Safety and Health: Criteria for a Recommended Standard . . . Occupational Exposure to Tetrachloroethylene (Perchloroethylene). DHEW (NIOSH) Pub No 76–185, pp 17–65. Washington, DC, US Government Printing Office, 1976
2. Hygienic Guide Series: Tetrachloroethylene (perchloroethylene). Am Ind Hyg Assoc J 26:640–643, 1965
3. Stewart RD, Erley DS, Schaffer AW, Gay HH: Accidental vapor exposure to anesthetic concentrations of a solvent containing tetrachloroethylene. Ind Med Surg 30:327–330, 1961
4. Stewart RD: Acute tetrachloroethylene intoxication. JAMA 208:1490–1492, 1969
5. von Oettingen WF: The Halogenated Aliphatic, Olefinic, Cyclic, Aromatic, and Aliphatic-Aromatic Hydrocarbons Including the Halogenated Insecticides, Their Toxicity and Potential Dangers. US Public Health Service Pub No 414, pp 227–235. Washington, DC, US Government Printing Office, 1955
6. Stewart RD, Baretta ED, Dodd HC, Torkelson TR: Experimental human exposure to tetrachloroethylene. Arch Environ Health 20:224–229, 1970

7. Environmental Health Criteria 31. Tetrachloroethylene, p 48. Geneva, World Health Organization, 1984
8. Reinhardt CF, Mullin LS, Maxfield ME: Epinephrine-induced cardiac arrhythmia potential of some common industrial solvents. J Occup Med 15:953–955, 1973
9. Schwetz BA et al: The effect of maternally inhaled trichloroethylene, perchloroethylene, methyl chloroform, and methylene chloride on embryonal and fetal development in mice and rats. Toxicol Appl Pharmacol 32:84–96, 1975
10. National Cancer Institute: Biosassay of Tetrachloroethylene for Possible Carcinogenicity. DHEW (NIH) Pub No 77–813. Washington, DC, US Government Printing Office, 1977
11. National Toxicology Program: Toxicology and Carcinogenesis Studies of Tetrachloroethylene (Perchloroethylene) (CAS No 127–18–4) in F344/N Rats and B6C3F1 Mice (Inhalation Studies). DHHS (NTP) TR-311, pp 1–197. Washington, DC, US Government Printing Office, August 1986
12. IARC Monographs on the Evaluation of the Carcinogenic Risks of Chemicals to Humans, Suppl 4, pp 243–245. Lyon, International Agency for Research on Cancer, October 1982

PERCHLOROMETHYL MERCAPTAN

CAS 594-42-3

CCl_3SCl 1987 TLV = 0.1 ppm

Synonyms: PCM; perchloromethanethiol; trichloromethanesulfenyl chloride

Physical Form. Yellow liquid

Uses. Production of fungicides; vulcanizing accelerator in rubber industry

Exposure. Inhalation

Toxicology. Perchloromethyl mercaptan is a severe pulmonary irritant and lacrimating agent; fatal exposure has also caused liver and kidney injury.

Humans can withstand exposures to 70 mg/m^3 (8.8 ppm); eye irritation begins at 10 mg/m^3 (1.3 ppm).[1] Intermediate concentrations caused marked irritation of the eyes, throat, and bronchi, as well as nausea.[2]

Of three chemical workers who were observed following accidental exposures to perchloromethyl mercaptan, two survived episodes of pulmonary edema, and the third died after 36 hours.[1] The fatality resulted from a spill of the liquid on the clothing and floor with exposure to the vapor. At autopsy, there was necrotizing tracheitis, massive hemorrhagic pulmonary edema, marked toxic nephrosis, and vacuolization of centrolobular hepatic cells.

The liquid splashed on the skin may be expected to cause irritation.

Mice and cats exposed for 15 minutes at 45 ppm died within 1 to 2 days from pulmonary edema; the LC_{50} for mice was 9 ppm for 3 hours. Repeated exposures over 3 months at 1 ppm resulted in the death of some of the mice tested.[1,2]

REFERENCES

1. Althoff H: Todliche Perchlormethylmercaptan—Intoxikation (fatal perchloromethyl mercaptan intoxication). Arch Fur Toxikol 31:121–135, 1973
2. Perchloromethyl mercaptan. Documentation of the TLVs and BEIs, 5th ed, p 466. Cincinnati, American Conference of Governmental Industrial Hygienists (ACGIH), 1986

PERCHLORYL FLUORIDE

CAS: 7616-94-6

ClO_3F 1987 TLV = 3 ppm

Synonyms: Chlorine fluoride oxide; chlorine oxyfluoride

Physical Form. Gas

Uses. In organic synthesis to introduce F atoms into organic molecules; oxidizing agent in rocket fuels; insulator for high-voltage systems

Exposure. Inhalation

Toxicology. Perchloryl fluoride is an irritant of mucous membranes; in animals, it causes methemoglobinemia and pulmonary edema.

One report states that workers suffered symptoms of upper respiratory irritation from brief exposure to unspecified concentrations.[1] There are no reports of methemoglobinemia in humans from exposure to perchloryl fluoride. However, severe exposure may be expected to cause the formation of methemoglobin and resultant anoxia with cyanosis (especially evident in the lips, nose, and earlobes), severe headache, weakness, and dizziness.[2] The liquid is stated to produce moderately severe burns with prolonged contact.[3]

Dogs exposed to 450 or 620 ppm for 4 hours developed hyperpnea, cyanosis, incoordination, and convulsions; methemoglobin levels were 29% and 71%, respectively.[4] In dogs that died from exposure, there was lung damage consisting of alveolar collapse and hemorrhage; pigment deposition in the liver, spleen, and bone marrow was observed.[5]

Repeated exposure of three species of animals to 185 ppm for 7 weeks caused the death of more than half of them—guinea pigs being the most susceptible. All the animals developed dyspnea, cyanosis, methemoglobinemia, alveolar edema, and pneumonitis.[4] With similar repeated exposure of animals to 104 ppm for 6 weeks, the normal fluoride levels were increased by a factor of 20 to 30 in the blood, 5 to 8 in the urine, and 12 in the rat femur.[4]

Diagnosis. Signs and symptoms include headache; signs of anoxia, including cyanosis of lips, nose, and earlobes; and anemia.

Differential Diagnosis: Other causes of cyanosis must be differentiated from methemoglobinemia owing to chemical exposure. These include hypoxia from lung disease, hypoventilation, and decreased cardiac output. Lung disease may be suspected from results of pulmonary function tests and arterial blood gas analysis. The arterial P_{O_2} may be normal in methemoglobinemia but tends to be decreased in cyanosis due to lung disease. Hypoventilation will cause elevation of arterial P_{CO_2}, which is not seen in chemical exposure. Decreased cardiac output states will cause cyanosis only when accompanied by arterial hypotension. If blood withdrawn from the vein shows the characteristic chocolate-brown coloration, the diagnosis of an abnormal pigment is almost certain, especially if the color remains after shaking the blood in air.[2]

Special Tests: Examine the urine for blood, and determine the methemoglobin concentration in the blood when chemical intoxication is suspected and at regular intervals until the methemoglobin has been fully reduced to normal hemoglobin.[6] Methemoglobin may be differentiated from sulfhemoglobin by the addition of a few drops of 10% potassium cyanide, which results in the rapid production of bright red cyanomethemoglobin but has no effect on the color of sulfhemoglobin.[2] Spectrophotometry is required for the precise identification of the pigment and its quantitation. Normal acid methemoglobin has a characteristic absorption spectrum with peaks at 502 and 632 nm, which disappear with the addition of cyanide, whereas sulfhemoglobin has a peak at 620 nm, which does not disappear with cyanide.[2]

Treatment. All the contaminant on the body must be removed. Immediately remove all clothing, and wash the entire body from head to foot with soap and water. Pay special attention to the hair and scalp, finger- and toenails, nostrils, and ear canals. Administer oxygen to alleviate the headache and general sense of weakness; confine the patient to bed. Determine the methemoglobin concentration in the blood, and repeat every 3 to 6 hours for 18 to 24 hours. Repeat skin cleansing if the methemoglobin concentration appears to rise after 3 to 4 hours. In general, patients will return to normal within 24 hours provided that all sources of further absorption are completely eliminated. The methemoglobin will be reduced spontaneously to ferrous hemoglobin in 2 to 3 days.[2]

Such therapy is not effective in subjects with glucose–6–dehydrogenase deficiency.[2]

The only justifiable use of methylene blue would be in cases of coma or stupor, usually at methemoglobin levels over 60%. In those patients in whom therapy is necessary, methylene blue, 1 to 2 mg/kg of a 1% solution in saline, may be given intravenously over a 10-minute period. If cyanosis has not disappeared within 1 hour, a second dose of 2 mg/kg may be administered.[2,6] The total dose should not exceed 7 mg/kg, because methylene blue may cause toxic effects such as dyspnea, precordial pain, restlessness, apprehension, red cell hemolysis, and changes in the electrocardiogram (reduction in the height or even reversal of the T wave, frequently with lowering of the R wave).[6]

REFERENCES

1. Perchloryl fluoride. Documentation of the TLVs and BEIs, 5th ed, p 466. Cincinnati, American Conference of Governmental Industrial Hygienists (ACGIH), 1986
2. Mangelsdorff AF: Treatment of methemoglobinemia. AMA Arch Ind Health 14:148–153, 1956
3. Boysen JE: Health hazards of selected rocket propellants. Arch Environ Health 7:77–81, 1963
4. Greene EA, Colbourn JL, Donati E, Weeks MH: The Inhalation Toxicity of Perchloryl Fluoride. US Army Chemical Research and Development Laboratories, Technical Report CRDLR 3010, pp 1–23. Army Chemical Center, Maryland, 1960
5. Greene EA, Brough R, Kunkel A, Rinehart W: Toxicity of Perchloryl Fluoride, An Interim Report. US Army Chemical Warfare Laboratories, CWL Technical Memorandum 26–5, pp 1–6. Army Chemical Center, Maryland, 1958
6. Rieder RF: Methemoglobinemia and sulfmethemoglobinemia. In Wyngaarden JB, Smith LH (eds): Cecil Textbook of Medicine, 16th ed, p 896. Philadelphia, WB Saunders, 1982

PHENOL

CAS: 108-95-2

C_6H_5OH 1987 TLV = 5 ppm; skin

Synonyms: Carbolic acid; phenic acid; phenylic acid; phenyl hydroxide; hydroxybenzene; oxybenzene

Physical Form. Colorless crystals or white, crystalline mass

Uses. Disinfectant; reagent in chemical analysis; manufacture of aromatic compounds

Exposure. Skin absorption; inhalation; ingestion

Toxicology. Phenol is an irritant of the eyes, mucous membranes, and skin; systemic absorption causes convulsions as well as liver and kidney damage.

Phenol does not frequently constitute a serious respiratory hazard in industry, due in large part to its low volatility.[1] The skin is a primary route of entry for the vapor, liquid, and solid. The vapor readily penetrates the skin with an absorption efficiency equal to that for inhalation. Skin absorption can occur at low vapor concentrations, apparently without discomfort. Signs and symptoms can develop rapidly with serious consequences, including shock, collapse, coma, convulsions, cyanosis, and death.

A laboratory technician repeatedly exposed to the vapor (unknown concentration) and to liquid spilled on the skin developed anorexia, weight loss, weakness, muscle pain, and dark urine.[2] During several months of nonexposure, there was gradual improvement in his condition, but, after brief reexposure, he suffered an immediate worsening of symptoms, with prompt darkening of the urine and tender enlargement of the liver.[2]

Brief intermittent industrial exposures to vapor concentrations of 48 ppm of phenol (accompanied by 8 ppm of formaldehyde) caused marked irritation of eyes, nose, and throat.[1] Workers at the same plant who were continuously exposed to an average concentration of 4 ppm experienced no respiratory irritation.[1]

Ingestion of lethal amounts (as little as 10 g) causes severe burns of the mouth and throat, marked abdominal pain, cyanosis, muscular weakness, collapse, coma, and death. Tremor, convulsions, and muscle twitching have also occurred.[1,3]

Concentrated phenol solutions are severely irritating to the human eye and cause conjunctival swelling; the cornea becomes white and hypesthetic. Loss of vision has occurred in some cases.[4]

In addition to systemic effects, contact with the solid or liquid can produce chemical burns.[1] Erythema, edema, tissue necrosis, and gangrene have been reported, and prolonged contact with dilute solutions may result in deposition of dark pigment in the skin (ochronosis).[1]

Rats and mice given doses of up to 120 mg/kg and 280 mg/kg, respectively, by gavage on days 6 to 15 of gestation showed dose-related signs of fetotoxicity with no evidence of teratogenic effects.[3] A Russian study demonstrated increased preimplantation loss and early postnatal death in the offspring of rats exposed to 0.13 and 1.3 ppm throughout pregnancy.[3]

Phenol was not considered carcinogenic to rats or mice receiving 2500 or 5000 ppm in drinking water for 103 weeks, although an increased incidence of leukemia and lymphomas was detected in the low-dose male rats.[5]

Mice were treated twice weekly for 42 weeks by application of one drop of a 10% solution of phenol in benzene to the shaved dorsal skin; after 52 weeks, there were papillomas in 5 of 14 mice, and a single fibrosarcoma appeared at 72 weeks. Phenol as a nonspecific irritant may promote development of tumors when applied in large amounts repeatedly to the skin.[1]

In bakelite factory workers, the urinary level of total phenol, free plus conjugated, was proportional to the air concentration of phenol up to 12.5 mg/m^3 of workroom air.[7]

Although phenol is a major metabolite of the leukemogen benzene, it does not exhibit any potential for myeloclastogenicity in animal tests.[8]

Phenol is detectable by odor at a threshold of 0.05 ppm.[9]

REFERENCES

1. National Institute for Occupational Safety and Health: Criteria for a Recommended Standard . . . Occupational Exposure to Phenol. DHEW (NIOSH) Pub No 76–1967, pp 23–69. Washington, DC, US Government Printing Office, 1976
2. Merliss RR: Phenol marasmus. J Occup Med 14:55–56, 1972
3. US Environmental Protection Agency (EPA): Summary Review of the Health Effects Associated with Phenol: Health Issue Assessment, p 37. Washington, DC, US Government Printing Office, January 1986
4. Grant WM: Toxicology of the Eye, 3rd ed, pp 720–721. Springfield, Illinois, Charles C Thomas, 1986
5. National Cancer Institute: Bioassay of Phenol

for Possible Carcinogenicity. CAS No 108-95-2, NCI-CG-TR-203, NTP-80-15. DHHS (NIH) Pub No 80-1759, p 123. Washington, DC, US Government Printing Office, August 1980
6. Boutwell RK, Bosch DK: The tumor-promoting action of phenol and related compounds for mouse skin. Cancer Res 19:413–424, 1959
7. Ohtsuji H, Ikeda M: Quantitative relationship between atmospheric phenol vapor and phenol in the urine of workers in bakelite factories. Br J Ind Med 29:70–73, 1972
8. Gad-el-Karim MM et al: Benzene myeloclastogenicity: A function of its metabolism. Am J Ind Med 7:475–484, 1985
9. Leonardos G et al: Odor threshold determination of 53 odorant chemicals. J Air Pollut Control Assoc 19:91–95, 1969

p-PHENYLENEDIAMINE

CAS: 106-50-3

$C_6H_4(NH_2)_2$ 1987 TLV = 0.1 mg/m^3; skin

Synonyms: *p*-Diaminobenzene; 1,4-benzenediamine

Physical Form. Colorless crystalline solid; with exposure to air, it turns red, brown, and, finally, black

Uses. Dyeing of furs; hair-dye formulations; in photographic developers; in antioxidants

Exposure. Inhalation; skin absorption

Toxicology. *p*-Phenylenediamine is a sensitizer of the skin and respiratory tract and may produce bronchial asthma.

Frequent inflammation of the pharynx and larynx has been reported in exposed workers.[1] Very small quantities of the dust have caused asthmatic attacks in workers after periods of exposure ranging from 3 months to 10 years. Sensitization dermatitis has been reported from its use in the fur-dyeing industry. In this process, oxidation products of *p*-phenylenediamine are generated that are also strong skin sensitizers. Many instances of inflammation and damage of periocular and ocular tissue have been reported from contact with hair dyes containing *p*-phenylenediamine, presumably in sensitized persons.[2,3]

Although *p*-phenylenediamine has been tested for carcinogenicity in mice by skin application and in rats by oral and subcutaneous administration, the IARC has determined that these studies are not adequate to evaluate carcinogenicity.[4] *p*-Phenylenediamine dihydrochloride was not carcinogenic in 2-year feeding studies with mice and rats.[5] In an analysis of structure and corresponding carcinogenicity, phenyldiamines appeared to be least active when the amine groups were para to one another and gained activity as they became ortho.[6]

REFERENCES

1. *p*-Phenylene diamine. Documentation of the TLVs and BEIs, 5th ed, p.474. Cincinnati, American Conference of Governmental Industrial Hygienists (ACGIH), 1986
2. Grant WM: Toxicology of the Eye, 3rd ed, pp 696–698. Springfield, Illinois, Charles C Thomas, 1986
3. Baer RL et al: The most common contact allergens. Arch Dermatol 108:74–78, 1973
4. IARC Monographs on the Evaluation of the Carcinogenic Risk of Chemicals to Man, Vol 16, pp 125–142. Lyon, International Agency for Research on Cancer, 1978
5. National Cancer Institute: Bioassay of *p*-Phenylenediamine Dihydrochloride for Possible Carcinogenicity, TR-174. DHEW (NIH) Pub No 79-1730. Washington, DC, US Government Printing Office, 1979
6. Milman HA, Peterson C: Apparent correlation between structure and carcinogenicity of phenylenediamines and related compounds. Environ Health Perspect 56:261–273, 1984

2-PHENYLETHANOL

CAS: 60-12-8

$C_6H_5CH_2CH_2OH$ 1987 TLV = none established

Synonyms: Benzyl carbinol; PEA; phenylethyl alcohol

Physical Form. Colorless, viscous liquid

Uses. Fragrance; antimicrobial agent; in organic synthesis; preservative, food additive

Exposure. Inhalation

Toxicology. Phenylethanol is an irritant of the eyes and a teratogen in rats.

An 8-hour exposure of rats to an essentially saturated atmosphere failed to cause any deaths.[1]

A solution containing 0.5% phenylethanol in 0.9% sodium chloride caused a sensation of smarting in human test subjects when dropped in the eye.[2] Application of a 1% solution to rabbit eyes caused irritation of the conjunctiva and transient clouding of corneal epithelium.[3]

The liquid on the skin of human test subjects was not irritating or sensitizing.[4]

Daily oral doses of 4.3, 43, or 432 mg/kg to rats on days 6 to 15 of gestation caused abnormalities in 50%, 93%, and 100% of the animals.[5] Major malformations, including micromelia, vertebral opening, and skull defects, were observed at the highest dose, whereas only skeletal variations occurred at 4.3 mg/kg.

REFERENCES

1. Carpenter CP et al: Range-finding toxicity data: List VIII. Toxicol Appl Pharmacol 28:313–319, 1974
2. Barkman R, Germanis M, Karpe G, Malmborg AS: Preservatives in drops. Acta Ophthalmol 47:461, 1969
3. Nakano M: Effect of various antifungal preparations on the conjunctiva and cornea of rabbits. Yakuzaigku 18:94, 1958
4. Greif N: Cutaneous safety of fragrance material as measured by the maximization test. Am Perfum Cosmet 82:54, 1967
5. Mankes RF et al: Effects of various exposure levels of 2-phenylethanol on fetal development and survival in Long-Evans rats. J Toxicol Environ Health 12:235–244, 1983

PHENYL ETHER

CAS: 101-84-8

$(C_6H_5)_2O$ 1987 TLV = 1 ppm

Synonyms: 1,1′-Oxybisbenzene; diphenyl ether; diphenyl oxide

Physical Form. Colorless liquid

Uses. Heat-transfer medium; in perfuming soaps; in organic syntheses

Exposure. Inhalation

Toxicology. Phenyl ether appears to be of relatively low toxicity.

There are no reported effects in humans, although complaints owing to the disagreeable odor may occur.

Twenty exposures at 10 ppm lasting 7 hours/day caused eye and nose irritation in rats and rabbits but not in dogs.[1] At 4.9 ppm, there were no signs of irritation or toxicity.[1] The acute lethal oral dose for rats and guinea pigs is 4 g/kg.[2] Rats receiving 2 g/kg and guinea pigs receiving 1 g/kg had liver and kidney injury at autopsy.[2] On the rabbit skin, the undiluted liquid is irritating if exposures are prolonged or repeated.[2]

The low vapor pressure of phenyl ether and its easily detectable odor should prevent exposure to hazardous concentrations.[2]

REFERENCES

1. Hefner RE Jr et al: Repeated inhalation toxicity of diphenyl oxide in experimental animals. Toxicol Appl Pharmacol 33:78–86, 1975

2. Kirwin CJ Jr, Sandmeyer EE: Ethers. In Clayton GD, Clayton FE (eds): Patty's Industrial Hygiene and Toxicology, 3rd ed, Vol 2A, Toxicology, pp 2541–2543. New York, Wiley–Interscience, 1981

PHENYL GLYCIDYL ETHER

CAS: 122-60-1

$C_6H_5OHC_2CHOCH_2$ 1987 TLV = 1 ppm

Synonyms: PGE; phenoxypropenoxide; 2,3–epoxypropyl phenyl ether

Physical Form. Colorless liquid

Uses. Chemical intermediate with high solvency for halogenated materials

Exposure. Inhalation

Toxicology. Phenyl glycidyl ether (PGE) is an irritant of mucous membranes and skin and causes sensitization.

Of 20 workers exposed to PGE, 13 had acute skin changes, including second-degree burns, vesicular rash, papules, and edema.[1] In another study of 15 workers with PGE-induced dermatitis, there was erythema with papules and vesicles.[2] Of these 15 workers, 8 reacted positively to patch tests.

During animal exposure studies, technicians experienced irritation of the eyes, nose, and respiratory tract.[1,2]

There are no reports describing systemic effects in humans, and the low vapor pressure should limit the risk of acute inhalation exposure.[2]

Rats exposed for 7 hours/day for 50 days to about 10 ppm showed no overt signs of toxicity and no deaths, although, when sacrificed, a few animals had mild pulmonary inflammation and nonspecific cellular changes in the liver.[1,2]

Exposure to 5 and 12 pm PGE for 30 hours/week for 13 weeks caused hair loss in rats; this was attributed to direct irritation of the skin rather than to systemic toxicity.[3]

Intragastric LD_{50} values were 1.40 and 3.85 g/kg, respectively, for mice and rats.[1] The predominant effect was central nervous system depression. Surviving animals exhibited a reversal of the depressant effect, with increased central nervous system activity manifested by hypersensitivity to sound, muscle twitching, and tremor. Rats given three daily injections of 400 mg/kg showed no signs of bone marrow depression.[4]

Exposure of pregnant rats to 11.5 ppm did not cause effects in mothers or their offspring.[2]

Direct application of PGE into rabbit eyes produced irritation ranging from mild to severe without permanent damage.[2]

REFERENCES

1. Hine CH et al: The toxicology of glycidol and some glycidyl ethers. AMA Arch Ind Health 14:250–264, 1956
2. National Institute for Occupational Safety and Health: Criteria for a Recommended Standard . . . Occupational Exposure to Glycidyl Ethers. DHEW (NIOSH) Pub No 78–166, pp 1–197. Washington, DC, US Government Printing Office, 1978
3. Terrill JB, Lee KP: The inhalation toxicity of phenylglycidyl ether. I. 90-day inhalation study. Toxicol Appl Pharmacol 42:263–269, 1977
4. Kodama JK et al: Some effects of epoxy compounds on the blood. Arch Environ Health 2:50–61, 1961

PHENYLHYDRAZINE

CAS: 100-63-0

$C_6H_5NHNH_2$ 1987 TLV = 5 ppm; skin

Synonym: Hydrazinobenzine

Physical Form. Pale-yellow crystal or an oily liquid; becomes reddish-brown when exposed to air and light

Uses. Chemical intermediate; manufacture of dyes

Exposure. Inhalation; skin absorption

Toxicology. Phenylhydrazine causes hemolytic anemia and is a skin sensitizer; in animals, it has caused liver and kidney injury secondary to hemolytic anemia, and it is carcinogenic.

Historically, phenylhydrazine hydrochloride was used to induce hemolysis in the treatment of polycythemia vera (a disease of abnormally high erythrocyte counts).[1] Oral doses totaling 3 to 4 g were administered; in a few cases, thrombosis occurred during excessive hemolysis, but, apparently, it was not caused by phenylhydrazine hydrochloride alone.[1] Several mild cases of hemolytic anemia from occupational exposure have been reported.[2] Symptoms of intoxication have included fatigue, headache, dizziness, and vertigo.[3]

Phenylhydrazine is a potent skin sensitizer that causes eczematous dermatitis with swelling and vesiculation in a high proportion of persons who have had repeated skin contact.[1,3] Based on results with other hydrazines, it is expected that phenylhydrazine could also be absorbed through the skin.[1]

The minimal lethal dose in mice by subcutaneous injection was 180 mg/kg; animals developed progressive cyanosis and dyspnea before death; at autopsy, there were degenerative lesions in the liver, kidneys, and other organs, with evidence of vascular damage.[4]

Hemoglobin concentration, hematocrit value, and erythrocyte count were significantly reduced in dogs receiving 20 mg/kg subcutaneously for 2 consecutive days.[5] At necropsy on day 5, the internal organs were dark brown, and the spleen, liver, and kidneys were severely congested. Large amounts of blood pigments were found in these organs, and the spleen was three to five times the normal size.[5]

Rats injected intraperitoneally (10 or 20 mg/kg) during pregnancy had offspring with severe jaundice, anemia, and reduced performance in certain areas of learning.[1] One milligram phenylhydrazine hydrochloride administered daily by gavage for 200 days to mice caused adenomas and adenocarcinomas of the lung in 53% compared with 13% in the control group.[1] Consumption of 0.6 to 0.8 mg/day in drinking water for life resulted in an increased incidence of blood vessel tumors.[1] Although other studies have reported negative carcinogenicity results, NIOSH recommends that phenylhydrazine be regulated as a carcinogen.[1]

Phenylhydrazine has a faint aromatic odor, which does not serve as an adequate warning property.[1]

REFERENCES

1. National Institute for Occupational Safety and Health: Criteria for a Recommended Standard . . . Occupational Exposure to Hydrazines, DHEW (NIOSH) Pub No 78–172, p 279. Washington, DC, US Government Printing Office, 1978
2. Schuckmann Von F: Beobachtungen zur Frage verschiedener Formen der Phenylhydrazine Intoxikation (Observations on the question of different forms of phenylhydrazine poisoning). Zbl Arbeitsmed 11:338–341, 1969
3. von Oettingen WF: The Aromatic Amino and Nitro Compounds, Their Toxicity and Potential Dangers. US Public Health Service Pub No 271,

pp 158–164. Washington, DC, US Government Printing Office, 1941
4. von Oettingen WF: Deichmann–Gruebler W: On the relation between the chemical constitution and pharmacological action of phenylhydrazine derivatives. J Ind Hyg Toxicol 18:1–16, 1936
5. Witchett CE: Exposure of dog erythrocytes in vivo to phenylhydrazine and monomethylhydrazine—a freeze-etch study of erythrocyte damage, p 33. Springfield, Virginia, US Department of Commerce, NTIS, 1975

PHOSGENE

CAS: 75-44-5

$Cl_2C{=}O$ 1987 TLV = 0.1 ppm

Synonyms: Carbonyl chloride; carbon oxychloride

Physical Form. Gas, liquefying at 8°C

Uses/Sources. Intermediate in organic synthesis, especially production of toluene diisocyanate and polymethylene polyphenylisocyanate; in metallurgy to separate ores by chlorination of the oxides and volatilization; occurs as a product of combustion whenever a volatile, chlorine compound comes in contact with a flame or very hot metal

Exposure. Inhalation

Toxicology. Phosgene gas is a severe respiratory irritant.

The least concentration capable of causing immediate irritation of the human throat is 3 ppm; 4 ppm causes immediate irritation of the eyes; 4.8 ppm causes cough; and exposure to 50 ppm may be rapidly fatal.[1]

The LC_{50} in humans is approximately 500 ppm/minute.[2] Prolonged exposure to low concentrations (e.g., 3 ppm for 170 minutes) is equally as fatal as acute exposure to higher concentrations (e.g., 30 ppm for 17 minutes). Exposure to lower concentrations, however, may not lead to noteworthy initial symptoms, whereas higher concentrations cause heavy lacrimation, coughing, nausea, and dyspnea.[2]

The onset of severe respiratory distress may be delayed for up to 72 hours, the latent interval depending upon the concentration and duration of exposure.[3] The delayed onset of pulmonary edema is characterized by cough, abundant quantities of foamy sputum, progressive dyspnea, and severe cyanosis. Pulmonary edema may progress to pneumonia, and cardiac failure may intervene. During the clinical latent period, phosgene reaches the terminal spaces of the lungs where hydrolysis occurs. Membrane function breaks down, fluid leaks from the capillaries into the interstitial space, and gradually increasing pulmonary edema ensues.[2] In time, air spaces are diminished, and the blood is thickened, leading to insufficient oxygen.[3,4] Death is due to asphyxiation or heart failure.[3,4]

Mortality experience among men occupationally exposed to phosgene in the years 1943 through 1945 was evaluated 30 years postexposure.[5] No excess overall mortality, or mortality from diseases of the respiratory tract, was found in a group of chemical workers chronically exposed to levels with daily excursions above 1 ppm. Another group of this cohort, 106 workers acutely exposed at some time to a concentration probably greater than 50 ppm, included one death from pulmonary edema, which occurred within 24 hours of exposure and three deaths versus 1.37 deaths expected owing to respiratory disease. No evidence of increased lung cancer mortality was found, but the small sample size was noted.

No chronic lung problems were found in 326 workers exposed to concentrations ranging from nondetectable to greater than 0.13 ppm.[4]

Forty-one per cent of animals exposed to 0.2 ppm for 5 hours per day for 5 consecutive days developed pulmonary edema.[6] At 1 ppm, lung lesions that would be likely to cause serious clinical symptoms in humans were observed.[6] Splashes of liquefied phosgene in the eye may produce severe irritation.[3] Skin contact with the liquefied material may cause severe burns.[3]

The irritant properties of phosgene are not sufficient to give warning of hazardous concentrations. A trained observer can recognize 0.5 ppm as being "sweet," and, at about 1 ppm, the odor becomes typical of the "musty or new-mown hay" smell usually ascribed to phosgene. Workers exposed to phosgene can lose their ability to detect low concentrations through olfactory fatigue.

Diagnosis. Signs and symptoms include eye irritation; dryness or burning sensation of the throat; vomiting; cough, foamy sputum, dyspnea, pain in the chest, and cyanosis; and severe skin or eye burns from splashes of liquefied material.

Special Tests: Diagnostic studies should include electrocardiogram, sputum gram stain and culture, differential white blood count, and arterial blood gas analysis.

Treatment. Since the extent of phosgene exposure is usually not known, any exposed person should be treated as though the exposure were life threatening.[2] Pulmonary edema may be forecast by x-ray well before clinical symptoms appear. If intensive therapy is delayed, rapidly developing edema may be fatal. There is no specific therapy for phosgene-induced pulmonary injury. Although methenamine (hexamethylene tetramin, Urotropin) has been recommended as a specific antidote, no justification has been documented.[7]

REFERENCES

1. Cucinell SA: Review of the toxicity of long-term phosgene exposure. Arch Environ Health 28:272–275, 1974
2. Diller WF: Medical phosgene problems and their possible solution. J Occup Med 20:189–193, 1978
3. Hygienic Guide Series: Phosgene. Am Ind Hyg Assoc J 29:308–311, 1968
4. National Institute for Occupational Safety and Health: Criteria for a Recommended Standard . . . Occupational Exposure to Phosgene. DHEW (NIOSH) Pub No 76–137, pp 43, 55. Washington DC, US Government Printing Office, 1976
5. Polednak AP: Mortality among men occupationally exposed to phosgene in 1943–1945. Environ Res 22:357–367, 1980
6. Cameron GR et al: First Report on Phosgene Poisoning: Part II. Ministry of Defense, UK Porton Report 2349, April 1942 (unclassified report)
7. Diller WF: The methenamine misunderstanding in the therapy of phosgene poisoning. Arch Toxicol 46:199–206, 1980

PHOSPHINE

CAS: 7803-51-2

PH_3 1987 TLV = 0.3 ppm

Synonyms: Hydrogen phosphide; phosphoretted hydrogen; phosphorus trihydride

Physical Form. Colorless gas

Uses. Insecticide used for fumigation; preparation of phosphonium halides; doping agent in semiconductor manufacture

Exposure. Inhalation

Toxicology. Phosphine is a severe pulmonary irritant.

Workers exposed intermittently to

concentrations up to 35 ppm, but averaging below 10 ppm, complained of nausea, vomiting, diarrhea, chest tightness and cough, headache, and dizziness; no evidence of cumulative effects was noted.[1] Single severe exposures cause similar signs and symptoms, as well as excessive thirst, muscle pain, chills, sensation of pressure in the chest, dyspnea, syncope, and stupor.[2] In a few cases of exposure, dizziness and staggering gait have also occurred.[1] From 1900 to 1958, there were 59 reported cases of phosphine poisoning with 26 deaths; the effect most frequently reported was marked pulmonary edema.[2]

Inhalation of phosphine released after fumigation using aluminium phosphide on a grain freighter resulted in acute illnesses among 29 of 31 crewmembers and two children, one of whom died. Air concentrations measured 2 days after onset of illness ranged from 0.5 ppm in some of the living quarters to 12 ppm at an air intake. The most common symptoms included headache, fatigue, nausea, vomiting, cough, and shortness of breath. Congestive heart failure with pulmonary edema and myocardial necrosis with inflammation were noted in the child who died. The other child had echocardiographic evidence of poor left ventricular function, an elevated MB (cardiac) isoenzyme fraction of creatine kinase, and an abnormal electrocardiogram, with resolution of abnormalities within 72 hours. No long-term clinical or laboratory abnormalities were observed in the survivors.[3]

Animals survived exposure to 5 ppm, 4 hours/day for 2 months, but seven similar exposures at 10 ppm were fatal.[4]

Phosphine has a fishy or garlic-like odor detectable at 2 ppm; the odor threshold does not provide sufficient warning of dangerous concentrations.

REFERENCES

1. Jones AT, Jones RC, Longley EO: Environmental and clinical aspects of bulk wheat fumigation with aluminum phosphide. Am Ind Hyg Assoc J 25:376–379, 1964
2. Harger PN, Spolyar LW: Toxicity of phosphine, with a possible fatality from this poison. AMA Arch Ind Health 18:497–504, 1958
3. Wilson R et al: Acute phosphine poisoning aboard a grain freighter. JAMA 244:148–150, 1980
4. Hygienic Guide Series: Phosphine. Am Ind Hyg Assoc J 25:314–316, 1964

PHOSPHORIC ACID

CAS: 7664-38-2

H_3PO_4 1987 TLV = 1 mg/m^3

Synonym: Orthophosphoric acid

Physical Form. Crystals or clear syrupy liquid

Uses. Manufacture of superphosphates; acid catalyst; in dental cements; rustproofing of metals

Exposure. Inhalation

Toxicology. Phosphoric acid mist is a mild irritant of the eyes, upper respiratory tract, and skin; the dust is especially irritating to skin in the presence of moisture.[1]

Unacclimated workers could not endure exposure to fumes of phosphorus pentoxide (the anhydride of phosphoric acid) at a concentration of 100 mg/m^3; exposure to concentrations between 3.6 and 11.3 mg/m^3 produced cough, whereas concentrations of 0.8 to 5.4 mg/m^3 were noticeable but not uncomfortable.[2]

There is no evidence that phosphorus poisoning can result from contact with phosphoric acid.[3] The risk of pulmonary edema resulting from the inhalation of mist or spray is remote.[3] A subcohort of workers from 16 phosphate companies who were occupationally exposed to unspecified amounts of phosphoric acid had no significant increase in cause-specific mortality.[4]

A dilute solution buffered to *p*H 2.5 caused a moderate brief stinging sensation but no injury when dropped in the human eye.[5] A 75% solution will cause severe skin burns.[1]

REFERENCES

1. Hygienic Guide Series: Phosphoric acid. Am Ind Hyg Assoc J 18:175–176, 1957
2. Phosphoric acid. Documentation of the TLVs and BEIs, 5th ed, p 483. Cincinnati, American Conference of Governmental Industrial Hygienists (ACGIH), 1986
3. Chemical Safety Data Sheet SD–70, Phosphoric Acid, pp 5–6, 12–13. Washington, DC, MCA, Inc, 1958
4. Checkoway H et al: Mortality among workers in the Florida phosphate industry. II. Cause-specific mortality relationships with work areas and exposures. J Occup Med 27:893–896, 1985
5. Grant WM: Toxicology of the Eye, 3rd ed, pp 733–734. Springfield, Illinois, Charles C Thomas, 1986

PHOSPHORUS (yellow)

CAS: 7723-14-0

P_4 1987 TLV = 0.1 mg/m^3

Synonym: Phosphorus (white)

Physical Form. Yellowish or colorless, transparent crystals that darken on exposure to light

Uses. Manufacture of rat poisons; for smoke screens; gas analysis; fireworks

Exposure. Inhalation

Toxicology. Yellow phosphorus fume is an irritant of the respiratory tract and eyes; the solid in contact with the skin produces deep thermal burns. Prolonged absorption of phosphorus causes necrosis of facial bones; it is a hepatotoxin.

Yellow phosphorus burns spontaneously in air, and the vapor released is irritating to the respiratory tract. The early signs of systemic intoxication by phosphorus are abdominal pain, jaundice, and a garlic odor of the breath. Prolonged intake may cause anemia, as well as cachexia and necrosis of bone, involving typically the maxilla and mandible (phossy jaw).[1–3]

The presenting complaints of overexposed workers may be toothache and excessive salivation. There may be a dull red appearance of the oral mucosa. One or more teeth may loosen, followed by pain and swelling of the jaw. Healing may be delayed following dental procedures such as extractions; with necrosis of bone, a sequestrum may develop with sinus tract formation.[2] In a series of 10 cases, the shortest period of exposure to phosphorus fume (concentrations not measured) that led to bone necrosis was 10 months (two cases) and the longest period of exposure was 18 years.[2]

Yellow phosphorus fume causes severe eye irritation with blepharospasm, photophobia, and lacrimation; the solid in the eye produces severe injury.[4] Phosphorus burns on the skin are deep and painful; a firm eschar is produced and is surrounded by vesiculation.[5]

Diagnosis Signs and symptoms include irritation of the eyes and respiratory tract; abdominal pain, nausea, and jaundice; anemia; cachexia; pain and loosening of the teeth, excessive salivation, pain and swelling of the jaw; and skin and eye burns.

Differential Diagnosis: Phossy jaw must be differentiated from other forms of osteomyelitis. With phossy jaw, a sequestrum forms in the bone and is released from weeks to months later. The sequestra are light in weight, yellow to brown, osteoporotic, and decalcified, whereas sequestra from acute staphylococcal osteomyelitis are sharp, white spicules of bone, dense and well calcified. In acute staphylococcal osteomyelitis, the radiographic picture changes rapidly and closely follows the clinical course, but, with phossy jaw, the diagnosis is sometimes clinically obvious before radiologic changes are discernible.

Special Tests: It is good dental practice to take routine x-ray films of jaws, but experience indicates that necrosis can occur in the absence of any pathology visible on the roentgenogram.[2]

Treatment. Remove the patient from exposure. Surgical intervention with appropriate therapy may be required in cases of phossy jaw. If particles of phosphorus come in contact with the skin, application of a 2% to 3% silver nitrate solution is recommended.[6] Further burning owing to phosphorus oxidation is prevented when silver granules precipitate onto the surface of the phosphorus, forming a thin layer that isolates the phosphorus from oxygen. (Copper sulfate solution has traditionally been used to treat phosphorus burns, but serious adverse effects from copper poisoning such as hemolysis, renal failure, and death have been reported.) Other symptomatic therapeutic measures, such as a wet compress with 3% to 5% sodium bicarbonate solution following silver nitrate application, should also be used.

Medical Control. Medical control involves preplacement and annual physical examination with emphasis on eyes, respiratory tract, liver, kidneys, and teeth; dental examination; and liver function tests.

REFERENCES

1. Rubitsky HJ, Myerson RM: Acute phosphorus poisoning. Arch Int Med 83:164–178, 1949
2. Hughes JPW et al: Phosphorus necrosis of the jaw: A present-day study. Br J Ind Med 19:83–99, 1962
3. Chemical Safety Data Sheet SD–16, Phosphorus, Elemental, pp 3, 9–13. Washington, DC, MCA, Inc, 1947
4. Grant WM: Toxicology of the Eye, 3rd ed, pp 734–735. Springfield, Illinois, Charles C Thomas, 1986
5. Summerlin WT, Walder AI, Moncrief JA: White phosphorus burns and massive hemolysis. J Trauma 7:476–484, 1967
6. Song ZY et al: Treatment of yellow phosphorus skin burns with silver nitrate instead of copper sulfate. Scand J Work Environ Health 11:33, 1985

PHOSPHORUS PENTACHLORIDE

CAS: 10026-13-8

PCl_5 1987 TLV = 0.1 ppm

Synonyms: Phosphoric chloride; phosphorus perchloride

Physical Form. White to pale yellow, fuming crystalline mass with pungent unpleasant odor

Uses. Catalyst in manufacture of acetylcellulose; chlorinating and dehydrating agent

Exposure. Inhalation

Toxicology. Phosphorus pentachloride fume is a severe irritant of the eyes and mucous membranes.

In humans, the fume causes irritation of the eyes and respiratory tract; cases of bronchitis have resulted from exposure.[1] Although not reported, delayed onset of pulmonary edema may occur. The material on the skin could be expected to cause dermatitis.

Exposure of mice to 120 ppm for 10 minutes was fatal.

The oral LD_{50} in rats is 660 mg/kg, and the inhalation LC_{50} for 4 hours is 205 mg/m^3.[2]

REFERENCES

1. Patty FA: As, P, Se, S, and Te. In Patty FA (ed): Industrial Hygiene and Toxicology, 2nd ed, Vol 2, Toxicology, p 885. New York, Wiley–Interscience, 1963
2. Lewis RJ Sr, Sweet DV (eds): Registry of Toxic Effects of Chemical Substances (RTECS). DHHS (NIOSH) Pub No 84–101–6, p 5477. Washington, DC, US Department of Health and Human Services, July 1986

PHOSPHORUS PENTASULFIDE

CAS: 1314-80-3

P_2S_5 1987 TLV = 1 mg/m^3

Synonyms: Phosphorus sulfide; phosphorus persulfide; thiophosphoric anhydride

Physical Form. Light yellow crystals

Uses. Manufacture of safety matches, ignition compounds; introducing sulfur into organic compounds

Exposure. Inhalation

Toxicology. Phosphorus pentasulfide is an irritant of the eyes and skin. In the presence of moisture, phosphorus pentasulfide is readily hydrolyzed to hydrogen sulfide gas (qv) and phosphoric acid (qv).[1]

The oral LD_{50} in rats was 389 mg/kg; 500 mg applied to rabbit skin for 24 hours was moderately irritating, and 20 mg instilled in rabbit eyes for 24 hours was severely irritating.[2]

REFERENCES

1. Phosphorus pentasulfide. Documentation of the TLVs and BEIs, 5th ed, p 486. Cincinnati, American Conference of Governmental Industrial Hygienists (ACGIH), 1986
2. Lewis RJ Sr, Sweet DV (eds): Registry of Toxic Effects of Chemical Substances (RTECS). DHHS (NIOSH) Pub No 84–101–6, p 5662. Washington, DC, US Department of Health and Human Services, July 1986

PHOSPHORUS TRICHLORIDE

CAS: 7719-12-2

PCl_3 1987 TLV = 0.2 ppm

Synonym: Phosphorus chloride

Physical Form. Colorless, clear, fuming liquid

Uses. As chlorinating agent; manufacture of other phosphorus chloride compounds; producing iridescent metallic deposits

Exposure. Inhalation

Toxicology. Phosphorus trichloride vapor is a severe irritant of the eyes, mucous membranes, and skin.

The irritant effects of phosphorus trichloride primarily result from the action of the strong acids (hydrochloric acid and acids of phosphorus) formed in contact with water.[1]

Inhalation by humans could be expected to cause injury ranging from mild bronchial spasm to severe pulmonary edema; the onset of severe respiratory symptoms may be delayed for 2 to 6 hours, and, after moderate exposure, the onset may not occur for another 12 to 24 hours.[1] Prolonged or repeated exposure to low concentrations may induce chronic cough and wheezing; pulmonary changes are nonfibrotic and nonprogressive.

Phosphorus trichloride causes severe burns in contact with the eyes, skin, or mucous membranes.[1] Although ingestion is unlikely to occur in industrial use, it will cause burns of the mouth, throat, esophagus, and stomach.[2]

Seventeen people exposed to phosphorus trichloride liquid and its hydration products following a tanker accident were evaluated.[3] Those closest to the spill experienced burning of the eyes, lacrimation, nausea, vomiting, dyspnea, and cough. Six patients had transient elevation of lactic dehydrogenase. Chest roentgenograms were normal. Pulmonary function tests showed statistically significant decreases in vital capacity, MBC, and FEV_1 in direct correlation with distance from the accident and duration of exposure. Of the 17 patients examined 1 month later, pulmonary function tests showed improvement in 7 patients, suggesting that acute effects were due to phosphorus trichloride toxicity.[3]

In rats, the LC_{50} was 104 ppm for 4 hours; at autopsy, the chief finding was nephrosis; pulmonary damage was negligible.[2]

REFERENCES

1. Chemical Safety Data Sheet SD–27, Phosphorus Trichloride, pp 5, 15–19. Washington, DC, MCA, Inc, 1972
2. Weeks MH: Acute vapor toxicity of phosphorus oxychloride, phosphorus trichloride and methyl phosphonic dichloride. Am Ind Hyg Assoc J 25:470–475, 1964
3. Wason S et al: Phosphorus trichloride toxicity, preliminary report. Am J Med 77:1039–1042, 1984

PHTHALIC ANHYDRIDE

CAS: 85-44-9

$C_6H_4(CO)_2O$ 1987 TLV = 1 ppm

Synonyms: Phthalic acid anhydride; phthalandione; 1,3–isobenzofurandione

Physical Form. White, crystalline solid

Uses. Production of plasticizers for vinyl, epoxy, acetate resins, alkyd resins, manufacture of dyes

Exposure. Inhalation

Toxicology. Phthalic anhydride is an irritant and sensitizer of the skin and respiratory tract and an irritant of the eyes.

In workers, air concentrations of 30 mg/m^3 (5 ppm) caused conjunctivitis; at 25 mg/m^3 (4 ppm), there were signs of mucous membrane irritation.[1] Workers exposed to undetermined concentrations of mixed vapors of phthalic acid and phthalic anhydride developed, in addition to conjunctivitis, bloody nasal discharge, atrophy of the nasal mucosa, hoarseness, cough, occasional bloody sputum, bronchitis, and emphysema.[2] Several cases of bronchial asthma resulted; there was also skin sensitization with occasional urticaria and eczematous response.[2]

Phthalic anhydride is a direct but delayed irritant of the skin; it is more severely irritating after contact with water, owing to the pronounced effects of the phthalic acid that is formed.[1] Prolonged or repeated exposure may also cause dermatitis.[1]

A worker who developed symptoms of rhinorrhea, lacrimation, and wheezing from exposure to phthalic anhydride over the period of a year was shown to have a positive patch test to the chemical and a high serum titer of specific IgE.[3] In another case, it was suggested that asthma was caused by the release of phthalic anhydride during the grinding of cured moldings. Unreacted phthalic anhydride may be trapped within cured resin and released during grinding, or, alternatively, heat generated during grinding may lead to disruption of bonds between resin and hardener and cause release of phthalic anhydride vapor.[4]

In 2-year feeding studies, phthalic anhydride was not carcinogenic to rats or mice.[5]

REFERENCES

1. Hygienic Guide Series: Phthalic anhydride. Am Ind Hyg Assoc J 28:395–398, 1967
2. Fassett DW: Organic acids and related compounds. In Patty FA (ed): Industrial Hygiene and Toxicology, 2nd ed, Vol 2, Toxicology, pp 1822–1823. New York: Interscience, 1963
3. Maccia CA, Bernstein IL, Emmett EA, Brooks SM: In vitro demonstration of specific IgE in phthalic anhydride hypersensitivity. Am Rev Respir Dis 113:701–704, 1976
4. Ward MJ, Davies D: Asthma due to grinding epoxy resin cured with phthalic anhydride. Clin Allergy 12:165–168, 1982
5. National Cancer Institute: Bioassay of Phthalic Anhydride for Possible Carcinogenicity. NCI–CG–TR–159. DHEW (NIH) Pub No 79–1715. Washington, DC, US Government Printing Office, 1979

PICRIC ACID

CAS: 88-89-1

$HOC_6H_2(NO_2)_3$ 1987 TLV = 0.1 mg/m^3; skin

Synonyms: 2,4,6–Trinitrophenol; carbazotic acid; picronitric acid

Physical Form. Yellow crystalline solid

Uses. High explosive; oxidant in rocket fuels; processing of leather; metal etching

Exposure. Inhalation; skin absorption

Toxicology. Picric acid causes sensitization dermatitis; absorption of large amounts causes liver and kidney damage.

The dermatitis usually occurs on the face, especially around the mouth and the sides of the nose; the condition progresses from edema, through the formation of papules and vesicles, to ultimate desquamation.[1,2] The skin and hair of workers handling picric acid may be stained yellow.[1]

Inhalation of high concentrations of the dust by one worker caused temporary coma followed by weakness, myalgia, anuria, and, later, polyuria.[3] Following ingestion of 2 to 5 g of picric acid, which has a bitter taste, there may be headache, vertigo, nausea, vomiting, diarrhea, yellow coloration of the skin, hematuria, and albuminuria; high doses cause destruction of erythrocytes, hemorrhagic nephritis, and hepatitis.[3,4]

High doses that cause systemic intoxication will color all tissues yellow, including the conjunctiva and aqueous humor, and cause apparent yellow vision. Corneal injury is stated to have resulted from a splash of a solution of picric acid in the eyes; dust or fume may cause eye irritation, which may be aggravated by sensitization.

REFERENCES

1. Schwartz L: Dermatitis from explosives. JAMA 125:186–190, 1944
2. Sunderman FW, Weidman FD, Batson OV: Studies of the effects of ammonium picrate on man and certain experimental animals. J Ind Hyg Toxicol 27:241–248, 1945
3. von Oettingen WF: The Halogenated Aliphatic, Olefinic, Cyclic, Aromatic, and Aliphatic-Aromatic Hydrocarbons Including the Halogenated Insecticides, Their Toxicity and Potential Dangers, US Public Health Service Pub No 414, pp 150–154. Washington, DC, US Government Printing Office, 1941
4. Harris AH, Binkley OF, Chenoweth BM Jr: Hematuria due to picric acid poisoning at a naval anchorage in Japan. Am J Public Health 36:727–733, 1946

PIVAL

CAS: 83-26-1

$C_{14}H_{14}O_3$ 1987 TLV = 0.1 mg/m^3

Synonyms: Pindone; Pivalyl Valone; Tri-Ban; 2–pivaloyl–1,3–indanedione

Physical Form. Yellow powder

Uses. Rodenticide

Exposure. Inhalation; ingestion

Toxicology. Pival is a vitamin K antagonist and causes inhibition of prothrombin formation, which results in hemorrhage.

There are no reports of effects in humans.

In rats, the ingestion of a single large dose of pival causes rapid death due to pulmonary and visceral congestion without hemorrhage and may not be related to vitamin K antagonism.[1] Death in animals from chronic exposure is due to multiple internal hemorrhage.

REFERENCE

1. US Public Health Service, US Department of Health, Education and Welfare: Operational Memoranda on Economic Poisons, pp 81–84. Atlanta, Georgia, Communicable Disease Center, 1956

PLATINUM (and soluble salts)

CAS: 7440-06-4

Pt 1987 TLV = 0.002 mg/m^3 soluble salts as Pt
= 1.0 mg/m^3—metal dust

Synonyms: Ammonium chloroplatinate; sodium chloroplatinate; platinic chloride; platinum chloride; sodium tetrachloroplatinate; potassium tetrachloroplatinate; ammonium tetrachloroplatinate; sodium hexachloroplatinate; potassium hexachloroplatinate; ammonium hexachloroplatinate

Physical Form. Crystalline solids

Uses. Jewelry; chemical and electrical industries; dentistry; windings of high temperature furnaces; electroplating; photography

Exposure. Inhalation

Toxicology. Exposure to the complex salts of platinum, especially ammonium hexachloroplatinate and ammonium tetrachloroplatinate, but not elemental platinum, causes a progressive allergic reaction that may lead to pronounced asthmatic symptoms. Skin sensitization and eye irritation may also occur.

A syndrome characterized by runny nose, sneezing, tightness of the chest, shortness of breath, cyanosis, wheezing, and cough has been described following exposure to soluble complex platinum salts and is referred to variously as platinum allergy, platinum asthma, and platinosis.[1,2] Of 91 men employed in four platinum refineries and exposed to the dust or spray of the complex platinum salts, 52 experienced these symptoms.[1] The severity of response was greatest in workers crushing platinum salts where airborne levels reached 1.7 mg/m^3. Thirteen of the men also complained of dermatitis. Contact dermatitis has also been said to occur from exposure to platinum oxides and chlorides, and occasionally from platinum itself.[3] Removal from platinum salt exposure results in almost immediate relief of asthma; the dermatitis usually clears in 1 to 2 days but may be persistent.[3]

The assumption that platinosis is due to an allergic response rather than to toxic or irritant effects is suggested by the following: (1) the appearance of sensitivity following previous exposure without apparent effect; (2) only a fraction of exposed persons exhibit a response; and (3) affected subjects show increasingly high degrees of sensitivity to small amounts.[4] Of 306 platinum refinery workers exposed to unspecified levels, 38 had a positive skin prick test to the platinum halide salts.[5]

The potent allergenicity of the divalent and tetravalent platinum compounds is thought to occur by conjugation with sulfhydryl-containing groups within proteins, thus forming immunogenic complexes.[5]

Soluble complex platinum salts have

been used as anticancer agents, and atopic hypersensitivity has been provoked following repeated injections.[6] In the eyes, the dusts cause a burning sensation, lacrimation, and conjunctival hyperemia, sometimes associated with photophobia.[7]

REFERENCES

1. Hunter D et al: Asthma caused by the complex salts of platinum. Br J Ind Med 2:92–98, 1945
2. Parrot JL, Herbert R, Saindelle A, Ruff F: Platinum and platinosis, allergy and histamine release due to some platinum salts. Arch Environ Health 19:685–691, 1969
3. Stokinger HE: The metals. In Clayton GD, Clayton FE (eds): Patty's Industrial Hygiene and Toxicology, 3rd ed, Vol 2A, Toxicology, pp 1853–1861. New York, Wiley–Interscience, 1981
4. National Research Council: Platinum Group Metals. Medical and Biologic Effects of Environmental Pollutants. Washington, DC, p 232. National Academy of Sciences, 1977
5. Murdoch RD, et al: IgE antibody responses to platinum group metals: A large scale refinery survey. Br J Ind Med 43:37–43, 1986
6. Orbaek P: Allergy to the complex salts of platinum. A review of the literature and three case reports. Scand J Work Environ Health 8:141–145, 1982
7. Grant WM: Toxicology of the Eye, 3rd ed, p 748. Springfield, Illinois, Charles C Thomas, 1986

POLYTETRAFLUOROETHYLENE DECOMPOSITION PRODUCTS

Perfluoroisobutylene
CAS: 382-21-8
Carbonyl fluoride
CAS: 353-50-4

1987 TLV = no value assigned (substance of variable composition; air concentrations should be minimal)

Synonyms: Teflon; Algoflon; Fluon; Tetran; PTFE

Physical Form. Grayish-white plastic

Exposure. Inhalation

Toxicology. Fumes of heated polytetrafluoroethylene cause polymer fume fever, an influenza-like syndrome.

When it is heated to between 315°C and 375°C, the fumes cause influenza-like effects, including chills, fever, and tightness of the chest, which last 24 to 48 hours.[1,2] Symptoms suggestive of pulmonary edema, including shortness of breath and chest discomfort, have been observed in a few instances.[3,4] Complete recovery has usually occurred within 12 to 48 hours after termination of exposure. The syndrome has been particularly associated with smoking of cigarettes contaminated with Teflon.

The decomposition products up to a temperature of 500°C are principally the monomer tetrafluoroethylene but also include perfluoropropene, other perfluoro compounds containing four or five carbon atoms, and an unidentified particulate waxy fume.[4] From 500°C to 800°C, the pyrolysis product is carbonyl fluoride, which can hydrolyze to form HF and CO_2.

In rats, the LC_{50} dose for polytetrafluoroethylene heated at 595°C was 45 mg/m^3 for a 30-minute exposure.[5] Conjunctival erythema and serous ocular and nasal discharge were observed immediately after exposure. Clinical signs included dyspnea, hunched posture, and lethargy. Pathologic findings included focal hemorrhages, edema, and fibrin deposition in the lungs. Disseminated intravascular coagulation developed in more than half the test animals, and its incidence and severity closely paralleled pulmonary damage.

Polytetrafluoroethylene implanted subcutaneously in animals has induced local sarcomas, suggesting a foreign body reaction rather than chemical carcinogenesis. The IARC has determined that there is insufficient evidence to assess the carcinogenic risk, especially with regard to occupational exposure in humans.[6]

REFERENCES

1. Zapp ZA Jr: "Polyfluorines." Encyclopaedia of Occupational Health and Safety, Vol II, pp 1095–1097. New York, McGraw-Hill, 1972
2. Harris DK: Polymer-fume fever. Lancet 7814:1008, 1951
3. Lewis CE, Kirby GR: An epidemic of polymer-fume fever. JAMA 191:375, 1965
4. National Institute for Occupational Safety and Health: Criteria for a Recommended Standard . . . Occupational Exposure to Decomposition Products of Fluorocarbon Polymers. DHEW (NIOSH) Pub No 77–193, pp 16, 63. Washington, DC, US Government Printing Office, 1977
5. Zook BC et al: Pathologic findings in rats following inhalation of combustion products of polytetrafluoroethylene (PTFE). Toxicology 26:25–36, 1983
6. IARC Monographs on the Evaluation of the Carcinogenic Risk of Chemicals to Humans, Vol 19, Some Monomers, Plastics and Synthetic Elastomers, and Acrolein, pp 285–297. Lyon, International Agency for Research on Cancer, 1979

PORTLAND CEMENT

CAS: 65997-15-1

Containing <1% quartz; no asbestos

1987 TLV = 10 mg/m³ total dust

Portland cement refers to a class of hydraulic cements in which the two essential constituents are tricalcium silicate ($3CaO \cdot SiO_2$) and dicalcium silicate ($2CaO \cdot SiO_2$) with varying amounts of alumina, tricalcium aluminate, and iron oxide. It is insoluble in water. The quartz content of most finished cements is below 1%. Chromium may be present.

Physical Form. Solid

Uses. Cement

Exposure. Inhalation

Toxicology. Portland cement is an irritant of the eyes and causes dermatitis.

Repeated and prolonged skin contact with cement can result in dermatitis of the hands, forearms, and feet; this is a primary irritant dermatitis and may be complicated in some instances by a secondary contact sensitivity to hexavalent chromium.[1] In a study of 95 cement workers, 15 had a mild dermatitis of the hands, which consisted of xerosis with erythema and mild scaling; of 20 workers who were patch tested with 0.25% potassium dichromate, one person had a mild reaction, and the others were negative.

In a survey of 2278 cement workers, it was concluded that exposure to the dust of finished Portland cement caused no significant findings on chest roentgenograms even after heavy and prolonged exposures.[2] However, in a follow-up study of 195 of these workers after further exposure of 17 to 20 years, 13 showed increases in lung markings on roentgenograms; an additional 6 workers who had been exposed largely to raw dusts that contained varying amounts of free silica had marked linear exaggeration with ill-defined micronodular shadows, but no symptoms referable to the chest.[3]

In contrast, a more recent study of 847 cement workers with at least 5 years of exposure to massive levels ranging up to 3020 mppcf in cement plants revealed that symptoms such as cough, expectoration, exertional dyspnea, wheezing, and chronic bronchitis syndromes were consistently more frequent than in a group of 460 control workers; a higher prevalence of these symptoms was also found in nonsmokers exposed to cement than in a control group of nonsmokers. It must be emphasized that these exposures were to cement dust not of Portland type.[3,4]

REFERENCES

1. Perone VB et al: The chromium, cobalt, and nickel contents of American cement and their relationship to cement dermatitis. Am Ind Hyg Assoc J 35:301–306, 1974

2. Sander OA: Roentgen resurvey of cement workers. AMA Arch Ind Health 17:96–103, 1958
3. Kalacic I: Chronic nonspecific lung disease in cement workers. Arch Environ Health 26:78–83, 1973
4. Kalacic I: Ventilatory lung function in cement workers. Arch Environ Health 26:84–85, 1973

POTASSIUM HYDROXIDE

CAS: 1310-58-3

KOH 1987 TLV = C 2 mg/m^3

Synonyms: Caustic potash; KOH

Physical Form. White solid, usually as lumps, rods, or pellets

Uses. Strong alkali; manufacture of soft and liquid soaps; manufacture of potassium carbonate for use in manufacture of glass

Exposure. Inhalation; skin/eye contact

Toxicology. Potassium hydroxide (KOH) is a severe irritant of the eyes, mucous membranes, and skin.

The effects of KOH are similar to those of other strong alkalis such as NaOH. Although inhalation of KOH is usually of secondary importance, the effects from the dust or mist will vary from mild irritation to severe pneumonitis, depending upon the severity of exposure.[1] The greatest industrial hazard is rapid tissue destruction of eyes or skin upon contact with either the solid or with concentrated solutions.

Contact with the eyes causes disintegration and sloughing of conjunctival and corneal epithelium, corneal opacification, marked edema, and ulceration.[2] After 7 to 13 days, either gradual recovery begins or there is progression of ulceration and corneal opacification, which may become permanent. If not removed from the skin, severe burns with deep ulceration will occur. Ingestion produces severe abdominal pain; corrosion of the lips, mouth, tongue, and pharynx; and the vomiting of large pieces of mucosa.

REFERENCE

1. National Institute for Occupational Safety and Health: Criteria for a Recommended Standard . . . Occupational Exposure to Sodium Hydroxide. DHEW (NIOSH) Pub No 76–105, pp 23–50. Washington, DC, US Government Printing Office, 1975
2. Grant WM: Toxicology of the Eye, 3rd ed, p 756. Springfield, Illinois, Charles C Thomas, 1986

PROPANE

CAS: 74-98-6

C_3H_8 1987 TLV = Appendix E; simple asphyxiant

Synonyms: Dimethylmethane; propyl hydride

Physical Form. Odorless gas

Uses. Fuel gas; refrigerant; in organic synthesis

Exposure. Inhalation

Toxicology. Propane is a simple asphyxiant. The determining factor in exposure is available oxygen. Minimal oxygen content of air in the workplace should be 18% by volume under normal atmospheric pressure, equivalent to Po_2 of 135 mm Hg.[1]

Exposure to 100,000 ppm propane for a few minutes produced slight dizziness but was not noticeably irritating to the eyes, nose, or respiratory tract.[2]

Direct contact with the liquefied product causes burns and frostbite.[3]

Propane is odorless, and atmospheres deficient in oxygen do not provide adequate warning.[1]

REFERENCES

1. Threshold Limit Values and Biological Exposure Indices for 1985–86, pp 6–7. Cincinnati, American Conference of Governmental Industrial Hygienists (ACGIH), 1985
2. Gerarde HW: The aliphatic (open chain, acyclic) hydrocarbons. In Fassett DW, Irish DD (eds): Patty's Industrial Hygiene and Toxicology, 2nd ed, Vol 2, Toxicology, pp 1195–1198. New York, Interscience, 1963
3. Sandmeyer EE: Aliphatic hydrocarbons. In Clayton GD, Clayton FE (eds): Patty's Industrial Hygiene and Toxicology, 3rd ed, Vol 2B, Toxicology, pp 3181–3182. New York, Wiley–Interscience, 1981

PROPANE SULTONE

CAS: 1120-71-4

$C_3H_6O_3S$ 1987 TLV = none; suspected human carcinogen

Synonyms: 1,3–Propane sultone; 3–hydroxy–1–propanesulphonic acid sultone

Physical Form. White crystals or colorless liquid

Uses. Chemical intermediate to confer water solubility and anionic properties

Exposure. Inhalation

Toxicology. Propane sultone is a carcinogen in experimental animals and a suspected human carcinogen. No human data are available.[1]

It is a carcinogen in rats when given orally, intravenously, or by prenatal exposure and a local carcinogen in mice and rats when given subcutaneously.[1]

In rats, twice-weekly oral doses by gavage of 56 mg/kg for 32 weeks or 28 mg/kg for 60 weeks resulted in several malignant manifestations, including tumors of the brain, ear duct, small intestine, and leukemia.[2,3]

In mice, weekly subcutaneous injection of 0.3 mg caused tumors at the injection site in 21 of 30 mice, compared with no tumors in 30 controls.[4] Weekly subcutaneous injection of 15 and 30 mg/kg into rats resulted in death of 7 of 12 and 11 of 11 animals, respectively, with local sarcomas.[2,5] A single subcutaneous dose of 100 mg/kg produced local sarcomas in all of 18 treated rats. A single intravenous dose of 150 mg/kg in 32 rats caused the death of 1 rat with a brain tumor after 235 days and the deaths of 9 others with malignant tumors of a variety of sites within 459 days. A single intravenous dose of 20 mg/kg given to pregnant rats on day 15 of gestation produced malignant neurogenic tumors in some of the offspring.[1]

REFERENCES

1. IARC Monographs on the Evaluation of the Carcinogenic Risk of Chemicals to Man, Vol 4, Some Aromatic Amines, Hydrazine and Related Substances, N-nitroso Compounds and Miscellaneous Alkylating Agents, pp 253–258. Lyon, International Agency for Research on Cancer, 1974
2. Druckery H et al: Carcerogene alkylierende Substanzen. IV 1,3–Propanesultone und 1,4–Butansulton. Z Krebsforsch 75:69, 1970
3. Ulland B et al: Carcinogenicity of the industrial chemicals propylene imine and propane sultone. Nature 230:460, 1971
4. Van Durren BL et al: Carcinogenicity of isoesters of epoxides and lactones: Aziridine ethanol, propane sultone and related compounds. J Natl Cancer Inst 46:143, 1971
5. Druckery H, Kruse H, Preussman R: Propane sultone, a potent carcinogen. Nautrwiessenschaften 55:449, 1968

β-PROPIOLACTONE

CAS: 57-57-8

$C_3H_4O_2$ 1987 TLV = 0.5 ppm (1.5 mg/m^3) suspected human carcinogen

Synonyms: BPL; 2–oxetanone

Physical Form. Colorless liquid

Uses. Vapor sterilant and disinfectant; intermediate in the production of acrylic acid and esters

Exposure. Inhalation; skin absorption

Toxicology. β-Propiolactone (BPL) is a skin irritant and a carcinogen in animals.

Although there is no epidemiologic evidence implicating BPL as a human carcinogen, the weight of experimental animal data suggests that BPL possesses a carcinogenic potential for humans.[1]

Skin Tumors—Mice: BPL applied to mouse skin one to seven times (over a period of 2 weeks) as undiluted BPL or in solutions of corn oil or acetone at doses of 0.8 to 100 mg caused skin irritation; the effects ranged from erythema to hair loss and scarring.[2] Lifetime painting (three times/week) using acetone and corn oil solutions showed that BPL produced both papillomas and cancer of the mouse skin; 0.25 mg in acetone caused papillomas in 12 of 30 animals and cancers in 3 animals, whereas 5 mg produced tumors in 21 of 30 animals and cancers in 11. In corn oil, 0.8 mg caused tumors in 27 of 30 mice; 12 of the tumors were malignant.[2] Papillomas developed in 11 of 90 and 14 of 80 of the acetone and corn oil control groups, respectively.[2]

Injection Site Tumors—Mice: Following weekly subcutaneous injections of 0.73 mg BPL in tricaprylin for 503 days in 30 female mice, 9 mice developed fibrosarcomas, 3 developed adenocarcinomas, 7 developed squamous cell carcinomas, and 3 developed squamous papillomas, all at the injection site. The number of months to first tumor was 7, and no local tumors developed in 110 controls treated with tricaprylin alone for up to 581 days.[3]

Injection Site Tumors—Rats: All of 10 rats injected biweekly for 44 weeks with 1 mg BPL in arachis oil developed injection-site sarcomas; no local sarcomas were observed in 7 controls given repeated injections of 0.5 mg arachis oil for 54 weeks.[4]

Squamous Cell Carcinomas—Rats: Repeated gastric administration of 10 mg BPL/0.5 ml tricaprylin/week for 70 weeks caused squamous cell carcinomas of the forestomach in three of five rats; there were no tumors in controls treated with tricaprylin alone.[3]

The 30-minute LC_{50} for rats was 250 ppm, whereas for 6 hours, the LC_{50} was 25 ppm. The undiluted liquid instilled in the rabbit eye caused permanent corneal opacification.[5]

Because of high acute toxicity and demonstrated skin tumor production in animals, human contact by all routes should be avoided.[5]

The IARC considers BPL only as showing evidence of producing carcinogenicity in experimental animals.[6]

REFERENCES

1. Department of Labor: Occupational Safety and Health Standards—Carcinogens. Federal Register 39:3757, 3786–3789, 1974
2. Palmes ED, Orris L, Nelson N: Skin irritation and skin tumor production by *beta*-propiolactone (BPL). Am Ind Hyg J 23:257–264, 1962
3. Van Duuren BL et al: Carcinogenicity of epoxides, lactones, and peroxy compounds. IV. Tumor response in epithelial and connective

tissue in mice and rats. J Natl Cancer Inst 37:825–834, 1966
4. Dickens F, Jones HEH: Carcinogenic activity of a series of reactive lactones and related substances. Br J Cancer 15:85–100, 1961
5. *beta*-Propiolactone. Documentation of the TLVs and BEIs, 5th ed, p 497. Cincinnati, American Conference of Governmental Industrial Hygienists (ACGIH), 1986
6. IARC Monographs on the Evaluation of the Carcinogenic Risk of Chemicals to Humans, Vol 4, Some Aromatic Amines, Hydrazines, and Related Substances, N-Nitroso Compounds and Miscellaneous Alkylating Agents, p 259. Geneva, International Agency for Research on Cancer, 1974

PROPYL ACETATE
CAS: 109-60-4
$CH_3C(O)OC_3H_7$ 1987 TLV = 200 ppm

Synonym: Propyl ester

Physical Form. Colorless liquid

Uses. Solvent; in flavoring agents and perfumes

Exposure. Inhalation

Toxicology. In animals, propyl acetate is an irritant of the eyes and mucous membranes. At high concentrations, it causes narcosis, and it is expected that severe exposure will produce the same effect in humans.

No chronic or systemic effects have been reported in humans.

In cats, 24,000 ppm for 30 minutes was lethal to some animals 4 days postexposure.[1] Exposure for 5 hours caused narcosis in cats at 9000 ppm, and in mice at 6000 ppm.[2] Exposures at 2600 ppm caused salivation and eye irritation in cats.[2]

Propyl acetate has a pear-like odor, but the odor threshold has not been determined.

REFERENCES

1. Sandmeyer EE, Kirwin CJ: Esters. In Clayton GD, Clayton FE (eds): Patty's Industrial Hygiene and Toxicology, 3rd ed, pp 2273–2277. New York, Wiley–Interscience, 1981
2. *n*-Propyl acetate. Documentation of the TLVs and BEIs, 5th ed, p 500. Cincinnati, American Conference of Governmental Industrial Hygienists (ACGIH), 1986

n-PROPYL ALCOHOL
CAS: 71-23-8
$CH_3CH_2CH_2OH$ 1987 TLV = 200 ppm; skin

Synonyms: 1–Propanol; ethyl carbinol

Physical Form. Clear liquid

Uses. Solvent; organic syntheses

Exposure. Inhalation; minor skin absorption

Toxicology. *n*-Propyl alcohol is an irritant of the eyes and mucous membranes. At high concentrations, it causes narcosis in animals, and it is expected that severe exposure in humans will produce the same effect.

Based on acute animal studies, *n*-propyl alcohol appears to be slightly more toxic than isopropyl alcohol. No chronic effects have been reported in humans, although a human fatality has been ascribed to ingestion.[1]

Mice exposed to 3250 ppm developed ataxia in 90 to 120 minutes, and prostration was evident in 165 to 180 minutes; deep narcosis was manifest in 240 minutes at 4100 ppm.[2] Exposure to 13,120 ppm for 160 minutes or 19,680 ppm for 120 minutes was lethal to mice.[2] Exposure of rats to 20,000 ppm for 1 hour resulted in no mortalities during a 14-day postexposure observation period.[2] *n*-Propanol is not appreciably irritating to the skin of rabbits even after prolonged contact, but it can be absorbed in significant amounts if confined to the skin. Application of 38 ml/kg/day for 30 days resulted in death of one third of the rabbits.[2]

Instilled in rabbit eyes, 0.1 ml produced marked conjunctivitis, corneal opacities, and ulcerations.[2]

In a limited study, lifetime administration of *n*-propanol by intubation or subcutaneous injection caused severe liver injury, hematopoietic effects, and a number of malignant tumors not found in controls.[1]

REFERENCES

1. Gosselin RE et al: Clinical toxicology of commercial products, 5th ed, p 218. Baltimore, Williams & Wilkins, 1984
2. Rowe VK, McCollister SB: Alcohols. In Clayton GD, Clayton FE (eds): Patty's Industrial Hygiene and Toxicology, 3rd ed, Vol 2C, Toxicology, pp 4557–4561. New York, Wiley–Interscience, 1982

PROPYLENE DICHLORIDE

CAS: 78-87-5

$CH_3CHClCH_2Cl$ 1987 TLV = 75 ppm

Synonym: 1,2–Dichloropropane

Physical Form. Colorless liquid

Uses. Solvent; stain remover; chemical intermediate; fumigant

Exposure. Inhalation

Toxicology. In animals, propylene dichloride is an eye irritant, and, at very high concentrations, it is a central nervous system depressant. It is expected that severe exposure in humans will produce the same effects.

Ingestion or inhalation of high levels caused severe liver damage, acute renal failure, hemolytic anemia, and disseminated intravascular coagulation in three reported cases.[1] Symptoms from inhalation included anorexia, abdominal pain, vomiting, ecchymoses, and hematuria. In all cases, more than 24 hours elapsed between exposure and onset of symptoms. Since 80% to 90% of propylene dichloride and its metabolites are eliminated within 24 hours, analysis of blood, urine, and feces for solvent is useless once symptoms appear.[1] Workmen tolerated short-term exposures to 400 to 500 ppm without apparent adverse effects. Repeated or prolonged skin contact with propylene dichloride may result in skin irritation due to defatting.[2]

Guinea pigs repeatedly exposed to 2200 ppm for 7 hours developed severe conjunctival swelling, as well as signs of respiratory irritation and incoordination; 11 of 16 animals died after daily exposure and had severe liver injury and some kidney injury.[3] Rats dying from repeated inhalation of 1000 ppm showed weakness, general debility, and signs of respiratory irritation a few days prior to death; mice died after a few hours of exposure to 1000 ppm.[3] In general, animals that survived 35 or more 7-hour exposures to 1000 to 2200 ppm showed no significant lesions at autopsy.[3]

At 400 ppm, rats, guinea pigs, and dogs exposed for up to 140 daily 7-hour exposures showed no adverse effects.[4] There was a high percentage of mortality among mice repeatedly exposed to 400 ppm. In mice of a susceptible strain, hepatomas were found that were similar histologically to those induced by carbon tetrachloride.[4]

Some skin absorption may occur; the dermal LD_{50} for rabbits was 8.75 ml/kg.[1]

The liquid is moderately irritating to the eye but does not cause serious or permanent injury.[2]

The liquid has a characteristic unpleasant, chloroform-like odor; human subjects described the odor as "strong" at 130 to 190 ppm and "not noticeable" at 15 to 23 ppm.[2]

REFERENCES

1. Pozzi C et al: Toxicity in man due to stain removers containing 1,2-dichloropropane. Br J Ind Med 42:770–772, 1985
2. Hygienic Guide Series: Propylene dichloride. Am Ind Hyg Assoc J 28:294–296, 1967
3. Heppel LA, Neal PA, Highman B, Porterfield VT: Toxicology of 1,2-dichloropropane (propylene dichloride). I. Studies on effects of daily inhalations. J Ind Hyg Toxicol 28:1–8, 1946
4. Heppel LA, Highman B, Peak EG: Toxicology of 1,2-dichloropropane (propylene dichloride). IV. Effects of repeated exposures to a low concentration of the vapor. J Ind Hyg Toxicol 30:189–191, 1948

PROPYLENE GLYCOL MONOMETHYL ETHER

CAS: 107-98-2

$CH_3OCH_2CHOHCH_3$

1987 TLV = 100 ppm (360 mg/m^3)

Synonyms: 1–Methoxy–2–propanol; Dowanol PM Glycol Ether; Propasol Solvent M; Poly-solv MPM Solvent

Physical Form. Colorless liquid

Uses. Solvent

Exposure. Inhalation

Toxicology. Propylene glycol monomethyl ether is low in toxicity but causes irritation of the eyes, nose, and throat, with discomfort from the objectionable odor.

In human studies, 100 ppm was reported as having a transient objectionable odor. At 1000 ppm, there was irritation of the eyes, nose, and throat and signs of central nervous system impairment.[1]

The LC_{50} in rats was 10,000 ppm for 5 to 6 hours, with death caused by central nervous system depression.[2] Rats and monkeys exposed for 132 daily exposures to 800 ppm over a period of 186 days showed no evidence of adverse effects.

Exposure of rats to 3000 ppm, 6 hours/day for a total of 9 days over an 11-day period caused central nervous system depression that was reversible. No other effects were observed, including no adverse testicular effects.[3] Exposure of pregnant rats and rabbits by inhalation to 500, 1500, or 3000 ppm for 6 hours/day on days 6 to 15 (rats) or 6 to 18 (rabbits) of gestation did not cause teratogenic or embryotoxic effects. Slight fetotoxicity in the form of delayed sternebral ossification was observed in the offspring of rats exposed at 3000 ppm—a dose that was also maternally toxic.[4]

The oral LD_{50} for rats was 6.6 g/kg and was on the order of 9.2 g/kg for dogs.[5,6] The dermal LD_{50} was in the range of 13 to 14 g/kg in rabbits, indicating minimal skin absorption.[2]

The liquid on the skin of rabbits caused only a very mild transient irritation after several weeks of constant application. In the rabbit eye, there was mild reversible irritation.

REFERENCES

1. Stewart RD, Baretta ED, Dodd HC, Torkelson TR: Experimental human exposure to vapor of propylene glycol monomethyl ether. Arch Environ Health 20:218, 1970
2. Rowe VK et al: Toxicology of mono-, di-, and tripropylene glycol methyl ethers. Arch Ind Hyg Occup Med 9:509, 1954
3. Miller RR et al: Comparative short-term inhalation toxicity of ethylene glycol monomethyl

ether and propylene glycol monomethyl ether in rats and mice. Toxicol Appl Pharmacol 61:368, 1981
4. Hanley TR Jr et al: Teratologic evaluation of inhaled propylene glycol monomethyl ether in rats and rabbits. Fund Appl Toxicol 4:784–794, 1984
5. Shideman FE, Procita L: Pharmacology of the monomethyl ethers of mono-, di-, and tripropylene glycol in the dog with observations of the auricular fibrillation produced by these compounds. J Pharmacol Exp Ther 102:79, 1951
6. Smyth HF Jr, Seaton J, Fisher L: Dose toxicity of some glycols and derivatives. J Ind Hyg Toxicol 23:259, 1941

PROPYLENE IMINE

CAS: 75-55-8

$NHCH_2CHCH_3$ 1987 TLV = 2 ppm; skin

Synonyms: 2–Methylaziridine; 1,2–propyleneimine

Physical Form. Flammable liquid

Uses. Intermediate in production of polymers, coatings, adhesives, textiles, and paper finishes

Exposure. Inhalation; skin absorption

Toxicology. Propylene imine vapor is an eye irritant and a carcinogen in animals.

There are no reports of systemic effects from industrial exposure.

Exposure of rats to 500 ppm for 4 hours was fatal, but inhalation for 2 hours resulted in no deaths.[1] Rats given 20 mg/kg by gavage twice weekly suffered from advanced flaccid paralysis after 18 weeks, and the mortality rate was high.[2] At 10 mg/kg, paralysis occurred to a lesser extent after 30 weeks. Granulocytic leukemia, squamous cell carcinoma of the ear duct, and brain tumors (glioma) were observed in different animals; females showed mammary adenocarcinomas, a number of which metastasized to the lung.[2]

Instilled in the eye of a rabbit, a 5% aqueous solution produced corneal damage.[1]

No information is available to assess the carcinogenic risk to humans.[3]

REFERENCES

1. Carpenter CP et al: The acute toxicity of ethylene imine to small animals. J Ind Hyg Toxicol 30:2–6, 1948
2. Ulland B et al: Carcinogenicity of industrial chemicals propylene imine and propane sultone. Nature 230:460–461, 1971
3. IARC Monographs on the Evaluation of the Carcinogenic Risk of Chemicals to Man, Vol 9, pp 61–65. Lyon, International Agency for Research on Cancer, 1975

PROPYLENE OXIDE

CAS: 75-56-9

CH_3CHOCH_2 1987 TLV = 20 ppm

Synonyms: 1,2–Epoxypropane; propene oxide; methyl oxirane

Physical Form. Colorless liquid

Uses. Production of polyether polyols to make polyurethane foams, propylene glycol, and other chemicals; fumigant; preservative

Exposure. Inhalation

Toxicology. Propylene oxide is an irritant of the eyes, mucous membranes, and skin. At high concentrations, it causes narcosis in animals, and it is expected that severe exposure will produce the same effect in humans. It is carcinogenic in experimental animals.

No chronic or systemic effects have been reported in humans, although three cases of corneal burns from the vapor have been described.[1]

The LC_{50} for rats exposed for 4 hours was 4000 ppm; for mice, it was 1740 ppm.[2] Rats and guinea pigs exhibited irritation, dyspnea, drowsiness, weakness, and some incoordination at concentrations of 2000 ppm or more.[3] Dogs exposed to 2030 ppm for 4 hours showed lacrimation, salivation, nasal discharge, and vomiting, and there were some deaths.[2]

Rats, guinea pigs, rabbits, and a monkey were given repeated (79 or more) 7-hour exposures to 457 ppm. Irritation of the eyes and respiratory passages was noted in the rats and guinea pigs; rats had increased mortality owing to pneumonia.[3] There were no adverse effects on the monkey or the rabbits.[3]

Rats exposed to 500 ppm 7 hours/day for 15 days 3 weeks prior to breeding and during gestation had a significant reduction in the numbers of corpora lutea, implants, and live fetuses.[4] Fetotoxicity was limited to minor skeletal abnormalities for exposed litters.

In an early study, propylene oxide produced local sarcomas in rats following subcutaneous injection.[5] In more recent studies, there was some evidence of carcinogenicity in rats exposed at 400 ppm, as indicated by an increased incidence of papillary adenomas of the nasal turbinates.[6] In mice, there was clear evidence of carcinogenicity at this dose, as indicated by increased incidence of hemangiomas and hemangiosarcomas of the nasal turbinates.[6] In the respiratory epithelium of the nasal turbinates, propylene oxide also caused suppurative inflammation, hyperplasia, and squamous metaplasia in rats and inflammation in mice. No case reports or epidemiologic studies are available to assess the carcinogenic risk to humans.[7]

Aqueous solutions of 10% and 20% propylene oxide applied to the skin of rabbits caused hyperemia and edema when the duration of skin contact was 6 minutes or longer; severe exposures resulted in scar formation.[3]

The odor has been described as sweet, alcoholic, and like natural gas, ether, or benzene. The median detectable concentration is 200 ppm, which does not provide sufficient warning for prolonged or repeated exposures.[2]

REFERENCES

1. McLaughlin RS: Chemical burns of the human cornea. Am J Ophthalmol 29:1355–1362, 1946
2. Jacobson KH, Hackley EB, Feinsilver L: The toxicity of inhaled ethylene oxide and propylene oxide vapors. AMA Arch Ind Health 13:237–244, 1956
3. Rowe VK et al: Toxicity of propylene oxide determined on experimental animals. AMA Arch Ind Health 13:228–236, 1956
4. Hardin BD et al: Reproductive-toxicologic assessment of the epoxides ethylene oxide, propylene oxide, butylene oxide, and styrene oxide. Scand J Work Environ Health 9:94–102, 1983
5. Walpole AL: Carcinogenic action of alkylating agents. Ann NY Acad Sci 68:750–761, 1958
6. National Toxicology Program: NTP Technical Report on the Toxicology and Carcinogenesis Studies of Propylene Oxide in F344/N Rats and B6C3F Mice (Inhalation Studies) NTP TR 26F NIH Pub No 85-252F, pp 1–168, 1985
7. IARC Monographs on the Evaluation of the Carcinogenic Risk of Chemicals to Man, Vol 11, Cadmium, Nickel, Some Epoxides, Miscellaneous Industrial Chemicals and General Considerations on Volatile Anesthetics, pp 191–199, Lyon, International Agency for Research on Cancer, 1976

n-PROPYL NITRATE

CAS: 627-13-4

$C_3H_7NO_3$ 1987 TLV = 25 ppm

Synonyms: Nitric acid *n*-propyl ester

Physical Form. Clear to yellow liquid

Uses. Fuel ignition promoter; rocket propellants; organic intermediate

Exposure. Inhalation

Toxicology. *n*-Propyl nitrate in animals causes anoxia owing to the formation of methemoglobin, as well as anemia and hypotension.

There have been no reports of human intoxication. It is speculated that in humans exposure severe enough to cause methemoglobin is unlikely, because lower concentrations produce sufficient warning in the form of irritation, headache, and nausea.[1]

Exposure of rats to 10,000 ppm for 4 hours caused nasal irritation, dyspnea, methemoglobinemia, weakness, cyanosis, and death.[2] In dogs repeatedly exposed to 260 ppm for 26 weeks, hemoglobinuria and mild anemia appeared during the first 2 weeks of exposure but then subsided; at 900 ppm for 6 days, effects were cyanosis, methemoglobinemia, hemolytic anemia, hemoglobinuria, collapse, and death.[1]

Anesthetized dogs given 50 to 250 mg/kg intravenously immediately showed hypotension, arrest of gut activity, respiratory paralysis, hyperpnea, and moderate methemoglobinemia. Since death was produced with methemoglobin levels of only 4%, *n*-propyl nitrate intoxication may be caused in part by a direct action on vascular smooth muscle.[3]

The liquid instilled into the eyes of rabbits caused mild transient inflammation with no evidence of corneal damage.[4] The liquid applied to the skin of rabbits daily for 10 days caused staining, inflammation, and thickening of the skin but no evidence of systemic toxicity.[2]

The odor of *n*-propyl nitrate is detectable at 50 ppm and above.[1]

Diagnosis. Signs and symptoms include headache; signs of anoxia, including cyanosis of lips, nose, and earlobes; and anemia.

Differential Diagnosis: Other causes of cyanosis must be differentiated from methemoglobinemia owing to chemical exposure. These causes include hypoxia resulting from lung disease, hypoventilation, and decreased cardiac output. Lung disease may be suspected from results of pulmonary function tests and arterial blood gas analysis. The arterial Po_2 may be normal in methemoglobinemia but tends to be decreased in cyanosis from lung disease. Hypoventilation will cause elevation of arterial Pco_2, which is not seen in chemical exposure. Decreased cardiac output states will cause cyanosis only when accompanied by arterial hypotension. If blood withdrawn from the vein shows the characteristic chocolate-brown coloration, the diagnosis of an abnormal pigment is almost certain, especially if the color remains after shaking the blood in air.[5]

Special Tests: Examine the urine for blood; determine the methemoglobin concentration in the blood when chemical intoxication is suspected and at regular intervals until the methemoglobin has been fully reduced to normal hemoglobin.[6] Methemoglobin may be differentiated from sulfhemoglobin by the addition of a few drops of 10% potassium cyanide, which results in the rapid production of bright red cyanomethemoglobin but has no effect on the color of sulfhemoglobin.[5] Spectrophotometry is required for the precise identification of the pigment and its quantitation. Normal acid methemoglobin has a characteristic absorption spectrum with peaks at 502 and 632 nm, which disappear with the addition of cyanide, whereas sulfhemoglobin has a peak at 620 nm, which does not disappear with cyanide.[5]

Treatment. All the contaminant on the body must be removed. Immediately remove all clothing, and wash the entire body from head to foot with soap and water. Pay special attention to the hair and scalp, finger- and toenails, nostrils, and ear canals. Administer oxygen to alleviate the headache and general sense of weakness; confine the patient to bed. Determine the methemoglobin concentration in the blood, and repeat every 3 to 6 hours for 18 to 24 hours. Repeat skin cleansing if the methemoglobin concentration appears to rise after 3 to 4 hours. In general, patients will return to normal within 24 hours provided that all sources of further absorption are completely eliminated. The methemoglobin will be reduced spontaneously to ferrous hemoglobin in 2 to 3 days.[5]

Such therapy is not effective in subjects with glucose–6–dehydrogenase deficiency.[5]

The only justifiable use of methylene blue would be in cases of coma or stupor, usually at methemoglobin levels over 60%. In those patients in whom therapy is necessary, methylene blue, 1 to 2 mg/kg of a 1% solution in saline, may be given intravenously over a 10-minute period. If cyanosis has not disappeared within 1 hour, a second dose of 2 mg/kg may be administered.[4,5] The total dose should not exceed 7 mg/kg, because methylene blue may cause toxic effects such as dyspnea, precordial pain, restlessness, apprehension, red cell hemolysis, and changes in the electrocardiogram (reduction in the height or even reversal of the T wave, frequently with lowering of the R wave).[6]

REFERENCES

1. Rinehart WE, Garbers RC, Greene EA, Stoufer RM: Studies on the toxicity of *n*-propyl nitrate vapor. Am Ind Hyg Assoc J 19:80–83, 1958
2. Hood DB: Toxicity of *n*-Propyl nitrate and isopropyl nitrate. Haskell Laboratory for Toxicology and Industrial Medicine, Report No 21–53. Wilmington, EI duPont de Nemours and Company, 1953
3. Murtha EF, Stabile DE, Wills JH: Some pharmacological effects of *n*-propyl nitrate. J Pharmacol Exp Ther 118:77–83, 1956
4. Sutton WL: Aliphatic nitro compounds, nitrates, nitrites. In Patty FA (ed): Industrial Hygiene and Toxicology, 2nd ed, Vol 2, Toxicology, pp 2090–2092. New York, Wiley–Interscience, 1963
5. Rieder RF: Methemoglobinemia and sulfmethemoglobinemia. In Wyngaarden JB, Smith LH (eds): Cecil Textbook on Medicine, 16th ed, p 896. Philadelphia, WB Saunders, 1982
6. Mangelsdorff AF: Treatment of methemoglobinemia. AMA Arch Ind Health 14:148–153, 1956

PYRETHRUM

CAS: 8003-34-7

$C_{21}H_{28}O_{31}$, or $C_{22}H_{28}O_5$

1987 TLV = 5 mg/m^3

Synonyms: Note: Pyrethrum flowers yield "pyrethrum extract," of which the insecticidal constituents are collectively the "pyrethrins" or the "natural pyrethrins."

Physical Form. Dust

Uses. Insecticide

Exposure. Inhalation

Toxicology. Pyrethrum dust causes dermatitis and occasionally sensitization.

Under practical conditions, pyrethrum and its derivatives are probably some of the least toxic to mammals of all insecticides currently in use.[1] It was used for many years as an anthelminthic agent at a suggested oral dose of 20 mg/day for 3 days with no apparent ill effects. However, ingestion of 14 mg was lethal to a 2-year-old child. Symptoms in an 11-month-old infant who ingested the powder included pallor, intermittent convulsions, vomiting, and bradycardia; there was extreme reddening of the lips and tongue and slight inflammation of the conjunctivae.[1]

Very young children are perhaps more susceptible to poisoning because they may not hydrolyze the pyrethrum esters efficiently.[1] Animal studies indicate that pyrethrum may undergo efficient destruction in the liver and/or be slowly absorbed from the gastrointestinal tract since oral LD_{50} values are several magnitudes of order higher than intravenous values.[1]

The chief effect from exposure to pyrethrum in humans is dermatitis.[2] The usual lesion is a mild erythematous dermatitis with vesicles, papules in moist areas, and intense pruritis; a bullous dermatitis may develop.[2]

In a study of workers engaged in processing pyrethrum powder, 30% had erythema, skin roughening, and pruritis, which subsided upon cessation of exposure.[3] One of these workers had an anaphylactic-type reaction. Shortly after entering a dust-laden room, the facial skin turned red and the person felt a sensation of burning and itching. The cheeks and eyes rapidly became swollen and pruritis became severe; the entire condition disappeared in 2 days after removal from exposure.[3]

Some persons exhibit sensitivity similar to pollinosis, with sneezing, nasal discharge, and nasal stuffiness.[2] A few cases of asthma due to pyrethrum mixtures have been reported; some of the people involved had a previous history of asthma with allergy to a wide spectrum of substances.[2]

Dogs fed pyrethrins at a dietary level of 5000 ppm for 90 days showed tremor, ataxia, labored respiration, and salivation during the first month of exposure.[1] Rats given up to 5000 ppm in their diets for 2 years suffered no significant effects on growth or survival but had slight liver damage.[1] A daily gavage dose of 50, 100, or 150 mg/kg on days 6 to 15 of pregnancy caused an increased incidence of resorptions in rats compared with controls.[4]

REFERENCES

1. Hayes WJ Jr: Pesticides Studies in Man, pp 75–80. Baltimore, Williams & Wilkins, 1982
2. Hayes WJ Jr: Clinical Handbook on Economic Poisons. Emergency Information for Treating Poisoning. US Public Health Service Pub No 476, pp 74–76. Washington, DC, US Government Printing Office, 1963
3. Casida JE (ed): Pyrethrum—The Natural Insecticide, pp 123–142. New York, Academic Press, 1973
4. Khera KS et al: Teratogenicity study on pyrethrum and rotenone (natural origin) and ronnel in pregnant rats. J Toxicol Environ Health 10:111–119, 1982

PYRIDINE

CAS: 110-86-1

NC_5H_5 1987 TLV = 5 ppm

Synonyms: Azabenzene; azine

Physical Form. Colorless liquid

Uses. Solvent; organic syntheses, especially agricultural chemicals

Exposure. Inhalation; skin absorption

Toxicology. Pyridine is an irritant and a central nervous system depressant; ingestion may cause liver and kidney damage.

Chemical plant workers chronically exposed to 6 to 12 ppm developed headache, vertigo, nervousness, sleeplessness, nausea, and vomiting.[1] Similar symptoms have occurred in workers repeatedly exposed to 125 ppm; in some cases, lower abdominal or back discomfort with urinary frequency was observed without associated evidence of liver or kidney damage.[2] Serious liver and kidney injury have been reported following oral administration of 1.8 to 2.5 ml of pyridine daily for 2 months in the treatment of epilepsy.[3] Skin irritation may result from prolonged or repeated contact.

In animals, the major effects from administration of large doses by any route include local irritation and narcosis, whereas repeated feeding results in kidney and liver injury.[2] Exposure of rats to 23,000 ppm was lethal in 1.5 hours, whereas exposure to 3,600 ppm for 6 hours was fatal to two of three rats tested.[2] The oral LD_{50} for rats was 1.58 g/kg; the dermal LD_{50} was 1 to 2 ml/kg in guinea pigs.[2] In the eye of a rabbit, a 40% solution caused corneal necrosis.[2]

No indications of real oncogenic potential have been found in chronic subcutaneous studies in animals.[2]

Pyridine has an unpleasant odor detectable at 1 ppm; the odor is objectionable to unacclimatized subjects at 10 ppm but does not provide sufficient warning of hazardous concentrations because olfactory fatigue occurs quickly.[4]

REFERENCES

1. Teisinger J: Mild chronic intoxication with pyridine. J Ind Hyg Toxicol 30:58, 1948
2. Reinhardt CF, Brittelli MR: Heterocyclic and miscellaneous nitrogen compounds. In Clayton GD, Clayton FE (eds): Patty's Industrial Hygiene and Toxicology, 3rd ed, rev, Vol 2A, pp 2727–2731. New York, Wiley–Interscience, 1981
3. Pollack LJ, Finkelman I, Arieff AJ: Toxicity of pyridine in man. Arch Intern Med 71:95–106, 1943
4. Santodonato J et al: Monograph on Human Exposure to Chemicals in the Workplace: Pyridine. Washington, DC, National Cancer Institute, 1985

QUINONE

CAS: 106-51-4

$C_6H_4O_2$ 1987 TLV = 0.1 ppm

Synonyms: *p*-Benzoquinone; 1,4–cyclohexadienedione; *p*-quinone

Physical Form. Yellow crystalline solid

Uses. Oxidizing agent; in photography; tanning hides; intermediate in the manufacturing of dyes, fungicides, and hydroquionone

Exposure. Inhalation

Toxicology. Quinone affects the eyes.

Acute exposure causes conjunctival irritation and, in some cases, corneal edema, ulceration, and scarring; transient eye irritation may be noted above 0.1 ppm and becomes marked at 1 to 2 ppm.[1] Chronic exposure causes the gradual development of changes characterized as: (1) brownish discoloration of the conjunctiva and cornea confined to the intrapalpebral fissure, (2) small opacities of the cornea, and (3) structural corneal changes that result in loss of visual acuity.[2,3] The pigmentary changes are reversible, but the more slowly developing structural changes in the cornea may progress. Although pigmentation may occur with less than 5 years of exposure, this is uncommon and usually is not associated with serious injury. Skin contact may cause irritation and staining.[1] Systemic effects from industrial exposure have not been reported.

The odor and irritant properties do not provide adequate protection from levels capable of producing chronic eye injury.[1]

REFERENCES

1. Hygienic Guide Series: Quinone. Am Ind Hyg Assoc J 24:194–195, 1963
2. Sterner JH, Oglesby FL, Anderson B: Quinone vapors and their harmful effects to corneal and

conjunctival injury. J Ind Hyg Toxicol 29:60–73, 1947
3. Anderson B, Oglesby F: Corneal changes from quinone—hydroquinone exposure. AMA Arch Ophthalmol 59:495–501, 1958

RESORCINOL

CAS: 108-46-3

$C_6H_4(OH)_2$ 1987 TLV = 10 ppm

Synonyms: m-Dihydroxybenzene; resorcin; 1,3–benzene diol; 1,3–dihydroxybenzene

Physical Form. White crystals that turn pink on exposure to air

Uses. Tanning; manufacture of resorcinol–formaldehyde resins, explosives, dyes, cosmetics, adhesives

Exposure. Inhalation

Toxicology. Resorcinol is an irritant of the eyes and skin.

Workers exposed to airborne levels of 10 ppm (45 mg/m^3) for periods of 30 minutes or more reported no irritation or discomfort.[1] Application to the skin of solutions or ointments containing from 3% to 25% resulted in hyperemia, itch, dermatitis, edema, and corrosion.[2] Systemic effects from skin absorption have included restlessness, methemoglobinemia, convulsions, and tachycardia.

No toxic signs were observed in rats exposed to 7800 mg/m^3 for 1 hour or 2800 for 8 hours.[1] When rats, rabbits, and guinea pigs were exposed to 34 mg/m^3 for 6 hours/day for 2 weeks, no toxic effects were observed.[1]

A 10% solution in rabbit eyes has caused pain, conjunctivitis, and corneal vascularization.[3] Dry, powdered resorcinol applied to rabbit eyes has caused necrosis and corneal perforation.

Resorcinol has been tested for teratogenicity in rats and mice and was negative in both studies.[4,5] In a dermal oncogenicity study, three groups of female Swiss mice were treated with 0.02 ml of 5%, 25%, and 50% solutions of resorcinol in acetone twice weekly for 100 weeks.[6] The percentage of tumor-bearing animals was similar in the resorcinol-treated, untreated, and acetone-treated groups. Under the conditions of the test, resorcinol was considered noncarcinogenic.

REFERENCES

1. Flickinger CW: The benzenediols: Catechol, resorcinol and hydroquinone—review of the industrial toxicology and current industrial exposure limits. Am Ind Hyg Assoc J 37:596–606, 1976
2. Strakosch EA: Studies on ointments: Ointments containing resorcinol. Arch Dermatol Syph 48:393, 1943
3. Estable JJ: The ocular effect of several irritant drugs applied directly to the conjunctiva. Am J Ophthalmol 31:837, 1948
4. Hazelton: Submission of Data by CTFA. Resorcin: Teratology Study in the Rat. Code No 2–23b–22, 1982
5. Hazelton: Submission of Data by CTFA. Resorcin: Teratology Study in the New Zealand White Rabbit. Code No 2–23b–22, 1982
6. Stenback F, Shubik P: Lack of toxicity and carcinogenicity of some commonly used cutaneous agents. Toxicol Appl Pharmacol 30:7, 1974

RHODIUM (and compounds)

CAS: 7440-16-6

Rh 1987 TLV = 1.0 mg/m^3 metal
= 1.0 mg/m^3 as Rh—insoluble compounds
= 0.01 mg/m^3 as Rh—soluble compounds

Principal Compounds: Rhodium trichloride, rhodium trioxide, rhodium (II)acetate, rhodium nitrate, rhodium potassium sulfate; rhodium sulfate; rhodium sulfite

Physical Form. Silver-white metal

Uses. Electroplating; manufacture of rhodium–platinum alloys; manufacture of high reflectivity mirrors

Exposure. Inhalation

Toxicology. There are no data demonstrating acute or chronic rhodium-related diseases, nor sensitization or irritation in humans. Solutions of insoluble salts splashed in the eye may cause mild irritation.

The LD_{50} for rhodium trichloride in rabbits by intravenous injection was 215 mg/kg; the clinical signs presented shortly after injection were increasing lethargy and waning respiration.[1] There were no abnormal findings at autopsy, but the rapid onset of death suggested central nervous system effects.

A solution of rhodium trichloride in the eye of a rabbit gave a delayed injurious reaction; 0.1 mg of solution adjusted to *p*H 7.2 with ammonium hydroxide was placed in a rabbit eye for 10 minutes after the corneal epithelium had been removed. An orange coloration of the cornea occurred, which faded to faint yellow within 8 weeks.[2] During the first 2 to 3 weeks, the cornea was slightly hazy; in the third week, white opacities gradually developed, and, finally, there was extensive opacification and vascularization.

Lifetime exposure to 5 ppm rhodium trichloride in the drinking water caused a minimally significant increase in malignant tumors in mice. Lymphomas, leukemias, and adenocarcinomas were most prevalent.[3] Chick embryos exposed to rhodium chloride on the eighth day of incubation were stunted; mild reduction of limb size and feather growth inhibition were also observed.[4] Possible relevance of these observations to occupational exposure has not been established.

A number of rhodium compounds have tested positive in bacterial assays for genetic altering capability.[5]

REFERENCES

1. Landholt RR, Berk HW, Russell HT: Studies on the toxicity of rhodium trichloride in rats and rabbits. Toxicol Appl Pharmacol 21:589–590, 1972
2. Grant WM: Toxicology of the Eye, 2nd ed, p 887. Springfield, Illinois, Charles C Thomas, 1974
3. Schroeder HA, Mitchener M: Scandium, chromium (VI) gallium, yttrium, rhodium, palladium, indium in mice: Effects on growth and life span. J Nutr 101:1431–1438, 1971
4. Ridgway LP, Karnofsky DA: The effects of metals on chick embryo: Toxicity and production of abnormalities in development. Ann NY Acad Scii 55:203–215, 1952
5. Warren G et al: Mutagenicity of a series of hexacordinate rhodium III compounds. Mutat Res 88:165–173, 1981

RONNEL

CAS: 299–84–3

$(CH_3O)_2P(S)OC_6H_2Cl_3$

1987 TLV = 10 mg/m^3

Synonym: *O,O*–Dimethyl–*O*–(2,4,5–trichlorophenyl) phosphorothioate

Physical Form. White, crystalline powder

Uses. Systemic insecticide in livestock

Exposure. Inhalation; ingestion

Toxicology. Ronnel is a weak cholinesterase inhibitor. It primarily affects the pseudocholinesterases of the blood plasma rather than the erythrocyte acetylcholinesterase.[1]

In an experiment on humans to evaluate the primary skin irritating and skin sensitizing potential of ronnel, 50 subjects received three applications per week for 3 weeks of gauze saturated with a 10% percent suspension of ronnel in sesame oil. There were no significant effects on the skin.[2]

In male rats, the oral LD_{50} was 1.7 g/kg; effects included salivation, tremor, diarrhea, miosis, and respiratory distress—all attributed to the anticholinesterase effect of ronnel.[2] Rats fed 50 mg/kg in the diet for 105 days developed slight liver and kidney damage.

Dogs fed 10 mg/kg/day for 2 years showed no overt clinical signs nor evidence of any effect upon urinalysis, hematologic analysis, organ weight measurement, or histologic evaluation of the tissues; depression of plasma cholinesterase was the only significant finding.[3]

When a small amount of ronnel powder was placed in the eye of a rabbit, effects were slight discomfort and transient conjunctival irritation, which subsided within 48 hours.[2]

Daily oral administration of 600 or 800 mg/kg ronnel to dams on days 6 through 15 of pregnancy caused a significant dose-related increase in fetuses with an extra rib.[4]

Ronnel has not been shown to potentiate the effect of other commonly used organophosphorus insecticides.[1]

Diagnosis: *Signs and Symptoms:* By analogy to other anticholinesterase agents, the following signs and symptoms may occur in extremely severe poisonings: Initial symptoms include headache, blurred vision, pallor, weakness, sweating, abdominal pain, nausea, vomiting, and diarrhea.

Moderate-to-severe intoxication includes miosis, lacrimation, excessive salivation, muscle fasciculations, dyspnea, cyanosis, convulsions, shock, cardiac arrhythmias, and coma.

Differential Diagnosis: Diagnosis is based primarily on a history of exposure and clinical evidence of diffuse parasympathetic stimulation. Careful observation of the effects of atropine and pralidoxime may be valuable. Patients with organophosphate poisoning are resistant to the action of atropine at moderate dosages; failure of 1 to 2 mg of atropine administered parenterally to produce signs of atropinization (flushing, mydriasis, tachycardia, or dryness of mouth) indicates organophosphate poisoning. Intravenous injections of 1 g pralidoxime generally result in some recovery from signs and symptoms.

Special Tests: Two types of cholinesterase are clinically significant: (1) true acetylcholinesterase, found principally in the nervous system and the red blood cell; and (2) pseudo- or butyrylcholinesterase, found in the plasma, liver, and nervous system. Whereas the action of both types is inhibited by most organophosphates, the level of depression of red blood cell cholinesterase is a better indicator of clinically significant reduction of cholinesterase activity in the nervous system. Ronnel may be an exception.

Laboratory evidence of depression of red blood cell cholinesterase to a level substantially below pre-exposure levels (at least 50% and usually much lower) is verification of poisoning. There is an imperfect correlation between the degree of depression of cholinestersase enzymes and the occurrence of symptoms. With a rapid drop in cholinesterase activity, generally reflecting an acute heavy exposure, there may be symptoms with only a 30% depression, whereas with slower drops to 70% depression, reflecting chronic low-level exposure, there may be no symptoms.[5]

If no pre-exposure baseline has been performed but symptoms are not sufficient to justify treatment with atropine, repeated testing during the recovery period demonstrating progressively increasing plasma and red blood cell cholinesterase levels over several days and weeks, respectively, suggests the diagnosis of anticholinesterase poisoning.

There are many different methods for estimation of cholinesterase content of blood, and associated with each method is a different set of normal values and a different set of reporting units. The laboratory report of a cholinesterase determination should state the units involved along with the appropriate normal range. Based on the Michel method, the normal range of red blood cell cholinesterase activity (delta *p*H per hour) is 0.39 to 1.02 for men, and 0.34 to 1.10 for women.[6] The normal range of the enzyme activity (delta *p*H per hour) of plasma is 0.44 to 1.63 for men, and 0.24 to 1.54 for women.

Treatment. Treatment of organophosphate poisoning ranges from simple removal from exposure in very mild cases to the provision of very rigorous supportive and antidotal measures in severe cases.[7–10] In moderate-to-severe cases, because of pulmonary involvement, there may be need for artificial respiration using a positive-pressure method. Careful attention must be paid to removal of secretions and to maintenance of a patent airway. Anticonvulsants such as thiopental sodium may be necessary. Maintenance of respiration is critical, because death usually results from weakness of the muscles of respiration and accumulation of excessive secretions in the respiratory tract.[8]

As soon as cyanosis has been overcome, 2 to 4 mg of atropine should be given intravenously. (Atropine may induce ventricular fibrillation in the presence of cyanosis.) *This dose of atropine is approximately 10 times the amount that is administered for other conditions in which atropine is considered therapeutic.* This dose should be repeated at 5- to 10-minute intervals until signs of atropinization appear (dry, flushed skin, tachycardia as high as 140 beats/minute, and pupillary dilatation). A mild degree of atropinization should be maintained for at least 48 hours.[8]

Pralidoxime (2–PAM, Protopam) chloride is a cholinesterase reactivator that complements the action of atropine. It has its greatest effect in reversing the nicotinic action of anticholinesterase agents at skeletal neuromuscular junctions but virtually no effect on central nervous system manifestations. In moderate-to-severe cases, the dose for adults is 1 to 2 g injected intravenously at a rate not in excess of 500 mg/minute. After 1 hour, a second dose of 1 g is indicated if muscle weakness has not been relieved. Treatment with pralidoxime chloride will be most effective if given within 24 hours after poisoning.[8] Morphine, aminophylline, and phenothiazines are contraindicated because of documented experience of adverse reactions in cases of organophosphate poisoning.[10]

It is of great importance to decontaminate the patient. Contaminated clothing should be removed at once, and the skin should be washed with generous amounts of soap or detergent and a flood of water, which is best accomplished under a shower or by submersion in a pond or other body of water if the exposure occurred in the field. Careful attention should be paid to cleansing of the skin and hair.

The patient should be attended and monitored continuously for not less than 24 hours, because serious and sometimes fatal relapses have occurred as a result of continuing absorption of the toxin or dissipation of the effects of the antidote.

Regeneration of cholinesterase is primarily by synthesis of new enzyme and takes place at the rate of approximately 1% per day.[10] A patient who has recovered from the acute phase of poisoning remains hypersusceptible to anticholinesterases for up to several weeks.

Medical Control. Medical control includes preplacement and annual physical examination with determination of pre-exposure red blood cell cholinesterase activity. A person whose red blood cell cholinesterase falls to or below 40% of the pre-exposure baseline should be removed from further exposure until the activity returns to within 80% of the pre-exposure baseline.

REFERENCES

1. Plapp FW, Casida JE: Bovine metabolism of organophosphorus insecticides. J Agric Food Chem 6:662–667, 1958
2. McCollister DD, Oyen F, Rowe VK: Toxicological studies of *O,O*-dimethyl-*O*-(2,4,5-trichlorophenyl) phosphorothioate (ronnel) in laboratory animals. J Agric Food Chem 7:689–693, 1959
3. Worden AN et al: Effect of ronnel after chronic feeding to dogs. Toxicol Appl Pharmacol 23:1–9, 1972
4. Khera KS et al: Teratogenicity study on pyrethrum and rotenone (natural origin) and ronnel in pregnant rats. J Toxicol Environ Health 10:111–119, 1982
5. Coye MJ, Lowe JA, Maddy KT: Biological monitoring of agricultural workers exposed to pesticides. Cholinesterase activity determinations. J Occup Med 28:619–627, 1986
6. Michel HO: Electrometric method for determination of red blood cell and plasma cholinesterase activity. J Lab Clin Med 34:1564–1568, 1949
7. Hayes WJ Jr: Organic phosphorus pesticides. In Pesticides Studied in Man, pp 284–435, Baltimore, Williams & Wilkins, 1982
8. Taylor P: Anticholinesterase agents. In Gilman AG et al (eds): Goodman and Gilman's The Pharmacological Basis of Therapeutics, 7th ed, pp 110–129. New York, Macmillan, 1985
9. Namba T, Nolte CT, Jackrel J, Grob D: Poisoning due to organophosphate insecticides. Am J Med 50:475–492, 1971
10. Milby TH: Prevention and management of organophosphate poisoning. JAMA 216:2131–2133, 1971

ROTENONE

CAS: 83-79-4

$C_{23}H_{22}O_6$ 1987 TLV = 5 mg/m^3

Synonyms: Derrin; nicouline; tubatoxin

Physical Form. Colorless crystals

Uses. Insecticide; lotion for chiggers; emulsion for scabies

Exposure. Inhalation

Toxicology. Rotenone affects the nervous system and causes convulsions in animals.

The lethal oral dose in humans is estimated to be 0.3 to 0.5 g/kg.[1] In humans, inhalation of dust is expected to cause pulmonary irritation.[1] Symptoms of absorption in humans (inferred mostly from animal studies) may include numbness of oral mucous membranes, nausea, vomiting, abdominal pain, muscle tremor, incoordination, clonic convulsions, and stupor.[1] The dust is irritating to the eyes and other mucous membranes and skin.[2]

Animals repeatedly fed derris powder (a botanical source containing 9.6% rotenone) at levels from 312 to 5000 ppm developed focal liver necrosis and mild kidney damage.[2] The oral LD_{50} values vary greatly, depending upon particle size, manner of dispersion, activity of sample, and species tested. Values ranging from 25 mg/kg in rats to more than 3000 mg/kg in rabbits have been reported.[3]

At the cellular level, rotenone inhibits cellular respiration by blocking electron transport between flavoprotein and ubinquinone. It also inhibits spindle microtubule assembly.[3]

Administered orally to rats on days 6 to 15 of pregnancy, 10 mg/kg was highly toxic to dams, killing 12 of 20. There was a significant decrease in the number of live fetuses per surviving dam and an increase in the proportion of resorptions.[4] In the 5-mg/kg group, there was an increased frequency of skeletal aberrations such as extra rib, delayed ossification of sternebrum, and missing sternebrae.[4]

Of 40 female rats given daily intraperitoneal injections of 1.7 mg/kg bodyweight rotenone in sunflower oil for 42 days, more than 60% developed mammary tumors 6 to 11 months after the end of treatment. Most of the tumors were mammary adenomas, and one was a differentiated adenocarcinoma. None of the control animals had tumors when examined 19 months after treatment.[5] Interestingly, the incidence of tumors was increased more by low dosages of rotenone and increased less by higher dosages that caused nonfatal inhibition of body growth.[3] Administered to two strains of mice at 0.4 mg/kg/day and higher for 18 months, rotenone was nontumorigenic.[3] It has been noted, however, that if mice respond similarly to rats, a lower dose may produce a positive result.[3]

A slightly increased incidence of parathyroid gland adenomas, 1 of 41 in controls vs 4 of 44 in treated animals, was found in male rats given 75 ppm rotenone in their diets for 2 years. Low-dose female rats given 38 ppm in their diets for 2 years had an increased incidence of subcutaneous tissue tumors. A *decreased* incidence of liver neoplasms, 12 of 47 in controls vs 1 of 50 in treated animals, in high-dose male mice receiving 1200 ppm for 2 years, was also related to administration of rotenone. Under conditions of this study, NTP determined that there was equivocal evidence of carcinogenicity.

REFERENCES

1. Gosselin RE et al: Clinical Toxicology of Commercial Products. Section III, 5th ed, pp 366–368. Baltimore, Williams & Wilkins, 1984
2. Negherbon WO (ed): Handbook of Toxicology, Vol III, p 665. Philadelphia, WB Saunders, 1957
3. Hayes WJ Jr: Pesticides Studied in Man, pp 82–86. Baltimore, Williams & Wilkins, 1982
4. Khera KS et al: Teratogenicity study on pyrethrum and rotenone (natural origin) and ronnel in pregnant rats. J Toxicol Environ Health 10:111–119, 1982
5. Gosalvez M, Merchan J: Induction of rat mammary adenomas with the respiratory inhibitor rotenone. Cancer Res 33:3047–3050, 1973

SELENIUM (and compounds)

CAS: 7782-49-2

Se — 1987 TLV = 0.2 mg/m^3

Synonyms: Selenium dioxide; selenium trioxide; selenium oxychloride; sodium selenite; sodium selenate; selenium anhydride

Physical Form. Powders or crystalline solids (red and grey)

Uses. Glass and ceramics manufacture; selenium rectifiers and photocells (exposure may also occur during smelting and refining of ores containing selenium)

Exposure. Inhalation

Toxicology. Elemental selenium and selenium compounds as dusts, vapors, and fumes are irritants of the eyes, mucous membranes, and skin. Chronic exposure may cause central nervous system effects, gastrointestinal disturbances, and loss of hair and fingernails.

A group of workers briefly exposed to unmeasured but high concentrations of selenium fume developed severe irritation of the eyes, nose, and throat, followed by headaches. Transient dyspnea occurred in one case.[1] A case of accidental hydrogen selenide poisoning also resulted in irritation of the mucous membranes; following a brief recovery period, pulmonary edema, bronchitis, and bronchial pneumonia occurred.[2] Workers exposed to an undetermined concentration of selenium oxide developed bronchospasm and dyspnea, followed within 12 hours by metal fume fever (chills, fever, and headache) and bronchitis leading to pneumonitis in a few cases; all workers were asymptomatic within 1 week.[3]

In a study of workers in a selenium plant, workroom air levels ranged from 0.2 to 3.6 mg/m^3, whereas urinary levels ranged from below 0.10 to 0.43 mg/liter of urine. The chief complaints included garlic odor of the breath, metallic taste, gastrointestinal disturbances, and skin eruptions.[4]

An endemic disease in China characterized by loss of hair and nails, skin lesions, and abnormalities of the nervous system, including some paralysis and hemiplegia, was attributed to chronic selenium poisoning.[5] The daily intake for six affected persons averaged 5.0 mg vs 0.1 mg for people from an unaffected area. Changing the diet resulted in recoveries. There have been no reports of disabling chronic disease or death from industrial exposure.

An accidental spray of selenium dioxide into the eyes of a chemist caused superficial burns of the skin and immediate irritation of the eyes. Within 16 hours, vision was blurred, and the lower portions of both corneas appeared dulled. Sixteen days after the accident, the corneas were normal.[6]

The element selenium is not particularly irritating, but various compounds such as selenium oxychloride and selenium dioxide are strong vesicants.[7] Skin contact with the fume of heated selenium dioxide caused an acute, weeping dermatitis, with the development of hypersensitivity in some cases.[8] Selenium dioxide forms selenious acid when in contact with water; if allowed to penetrate beneath the fingernails, it causes an especially painful inflammatory reaction.[8] Compounds of selenium can be absorbed through the unbroken skin. Selenium sulfide, which is found in some shampoos, can penetrate through the scalp and cause generalized toxic responses.[7]

Prolonged feeding of animals with diets containing selenium in amounts of 5 to 15 ppm caused hepatic necrosis, hemorrhage, and cirrhosis; marked and progressive anemia occurred in some species.[8]

High doses of selenium sulfide administered by gavage caused liver tumors in rats, and lung and liver tumors in female mice.[9] A number of other studies, however, have shown selenium compounds to have anticarcinogenic or preventive effects against carcinogen-induced breast, colon, liver, and skin cancers in animals.[10] Mutagenic and antimutagenic effects of selenium have also been reported.[10] In humans, no excess death rates due to malignant neoplasms have been found in selenium workers, nor have dietary intake levels been shown to correlate with cancer deaths.[11] The role of selenium in cancer induction and/or prevention requires further investigation.

See separate monograph on selenium hexafluoride.

REFERENCES

1. Clinton M Jr: Selenium fume exposure. J Ind Hyg Toxicol 29:225–226, 1947
2. Olson OE: Selenium toxicity in animals with emphasis on man. J Am Coll Toxicol 5:45–70, 1986
3. Wilson HM: Selenium oxide poisoning. JAMA 180(8):173–174, 1962
4. Glover JR: Selenium and its industrial toxicology. Ind Med Surg 39:50–54, 1970
5. Yang G et al: Endemic selenium intoxication of humans in China. Am J Clin Nutr 37:872–881, 1983
6. Middleton JM: Selenium burn of the eye. AMA Arch Ophthalmol 38:806–811, 1947

7. Wilber CG: Toxicology of selenium: A review. Clin Toxicol 17:171–230, 1980
8. Committee on Medical and Biologic Effects of Environmental Pollutants, National Research Council: Selenium, pp 116–118. Washington, DC, National Academy of Sciences, 1976
9. National Cancer Institute: Bioassay of Selenium Sulfide (Gavage) for Possible Carcinogenicity, DHHS (NIH) Pub No 80–1750, p 130. Washington, DC, US Government Printing Office, 1980
10. Shamberger RJ: The genotoxicity of selenium. Mutat Res 154:29–48, 1985
11. Diplock AT: Biological effects of selenium and relationships with carcinogenesis. Toxicol Environ Chem 8:305–311, 1984

SELENIUM HEXAFLUORIDE

CAS: 7783-79-1

SeF_6 1987 TLV = 0.05 ppm

Synonym: Selenium fluoride

Physical Form. Colorless gas

Uses. Gaseous electric insulator

Exposure. Inhalation

Toxicology. Selenium hexafluoride is a severe pulmonary irritant in animals; heavy exposure is expected to cause the same effect in humans.

Exposure of four animal species to 10 ppm for 4 hours was fatal; 5 ppm for 5 hours was not fatal but caused pulmonary edema, whereas 1 ppm produced no effects.[1] Animals exposed to 5 ppm for 1 hour daily for 5 days developed signs of pulmonary injury; 1 ppm for the same time period caused no effects.

REFERENCE

1. Selenium hexafluoride. Documentation of the TLVs and BEIs, 5th ed, p 518. Cincinnati, American Conference of Governmental Industrial Hygienists (ACGIH), 1986

SILICA (amorphous–diatomaceous earth)

CAS: 68855-54-9

SiO_2 Uncalcined; containing <1% quartz
1987 TLV = 10 mg/m^3, total dust

Synonyms: Diatomite; diatomaceous silica; infusorial earth; tripoli; kiesselguhr

Physical Form. Solid; soft, chalky powder

Uses. Production of filters, polishes, absorbents, insulators

Exposure. Inhalation

Toxicology. Amorphous silica, natural diatomaceous earth, is usually considered to be of low toxicity; however, pure amorphous silica is rarely found, and diatomaceous earth usually contains some amount of crystalline silica. Processing of amorphous silica by high temperature calcining alters the silica from the benign amorphous to the pathogenic crystalline form, which causes fibrosis.[1]

In a study of diatomaceous earth workers, those employed in the quarry for more than 5 years and exposed only to natural diatomaceous earth had no significant roentgenologic changes. Of other workers employed for more than 5 years in the milling process and exposed to calcined material, 17% had simple pneumoconiosis, and 23% had the confluent form, probably the result of fibrogenic action of the crystalline silica formed by calcination of the naturally occurring mineral.[1,2]

In humans, calcined diatomaceous earth pneumoconiosis is characterized roentgenographically by fine linear and/or minute nodular shadows, either or both of which may be accompanied by conglomerate fibrosis. In the simple phase of the disease, the upper lobes are affected more than the lower lobes, and the condition progresses by an increase in the apparent number of the nodules, which rarely attain the density or size of nodules often seen in quartz silicosis.[2] In the early confluent stage of the disease, the linear and nodular changes in the upper lung fields become more circumscribed and homogeneous. Histologically, there is an absence of the focal, discrete, hyaline nodules or the whorled pattern of collagenous fibers of typical silicosis.[2,3]

Repeated exposure of guinea pigs to natural diatomaceous earth for periods of up to 50 weeks to average concentrations ranging from 60 to 124 mg/m^3 caused thickening of the alveolar septa by infiltration of macrophages, accumulation of large numbers of multinuclear cells containing dust particles, and lymphadenopathy, but no proliferation of connective tissue.[4]

REFERENCES

1. Dutra FR: Diatomaceous earth pneumoconiosis. Arch Environ Health 11:613–619, 1965
2. Dechsli WR, Jacobson G, Brodeur AE: Diatomite pneumoconiosis: Roentgen characteristics and classification. Am J Roentgenol Radium Ther Nucl Med 85:263–270, 1961
3. Smart RH, Anderson WM: Pneumoconiosis due to diatomaceous earth–clinical and x-ray aspects. Ind Med Surg 21:509–518, 1952
4. Tebbens BD, Beard RR: Experiments on diatomaceous earth pneumoconiosis. I. Natural diatomaceous earth in guinea pigs. AMA Arch Ind Health 16:55–63, 1957

SILICA (crystalline [quartz])

CAS: 14808-60-7

SiO_2 1987 TLV = 0.1 mg/m^3, respirable dust

Synonyms: Silicon dioxide; silicic anhydride

Physical Form. Colorless crystals

Uses. Manufacture of glass, porcelain, and pottery; metal casting; sandblasting; granite cutting; manufacture of refractory, grinding, and scouring compounds

Exposure. Inhalation

Toxicology. Silica causes silicosis, a form of disabling, progressive, and sometimes fatal pulmonary fibrosis characterized by the presence of typical nodulation in the lungs.[1]

The earliest lesions are seen in the region of the respiratory bronchioles. Lymphatics become obliterated by infiltration with dust-laden macrophages and granulation tissue. Morphologically, the typical lesion of silicosis is a firm nodule composed of concentrically arranged bundles of collagen; these nodules usually measure between 1 and 10 mm in diameter and appear around blood vessels and beneath the pleura, as well as in mediastinal lymph nodes. There may be conglomeration of nodules as the disease progresses, leading to massive fibrosis.[1] The pulmonary pleura is usually thickened owing to fibrosis and is often adherent to the parietal pleura, especially over the upper lobes and in the vicinity of underlying conglomerate lesions.[2]

Histologically, the silicotic nodule consists of a relatively acellular, avascular core of hyalinized reticulin fibers arranged concentrically and blending with collagen fibers toward the periphery, which has well-defined borders.[3] The particles of silica responsible for the reaction are birefringent and can be visualized under polarized light if they exceed 1 micron in diameter. Silica in the lungs can be identified by x-ray diffraction studies and incinerating a portion of the lung, with subsequent analysis of the ash. The silica content of the normal lung should not exceed 0.2% dry weight.

The clinical signs and symptoms of silicosis tend to be progressive with continued exposure to quantities of dust containing free silica, with advancing age, and with continued smoking habits.[1] Symptoms may also be exacerbated by pulmonary infections and cardiac decompensation. Symptoms include cough, dyspnea, wheezing, and repeated nonspecific chest illnesses. Impairment of pulmonary function may be progressive. In individual cases, there may be little or no decrement when simple discrete nodular silicosis is present, but when nodulations become larger or when conglomeration occurs, recognizable cardiopulmonary impairment tends to occur.

Progression of symptoms may continue after dust exposure ceases. Although there may be a factor of individual susceptibility to a given exposure to silica dust, the risk of onset and the rate of progression of the pulmonary lesion is clearly related to the character of the exposure (dust concentration and duration).[1] The disease tends to occur after an exposure measured in years rather than in months. It is generally accepted that silicosis predisposes to active tuberculosis and that the combined disease tends to be more rapidly progressive than uncomplicated silicosis.

The earliest radiographic evidence of nodular silicosis consists of small discrete opacities of 1 to 3 mm in diameter appearing in the upper lung fields. As the disease advances, discrete opacities increase in number and size and are seen in the lower as well as the other zones of the lung fields. Small conglomerations may then appear, subsequently developing into large, irregular, and sometimes massive opacities occupying the greater part of both lung fields. Bullae may be seen in the vicinity of conglomerations.[2]

A group of 972 granite shed workers were studied to relate exposure levels to incidence of silicosis.[4] The workers were grouped according to four average exposure levels: (1) 37 to 60, (2) 27 to 44, (3) 20, and (4) 3 to 9 mppcf. Those with the highest dust exposure showed development of early silicosis in 40% of the workers after 2 years, and 100% after 4 years of exposure. The development of silicosis in the remaining workers appeared to be proportional to the dust exposure. At the second highest exposure level (27 to 444 mppcf), early stages of silicosis appeared after 4 years of exposure, and more advanced stages developed by the seventh year. In the group exposed at an average of 20 mppcf, there was little indication of severe effects upon the health of the workers. In the lowest exposure group where the average dust concentration was 6 mppcf (range 3 to 9 mppcf), there was no indication of any untoward effects of dust exposure on workers.

In some occupations, such as sandblasting and production of silica flour, exposure to high concentrations of silica over only a few years has produced a more rapidly progressive form of the disease termed *accelerated silicosis.* The symptoms are those of the more chronic disease, but clinical and radiologic progression is rapid.[5]

An acute form of silicosis has occurred in a few workers exposed to very high concentrations of silica over periods of as little as a few weeks. The history is one of progressive dyspnea, fever, cough, and weight loss. In acute silicosis, the nodular pattern is absent—the lungs showing a diffuse ground-glass appearance, similar to pulmonary edema.[5]

REFERENCES

1. National Institute for Occupational Safety and Health: Criteria for a Recommended Standard . . . Occupational Exposure to Crystalline Sil-

ica. DHEW (NIOSH) Pub No 75–120. Washington, DC, US Government Printing Office, 1974
2. Parkes WR: Occupational Lung Disorders, 2nd ed, pp 142, 147–148. London, Butterworths, 1982
3. Levy SA: Occupational pulmonary diseases. In Zenz C (ed): Occupational Medicine–Principles and Practical Applications, pp 117, 129–134. Chicago, Year Book Medical Publishers, 1975
4. Russell AE, Britten RH, Thompson LR, Bloomfield JJ: The Health of Workers in Dusty Trades—II. Exposure to Siliceous Dust (Granite Industry). US Public Health Service Bull No 187. Washington, DC, US Government Printing Office, 1929
5. Seaton A. In Morgan WKC, Seaton A: Occupational Lung Diseases, 2nd ed. Philadelphia, WB Saunders, 1984

SILVER (and compounds)

CAS: 7440-22-4

Ag 1987 TLV = 0.01 mg/m^3—soluble compounds, as Ag
= 0.1 mg/m^3—metal dust and fume

Synonyms: Silver nitrate; silver chloride; silver sulfide

Physical Form. Crystals

Uses. Oxidizing agent; in photography; in silverplating

Exposure. Inhalation

Toxicology. The dust of silver and its soluble compounds causes local or generalized impregnation of the mucous membranes, skin, and eyes with silver, a condition termed *argyria.*

Localized argyria occurs in the skin and eyes where grey-blue patches of pigmentation are formed without evidence of tissue reaction.[1] Generalized argyria is recognized by the widespread pigmentation of the skin and may be seen first in the conjunctiva, with some localization in the inner canthus. Argyrosis of the respiratory tract has been described in two workers involved in the manufacture of silver nitrate. Their only symptom was mild chronic bronchitis. Bronchoscopy revealed tracheobronchial pigmentation. Biopsy of the nasal mucous membrane showed silver deposition in the subepithelial area.[1] It has been estimated that gradually accumulated intake of from 1 to 5 g of silver will lead to generalized argyria.[1]

Massive exposure to heated vapor of metallic silver for 4 hours by a workman caused lung damage with pulmonary edema.[2] Ingestion of 10 g silver nitrate is usually fatal. Large oral doses of the compound cause abdominal pain and rigidity, vomiting, convulsions, and shock.[3] Patients dying after intravenous administration of Collargol (silver plus silver oxide) showed necrosis and hemorrhage in the bone marrow, liver, and kidney.[3]

REFERENCES

1. Browning E: Toxicity of Industrial Metals, 2nd ed, pp 296–301. London, Butterworths, 1969
2. Forycki Z et al: Acute silver poisoning through inhalation. Bull Inst Maritime Trop Med in Gydnia 34:199–202, 1983
3. US Environmental Protection Agency: Ambient Water Quality, Criteria for Silver. Springfield, Virginia, National Technical Information Service, PB81–117822, October 1980

SOAPSTONE

$3MgO\text{-}4SiO_2\text{-}H_2O$

containing <1% quartz

1987 TLV = 3 mg/m^3 respirable dust
= 6 mg/m^3 total dust

Soapstone does not have a precise mineralogic definition but is of variable composition dependent upon its source; talc is mined as soapstone, but some forms of soapstone have as little as 50% talc.

Physical Form. Talc-like material of varying composition, but generally greyish-white, fine, odorless powder. It is noncombustible and insoluble in water.

Uses. Pigment in paint, varnishes; filler for paper, rubber, soap; lubricating molds and machinery; heat insulator

Exposure. Inhalation

Toxicology. The fibrous talc in soapstone dust causes fibrotic pneumoconiosis and an increased incidence of cancer of the lungs and pleura.[1-4]

In the development of talc pneumoconiosis or talcosis, the subject is initially symptom-free, but cough and dyspnea develop as the disease progresses; cyanosis, digital clubbing, and cor pulmonale occur in advanced cases. The disease progresses slowly, even in the absence of continued exposure; occasionally, the disease may progress rapidly, with death occurring within a few years of a very heavy exposure.

An epidemiologic study of 260 workers with 15 or more years of exposure to commercial talc dust, containing talc, tremolite, anthophyllite, carbonate dusts, and a small amount of free silica revealed a fourfold greater than expected mortality rate from cancer of the lungs and pleura; in addition, a major cause of death among these workers was cor pulmonale—a result of the pneumoconiosis.[5]

REFERENCES

1. Soapstone. Documentation of the TLVs and BEIs, 5th ed, p 532. Cincinnati, American Conference of Governmental Industrial Hygienists (ACGIH), 1986
2. Spiegel RM: Medical aspects of talc. In Goodwin A (ed): Proceedings of the Symposium on Talc, Bureau of Mines Report No 8639, pp 97–102. Washington, DC, US Government Printing Office, 1973
3. Kleinfeld M, Messite J, Kooyman O, Zaki MH: Mortality among talc miners and millers in New York State. Arch Environ Health 14:663–667, 1967
4. Blejer HP, Arlon R: Talc: A possible occupational and environmental carcinogen. J Occup Med 15:92–97, 1973
5. Kleinfeld M, Messite J, Zaki MH: Mortality experiences among talc workers: A follow-up study. J Occup Med 16:345–349, 1974

SODIUM FLUOROACETATE

CAS: 62-74-8

$CH_2FCOONa$ 1987 TLV = 0.05 mg/m^3; skin

Synonyms: Compound 1080; fluoroacetic acid, sodium salt; Fratol; sodium monofluoroacetate

Physical Form. Fine, white powder

Uses. Rodenticide

Exposure. Inhalation; ingestion

Toxicology. Sodium fluoroacetate is highly toxic and causes convulsions and ventricular fibrillation.

Fluoroacetate produces its toxic action by inhibiting the citric acid cycle.[1] The fluorine-substituted acetate is metabolized to fluorocitrate, which inhibits the conversion of citrate to isocitrate. There is an accumulation of large quantities of citrate in the tissue, and the cycle is blocked. The heart and central nervous system are the most critical tissues involved in poisoning by a general inhibition of oxidative energy metabolism.[1]

Onset of symptoms after ingestion is frequently delayed for 30 minutes to 2 hours; effects are vomiting, apprehension, auditory hallucinations, nystagmus, tingling sensation of the nose, numbness of face, facial twitching, and epileptiform convulsions.[2,3] After a period of several hours, there may be pulsus alterans, long sequences of ectopic heartbeats (often multifocal), tachycardia, ventricular fibrillation, and death.[3,4] The lethal oral dose in humans is estimated to be approximately 5.0 mg/kg.[4,5] In a fatal case of ingestion, autopsy findings included hemorrhagic pulmonary edema and degeneration of renal tubules.[5]

In the only alleged case of chronic human poisoning, an exterminator repeatedly exposed over a period of 10 years presented with severe and progressive lesions of the renal tubular epithelium and with milder hepatic, neurologic, and thyroid dysfunctions.[6]

Treatment. Induce vomiting if convulsions are not imminent; administer sodium or magnesium sulfate in water (15 to 30 g).[4] Although the clinical efficacy of monoacetin (glyceral monoacetate) is not established, it should probably be administered.[4] Glyceral monoacetate appears to serve as an acetate donor to block fluoroacetate metabolism in a competitive manner.[4] If monoacetin is not available, acetamide or ethyl alcohol may be given in the same doses.[4]

REFERENCES

1. Murphy SD: Toxic effects of pesticides. In Klaassen CD et al (eds): Casarett and Doull's Toxicology, the Basic Science of Poisons, 3rd ed, p 565. New York, Macmillan, 1986
2. Harrisson JWE et al: Acute poisoning with sodium fluoroacetate (compound 1080). JAMA 149:1522, 1952
3. Hayes WJ Jr: Clinical Handbook on Economic Poisons. Emergency Information for Treating Poisoning. US Public Health Service Pub No 476, pp 79–82. Washington, DC, US Government Printing Office, 1963
4. Gosselin RE et al: Clinical Toxicology of Commercial Products, Section III, 5th ed, pp 193–196. Baltimore, Williams & Wilkins, 1984
5. Harrisson JWE, Ambrus JL, Ambrus CM: Fluoroacetate (1080) poisoning. Ind Med Surg 21:440–442, 1952
6. Parkin PJ: Chronic sodium monofluoroacetate (compound 1080) intoxication in a rabbiter. NZ Med J 85:93–99, 1977

SODIUM HYDROXIDE

CAS: 1310-73-2

NaOH 1987 TLV = C 2 mg/m^3

Synonyms: Caustic soda; caustic flake; lye; caustic; liquid caustic

Physical Form. Solid

Uses. Manufacture of rayon, mercerized cotton, soap, paper, aluminum, petroleum products; metal cleaning; electrolytic extraction of zinc; tin plating; oxide coating

Exposure. Inhalation

Toxicology. Sodium hydroxide is a severe irritant of the eyes, mucous membranes, and skin.

Although inhalation of sodium hydroxide is usually of secondary importance in industrial exposures, the effects from the dust or mist will vary from mild irritation of the nose at 2 mg/m^3 to severe pneumonitis, depending upon the severity of exposure.[1,2] The greatest industrial hazard is rapid tissue destruction of eyes or skin upon contact with either the solid or with concentrated solutions.

Contact with the eyes causes disintegration and sloughing of conjunctival and corneal epithelium, corneal opacification, marked edema, and ulceration; after 7 to 13 days, either gradual recovery begins, or there is progression of ulceration and corneal opacification.[3] Complications of severe eye burns are symblepharon with overgrowth of the cornea by a vascularized membrane, progressive or recurrent corneal ulceration, and permanent corneal opacification.[1]

On the skin, solutions of 25% to 50% cause the sensation of irritation within about 3 minutes; with solutions of 4%, this does not occur until after several hours.[1] If not removed from the skin, severe burns with deep ulceration will occur. Exposure to the dust or mist may cause multiple small burns with temporary loss of hair.[2]

Ingestion produces severe abdominal pain; corrosion of the lips, mouth, tongue, and pharynx; and the vomiting of large pieces of mucosa. Cases of squamous cell carcinoma of the esophagus have occurred with latent periods of 12 to 42 years after ingestion. These cancers were undoubtedly sequelae of tissue destruction and possibly scar formation, rather than from a direct carcinogenic action of sodium hydroxide itself.[1]

REFERENCES

1. National Institute for Occupational Safety and Health: Criteria for a Recommended Standard . . . Occupational Exposure to Sodium Hydroxide. DHEW (NIOSH) Pub No 76–105, pp 23–50. Washington, DC, US Government Printing Office, 1975
2. Chemical Safety Data Sheet SD–9, Caustic Soda, pp 5, 16–17. Washington, DC, MCA, Inc, 1968
3. Patty FA: Alkaline materials. In Fassett DW, Irish DD (eds): Patty's Industrial Hygiene and Toxicology, 2nd ed, Vol 2, Toxicology, pp 867–868. New York, Wiley–Interscience, 1963

STIBINE

CAS: 7803-52-3

SbH_3 1987 TLV = 0.1 ppm

Synonyms: Antimony hydride; hydrogen antimonide

Physical Form. Colorless gas

Sources. Produced accidentally as a result of the generation of nascent hydrogen in the presence of antimony; may be liberated when certain drosses are treated with water or acid

Exposure. Inhalation

Toxicology. Stibine is a hemolytic agent in animals; it is expected that the same effect will occur in humans.

No clear-cut case of fatal stibine poisoning in humans has been reported.[1–4] Workers exposed to a mixture of gases (concentrations unmeasured) of stibine, arsine, and hydrogen sulfide developed headache, weakness, nausea, abdominal and lumbar pain, hemoglobinuria, hematuria, and anemia.[1] Although these signs and symptoms are clearly manifestations of acute hemolytic anemia, it is not possible to determine the relative contribution of arsine, which is also a hemolytic agent. By analogy to other effects caused by arsine, additional signs of stibine poisoning may be leukocytosis and jaundice.

Guinea pigs exposed to 65 ppm of stibine for 1 hour developed hemoglobinuria followed within a few days by profound anemia.[5] Stibine is also a pulmonary irritant in animals, causing pulmonary congestion and edema and, ultimately, death in cats and dogs following a 1-hour exposure at 40 to 45 ppm.[6]

REFERENCES

1. Dernehl C, Stead FM, Nau CA: Arsine, stibine and H_2S: Accidental generation in a metal refinery. Ind Med Surg 13:361, 1944
2. Stokinger HE: The metals. In Clayton GD, Clayton FE (eds): Patty's Industrial Hygiene and Toxicology, 3rd ed, Vol 2, Toxicology, p 1511. New York, Wiley–Interscience, 1981
3. Hygienic Guide Series: Stibine. Am Ind Hyg Assoc J 21:529–530, 1960
4. Pinto SS: Arsine poisoning: Evaluation of the acute phase. J Occup Med 18:633–635, 1976
5. Webster SH: Volatile hydrides of toxicological importance. J Ind Hyg Toxicol 28:167–182, 1946
6. Stibine. Documentation of the TLVS and BEIs, p 536. 5th ed, Cincinnati, American Conference of Governmental Industrial Hygienists (ACGIH), 1986

STODDARD SOLVENT

CAS: 8052-41-3
15% to 20% aromatic hydrocarbons
80% to 85% paraffin and naphthenic hydrocarbons 1987 TLV = 100 ppm

Synonyms: White spirits; safety solvent; varnoline

Physical Form. Colorless liquid

Uses. Dry cleaning; degreasing; paint thinner

Exposure. Inhalation

Toxicology. Stoddard solvent is a mild central nervous system depressant and a mucous membrane irritant.

Stoddard solvent is a mixture of predominantly C9 to C11 hydrocarbons of which 30% to 50% are straight and branched chain paraffins, 30% to 40% naphthenes, and 10% to 20% aromatic hydrocarbons.[1] Although uses may differ, Stoddard solvent is chemically similar to mineral spirits, and the terms have been used interchangeably.[1]

One of six volunteers exposed to 150 ppm of Stoddard solvent for 15 minutes had transitory eye irritation; at 470 ppm, all subjects had eye irritation and two had slight dizziness.[2] Reports of Stoddard solvent as an etiologic agent in the development of aplastic anemia are of questionable validity.[1] Skin exposure may cause dermatitis and sensitization.[1]

Rats exposed to Stoddard solvent at a level of 1400 ppm for 8 hours exhibited eye irritation, bloody exudate around the nostrils, and slight loss of coordination. Exposure of a dog resulted in increased salivation at 3 hours, tremor at 4 hours, and clonic spasms after 5 hours; 1700 ppm caused tremors, convulsions, and finally, death in cats after 2.5 to 7.5 hours.[2] No significant effects were observed in dogs exposed to 6 hours/day for 65 days to 330 ppm; there was elevated blood urea nitrogen levels and marked tubular degeneration in the kidneys of rats similarly exposed.[2]

The odor threshold is 0.9 ppm; the odor and irritative properties probably do not provide adequate warning of dangerous concentrations.[2]

REFERENCES

1. National Institute for Occupational Safety and Health: Criteria for a Recommended Standard . . . Occupational Exposure to Refined Petroleum Solvent. DHEW (NIOSH) Pub No 77–192. Washington, DC, US Government Printing Office, 1977
2. Carpenter CP et al: Petroleum hydrocarbon toxicity studies, III. Animal and human response to vapors of Stoddard solvent. Toxicol Appl Pharmacol 32:282–297, 1975

STRYCHNINE

CAS: 57-24-9
$C_{21}H_{22}N_2O_2$ 1987 TLV = 0.15 mg/m^3

Synonym: Stricnina

Physical Form. White, crystalline powder

Uses. Rodenticide

Exposure. Inhalation; ingestion

Toxicology. Strychnine is a potent convulsant.

Strychnine poisoning occurs from accidental and intentional ingestion and from misuse as a therapeutic agent.[1] Doses of 5 to 7 mg cause muscle tightness, especially in the neck and jaws, and twitching of individual muscles, especially in the little fingers.[1]

The lethal oral dose is approximately 100 to 120 mg, but doses of up to 13,000 mg have reportedly been survived following treatment.[1,2] After ingestion, effects usually occur within 10 to 30 minutes and include stiffness of the face and neck muscles and increased reflex excitability.[3] Any sensory stimulus may produce a violent motor response, which, in the early stages of intoxication, tends to be a coordinated extensor thrust and, in later stages, may be a tetanic convulsion with opisthotonos; anoxia and cyanosis develop rapidly. Between convulsions, muscular relaxation is complete, breathing is resumed, and cyanosis lessens.[1] Because sensation is unaffected, the convulsions are painful and lead to overwhelming fear. As many as 10 convulsions separated by intervals of 10 to 15 minutes may be experienced, but death often occurs after the second to fifth convulsion, and even the first convulsion may be fatal if sustained; death is commonly due to asphyxia.[2,3] If recovery occurs, it is remarkably prompt and complete in spite of the violence of the illness; muscle soreness may persist for a number of days.[1]

In fatal cases, the pathological findings are entirely nonspecific. They usually consist of petechial hemorrhages and congestion of the organs, indicating combined action of severe convulsions and anoxia.[1] Compression fractures and related injury may be found in cases with violent tetany.[1]

Diagnosis. Signs and symptoms include stiffness of the neck and facial muscles, increased reflex excitability, tetanic convulsions with opisthotonos, and cyanosis.

Differential Diagnosis: Other causes of convulsions must be differentiated from strychnine exposure. These include idiopathic epilepsy; hypertensive encephalopathy; metabolic disturbances such as hypoglycemia, hypocalcemia, uremia, and porphyria; hypoxic encephalopathy; infections such as viral encephalitis; and exposure to other convulsants. Fully developed tetanus closely resembles strychnine poisoning with generalized increased rigidity and convulsive spasms of skeletal muscles.[4]

Special Tests: Except for the demonstration of the poison itself, laboratory findings are not helpful in diagnosis or treatment.[1]

Treatment. Because of the very rapid absorption of strychnine and the danger of precipitating convulsions, emptying of the stomach should not be attempted unless it can be done immediately after ingestion or after the patient is completely protected against convulsions.[1] Potassium permanganate or charcoal may be given to reduce and delay absorption if symptoms are minimal or absent. Tea and coffee should be avoided because of the synergistic stimulation of caffeine. The control of convulsions calls for the intravenous administration of a short-acting barbiturate or the use of inhalation anesthesia.[5] Endotracheal intubation is an important safeguard.[5] The duration of poisoning can be substantially shortened by peritoneal dialysis and forced diuresis.[1] Morphine is contraindicated because of respiratory depression.

In the past, it was recommended that the patient should be placed in a quiet, darkened room and protected from sudden, unexpected stimuli.[1] The patient may often tolerate manipulation without convulsions provided that all motions are expected and gentle. Although avoidance of unexpected stimuli remains a useful element in management of the patient, the availability of effective drugs has caused greater attention to be given to treating the patient in such a way that he reacts normally to stimuli.[1]

REFERENCES

1. Hayes WJ: Pesticides Studied in Man, pp 96–101. Baltimore, Williams & Wilkins, 1982
2. Gosselin RE et al: Clinical Toxicology of Commercial Products, Section III, 5th ed, pp 375–379. Baltimore, Williams & Wilkins, 1984
3. Franz DN: Central nervous system stimulants. In Goodman LS, Gilman A (eds): The Pharmacological Basis of Therapeutics, 7th ed, pp 582–584. New York, Macmillan, 1985
4. Beaty HN: Tetanus. In Petersdorf RG et al (eds): Harrison's Principles of Internal Medicine, 10th ed, p 1004. New York, McGraw-Hill, 1983
5. Victor M, Adams R: Sedatives, stimulants, and psychotropic drugs. In Petersdorf RG et al (eds): Harrison's Principles of Internal Medicine, 10th ed, p 1301. New York, McGraw-Hill, 1983

STYRENE

CAS: 100-42-5

$C_6H_5CHCH_2$ 1987 TLV = 50 ppm

Synonyms: Vinylbenzene; phenylethylene; styrene monomer; cinnamene

Physical Form. Colorless liquid

Uses. Solvent for synthetic rubber and resins; intermediate in chemical synthesis; manufacture of polymerized synthetic materials

Exposure. Inhalation; skin absorption

Toxicology. Styrene is an irritant of the eyes and mucous membranes and a central nervous system depressant.

Humans exposed to 376 ppm experienced eye and nasal irritation within 15 minutes; after 1 hour at 376 ppm, effects included headache, nausea, decreased dexterity and coordination, and other signs of transient neurologic impairment.[1] Subjective complaints, including headache, fatigue, and concentration difficulty, have been reported following 90-minute experimental exposures at concentrations as low as 50 ppm.[2] Some acute effects on neuropsychological tests of verbal learning skills and visuoconstructive abilities have been demonstrated among workers exposed to mean concentrations of about 50 ppm.[3]

Although upper respiratory and eye irritation have been reported at concentrations as low as 50 ppm, there is no convincing evidence of lower respiratory effects or changes in pulmonary function.[2]

The rate of absorption of the liquid through the skin of the hand and the forearm in man was 9 to 15 mg/cm^2/hour.[4] Prolonged or repeated exposure may lead to dermatitis due to defatting action on the skin.

Although high-level experimental exposure to animals (1300 ppm or higher) has resulted in evidence of liver damage, there is no clear-cut evidence of human liver toxicity from industrial exposures.[2] Liver enzymes and serum bile acid concentrations among 34 workers with average 30 to 40 ppm styrene exposures for a mean of 5.1 years did not differ significantly from a control group of unexposed workers.[5]

Rats and guinea pigs exposed to 10,000 ppm became comatose in a few minutes and died after 30 to 60 minutes of exposure.[6] Animals exposed to 2500 ppm showed weakness and stupor, followed by incoordination, tremor, coma, and death in 8 hours.[6] Rats and guinea pigs showed signs of eye and nasal irritation after exposure to 1300 ppm for 8 hours/day, 5 days/week for 6 months.[3]

Although some excretion of unchanged styrene occurs through the lungs in humans, most absorbed styrene is excreted in the urine after metabolism to mandelic acid and phenylglyoxylic acid.[2] The sum of these two metabolites in the urine correlates with the TWA exposure to styrene, and this parameter may prove useful in biologic monitoring.[7]

There is limited evidence for carcinogenicity of styrene to experimental animals. Epidemiologic studies of styrene-exposed workers have not revealed an excess in overall cancer incidence. Although some mortality studies have identified several cases of lymphoma or leukemia among styrene workers that were greater than the expected occurrence, the numbers are small; this finding has not been observed in all studies.[2,8,9] Human studies of mutagenic and reproductive effects among styrene workers are limited and have revealed conflicting results, without consistent evidence of adverse effects.[2]

The odor threshold is 0.1 ppm; the disagreeable odor and the eye and nose irritation make the inhalation of seriously acute toxic quantities unlikely, although the warning properties may not be sufficient for prolonged exposures.

REFERENCES

1. Stewart RD, Dodd HC, Baretta ED, Schaffer AW: Human exposure to styrene vapor Arch Environ Health 16:656–662, 1968
2. National Institute for Occupational Safety and Health: Criteria for a Recommended Standard. . . Occupational Exposure to Styrene. DHHS (NIOSH) Pub No 83-119, pp. 121–130. Washington, DC, US Government Printing Office, 1983
3. Mutti A et al: Exposure-effect and exposure-response relationships between occupational exposure to styrene and neuropsychological functions. Am J Ind Med 5:275–286, 1984
4. Dutkiewicz T, Tyras H: Skin absorption of toluene, styrene, and xylene by man. Br J Ind Med 25:243, 1968
5. Harkonen H, Lehtniewi A, Aitio A: Styrene exposure and the liver. Scand J Work Environ Health 10:59–61, 1984
6. Sandmeyer EE: The aromatic hydrocarbons. In Clayton GD, Clayton FE (eds): Patty's Industrial Hygiene and Toxicology, 3rd ed, Vol 2A, Toxicology, pp 3312–3319. New York, Wiley–Interscience, 1981
7. Bartolucci G et al: Biomonitoring of occupational exposure to styrene. Appl Ind Hyg 3:125–131, 1986
8. Hodgson J, Jones R: Mortality of styrene production, polymerization and processing workers of a site in northwest England. Scand J Work Environ Health 11:347–352, 1985
9. Okun A et al: Mortality patterns among styrene-exposed boatbuilders. Am J Ind Med 8:193–205, 1985

SULFOLANE

CAS: 126-33-0

$C_4H_8SO_2$ TLV = none established

Synonyms: 1,1–Dioxidetetrahydrothiofuran; 1,1–dioxothiolan; cyclotetramethylene sulfone; dioxothiolan; sulphoxaline; tetrahydrothiophene 1,1–dioxide; tetramethylene sulfone; thiocyclopentane dioxide; thiophane dioxide

Physical Form. Colorless, oily liquid (solid at 15°C)

Uses. Process solvent for extractions of aromatics and for purification of acid gases

Exposure. Inhalation

Toxicology. Sulfolane is a convulsant in animals.

Sulfolane is not highly acutely toxic. Oral LD_{50} values in the rat range from 1846 to 2500 mg/kg.[1] Symptoms of neurotoxicity have been observed in rats, dogs, and monkeys after ingestion, injection, inhalation, or dermal application. Effects included convulsions, hyperactivity, tremor, and ataxia.[2] In acute inhalation studies, no rats died in the 2 weeks following 4-four exposure to levels as high as 12,000 mg/m^3.[2] Dogs exposed continuously to 200 mg/m^3 for 7 days experienced convulsions.

The liquid is not irritating to the skin and is mildly irritating to the eyes.[3] It was not a sensitizer in the guinea pig.[4,5]

REFERENCES

1. Anderson ME et al: Sulfolane-induced convulsions in rodents. Res Commun Chem Pathol Pharmacol 15:571, 1976
2. Anderson ME et al: The inhalation toxicity of sulfolane (tetrahydrothiopene–1,1–dioxide). Toxicol Appl Pharmacol 40:463, 1977
3. Weiss G (ed): Hazardous Chemicals Data Book, p 840. Park Ridge, New Jersey, Noyes Data Corporation, 1980
4. Brown VKH, Ferrigan LW, Stevenson DE: Acute toxicity and skin irritation properties of sulfolane. Br J Ind Med 23:302, 1966
5. Phillips Petroleum Co: FYI–OTS–0484–034 Supplement Sequence D. Summary of Toxicity of Sulfolane. Washington, DC, US Environmental Protection Agency, Office of Toxic Substances, June 6, 1983

SULFUR DIOXIDE

CAS: 7446-09-5

SO_2 1987 TLV = 2 ppm

Synonyms: Sulfurous anhydride; sulfurous oxide

Physical Form. Colorless gas

Uses. Manufacture of sulfuric acid; as a bleach; casting of nonferrous metal; food processing, manufacture of sodium sulfite, pulp/paper manufacture; combustion product from burning of materials containing sulfur

Exposure. Inhalation

Toxicology. Sulfur dioxide is a severe irritant of the eyes, mucous membranes, and skin.

The irritant effects of SO_2 are due to the rapidity with which it forms sulfurous acid on contact with moist membranes.[1,2] Approximately 90% of all SO_2 inhaled is absorbed in the upper respiratory passages, where most effects occur; however, it may produce respiratory paralysis and may also cause pulmonary edema.[2] Exposure to concentrations of 10 to 50 ppm for 5 to 15 minutes causes irritation of the eyes, nose, and throat; rhinorrhea; choking; cough; and, in some instances, reflex bronchoconstriction with increased pulmonary resistance.[2]

The phenomenon of adaptation to irritating concentrations is a recognized occurrence in experienced workers.[2] Workers repeatedly exposed to 10 ppm experienced upper respiratory irritation and some nosebleeds, but the symptoms did not occur at 5 ppm. In another study, initial cough and irritation did occur at 5 ppm and 13 ppm, but subsided after 5 minutes of exposure.[2,4]

In a human experimental study with the subjects breathing through the mouth, brief exposure to 13 ppm caused a 73% increase in pulmonary flow resistance, 5 ppm resulted in a 40% increase; and 1 ppm produced no effects.[3]

Studies of persons with mild asthma have demonstrated much greater sensitivity to low levels of sulfur dioxide exposure, particularly during exercise. Exposures to concentrations of 0.5 to 0.1 ppm during exercise resulted in significant increases in airway resistance in these subjects.[5] At rest, exposures to 1 ppm resulted in a significant increase in airway resistance in mild asthmatics.[6]

Epidemiologic studies of workers chronically exposed to sulfur dioxide, as in copper smelters, have yielded conflicting results regarding excessive occurrence of chronic respiratory disease, chronic bronchitis, or decrements in pulmonary function. Such studies are plagued by the confounding effect of smoking and difficulties in exposure assessment. Overall, the evidence for chronic effects in humans is quite limited.[7]

Exposure of the eyes to liquid SO_2 from pressurized containers causes corneal burns and opacification resulting in a loss of vision.[2] The liquid on the skin produces skin burns from the freezing effect of rapid evaporation.[2]

REFERENCES

1. National Institute for Occupational Safety and Health: Criteria for a Recommended Standard . . . Occupational Exposure to Sulfur Dioxide, pp 16–54. Washington, DC, US Government Printing Office, 1974
2. Department of Labor: Occupational exposure to sulfur dioxide. Federal Register 40:54520–54534, 1975
3. Whittenberger JL, Frank RN: Human exposures to sulfur dioxide. Arch Environ Health 7:244–245, 1963
4. Sulfur dioxide. Documentation of the TLVs and BEIs, 5th ed, pp 542–543. Cincinnati, American Conference of Governmental Industrial Hygienists (ACGIH), 1986
5. Sheppard D et al: Exercise increases sulfur dioxide-induced bronchoconstriction in asthmatic subjects. Am Rev Respir Dis 123:486–491, 1981
6. Sheppard D et al: Lower threshold and greater bronchomotor responsiveness of asthmatic subjects to sulfur dioxide. Am Rev Respir Dis 122:873–878, 1980
7. Federspiel C et al: Lung function among employees of a copper mine smelter: Lack of effect of chronic sulfur dioxide exposure. J Occup Med 22:438–444, 1980

SULFUR HEXAFLUORIDE

CAS: 2551-62-4

SF_6 1987 TLV = 1000 ppm

Synonym: Sulfur fluoride

Physical Form. Colorless, odorless gas

Uses. Dielectric for high-voltage equipment

Exposure. Inhalation

Toxicology. Sulfur hexafluoride is an agent of low toxicity; at extremely high levels, it has a mild effect on the nervous system.

In humans, inhalation of 80% sulfur hexofluoride and 20% oxygen for 5 minutes produced peripheral tingling and a mild excitement stage, with some altered hearing in most subjects.[1] According to the ACGIH, the chief hazard, as with other inert gases, would be asphyxiation as a result of the displacement of air by this heavy gas.[2]

Rats exposed for many hours to an atmosphere containing 80% sulfur hexafluoride and 20% oxygen gave no perceptible indications of intoxication, irritation, or other toxicologic effects.[3]

Electrical discharges and high temperatures will cause sulfur hexafluoride decomposition.[2,4] Although some decomposition products are highly toxic, the concentrations produced and practical significance under usual working conditions are undetermined.

REFERENCES

1. Glauser SC, Glauser EM: Sulfur hexafluoride—a gas not certified for human use. Arch Environ Health 13:467, 1966
2. Sulfur hexafluoride. Documentation of the TLVs and BEIs, 5th ed, p 543. Cincinnati, American Conference of Governmental Industrial Hygienists (ACGIH), 1986
3. Lester D, Greenberg LA: The toxicity of sulfur hexafluoride. Arch Ind Hyg Occup Med 2:348–349, 1950
4. Griffin GD et al: On the Toxicity of Sparked SF_6. IEEE Transactions on Electrical Insulation, Vol EI-18 No 5, pp 551–552. Washington, DC, Institute of Electrical and Electronic Engineers, 1983

SULFURIC ACID

CAS: 7664-93-9

H_2SO_4 1987 TLV = 1 mg/m³

Synonym: Oil of vitriol

Physical Form. Liquid

Uses. Fertilizer manufacturing; metal cleaning; manufacture of chemicals, plastics, and explosives; petroleum refining; pickling of metal

Exposure. Inhalation

Toxicology. Sulfuric acid is a severe irritant of the respiratory tract, eyes, and skin. Exposure also causes dental erosion.

Concentrated sulfuric acid destroys tissue as a result of its severe dehydrating action, whereas the dilute form acts as a milder irritant owing to acid properties. A worker sprayed in the face with liquid fuming sulfuric acid suffered skin burns of the face and body, as well as pulmonary edema from inhalation.[1] Sequelae were pulmonary fibrosis, residual bronchitis, and pulmonary emphysema; in addition, necrosis of the skin resulted in marked scarring.[1]

In human subjects, concentrations of about 5 mg/m^3 were objectionable, usually causing cough, with an increase in respiratory rate and impairment of ventilatory capacity.[2]

In a recent study of 248 workers, no significant association was found with exposure to vapor concentrations of up to 0.42 mg/m^3 (2.6 to 10 μm mass median diameter) and symptoms of cough, phlegm, dyspnea, and wheezing.[3] However, the FVC in the highest exposure group was reduced compared with that of a low exposure group. Repeated exposure of workers to unspecified concentrations has reportedly caused chronic conjunctivitis, tracheobronchitis, stomatitis, and dermatitis.[1]

The dose–effect relationship for chronic exposure is difficult to determine because of the number of factors that influence toxicity,[4] including the particle size of the mist, presence of particulates, synergistic and protective agents, and humidity. In regard to particle size, the smallest aerosol particles appear to cause the greatest alteration in pulmonary function and in microscopic lesions because of their ability to penetrate deeply into the lungs.[1] Larger particles that deposit in the upper lung may be more acutely harmful because reflexive bronchoconstriction occurs. Very large particles that penetrate only the nasal passages and upper respiratory tract would not lead to either effect. Sulfuric acid may also be carried deeper into the respiratory tract if it is absorbed on other particulates.[4] Synergism has been demonstrated between sulfuric acid and ozone, sulfur dioxide and metallic aerosols.[4] Increased ammonia concentrations in expired air afford protection. Because of the hygroscopic nature of sulfuric acid, humidity directly affects particle size and, hence, toxicity.[4]

The corrosive effects upon teeth with chronic exposure are well established.[1] The damage, etching of dental enamel, followed by erosion of enamel and dentine with loss of tooth substance is limited to the parts of the teeth that are exposed to direct impingement of acid mist upon the surface. Although etching typically occurs after years of occupational exposure, in one case, exposure to an average of 0.23 mg/m^3 for 4 months was sufficient to initiate erosion.[3]

A positive association between the development of upper respiratory cancer, especially laryngeal cancer, and exposure to sulfuric acid was found in workers at a chemical refinery.[5] It is not known whether sulfuric acid can act as a direct carcinogen, as a promoter, or in combination with other substances.

Splashed in the eye, the concentrated acid causes extremely severe damage often leading to blindness, whereas dilute acid produces more transient effects from which recovery may be complete.[6] Although ingestion of the liquid is unlikely in ordinary industrial use, the highly corrosive nature of the substance will produce serious burns of the mouth and esophagus.[1]

In guinea pigs, aerosols of larger, but still respirable, size were more lethal than those of smaller size.[7] For 8-hour exposures, the LC_{50} was 30 mg/m^3 for mist of 0.8 μm and greater than 109 mg/m^3 for 0.4 μm mists.[7] Animals that died from exposure to the larger mists had hyperinflated lungs, whereas those that died from the finer mists also had hemorrhage and transudation. Changes in pulmonary function, however, were more severe for aerosols of smaller diameter.[8] The concentration producing 50% increase in pulmonary flow resistance is 0.3 mg/m^3 for 0.3 μm particles, 0.7 mg/m^3 for 1 μm, 6 mg/m^3 for 3.5 μm, and 30 mg/m^3 for 7 μm.[8] Long-term exposure of monkeys at concentrations between 0.1 and 1 mg/m^3, regardless of particle size, produced slight but increasingly severe microscopic pulmonary lesions.[9] Impairment of pulmonary ventilation occurred above 2.5 mg/m^3.[9]

Sulfuric acid mist was not teratogenic in mice or rabbits exposed 7 hours/day to 20 mg/m^3 during the period of major organogenesis.[10]

REFERENCES

1. National Institute for Occupational Safety and Health: Criteria for a Recommended Standard . . . Occupational Exposure to Sulfuric Acid. DHEW (NIOSH) Pub No 74–128, pp 19–49. Washington, DC, US Government Printing Office, 1974
2. Amdur MO, Silverman L, Drinker P: Inhalation of sulfuric acid mist by human subjects. AMA Arch Ind Hyg Occup Med 6:305–313, 1952
3. Gamble J et al: Epidemiological–environmental study of lead acid battery workers. III. Chronic effects of sulfuric acid on the respiratory system and teeth. Environ Res 35:30–52, 1984
4. Health Effects Assessment for Sulfuric Acid. Report No EPA/540/1-86/031, p 33. Washington, DC, US Environmental Protection Agency: Environmental Criteria and Assessment Office, 1984
5. Soskolne CL et al: Laryngeal cancer and occupational exposure to sulfuric acid. Am J of Epidemiol 120:358–369, 1984
6. Grant WM: Toxicology of the Eye, 3rd ed, pp 866–868. Springfield, Illinois, Charles C Thomas, 1986
7. Wolff RK et al: Toxicity of 0.4- and 0.8-μm sulfuric acid aerosols in the guinea pig. J Toxicol Environ Health 5:1037–1047, 1979
8. Amdur MO et al: Respiratory response of guinea pigs to low levels of sulfuric acid. Environ Res 5:418–423, 1978
9. Alarie YC et al: Long-term exposure to sulfur dioxide, sulfuric acid mist, fly ash, and their mixtures—results of studies in monkeys and guinea pigs. Arch Environ Health 30:254–263, 1975
10. Murray FJ et al: Embryotoxicity of inhaled sulfuric acid aerosol in mice and rabbits. J Environ Sci Health 13:251–266, 1979

SULFUR MONOCHLORIDE

CAS: 10025-67-9

S_2Cl_2 1987 TLV = C 1 ppm

Synonyms: Sulfur chloride; sulfur subchloride; disulfur dichloride

Physical Form. Nonflammable, light amber to yellowish-red fuming, oily liquid

Uses. Intermediate and chlorinating agent in manufacture of organics, sulfur dyes, insecticides, and synthetic rubber

Exposure. Inhalation

Toxicology. Sulfur monochloride is an irritant of the eyes, mucous membranes, and skin.

On contact with water, it decomposes to form hydrogen chloride and sulfur dioxide; because this occurs rapidly, it acts primarily as an upper respiratory irritant and does not ordinarily reach the lungs.[1] In humans, exposure to the vapor causes lacrimation and cough; exposure to high concentrations may cause pulmonary edema.[2] Concentrations of 2 to 9 ppm are reported to be mildly irritating.[3] Splashes of the liquid in the eyes will produce severe immediate damage, which may result in permanent scarring.[2] The liquid on the skin will produce irritation and burns if not removed.[2]

Exposure of mice to 150 ppm for 1 minute is fatal.[1] In cats, some deaths occurred following a 15-minute exposure to 60 ppm.

REFERENCES

1. Patty FA: As, P, Se, S, and Te. In Patty FA (ed): Industrial Hygiene and Toxicology, 2nd ed, Vol 2, Toxicology, pp 905–906. New York, Wiley–Interscience, 1963
2. Chemical Safety Data Sheet SD–77, Sulfur Chlorides, pp 5, 11–13. Washington, DC, MCA, Inc, 1960
3. Sulfur monochloride. Documentation of the TLVs for Substances in Workroom Air, 5th ed, p 545. Cincinnati, American Conference of Governmental Industrial Hygienists (ACGIH), 1986

SULFUR PENTAFLUORIDE

CAS: 5714-22-7

S_2F_{10} 1987 TLV = C 0.01 ppm

Synonym: Disulfur decafluoride

Physical Form. Colorless liquid

Source. Production by-product of synthesis of sulfur hexafluoride

Exposure. Inhalation

Toxicology. Sulfur pentafluoride is a severe pulmonary irritant in animals; severe exposure is expected to cause the same effect in humans.

Exposure of rats to 1 ppm for 16 to 18 hours was fatal; 0.5 ppm caused pulmonary edema and hemorrhage; 0.1 ppm caused irritation of the lungs; 0.01 ppm had no effect.[1] Nonfatal exposure of rats to 10 ppm for 1 hour caused pulmonary hemorrhage.[1]

REFERENCE

1. Greenberg LA, Lester D: The toxicity of sulfur pentafluoride. AMA Arch Ind Hyg Occup Med 2:350–352, 1950

SULFURYL FLUORIDE

CAS: 2699-79-8

SO_2F_2 1987 TLV = 5 ppm

Synonyms: Sulfuric oxyfluoride; Vikane

Physical Form. Colorless gas

Uses. Insect fumigant

Exposure. Inhalation

Toxicology. Sulfuryl fluoride is a central nervous system depressant and a pulmonary irritant in animals.

A worker exposed to an undetermined concentration of a mixture of sulfuryl fluoride and 1% chloropicrin for 4 hours developed nausea, vomiting, abdominal pain, and pruritis. Physical examination revealed conjunctivitis, rhinitis, pharyngitis, diffuse rhonchi, and paresthesia of the right leg, all of which rapidly subsided.[1] The role of sulfuryl fluoride in this case is not known, but the signs and symptoms are those expected of chloropicrin overexposure.

Exposure of animals to an unspecified but high concentration caused signs of narcosis and, in some instances, tremor, convulsions, and pulmonary edema; repeated exposure caused lung and kidney injury.[2] Exposure of animals to 1000 ppm for 3 hours or to 15,000 ppm for 6 minutes was fatal to less than 5% of animals tested.[1] The inhalation LC_{50} for 1 hour was 3020 ppm for female rats and 3730 ppm for male rats.[3] The oral LD_{50} for rats and guinea pigs was 100 mg/kg.[1]

Exposure of rabbits to 225 ppm during days 6 to 18 of gestation for 6 hours/day was slightly fetotoxic, causing decreased bodyweight and decreased rump–crown length.[4]

There are no warning properties of overexposure, because the gas is odorless and colorless.

REFERENCES

1. Taxay EP: Vikane inhalation. J Occup Med 8:425–426, 1966
2. Sulfuryl fluoride. Documentation of the TLVs and BEIs, 5th ed, p 546. Cincinnati, American Conference of Governmental Industrial Hygienists (ACGIH), 1986
3. Vernot EH et al: Acute toxicity and skin corrosion data for some organic and inorganic compounds and aqueous solutions. Toxicol Appl Pharmacol 42:417–423, 1977
4. Chemical Fact Sheet For Sulfuryl Fluoride. Fact Sheet No 51. Washington, DC, US Environmental Protection Agency, June 30, 1985

TALC (nonasbestos form)

CAS: 14807-96-6

$Mg_3Si_4O_{10}(OH)_2$ 1987 TLV = 2 mg/m^3 respirable dust

Synonym: Nonfibrous talc

Physical Form. Talc as a pure chemical compound is hydrous magnesium silicate. Talc is usually crystalline, flexible and soft. The purity and physical form of any sample of talc depends upon the source of the talc and on the minerals found in the ore body from which it is refined.

Uses. For clarifying liquid by filtration; pigment; for lubricating molds and machinery; electric and heat insulator; cosmetics

Exposure. Inhalation

Toxicology. The nonasbestos form of talc, also termed *nonfibrous*, has not been proved to cause the effects produced by exposure to fibrous talc: fibrotic pneumoconiosis and an increased incidence of cancer of the lungs and pleura.

Although there are a number of contradictory reports regarding the effects of talc, the contradiction has been ascribed to the differences in the mineral composition of the various talcs.

In a study of 20 workers exposed for 10 to 36 years to talc described as "pure," at levels ranging from 15 to 35 mppcf, no evidence of pneumoconiosis was found.[1,2] In another study that compared the pulmonary function of workers exposed to either fibrous or nonfibrous talc, it was concluded that although the fibrous form was the more pathogenic type, both talcs produced pulmonary fibrosis; no data were presented to document the types of talc involved.[3]

A study of 260 workers with 15 or more years of exposure to commercial talc dust (containing not only talc but also tremolite, anthophyllite, carbonate dusts, and a small amount of free silica) revealed a 40-fold greater than expected proportional mortality from cancer of the lungs and pleura. In addition, a major cause of death was cor pulmonale, a result of the pneumoconiosis; the effects were likely due to the asbestosform contaminants.[4,5] The role of nonfibrous talc in these disease states could not be assessed.

In a study of 80 talc workers, there was an excess prevalence of productive cough and of criteria of chronic obstructive lung disease (COLD) when compared with 189 nonexposed workers.[6] The increase in COLD and wheezing occurred only among smokers. Those talc workers with more than 10 years of exposure had significantly decreased FEV_1; none of the talc workers had chest x-rays definitely consistent with classical talc pneumoconiosis.[6] Exposure had been to talc of industrial grade with less than 1% silica and less than two fibers asbestos/ml at levels of 0.51 to 3.55 mg/m^3, with most of the workers being exposed to less than 1 mg/m^3 (or 2 mppcf).

A mortality study of 392 miners and millers of nonasbestos talc in Vermont showed an excess of deaths due to nonmalignant respiratory disease among millers and an excess of lung cancer mortality among miners.[7] The fact that the excess lung cancer mortality was observed for miners and not millers, despite probable higher dust exposure, led the investigators to conclude that other etiologic agents either alone or in combination with talc dust affected the miners.[7]

Another historical cohort study of 655 workers in a New York talc mine and mill revealed no significant differences in death rates from all causes, from cancer of the respiratory system, and from nonmalignant respiratory disease for the period ranging from 1948 to 1978.[8] However, workers with previous occupational histories were found to have excessive mortality from lung cancer and from nonmalignant respiratory tract disease, again suggesting another etiologic agent.

A 1-year follow-up of 103 miners and millers of talc ore free from asbestos and silica showed an association between exposure and small opacities on chest radiographs; the annual loss in FEV_1 and FVC was greater than expected and could not be wholly attributed to cigarette smoking.[9] However, effects on pulmonary function in nonsmokers was not associated with lifetime or current talc exposure.[9]

In inhalation studies with hamsters exposed to 8 mg/m^3 at a cumulative dust dose ranging from 15 to 6000 mg/m^3/hour, no talc-induced lung lesions were found[10] However, Italian talc, containing some quartz, was fibrogenic in specified pathogen-free rats exposed to a respirable dust concentration of 10.8 mg/m^3—the cumulative dust doses being approximately 4,100, 8,200, and 16,400 mg/m^3/hour for 3-, 6-, and 12-month exposures.[11] There was some evidence of progression of the fibrosis after exposure to talc had been discontinued in the animals exposed for the longest period of time.

In vitro assay of a number of respirable talc specimens of high purity demonstrated a modest but consistent cytotoxicity to macrophages; the investigators conclude that the talcs would be expected to be slightly fibrogenic in vivo.[12]

REFERENCES

1. Spiegel RM: Medical aspects of talc. In Goodwin A (ed): Proceedings of the Symposium on Talc, Bureau of Mines Report No 8639, pp 97–102. Washington, DC, US Government Printing Office, 1973
2. Hogue WL, Mallette FS: A study of workers exposed to talc and other dusting compounds in the rubber industry. J Ind Hyg Toxicol 31:359–364, 1949
3. Kleinfeld M et al: Lung function in talc workers—a comparative physiologic study of workers exposed to fibrous and granular talc dusts. Arch Environ Health 9:559–566, 1964
4. Kleinfeld M, Messite J, Kooyman O, Zaki MH: Mortality among talc miners and millers in New York State. Arch Environ Health 14:663–667, 1967
5. Kleinfeld M, Messitte J, Zaki MH: Mortality experiences among talc workers: A follow-up study. J Occup Med 16:345–349, 1974
6. Fine LJ, Peters JM, Burgess WA, Di Berardinis LJ: Studies of respiratory morbidity in rubber workers, Part IV. Respiratory morbidity in talc workers. Arch Environ Health 31:195–200, 1976
7. Selevan SG et al: Mortality patterns among miners and millers of non-asbestiform talc: preliminary report. J Environ Pathol Toxicol 2:273–284, 1979
8. Stille WT, Tabershaw IR: The mortality experience of upstate New York talc workers. J Occup Med 24:480–484, 1982

9. Wegman DH et al: Evaluation of respiratory effects in miners and millers exposed to talc free of asbestos and silica. Br J Ind Med 39:233–238, 1982
10. Wehner AP et al: Inhalation of talc baby powder by hamsters. Fd Cosmet Toxicol 15:121, 1977
11. Wagner JC et al: An animal model for inhalation exposure to talc. In Lemen R, Dement JM (eds): Dusts and Disease, p 389. Proceedings of the Conference on Occupational Exposure to Fibrous and Particulate Dust and Their Extension into the Environment. Park Forest South, Illinois, Pathotox Publishers, 1979
12. Davies R et al: Cytotoxicity of talc for macrophages in vitro. Fd Chem Toxicol 21:201–207, 1983

TANTALUM

CAS: 7440-25-7
Ta 1987 TLV = 5 mg/m^3

Synonyms: None

Physical Form. Solid (powder)

Uses. Manufacture of capacitors and other electronic components; chemical equipment and corrosion-resistant tools

Exposure. Inhalation

Toxicology. Tantalum has a low order of toxicity but has produced transient inflammatory lesions in the lungs of animals.

Surgical implantation of tantalum metal products as plates and screws has not shown any adverse tissue reaction, thus demonstrating its physiologic inertness.[1]

Intratracheal administration to guinea pigs of 100 mg tantalum oxide produced transient bronchitis, interstitial pneumonitis, and hyperemia, but it was not fibrogenic.[2] There were some slight residual sequelae in the form of focal hypertrophic emphysema and organizing pneumonitis around metallic deposits; and there was slight epithelial hyperplasia in the bronchi and bronchioles. Doses as high as 8000 mg/kg given orally produced no untoward effects in rats.[3]

REFERENCES

1. Stokinger HE: The metals. In Clayton GD, Clayton FE (eds): Patty's Industrial Hygiene and Toxicology, 3rd ed, rev, Vol 2A, Toxicology, pp 1493–2060, New York, Wiley-Interscience, 1981
2. Schepers GWH: The biologic action of tantalum oxide. AMA Arch Ind Health 12:121–123, 1955
3. Cochran KW et al: Acute toxicity of zirconium, columbium, strontium, lanthanum, cesium, tantalum and yttrium. Arch Ind Hyg Occup Med 1:637–650, 1950

TELLURIUM

CAS: 13494-80-9
Te 1987 TLV = 0.1 mg/m^3

Synonyms: None

Physical Form. Greyish-white, lustrous, brittle, crystalline solid; or dark grey-to-brown amorphous powder

Uses. Coloring agent in chinaware, porcelains, glass; reagent in producing black finish on silverware

Exposure. Inhalation

Toxicology. Tellurium causes garlic odor of the breath and malaise in humans.

Serious cases of tellurium intoxication have not been reported from industrial exposure. Iron foundry workers exposed to concentrations between 0.01 and 0.1 mg/m^3 complained of garlic odor of the breath and sweat, dryness of the mouth and metallic taste, somnolence, anorexia, and occasional nausea; urinary concentrations ranged from 0 to 0.6 mg/liter. Somnolence and metallic taste in the mouth did not appear with regularity until the level of tellurium in the urine was at least 0.01 mg/liter.[1] Skin lesions in the form of scaly itching patches and loss of sweat function occurred in workers exposed to tellurium dioxide in an electrolytic lead refinery.[2]

Hydrogen telluride has caused pulmonary irritation and hemolysis of red blood cells in animals; this gas is very unstable, however, and its occurrence as an actual industrial hazard is unlikely.[1,3]

In animals, acute tellurium intoxication results in restlessness, tremor, diminished reflexes, paralysis, convulsions, somnolence, coma, and death.[4] Administration of 500 to 3000 ppm tellurium in the diet of pregnant rats resulted in a high incidence of hydrocephalic offspring.[5] Weanling rats fed elemental tellurium at a level of 1% (10,000 ppm) in the diet developed a neuropathy characterized by segment demyelination; remyelination and functional recovery occurred despite continued administration of tellurium.[6]

See separate monograph on tellurium hexafluoride.

REFERENCES

1. Hygienic Guide Series: Tellurium. Am Ind Hyg Assoc J 25:198–201, 1964
2. Browning E: Toxicity of Industrial Metals, 2nd ed., pp 310–316. London, Butterworths, 1969
3. Cerwenka EA, Cooper WC: Toxicology of selenium and tellurium and their compounds. Arch Environ Health 3:189–200, 1961
4. Cooper WC (ed): Tellurium, pp 313–321. New York, Van Nostrand Reinhold, 1971
5. Duckett S: Fetal encephalopathy following ingestion of tellurium. Experientia 26:1239–1241, 1970
6. Lampert P, Garro F, Pentschew A: Tellurium neuropathy. Acta Neuropathol 15:308–317, 1970

TELLURIUM HEXAFLUORIDE

CAS: 7783-80-4

TeF_6 1987 TLV = 0.02 ppm

Synonyms: None

Physical Form. Gas

Source. By-product of ore refining

Exposure. Inhalation

Toxicology. Tellurium hexafluoride is a pulmonary irritant in animals; severe exposure is expected to cause the same effect in humans.

Human exposure has caused headache and dyspnea.[1,2] Two subjects accidentally exposed to tellurium hexafluoride following leakage of 50 g into a small laboratory experienced garlic breath, fatigue, a bluish-black discoloration of the webs of the fingers, and streaks on the neck and face.[3] Complete recovery occurred without treatment.

Rodents exposed to 1 ppm for 1 hour had increased respiratory rates, whereas a 4-hour exposure at this concentration caused pulmonary edema.[4] However, repeated exposure at 1 ppm for 5 days produced no effect; 5 ppm for 4 hours was fatal.

REFERENCES

1. Cooper WC: Tellurium, pp 317, 320–321. New York, Van Nostrand Reinhold Co, 1971
2. Cerwenka EA Jr, Cooper WC: Toxicology of selenium and tellurium and their compounds. Arch Environ Health 3:71–82, 1961
3. Blackadder ES, Manderson WG: Occupational absorption of tellurium: A report of two cases. Br J Ind Med 32:59–61, 1975
4. Tellurium hexafluoride. Documentation of the TLVs and BEIs, 5th ed, p 556. Cincinnati, American Conference of Governmental Industrial Hygienists (ACGIH), 1986

TERPHENYLS

CAS: 26140-60-3

$C_6H_5C_6H_4C_6H_5$ 1987 TLV = C 5 mg/m^3

Synonyms: Phenylbiphenyls; diphenylbenzenes; triphenyls; *o*-terphenyl; *m*-terphenyl; *p*-terphenyl

Physical Form. Colorless or light-yellow solids

Uses. Coolant for heat exchange in nuclear reactors

Exposure. Inhalation

Toxicology. Terphenyls are irritants of the eyes, mucous membranes, and skin.

There are no well-documented studies showing the effects of terphenyls on humans. Clinical studies of an exposed group of workers showed no ill effects from prolonged exposure to 0.1 to 0.9 mg/m^3.[1] Workers have experienced eye and respiratory irritation at levels above 10 mg/m^3.[2] As a class of compounds, organic coolants (including terphenyls) have caused transient headache and sore throat.[1] In addition, there have been cases of dermatitis attributed to skin contact with organic coolant compounds.[1]

Inhalation by rats of relatively high concentrations (660 to 3390 mg/m^3) of mixed and single isomers for periods of 1 hour for up to 14 days caused tracheobronchitis, pulmonary edema, and death at the higher concentrations.[3] In rats, the oral LD_{50} for *o*-terphenyl was 1.9 g/kg; for *m*-terphenyl, it was 2.4 g/kg; and, for *p*-terphenyl, it was greater than 10 g/kg.[4]

Transient morphologic changes in mitochondria of pulmonary cells were found in rats exposed to 50 mg/m^3 terphenyls 7 hours/day for up to 8 days.[5] The number of vacuolated mitrochondria increased with days of exposure.[5]

REFERENCES

1. Weeks JL, Lentle BC: Health considerations in the use of organic reactor coolants. J Occup Med 12:246–252, 1970
2. Testa C, Masi G: Determination of polyphenyls in working environments of organic reactors by spectrophotometric methods. Anal Chem 36:2284–2287, 1964
3. Haley TJ et al: Toxicological studies on polyphenyl compounds used in atomic reactor moderator-coolants. Toxicol Appl Pharmacol 1:515–523, 1959
4. Cornish HH, Bahor RE, Ryan RC: Toxicity and metabolism of ortho-, meta-, and para-terphenyls. Am Ind Hyg Assoc J 23:372–378, 1962
5. Adamson IYR et al: The acute toxicity of reactor polyphenyls on the lung. Arch Environ Health 19:499–504, 1969

1,1,2,2-TETRACHLORO-1,2-DIFLUOROETHANE

CAS: 76-12-0

CCl_2FCCl_2F 1987 TLV = 500 ppm

Synonyms: TCDF; Refrigerant 112

Physical Form. Colorless solid; melting point at 23.8°C

Uses. Solvent

Exposure. Inhalation

Toxicology. At high concentrations, 1,1,2,2-tetrachloro-1,2-difluoroethane affects the nervous system and causes pulmonary edema in animals; it is expected that severe exposure in humans will produce the same effects.

There are no reports of adverse effects in humans.

Rats exposed to 30,000 ppm died within 1 hour after onset of exposure with severe pulmonary hemorrhage.[1] At 15,000 ppm, rats exhibited excitability, incoordination, coma, rapid respiration, tremor, and convulsions; three of four rats died in 3 hours with pulmonary edema and hyperemia of the lungs and liver.[2] Exposure at 5000 ppm for 18 hours caused coma, pulmonary damage, and death.[1] Rats survived 10 exposures of 4 hours each at 3000 ppm with rapid, shallow respiration, hyperresponsiveness, and slight incoordination; recovery was immediate after exposure.[2] Decreased leukocyte count occurred in female rats exposed to 1000 ppm 6 hours/day for 31 days.[2]

1,1,2,2-Tetrachloro-1,2-difluoroethane was mildly irritating to rabbit eyes and guinea pig skin.

REFERENCES

1. Greenberg LA, Lester D: Toxicity of the tetrachlorodifluoroethanes. Arch Ind Hyg Occup Med 2:345, 1950
2. Clayton JW Jr et al: Toxicity studies on 1,1,2,2-tetrachloro-1,2-difluoroethane and 1,1,1,2-tetrachloro-2,2-difluoroethane. Am Ind Hyg Assoc J 27:332, 1966

1,1,2,2-TETRACHLOROETHANE

CAS: 79-34-5

$CHCl_2CHCl_2$ 1987 TLV = 1 ppm; skin

Synonyms: Acetylene tetrachloride; sym-tetrachloroethane

Physical Form. Heavy, clear liquid

Uses. Intermediate in the production of trichloroethylene from acetylene; insecticide; solvent

Exposure. Inhalation; skin absorption

Toxicology. Tetrachloroethane is toxic to the liver and causes central nervous system depression and gastrointestinal effects. There is limited evidence that it is carcinogenic in experimental animals.

Reports of industrial experience indicate that cases of intoxication most commonly have presented symptoms of gastrointestinal irritation (nausea, vomiting, abdominal pain, and anorexia) and liver involvement (liver enlarged and tender, jaundice, and bilirubinuria).[1,2] Jaundice sometimes progressed to cirrhosis and was often accompanied by delirium, convulsions, coma, and death. Other cases have primarily been characterized by central nervous system effects (dizziness, headache, irritability, nervousness, insomnia, paresthesia, and tremors).[1,2]

In one study, exposure of two men at 116 ppm for 20 minutes caused dizziness and mild vomiting; at 146 ppm, dizziness occurred after 10 minutes, mucosal irritation occurred at 12 minutes, and fatigue occurred within 20 minutes.[2] Concentrations up to 335 ppm produced the same symptoms with shorter exposure times. Occupational exposure to concentrations ranging from 1.5 to 247 ppm caused signs of liver injury such as hepatomegaly and increased serum bilirubin. These signs were still found after air concentrations had been reduced below 36 ppm.[2] Among a group of workers in India exposed to 20 to 65 ppm, effects included nausea, vomiting, and abdominal pain and a high incidence of tremor of the fingers.[3]

Oral ingestion of 3 ml caused coma or impaired consciousness in eight adult patients mistakenly administered tetrachlorethane.[2] Dermal absorption has been suspected in some poisoning cases.[2] Skin exposure may also produce dermatitis due to defatting action; in rare cases, the dermatitis may be caused by hypersensitivity to the substance.[4]

Treatment of mice during gestation caused embryotoxic effects and a low incidence of malformations.[5] 1,1,2,2-Tetrachloroethane administered by gavage produced an increased incidence of hepatocellular carcinomas in mice but not in rats.[6] There is limited evidence of carcinogenicity in experimental animals according to the IARC.[5]

Tetrachloroethane has a mild, sweetish odor detectable at 3 ppm that may not provide sufficient warning of dangerous levels owing to olfactory fatigue.

REFERENCES

1. von Oettingen WF: The Halogenated Aliphatic, Olefinic, Aromatic, and Aliphatic-Aromatic Hydrocarbons Including the Halogenated Insecticides, Their Toxicity and Potential Dangers. US Public Health Service Pub No. 414, pp 158–164. Washington, DC, US Government Printing Office, 1955
2. National Institute for Occupational Safety and Health: Criteria for a Recommended Standard . . . Occupational Exposure to 1,1,2,2–Tetrachloroethane. DHEW (NIOSH) Pub No 77–121, p 143. Washington, DC, US Government Printing Office, 1976
3. Lobo–Mendonca R: Tetrachloroethane—a survey. Br J Ind Med 20:50–56, 1963
4. Chemical Safety Data Sheet SD–34, Tetrachloroethane, p 12. Washington, DC, MCA, Inc, 1949
5. IARC Monographs on the Evaluation of the Carcinogenic Risk of Chemicals to Humans. Some Halogenated Hydrocarbons, Vol 20, pp 477–489. Lyon, International Agency for Research on Cancer, 1979

TETRACHLORONAPHTHALENE

CAS: 1335-88-2

$C_{10}H_4Cl_4$ 1987 TLV = 2 mg/m^3

Synonym: Halowax

Physical Form. Solid

Uses. Synthetic wax; dielectrics in capacitors; wire insulation

Exposure. Inhalation; skin absorption

Toxicology. Tetrachloronaphthalene may cause liver injury.

Experiments on human volunteers showed tetrachloronaphthalene to be nonacneigenic as opposed to the penta- and hexachloro derivatives that produce very severe chloracne.[1]

Rats exposed 16 hours/day to 10.97 mg/m^3 of tri- and tetrachloronaphthalene vapor for up to 4.5 months had slight liver injury.[2] When a mixture of tetra- and pentachloronaphthalene was fed to rats at a dose of 0.5 mg/day for 2 months, definite liver injury and some mortality occurred.[2]

REFERENCES

1. Shelley WB, Kligman AM: The experimental production of acne by penta- and hexachloronaphthalenes. Arch Dermatol 75:689–695, 1957
2. Deichmann WB: Halogenated cyclic hydrocarbons. In Clayton GD, Clayton FE (eds): Patty's Industrial Hygiene and Toxicology, 3rd ed, rev, Vol 2B, Toxicology, pp 3669–3675. New York, Wiley–Interscience, 1981

TETRAETHYL LEAD

CAS: 78-00-2

$Pb(C_2H_5)_4$ 1987 TLV = 0.1 mg/m^3, as Pb; skin

Synonyms: Lead tetraethyl; TEL; tetraethylplumbane

Physical Form. Colorless liquid

Uses. Gasoline additive to prevent "knocking" in motors

Exposure. Inhalation; skin absorption; ingestion

Toxicology. Tetraethyl lead (TEL) affects the nervous system and causes mental aberrations, including psychosis, mania, and convulsions. Of approximately 150 reported fatal cases of TEL poisoning, most have been related to early production methods, to cleaning of leaded gasoline storage tanks without protective equipment, and to suicidal or accidental ingestion.[1] Milder cases of intoxication have been caused by exposures to leaded gasoline in the workplace.[1]

The signs and symptoms of TEL intoxication differ in many respects from those owing to inorganic lead intoxication and are often vague and easily undetected. The absorption by humans of a sufficient quantity of tetraethyl lead, either briefly at a high rate (100 mg/m^3 for 1 hour) or for prolonged periods at a lower rate, causes intoxication.[2] The interval between exposure and the onset of symptoms varies inversely with dose and may last from 1 hour to several days.[1] This clinical latency is related to the time it takes for TEL to be absorbed, distributed, and metabolized to triethyl lead before toxic action develops.[1]

The initial or prodromal symptoms are nonspecific and include asthenia, weakness, fatigue, headache, nausea, vomiting, diarrhea, and anorexia.[1] Insomnia is usually present, and any sleep is light with nightmares. Signs of nervous system involvement (ataxia, tremor, hypotonia), as well as bradycardia and hypothermia, referred to as the TEL triad, may then develop.[1]

More severe intoxication causes recurrent or nearly continuous episodes of disorientation, hallucinations, facial contortions, and intense hyperactivity, which requires that the individual be restrained. Such episodes may convert abruptly into maniacal or violent convulsive seizures, which may terminate in coma and death.[2] Autopsy reports from humans who succumbed to TEL poisoning confirm that the brain is the critical target organ, and both focal and generalized damage have been described. Unless death occurs, recovery may take many weeks or months.[1] There is some question as to whether or not all changes are reversible following heavy or long-term exposures.[1]

During intoxication, there is a striking elevation of the rate of excretion of lead in the urine, but only a negligible or slight elevation of the concentration of lead in the blood.[2,3] In severe intoxication, the urine lead is rarely less than 350 μg/liter of urine, whereas the blood lead is rarely more than 50 μg/100 g of blood.[2,3] There is also a total absence of morphologic or chemical abnormalities in the erythrocytes—in sharp contrast to intoxication caused by inorganic lead.[2]

In a mortality study of 592 workers, the mean exposure time to TEL was 17.9 years, and urinary lead levels during this period did not exceed 180 μg/liter. The incidence of death in this group and in a control group of employees was less than that expected in the general population, and there were no peculiarities in the specific causes of death in either group.[4] In a similar study of a different cohort of these exposed workers, there were no significant health differences when compared with a control group.[5]

Tetraethyl lead is not an irritant, and no unpleasant sensations are related to skin contact or inhalation.[1] The ability to penetrate skin makes reliance on airborne concentrations impractical.[2]

Of 41 female Swiss mice that survived for 36 weeks after a single subcutaneous injection of 0.6 mg, 5 developed malignant lymphomas during the next 15 weeks; the significance of the study cannot be evaluated, because this tumor occurs spontaneously with a variable incidence in the mouse strain used.[6,7]

Teratogenic effects have not been observed after exposure to maximally tolerated doses in mice or rats.[1] Rodent embryos may serve as a poor model for human fetuses because the hepatic microsomal metabolizing enzymes do not develop until after birth in rodents, whereas these enzymes develop early in humans.

Diagnosis. Signs and symptoms include insomnia, lassitude, lurid dreams, anxiety; tremor, hyperreflexia, spasmodic muscular contractions; bradycardia, hypotension, hypothermia, pallor; nausea, anorexia, weight loss; disorientation, hallucinations, psychosis, mania, convulsions, coma; and eye irritation.

Differential Diagnosis: Other causes of convulsions must be differentiated from TEL exposure. These include idiopathic epilepsy; hypertensive encephalopathy; metabolic disturbances such as hypoglycemia, hypocalcemia, uremia, and porphyria; hypoxic encephalopathy; infections such as viral encephalitis; and exposure to other convulsants.

A history of exposure to TEL and the presence of the aforementioned signs and symptoms should confirm the diagnosis.

Special Tests: An analysis of the urinary concentration of lead is useful in evaluating the amount of TEL absorption. Blood lead concentration is an unreliable index of TEL absorption. A urine concentration of lead of 150 μg/liter corrected to a specific gravity of 1.024 indicates a dangerous degree of absorption; if the level rises to 180 μg/liter, the worker should be removed from exposure. TEL poisoning is typically associated with levels of 300 μg/liter or more; if the level is less than 100 μg/liter at the time of symptoms, TEL absorption is not the cause.[8]

Treatment. If TEL gets on the skin, immediately remove it by rinsing first with kerosene or similar petroleum distillate product and then wash with soap and water. Chelating agents are not useful for organolead poisoning.[9] Heavy and prolonged sedation with short-acting barbiturates in a hospital provides the most effective therapy available. Drugs with a cortical effect, such as morphine, are contraindicated, because they may worsen symptoms.[1] Fluid and electrolyte balance must be carefully maintained; this may be difficult owing to the patient's hyperactivity.

Persons suspected of having TEL intoxication should be kept under close surveillance, because personality changes may occur and be manifested in suicidal atttempt. Relapses during recovery are common.

Medical Control. Medical control involves preplacement and annual physical examination with emphasis on the central nervous system and the cardiovascular system and urinalysis. Biologically monitor the lead in urine every 3 months; urine specimens with a specific gravity of less than 1.020 should be discarded, and another sample of urine should be obtained.

REFERENCES

1. Grandjean P: Organolead exposures and intoxications. In Grandjean P (ed): Biological Effects of Organolead Compounds. pp 1–278. Boca Raton, Florida, CRC Press, 1984
2. Kehoe R: Lead, alkyl compounds. In International Labour Office: Encyclopaedia of Occupational Health and Safety, Vol II, L–Z, pp 1197–1199. Geneva, Switzerland, 1983
3. Fleming AJ: Industrial hygiene and medical control procedures—manufacture and handling of organic lead compounds. Arch Environ Health 8:266–270, 1964
4. Robinson TR: 20-year mortality of tetraethyl lead workers. J Occup Med 17:601–605, 1974
5. Robinson TR: The health of long service tetraethyl lead workers. J Occup Med 18:31–40, 1976
6. IARC Monographs on the Evaluation of the Carcinogenic Risk of Chemicals to Man, Vol 2, Some Inorganic and Organometallic Compounds, pp 150–160. Lyon, International Agency for Research on Cancer, 1973
7. Epstein SS, Mantel N: Carcinogenicity of tetraethyl lead. Experientia 24:580–581, 1968
8. Tsuchiya K: Lead. In Friberg L, Norsberg G, Vouk V (eds): Handbook on the Toxicology of Metals, 2nd ed, Vol II, Specific Metals, pp 340–342. Amsterdam, Elsevier, 1986
9. Chisholm JJ Jr: Treatment of lead poisoning. Mod Treatment 8:593–611, 1971

TETRAETHYL PYROPHOSPHATE

CAS: 107-49-3

$(C_2H_5)_4P_2O_7$ 1987 TLV = 0.05 mg/m^3; skin

Synonyms: TEPP; Tetron; NIFOS; TEP

Physical Form. Colorless, odorless liquid (pure); amber liquid (crude)

Uses. Insecticide

Exposure. Inhalation; skin absorption; ingestion

Toxicology. Tetraethyl pyrophosphate is an anticholinesterase agent.

Signs and symptoms of overexposure are caused by the inactivation of the enzyme cholinesterase, which results in the accumulation of acetylcholine at synapses in the nervous system, skeletal and smooth muscle, and secretory glands.[1–3] The sequence of the development of systemic effects varies with the route of entry.[2] The onset of signs and symptoms is usually prompt but may be delayed for up to 12 hours.[1,2] After inhalation, respiratory and ocular effects are the first to appear, often within a few minutes of exposure. Respiratory effects include tightness in the chest and wheezing due to bronchoconstriction and excessive bronchial secretion; laryngeal spasms and excessive salivation may add to the respiratory distress; cyanosis may also occur. Ocular effects include miosis, blurring of distant vision, tearing, rhinorrhea, and frontal headache.

After ingestion, gastrointestinal effects such as anorexia, nausea, vomiting, abdominal cramps, and diarrhea appear within 15 minutes, and muscular fasciculations in the immediate area usually occur within 15 minutes to 4 hours. Skin absorption is somewhat greater at higher ambient temperatures and is increased by the presence of dermatitis.[1,2]

With severe intoxication by all routes, an excess of acetylcholine at the neuromuscular junctions of skeletal muscle causes weakness aggravated by exertion, involuntary twitchings, fasciculations, and, eventually, paralysis.[2] The most serious consequence is paralysis of the respiratory muscles; in fatal cases, death usually occurs within 24 hours.[2] Effects on the central nervous system include giddiness, confusion, ataxia, slurred speech, Cheyne–Stokes respiration, convulsions, coma, and loss of reflexes. The blood pressure may fall to low levels, and cardiac irregularities, including complete heart block, may occur.[1]

In nonfatal cases, recovery usually occurs within 1 week, but increased susceptibility to anticholinesterase agents persists for up to several weeks after exposure.[2] Daily exposure to concentrations that are insufficient to produce symptoms following a single exposure may result in the onset of symptoms. Continued daily exposure may be followed by increasingly severe effects.[1]

Mild intoxication was reported in 15 people exposed to a dust of 1% TEPP; the predominant symptom was shortness of breath, which occurred after breathing the dust-laden air for 30 minutes. Symptoms rapidly abated after exposure was terminated.[4]

Eye exposure can produce visual disturbances without affecting blood cholinesterase levels. Exposed crop duster pilots, unable to judge distances, have been involved in accidents. Volunteers instilled with 2 drops of 0.1% TEPP 30 minutes apart experienced maximal miosis without any change in blood cholinesterase.[5]

TEPP inactivates cholinesterase by phosphorylation of the active site of the enzyme to form the diethylphosphoryl enzyme. Over the following 24 to 48 hours, there is a process, termed *aging*, of conversion to the monoethylphosphoryl enzyme. Aging is of clinical interest in the treatment of poisoning because cholinesterase reactivators such as pralidoxime (2–PAM, Protopam) chloride are ineffective after aging has occurred.

Diagnosis. *Signs and Symptoms:* Initial signs include headache, blurred vision, pallor, weakness, sweating, abdominal pain, nausea, vomiting, and diarrhea.

Moderate-to-severe intoxication includes miosis, lacrimation, excessive salivation, muscle fasciculations, dyspnea, cyanosis, convulsions, shock, cardiac arrhythmias, and coma.

Differential Diagnosis: Diagnosis is based primarily on a history of exposure and clinical evidence of diffuse parasympathetic stimulation. Careful observation of the effects of atropine and pralidoxime may be valuable. Patients with organophosphate poisoning are resistant to the action of atropine at moderate dosages; failure of 1 to 2 mg of atropine administered parenterally to produce signs of atropinization (flushing, mydriasis, tachycardia, or dryness of mouth) indicates anticholinesterase poisoning. Intravenous injection of 1 g pralidoxime generally causes some recovery from signs and symptoms.

Special Tests: Two types of cholinesterase are clinically significant: (1) true acetylcholinesterase, found principally in the nervous system and the red blood cell; and (2) pseudo- or butyrylcholinesterase, found in the plasma, liver, and nervous system. Whereas the action of both types is inhibited by organophosphates, the level of depression of red blood cell cholinesterase is a better indicator of clinically significant reduction of cholinesterase activity in the nervous system.

Laboratory evidence of depression of red blood cell cholinesterase to a level substantially below pre-exposure levels (at least 50% and usually much lower) is verification of organophosphate poisoning. There is an imperfect correlation between the degree of depression of cholinesterase enzymes and the occurrence of symptoms. With a rapid drop in cholinesterase activity, generally reflecting an acute heavy exposure, there may be symptoms with only a 30% depression, whereas with slower drops to 70% depression, reflecting chronic low-level exposure, there may be no symptoms.[6]

If no pre-exposure baseline has been performed but symptoms are not sufficient to justify treatment with atropine, repeated testing during the recovery period demonstrating progressively increasing plasma and red blood cell cholinesterase levels over several days and weeks, respectively, suggests the diagnosis of anticholinesterase poisoning.

There are many different methods for estimation of cholinesterase content of blood, and associated with each method is a different set of normal values and a different set of reporting units. The laboratory report of a cholinesterase determination should state the units involved along with the appropriate normal range. Based on the Michel method, the normal range of red blood cell cholinesterase activity (delta *p*H per hour) is 0.39 to 1.02 for men, and 0.34 to 1.10 for women.[7] The normal range of the enzyme activity (delta *p*H per hour) of plasma is 0.44 to 1.63 for men, and 0.24 to 1.54 for women.

Treatment. Treatment of organophosphate poisoning ranges from simple removal from exposure in very mild cases to the provision of rigorous supportive and antidotal measures in severe cases.[2,3,5,8,9] In moderate-to-severe cases, because of pulmonary involvement, there may be need for artificial respiration using a positive-pressure method. Careful attention must be paid to removal of secretions and to maintenance of a patent airway. Anticonvulsants such as diazepam or thiopental sodium may be necessary. Maintenance of respiration is critical because death usually results from weakness of the muscles of respiration and accumulation of excessive secretions in the respiratory tract.[3]

As soon as cyanosis has been overcome, 2 to 4 mg of atropine should be given intravenously. (Atropine may induce ventricular fibrillation in the presence of cyanosis.) *This dose of atropine is approximately 10 times the amount that is administered for other conditions in which atropine is considered therapeutic.* This dose should be repeated at 5- to 10-minute intervals until signs of atropinization appear (dry, flushed skin, tachycardia as high as 140 beats/minute, and pupillary dilatation). If muscarinic symptoms reappear, the dose of atropine should be repeated. A mild degree of atropinization should be maintained for at least 48 hours.[2]

Pralidoxime (2–PAM, Protopam) chloride is a cholinesterase reactivator that complements the action of atropine. It has its greatest effect in reversing the nicotinic action of anticholinesterase agents at skeletal neuromuscular junctions but virtually no effect on central nervous system manifestations. In moderate-to-severe cases, the dose for adults is 1 to 2 g injected intravenously at a rate not in excess of 500 mg/minute. If muscle weakness has not been relieved or if it recurs within 20 minutes, a second dose of 1 g is indicated. Treatment with pralidoxime chloride will be most effective if given within 24 hours after poisoning.[2] Morphine, aminophylline, and phenothiazines are contraindicated because of documented experience of adverse reactions in cases of organophosphate poisoning.[9]

It is of great importance to decontaminate the patient. Contaminated clothing should be removed at once, and the skin should be washed with generous amounts of soap or detergent and a flood of water, which is best accomplished under a shower or by submersion in a pond or other body of water if the exposure occurred in the field. Careful attention should be paid to cleansing of the skin and the hair.

The patient should be attended to and monitored continuously for not less than 24 hours, because serious and sometimes fatal relapses have occurred as a result of continuing absorption of the toxin or dissipation of the effects of the antidote.

Regeneration of cholinesterase is primarily by synthesis of new enzyme and takes place at the rate of approximately 1% per day.[9] A patient who has recovered from the acute phase of poisoning remains hypersusceptible to anticholinesterases for up to several weeks.

Medical Control. Medical control involves preplacement and annual physical examination with determination of pre-exposure red blood cell cholinesterase activity. A person whose red blood cell cholinesterase falls to or below 40% of the pre-exposure baseline should be removed from further exposure until the activity returns to within 80% of the pre-exposure baseline.

REFERENCES

1. Koelle GB (ed): Cholinesterases and Anticholinesterase Agents. Handbuch der Experimentallen Pharmakologie, Vol 15, pp 989–1027. Berlin, Springer-Verlag, 1963

2. Taylor P: Anticholinesterase agents. In Gilman AG et al (eds): Goodman and Gilman's The Pharmacological Basis of Therapeutics, 7th ed, pp 110–129. New York, Macmillan, 1985
3. Hayes WJ Jr: Clinical Handbook on Economic Poisons, Emergency Information for Treating Poisoning. US Public Health Service Pub No 476, pp 12–23, 40–42. Washington, DC, US Government Printing Office, 1963
4. Quinby GE, Doornink GM: Tetraethyl pyrophosphate poisoning following airplane dusting. JAMA 191:95–100, 1965
5. Hayes WJ Jr: Pesticides Studied in Man, pp 391–394. Baltimore, Williams & Wilkins, 1982
6. Coye MJ, Lowe JA, Maddy KT: Biological monitoring of agricultural workers exposed to pesticides. I. Cholinesterase activity determinations. J Occup Med 28:619–627, 1986
7. Michel HO: Electrometric method for determination of red blood cell and plasma cholinesterase activity. J Lab Clin Med 34:1564–1568, 1949
8. Namba T, Nolte CT, Jackrel J, Grob D: Poisoning due to organophosphate insecticides. Am J Med 50:475–492, 1971
9. Milby TH: Prevention and management of organophosphate poisoning. JAMA 216:2131–2133, 1971

TETRAHYDROFURAN

CAS: 109-99-9

$(C_2H_4)_2O$ 1987 TLV = 200 ppm

Synonyms: Cyclotetramethylene oxide; diethylene oxide; THF; tetramethylene oxide

Physical Form. Colorless liquid

Uses. Solvent for resins; manufacture of lacquers

Exposure. Inhalation

Toxicology. Tetrahydrofuran is an upper respiratory tract irritant; at high concentrations, it is a central nervous system depressant in humans.

No chronic systemic effects have been reported in humans, although nausea, dizziness, and headaches are said to occur with overexposure but are readily reversible in fresh air.[1]

In dogs and mice, concentrations above 25,000 ppm were required to produce anesthesia.[2] Exposure of animals to vapor above 3000 ppm for 8 hours/day for 20 days produced irritation of the upper respiratory tract.[2] Some injury to the liver and kidneys was observed; this was possibly due to impurities, because other studies have not confirmed these findings.[1,2] Daily exposure of dogs for 6 hours to 200 to 400 ppm for 12 weeks produced no significant effects.[2] In recent studies, it was not found to be a skin irritant or sensitizer.[2]

The liquid has an etheral odor similar to acetone and a pungent taste.

REFERENCES

1. Hygienic Guide Series: Tetrahydrofuran. Am Ind Hyg Assoc J 20:250–251, 1959
2. Tetrahydrofuran. Documentation of the TLVs and BEIs, 5th ed, p 564. Cinncinnati, American Conference of Governmental Industrial Hygienists (ACGIH), 1986

TETRAMETHYL LEAD

CAS: 75-74-1

$Pb(CH_3)_4$ 1987 TLV = 0.15 mg/m^3; skin

Synonyms: Lead tetramethyl; TML

Physical Form. Colorless liquid

Uses. Gasoline additive, especially to aviation and premium grades with high aromatic content

Exposure. Inhalation; skin absorption; ingestion

Toxicology. Tetramethyl lead (TML) affects the nervous system in animals.

Accidental human exposure to a high

level of TML liquid for approximately 5 minutes caused no signs or symptoms of lead poisoning. Significant exposure was corroborated by high levels of urinary lead, averaging almost 1000 μg/24 hours for the first 4 days postexposure.[1] By comparison, urinary lead levels less than 750 μg/24 hours following tetraethyl lead (TEL) exposure have been associated witth confusion, agitation, and acute toxic delirium.[1]

In a plant, 21 workers were exposed at different times to TEL and then to TML under similar conditions for similar periods of time. TML had three times the airborne level found during TEL production, yet the urinary lead levels were nearly the same in both cases; this suggests that TML is absorbed more slowly than TEL.[2] No signs or symptoms of toxicity were noted.

In rats, the approximate oral LD_{50} for TML is 108 mg/kg vs 17 mg/kg for TEL. Effects included tremor, hyperactivity, and convulsions.[3] Inhalation studies on rats showed TML to have less than one tenth the toxicity of TEL.[4] In dogs and mice, however, the reverse is true, with TML being more potent than TEL.[5]

Prudent practice suggests that TML be treated as though it were TEL.[5] Further caution is indicated by recent reports that the degradation product of TML, trimethyllead, acts differently from higher trialkylated compounds, inducing lipid peroxidation.[6] This difference indicates a potential for more severe chronic toxicity from TML exposure.

REFERENCES

1. Gething J: Tetramethyl lead absorption: A report of human exposure to a high level of tetramethyl lead. Br J Ind Med 32:329–333, 1975
2. deTreville RTP, Wheeler HW, Sterling T: Occupational exposure to organic lead compounds—the relative degree of hazard in occupational exposure to air-borne tetraethyllead and tetramethyllead. Arch Environ Health 5:532–536, 1962
3. Schepers GWH: Tetraethyllead and tetramethyllead—comparative experimental pathology. Part I. Lead absorption and pathology. Arch Environ Health 8:277–295, 1964
4. Cremer JE, Callaway S: Further studies on the toxicity of some tetra and trialkyl lead compounds. Br J Ind Med 18:277–282, 1961
5. Grandjean P: Biological Effects of Organolead Compounds, p 278. Boca Raton, Florida, CRC Press, Inc, 1983
6. Ramstock ER et al: Trialkyllead metabolism and lipid peroxidation in vivo in vitamin E—and selenium deficient rats as measured by ethane production. Toxicol Appl Pharmacol 54:251–257, 1980

TETRAMETHYLSUCCINONITRILE

CAS: 3333-52-6

$(CH_3)_2C(CN)C(CN)(CH_3)_2$

1987 TLV = 0.5 ppm; skin

Synonyms: TMSN; succinonitrile, tetramethyl

Physical Form. Crystalline solid

Source. Breakdown product of azobisisobutyronitrile used as a blowing agent for the production of vinyl foam

Exposure. Inhalation, skin absorption

Toxicology. Tetramethylsuccinonitrile is a convulsant.

A NIOSH literature search resulted in only one European report of toxic effects in the workplace. This occurred in a group of 16 workers using azoisobutyronitrile over an 18-month period in the production of polyvinyl chloride foam.[1] There were five cases of convulsions and unconsciousness. Other symptoms reported included headache, dizziness, nausea, and vomiting. Although an unknown concentration of tetramethylsuccinonitrile was the suspected etiologic agent, it was noted that exposure to a number of other substances also occurred. All symptoms subsided following installation of improved ventilation in the work area.

Exposure of rats to the vapor at 60 ppm for 2 to 3 hours, or to 6 ppm for 30 hours, caused death.[1] Mice exposed to 22 ppm had muscle spasms and died within 2 to 3 hours.[1,2] Rats, guinea pigs, rabbits, and dogs administered tetramethylsuccinonitrile by a variety of routes developed violent convulsions and asphyxia, which eventually led to death from 1 minute to 5 hours following convulsions.[1] In a variety of species, LD_{50} values for intravenous, intraperitoneal, subcutaneous, and oral administration ranged from 17.5 to 30 mg/kg.[1] Administration of a quick-acting barbiturate followed by phenobarbital reduced the toxicity of tetramethylsuccinonitrile given in doses up to 50 mg/kg.[1,2]

REFERENCES

1. National Institute for Occupational Safety and Health: Criteria for a Recommended Standard . . . Occupational Exposure to Nitriles. DHEW (NIOSH) Pub No 78–212, p 155. Washington, DC, US Government Printing Office, 1978
2. Harger RN, Hulpieu HR: Toxicity of tetramethyl succinonitrile and the antidotal effects of thiosulphate, nitrile and barbiturates. Fed Proc 8(abstr):205, 1949

TETRANITROMETHANE

CAS: 509-14-8

$C(NO_2)_4$ 1987 TLV = 1 ppm

Synonyms: None

Physical Form. Pale yellow liquid

Uses. Oxidizer in rocket propellants; explosive in admixture with toluene; reagent for detecting presence of double bonds in organic compounds

Exposure. Inhalation

Toxicology. Tetranitromethane vapor is a severe irritant of the eyes and respiratory tract; it can cause methemoglobinemia.

In workers, various studies showed that exposure caused irritation of the eyes, nose, and throat; dizziness; headache; chest pain; dyspnea; and rarely, skin irritation.[1]

Severe exposure may be expected to cause the formation of methemoglobin and resultant anoxia with cyanosis (especially evident in the lips, nose, and earlobes); other effects include weakness, dizziness, and severe headache.[2–4] Concentrations in excess of 1 ppm cause lacrimation and upper respiratory irritation, whereas 0.4 ppm may cause mild irritation.[2] The liquid on the skin may cause mild burns.[2]

The LC_{50} for rats was 1230 ppm for 36 minutes; effects included lacrimation, rhinorrhea, gasping, and cyanosis. Pulmonary edema was present at autopsy.[1] In three species of animals, intravenous injection caused methemoglobinema, anemia, damage to the central nervous system, and pulmonary edema.[1]

Diagnosis. Signs and symptoms include headache; signs of anoxia, including cyanosis of lips, nose, and earlobes; anemia; and hematuria.

Differential Diagnosis: Other causes of cyanosis must be differentiated from methemoglobinemia resulting from chemical exposure. These include hypoxia due to lung disease, hypoventilation, and decreased cardiac output. Lung disease may be suspected from results of pulmonary function tests and arterial blood gas analysis. The arterial P_{O_2} may be normal in methemoglobinemia but tends to be decreased in cyanosis due to lung disease. Hypoventilation will cause elevation of arterial P_{CO_2}, which is not seen in chemical exposure. Decreased cardiac output states will cause cyanosis only when accompanied by arterial hypotension. If blood withdrawn from the vein shows the characteristic chocolate-brown coloration, the diagnosis of an abnormal pigment is almost certain, especially if the color remains after shaking the blood in air.[5]

Special Tests: Examine the urine for blood, and determine the methemoglobin concentration in the blood when chemical intoxication is suspected and at regular intervals until the methemoglobin has been fully reduced to normal hemoglobin.[5] Methemoglobin may be differentiated from sulfhemoglobin by the addition of a few drops of 10% potassium cyanide, which results in the rapid production of bright red cyanomethemoglobin but has no effect on the color of sulfhemoglobin.[4] Spectrophotometry is required for the precise identification of the pigment and its quantitation. Normal acid methemoglobin has a characteristic absorption spectrum with peaks at 502 and 632 nm, which disappear with the addition of cyanide, whereas sulfhemoglobin has a peak at 620 nm, which does not disappear with cyanide.[4]

Treatment. All the contaminant on the body must be removed. Immediately remove all clothing, and wash the entire body from head to foot with soap and water. Pay special attention to the hair and scalp, finger- and toenails, nostrils, and ear canals. Administer oxygen to alleviate the headache and general sense of weakness; confine the patient to bed. Determine the methemoglobin concentration in the blood, and repeat every 3 to 6 hours for 18 to 24 hours. Repeat skin cleansing if the methemoglobin concentration appears to rise after 3 to 4 hours. In general, patients will return to normal within 24 hours provided that all sources of further absorption are completely eliminated. The methemoglobin will be reduced spontaneously to ferrous hemoglobin in 2 to 3 days.[4]

Such therapy is not effective in subjects with glucose–6–dehydrogenase deficiency.[4]

The only justifiable use of methylene blue would be in cases of coma or stupor, usually at methemoglobin levels above 60%. In those patients in whom therapy is necessary, methylene blue, 1 to 2 mg/kg of a 1% solution in saline may be given intravenously over a 10-minute period. If cyanosis has not disappeared within 1 hour, a second dose of 2 mg/kg may be administered.[4,5] The total dose should not exceed 7 mg/kg, because methylene blue may cause toxic effects such as dyspnea, precordial pain, restlessness, apprehension, red cell hemolysis, and changes in the electrocardiogram (reduction in the height or even reversal of the T wave, frequently with lowering of the R wave).[4]

REFERENCES

1. Horn HJ: Inhalation toxicology of tetranitromethane. AMA Arch Ind Hyg Occup Med 10:213–222, 1954
2. Hygienic Guide Series: Tetranitromethane. Am Ind Hyg Assoc J 25:513–515, 1964
3. Hager KF: Tetranitromethane. Ind Engr Chem 41:2168–2172, 1949
4. Rieder RF: Methemoglobinemia and sulfmethemoglobinemia. In Wyngaarden JB, Smith LH (eds): Cecil Textbook of Medicine, 16th ed, p 896. Philadelphia, WB Saunders, 1982
5. Mangelsdorff AF: Treatment of methemoglobinemia. AMA Arch Ind Hyg Occup Med 14:148–153, 1956

TETRYL

CAS: 479-45-8

$(NO_2)_3C_6H_2N(NO_2)CH_3$

1987 TLV = 1.5 mg/m^3; skin

Synonyms: 2,4,6–Trinitrophenylmethylnitramine; tetralite; nitramine; n-methyl-n,2,4,6–tetranitroaniline

Physical form. Yellow crystals

Uses. Explosives

Exposure. Inhalation; skin absorption

Toxicology. Tetryl causes contact and sensitization dermatitis and irritation of the upper respiratory tract.

Dermatitis in workers appears as early as the first week of exposure to the dust, with itching of and around the eyes. There is a progression to erythema and edema occurring most often on the nasal folds, cheeks, and neck; papules and vesicles may develop. The remainder of the body is rarely affected.[1] The severest forms show massive generalized edema with partial obstruction of the trachea owing to swelling of the tongue and require hospitalization; exfoliation usually occurs after the edema subsides.[1] The majority of these effects occur between the 10th and 20th days of exposure; upon cessation of exposure, there is rapid abatement of the mild symptoms and, after 3 to 10 days, disappearance of physical signs.[1]

Contact with tetryl causes a bright yellow staining, most often seen on the palms, face, and neck and in the hair.[1] The irritant effects on the upper respiratory tract are variously localized from the nostrils to the bronchi and cause burning, itching, sneezing, coryza, epistaxis, and cough. The symptoms may begin the first day of exposure or as late as the third month; upon removal from exposure, the symptoms regress over 2 to 4 weeks.[1]

Other effects reported in tetryl workers include irritability, fatigue, malaise, headache, lassitude, insomnia, nausea, and vomiting.[1] Anemia, of either the marrow depression or deficiency type, has been observed among tetryl workers.[1] Conjunctivitis may be caused by rubbing the eyes with contaminated hands or by airborne dust; keratitis and iridocyclitis have occurred.[2] Tetryl has been reported to cause irreversible liver damage and death following heavy exposure; no cases of systemic poisoning have been reported at concentrations below 1.5 mg/m^3.[1,3]

REFERENCES

1. Bergman BB: Tetryl toxicity: A summary of ten years' experience. AMA Arch Ind Hyg Occup Med 5:10–20, 1952
2. Troup HE: Clinical effects of tetryl (CE powder). Br J Ind Med 3:20–23, 1946
3. Hardy HL, Maloof CC: Evidence of systemic effect of tetryl, with summary of available literature. AMA Arch Ind Hyg Occup Med 1:545–555, 1950

THALLIUM (soluble compounds)

CAS: 7440-28-0

Tl   1987 TLV = 0.1 mg/m^3; skin

Synonym: Thallium salts

Physical Form. Bluish-white, very soft, inelastic, easily fusible, heavy metal

Uses. Poison for rodents

Exposure. Inhalation; skin absorption; ingestion

Toxicology. Soluble thallium compounds are extremely toxic, and intoxication is cumulative; they primarily affect the nervous system and the body hair.

Many deaths have resulted from ingestion; poisonings from industrial exposures have been reported rarely and have not been fatal. The lethal oral dose of thallium acetate for humans is estimated to be about 12 mg/kg bodyweight.[1] Symptoms are usually nonspecific owing to multiorgan involvement.[2] Ingestion causes nausea, vomiting, diarrhea, abdominal pain, and gastrointestinal hemorrhage, which usually occur within 1 to 3 days.[3] These symptoms are followed or accompanied by ptosis and strabismus; peripheral neuritis; pain, weakness, and paresthesias in the legs; tremor; and retrosternal tightness and chest pain.[1,3] Severe and abrupt alopecia is pathognomonic of the toxic effects of thallium and usually, but not always, occurs after 2 to 3 weeks.[3,4]

Severe intoxication has resulted in prostration, tachycardia, blood pressure fluctuations, convulsive seizures, choreiform movements, and psychosis. Recovery may be complete, but permanent residual effects such as ataxia, optic atrophy, tremor, mental abnormalities, and footdrop have been reported.[3] In cases of fatal intoxication, typical autopsy findings include pulmonary edema, necrosis of the liver, nephritis, and degenerative changes in peripheral axons.[1]

Prolonged ingestion of thallium produces a variable clinical picture, which includes stomatitis, tremor, cachexia, polyneuropathy, alopecia, and emotional disturbance.[3]

In a study of 15 workers who had handled solutions of organic thallium salts over a 7½-year period, six workers suffered thallium intoxication. Chief complaints included abdominal pain, fatigue, weight loss, pain in the legs, and nervous irritability; three of the workers had albuminuria, and one had hematuria.[5]

In another cohort study, no statistically significant clinical effects were found, even though urinary concentrations ranging up to 236 μg/liter indicated exposures above the TLV of 0.1 mg/m^3.[6] A urine thallium concentration of 100 μg/liter corresponds approximately to a 40 hour/week exposure at 0.1 mg/m^3, and normal values range between 0.6 and 2.0 μg/liter.[6]

In six cases of thallium intoxication of pregnant women during their first trimester, no congenital abnormalities were observed.[7]

Administration to pregnant mice, rabbits, or rats produced slight embryotoxic effects at maternally toxic doses.[7]

Diagnosis. Signs and symptoms include nausea, vomiting, diarrhea, abdominal pain; ptosis, strabismus; peripheral neuritis, tremor; pain, weakness, and paresthesias in the legs; retrosternal tightness, chest pain, pulmonary edema; convulsions, chorea, psychosis; liver and kidney damage; and alopecia.

Differential Diagnosis: Other causes of convulsions must be differentiated from convulsions due to thallium exposure. These include idiopathic epilepsy; hypertensive encephalopathy; metabolic disturbances such as hypoglycemia, hypocalcemia, uremia, and porphyria; hypoxic encephalopathy; infections such as viral encephalitis; and exposure to other convulsants.

Differentiate from a mononeuropathy or myositis; other causes of polyneuropathy include (1) chronic intoxication—alcohol, other neurotoxic agents; (2) vascular and metabolic conditions—atherosclerosis, polyarteritis nodosa, diabetes mellitus, pregnancy, acute intermittent porphyria, uremia; (3) infections—meningitis, syphilis, pneumonia, Guillain–Barré syndrome, mumps; (4) nutritional causes–vitamin B_{12} or B complex deficiency; and (5) malignancies.

Special Tests: If convulsions occur, diagnostic studies should include analysis of blood for glucose, calcium, urea, nitrogen, and CO_2. Thallium in the urine indicates that systemic absorption has occurred. Urinary thallium in normal unexposed subjects is less than 1.5 μg/liter.[8] No relationship between excretion rate of thallium and either exposure or appearance of signs and symptoms has been established.

Treatment. The therapeutic management of thallium poisoning is controversial.[9–11] Treatment is aimed at gastrointestinal decontamination, diminishing absorption, and increasing the excretion of the compound. Prussian Blue (potassium ferric ferrocyanide) has been used to bind thallium in the intestine and prevent its absorption, thereby increasing fecal excretion. The human therapeutic dose is two 500-mg capsules three times daily (t.i.d.) for 2 to 3 weeks, depending upon the severity of the case.[2] In severe poisoning, Prussian Blue solution should be administered by a duodenal tube in a dose of 1000 mg t.i.d. because of the pyloric spasm and gastric dilatation that can accompany thallotoxicosis.[2] The use of Prussian Blue is of no value if the patient is suffering from constipation due to absence of intestinal motility.[9–11] Prussian Blue adsorbs thallium ions during their entero-enteral cycling through the gut, exchanging these for potassium ions. Neither Prussian Blue nor its complex with thallium are absorbed systemically, and side effects have not been reported.[12]

Potassium chloride in large doses has been used to mobilize thallium from tissue storage sites and to enhance renal excretion.[2] A paradoxical transient worsening of clinical symptoms may follow KCl use owing to the redistribution of intracellular thallium into the serum. The aggravation of symptoms does not contraindicate further KCl therapy, because the ultimate outcome after several days is usually beneficial.

Hemodialysis and hemoperfusion therapy have been performed for thallium poisoning; some practitioners found the procedure very effective, whereas others found it equal to forced diuresis.[2,9,11] The procedures are probably indicated in severe cases in which renal failure and paralytic bowel complicate other elimination techniques.

Medical Control. Medical control involves preplacement and annual physical examination, with emphasis on the nervous system, eyes, lungs, liver, kidneys, and body hair; urinalysis is also suggested.

REFERENCES

1. Browning E: Toxicity of Industrial Metals, 2nd ed, pp 317–322. London, Butterworths, 1969
2. Saddique A, Peterson CD: Thallium poisoning: A review. Vet Hum Toxicol 25:16–22, 1983
3. Paulson G, Vergara G, Young J, Bird M: Thallium intoxication treated with dithizone and hemodialysis. Arch Intern Med 129:100–103, 1972
4. Bank WJ et al: Thallium poisoning. Arch Neurol 26:456–464, 1972
5. Richeson EM: Industrial thallium intoxication. Ind Med Surg 27:607–619, 1958
6. Marcus RL: Investigation of a working population exposed to thallium. J Soc Occup Med 35:4–9, 1985
7. Dolgner R et al: Repeated surveillance of exposure to thallium in a population living in the vicinity (SIC) of a cement plant emitting dust containing thallium. Int Arch Occup Environ Health 52:69–94, 1983
8. Lauwerys RR: Industrial Chemical Exposure: Guidelines for Biological Monitoring, pp 48–49. Davis, California, Biomedical Publications, 1983
9. Heath A et al: Thallium poisoning—toxin elimination and therapy in three cases. J Toxicol Clin Toxicol 20:451–463, 1983
10. De Groot G, van Heijst ANP: Thallium concentrations in body fluids and tissues in a fatal case of thallium poisoning. Vet Hum Toxicol 27:115–119, 1985
11. Nogue S et al: Acute thallium poisoning: An evaluation of different forms of treatment. J Toxicol Clin Toxicol 19:1015–1021, 1983
12. Kazantzis G: Thallium. In Friberg L, Nordberg G, Vouk V. (eds): Handbook on the Toxicology of Metals, 2nd ed, Vol II: Specific Metals, pp 549–567. Amsterdam, Elsevier, 1986

THIRAM

CAS:137-26-8

$([CH_3]_2NCS)_2S_2$ 1987 TLV = 5 mg/m^3

Synonyms: Tetramethylthiuram disulfide; TMTD

Physical Form. White powder

Uses. Agricultural fungicide; rubber accelerator

Exposure. Inhalation

Toxicology. Thiram is an irritant of the eyes, mucous membranes, and skin and causes sensitization dermatitis.

Thiram is the methyl analogue of disulfiram or Antabuse, a drug used to establish a conditioned reflex of fear of alcohol in the treatment of alcoholism.[1] Ingestion of even a small amount of alcohol while undergoing Antabuse therapy is followed by distressing and occasionally dangerous symptoms, including flushing, palpitations, headache, nausea, vomiting, and dyspnea. The systemic "Antabuse–alcohol" syndrome is apparently rare in thiram-exposed workers, but it has been reported.[2] In one case, a man became ill and died 4 days after treating seed with thiram. Although he received substantial exposure over 10 hours, it is unclear whether he received enough thiram to produce death without associated alcohol ingestion.[2] A skin reaction, without other systemic effects, is said to occur in chronically exposed workers after ingestion of alcohol. The response of the skin is rapid and takes the form of flushing, erythema, pruritis, and urticaria.[1] Thiram without alcohol can produce dermatitis but only in a few susceptible people. Sensitization dermatitis in the form of eczema has occurred on the hands, forearms, and feet.[1]

In mice and male rats, the oral LD_{50} was approximately 4 g/kg; symptoms of toxicity were ataxia and hyperactivity followed by inactivity, loss of muscular tone, labored breathing, clonic convulsions, and death within 2 to 7 days.[3]

A dietary level of 1000 ppm for 2 years produced weakness, ataxia, and varying degrees of paralysis of the hind legs of rats.[2]

Thiram was teratogenic (skeletal malformations) in hamsters given a single oral dose of 250 mg/kg during the period of organogenesis and in mice given oral doses of 5 to 30 mg per animal daily between days 6 and 17 of pregnancy.[4,5]

Thiram was not carcinogenic in rats in chronic feeding studies or by gavage, or in mice by single subcutaneous injection.[2,6] The IARC has noted, however, that thiram can react with nitrite under mildly acidic conditions, simulating those in the human stomach, to form N-nitrosodimethylamine, which is carcinogenic in a number of species.[6] In a recent study, dietary administration of 500 ppm thiram plus 2000 ppm sodium nitrite for 2 years caused a high incidence of nasal cavity tumors in rats vs no tumors in controls or in animals given only one compound.[7] The carcinogenic risk to humans cannot be evaluated at this time.

REFERENCES

1. Shelley WB: Golf-course dermatitis due to thiram fungicide. JAMA 188:415–417, 1964
2. Hayes WJ Jr: Pesticides Studies in Man, pp 603–606. Baltimore, Williams & Wilkins, 1982
3. Lee CC et al: Oral toxicity of ferric dimethyldithiocarbamate (ferbam) and tetramethylthiram disulfide (thiram) in rodents. J Toxicol Environ Health 4:93–106, 1978
4. Robens JF: Teratologic studies of carbaryl, diazinon, norea, disulfiram, and thiram in small laboratory animals. Toxicol Appl Pharmacol 15:152–173, 1969
5. Roll R: Teratologische Untersuchungen mit Thiram (TMTD) an zwei Mausestammen. Arch Toxicol 27:163–186, 1971
6. IARC Monographs on the Evaluation of Car-

cinogenic Risk of Chemicals to Man. Some Carbamates, Thiocarbamates and Carbazides. Vol 12, pp 225–236. Lyon, International Agency for Research on Cancer, 1976
7. Lijinsky W: Induction of tumors of the nasal cavity in rats by concurrent feeding of thiram and sodium nitrite. J Toxicol Environ Health 13:609–614, 1984

TIN (inorganic compounds, except SnH_4)

CAS: 7440-31-5

Sn 1987 TLV = 2 mg/m^3 as Sn (except SnH_4)

Synonyms: Stannic oxide; tin tetrachloride; stannic chloride; stannous chloride; stannous sulfate; sodium stannate; potassium stannate

Physical Form. Metal; powders

Uses. Protective coatings and alloys; glass bottle manufacture

Exposure. Inhalation

Toxicology. Inorganic tin salts are irritants of the eyes and skin.

No systemic effects have been reported from industrial exposure. Some inorganic tin compounds can cause skin or eye irritation because of acid or alkaline reaction produced with water. Tin tetrachloride, stannous chloride, and stannous sulfate are strong acids; sodium and potassium stannate are strong alkalis.[1] Glass bottle makers exposed to a hot mist of stannic chloride (0.10 to 0.18 mg/m^3) and hydrogen chloride (5 ppm) had an excess of symptoms of respiratory irritation over workers predominantly exposed to hydrogen chloride in the same plant.[2] Exposure to dust and fume of tin oxide results in stannosis, a rare benign pneumoconiosis.[3]

Ingested inorganic tin exhibits only moderate toxicity—probably because of poor absorption and rapid tissue turnover. However, consumption of food and fruit juices heavily contaminated with tin compounds in the range of 1400 ppm or more results in symptoms of gastrointestinal irritation, including nausea, abdominal cramps, vomiting, and diarrhea.[4]

In animals, high doses of soluble tin salts induce neurologic disturbances.[4] Subcutaneous injection of animals with sodium stannous tartrate at a daily dose of 12.5 mg/kg was fatal. Death was preceded by vomiting, diarrhea, and paralysis with twitching of the limbs.[5] Daily administration to a dog of stannous chloride in milk at a level of 500 mg/kg produced paralysis after 14 months.[1]

Administration of 1 and 3 mg Sn/kg to rats resulted in inhibition of various enzymes, including hepatic succinate dehydrogenase and the acid phosphatase of the femural epiphysis. Tin also appears to interact with the absorption and metabolism of biological essential metals such as copper, zinc, and iron and to influence heme metabolism.[4]

REFERENCES

1. Stauden A (ed): Kirk-Othmer Encyclopedia of Chemical Technology, 2nd ed, Vol 20, pp 323–325. New York, Interscience, 1972
2. Levy BS, Davis F, Johnson B: Respiratory symptoms among glass bottle makers exposed to stannic chloride solution and other potentially hazardous substances. J Occup Med 27:277–282, 1985
3. Tin (metal, tin oxide and inorganic compounds, except SnH_4). Documentation of the TLVs and BEIs, 5th ed, p 574. Cincinnati, American Conference of Governmental Industrial Hygienists (ACGIH), 1986
4. Schafer SG, Femfurt U: Tin—a toxic heavy metal? A review of the literature. Regul Toxicol Pharmacol 4:57–69, 1984
5. Barnes JM, Stoner HB: The toxicology of tin compounds. Pharmacol Rev 11:214–216, 1959

TIN (organic compounds)
Sn 1987 TLV = 0.1 mg/m^3; skin

Synonyms: Triethyltin iodide; dibutyltin chloride; tributyltin chloride; triphenyltin acetate; bis(tributyltin) oxide

Physical Form. Solids and liquids

Uses. Stabilizers in polymers; biocides, catalysts

Exposure. Inhalation; skin absorption

Toxicology. Organotin compounds cause irritation of the eyes, mucous membranes, and skin; some produce cerebral edema, and others cause hepatic necrosis.

The most toxic of the organotin compounds are the trialkyltins, followed by the dialkyltins and monoalkyltins.[1] The tetraalkyltins are metabolized to their trialkyltin homologs; their effects are those of the trialkyltins, with severity of effects dependent upon the rate of metabolic conversion. In each major organotin group, the ethyl derivative is the most toxic.[1]

Triethyltin: Oral administration of a French medication (Stalinon, containing diethyltin diiodide and isolinoleic esters) for treatment of human furunculosis resulted in 217 cases of poisoning, of which 102 were fatal.[1,2] The capsules were found to be contaminated with triethyltin and other organotin compounds. After a latent period of 4 days, effects included severe, persistent headache, vertigo, visual disturbances (including photophobia), abdominal pain, vomiting, and urinary retention. The more severe cases showed transient or permanent paralysis and psychic disturbances. Residual symptoms in those who recovered included persistent headache, diminished visual acuity, paresis, focal anesthesia, and, in four severe cases, flaccid paraplegia with incontinence. The most significant lesion found at autopsy was cerebral edema.

Tributyltin: Workers exposed to the vapor or fume of tributyltin compounds developed sore throat and cough several hours after exposure.[3] When a worker was splashed in the face with a tributyltin compound, lacrimation and severe conjunctivitis appeared within minutes, despite immediate lavage, and persisted for 4 days. At the end of 7 days, the eyes appeared normal.[3] Chemical burns may result after only brief contact with the skin. Pain is usually moderate, and itching is the chief symptom. Healing is usually complete within 7 to 10 days.[3]

Triphenyltin Acetate: Liver damage has occurred from occupational exposure to triphenyltin acetate.[1] In two cases, both workers developed hepatomegaly; one worker had slightly elevated SGPT and SGOT activity. Occupational exposure to a 20% solution produced skin irritation 2 to 3 days after prolonged contact with contaminated clothing. Other nonspecific effects of exposure have included headache, nausea, vomiting, diarrhea, and blurred vision.[1]

Trimethyltin: Neurologic and behavioral changes in rodents exposed to trimethyltin compounds, including aggression, hyperexcitability, tremor, spontaneous seizures, and hyperreactivity, are well documented.[4]

Bis (tributyltin) oxide (TBTO): TBTO is an irritant of the eyes and respiratory tract.[1]

REFERENCES

1. National Institute for Occupational Safety and Health: Criteria for a Recommended Standard . . . Occupation Exposure to Organotin Compounds. DHEW (NIOSH) Pub No 77–115, pp

26–105. Washington, DC, US Government Printing Office, 1976
2. Barnes JM, Stoner HB: The toxicology of tin compounds. Pharmacol Rev 11:211–231, 1959
3. Lyle WH: Lesions of the skin in process workers caused by contact with butyl tin compounds. Br J Ind Med 15:193–196, 1958
4. Chang LW: Neuropathology of trimethyltin: A proposed pathogenic mechanism. Fund Appl Toxicol 6:217–232, 1986

TITANIUM DIOXIDE

CAS: 13463-67-7

TiO_2 1987 TLV = 10 mg/m^3

Synonyms: Unitane; rutile; anatase; octahedrite; brookite

Physical Form. White powder

Uses. Welding rod coating; acid-resistant vitreous enamels; white pigment for paints; acetate rayon; ceramics

Exposure. Inhalation

Toxicology. Titanium dioxide is a mild pulmonary irritant and is generally regarded as a nuisance dust.

Of 15 workers who had been exposed to titanium dioxide dust three showed radiographic signs in the lungs resembling "slight fibrosis," but disabling injury did not occur. The magnitude and duration of exposure were not specified.[1,2] In the lungs of three workers involved in processing titanium dioxide pigments, deposits of the dust in the pulmonary interstitium were associated with cell destruction and slight fibrosis; the findings indicated that titanium dioxide is a mild pulmonary irritant.[3]

Rats repeatedly exposed to concentrations of 10 to 328 mppcf of air for as long as 13 months showed small focal areas of emphysema, which were attributed to large deposits of dust. There was no evidence of any specific lesion being produced by titanium dioxide.[4] Exposure of rats to 250 mg/m^3 titanium dioxide in a 2-year inhalation bioassay resulted in the development of squamous cell carcinomas in 13 of 74 female rats and in 1 of 77 male rats, as well as an increase in bronchoalveolar adenomas. No excess tumor incidence was observed at 50 mg/m^3. Given the extremely high concentration exposures, the unusual histology and location of the tumors, and the absence of metastases, the authors questioned the biological relevance of these tumors to humans.[5]

REFERENCES

1. Browning E: Toxicity of Industrial Metals, 2nd ed, pp 331–335. London, Butterworths, 1969
2. AIHA Hygienic Guide Series: Titanium Dioxide. Akron, Ohio, American Industrial Hygiene Association, 1978
3. Elo R, Maatta K, Uksila E, Arstila AU: Pulmonary deposits of titanium dioxide in man. Arch Pathol 94:417–424, 1972
4. Christie H, Mackay RJ, Fisher AM: Pulmonary effects of inhalation of titanium dioxide by rats. Am Ind Hyp Assoc J 24:42–46, 1963
5. Lee K et al: Pulmonary response of rats exposed to titanium dioxide (TiO_2) by inhalation for two years. Toxicol Appl Pharmacol 79:179–192, 1985

TOLUENE

CAS: 108-88-3

$C_6H_5CH_3$ 1987 TLV = 100 ppm

Synonyms: Toluol; methyl benzene

Physical Form. Colorless liquid

Uses. Manufacturing of benzene and other chemicals; solvent for paints and coatings; component of gasoline

Exposure. Inhalation; skin absorption

Toxicology. Toluene causes central nervous system depression.

Controlled exposure of human subjects to 200 ppm for 8 hours produced mild fatigue, weakness, confusion, lacrimation, and paresthesias of the skin. At 600 ppm for 8 hours, other effects included euphoria, headache, dizziness, dilated pupils, and nausea. At 800 ppm for 8 hours, symptoms were more pronounced, and aftereffects included nervousness, muscular fatigue, and insomnia that persisted for several days.[1–4]

Subjects exposed to 100 ppm of toluene for 6 hours in an environmental chamber complained of eyes and nose irritation and, in some cases, headache, dizziness, and a feeling of intoxication. However, no significant differences were noted in performance on a variety of neurobehavioral tests. No symptoms were noted at 10 or 40 ppm.[5]

Chronic organic brain dysfunction, associated with cerebral and cerebellar atrophy, has been described following long-term inhalational abuse of toluene among glue-sniffers exposed to very high concentrations. Several studies of workers chronically exposed to toluene or mixtures of toluene and other solvents have suggested minor abnormalities on neuropsychological testing or differences in performance on such testing when compared with unexposed controls.[6]

However, a recent study of 43 rotogravure printers exposed to estimated mean levels of 117 ppm for a mean of 22 years failed to demonstrate significant clinical neuroradiologic, neurophysiologic, or neuropsychological differences when compared with a control group of 31 unexposed printers.[7]

Severe but reversible liver and kidney injury occurred in a person who was a glue-sniffer for 3 years. The chief component of the inhaled solvent was toluene (80% V/V); other ingredients were not listed.[3]

In workers exposed for many years to concentrations in the range of 80 to 300 ppm, there was no clinical or laboratory evidence of altered liver function.[3]

Toluene exposure does not result in the hematopoietic effects caused by benzene. The myelotoxic effects previously attributed to toluene are judged by more recent investigations to be the result of concurrent exposure to benzene present as a contaminant in toluene solutions.[3] Most of the toluene absorbed from inhalation is metabolized to benzoic acid, conjugated with glycine in the liver to form hippuric acid, and excreted in the urine. The average amount of hippuric acid excreted in the urine by persons not exposed to toluene is approximately 0.7 to 1.0 g/liter of urine.[3]

The liquid splashed in the eyes of two workers caused transient corneal damage and conjunctival irritation; complete recovery occurred within 48 hours.[3] Repeated or prolonged skin contact with liquid toluene has a defatting action, causing drying, fissuring, and dermatitis.[3]

A chronic inhalation bioassay in rats, with exposures up to 300 ppm for 24 months, failed to demonstrate any evidence of carcinogenicity, but there were some deficiencies in the study.[6] Further studies are being conducted currently by the National Toxicology Program. Some evidence of embryo/fetotoxicity with reduced fetal weight and retarded skeletal development but without teratogenicity has been observed in mice and rats exposed to 133 ppm and 266 to 399 ppm, respectively.[6]

REFERENCES

1. Department of Labor: Occupational exposure to toluene, Federal Register 40:46206–46219, 1975
2. von Oettingen WF, Neal PA, Donahue DD: The toxicity and potential dangers of toluene preliminary report. JAMA 113:578–584, 1942
3. National Institute for Occupational Safety and Health: Criteria for a Recommended Standard . . . Occupational Exposure to Toluene. DHEW (NIOSH) Pub No (HSM) 7311023, pp 14–45.

Washington, DC, US Government Printing Office, 1973
4. Toluene. Documentation of the TLVs and BEIs, 5th ed, pp 578–579. Cincinnati, American Conference of Governmental Industrial Hygienists (ACGIH), 1986
5. Anderson I et al: Human response to controlled levels of toluene in six-hour exposures. Scand J Work Environ Health 9:405–418, 1983
6. Environmental Protection Agency: Health Assessment Document for Toluene, NTIS, 1983
7. Juntunen J et al: Nervous system effects of long-term occupational exposure to toluene. Acta Neurol Scand 75:512–517, 1985

TOLUENE–2,4–DIISOCYANATE

CAS: 584-84-9

$CH_3C_6H_3(NCO)_2$ 1987 TLV = 0.005 ppm

Synonyms: TDI; toluene diisocyanate

Physical Form. Colorless liquid; aerosol

Uses. Production of polyurethane foams and plastics; used in polyurethane paints and wire coatings; the most commonly used material is a mixture of 80% 2,4 isomer and 20% 2,6 isomers

Exposure. Inhalation

Toxicology. Toluene–2,4–diisocyanate (TDI) is a strong irritant of the eyes, mucous membranes, and skin and is a potent sensitizer of the respiratory tract.

Exposure of humans to sufficient concentrations causes irritation of the eyes, nose, and throat; a choking sensation; and a productive cough of paroxysmal type, often with retrosternal soreness and chest pain.[1,2] If the breathing zone concentration reaches 0.5 ppm, the possibility of respiratory response is imminent.[3] Depending upon length of exposure and level of concentration above 0.5 ppm, respiratory symptoms will develop with a latent period of 4 to 8 hours.[3]

Higher concentrations produce a sensation of oppression or constriction of the chest. There may be bronchitis and severe bronchospasm; pulmonary edema may also occur. Nausea, vomiting, and abdominal pain may complicate the presenting symptoms. Upon, removal from exposure, the symptoms may persist for 3 to 7 days.[3]

Although the acute effects may be severe, their importance is overshadowed by respiratory sensitization in susceptible persons; this has occurred after repeated exposure to TDI at levels of 0.02 ppm and below.[2] The onset of symptoms of sensitization may be insidious, becoming progressively more pronounced with continued exposure over a period of days to months. Initial symptoms are often nocturnal dyspnea and/or nocturnal cough with progression to asthmatic bronchitis.[1] Immediate, late, and dual patterns of bronchospastic response to laboratory exposure to TDI in sensitized persons have been observed, confirming the clinical findings of nocturnal symptoms in some exposed workers. The time from initial employment to the development of symptoms suggestive of asthma has been reported to vary from 6 months to 20 years.[4,5]

In another pattern of sensitization response, a worker who has had only minimal upper respiratory symptoms or no apparent effects from several weeks of low-level exposure may suddenly develop an acute asthmatic reaction to the same or slightly higher level. The asthmatic reaction may be severe, sometimes resulting in status asthmaticus, which may be fatal if exposure continues.[1]

Susceptibility to TDI-induced asthma does not require a prior history of atopy or allergic conditions, and sensitization, may not be any more common in atopics.[6] Given sufficient exposure, it appears that virtually any person may become sensitized. The proportion of people with TDI asthma in working populations has varied from 4.3% to

25%.[7] There is some evidence that this percentage decreases with decreasing air concentrations. Exposure to spills of TDI appear to increase the risks of sensitization. The pathophysiology of TDI-induced asthma is unknown; both immunologic and nonimmunologic pharmacologic mechanisms have been postulated. It is clear, however, that TDI-induced asthma is not solely mediated by a type I hypersensitivity response associated with IgE antibody.[6]

Several studies have provided evidence of cross-shift and progressive annual declines in FEV_1 and FEF 25% to 75% among asymptomatic workers without evidence of TDI asthma exposed to low levels of TDI (below 0.02 ppm and as low as 0.003 ppm). The annual declines were two- to threefold greater than expected, appeared dose related, and correlated with observed cross-shift declines. Workers, in general, exhibited no acute or chronic symptoms related to these exposures or pulmonary function decrements.[8,9]

The diagnosis of TDI-induced asthma relies primarily on the clinical history in a worker with known exposure, recognizing that symptoms (wheezing, dyspnea, cough) may develop at night long after the end of the shift. Serial measurement of peak flow rates by the workers may aid in making the diagnosis.[10] Nonspecific bronchial hyperreactivity to histamine or methacholine is frequently, but not invariably, present in patients with TDI-induced asthma. Its absence may reflect that the asthma is quiescent owing to no recent exposure, and re-exposure may lead to hyperreactivity. Failure to demonstrate nonspecific hyperreactivity on a single test does not exclude the diagnosis of TDI-induced asthma.[11] RAST testing for IgE antibodies against p-tolyl monoisocyanate antigens is probably not useful because of the occurrence of false-positive (in exposed but asymptomatic workers) and false-negative results.[12] Specific bronchoprovocation challenge with TDI is a definitive way to make the diagnosis but is often not practical because of the need for prolonged observation for late reactions and the risk of severe reactions.

Following removal from exposure, some patients have had resolution of symptoms and findings suggestive of asthma. However, there is evidence from several studies that persons with TDI-induced asthma may continue to experience symptoms of dyspnea and wheezing and bronchial hyperreactivity for 2 or more years following cessation of exposure.[13,14] In one study, patients with TDI-induced asthma who continued to have exposure to TDI for 2 more years had, as a rule, marked abnormal decreases in spirometric parameters and increases in nonspecific hyperreactivity.[13] In another study, 6 of 12 workers with a convincing history of TDI-induced asthma had positive responses to specific bronchial provocation testing with low concentrations of TDI (up to 0.02 ppm) at a mean of 4.5 years after cessation of exposure. These persons had experienced persistent respiratory symptoms requiring daily treatment for asthma and persistent airway hyperreactivity.[14] Once sensitized, it is clear that patients can react to concentrations of 0.005 ppm or less.[7]

Splashes of TDI liquid in the eye cause severe conjunctival irritation and lacrimation. On the skin, the liquid produces a marked inflammatory reaction. Sensitization of the skin occurs but is uncommon owing to proper work practices. There seems to be little relation between skin sensitivity and respiratory sensitivity to TDI.[1]

REFERENCES

1. National Institute for Occupational Safety and Health: Criteria for a Recommended Standard . . . Occupational Exposure to Toluene Diisocyanate. DHEW (NIOSH) Pub No (HSM) 73-11022. Washington, DC, US Government Printing Office, 1973
2. Elkins HB, McCarl GW, Brugsch HG, Fahy JP: Massachusetts experience with toluene diisocyanate. Am Ind Hyg Assoc J 23:265–272, 1962
3. Rye WA: Human responses to isocyanate exposure. J Occup Med 15:306–307, 1973
4. O'Brien I, Harris M, Burge P, Pepys J: Toluene di-isocyanate-induced asthma. Clin Allergy 9:1, 1979
5. Chester E et al: Patterns of airway reactivity to asthma produced by exposure to toluene diisocyanate. Chest 75:229, 1979
6. Bernstein I: Isocyanate-induced pulmonary diseases: A current perspective. J Allergy Clin Immunol 70:24–31, 1982
7. Toluene-2,4-diisocyanate. Documentation of the TLVs and BEIs, 5th ed, pp 580–585. Cincinnati, American Conference of Governmental Industrial Hygienists (ACGIH), 1986
8. Diem JE et al: Five-year longitudinal study of workers employed in a new toluene diisocyanate manufacturing plant. Am Rev Respir Dis 126:420–428, 1982
9. Wegman D et al: Accelerated loss of FEV-1 in polyurethane production workers: A four-year prospective study. Am J Ind Med 3:209–215, 1982
10. Burge P, O'Brien I, Harris M: Peak flow rate record in the diagnosis of occupational asthma due to isocyanates. Thorax 34:317, 1979
11. Burge P: Nonspecific bronchial hyperreactivity in workers exposed to toluene diisocyanate diphenylmethane diisocyanate and colophony. Eur J Respir Dis 63(suppl 123):91–96, 1982
12. Butcher B et al: Radioallergosorbent testing with p-tolyl monoisocyanate in toluene diisocyanate workers. Clin Allergy 13:31–34, 1983
13. Paggiaro P et al: Follow-up study of patient with respiratory disease due to toluene diisocyanate (TDI). Clin Allergy 14:463–469, 1984
14. Moller D et al: Chronic asthma due to toluene di-isocyanate. Chest 90:494–499, 1986

o-TOLUIDINE

CAS: 95-53-4

$CH_3C_6H_4NH_2$ — 1987 TLV = 2 ppm; skin

Synonyms: *ortho*-Aminotoluene; 1 methyl–2–aminobenzene; 2–methylaniline

Physical Form. Clear to light yellow liquid

Uses. Dye intermediate

Exposure. Inhalation; skin absorption

Toxicology. *o*-Toluidine causes anoxia owing to the formation of methemoglobin, and hematuria; *ortho*-toluidine hydrochloride is carcinogenic in experimental animals.

Signs and symptoms of overexposure are due to the loss of oxygen-carrying capacity of the blood. The earliest manifestations of poisoning in humans are headache and cyanosis of the lips, mucous membranes, fingernail beds, and tongue.[1] Minor degrees of hypoxia may lead to a temporary sense of well-being and exhilaration. As the lack of oxygen increases, however, there is growing weakness, dizziness, and drowsiness, leading to stupor, unconsciousness, and even death if treatment is not prompt. Exposure to 10 ppm for more than a short period of time may lead to symptoms of illness, and 40 ppm for 60 minutes may cause severe toxic effects.[2] Transient microscopic hematuria has been observed in *o*-toluidine workers and is presumably of renal origin, since no alterations in the bladder mucosa were observed by cystoscopy.[3]

In general, higher ambient temperatures increase susceptibility to cyanosis from exposure to methemoglobin-forming agents.[4]

Rats survived an 8-hour exposure to

concentrated vapor.[5] Animals exposed from 6 to 23 ppm for several hours developed mild methemoglobinemia.[6] In the eye of a rabbit, the liquid caused a severe burn.[5] Excessive drying of the skin may result from repeated or prolonged contact.[1]

Ortho-toluidine hydrochloride was carcinogenic in mice fed diets containing 1000 or 3000 mg/kg for 2 years, producing hepatocellular carcinomas or adenomas in females and hemangiosarcomas at multiple sites in males.[7] In another strain of mice fed diets of 16,000 ppm for 3 months and then 8,000 ppm for an additional 15 months or 32,000 ppm for 3 months followed by 16,000 ppm for 15 months, there were significant dose-dependent increases in the incidences of vascular tumors.[8] It was also carcinogenic in rats fed 0.028 mol/kg diet for 72 weeks, producing tumors of multiple organs.[9]

Epidemiologic studies have dealt only with workers exposed to *o*-toluidine in combination with other chemicals. The role of *o*-toluidine in human carcinogenesis cannot be evaluated.[10] The IARC has determined that there is sufficient evidence for carcinogenicity of *o*-toluidine hydrochloride in animals and that it should be regarded as though it presented a carcinogenic risk to humans.[10]

Diagnosis. *Signs and Symptoms:* Signs and symptoms include headache; signs of anoxia, including cyanosis of lips, nose, and earlobes; eye irritation; anemia; and hematuria.

Differential Diagnosis: Other causes of cyanosis must be differentiated from methemoglobinemia due to *o*-toluidine exposure. These include hypoxia resulting from lung disease, hypoventilation, and decreased cardiac output. Lung disease may be suspected from results of pulmonary function tests and arterial blood gas analysis. The arterial P_{O_2} may be normal in methemoglobinemia but tends to be decreased in cyanosis due to lung disease. Hypoventilation will cause elevation of arterial P_{CO_2}, which is not seen in *o*-toluidine exposure. Decreased cardiac output states will cause cyanosis only when accompanied by arterial hypotension. If blood withdrawn from the vein shows the characteristic chocolate-brown coloration, the diagnosis of an abnormal pigment is almost certain, especially if the color remains after shaking the blood in air.[11]

Special Tests: Examine the urine for blood; determine the methemoglobin concentration in the blood when *o*-toluidine intoxication is suspected and at regular intervals until the methemoglobin has been fully reduced to normal hemoglobin.[12] Methemoglobin may be differentiated from sulfhemoglobin by the addition of a few drops of 10% potassium cyanide, which results in the rapid production of bright red cyanomethemoglobin but has no effect on the color of sulfhemoglobin. Spectrophotometry is required for the precise identification of the pigment and its quantitation. Normal acid methemoglobin has a characteristic absorption spectrum with peaks at 502 and 632 nm, which disappear with the addition of cyanide, whereas sulfhemoglobin has a peak at 620 nm, which does not disappear with cyanide.[11]

Treatment. All *o*-toluidine on the body must be removed. Immediately remove all clothing, and wash the entire body from head to foot with soap and water. Pay special attention to the hair and scalp, finger- and toenails, nostrils, and ear canals. Administer oxygen to alleviate the headache and general sense of weakness; confine the patient to bed. Determine the methemoglobin concentration in the blood, and repeat every 3 to 6 hours for 18 to 24 hours. Repeat skin cleansing if the methemoglobin concentration appears to rise after 3 to 4 hours. In general, patients will return to normal within 24 hours provided that all sources of further absorption have been completely eliminated.

The only justifiable use of methylene blue would be in cases of coma or stupor, usually at methemoglobin levels above 60%. In those patients in whom therapy is considered necessary, methylene blue, 1 to 2 mg/kg, may be given intravenously over a 5-minute period as a 1% solution. If cyanosis has not disappeared within 1 hour, a second dose of 2 mg/kg should be administered.[11,12] The total dose should not exceed 7 mg/kg, because methylene blue may cause toxic effects such as dyspnea, precordial pain, restlessness, apprehension, red cell hemolysis, and changes in the electrocardiogram (reduction in the height or even reversal of the T wave, frequently with lowering of the R wave).[11,12]

REFERENCES

1. Chemical Safety Data Sheet SD–82, Toluidine, pp 13–14. Washington, DC MCA, Inc, 1961
2. Goldblatt MW: Research in industrial health in the chemical industry. Br J Ind Med 12:1–20, 1955
3. Hamblin DO: Aromatic nitro and amino compounds. In Patty FA (ed): Industrial Hygiene and Toxicology, 2nd ed, Vol 2, Toxicology, pp 2123, 2155. New York, Wiley–Interscience, 1963
4. Linch AL: Biological monitoring for industrial exposure to cyanogenic aromatic nitro and amino compounds. Am Ind Hyg Assoc J 35:426–432, 1974
5. Smyth HF Jr et al: Range-finding toxicity data: List VI. Am Ind Hyg Assoc J 23:95–96, 103, 1962
6. Henderson Y, Haggard HW: Noxious Gases, 2nd ed, p 228. New York, Reinhold, 1943
7. National Cancer Institute: Bioassay of *o*-Toluidine Hydrochloride for Possible Carcinogenicity, TR-153. DHEW (NIH) Pub No 79-1709. Washington, DC, US Government Printing Office, 1979
8. Weisburger EK et al: Testing of twenty-one environmental aromatic amines or derivatives for long-term toxicity or carcinogenicity. J Environ Pathol Toxicol 2:325–356, 1978
9. Hecht SS et al: Comparative carcinogenicity of *o*-toluidine hydrochloride and nitrosotoluene in F–344 Rats. Cancer Lett 16:103–108, 1982
10. IARC Monographs on the Evaluation of the Carcinogenic Risk of Chemicals to Humans, Vol 27, pp 155–175. Lyon, International Agency for Research on Cancer, 1982
11. Rieder RF: Methemoglobinemia and sulfhemoglobinemia. In Wyngaarden JB, Smith LH (eds): Cecil Textbook of Medicine, 16th ed, p 896. Philadelphia, WB Saunders, 1982
12. Mangelsdorff AF: Treatment of methemoglobinemia. AMA Arch Ind Health 14:148–153, 1956

TOXAPHENE

CAS: 8001–35–2

$C_{10}H_{10}Cl_8$ 1987 TLV = 0.5 mg/m^3; skin

Synonyms: Chlorinated camphene; polychlorocamphene; octachlorocamphene

Physical Form. Yellow waxy solid

Uses. Insecticide

Exposure. Inhalation; skin absorption; ingestion

Toxicology. Toxaphene is a convulsant; it is carcinogenic in experimental animals.

Most fatal cases of poisoning have been due to accidental ingestion, resulting in convulsions, coma, and death.[1–3] The lethal oral dose for humans is estimated to be 2 to 7 g.[1]

Symptoms of acute poisoning are those of diffuse stimulation of the central nervous system.[1] Convulsions may be preceded by nausea, vomiting, and muscle spasms or may begin without antecedent symptoms.[1] Onset of symptoms occurs within 4 hours, with death occurring from 4 to 24 hours postexposure.

Nonfatal poisoning has been characterized by nausea, mental confusion, jerking of the arms and legs, and convulsions.[2,3] Few cases of intoxication owing to occupational exposure have been reported, and, of these, two cases of pneumonitis in insecticide sprayers are of dubious validity.[4] In one acute study, 25 volunteers were exposed to 500 mg/m^3 for 30 minutes for 10 days.[5] Following a 3-week respite, the exposure was repeated for 3 days. Each subject was thought to have absorbed 1 mg/kg/day. Physical examination and blood and urine tests revealed no toxic manifestations. Chronic human intoxication by toxaphene is not reported in the literature.

Toxaphene caused a dose-related increase of hepatocellular carcinomas in mice fed 98 or 198 ppm for 80 weeks. In rats, there was a significantly increased incidence of neoplastic thyroid lesions at the high dose.[6] The IARC has determined that sufficient evidence exists in rodents to regard toxaphene as though it presented a carcinogenic risk to humans.[7]

REFERENCES

1. Starmont RT, Conley BE: Pharmacologic properties of toxaphene, a chlorinated hydrocarbon insecticide. JAMA 149:1135–1137, 1952
2. McGee LC, Reed HL, Fleming JP: Accidental poisoning by toxaphene. JAMA 149:1124–1126, 1952
3. Hayes WJ JR: Clinical Handbook on Economic Poisons, Emergency Information for Treating Poisoning. US Public Health Service Pub No 476, pp 47–50, 71–73. Washington, DC, US Government Printing Office, 1963
4. Warraki S: Respiratory hazards of chlorinated camphene. Arch Environ Health 7:137–140, 1963
5. Chlorinated camphene (60%). Documentation of the TLVs and BEIs, 5th ed, p 115. Cincinnati, American Conference of Governmental Industrial Hygienists (ACGIH), 1986
6. National Cancer Institute. Bioassay of Toxaphene for Possible Carcinogenicity. DHEW (NIH) Pub No 79-837. Bethesda, Maryland, Carcinogenesis Testing Program, Division of Cancer Cause and Prevention. 1979
7. IARC Monographs on the Evaluation of the Carcinogenic Risk of Chemicals to Man, Vol 20, Some Halogenated Hydrocarbons, pp 327–438. Lyon, International Agency for Research on Cancer, 1978

TRIBUTYL PHOSPHATE

CAS: 126-73-8

$(C_4H_9)_3PO_4$ 1987 TVL = 0.2 ppm

Synonym: TBP

Physical Form. Colorless liquid

Uses. Antifoaming agent; plasticizer for cellulose esters, lacquers, plastic and vinyl resins

Exposure. Inhalation

Toxicology. Tributyl phosphate is an irritant of the eyes, mucous membranes, and skin; it causes pulmonary edema in animals, and severe exposure is expected to cause the same effect in humans.

Workers exposed to unspecified concentrations of vapor complained of headache and nausea; hot vapor was severely irritating to the eyes and throat.[1] The liquid on the skin is said to be irritating.[2]

In rats, 123 ppm for 6 hours caused respiratory irritation.[2] The oral LD_{50} for rats was 3 g/kg; effects included weakness, dyspnea, pulmonary edema, and muscle twitching.[2] The liquid dropped on the eye of a rabbit caused temporary epithelial injury and discomfort.[3] In vitro, tributyl phosphate caused weak inhibition of cholinesterase in human erythrocytes and plasma.[4]

REFERENCES

1. Tributyl phosphate. Documentation of the TLVs and BEIs, 5th ed, p 591. Cincinnati, American Conference of Governmental Industrial Hygienists (ACGIH), 1986
2. Sandmeyer EE, Kirwin CJ Jr: Ethers. In Clayton GD, Clayton FE (eds): Patty's Industrial Hygiene and Toxicology, 3rd ed, rev, Vol 2A Toxicology, pp 2370, 2379, New York, Wiley–Interscience, 1981
3. Grant WM: Toxicology of the Eye, 2nd ed, p 1032. Springfield, IL, Charles C Thomas, 1974
4. Sabine JC, Hayes FN: Anticholinesterase activity of tributyl phosphate. AMA Arch Ind Hyg Occup Med 6:174–177, 1952

1,1,1-TRICHLOROETHANE

See Addendum, p. 552.

1,1,2-TRICHLOROETHANE

CAS:79-00-5

$CH_2ClCHCl_2$ 1987 TLV = 10 ppm; skin

Synonyms: Vinyl trichloride; ethane trichloride; β-trichloroethane

Physical Form. Colorless liquid

Uses. Chemical intermediate; solvent

Exposure. Inhalation; skin absorption

Toxicology. In animals, 1,1,2–trichloroethane is a central nervous system depressant and causes liver and kidney damage; it is expected that severe exposure will produce the same effects in humans.

No cases of human intoxication or systemic effects from industrial exposure have been reported.[1]

The lethal concentration for rats was 2000 ppm for 4 hours, with the deaths occurring during a 14-day observation period.[2] An 8-hour exposure to 500 ppm was also lethal to about half of the exposed rats.[3] Rats exposed to 250 ppm for 4 hours survived but showed liver and kidney necrosis.[4] Repeated exposure to 30 ppm resulted in minor liver changes in female rats.

Application of 0.5 ml to the skin of guinea pigs was lethal to all animals within 3 days, whereas 0.25 ml was fatal to 5 of 20 animals.[5]

Mice treated by intraperitoneal injection with anesthetic doses showed moderate hepatic dysfunction and renal dysfunction. At autopsy, findings included centrolobular necrosis of the liver and tubular necrosis of the kidneys; the 24-hour LD_{50} for intraperitoneal injection was 0.35 mg/kg.[6] The LD_{50}s for 1,1,2–trichloroethane administered by a single gavage dose to male and female mice were 378 and 491 mg/kg, respectively.[7] Above 450 mg/kg, animals became sedated within 1 hour, and deaths from central nervous system depression occurred within 24 hours. Necropsies showed irritation of the upper gastrointestinal tract, pale livers, and some lung damage. Dose-dependent alterations in hepatic microsomal enzyme activities and serum enzyme levels were found in mice given 1,1,2–trichloroethane in their drinking water for 90 days.[7]

A significant increase in hepatocellular carcinomas occurred in mice given 195 or 390 mg/kg/day by gavage for 78 weeks.[8] Adrenal pheochromocytomas were also increased for the high-dose female mice. No neoplasms were observed at statistically significant incidences in rats given up to 92 mg/kg/day. The IARC has determined that there is limited evidence that 1,1,2–trichloroethane is carcinogenic in mice.[9]

1,1,2–Trichloroethane is not highly irritating to the skin and eyes but may injure the skin by defatting.[4]

REFERENCES

1. National Institute for Occupational Safety and Health: Current Intelligence Bulletin 27, Chloroethanes: Review of Toxicity. DHEW (NIOSH) Pub No 78–181, pp 22, 1978

2. Carpenter CP, Smyth HF Jr, Pozzani UC: The assay of acute vapor toxicity, and the grading and interpretation of results on 96 chemical compounds. J Ind Hyg Toxicol 31:343–346, 1949
3. Smyth HF Jr et al: Range-finding toxicity data: List VII. Am Ind Hyg Assoc J 30:470–476, 1969
4. Torkelson TR, Rowe VK: Halogenated aliphatic hydrocarbons. In Clayton GD, Clayton FE (eds): Patty's Industrial Hygiene and Toxicology, 3rd ed, rev, Vol 2B, Toxicology, pp 3510–3513. New York, Interscience, 1981
5. Wahlberg JE: Percutaneous toxicity of solvents. A comparative investigation in the guinea pig with benzene, toluene, and 1,1,2-trichloroethane. Ann Occup Hyg 19:226–229, 1976
6. Klaassen CD, Plaa GL: Relative effects of various chlorinated hydrocarbons on liver and kidney function in mice. Toxicol Appl Pharmacol 9:139–151, 1966
7. White KL et al: Toxicology of 1,1,2-trichloroethane in the mouse. Drug Chem Toxicol 8:333–335, 1985
8. National Cancer Institute: Carcinogenesis Technical Report Series No 74. Bioassay of 1,1,2-Trichloroethane for Possible Carcinogenicity. NCI-CG-TR-74. Washington, DC, US Department of Health, Education and Welfare, 1978
9. IARC Monographs on the Evaluation of the Carcinogenic Risks of Chemicals to Humans, Vol 20, Some Halogenated Hydrocarbons, pp 533–543. Lyon, International Agency for Research on Cancer, 1979

TRICHLOROETHYLENE

CAS: 79-01-6

$CHCl{=}CCl_2$ 1987 TLV = 50 ppm

Synonyms: TCE; 1,1,2-trichloroethylene; trichloroethene; 1,1-dichloro-2-chloroethylene; acetylene trichloride; ethylene trichloride

Physical Form. Nonflammable, mobile liquid

Uses. Degreasing solvent; dry cleaning and extraction; chemical intermediate; limited use as an anesthetic and analgesic

Exposure. Inhalation

Toxicology. Trichloroethylene (TCE) is primarily a central nervous system (CNS) depressant; it is carcinogenic in experimental animals.

Inhalation of concentrations in the range of 5,000 to 20,000 ppm have been used to produce light anesthesia.[1] Recovery from unconsciousness is usually uneventful, but ventricular arrhythmias and death from cardiac arrest have occurred rarely. Exposure of volunteers to 500 to 1000 ppm has resulted in some symptoms of CNS disturbance such as dizziness, light-headedness, lethargy, and impairment in visual–motor response tests. In general, no significant signs of toxicity or impaired performance have been noted in subjects acutely exposed to 300 ppm or less.

Prenarcotic symptoms, including visual disturbances and feelings of inebriation, occurred in workers exposed to mean levels of 200 to 300 ppm. Some evidence of mild liver dysfunction has occurred in workers exposed to levels sufficient to produce marked CNS effects. Prolonged exposure at toxic levels may also result in hearing defects.

Workers exposed to average levels of TCE, estimated to be 100 to 200 ppm, have reported increased incidence of fatigue, vertigo, dizziness, headaches, memory loss, and impaired ability to concentrate. Other effects noted at about 100 ppm and above include paresthesia, muscular pains, and gastrointestinal disturbances.

Intolerance to alcohol, presenting as a transient redness affecting mainly the face and neck (trichloroethylene flush), has frequently been observed following repeated exposure to TCE and alcohol ingestion. It has been suggested that ingestion of alcohol may potentiate the effect of TCE intoxication.[2]

TCE is mildly irritating to the skin; repeated contact may cause chapping and erythema due to defatting.[1] Direct eye contact produces injury to the corneal epithelium; recovery usually occurs within a few days.[1]

Breath analysis for TCE has provided a more accurate index of exposure than the measurement of metabolites (trichloroethanol and trichloroacetic acid) in the urine.[3]

Technical grade TCE (later shown to be contaminated with other chemicals) was found in an NCI study to cause liver cancer in B6C3F1 mice but not in Osborne–Mendel rats.[4] Intragastric administration of 2.4 g/kg, five times per week for 78 weeks resulted in hepatocellular carcinomas in 31 of 48 male mice. At 1.2 g/kg, 26 of 50 males were affected, whereas male controls had a 5% liver cancer rate. Among female mice, 11 of 47 developed liver hepatocellular carcinomas, whereas only 1 of 80 control animals did.[4] In a second gavage bioassay using epichlorohydrin-free reagent grade TCE, results paralleled the NCI study; significantly elevated incidences of hepatocellular adenomas and carcinomas occurred in mice administered 1.0 g/kg for 2 years.[5] An increase in renal adenocarcinomas was also found in male rats.[5]

Mice, rats, and hamsters inhaling up to 500 ppm 6 hours/day 5 days/weeks for 18 months showed no increase in tumor formation except for an increased incidence of malignant lymphomas in female MRI mice.[6] This strain normally has a high spontaneous incidence of lymphomas, and the significance of TCE exposure is unclear. ICR mice exposed at 150 and 450 ppm for 107 weeks developed a 16% and 15% incidence of adenocarcinomas of the lungs vs 2% for controls.[7] Rats did not show a higher incidence at any site.

Limited epidemiologic studies have not shown a potent carcinogenic effect in TCE-exposed populations.[8–10] A mortality study of 2117 workers exposed at some time between 1963 and 1976 showed no increase in overall mortality or in cancer deaths.[10] Limitations of the study include short latency period, young age of cohort, no direct data on exposure levels, exposure to other chemicals, and possible inclusion of unexposed workers.

No evidence of teratogenic effects have been seen in rodent assays.[1] At 1800 ppm, 6 hours/day on days 0 to 20 of gestation, there were some fetotoxic effects, including incomplete ossification of the sternum in rats.[11] At 300 ppm, on days 6 to 15 of gestation, there were slight fetotoxic effects in mice but not in rats.[12] In humans, there is no evidence of an increased incidence of adverse effects in the offspring of female TCE-exposed workers. An increased incidence of menstrual disorders in female workers and decreased libido in male workers has been reported in workers exposed to levels sufficient to produce marked CNS disturbances.[1]

REFERENCES

1. Fielder RJ et al: Toxicity Review 6. Trichloroethylene. Health and Safety Executive, pp 1–70. London, Her Majesty's Stationery Office, 1982
2. National Institute for Occupational Safety and health: Criteria for a Recommended Standard . . . Occupational Exposure to Trichloroethylene. DHEW (NIOSH) Pub No (HSM) 73-11025, pp 15–40. Washington, DC, US Government Printing Office, 1976
3. Stewart RD, Hake CL, Peterson JE: Use of breath analysis to monitor trichloroethylene exposures. Arch Environ Health 29:6–13, 1974
4. National Cancer Institute: Carcinogenesis Bioassay of Trichloroethylene, TR-2. DHEW (NIH) Pub No 76–802. Washington, DC, US Government Printing Office, 1976
5. Kimbrough RD et al: Trichlorethylene: An update. J Toxicol Environ Health 15:369–383, 1985
6. Henschler D et al: Carcinogenicity study of trichloroethylene by long-term inhalation in three animal species. Arch Toxicol 43:237–248, 1980
7. Fukuda K et al: Inhalation carcinogenicity of trichloroethylene in mice and rats. Ind Health 21:243–254, 1983
8. Axelson O et al: A cohort study on trichloro-

ethylene exposure and cancer mortality. J Occup Med 20:194–196, 1978
9. Shindell S, Ulrich S: A cohort study of employees of a manufacturing plant using trichloroethylene. J Occup Med 27:577–579, 1985
10. Tola S et al: A cohort study on workers exposed to trichloroethylene. J Occup Med 22:737–740, 1980
11. Dorfmueller MA et al: Evaluation of teratogenicity and behavioural toxicity with inhalation exposure of maternal rats to trichloroethylene. Toxicology 14:153–166, 1979
12. Schwetz BA et al: The effect of maternally inhaled trichloroethylene, perchloroethylene, methyl chloroform and methylene chloride on embryonal and fetal development in mice and rats. Toxicol Appl Pharmacol 32:84–96, 1975

TRICHLOROFLUOROMETHANE

CAS: 75-69-4

$FCCl_3$ 1987 TLV = C 1000 ppm

Synonyms: Freon 11; fluorotrichloromethane; fluorocarbon 11

Physical Form. Colorless liquid

Uses. Aerosol propellant; refrigerant and blowing agent; solvent for cleaning and degreasing

Exposure. Inhalation

Toxicology. Trichlorofluoromethane causes narcosis and death from respiratory depression at extremely high concentrations. Death can also occur following sensitization of the heart to the arrhythmogenic actions of adrenalin.

Exposure of volunteers to 250, 500, or 1000 ppm for up to 8 hours did not produce adverse effects.[1] Chronic exposure 6 hours/day for 20 days to 1000 ppm caused a slight, but insignificant, decrement in cognitive tests; there were no changes in pulmonary function or cardiac rhythm.[1] Workmen near a large area of spilled trichlorofluoromethane experienced narcotic effects, including loss of consciousness; prolonged tachycardia was also observed in one case.[2] Accidental ingestion caused necrosis and multiple perforations of the stomach.[2]

Sudden deaths from "sniffing" aerosols have been associated with a number of chlorofluorocarbons. The deaths are thought to be due to ventricular fibrillation following cardiac sensitization.[3]

Exposure of rats to 500,000 ppm for 1 minute, 150,000 ppm for 8 minutes, or 100,000 ppm for 30 minutes was always fatal.[4] At 66,000 ppm, one of four rats died within 2 hours, but all survived 4 hours at 36,000 ppm.[2] Toxicity symptoms at the higher dose levels included rapid or labored breathing, twitching, unresponsiveness, or unconsciousness.

No symptoms were observed in rats, guinea pigs, monkeys, and dogs continuously exposed to 1000 ppm for 90 days or exposed to 10,250 ppm 8 hours/day for 6 weeks.[5]

Cardiac arrhythmias have been provoked in a number of species. Inhalation of 3500 to 6100 ppm by dogs for 5 minutes caused ventricular fibrillation and cardiac arrest following injection of epinephrine.[3] The minimal concentration that elicited cardiac arrhythmias in the anesthetized monkey was 50,000 ppm.[6]

Cardiac sensitization is unlikely to occur in humans in the absence of any effects on the central nervous system, and dizziness should act as an early warning that a dangerous concentration is being reached.[7]

Administered by gavage, 3925 mg/kg/day trichlorofluoromethane for 78 weeks was not carcinogenic to mice; results from rats were inconclusive because of poor survival rates.[8]

REFERENCES

1. Stewart RD et al: Physiological response to aerosol propellants. J Environ Health Perspect 26:275–285, 1978
2. Du Pont Company Haskell Laboratory—Tox-

icity Review: Freon 11, p. 25, 1982 (unpublished)
3. Reinhardt CF et al: Cardiac arrhythmias and aerosol "sniffing." Arch Environ Health 22:265–279, 1971
4. Lester D, Greenburg LA: Acute and chronic toxicity of some halogenated derivatives of methane and ethane. Arch Ind Hyg Occup Med 2:335–344, 1950
5. Jenkins LJ et al: Repeated and continuous exposures of laboratory animals to trichlorofluoromethane. Toxicol Appl Pharmacol 16:133–142, 1970
6. Belej MA et al: Toxicity of aerosol propellants in the respiratory and circulatory systems. IV. Cardiotoxicity in the monkey. Toxicology 2:381–395, 1974
7. Clark DG, Tinston DJ: Acute inhalation toxicity of some halogenated and non-halogenated hydrocarbons. Hum Toxicol 1:239–247, 1982
8. National Cancer Institute: Bioassay of Trichlorofluoromethane for Possible Carcinogenicity. CAS No 75-69-4. NCI-CG-TR-106, p 46. Washington, DC, US Department of Health, Education and Welfare, 1978

TRICHLORONAPHTHALENE

CAS: 1321-65-9

$C_{10}H_5Cl_3$ 1987 TLV = 5 mg/m^3; skin

Synonyms: 1,4,5–Trichloronaphthalene; 1,4,6–trichloronaphthalene

Physical Form. White solid

Uses. Electric wire insulation; lubricants

Exposure. Inhalation; skin absorption

Toxicology. Trichloronaphthalene is moderately toxic to the liver.

Industrial exposure to trichloronaphthalene (usually mixed with tetrachloronaphthalene) has been relatively free of untoward effects compared with the more highly chlorinated naphthalenes.[1] No fatal cases of liver injury have been reported, but one instance of toxic hepatitis supposedly resulted from exposure to 3 mg/m^3.[1] Although there are several reports of chloracne from exposure to trichloronaphthalene, they do not stand up well to critical analysis.[2] Experiments on human volunteers showed that the mist was entirely nonacneigenic as opposed to the penta- and hexachloro derivatives, which produce severe chloracne.[3]

Rats exposed to 11 mg/m^3 of trichloronaphthalene, containing some tetrachloronaphthalene, 16 hours/day for 2.5 months showed slightly swollen liver cells with granular cytoplasm.[1] No effects were found in calves fed 26 mg/kg for 1 week.[1] The higher chlorinated naphthalenes show a much greater toxicity.[2]

REFERENCES

1. Trichloronaphthalene. Documentation of the TLVs and BEIs, 5th ed, p 600. Cincinnati, American Conference of Governmental Industrial Hygienists (ACGIH), 1986
2. Deichmann WB: Halogenated cyclic hydrocarbons. In Clayton GD, Clayton FE (eds): Patty's Industrial Hygiene and Toxicology, 3rd ed, rev, Vol 2B, Toxicology, pp 3669–3675. New York, Wiley–Interscience, 1981
3. Shelley WB, Kligman AM: The experimental production of acne by penta- and hexachloronaphthalenes. Arch Dermatol 75:689–695, 1957

2,4,5–TRICHLOROPHENOXYACETIC ACID

CAS: 93-76-5

$C_8H_5Cl_3O$ 1987 TLV = 10 mg/m^3

Synonym: 2,4,5–T

Physical Form. Solid

Uses. Herbicide in brush control

Exposure. Inhalation

Toxicology. 2,4,5–T is of low-order acute toxicity; at high doses, it is teratogenic in experimental animals.

Eleven men in two separate experiments experienced no clinical effects after ingestion of 5 mg/kg of 2,4,5–T. Most of the men reported a metallic taste lasting 1 to 2 hours after ingestion.[1]

Most if not all occupational illness associated with 2,4,5–T has been found to be the result of product contamination with 2,3,7,8–tetrachlorodibenzo-*p*-dioxin (TCDD)[1]—for example, chloracne.[2] TCDD is extremely toxic to animals, and exposure has also been associated with liver function impairment, peripheral neuropathy, personality changes, porphyria cutanea, hypertrichosis, and hyperpigmentation.[3] The role of dioxin contaminants must always be considered in the discussion of 2,4,5–T toxicology.

A study of 204 workers exposed from 1 month to 20 years to 2,4,5–T and its contaminants (concentrations unspecified) showed no evidence of increased risk for cardiovascular disease, hepatic disease, renal damage, central or peripheral nervous system effects, reproductive problems, or birth defects.[3] Clinical evidence of chloracne persisted in 55.7%%, and an association between exposure and history of upper gastrointestinal tract ulcer was found.

The oral LD_{50} for dogs was in the range of 100 mg/kg; effects were limited to a slight or moderate stiffness in the hind legs with development of ataxia.[4] Dogs survived 10 mg/kg/day for 90 days without illness. In rats fed diets containing 2000 ppm of 2,4,5–T (<0.05 TCDD), the minimal cumulative fatal dose was approximately 900 mg/kg.[5]

Concern about the toxicology of 2,4,5–T has centered on its teratogenic action in experimental animals.[2] Although the first studies were carried out with 2,4,5–T contaminated by 30 ppm TCDD, subsequent experiments using analytical grade 2,4,5–T (<0.05% TCDD) showed that 100 mg/kg/day administered subcutaneously to mice on days 6 through 15 of gestation caused an increased incidence of cleft palates.[6] 2,4,5–T containing no detectable TCDD was feticidal and teratogenic to hamsters when administered orally on days 6 to 10 of gestation at a dosage of 100 mg/kg/day.[7] At 80 mg/kg/day, there was a reduction in the number of pups per litter, in fetal weight, and in survival.[7] Rats, rabbits, and monkeys have appeared relatively resistant to teratogenic effects in a number of studies.[1,2]

An epidemiologic investigation of New Zealand chemical applicators using 2,4,5–T found no significant differences in the rates of congenital defects, stillbirths, or miscarriages compared with controls.[8]

Several epidemiologic studies in Sweden suggested an association between exposure to phenoxyherbicides (and/or their contaminants) and soft-tissue sarcomas.[2] There has also been widespread concern among Vietnam veterans that exposure to the defoliant Agent Orange, which contains equal quantities of 2,4–D and 2,4,5–T (with its contaminant TCDD), might increase their risk of adverse health effects, particularly various forms of cancer.[2] Information to support or refute these claims is fragmentary, and no conclusions as to the carcinogenicity of 2,4,5–T can be made at this time.[2] Animal studies do not support the notion that 2,4,5–T itself is carcinogenic. Chronic feeding studies in rats did not produce an increased tumor incidence, even at doses of 30 mg/kg/day, which produced toxic effects.[9] The IARC has determined that there is inadequate evidence for carcinogenicity in both animals and humans.[10]

REFERENCES

1. Hayes WJ Jr: Pesticides Studied in Man, pp 526–533. Baltimore, Williams & Wilkins, 1982
2. Murphy SD: Toxic effects of pesticides. In

Klaassen CD et al (eds): Casarett and Doull's Toxicology. The Basic Science of Poisons, 3rd ed, pp 554–555. New York, Macmillan, 1986
3. Suskind RR, Hertzberg VS: Human health effects of 2,4,5-T and its toxic contaminants. JAMA 251:2372–2380, 1974
4. Drill VA, Hiratzka T: Toxicity of 2,4-dichlorophenoxyacetic acid and 2,4,5-trichlorophenoxyacetic acid. AMA Arch Ind Hyg Occup Med 7:61–67, 1953
5. Chang H et al: Effects of phenoxyacetic acids on rat liver tissues. J Agric Food Chem 22:62–65, 1974
6. Moore JA, Courtney KD: Teratology studies with the trichlorophenoxyacid herbicides, 2,4,5-T and silvex. Teratology 4(abstr):236, 1971
7. Collins TFX, Williams CH: Teratogenic studies with 2,4,5-T and 2,4-D in the hamster. Bull Environ Contam Toxicol 6:559–567, 1971
8. Smith AH et al: Preliminary report of reproductive outcomes among pesticide applicators using 2,4,5-T. NZ Med J 93:177–179, 1981
9. Kociba RJ et al: Results of a two-year chronic toxicity and oncogenic study of rats ingesting diets containing 2,4,5-trichlorophenoxyacetic acid (2,4,5-T). Food Cosmet Toxicol 17:205–221, 1979
10. IARC Monographs on the Evaluation of the Carcinogenic Risk of Chemicals to Humans, suppl 4, pp 235–238. Lyon, International Agency for Research on Cancer, 1982

1,2,3-TRICHLOROPROPANE

CAS: 96-18-4

$CH_2ClCHClCH_2Cl$ 1987 TLV = 10 ppm; skin

Synonyms: Glycerol trichlorohydrin; allyl trichloride; trichlorohydrin

Physical Form. Colorless liquid

Uses. Intermediate in the manufacture of pesticides and polysulfide rubbers

Exposure. Inhalation; skin absorption

Toxicology. 1,2,3-Trichloropropane is an irritant of the eyes and mucous membranes; at high concentrations, it causes narcosis in animals, and it is expected that severe exposure will produce the same effect in humans.

Human subjects exposed to 100 ppm for 15 minutes noted eye and throat irritation and objected to the unpleasant odor.[1] Ingestion of 3 g caused drowsiness, headache, unsteady gait, and lumbar pain.[2]

In rats, 1000 ppm caused death in five of six animals after 4 hours of exposure.[3] Eight of 15 mice did not survive exposure to 5000 ppm for 20 minutes; liver damage accounted for four additional deaths after 7 to 10 days.[2] Daily 10-minute exposures to 2500 ppm for 10 days resulted in the death of 7 of 10 mice tested.[2]

The liquid was not irritating to the skin of rabbits but was extremely irritating to the eyes.[3] The dermal LD_{50} was 2.5 g/kg.[3]

Intraperitoneal doses causing maternal toxicity in rats were not fetotoxic or teratogenic.[4] Male rats administered 80 mg/kg/day by gavage for 5 days and then mated with an untreated female did not produce any meaningful changes in indices such as number of implants and number of live embryos compared with controls.[5]

REFERENCES

1. Silverman L, Schulte HF, First MW: Further studies on sensory response to certain industrial solvent vapors. J Ind Hyg Toxicol 28:262–266, 1946
2. McOmie WA, Barnes TR: Acute and subacute toxicity of 1,2,3-trichloropropane in mice and rabbits. Fed Proc 8:319, 1948
3. Smyth HF Jr et al: Range-finding toxicity data: List VI. Am Ind Hyg Assoc J 23:95–107, 1962
4. Hardin BD et al: Testing of selected workplace chemicals for teratogenic potential. Scand Work Environ Health 7(suppl 4): 66–75, 1981
5. Saito-Suzuki R et al: Dominant lethal studies in rats with 1,2-dibromo-3-chloropropane and its structurally related compounds. Mutat Res 191:321–327, 1982

1,1,2–TRICHLORO–1,2,2–TRIFLUOROETHANE

CAS: 76-13-1

CCl_3CF_3 1987 TLV = 1000 ppm

Synonyms: Refrigerant 113; fluorocarbon 113; Freon 113, TCTFE

Physical Form. Colorless gas

Uses. Solvent for cleaning electronic equipment and degreasing of machinery; refrigerant

Exposure. Inhalation

Toxicology. 1,1,2–Trichloro–1,2,2–trifluoroethane is a central nervous system depressant, a cardiac sensitizer, and a mild mucous membrane irritant.

In experimental human studies, exposure to 4500 ppm for 30 to 100 minutes resulted in significant impairment of manual dexterity and vigilance. Subjects reported loss of concentration and a tendency to somnolence, which disappeared 15 minutes after the exposure ended; at 1500 ppm, no effects were observed.[1] More prolonged human exposures of 6 hours daily, 5 days/week for 2 weeks at concentrations of approximately 500 and 1000 ppm caused mild throat irritation on the first day; there was no decrement in performance of complex mental tasks.[2] No signs or symptoms of adverse effects were found among 50 workers exposed to levels ranging from 46 to 4700 ppm for an average duration of 2.8 years.[3]

The liquid dissolves the natural oils of the skin, and dermatitis may occur as a result of repeated contact; one worker experienced drying of the skin attributed to contact with 1,1,2–trichloro–1,2,2–trifluoroethane.[3,4]

The LC_{50} for 2-hour exposures of experimental animals ranged from 50,000 to 120,000 ppm.[5] Dogs exposed at 11,000 to 13,000 ppm for 6 hours experienced vomiting, lethargy, nervousness, and tremors—all reversible within 15 minutes after exposure. Chronic exposure of rats and rabbits to 12,000 ppm for up to 2 years caused no adverse effects.

In dogs, cardiac sensitization to intravenously administered epinephrine occurred at concentrations of 5,000 to 10,000 ppm.[6] Concentrations greater than 25,000 ppm were necessary to produce arrhythmias in animals under anesthesia.[7]

Occluded contact with rabbit skin of 5 g/kg/day for 5 days caused local necrosis of skin and enlargement of liver cells; no effects were observed after 20 weeks of applications to uncovered skin.[5] The liquid produced no significant irritation in a rabbit eye test.[5]

REFERENCES

1. Stopps GJ, McLaughlin M: Psychophysiological testing of human subjects exposed to solvent vapors. Am Ind Hyg Assoc J 28:43–50, 1967
2. Reinhardt CF et al: Human exposures of fluorocarbon 113. Am Ind Hyg Assoc J 32:143–152, 1971
3. Imbus HR, Adkins C: Physical examination of workers exposed to trichlorotrifluoroethane. Arch Environ Health 24:257–261, 1972
4. Hygienic Guide Series: 1,1,2–Trichloro–1,2,2–trifluoroethane. Am Ind Hyg Assoc J 29:521–525, 1968
5. 1,1,2–Tricholoro–1,2,2–trifluoroethane. Documentation of the TLVs and BEIs, 5th ed, pp 603–604. Cincinnati, American Conference of Governmental Industrial Hygienists (ACGIH), 1986
6. Reinhardt CF, Mullin LS, Maxfield ME: Cardiac sensitization potential of some common industrial solvents. Ind Hyg News Rep 15:3–4, 1972
7. Aviado DM: Toxicity of aerosol propellants in the respiratory and circulatory systems. X. Proposed classification. Toxicology 3:321–332, 1975

TRIETHANOLAMINE

CAS: 102–71–6

$N(CH_2CH_2OH)_3$ TLV = none established

Synonyms: TEA, 2,2′,2″,–nitrilotriethanol

Physical Form. Clear, colorless, viscous liquid with ammonia odor

Uses. Manufacture of emulsifiers and dispersing agents; cosmetic formulations; household and commercial cleaners and detergents

Exposure. Inhalation

Toxicology. TEA is a moderate irritant to the eyes and skin; it is a carcinogen in mice.

The acute toxicity of TEA is low, as reflected in the high values for the oral LD_{50} in rats of 4.2 to 11.3 g/kg.[1–3]

In rats fed 0.73 g/kg daily for 90 days, the only major effect was fatty degeneration of the liver.[2,4] There were no effects at 0.08 g/kg.

When TEA was applied to the skin of rabbits for 72 hours, there was moderate hyperemia, edema, and necrosis.[5] In a guinea pig sensitization test, there was no evidence of sensitization.[6]

In the eyes of rabbits, one drop caused moderate, transient injury at 24 hours.[7]

TEA in the diet of mice at levels of 0.03% or 0.3% caused a significant increase in the occurrence of thymic and nonthymic tumors in lymphoid tissues of females.[8] TEA had no carcinogenic or cocarcinogenic activity when dermally applied to mice for 18 months.[9]

In the presence of N-nitrosating agents, TEA may give rise to N-nitrosodiethanolamine, a known animal carcinogen.[9]

REFERENCES

1. CTFA. Submission of Data by CTFA. (2–9–59). Acute Oral Toxicity of Triethanolamine, 1973
2. Mellon Institute. Submission of Data by FDA. Mellon Institute of Industrial Research, University of Pittsburgh, Special Report on the Acute and Subacute Toxicity of Mono-, Di-, and Triethanolamine, Carbide and Carbon Chem Div, UCC Industrial Fellowship No 274–13 (Report 13–67), August 18, 1950
3. Mellon Institute. Submission of Data by FDA. Letter from HF Smyth Jr to ER Weidlein Jr, Union Carbide Chemicals Co UCC, Acute Oral Toxicity of Triethanolamine, June 16, 1961
4. Smyth HF Jr et al: Range-finding toxicity data: List IV. AMA Arch Ind Hyg Occup Med 4:119, 1951
5. CTFA. Submission of Data by CTFA. CIR Safety Data Test Summary, Primary Skin Irritation and Eye Irritation of Triethanolamine, 1959
6. Life Science Research: Submission of Data by CTFA (2–5–50). Dermal Sensitization Test in Guinea Pigs (TEA), 1975
7. Carpenter CP, Smyth HF Jr: Chemical burns of the rabbit cornea. Am J Ophthalmol 29:1363, 1946
8. Hoshino H, Tanooka H: Carcinogenicity of triethanolamine in mice and its mutagenicity after reaction with sodium nitrite in bacteria. Cancer Res 38:3918, 1978
9. Beyer KH Jr et al: Final report on the safety assessment of triethanolamine, diethanolamine, and monoethanolamine. J Am Coll Toxicol 1:183–235, 1983

TRIETHYLAMINE

CAS: 121-44-8

$(C_2H_5)_3N$ 1987 TLV = 10 ppm

Synonym: N,N-Diethyl-ethanamine

Physical Form. Colorless liquid

Uses. Catalyst in polyurethane foam

Exposure. Inhalation

Toxicology. Triethylamine is an irritant of the eyes and mucous membranes.

Two volunteers exposed to approximately 4.5 ppm for 8 hours experienced slight subjective visual disturbances.[2] At 12 ppm for 1 hour, subjects experienced heavy hazing of the visual field, an inability to distinguish outlines of objects 100 m or more away, and bluish halos around lights. There was pronounced increase in corneal thickness. The investigators suggest that the decrease in visual acuity may be severe enough to cause accidents in the workplace or in traffic at the end of work. Effects are reversible, and it appears that even repeated bouts of edema do not cause permanent damage to the cornea.

Among 19 workers repeatedly exposed to time-weighted average levels of 3 ppm with brief excursions to higher levels, five workers reported foggy vision, blue haze, and halo phenonema on 47 occasions over an 11-week period.[2]

Exposure of six rats to 1000 ppm for 4 hours was lethal to one.[3] Rabbits survived exposures to 100 ppm daily for 6 weeks but showed pulmonary irritation, myocardial degeneration, and cellular necrosis of liver and kidneys; at 50 ppm, the effects on lung, liver, and kidneys were less severe, but there was also damage to the cornea.[4]

In rabbits, skin contact caused irritation.[3]

REFERENCES

1. Akesson B et al: Visual disturbances after experimental human exposure to triethylamine. Br J Ind Med 42:848–850, 1985
2. Akesson B et al: Visual disturbances after industrial triethylamine exposure. Int Arch Occup Environ Health 47:297–302, 1986
3. Smyth HF Jr et al: Range-finding toxicity data: List IV. AMA Arch Ind Hyg Occup Med 4:109–122, 1951
4. Brieger H, Hodes WA: Toxic effects of exposure to vapors of aliphatic amines. AMA Arch Ind Hyg Occup Med 3:287–291, 1951

TRIETHYLENE TETRAMINE

CAS: 112-24-3

$H_2N(CH_2)_2–N(CH_2)_2NH–(CH_2)_2NH_2$

1987 TLV = none established

Synonyms: TETA; Araldite hardener HY 951; DEH 24; TECZA; 1,3,7,10–Tetra-azadecane; Trien

Physical Form. Slightly viscous, yellow liquid; commercially available form is 95% to 98% pure, and impurities include linear, branched, and cyclic isomers

Uses. Hardener/cross-linker for epoxy resins; metal chelator; constituent of wet strength paper resins; copolymer with fatty acids in metal spray coatings; constituent of synthetic elastomer formulations

Exposure. Inhalation

Toxicology. Triethylene tetramine (TETA) is a strong irritant of the eyes, mucous membranes, and skin and is a sensitizer of the respiratory tract and skin.

Exposure to the vapor causes irritation of the eyes, nose, throat, and respiratory tract.[1] Exposure to hot vapor causes itching of the face with erythema and edema.[2]

Sensitization of the respiratory tract has followed chronic exposure to fumes or dust of TETA, manifested by bronchial asthma.[3–5] One worker developed asthma after working with an epoxy resin/TETA formulation for 6 months in a job laminating aircraft windows.[5] In an environmental chamber, the worker developed flu-like symptoms and asthmatic breathing after simulating the job conditions for 2 hours with the resin/TETA mixture. Similar exposure to the resin alone did not produce the symptoms.

TETA on the skin causes irritation and dermatitis, and continued exposure can induce allergic contact dermatitis.[3] Cross-sensitization to other amines has occurred.

TETA was teratogenic when fed to rats at 1.67% in the diet.[6]

REFERENCES

1. Spitz RD: Diamines and higher amines, aliphatic. In Grayson M, Eckroth D (eds): Kirk–Othmer Encyclopedia of Chemical Technology, 3rd ed, New York, Wiley–Interscience, 1979
2. Beard RR, Noe JT: Aliphatic and alicyclic amines. In Patty's Industrial Hygiene and Toxicology, Vol 2B, pp 3235–3273. New York, John Wiley & Sons, 1971
3. Eckardt RE, Hindin R: The health hazards of plastics. J Occup Med 15:808–819, 1973
4. Eckardt RE: Occupational and environmental health hazards in the plastics industry. Environ Health Perspect 17:103–196, 1976
5. Fawcett IW, Taylor AJN, Pepys J: Asthma due to inhaled chemical agents—epoxy resin systems containing phthalic acid anhydride, trimellitic acid anhydride and triethylene tetramine. Clin Allergy 7:1–14, 1977
6. Cohen NL, Keen CL, Lonnerdal B, Hurley LS: Low tissue copper and teratogenesis in triethylenetetramine-treated rats. Fed Proc 41:944, 1982

TRIFLUOROMONOBROMOMETHANE

CAS: 75-63-8

CF_3Br 1987 TLV = 1000 ppm

Synonyms: Bromotrifluoromethane; Freon 13B1; Halon 1301

Physical Form. Colorless gas

Uses. Fire-extinguishing agent; refrigerant

Exposure. Inhalation

Toxicology. Trifluoromonobromomethane in animals causes sensitization of the myocardium to epinephrine and central nervous system effects.

Human exposure to 70,000 ppm for 3 minutes caused no adverse effects.[1] Light-headedness, paresthesia, and diminished performance were reported during exposures up to 100,000 ppm; at 150,000, a feeling of impending unconsciousness developed.[2]

In dogs and rats repeatedly exposed to 23,000 ppm, there were no toxic signs or pathologic changes.[2] Monkeys exposed to concentrations of 200,000 ppm were lethargic and suffered spontaneous cardiac arrhythmias within 5 to 40 seconds of exposure.[3] Dogs exposed to 200,000 ppm or greater became agitated within 1 to 2 minutes, and tremor occurred within 3 minutes.[3] Epileptiform convulsions characterized by generalized rigidity, apnea, and cyanosis of the tongue were observed in about half the dogs exposed to 500,000 to 800,000 ppm. Intravenous injection of a pressor dose of epinephrine produced arrhythmias in all animals exposed to 400,000 ppm; larger doses of epinephrine (5 to 10 μg/kg) caused ventricular fibrillation with cardiac arrest in dogs and spontaneous defibrillation in monkeys.

REFERENCES

1. Smith DG, Harris DJ: Human exposure to Halon 1301 ($CBrF_3$) during simulated aircraft cabin fires. Aerosp Med 44:198–201, 1973
2. Trifluoromonobromomethane. Documentation of the TLVs and BEIs, 5th ed, p 605. Cincinnati, American Conference of Governmental Industrial Hygienists (ACGIH), 1986
3. Van Stee EW, Back KC: Short-term inhalation of bromotrifluoromethane. Toxicol Appl Pharmacol 15:164–174, 1969

2,4,6–TRINITROTOLUENE

CAS: 118-96-7

$CH_3C_6H_2(NO_2)_3$ 1987 TLV = 0.5 mg/m^3; skin

Synonyms: TNT; *a*-trinitrotoluol; *sym*-trinitrotoluene; 1–methyl–2,4,6–trinitrobenzene

Physical Form. Colorless, monoclinic prisms, crystals; commercial crystals are yellow

Uses. Explosives

Exposure. Inhalation; skin absorption

Toxicology. 2,4,6–Trinitrotoluene (TNT) causes liver damage and aplastic anemia.

Deaths from aplastic anemia and toxic hepatitis were reported in TNT workers prior to the 1950s; with improved industrial practices, there have been few reports of fatalities or serious health problems related to its use.[1]

Exposures exceeding 0.5 mg/m^3 cause destruction of red blood cells.[2] Hemolysis is partially compensated for by enhanced regeneration of red blood cells in the bone marrow, which is manifest as an increased percentage of reticulocytes in peripheral blood.[2] Among some groups of workers, there is a reduction in average hemoglobin and hematocrit values.[2] Workers deficient in glucose–6–phosphate dehydrogenase may be particularly at risk of acute hemolytic disease.[3] Three such cases occurred after a latent period of 2 to 4 days and were characterized by weakness, vertigo, headache, nausea, paleness, enlarged liver and spleen, dark urine, decreased hemoglobin levels, and reticulocytosis.[3] Although no simultaneous measurements of atmospheric levels were available, measurement on other occasions showed levels up to 3.0 mg/m^3.[3]

Above 1.0 mg/m^3, the liver is unable to handle the increased amounts of red blood cell breakdown products, and indirect bilirubin levels rise.[2] Elevations of liver function enzymes may occur, particularly in new employees or those recently exposed to higher levels. There are suggestions of marked individual susceptibility to liver damage, with most cases not showing effects unless exposures considerably exceed 1.0 mg/m^3.[2]

A characteristic TNT cataract is reportedly produced with exposures regularly exceeding 1.0 mg/m^3 for more than 5 years.[2] In one study, 6 to 12 workers had bilateral peripheral cataracts, visible only with maximal dilation.[4] The opacities did not interfere with visual acuity or visual fields. The induced cataracts may not regress once exposure ceases, although progression is arrested.

The vapor or dust can cause irritation of mucous membranes resulting in sneezing, cough, and sore throat.[5] Although intense or prolonged exposure to TNT may cause some cyanosis, it is not regarded as a strong producer of methemoglobin.[6] Other occasional effects include leukocytosis or leukopenia, peripheral neuritis, muscular pains, cardiac irregularities, and renal irritation.[2]

Trinitrololuene is absorbed through skin fairly rapidly, and reference to airborne levels of vapor or dust may underestimate total systemic exposure if skin exposure also occurs.[2] Apparent differences in dose–response relationships based only on airborne levels may be explained by differences in skin contact.[2] TNT causes sensitization dermatitis; the hands, wrist, and forearms are most commonly affected, but skin at friction points such as the collar line, belt line, and ankles is also often involved. Erythema, papules, and an itchy eczema can be severe.[7]

The skin, hair, and nails of exposed workers may be stained yellow.[5]

Rats administered 50 mg/kg/day in their diets had anemia, splenic lesions, and liver and kidney damage.[8] Hyperplasia and carcinoma of the urinary bladder were also observed in females.

REFERENCES

1. Woollen BH et al: Trinitrotoluene: Assessment of occupational absorption during manufacture of explosives. Br J Ind Med 43:465–473, 1986
2. Hathaway JA: Subclinical effects of trinitrotoluene: A review of epidemiology studies. In Richert DE (ed): Toxicity of Nitroaromatic Compounds, pp 255–274. New York, Hemisphere Publishing Co, 1985
3. Djerassi LS, Vitany L: Haemolytic episode in G6–PD deficient workers exposed to TNT. Br J Ind Med 32:54–58, 1975
4. Härkonen H et al: Early equatorial cataracts in workers exposed to trinitrotoluene. Am J Ophthalmol 95:807–810, 1983
5. Hygienic Guide Series: 2,4,6–Trinitrotoluene (TNT). Am Ind Hyg Assoc J 25:516–519, 1964
6. Goodwin JW: Twenty years handling TNT in a shell loading plant. Am Ind Hyg Assoc J 33:41–44, 1972
7. Schwartz L: Dermatitis from explosives. JAMA 125:186–190, 1944
8. Levine BS et al: Two-year chronic oral toxicity/carcinogenicity study on the munitions compound trinitrotoluene (TNT) in rats. Toxicologist 5(abstr 697):175, 1985

TRI-*o*-CRESYL PHOSPHATE

CAS: 78-30-8

$(CH_3C_6H_4)_3PO_4$ 1987 TLV = 0.1 mg/m^3; skin

Synonyms: TOCP; tri-*o*-tolyl phosphate; phosphoric acid, tri-*o*-tolyl ester

Physical Form. Colorless or pale yellow liquid

Uses. Plasticizer in lacquers and varnishes; production of heat-stable lubricating oils

Exposure. Inhalation; skin absorption; ingestion

Toxicology. Tri-*o*-cresyl phosphate (TOCP) causes peripheral neuropathy with flaccid paralysis of the distal muscles of the upper and lower extremities, followed in some cases by spastic paralysis.

Thousands of people have been poisoned by the accidental ingestion of TOCP in contaminated foods and beverages.[1] The most notable example was the consumption of an adulterated Jamaica ginger extract ("Jake").[2] However, reports of intoxication from occupational exposure are rare.[1] Shortly after ingestion, there may be nausea, vomiting, diarrhea, and abdominal pain.[1] After a symptom-free interval of 3 to 28 days, most patients complain of sharp, cramplike pains in the calf muscles; some patients complain of numbness and tingling in the feet and sometimes the hands.[1,3] Within a few hours, there is increasing weakness of the legs and feet, progressing to bilateral footdrop.[3] After an interval of another 10 days, weakness of the fingers and wristdrop develop, but the paralysis is not usually as severe as occurs in the feet and legs. This process does not extend above the elbows; the thigh muscles are infrequently involved. Sensory changes, if they occur, are minor.[4,5]

With severe intoxication, lesions of the anterior horn cells and the pyramidal tracts may also occur.[5,6] Muscular weakness may increase over a period of several weeks or months; recovery may take months or years, and, in 25% to 30% of cases, permanent residual effects remain, usually confined to the lower limbs.[3,5] Gait impairment, permanent in some people, was called "Jake Walk."[2]

Fatalities are rare and occur principally in those people who have taken large quantities in a short period of time; autopsy of six human cases revealed involvement of anterior horn cells and demyelination of nerve cells.[4] The lethal dose for humans by ingestion is about 1.0 g/kg; severe paralysis has been produced by ingestion of 6 to 7 mg/kg.[4]

In workers engaged in the manufacture of aryl phosphates (including up to 20% TOCP) and exposed to concentrations of aryl phosphates at 0.2 to 3.4 mg/m^3, there was some inhibition of plasma cholinesterase, but there was no correlation of this effect with degree of exposure or with minor gastrointestinal or neuromuscular symptoms.[7,8] No effects on the eyes or skin have been reported; TOCP is readily absorbed through the skin without local irritant effects.

In affected cats and hens, extensive damage is observed in the spinal cord and sciatic nerves; damage to the myelin sheath and Schwann cells is secondary to the destructive lesion in the axon, which starts at the distal end of the longer axons.[9]

REFERENCES

1. Hygienic Guide Series: Triorthocresylphosphate. Am Ind Hyg Assoc J 24:534–536, 1963
2. Morgan JP, Tulloss TC: The Jake Walk blues—a toxicologic tragedy mirrored in American popular music. Ann Intern Med 85:804–808, 1976
3. Susser M, Stein Z: An outbreak of tri-*ortho*-cresyl phosphate (TOCP) poisoning in Durban. Br J Ind Med 14:111–120, 1957
4. Fassett DW: Esters. In Patty FA (ed): Industrial Hygiene and Toxicology, 2nd ed, Vol 2, Toxicology, pp 1853, 1914–1925, 1935–1937. New York, Wiley–Interscience, 1963
5. Hunter D, Perry KMA, Evans RB: Toxic polyneuritis arising during the manufacture of tricresyl phosphate. Br J Ind Med 1:227–231, 1944
6. Vora DD: Toxic polyneuritis in Bombay due to *ortho*-cresyl-phosphate poisoning. J Neurol Neurosurg Psychiatry 25:234–242, 1962
7. Tabershaw IR, Kleinfeld M, Feiner B: Manufacture of tricresyl phosphate and other alkyl phenyl phosphates: An industrial hygiene study. I. Environmental factors. AMA Arch Ind Health 15:537–540, 1957
8. Tabershaw IR, Kleinfeld M: Manufacture of tricresyl phosphate and other alkyl phenyl phosphates: An industrial hygiene study. II. Clinical effects of tricresyl phosphate. AMA Arch Ind Health 14:541–544, 1957
9. Johnson MK: Delayed neurotoxic action of some organophosphorus compounds. Br Med Bull 25:231–235, 1969

TRIPHENYL PHOSPHATE

CAS: 115-86-6

$(C_6H_5O)_3PO_4$ 1987 TLV = 3 mg/m^3

Synonyms: TPP; Celluflex TPP

Physical Form. Solid

Uses. Noncombustible substitute for camphor in celluloid; impregnating roofing paper; plasticizer in lacquers and varnishes

Exposure. Inhalation

Toxicology. Triphenyl phosphate is of low toxicity in humans.

A group of 16 workers exposed to vapor, mist, or dust at an average concentration of 3.5 mg/m^3, and occasionally as high as 40 mg/m^3, for 8 to 10 years exhibited no signs of illness; the only positive finding was a slight but statistically significant reduction in erythrocyte cholinesterase activity.[1] In workers engaged in the manufacture of aryl phosphates (including triphenyl phosphates and up to 20% tri-*o*-cresyl phosphate) and exposed to concentrations of aryl phosphates of 0.2 to 3.4 mg/m^3, there was some inhibition of plasma cholinesterase but no correlation of this effect with degree of exposure or with minor gastrointestinal or neuromuscular symptoms.[2,3]

Two of six cats given a single intraperitoneal injection of triphenyl phosphate at 0.1 to 0.5 g/kg developed paralysis after 16 to 18 days.[1] The effects of triphenyl phosphate on the eye have not been reported; application in ethanol to the skin of mice produced no more irritation than was expected from the solvent.[1]

REFERENCES

1. Sutton WL et al: Studies on the industrial hygiene and toxicology of triphenyl phosphate. Arch Environ Health 1:45–48, 1960
2. Tabershaw IR, Kleinfeld M, Feiner B: Manufacture of tricresyl phosphate and other alkyl phenyl phosphates: An industrial hygiene study. I. Environmental factors. AMA Arch Ind Health 15:537–540, 1957
3. Tabershaw IR, Kleinfeld M, Feiner B: Manufacture of tricresyl phosphate and other alkyl phenyl phosphates: An industrial hygiene study. II. Clinical effects of tricresyl phosphate. AMA Arch Ind Health 15:541–544, 1957

TRIPHENYL PHOSPHITE

CAS: 101-02-0

$P(O\text{–}C_6H_5)_3$ 1987 TLV = none established

Synonyms: Phenyl phosphite; triphenoxyphosphine

Physical Form. Water-white to pale yellow; solid (below 22°C) or oily liquid

Uses. Stabilizer/antioxidant for vinyl plastics and polyethylene, polypropylene, styrene copolymers, and rubber

Exposure. Inhalation

Toxicology. Triphenyl phosphite is a skin irritant and sensitizer in humans and is neurotoxic in laboratory animals.

In an early study in rats, subcutaneous injections of triphenyl phosphite caused two distinct stages of neurotoxic action.[1] The early, rapidly developing stage was characterized by fine or coarse tremor, usually involving the large muscle groups. The tremor disappeared in surviving animals within a few hours. The later stage occurred several days after treatment and was characterized by hyperexcitability, some spasticity, and incoordination, followed by partial flaccid paralysis of the extremities. The posterior extremities were usually more affected.

In the same study, cats received a one-time subcutaneous injection of 0.1, 0.2, 0.3, or 0.5 ml/kg. At the lower doses, the compound produced ataxia and paresis of the extremities after several days. The intermediate dose (0.3 ml/kg) was eventually lethal in two animals and produced rapidly progressing ataxis on day 6 followed in 1 to 2 days by extensor rigidity. The highest dose (0.5 ml/kg) produced death within 30 hours.

In a more recent study, rats were injected with two 1.0 ml/kg (1184 mg/kg) subcutaneous injections spaced 1 week apart, and were killed 1 week after the second injection.[2] Dysfunctional changes, including tail rigidity, circling, and hindlimb paralysis, were noted. However, the pattern of TPP-induced spinal cord damage in conjunction with marginal neurotoxic esterase inhibition suggested that this toxic neuropathy differed from those previously described for organophosphorus-induced delayed neuropathy.

Applied to human skin in patch tests, triphenyl phosphite diluted 1:3 with cold cream produced slight irritation in two thirds of volunteers tested after a 48-hour contact time. A challenge with the compound 14 days later produced a moderate sensitization reaction.[3] When applied to the skin of laboratory animals, the undiluted chemical was severely irritating and produced moderate sensitization. Instillation of 0.1 ml of triphenyl phosphite into the eyes of rabbits did not produce primary eye irritation.[4] However, another report lists triphenyl phosphite as an eye irritant.[5]

REFERENCES

1. Smith MI, Lillie RD, Elvove E, Stohlman EF: The pharmacologic action of phosphorous acid esters of the phenols. J Pharmacol Exp Ther 49:78–79, 1983
2. Veronesi B, Padilla S, Newland D: Biochemical and neuropathological assessment of triphenyl phosphite in rats. Toxicol Appl Pharmacol 83:203–210, 1986
3. Mallette FS, VonHaam E: Studies on the toxicity and skin effects of compounds used in the rubber and plastics industries. Arch Ind Hyg Occup Med 5:311–317, 1952
4. Borg–Warner Chemicals. FYI–OTS–0785–0422 FLWP. Sequence D. Primary Eye Irritation Tests of Triphenyl Phosphite in Rabbits. Washington, DC, US Environmental Protection Agency, Office of Toxic Substances, 1980
5. Sandmeyer EE, Kirwin CJ: Esters. In Clayton GD, Clayton FE (eds): Patty's Industrial Hygiene and Toxicology, 3rd ed, rev, Vol 2A, Toxicology, pp 2362—2377. New York, Wiley–Interscience, 1981

TURPENTINE

CAS: 8006-64-2

$C_{10}H_{16}$ — 1987 TLV = 100 ppm

Synonyms: Spirit of turpentine; oil of turpentine; wood turpentine

Physical Form. Volatile liquid, colorless or yellow, which varies in composition according to its source and method of production. A typical analysis of turpentine is as follows: α-pinene, 82.5%; camphene, 8.7%; β-pinene, 2.1%; unidentified natural turpenes, 6.8%.

Uses. Solvent; paint thinner

Exposure. Inhalation; skin absorption; ingestion

Toxicology. Turpentine is a skin and mucous membrane irritant and a central nervous system depressant.

Several human subjects had nose and throat irritation at exposures of 75 ppm for 3 to 5 minutes; 175 ppm was intolerable to the majority.[1] Although often reported in the older literature, toxic nephrosis characterized by albuminuria, dysuria, hematuria, and glycosuria is seldom seen today with turpentine overexposure.[2] The apparent rarity of renal lesions in current poisonings may be related to the change in composition of domestic turpentine; turpentine is now more "pure" because of the removal of a hydroperoxide of δ 3-carene.[2]

By ingestion, the mean lethal dose for humans probably lies between 120 and 180 ml.[2] Symptoms include a burning pain in the mouth and throat, abdominal pain, nausea, vomiting, and occasionally diarrhea.[2] Central nervous system effects include excitement, ataxia, confusion, and stupor. Convulsions may occur several hours after ingestion. Fever and tachycardia commonly occur, and death is usually attributed to respiratory failure.[2]

The liquid may cause conjunctivitis and corneal burns.[3] Turpentine from any source is a skin irritant if allowed to remain in contact for a sufficient length of time; hypersensitivity occurs in some persons.[3] The liquid can be absorbed by the skin and mucous membranes, and intoxication by this route has been reported.[2]

The LC_{50} for rats was 3590 ppm for 1 hour and 2150 ppm for 6 hours; hyperpnea, ataxia, tremor, and convulsions were noted.[4] Mucous membrane irritation, particularly of the eyes, and mild convulsions were observed in cats exposed to 540 to 720 ppm for a few hours.[2]

REFERENCES

1. Nelson KW et al: Sensory response to certain industrial solvent vapors. J Ind Hyg Toxicol 25:282–285, 1943

2. Gosselin RE, Smith RP, Hodge HC: Clinical Toxicology of Commercial Products, Section III, 5th ed, pp 393–394. Baltimore, Williams & Wilkins, 1984
3. Hygienic Guide Series: Turpentine. Am Ind Hyg Assoc J 28:297–300, 1967
4. Sperling F, Marcus WL, Collins C: Acute effects of turpentine vapors on rats and mice. Toxicol Appl Pharmacol 10:8–20, 1967

URANIUM (natural soluble and insoluble compounds)

CAS: 7440-61-1

U   1987 TLV = 0.2 mg/m^3

Synonyms:
Soluble: Uranyl nitrate, $UO_2(NO_3)$; uranyl fluoride, UO_2F_2; uranium hexafluoride, UF_6
Insoluble: Uranium dioxide, UO_2; uranium tetrafluoride, UF_4

Physical Form. Solids

Uses. Intensifier in photography; nuclear fuel

Exposure. Inhalation

Toxicology. Insoluble compounds of uranium are respiratory irritants, whereas soluble compounds are also toxic to the kidneys.

Soluble Compounds: Animals repeatedly exposed to dusts of soluble uranium compounds in concentrations from 3 to 20 mg/m^3 died of pulmonary and renal damage; both feeding and percutaneous toxicity studies on animals indicated that the more soluble compounds are the most toxic.[1] In animals, effects on the liver are a consequence of the acidosis and azotemia induced by renal dysfunction.[1]

Animal studies indicate that the primary toxic effect of uranium exposure is on the kidney, with particular damage to the proximal tubules. Functionally, this may result in increased excretion of glucose and amino acids. Structurally, the necrosis of tubular epithelium leads to formation of cellular casts in the urine. If exposure is insufficient to cause death from renal failure, the tubular lesion is reversible with epithelial regeneration. Although bone is the other major site of deposition, there is no evidence of toxic or radiocarcinogenic effects to bone or bone marrow from experimental studies.[2]

Insoluble Compounds: Repeated exposures of three animal species to uranium dioxide dust at a concentration of 5 mg uranium/m^3 for periods of up to 5 years resulted in no kidney injury. More than 90% of the uranium found in the body was in the lungs and tracheobronchial lymph nodes (TLN).[2] Fibrotic changes suggestive of radiation injury were occasionally seen in the TLN of dogs and monkeys and in the lungs of monkeys after exposure periods of more than 3 years; the estimated alpha dose to tissues was greater than 500 rads for lungs and 7000 rads for TLN.[3]

Uranium: Rats injected with metallic uranium in the femoral marrow and in the chest wall developed sarcomas; whether this was due to metallocarcinogenic or radiocarcinogenic action could not be determined.[1] The increased incidence of lung cancer reported among uranium miners is probably the result of exposure to radon gas and its particulate daughters, rather than to uranium dust.[4] In a group of uranium mill workers, there was an excess of deaths from malignant disease of lymphatic and hematopoietic tissue; data from animal experiments suggested that this excess may have resulted from irradiation of lymph nodes by thorium-230, a disintegration product of uranium.[4] Some absorbed uranium is deposited in bone. A potential risk of radiation effects on bone marrow has been postulated, but extensive clinical studies on exposed workers have disclosed no hematologic abnormalities.[5,6]

Accidental exposure of workers to a mixture of uranium hexafluoride, uranyl fluoride, hydrofluoric acid, and live steam caused lacrimation, conjunctivitis, shortness of breath, paroxysmal cough, rales in the chest, nausea, vomiting, skin burns, transitory albuminuria, and elevation of blood urea nitrogen.[5] Two deaths occurred among the most heavily exposed workers shortly after exposure. The persons having the greatest exposure showed the highest urinary uranium levels. In addition, their urinary abnormalities were the most severe, including albuminuria plus red cells and casts in the urinary sediment, and blood urea nitrogen remained elevated for several weeks. The injurious effects observed on the skin, eyes, and respiratory tract were apparently caused by the irritant action of the hydrofluoric acid, whereas the uranium was believed to be responsible for the transient renal changes.

No evidence of chronic toxicity, either chemical or radiation, was observed for any uranium compound during the first 6 years of the atomic energy program; all exposed workers were under very close medical surveillance.[1] Several uranium compounds tested on the eyes of animals caused severe eye damage as well as systemic poisoning. The anion and its hydrolysis products determine the degree of injury.[7] A hot nitric acid solution of uranyl nitrate spilled on the skin caused skin burns, nephritis, and heavy metal encephalopathy.[7] Prolonged skin contact with uranium compounds should be avoided because of potential radiation damage to basal cells. Dermatitis has occurred as a result of handling uranium hemofluoride.[7]

REFERENCES

1. Stokinger HE: The metals. In Clayton GD, Clayton FE (eds): Patty's Industrial Hygiene and Toxicology, 3rd ed, Vol 2, Toxicology, pp 1995–2012. New York, Wiley–Interscience, 1981
2. US Environmental Protection Agency: Drinking Water Criteria Document for Uranium. Washington, DC, Office of Drinking Water, F-198, 1985
3. Leach LJ et al: A five-year inhalation study with natural uranium dixoide (UO_2) dust. I. Retention and biologic effect in the monkey, dog and rat. Health Phys 18:599–612, 1970
4. Archer VE, Wagoner JK, Lundin FE Jr: Cancer mortality among uranium mill workers. J Occup Med 15:11–14, 1973
5. Voegtlin C, Hodge HC: Pharmacology and Toxicology of Uranium Compounds, Vol 1, pp 413–414. New York, McGraw-Hill, 1949
6. Voegtlin C, Hodge HC: Pharmacology and Toxicology of Uranium Compounds, Vol 2, pp 687–689, 993–1017. New York, McGraw-Hill, 1949
7. Hygienic Guide Series: Uranium (natural) and its compounds. Am Ind Hyg Assoc J 30:313–317, 1969

VANADIUM PENTOXIDE

CAS: 1314-62-1

V_2O_5 — 1987 TLV = 0.05 mg/m^3; respirable dust and fume

Synonym: None

Physical Form. Dust or fume

Uses. In welding electrode coatings; additive to special steels; catalyst in glass, ceramic glazes, and oxidation of sulfur dioxide; present in some fuel oils

Exposure. Inhalation

Toxicology. Vanadium pentoxide is an irritant of the eyes and respiratory tract. Fume is recognized as being generally more toxic than dust owing to the smaller particle size of fume, which allows more complete penetration to the small airways of the lungs.

Sixteen workers exposed to concentrations of dust (and possibly some fume) in excess of 0.5 mg/m^3 with particle sizes ranging from 0.1 to 10 μ developed conjunctivitis, nasopharyngitis, hacking cough, fine rales, and wheezing. In three workers exposed to the highest concentrations, the onset of symptoms occurred at the end of the first workday.[1] The bronchospastic element in the more seriously ill persisted for 48 hours after removal from exposure; rales lasted for 3 to 7 days, and, in several cases, cough lasted for up to 14 days.[1] Among those workers with acute intoxication, there was dramatically increased severity of symptoms from repeated exposures of lesser time and intensity. Absorbed vanadium is primarily excreted in the urine, and it was detectable in 12 of the workers for periods of up to 2 weeks. Urinary concentrations in unexposed subjects average 8 mg/24 hours and range as high as 22 mg/24 hours.[6] Urinary vanadium concentrations were elevated in workers exposed to mean air concentrations of 0.1 to 0.28 mg/m^3, but there was no correlation between the air and urinary concentrations. Although most absorbed vanadium was excreted within 1 day after cessation of exposure, increased excretion relative to unexposed controls continued for more than 2 weeks among chronically exposed workers.[7]

Workers exposed to a mixture of ammonium metavanadate and vanadium pentoxide at concentrations near 0.25 mg/m^3 developed green tongue, metallic taste, throat irritation, and cough.[4] Of 36 workers examined 8 years after their original exposure to vanadium pentoxide, there was no evidence of either pneumoconiosis or emphysema, although 6 of the workers still had bronchitis with rhonchi resembling asthma and bouts of dyspnea.[5]

Two volunteers exposed to a concentration of 1 mg/m^3 for 8 hours developed a persistent cough, which lasted for 8 days; 21 days after the original exposure, re-exposure for 5 minutes to a heavy cloud of vanadium pentoxide dust occurred, and, within 16 hours, marked cough developed. The following day, rales and expiratory wheezes were present throughout the entire lung field, but pulmonary function was normal.[6] Subjects exposed to a concentration of 0.2 mg/m^3 for 8 hours developed a loose cough the following morning; other subjects exposed for 8 hours to 0.1 mg/m^3 developed slight cough with increased mucus, which lasted 3 to 4 days.[6]

Although there have been some cases of emphysema observed among workers with exposure to vanadium pentoxide, other possible causes, such as smoking, were not excluded. Cases of asthma have occurred more frequently, suggesting that this may be an effect of chronic exposure.[4]

Eyes and skin are irritated by the dust or by contact with an acid solution of vanadium pentoxide. Eczematous lesions have occurred, and, in three cases, there was an allergic response to patch tests with sodium vanadate.[7]

REFERENCES

1. Zenz C, Bartlett JP, Thiede WH: Acute vanadium pentoxide intoxication. Arch Environ Health 5:542–546, 1962
2. Committee on Occupational Medical Practice: Vanadium pentoxide exposure. J Occup Med 29:14–17, 1987
3. Kiviluoto M: Serum and urinary vanadium of workers processing vanadium pentoxide. Int Arch Occup Environ Health 48:251–256, 1981
4. National Institute for Occupational Safety and Health: Criteria for a Recommended Standard . . . Occupational Exposure to Vanadium. DHEW (NIOSH) Pub No 77–222, p 142. Washington, DC, US Government Printing Office, 1977

5. Sjoberg SG: Follow-up investigation of workers at a vanadium factory. Acta Med Scand 154:381, 1956
6. Zenz C, Berg BA: Human responses to controlled vanadium pentoxide exposure. Arch Environ Health 14:709–712, 1967
7. Sjoberg SG: Health hazards in the production and handling of vanadium pentoxide. Arch Ind Health 3:631–646, 1951

VINYL ACETATE

CAS: 108-05-4

$CH_3COOCH{=}CH_2$ 1987 TLV = 10 ppm

Synonyms: 1–Acetoxyethylene; acetic acid ethenyl ester

Physical Form. Colorless liquid that polymerizes to a transparent solid on exposure to light

Uses. Production of vinyl acetate polymers

Exposure. Inhalation

Toxicology. Vinyl acetate is an irritant of the eyes, nose, and throat.

Volunteers exposed to vinyl acetate showed a wide variation in individual sensitivity to its irritant effects; one of three volunteers had throat irritation at 20 ppm for 4 hours, whereas 72 ppm for 30 minutes produced eye irritation in three of four participants.[1] All subjects agreed that they could not work at 72 ppm for 8 hours.

From a study of 21 workers exposed for an average of 15 years at concentrations between 5 and 10 ppm (with occasional excursions above 300 ppm), vinyl acetate produced no serious chronic effects.[2] Some subjects were sensitive at concentrations of about 6 ppm, and concentrations above 20 ppm produced irritation in most persons.

Prolonged dermal contact, such as that afforded by clothing wet with vinyl acetate, may result in severe irritation or blistering of the skin in some persons.[1]

The LC_{50} for 4 hours in rats was 14,000 mg/m^3 (about 4667 ppm).[3] Dogs exposed 6 hours daily for several weeks starting at 91 ppm and ending after 11 weeks at 186 ppm exhibited eye irritation and lacrimation.[4] Rats exposed repeatedly to 100 ppm showed no effects.[5]

Two carcinogenicity/chronic toxicity studies for vinyl acetate have been performed in which animals were treated for a year or more with the chemical. In one study, rats were exposed to 2500 ppm of the chemical by inhalation for 52 weeks and were observed for tumorigenic effects for 83 weeks after exposure.[6,7] None of the treated rats developed tumors, and although there was increased mortality among the exposed rats, no other toxic effects were reported. In the second study, rats were fed a nominal dose of approximately 20 or 50 mg/day in drinking water for 100 weeks.[8] However, the investigators stated that the vinyl acetate was degraded in the drinking water and estimated that the animals received at least half the nominal dose. The total numbers of tumor-bearing animals in this study were similar for treated and control groups.

Vinyl acetate has been tested for teratogenicity in inhalation and oral assays.[9] Pregnant rats exposed to levels as high as 1000 ppm by inhalation or 5000 ppm in drinking water on gestation days 6 to 15 had significantly reduced weight gain during exposure. The fetuses of the rats exposed by inhalation were also significantly smaller than control fetuses and had an increased incidence of minor skeletal defects. However, the investigators thought that the fetal effects were a consequence of the maternal growth retardation and not of vinyl acetate treatment. In the drinking water study, there were no significant effects on the fetuses, and the investigators concluded that vinyl acetate did not elicit embryolethality, embryotoxicity, or teratogenicity.

REFERENCES

1. National Institute for Occupational Safety and Health: Criteria for a Recommended Standard . . . Occupational Exposure to Vinyl Acetate. DHEW (NIOSH) Pub No 78–205, p 78. Washington, DC, US Government Printing Office, 1978
2. Deese DE, Joyner RE: Vinyl acetate—a study of chronic human exposure. Am Ind Hyg Assoc J 30:449, 1969
3. Carpenter CP et al: The assay of acute vapor toxicity, and the grading and interpretation of results on 96 chemical compounds. J Ind Hyg Toxicol 31:343–346, 1949
4. Haskell Laboratory: Report of Toxicity of Vinyl Acetate. Wilmington, Delaware, EI DePont de Nemours & Co, January 1967
5. Gage JC: The subacute inhalation toxicity of 109 industrial chemicals. Br J Ind Med 27:1–18, 1970
6. Maltoni C, Lefemine G: Carcinogenicity bioassays of vinyl chloride. I. Research plan and early results. Environ Res 7:387–405, 1974
7. Maltoni C: Carcinogenicity of vinyl chloride—current results—experimental evidence. Adv Tumor Prev Detect Charac 3:216–237, 1976
8. Lijinsky W, Rueber MD: Chronic toxicity studies of vinyl acetate in Fischer rats. Toxicol Appl Pharmacol 68:43–53, 1983
9. The Society of the Plastics Industry, Inc. FYI-AX-1283-0278 (Seq C). Protocols for Toxicological Studies on Vinyl Acetate. Washington, DC, US Environmental Protection Agency, 1983

VINYL BROMIDE

CAS: 593-60-2

$H_2C{=}CHBr$ 1987 TLV = 5 ppm

Synonyms: Bromoethene; bromoethylene

Physical Form. Gas

Uses. Production of flame-resistant plastics or thermoplastic resins

Exposure. Inhalation

Toxicology. Vinyl bromide causes central nervous system (CNS) depression in animals at high levels and has a carcinogenic action in rats.

There are no data on human exposures.

Exposure of rats to 100,000 ppm for 15 minutes resulted in deep anesthesia and death.[1] Exposure to 50,000 ppm caused anesthesia in 25 minutes and was lethal after exposure for 7 hours.

A significant decline in animal body weights was the only treatment-related effect following exposure at 10,000 ppm, 7 hours/day for 4 weeks. In a 6-month inhalation study in a number of species, serum bromide levels increased following exposure to 250 and 500 ppm.

It was carcinogenic in male and female rats exposed to 10, 50, 250, or 1250 ppm in a lifetime inhalation study and caused angiosarcomas of the liver in both sexes at all levels.[2] A significant increase in hepatocellular neoplasms was also seen in male rats exposed at 250 ppm and in female rats exposed at 10, 50, and 250 ppm. The lack of increase in hepatocellular neoplasms in rats at the 1250 level was probably due to their early mortality and termination at 72 weeks. It is regarded as suspect of carcinogenic potential to humans.[3]

The liquid was moderately irritating to the rabbit eye but essentially nonirritating to the skin.

REFERENCES

1. Leong BKJ, Torkelson TR: Effects of repeated inhalation of vinyl bromide in laboratory animals with recommendations for industrial handling. Am Ind Hyg Assoc J 31:1–11, 1970.
2. Benya TJ et al: Inhalation carcinogenicity bioassay of vinyl bromide in rats. Toxicol Appl Pharmacol 64:367–379, 1982
3. Vinyl bromide. Documentation of the TLVs and BEIs, 5th ed, p 622. Cincinnati, American Conference of Governmental Industrial Hygienists (ACGIH), 1986

VINYL CHLORIDE
CAS: 75-01-4
$H_2C{=}CHCl$
1987 TLV = 5 ppm (10 mg/m^3); recognized human carcinogen

Synonyms: Chloroethene; chloroethylene; ethylene monochloride

Physical Form. A colorless gas, but usually handled as a liquid under pressure

Uses. Production of polyvinyl chloride resins; organic synthesis

Exposure. Inhalation

Toxicology. Occupational exposure to vinyl chloride is associated with an increased incidence of angiosarcoma of the liver and other malignant tumors, acroosteolysis, Raynaud's syndrome, scleroderma, thrombocytopenia, circulatory disturbances, and impaired liver function. Very high concentrations cause central nervous system (CNS) depression.

Humans exposed to 20,000 ppm for 5 minutes experienced dizziness, lightheadedness, nausea, and dulling of vision and auditory cues.[1] No clinical changes nor abnormal neurologic responses were found in 13 volunteers exposed 7.5 hours to 500 ppm, although exposure to 300 ppm caused impairment in liver function tests without overt clinical disease.[2]

Twenty-five cases of acroosteolysis (degeneration of terminal phalanges) have been reported in vinyl chloride workers; Raynaud's phenomenon was the first manifestation noted by a majority of subjects, suggesting that the vascular lesion antecedes the bone changes in most cases.[3,4] Radiologic findings in patients with acroosteolysis included lytic lesions in the distal phalanges of the hands, in the styloid processes of the ulna and radius, and in the sacroiliac joints.[4] Of 20 autoclave cleaners with exposure to vinyl chloride, 16 had thrombocytopenia, 7 had splenomegaly, 6 had hepatomegaly, 14 had fibrosis of the liver capsule, and 4 had signs of acroosteolysis.[5]

In four facilities engaged in the polymerization of vinyl chloride for at least 15 years, a study of workers exposed for at least 5 years revealed a significant number of excess deaths due to malignant neoplasms (35 deaths observed, 23.5 expected).[6] The excesses were found for four organ systems: CNS (3 observed, 0.9 expected); respiratory system (12 observed, 7.7 expected); hepatic system (7 observed, 0.6 expected); and lymphatic and hematopoietic systems (4 observed, 2.5 expected).

By 1975, more than 30 cases of angiosarcoma of the liver had been reported among vinyl chloride polymerization workers in the United States and nine other nations.[7] Because this tumor is rare, the occurrence of these cases under similar occupational conditions strongly suggests a causal relationship to some phase of vinyl chloride production.[8] Clinical features of seven patients with the malignancy varied from no signs or symptoms to weakness, pleuritic pain, abdominal pain, weight loss, gastrointestinal bleeding, and hepatosplenomegaly. Liver function abnormalities were present in all subjects, but there was no consistent pattern.[8] In addition to the malignant tumors, four cases of nonmalignant hepatic disease characterized by portal fibrosis and portal hypertension have been attributed to vinyl chloride exposure.[8]

In animal experiments, vinyl chloride has been shown to be oncogenic; Zymbal gland carcinomas, nephroblastomas, and angiosarcomas were the prevailing tumors in rats. Results ranged from a 16% tumor incidence at an exposure level of 250 ppm to a 39% incidence at 10,000 ppm. In mice, liver angiosarcomas, pulmonary adenomas, and mammary carcinomas were observed after exposures ranging from 50 to 10,000 ppm.[9] The development of some tumors was more dependent upon duration of exposure than upon concentration of vinyl chloride.[9]

When vinyl chloride was used as an anesthetic agent in dogs, muscular incoordination, sensitization of the myocardium, and serious cardiac arrhythmias were observed.[10] Mice exposed to 80,000 to 120,000 ppm were anesthetized, whereas 250,000 to 300,000 ppm was lethal in 10 minutes. Guinea pigs were killed in a short time at 200,000 to 400,000 ppm; pulmonary edema and hyperemia of the liver and kidneys were observed.[10]

Contact of the skin or eyes with the liquefied gas can produce freezing and frostbite.[11]

Vinyl chloride is now designated a recognized human carcinogen.[11]

REFERENCES

1. Lester D, Greenberg LA, Adams WR: Effects of single and repeated exposures of humans and rats to vinyl chloride. Am Ind Hyg Assoc J 24:265–275, 1963
2. Kramer CG, Mutchler JE: The correlation of clinical and environmental measurements for workers exposed to vinyl chloride. Am Ind Hyg Assoc J 33:19–30, 1972
3. Dinman BD et al: Occupational acroosteolysis. I. An epidemiological study. Arch Environ Health 22:61–73, 1971
4. Dodson VN et al: Occupational acroosteolysis. III. A clinical study. Arch Environ Health 22:83–91, 1971
5. Marstellar HJ et al: Chronisch-toxische Leberschaden bei Arbeitern in der PVC Produktion. Dtsch Med Wochenschr 98:2311–2314, 1973
6. Waxweiler RJ et al: Neoplastic risk among workers exposed to vinyl chloride. Ann NY Acad Sci 271:40–48, 1976
7. Lloyd JW: Angiosarcoma of the liver in vinyl chloride/polyvinyl chloride workers. J Occup Med 17:333–334, 1975
8. Falk H et al: Hepatic disease among workers at a vinyl chloride polymerization plant. JAMA 230:59–63, 1974
9. Maltoni C, Lefmine G: Carcinogenicity bioassays of vinyl chloride. I. Research plan and early results. Environ Res. 7:387–405, 1974
10. Mastromatteo E, Fischer AM, Christie H, Danziger H: Acute inhalation toxicity of vinyl chloride to laboratory animals. Am Ind Hyg Assoc J 21:394–398, 1960
11. Vinyl chloride. Documentation of TLVs and BEIs, 5th ed, pp 623–626. Cincinnati, American Conference of Governmental Industrial Hygienists (ACGIH), 1986

VINYLIDENE CHLORIDE

CAS: 75-35-4

$CH_2{=}CCl_2$ 1987 TLV = 5 ppm

Synonyms: 1,1–Dichloroethylene; VDC; asym-dichloroethylene; 1,1–dichloroethene; 1,1–DCE

Physical Form. Clear liquid that is highly flammable and reactive and in the presence of air can form complex peroxides in the absence of chemical inhibitors

Uses. Production of copolymers of high vinylidene chloride content, the other major monomer usually being vinyl chloride such as Saran and VELON for films and coatings

Exposure. Inhalation

Toxicology. Vinylidene chloride causes CNS depression at high levels, and repeated exposure to lower concentrations results in liver and kidney damage in experimental animals.

In male rats, the 4-hour LC_{50} was 6350 ppm.[1] Rats exposed 6 hours/day for 20 days to 200 ppm exhibited only slight nasal irritation.[2] Animals exposed 8 hours/day, 5 days/week for 6 months to 50 or 100 ppm had liver and kidney injury.[3] The oral LD_{50} of vinylidene chloride in corn oil in male rats was 1500 mg/kg.[4] The liquid is moderately irritating to the eyes and irritating to the skin of rabbits.

Vinylidene chloride affects several liver enzymes: It decreases the activity of hepatic glucose-6-phosphatase and the content of glutathione, and increases serum alanine alpha ketoglutanate transaminase activity and liver content of triglycerides.[5]

In rats, ingestion of drinking water containing up to 200 ppm vinylidene chloride caused mild, dose-related changes in the liver but did not affect the reproductive capacity through three generations that produced six sets of litters.[6]

In a carcinogenicity study, Swiss mice were exposed to 10 or 25 ppm for 4 hours/day, 5 days/week for 52 weeks.[7,8] After 98 weeks, 25 ppm had caused kidney adenocarcinomas in 24 of 150 males and in 1 of 150 females, whereas none were seen in the control group. Rats exposed to 75 ppm 6 hours/day, 5 days/week for 18 months then held until 24 months showed a reversible hepatocellular fatty change but no increase in tumor incidence that could be attributed to vinylidene chloride exposure.[9] Several other studies in other strains of mice, rats, and hamsters did not produce carcinogenic effects.[5]

In one epidemiologic study of 138 exposed workers, no excess of cancer cases was found, but follow-up was incomplete; nearly 40% of the workers had less than 15 years latency since first exposure, and only five deaths were observed.[10] According to the IARC, there is no adequate information to permit assessment of human carcinogenicity.[5]

REFERENCES

1. Siegel J, Jones RA, Con RA, Lyon JP: Effects on experimental animals of acute, repeated and continuous inhalation exposures to dichloroactylene mixture. Toxicol Appl Pharmacol 18:168, 1971
2. Gage JC: The subacute inhalation toxicity of 109 industrial chemicals. Br J Ind Med 27:1, 1970
3. Unpublished data. The Dow Chemical Company, Midland, Michigan
4. Jenkins LJ Jr, Trabulus MJ, Murphy SD: Biochemical effects of 1,1-dichloroethylene. Toxicol Appl Pharmacol 23:501, 1972
5. IARC Monographs on the Evaluation of the Carcinogenic Risk of Chemicals to Man, Vol 19, Some Monomers, Plastics and Synthetic Elastomers, and Acrolein, pp 439–447. Lyon, International Agency for Research on Cancer, 1979
6. Reitz RH et al: Effects of vinylidene chloride on DNA synthesis and DNA repair in the rat and mouse: A comparative study with dimethyl nitrosamine. Toxicol Appl Pharmacol 2:357–370, 1980
7. Maltoni C: Recent findings on the carcinogenicity of chlorinated olefins. Environ Health Perspect 21:1–5, 1977
8. Maltoni C, Cotti G, Morisi L, Chieco P: Carcinogenicity bioassays of vinylidene chloride. Research plans and early results. Med Lav 58:241–262, 1977
9. Quast JF et al: Chronic toxicity and oncogenicity study on inhaled vinylidene chloride in rats. Fund Appl Toxicol 6:105–144, 1986
10. Ott MG et al: A health study of employees exposed to vinylidene chloride. J Occup Med 18:735–738, 1976

VINYL TOLUENE

CAS: 25013-15-4

$CH_3C_6H_4CHCH_2$ 1987 TLV = 50 ppm

Synonyms: meta- and *para-*Vinyl toluene

Physical Form. Colorless liquid

Uses. Chemical synthesis; pharmaceuticals

Exposure. Inhalation

Toxicology. Vinyl toluene is an irritant of the eyes and mucous membranes; at high concentrations, it causes narcosis in animals, and it is expected that severe exposure will produce the same effect in humans.

Human subjects exposed to 200 ppm detected a strong odor, but excessive discomfort was not experienced; at 400 ppm, there was strong eye and nasal irritation.[1]

Exposure of rats and guinea pigs to 1350 ppm for 7 hours/day for 100 days caused the death of some of the rats and slight liver damage in guinea pigs; there were no effects in female monkeys at this concentration.[1]

Rats tolerated exposure to 300 ppm for 60 hours without clinical symptoms, although they appeared relatively inactive.[2] At this concentration, vinyl toluene was found to accumulate in perirenal fat and was more effective than styrene, xylene, or toluene in producing neurochemical effects as determined by enzyme assays.

The liquid dropped in the eyes of rabbits caused slight conjunctival irritation.[1] Applied to rabbit skin, vinyl toluene produced erythema with the development of edema and superficial necrosis.[1]

Vinyl toluene has a disagreeable odor detectable at 50 ppm.[1]

REFERENCES

1. Wolf MA et al: Toxicological studies of certain alkylated benzenes and benzene. AMA Arch Ind Health 14:387–398, 1956
2. Savolainen H, Pfäffli P: Neurochemical effects of short-term inhalation exposure to vinyltoluene vapor. Arch Environ Contam Toxicol 10:511–517, 1981

VM&P NAPHTHA

CAS: 8030-30-6 — 1987 TLV = 300 ppm

Synonyms: Varnish Makers' and Printers' Naphtha; light naphtha; dry-cleaners' naphtha; spotting naphtha

Physical Form. Clear colorless to yellow liquid; petroleum distillate containing C5 to C11 hydrocarbons; a typical composition is paraffins 55.4%, naphthenes 30.3%, alkyl benzene 11.7%, dicycloparaffins 2.4%, benzene less than 0.1%.

Uses. Dilutent for paints, coatings, resins, printing inks, rubbers, and cements; solvent

Exposure. Inhalation

Toxicology. VM&P Naphtha vapor is a central nervous system (CNS) depressant and a mild irritant of the eyes and upper respiratory tract.

In human tests, exposure to 880 ppm (4100 mg/m^3) for 15 minutes resulted in eye and throat irritation with olfactory fatigue.[1] The chief effect of exposure to high levels of the vapor is reported to be CNS depression.[2,3] However, in an accidental brief exposure of 19 workers from an overheated solvent tank, the chief effect was dyspnea, which lasted for several minutes after the exposure.[2] Two of the workers were cyanotic with tremor and nausea, but these symptoms were of brief duration. The absence of CNS depression was noteworthy.

The LC_{50} in rats was 3400 ppm for 4 hours; incoordination was observed.[4] In rats and beagle dogs exposed to 500 ppm for 30 hours weekly for 13 weeks, there was no evidence of latent or chronic effects.

REFERENCES

1. Carpenter CP et al: Petroleum hydrocarbon toxicity series. IV. Animal and human response to vapors of rubber solvent. Toxicol Appl Pharmacol 33:526, 1975
2. Wilson WF: Toxicology of petroleum naphtha distillate vapors. J Occup Med 18:821, 1976
3. National Institute for Occupational Safety and Health: Criteria for a Recommended Standard . . . Occupational Exposure to Refined Petroleum Solvents. DHEW (NIOSH) Pub No 77-192. Washington, DC, US Government Printing Office, 1977
4. Carpenter CP et al: Petroleum hydrocarbon toxicity series. II. Animal and human response to vapors of varnish makers' and printers' naphtha. Toxicol Appl Pharmacol 32:263, 1975

WARFARIN

CAS: 81-81-2

$C_{19}H_{16}O_4$ 1987 TLV = 0.1 mg/m^3

Synonyms: 3–(a-Acetonylbenzyl)–4,hydroxycoumarin; coumadin; compound 42

Physical Form. Colorless, odorless, tasteless crystals

Uses. Rodenticide

Exposure. Inhalation; skin absorption

Toxicology. Warfarin causes hypoprothrombinemia and vascular injury, which results in hemorrhage.

Warfarin suppresses the hepatic formation of prothrombin and of factors VII, IX, and X, causing a markedly reduced prothrombin activity of the blood. Warfarin also causes dilatation and engorgement of blood vessels and an increase in capillary fragility.[1] The inhibition of prothrombin formation does not become apparent until the prothrombin reserves are depleted, which usually requires exposure for a number of days. A single large exposure may cause intoxication after a latency period of several days, but, in a series of acute ingestion episodes, there were no signs of hemorrhage or depression of plasma prothrombin.

A farmer whose hands were intermittently wetted with an 0.5% solution of warfarin over a period of 24 days developed gross hematuria 2 days after the last contact with the solution; the following day, spontaneous hematomas appeared on the arms and legs.[2] Within 4 days, other effects included epistaxis, punctate hemorrhages of the palate and mouth, and bleeding from the lower lip. Four days later, after treatment for 2 days with phytonadione, hematologic indices had returned to the normal range. Other effects of warfarin intoxication have included back pain, abdominal pain, vomiting, and petechiae of the skin.[1,3]

REFERENCES

1. Gosselin RE, Hodge HC, Smith RP: Clinical Toxicology of Commercial Products, Section III, 5th ed, pp 395–397. Baltimore, Williams & Wilkins, 1984
2. Fristedt B, Sterner N: Warfarin intoxication from percutaneous absorption. Arch Environ Health 11:205, 1965
3. Hayes WJ Jr: Pesticides Studied in Man, pp 508–512. Baltimore, Williams & Wilkins, 1982

XYLENE

CAS: 133-20-7

o-Xylene—95-47-6

m-Xylene—108-38-3

p-Xylene—106-42-3

$C_6H_4(CH_3)_2$ 1987 TLV = 100 ppm

Synonyms: Xylol; dimethylbenzene

Physical Form. Colorless liquid

Uses. Solvent; manufacture of certain organic compounds

Exposure. Inhalation; skin absorption

Toxicology. Xylene vapor is an irritant of the eyes, mucous membranes, and skin; at high concentrations, it causes narcosis.

Three painters working in the confined space of a fuel tank were overcome by xylene vapor estimated to be 10,000 ppm. They were not found until 18.5 hours after entering the tank, and one died from pulmonary edema shortly thereafter. The other two workers recovered completely in 2 days; both had temporary hepatic impairment (inferred from elevated serum transaminase levels), and one had evidence of temporary renal impairment (increased blood urea and reduced creatinine clearance).[1]

Giddiness, anorexia, and vomiting occurred in a worker exposed to a solvent containing 75% xylene at levels of 60 to 350 ppm, with possible higher excursions.[2] In another report, eight painters exposed to a solvent consisting of 80% xylene and 20% methylglycolacetate experienced headache, vertigo, gastric discomfort, dryness of the throat, and signs of slight drunkenness.[3] Volunteers exposed to 460 ppm for 15 minutes had slight tearing and light-headedness.[4] A level of 230 ppm was not considered to be objectionable to most of these subjects. However, in an earlier study, the majority of subjects found 200 ppm irritating to the eyes, nose, and throat and judged 100 ppm to be the highest concentration subjectively satisfactory for an 8-hour exposure.[5]

Prior to 1940, most reports on the possible chronic toxicity of xylene also involved exposure to solvents that also contained high percentages of benzene or toluene as well as other compounds. Consequently, the effects attributed to xylene in these reports are questionable.[6] Blood dyscrasias such as those reportedly caused by benzene exposure have not been associated with the xylenes.[6]

After 24-hour exposures, some mice died at 2010 ppm (*m*-xylene) and 3062 ppm (*o*-xylene) but survived 4912 ppm (*p*-xylene).[5] Exposure of rats to 1600 ppm for 2 or 4 days produced mucous membrane irritation, incoordination, narcosis, weight loss, increased erythrocyte count, and death. Exposure to 980 ppm for 7 days caused leukopenia, kidney congestion, and hyperplasia of the bone and spleen.[5]

The liquid is a skin irritant and causes erythema, dryness, and defatting; prolonged contact may cause the formation of vesicles.[3]

Repeated exposure of rabbits to 1150 ppm of a mixture of isomers of xylene for 40 to 55 days caused a reversible decrease in red and white cell count and an increase in thrombocytes; exposure to 690 ppm for the same time period caused only a slight decrease in the white cell count.[7]

Repeated application of 95% xylene to rabbit skin caused erythema and slight necrosis. Instilled in rabbit eyes, it produced conjunctival irritation and temporary corneal injury. Exposure to the vapors produced reversible vacuoles in the corneas of cats. Exposure of rats to 100 to 400 ppm on days 6 to 15 of pregnancy for 6 to 24 hours caused dose-dependent retardation of fetal development at levels that were also maternally toxic.[8]

In 2-year gavage studies, there was no evidence of carcinogenicity of mixed xylenes for male or female rats given 250 or 500 mg/kg for male or female mice given 500 or 1000 ml/kg.[9]

The odor threshold has been reported as 1 ppm.[6]

REFERENCES

1. Morley R et al: Xylene poisoning—a report on one fatal case and two cases of recovery after prolonged unconsciousness. Br Med J 3:442–443, 1970
2. Glass WI: A case of suspected xylol poisoning. NZ Med J 60:113, 1961
3. Goldie I: Can xylene (xylol) provoke convulsive seizures? Ind Med Surg 29:33–35, 1960
4. Carpenter CP et al: Petroleum hydrocarbon toxicity studies. V. Animal and human responses to vapors of mixed xylenes. Toxicol Appl Pharmacol 33:543–558, 1975
5. National Institute for Occupational Safety and Health: Criteria for a Recommended Standard . . . Occupational Exposure to Xylene. DHEW (NIOSH) Pub No 75–168. Washintgon, DC, US Government Printing Office, 1975

6. Von Burg R: Toxicology updates. Xylene. J Appl Toxicol 2:269–271, 1982
7. Fabre R et al: Toxicological research on replacement solvents for benzene. IV. Study of xylenes. Arch Mal Prof Med Trav Secur Soc 21:301, 1960
8. Ungvary G et al: Studies on the embryotoxic effects of *ortho, meta,* and *para*-xylene. Toxicology 18:61–74, 1980

XYLIDINE

CAS: 1300-73-8

$(CH_3)_2C_6H_3NH_2$ — 1987 TLV = 2 ppm; skin

Synonyms: None

Physical Form. Liquid, except *o*–4–xylidine is a solid

Uses. Manufacture of dyes

Exposure. Inhalation; skin absorption

Toxicology. Xylidine causes anoxia due to the formation of methemoglobin. In experimental animals, it also causes damage to the lungs, liver, and kidneys.

In humans, the onset of xylidine intoxication may be insidious, in that early warning signs of methemoglobinemia such as headache and dizziness are not always interpreted as being a result of overexposure, and even mild cyanosis may not be recognized.[1] The effects of methemoglobinemia include cyanosis (especially in the lips, nose, and earlobes), weakness, dizziness, and severe headache. These symptoms occur at methemoglobin levels of about 35%, whereas levels in excess of 80% are probably incompatible with life.[1] In general, higher ambient temperatures increase susceptibility to cyanosis from exposure to methemoglobin-forming agents.[2]

Repeated exposure of cats to 138 ppm caused loss of coordination, cyanosis, prostration, and death; at autopsy, there was pulmonary edema and lobular pneumonia, necrosis of the liver, and toxic nephrosis.[3] In cats exposed to 132 ppm for 3 days, the methemoglobin concentration was 55%; repeated exposure to cats to 17.4 ppm caused toxic hepatitis and some deaths, whereas repeated exposure to 7.8 ppm caused no adverse effects.[3] Liquid xylidine penetrates the skin of rabbits in sufficient quantity to cause cyanosis and death, but with no local effects on the skin.

Diagnosis. *Signs and Symptoms:* Signs and symptoms include headache; signs of anoxia, including cyanosis of lips, nose, and earlobes; anemia; and hematuria.

Differential Diagnosis: Other causes of cyanosis must be differentiated from methemoglobinemia due to chemical exposure. These include hypoxia resulting from lung disease, hypoventilation, and decreased cardiac output. Lung disease may be suspected from results of pulmonary function tests and arterial blood gas analysis. The arterial Po_2 may be normal in methemoglobinemia but tends to be decreased in cyanosis due to lung disease. Hypoventilation will cause elevation of arterial Pco_2, which is not seen in chemical exposure. Decreased cardiac output states will cause cyanosis only when accompanied by arterial hypotension. If blood withdrawn from the vein shows the characteristic chocolate-brown coloration, the diagnosis of an abnormal pigment is almost certain, especially if the color remains after shaking the blood in air.[4]

Special Tests: Examine the urine for blood, and determine the methemoglobin concentration in the blood when chemical intoxication is suspected and at regular intervals until the methemoglobin has been fully reduced to normal hemoglobin.[5] Methemoglobin may be differentiated from sulfhemoglobin by the addition of a few drops of 10% potassium cyanide, which results in the rapid production of bright red cyanomethemoglobin but has no effect on the color of sulfhemoglobin.[4] Spectrophotometry is required for the precise identification of the pigment and its quantitation. Normal acid methemoglobin has a characteristic absorption spectrum with peaks at 502 and 632 nm, which disappear with the addition of cyanide, whereas sulfhemoglobin has a peak at 620 nm, which does not disappear with cyanide.[4]

Treatment. All the contaminant on the body must be removed. Immediately remove all clothing, and wash the entire body from head to foot with soap and water. Pay special attention to the hair and scalp, finger- and toenails, nostrils, and ear canals. Administer oxygen to alleviate the headache and general sense of weakness; confine the patient to bed. Determine the methemoglobin concentration in the blood, and repeat every 3 to 6 hours for 18 to 24 hours. Repeat skin cleansing if the methemoglobin concentration appears to rise after 3 to 4 hours. In general, patients will return to normal within 24 hours provided that all sources of further absorption are completely eliminated. The methemoglobin will be spontaneously changed to ferrous hemoglobin in 2 to 3 days.[4]

Such therapy is not effective in subjects with glucose-6-phosphate dehydrogenase deficiency.[4]

The only justifiable use of methylene blue would be in cases of coma or stupor, usually at methemoglobin levels over 60%. In those patients in whom therapy is necessary, methylene blue, 1 to 2 mg/kg of a 1% solution, may be given intravenously over a 10-minute period. If cyanosis has not disappeared within 1 hour, a second dose of 2 mg/kg may be administered.[4,5] The total dose should not exceed 7 mg/kg, because methylene blue may cause toxic effects such as dyspnea, precordial pain, restlessness, apprehension, red cell hemolysis, and changes in the electrocardiogram (reduction in the height or even reversal of the T wave, frequently with lowering of the R wave).[5]

REFERENCES

1. Bunn HF: Disorders of hemoglobin structure, function and synthesis. In Petersdorf RG et al (eds): Harrison's Principles of Internal Medicine, 10th ed, pp 1881–1882. New York, McGraw-Hill, 1983
2. Linch AL: Biological monitoring for industrial exposure to cyanogenic aromatic nitro and amino compounds. Am Ind Hyg Assoc J 35:426–432, 1974
3. Treon JF et al: The toxic properties of xylidine and monomethylaniline. Arch Ind Hyg Occup Med 1:506–524, 1950
4. Rieder RF: Methemoglobinemia and sulfhemoglobinemia. In Wyngaarden JB, Smith LH (eds): Cecil Textbook of Medicine, 16th ed, p 896. Philadelphia, WB Saunders, 1982
5. Mangelsdorff AF: Treatment of methemoglobinemia. AMA Arch Ind Health 14:148–153, 1956

YTTRIUM

CAS: 7440-65-5

Y 1987 TLV = 1 mg/m^3

Compounds: Yttrium chloride; yttrium nitrate; yttrium oxide; yttrium phosphate

Physical Form. White powder

Uses. Yttrium is mixed with rare earths as phosphors for color television receivers; oxide for mantles in gas and acetylene lights

Exposure. Inhalation

Toxicology. Yttrium compounds cause pulmonary irritation in animals.

No effects in humans have been reported.

Intratracheal administration of 50 mg yttrium oxide in rats caused granulomatous nodules to develop in the lungs by 8 months.[1] Nodules in the peribronchial tissue compressed and deformed several bronchi; and the surrounding lung areas were emphysematous, the interalveolar walls were thin and sclerotic, and the alveolar cavities were dilated. Intraperitoneal injection of yttrium chloride in animals caused peritonitis with serous or hemorrhagic ascites.[2] It was speculated that the development of ascites may have been related to the acidity of the administered solution rather than to the yttrium.[2]

Application of a 0.1 M solution of yttrium chloride to the eyes of rabbits caused no injury; similar exposure of eyes from which the corneal epithelium had been removed resulted in immediate slight haziness of the cornea, which subsequently became opaque and vascularized.[3]

REFERENCES

1. Stokinger HE: The metals. In Clayton GD, Clayton FE (eds): Patty's Industrial Hygiene and Toxicology, 3rd ed, rev, Vol 2A, Toxicology, p 1682. New York, Wiley–Interscience, 1981
2. Steffee CH: Histopathologic effects of rare earths administered intraperitoneally to rats. AMA Arch Ind Health 20:414–419, 1959
3. Grant WM: Toxicology of the Eye, 3rd ed, p 986. Springfield, Illinois, Charles C Thomas, 1986

ZINC CHLORIDE FUME

CAS: 7646-85-7

$ZnCl_2$ 1987 TLV = 1 mg/m^3

Synonym: Zinc dichloride fume

Physical Form. Fume

Uses. Smoke generators; flux in soldering

Exposure. Inhalation

Toxicology. Zinc chloride fume is an irritant of the eyes, mucous membranes, and skin, and, with very high concentrations, it causes pulmonary edema.

Ten deaths and 25 cases of nonfatal injury occurred among 70 persons exposed to a high concentration of zinc chloride released from smoke generators. Presenting symptoms included conjunctivitis (two cases with burns of the corneas), irritation of the nose and throat, cough with copious sputum, dyspnea, constrictive sensation in the chest, stridor, retrosternal pain, nausea, epigastric pain, and cyanosis.[1]

Of the 10 fatalities, a few persons died immediately or in a few hours with pulmonary edema, whereas those who survived longer developed bronchopneumonia.[1] Between the second and fourth days after exposure, almost all cases developed moist adventitious sounds in the lungs, and the majority continued to present a pale, cyanotic color. A prominent feature was the disparity between the severe symptoms and the paucity of physical signs in the lungs. Recovery occurred within 1 to 6 weeks after the incident.[1]

In a firefighter who was fatally exposed to a high but undetermined concentration of zinc chloride fume from a smoke generator, presenting symptoms were nausea, sore throat, and chest tightness aggravated by deep inspiration.[2] The patient improved initially but developed tachypnea, substernal soreness, fever, cyanosis, and coma. The lung fields were clear on auscultation despite diffuse pulmonary infiltrations seen on the chest roentgenogram. Death occurred 18 days after exposure, and autopsy revealed active fibroblastic proliferation of lung tissue and cor pulmonale.

An outdoor exposure to zinc chloride aerosol following the detonation of a smoke bomb in an airport disaster drill resulted in upper respiratory tract irritative symptoms in the victims, correlating with the presumed intensity and duration of exposure.[3] Questionnaire responses from 81 exposed persons most commonly reported cough, hoarseness, and sore throat, with onset primarily at the time of exposure. Other common symptoms among people with self-reported moderate and heavy exposures included listlessness, metallic taste, light-headedness, chest tightness, and soreness in the chest. Wheezing was relatively uncommon, and, by spirometry, 1 to 2 days after exposure, the mean results for FEV_1 and FVC as a percentage of predicted were normal. The predominance of upper respiratory symptoms was attributed to the solubility and hygroscopic tendency of zinc chloride, resulting in upper respiratory tract deposition. Upon dissolution of zinc chloride, both hydrochloric acid and zinc oxychloride are formed, contributing to the corrosive action. Most of the exposed victims became asymptomatic within 48 hours, but symptoms persisted in a few patients for up to several weeks.[3]

Accidental installation in a human eye of one drop of a 50% zinc chloride solution caused immediate severe pain, which persisted despite immediate irrigation with water. The corneal epithelium was burned, and corneal vascularization followed. After many weeks, areas of opacification and vascularization remained in the cornea.[4] Zinc chloride has caused ulceration of the fingers, hands, and forearms of workers who used it as flux in soldering.[5]

Zinc is an essential element in human metabolism; the normal intake of zinc in food is 10 to 15 mg/day, and urinary excretion is 0.3 to 0.4 mg/24 hours.[6]

Injection of zinc chloride solution into the testes of 49 Syrian hamsters resulted in areas of necrosis occupying approximately 25% of each testis; two embryonal carcinomas of the testis were found 10 weeks later at necropsy.[7]

Treatment. If eyes are irritated, immediately flush with water and then further irrigate with 1.7% EDTA for 15 minutes, starting the EDTA within 2 minutes after injury if possible.[8] Flush skin with water if it is contaminated.

REFERENCES

1. Evans EH: Casualties following exposure to zinc chloride smoke. Lancet 2:368–370, 1945
2. Milliken JA, Waugh D, Kadish ME: Acute interstitial pulmonary fibrosis caused by a smoke bomb. Can Med Assoc J 88:36–39, 1963
3. Schenker M et al: Acute upper respiratory symptoms resulting from exposure to zinc chloride aerosol. Environ Res 25:317–324, 1981
4. Grant WM: Toxicology of the Eye, 3rd ed, pp 986–987. Springfield, Illinois, Charles C Thomas, 1986
5. Stokinger HE: The metals (excluding lead). In Fassett DW, Irish DD (eds): Patty's Industrial Hygiene and Toxicology, 2nd ed, Vol 2, Toxicology, pp 1182–1188. New York, Wiley–Interscience, 1963
6. Vallee BL: Zinc and its biological significance. AMA Arch Ind Health 16:147–154, 1957
7. Guthrie J, Guthrie OA: Embryonal carcinomas in Syrian hamsters after intratesticular inoculation of zinc chloride during seasonal testicular growth. Cancer Res 34:2612–2613, 1974
8. Johnstone MA, Sullivan WR, Grant WM: Experimental zinc chloride ocular injury and treatment with disodium edetate. Am J Ophthalmol 76:137–142, 1973

ZINC OXIDE

CAS: 1314-13-2

ZnO 1987 TLV = 5 mg/m^3—fume
= 10 mg/m^3—total dust

Synonyms: None

Physical Form. Fume; dust

Uses. Metallic zinc in galvanizing, electroplating, in dry cells, alloying; zinc oxide in pigments

Exposure. Inhalation

Toxicology. Inhalation of zinc oxide fume causes an influenza-like illness termed *metal fume fever.*[1,2]

During human exposure to zinc oxide fume, effects include dryness and irritation of the throat, a sweet or metallic taste, substernal tightness and constriction in the chest, and a dry cough.[3] Several hours following exposure, the subject develops chills, lassitude, malaise, fatigue, frontal headache, low back pain, muscle cramps, and, occasionally, blurred vision, nausea, and vomiting.[3] Physical signs include fever, perspiration, dyspnea, rales through the chest, and tachycardia; in some instances, there has been a reversible reduction in pulmonary vital capacity. There is usually leukocytosis, which may reach 12,000 to 16,000/cmm.[2]

An attack usually subsides after 6 to 12 hours but may last for up to 24 hours; recovery is usually complete.[3] Most workers develop an immunity to these attacks, but it is quickly lost; attacks tend to be more severe on the first day of the workweek.[3] Only freshly formed fume causes the illness, presumably because flocculation occurs in the air with formation of larger particles that are deposited in the upper respiratory tract and do not penetrate deeply into the lungs.[4] Chills have been reported in workers from exposure to concentrations of zinc oxide fume below 5 mg/m^3.[3]

Zinc is an essential element in human metabolism; the normal intake of zinc in food is 10 to 15 mg/day, and the average urinary excretion is 0.3 to 0.4 mg/24 hours; 0.6 to 0.7 mg zinc/liter of urine have been found in workers exposed to zinc oxide fume in concentrations between 3 and 5 mg/m^3.[5]

A short-term study of guinea pigs exposed to zinc oxide fume for 3 hours/day for 6 days at the TLV of 5 mg/m^3 revealed pulmonary function changes and morphologic evidence of small airway inflammation and edema. Pulmonary flow resistance increased, compliance decreased, and lung volume and carbon monoxide diffusing capacity decreased. Some of these changes persisted for the 72-hour duration of postexposure follow-up.[6]

The dust of zinc oxide is considered a nuisance dust that has little adverse effect on the lung and does not produce significant organic disease when exposures are kept under reasonable control.[7]

REFERENCES

1. National Institute for Occupational Safety and Health: Criteria for a Recommended Standard . . . Occupational Exposure to Zinc Oxide. DHEW (NIOSH) Pub No 76–104, pp 36–38. Washington, DC, US Government Printing Office, 1975
2. McCord CP: Metal-fume fever as an immunological disease. Ind Med Surg 29:101–107, 1960
3. Rohrs LC: Metal-fume fever from inhaling zinc oxide. AMA Arch Ind Health 16:42–47, 1957
4. Hygienic Guide Series: Zinc oxide. Am Ind Hyg Assoc J 30:422–424, 1969
5. Vallee BL: Zinc and its biological significance. AMA Arch Ind Health 16:147–154, 1957
6. Lam H et al: Functional and morphologic changes in the lungs of guinea pigs exposed to freshly generated ultrafine zinc oxide. Toxicol Appl Pharmacol 78:29–38, 1985
7. Methyl hydrazine. Documentation of the TLVs and BEIs, 5th ed, pp 645–646. Cincinnati, American Conference of Government Industrial Hygienists (ACGIH), 1986

ZIRCONIUM COMPOUNDS

CAS: 7440-67-2

Zr 1987 TLV = 5 mg/m^3

Synonyms: Zirconium dioxide; zirconium silicate; zirconium tetrachloride

Physical Form. Solids

Uses. Structural material for atomic reactors; ingredient of priming and explosive mixtures; reducing agents; pigment; textile water repellent

Exposure. Inhalation

Toxicology. Zirconium compounds are of low toxicity, although granulomas have been produced by repeated topical applications of zirconium salts to human skin.

A study of 22 workers exposed to fume from a zirconium reduction process for 1 to 5 years revealed no abnormalities related to the exposure.[1] There are no well-documented cases of toxic effects from industrial exposure. Two persons given zirconium malate in 50-mg intravenous injections developed vertigo.[2] Granulomas of the human axillary skin have occurred from use of deodorants or poison-ivy remedies containing zirconium.[3]

In rats, the oral LD_{50} of several zirconium compounds ranged from 1.7 to 10 g/kg.[4] Animals acutely poisoned by zirconium compounds show progressive depression and decrease in activity until death.[2] Repeated inhalation of zirconium tetrachloride mist by dogs for 2 months at 6 mg zirconium/m^3 caused slight decreases in hemoglobin and in erythrocyte counts, with some increased mortality over that of controls. These effects may have been due to the liberation of hydrogen chloride.[4] Animals exposed to zirconium dioxide dust for 1 month at 75 mg zirconium/m^3 showed no detectable effects. Rats exposed to high concentrations of zirconium silicate dust for 7 months developed radiographic shadows in the lungs; these were attributed solely to the deposition of the radiopaque particles, since histologic examination showed no cellular reaction. The addition of 5 ppm of zirconium as the sulfate to the drinking water of mice for their lifetime did not increase the incidence of tumors.[5]

REFERENCES

1. Reed CE: A study of the effects on the lung of industrial exposure to zirconium dusts. AMA Arch Ind Health 13:578–580, 1956
2. Smith EC, Carson BL: Trace Metals in the Environment, Vol 3, Zirconium, p 405. Ann Arbor, Michigan, Ann Arbor Science Publishers, 1978
3. Shelley WB, Hurley HJ: The allergic origin of zirconium deodorant granulomas. Br J Dermatol 70:75–101, 1958
4. Stokinger HE: The metals. In Clayton GD, Clayton FE (eds): Patty's Industrial Hygiene and Toxicology, 3rd ed, rev, Vol 2A, Toxicology, pp 2049–2059. New York, Wiley–Interscience, 1981
5. Kanisawa M, Schroeder HA: Life term studies on the effect of trace elements on spontaneous tumors in mice and rats. Cancer Res 29:892–895, 1969

Appendix

NIOSH Recommendations for Occupational Safety and Health Standards (September 1986)*

INTRODUCTION

Acting under the authority of the Occupational Safety and Health Act of 1970 (Public Law 91-596), the National Institute for Occupational Safety and Health (NIOSH) develops, and periodically revises, recommendations for limits of exposure to potentially hazardous substances or conditions in the workplace. It also recommends preventive measures designed to reduce or eliminate adverse health effects associated with these hazards. In formulating these recommendations, NIOSH evaluates all known and available scientific information relevant to the potential hazard. These recommendations are then published and transmitted to the Department of Labor, Occupational Safety and Health Administration (OSHA) or Mine Safety and Health Administration (MSHA) for use in promulgating legal standards.

NIOSH recommendations are published in a variety of documents. Criteria Documents specify a NIOSH Recommended Exposure Limit (REL) and appropriate preventive measures designed to reduce or eliminate adverse health effects. Special Hazard Reviews, Occupational Hazard Assessments, and Technical Guidelines are other types of documents published by NIOSH to complement the Institute's recommendations for standards. These documents provide assessments, from a safety and health standpoint, of specific problems associated with a given agent or hazard and recommend appropriate control and surveillance methods.

Although these documents do not supplant the more comprehensive Criteria Document, they are prepared in such a way as to assist OSHA in the formulation of regulations. NIOSH also periodically presents testimony before various Congressional committees and at regulatory hearings convened by

* U.S. Department of Health and Human Services, Public Health Service, National Institute for Occupational Safety and Health, Centers for Disease Control, Atlanta, Georgia 30333

OSHA. The testimony presented always includes the current NIOSH policy concerning the particular hazard in question.

NIOSH also publishes documents known as Current Intelligence Bulletins (CIB) which review and evaluate emerging information on occupational hazards. These bulletins are based on the rapid evaluation of new and changing information on a particular hazard in light of existing knowledge.

The "NIOSH Recommendations for Occupational Safety and Health Standards" are based on existing NIOSH policy as previously published in any of the forms listed above. The intent of this table is to provide, in rapid-reference form, the most recent NIOSH REL or other recommendation for each potential hazard. The current OSHA Permissible Exposure Limit (PEL) or standard is also presented. Unless otherwise noted in the table, the NIOSH recommendations were originally published in Criteria Documents.

Definitions of abbreviations and terms used in this publication:

Action level	the exposure concentration at which certain provisions of the NIOSH recommended standard must be initiated, such as periodic measurements of worker exposure, training of workers, and medical surveillance
Ca	NIOSH recommends that the substance be treated as a potential human carcinogen
CFR	Code of Federal Regulations
CIB	Current Intelligence Bulletin
dBA	decibel, weighted according to the A scale, which approximates the response of the human ear
ECG	electrocardiogram
J/cm^2	joules per square centimeter
μm	micrometer
$\mu g/m^3$	micrograms per cubic meter
mg/m^3	milligrams per cubic meter
mppcf	millions of particles per cubic foot
mW/cm^2	milliwatts per square centimeter
nm	nanometer
NIOSH	National Institute for Occupational Safety and Health
OSHA	Occupational Safety and Health Administration
PCB's	polychlorinated biphenyls
PCDD's	polychlorinated dibenzo-*p*-dioxins
PCDF's	polychlorinated dibenzofurans
PEL	Permissible Exposure Limit (OSHA)
ppb	parts per billion
ppm	parts per million
REL	Recommended Exposure Limit (NIOSH)
(Skin)	potential contribution to overall exposure by the cutaneous route including mucous membranes and eyes
TCDD	2,3,7,8-tetrachlorodibenzo-*p*-dioxin
TWA	time-weighted average

SUMMARY OF OSHA REGULATIONS AND NIOSH RECOMMENDATIONS FOR OCCUPATIONAL SAFETY AND HEALTH STANDARDS, 1986

		NIOSH		
Potential Hazard*	**OSHA PEL'S/ Standard**	**REL'S†/Other Recommendations**	**Health Effect(s) Considered§**	**Comments**
2-Acetylamino-fluorene (NIOSH testimony at OSHA hearing September 1973)	No PEL Cancer-suspect agent Stringent workplace controls, recordkeeping, and medical surveillance required 29 CFR 1910.1014	Ca Use 29 CFR 1910.1014	Potential for cancer in humans; produced tumors of the liver, bladder, lungs, pancreas, and skin in animals	None
Acetylene (July 1976)	2,500 ppm (10% of lower explosive limit)	No exposure > 2,500 ppm (2,662 mg/m^3)	Asphyxia	Employers to check for and inform workers of contaminants such as arsine and phosphine
Acrylamide (October 1976)	0.3 mg/m^3, 8-hr TWA (Skin)	0.3 mg/m^3 TWA	Skin, eye, and nervous system effects	Prevent skin and eye contact
Acrylonitrile (January 1978; revised March 1978 as part of NIOSH testimony at OSHA hearing)	2 ppm, 8-hr TWA; 10 ppm ceiling (15 min) (Skin) 29 CFR 1910.1045	Ca 1 ppm, 8-hr TWA; 10 ppm ceiling (15 min) (Skin)	Brain tumors; lung and bowel cancer	Chest x ray required; first-aid and medical kits to be available during use; prevent skin contact
Aldrin/dieldrin (Special Hazard Review September 1978)	0.25 mg/m^3, 8-hr TWA (Skin)	Ca Lowest reliably detectable level	Potential for cancer in humans; produced tumors of the lungs, liver, thyroid, and adrenal glands in animals	Aldrin/dieldrin no longer produced in U.S.; prevent skin contact

*Date recommendation was published or testimony presented is in parentheses.

†NIOSH TWA recommendations are based on exposures up to 10 hours unless otherwise noted.

§Unless otherwise noted health effects cited are for humans.

Potential Hazard*	OSHA PEL'S/ Standard	NIOSH		
		REL'S[†]/Other Recommendations	Health Effect(s) Considered[§]	Comments
Alkanes (C5-C8) (March 1977)	All are 8-hr TWA values: Pentane: 1,000 ppm (2,950 mg/m^3); n-hexane: 500 ppm (1,800 mg/m^3); n-heptane: 500 ppm (2,000 mg/m^3); octane: 500 ppm (2,350 mg/m^3)	All are TWA values: Pentane: 120 ppm (350 mg/m^3); hexane: 100 ppm (350 mg/m^3); heptane: 85 ppm (350 mg/m^3); octane: 75 ppm (350 mg/m^3); Mixtures not to exceed 350 mg/m^3 TWA; All are ceiling values (15 min) singly or mixtures: pentane: 610 ppm (1,800 mg/m^3); hexane: 510 ppm (1,800 mg/m^3); heptane: 440 ppm (1,800 mg/m^3); octane: 385 ppm (1,800 mg/m^3); Action level set at 200 mg/m^3 for these substances	Skin and nervous system effects	None
Allyl chloride (September 1976)	1 ppm (3 mg/m^3), 8-hr TWA	1 ppm (3.1 mg/m^3) TWA; 3 ppm (9.3 mg/m^3) ceiling (15 min)	Liver, kidney, and lung effects	Urine, blood, and pulmonary function testing required
4-Aminodiphenyl (NIOSH testimony at OSHA hearing September 1973)	No PEL Cancer-suspect agent 29 CFR 1910.1011 Stringent workplace controls, recordkeeping, and medical surveillance required	Ca Use 29 CFR 1910.1011	Bladder cancer	None
Ammonia (July 1974)	50 ppm (35 mg/m^3), 8-hr TWA	50 ppm (34.8 mg/m^3) ceiling (5 min)	Respiratory irritation	Prevent eye contact
Anesthetic gases (see Waste anesthetic gases)				

Animal rendering processes (Occupational Hazard Assessment March 1981)	OSHA PEL's or NIOSH REL's for specific hazards are applicable		Mechanical injury; burns; heat stress; infections from biologic agents; chemical hazards	Guidelines for engineering controls and work practices to reduce injury and illness presented
Antimony (September 1978)	0.5 mg Sb/m^3, 8-hr TWA	0.5 mg Sb/m^3 TWA	Irritation; cardiovascular and lung effects	Chest x ray, pulmonary function testing, and electrocardiogram required
Arsenic, inorganic (September 1974; revised June 1975; reaffirmed July 1982 as part of NIOSH testimony at OSHA hearing)	10 μg As/m^3, 8-hr TWA 29 CFR 1910.1018	Ca 2 μg As/m^3 ceiling (15 min)	Lung and lymphatic cancer; dermatitis	Chest x ray required
Arsine (CIB August 1979)	0.2 mg/m^3 (0.05 ppm), 8-hr TWA	Ca 2 μg As/m^3 (0.002 mg As/m^3) ceiling (15 min)	Sudden extensive hemolysis	Workers to be warned of working with arsenic compounds in presence of freshly formed hydrogen
Asbestos (January 1972; revised December 1976; revised March 1984 as part of NIOSH testimony at Congressional hearing; [Continued on next page]	200,000 $fibers/m^3$, over 5 μm in length, 8-hr TWA; Action level of 100,000 $fibers/m^3$, 8-hr TWA 29 CFR 1910.1001	Ca 100,000 $fibers/m^3$, over 5 μm in length, 8-hr TWA in a 400-liter air sample	Lung cancer; mesothelioma; asbestosis	Chest x ray and pulmonary function testing required

*Date recommendation was published or testimony presented is in parentheses.

†NIOSH TWA recommendations are based on exposures up to 10 hours unless otherwise noted.

§Unless otherwise noted health effects cited are for humans.

		NIOSH		
Potential Hazard*	OSHA PEL'S/ Standard	REL'S[†]/Other Recommendations	Health Effect(s) Considered[§]	Comments
reaffirmed June 1984 as NIOSH testimony at OSHA hearing)				
Asphalt fumes (September 1977)	None	5 mg/m^3 ceiling, measured as total particulate (15 min)	Eye and respiratory irritation	Medical surveillance required; prevent skin contact
Benzene (July 1974; revised August 1976; revised July 1977 as part of NIOSH testimony at OSHA hearing; revised March 1986 as part of NIOSH testimony at OSHA hearing)	10 ppm, 8-hr TWA; 25 ppm acceptable ceiling; 50 ppm maximum ceiling (10 min)	Ca 0.1 ppm (0.32 mg/m^3) 8-hr TWA; 1 ppm (3.2 mg/m^3) ceiling (15 min)	Cancer (leukemia)	Prevent skin contact
Benzidine (NIOSH testimony at OSHA hearing September 1973)	No PEL Cancer-suspect agent 29 CFR 1910.1010 Stringent workplace controls, recordkeeping, and medical surveillance required	Ca Use 29 CFR 1910.1010	Bladder, liver, and kidney cancer	None

Benzidine-based dyes (Special Hazard Review January 1980; revised in "Preventing health hazards from . . . benzidine congener dyes" January 1983)	No PEL for benzidine-based dyes	Ca Reduce exposure to lowest feasible level; replace with less toxic materials	Bladder cancer	Stringent workplace controls and medical surveillance required. Urine monitoring for benzidine suggested
Benzoyl peroxide (June 1977)	5 mg/m^3, 8-hr TWA	5 mg/m^3 TWA	Respiratory and eye irritation; skin effects	None
Benzyl chloride (August 1978)	5 mg/m^3 (1 ppm), 8-hr TWA	5 mg/m^3 ceiling (15 min)	Irritation; skin and eye effects	Chest x ray and pulmonary function testing required
Beryllium (June 1972; revised August 1977 as part of NIOSH testimony at OSHA hearing)	2 $\mu g/m^3$, 8-hr TWA; 5 $\mu g/m^3$ acceptable ceiling; 25 $\mu g/m^3$ maximum ceiling (30 min)	Ca Not to exceed 0.5 μg Be/m^3	Lung cancer	Pulmonary function testing and chest x ray required
Boron trifluoride (December 1976)	1 ppm (3 mg/m^3) ceiling	No exposure limit recommended due to the absence of a reliable monitoring method	Respiratory effects	Appropriate engineering and work-practice controls to reduce exposure to lowest feasible level; pulmonary function testing required
1,3-Butadiene (CIB February 1984)	1,000 ppm (2,200 mg/m^3), 8-hr TWA	Ca Reduce exposure to lowest feasible level	Hematopoietic cancer; teratogenicity; reproductive system effects	Appropriate engineering and work-practice controls; restrict access to areas where 1,3-butadiene is used

*Date recommendation was published or testimony presented is in parentheses.

†NIOSH TWA recommendations are based on exposures up to 10 hours unless otherwise noted.

§Unless otherwise noted health effects cited are for humans.

Potential Hazard*	OSHA PEL'S/ Standard	NIOSH		
		REL'S[†]/Other Recommendations	Health Effect(s) Considered[§]	Comments
Cadmium (August 1976; revised in CIB September 1984)	Fume: 0.1 mg/m^3, 8-hr TWA; 0.3 mg/m^3 ceiling; dust: 0.2 mg/m^3, 8-hr TWA; 0.6 mg/m^3 ceiling	Ca Reduce exposure to lowest feasible level	Lung cancer, prostatic cancer, renal system effects	None
Carbaryl (September 1976)	5 mg/m^3, 8-hr TWA	5 mg/m^3 TWA	Central nervous system and reproductive system effects	Workers to be warned of possible effects on reproductive system and to have only minimum exposure during pregnancy; prevent skin and eye contact
Carbon black (September 1978)	3.5 mg/m^3, 8-hr TWA	3.5 mg/m^3 TWA; Ca 0.1 mg/m^3 TWA in presence of polycyclic aromatic hydrocarbons	Lung, cardiovascular and skin effects; cancer of the lymphatic-bone marrow complex when exposed to carbon black in the presence of polycyclic aromatic hydrocarbons	Chest x rays, pulmonary function testing, and ECG required
Carbon dioxide (August 1976)	5,000 ppm (9,000 mg/m^3), 8-hr TWA	10,000 ppm (18,000 mg/m^3) TWA; 30,000 ppm (54,000 mg/m^3) ceiling (10 min)	Respiratory effects	None
Carbon disulfide (May 1977)	20 ppm, 8-hr TWA; 30 ppm acceptable ceiling; 100 ppm maximum ceiling (30 min)	1 ppm (3 mg/m^3) TWA; 10 ppm (30 mg/m^3) ceiling (15 min)	Cardiovascular, central nervous system, and reproductive system effects	Workers to be advised of potential effects on reproductive system
Carbon monoxide (August 1972)	50 ppm (55 mg/m^3), 8-hr TWA	35 ppm (40 mg/m^3), 8-hr TWA; 200 ppm (229 mg/m^3) ceiling (no defined time)	Cardiovascular effects	None

Carbon tetrachloride (December 1975; revised June 1976)	10 ppm, 8-hr TWA; 25 ppm acceptable ceiling; 200 ppm maximum ceiling (5 min in 4 hr)	Ca 2 ppm (12.6 mg/m^3) ceiling in a 45-liter sample (60 min)	Liver cancer	REL based on lower limit of detection at time of document publication
Chlorine (May 1976)	1 ppm (3 mg/m^3) ceiling	0.5 ppm (1.45 mg/m^3) ceiling (15 min)	Eye and respiratory irritation	Chest x rays required
Chloroethane (CIB August 1978)	1,000 ppm (2,600 mg/m^3), 8-hr TWA	To be handled in the workplace with caution	Central nervous system effects; possible liver and/or kidney effects	Exposures should be minimized due to the structural similiarity to the carcinogenic chloroethanes
Chloroform (September 1974; revised June 1976)	50 ppm (240 mg/m^3) ceiling	Ca 2 ppm (9.78 mg/m^3) ceiling in a 45-liter sample (60 min)	Liver or kidney tumors and central nervous system effects	None
bis-Chloromethyl ether (NIOSH testimony at OSHA hearing September 1973)	No PEL Cancer-suspect agent Stringent workplace controls, recordkeeping, and medical surveillance required 29 CFR 1910.1008	Ca Use 29 CFR 1910.1008	Lung cancer	None
Chloroprene (August 1977)	25 ppm (90 mg/m^3), 8-hr TWA	Ca 1 ppm (3.6 mg/m^3) ceiling (15 min)	Lung and skin cancer; reproductive effects	Chest x ray and pulmonary function testing required; pregnant workers to be counseled about continuing work with chloroprene
Chromic acid (July 1973; revised—see Chromium (VI) December 1975)	1 mg/10 m^3 (100 μg/m^3) ceiling	25 μg/m^3 (0.025 mg/m^3) TWA; 50 μg/m^3 (0.05 mg/m^3) ceiling (15 min) as noncarcinogenic Cr (VI)	Nasal ulceration	None

*Date recommendation was published or testimony presented is in parentheses.

†NIOSH TWA recommendations are based on exposures up to 10 hours unless otherwise noted.

§Unless otherwise noted health effects cited are for humans.

		NIOSH		
Potential Hazard*	OSHA PEL'S/ Standard	REL'S†/Other Recommendations	Health Effect(s) Considered§	Comments
Chromium (VI) (December 1975)	1 mg/10 m^3 (100 $\mu g/m^3$) ceiling	Ca Carcinogenic Cr (VI): 1 $\mu g/m^3$ TWA; other Cr (VI): 25 $\mu g/m^3$ TWA; 50 $\mu g/m^3$ ceiling (15 min)	Lung cancer, skin ulcers, and lung irritation	Employer must demonstrate absence of carcinogenic Cr (VI); x ray required
Chrysene (Special Hazard Review June 1978)	0.2 mg/m^3, 8-hr TWA	Ca To be controlled as an occupational carcinogen	Potential for cancer in humans; produced liver and skin tumors in animals	Document also contains control recommendations for polycyclic aromatic hydrocarbons
Coal gasification plants (September 1978)	OSHA PEL's or NIOSH REL's for specific hazards are applicable		Various effects depending on substances present; potential for skin cancer	Extensive work-practice and control procedures recommended
Coal liquefaction, volumes I and II (Occupational Hazard Assessment March 1981)	OSHA PEL's or NIOSH REL's for specific hazards are applicable		Various effects depending on substances present; potential for skin cancer	Extensive work-practice and control procedures recommended
Coal tar products (September 1977)	0.2 mg/m^3, 8-hr TWA (benzene-soluble fraction) 29 CFR 1910.1002 (coal tar pitch volatiles)	Ca 0.1 mg/m^3 TWA (cyclohexane-extractable fraction)	Lung and skin cancer	Includes coal tar, creosote, and coal-tar pitch; pulmonary function testing and chest x rays required

Cobalt (Occupational Hazard Assessment October 1981)	0.1 mg/m^3, 8-hr TWA	NIOSH has concluded that there is insufficient evidence to warrant recommending a new exposure limit	Dermatitis; potential for pulmonary fibrosis	Includes recommendations for engineering controls, work practices, protective equipment, worker education, monitoring, and medical surveillance
Coke oven emissions (February 1973; revised November 1975 as part of NIOSH testimony at OSHA hearing)	150 $\mu g/m^3$, 8-hr TWA 29 CFR 1910.1029	Ca 0.5-0.7 mg/m^3 (500-700 $\mu g/m^3$) (total particulates) as screening level	Lung cancer	Chest x ray required; work practices to minimize exposure to emissions
Confined spaces, working in (December 1979)	Covered under numerous OSHA regulations for General Industry (29 CFR 1910)	Various recommendations including a permit system to prevent worker injury and death	Injury and death	None
Cotton dust (September 1974; reaffirmed September 1983 as part of NIOSH testimony at OSHA hearing)	Yarn manufacturing: 200 $\mu g/m^3$, 8-hr TWA; slashing and weaving operations: 750 $\mu g/m^3$, 8-hr TWA; all other operations: 500 $\mu g/m^3$, 8-hr TWA 29 CFR 1910.1043	<200 $\mu g/m^3$ lint-free cotton dust	Pulmonary disease (byssinosis)	Pulmonary function testing required
Cresol (February 1978)	5 ppm (22 mg/m^3), 8-hr TWA (Skin)	2.3 ppm (10 mg/m^3) TWA	Skin, liver, kidney, and pancreas effects	Applies to mixtures of cresols and cresylic acid; prevent skin and eye contact; possible delayed effects

*Date recommendation was published or testimony presented is in parentheses.

†NIOSH TWA recommendations are based on exposures up to 10 hours unless otherwise noted.

§Unless otherwise noted health effects cited are for humans.

Potential Hazard*	OSHA PEL'S/ Standard	NIOSH REL'S[†]/Other Recommendations	NIOSH Health Effect(s) Considered[§]	NIOSH Comments
DDT (Special Hazard Review September 1978)	1 mg/m^3, 8-hr TWA (Skin)	Ca Lowest reliably detectable level; 0.5 mg/m^3 TWA by NIOSH-validated method	Potential for cancer in humans; produced tumors of the liver, lungs, and lymphatic system in animals	Prevent skin contact
2,4-Diaminoanisole and its salts (CIB January 1978)	None	Ca Reduce exposure to lowest feasible level	Potential for cancer in humans; produced tumors of the thyroid, skin, and lymphatic system in animals	Prevent skin contact; engineering and work-practice controls are recommended
o-Dianisidine-based dyes (Joint NIOSH/OSHA Health Hazard Alert December 1980)	None	Ca Should be handled in the workplace with caution; exposures should be minimized	Potential for cancer in humans; produced tumors of the bladder, stomach, and mammary glands in animals	Substitute less toxic dyes wherever possible
Dibromochloropropane (January 1978)	1 ppb, 8-hr TWA; eye and skin contact to be avoided 29 CFR 1910.1044	10 ppb (0.1 mg/m^3) TWA (NIOSH recommendation superseded by OSHA standard promulgated in 1978)	Sterility; renal and liver effects	Regulated by OSHA as a carcinogen
3,3'-Dichlorobenzidine (NIOSH testimony at OSHA hearing September 1973)	No PEL Cancer-suspect agent Stringent workplace controls, recordkeeping, and medical surveillance required 29 CFR 1910.1007	Ca Use 29 CFR 1910.1007	Potential for cancer in humans; produced tumors of the liver, bladder, and lungs in animals	None

1,1-Dichloroethane (CIB August 1978)	100 ppm (400 mg/m^3), 8-hr TWA	To be handled in the workplace with caution	Central nervous system effects; possible liver and/or kidney damage	Exposures should be minimized due to the structural similiarity to the carcinogenic chloroethanes
Dieldrin (see Aldrin/dieldrin)				
Di-2-Ethylhexyl Phthalate (DEHP) (Special Hazard Review March 1983)	5 mg/m^3, 8-hr TWA	Ca Reduce exposure to lowest feasible level	Potential for cancer in humans; produced liver tumors in animals	DEHP, widely used in the quantitative fit testing of respirators, should be replaced with less toxic material such as refined corn oil
Diisocyanates (September 1978)	Toluene diisocyanate (TDI): 0.02 ppm (0.14 mg/m^3) ceiling; diphenylmethane diisocyanate (MDI): 0.02 ppm (0.2 mg/m^3) ceiling	All values given in $\mu g/m^3$ and all ceiling values for 10 min (each equivalent to 5 ppb TWA and 20 ppb ceiling): TDI: 35 TWA, 140 ceiling; MDI: 50 TWA, 200 ceiling; hexamethylene diisocyanate (HDI): 35 TWA, 140 ceiling; naphthalene diisocyanate (NDI): 40 TWA, 170 ceiling; isophorone diisocyanate (IPDI): 45 TWA, 180 ceiling; dicyclohexylmethane-4,4'-diisocyanate (hydrogenated MDI): 55 TWA, 210 ceiling; other diisocyanates to be controlled to 20 ppb ceiling and 5 ppb TWA	Respiratory effects and sensitization; pulmonary irritation	Chest x ray and pulmonary function testing required

*Date recommendation was published or testimony presented is in parentheses.
†NIOSH TWA recommendations are based on exposures up to 10 hours unless otherwise noted.
§Unless otherwise noted health effects cited are for humans.

Potential Hazard*	OSHA PEL'S/ Standard	NIOSH REL'S[†]/Other Recommendations	NIOSH Health Effect(s) Considered[§]	NIOSH Comments
4-Dimethylamino-azobenzene (NIOSH testimony at OSHA hearing September 1973)	No PEL Cancer-suspect agent Stringent workplace controls, recordkeeping, and medical surveillance required 29 CFR 1910.1015	Ca Use 29 CFR 1910.1015	Potential for cancer in humans; produced tumors of the liver and bladder in animals	None
Dinitro-ortho-cresol (February 1978)	0.2 mg/m^3, 8-hr TWA (Skin)	0.2 mg/m^3 TWA	Central nervous system and metabolic effects	Blood and urine monitoring required; prevent skin and eye contact; possible delayed effects
Dinitro-toluenes (CIB July 1985)	1.5 mg/m^3, 8-hr TWA (Skin)	Ca Reduce exposure to lowest feasible level	Potential for cancer in humans; produced tumors of the liver, skin, and kidneys in animals; reproductive system effects	Prevent skin contact
Dioxane (September 1977)	100 ppm (360 mg/m^3), 8-hr TWA (Skin)	Ca 1 ppm (3.6 mg/m^3) ceiling (30 min)	Potential for cancer in humans; produced tumors of liver, lungs, and nasal cavity in animals; effects on liver and kidney	Blood and urine testing required; prevent skin contact
Dioxin (see 2,3,7,8-Tetrachloro-dibenzo-*p*-dioxin)				
Elevated workstations, emergency egress from (December 1975)	Sections under Subpart E, Means of Egress, General Industry Standards, and Subpart R, Special Industries (29 CFR 1910.261)	Various recommendations concerning means and availability of egress	Trauma and injury	None

Epichlorohydrin (September 1976; revised in CIB October 1978)	5 ppm (19 mg/m^3), 8-hr TWA	Ca Occupational exposure to epichlorohydrin to be minimized	Respiratory cancer; mutagenesis; reproductive effects; skin, kidney, liver, and respiratory effects	Prevent skin contact
2-Ethoxy-ethanol (see Glycol ethers)				
Ethyl chloride (see Chloroethane)				
Ethylene dibromide (August 1977; revised November 1983; reaffirmed February 1984 as part of NIOSH testimony at OSHA hearing)	20 ppm, 8-hr TWA; 30 ppm acceptable ceiling; 50 ppm maximum peak (5 min)	Ca 0.045 ppm (0.38 mg/m^3), 8-hr TWA; 0.13 ppm (1 mg/m^3) ceiling (15 min)	Potential for cancer in humans; mutagenesis; damage to skin, eyes, cardiovascular, liver, spleen, reproductive, respiratory, and central nervous systems	Workers to be warned of potential for reproductive abnormalities and cancer; hazardous liquid; prevent skin contact
Ethylene dichloride (March 1976; revised in CIB April 1978; revised September 1978)	50 ppm, 8-hr TWA; 100 ppm acceptable ceiling; 200 ppm maximum ceiling (5 min in 3 hr)	Ca 1 ppm (4 mg/m^3) TWA; 2 ppm (8 mg/m^3) ceiling (15 min)	Potential for cancer in humans; nervous system, respiratory, cardiovascular, and liver effects	Nursing infants of exposed mothers at risk

*Date recommendation was published or testimony presented is in parentheses.

†NIOSH TWA recommendations are based on exposures up to 10 hours unless otherwise noted.

§Unless otherwise noted health effects cited are for humans.

Potential Hazard*	OSHA PEL'S/ Standard	NIOSH REL'S[†]/Other Recommendations	NIOSH Health Effect(s) Considered[§]	NIOSH Comments
Ethyleneimine (NIOSH testimony at OSHA hearing September 1973)	0.5 ppm (1 mg/m^3), 8-hr TWA (Skin) 29 CFR 1910.1012	Ca Use 29 CFR 1910.1012	Potential for cancer in humans; produced tumors of the liver and lung in animals	Stringent workplace controls and medical surveillance required
Ethylene oxide (Special Hazard Review September 1977; revised July 1983 as part of NIOSH testimony at OSHA hearing)	1 ppm (1.8 mg/m^3), 8-hr TWA 29 CFR 1910.1047	Ca 5 ppm (9 mg/m^3) ceiling (10 min/day); <0.1 ppm (0.18 mg/m^3), 8-hr TWA	Peritoneal cancer; leukemia; mutagenesis; reproductive effects	Blood monitoring and medical counseling recommended
Ethylene thiourea (Special Hazard Review October 1978)	None	Ca Should be used in encapsulated form in industry; worker exposure to be minimized	Potential for cancer and teratogenicity in humans; produced tumors of the liver, thyroid, and lymphatic system in animals	Workers to be informed of carcinogenic and teratogenic hazards; special attention to be given to thyroid function tests
Excavations, development of draft construction safety standards for (Technical Guideline May 1983)	Many aspects covered under OSHA regulations governing excavations, trenching, and shoring practices in the construction industry (29 CFR 1926, Subpart P)	Many work-practice recommendations concerning safety standards for excavations	Injury and death	None
Fibrous glass (April 1977)	Nuisance dust PEL applies, 15 mg/m^3 total dust; 5 mg/m^3 respirable fraction	3 million fibers/m^3 TWA (fibers $\leqslant$ 3.5 μm diameter and $\geqslant$ 10 μm length); 5 mg/m^3 TWA (total fibrous glass)	Eye, skin, and respiratory effects	NIOSH recommends that this REL also apply to other synthetic fibers

Fluorides, inorganic (June 1975)	2.5 mg F/m^3, 8-hr TWA	2.5 mg F/m^3 TWA	Kidney and bone effects	Urine monitoring required
Fluorocarbon polymers, decomposition products of (September 1977)	None	Various recommendations emphasizing good work practices, engineering controls, and medical surveillance	Lung effects; polymer fume fever	Workroom air to be monitored for inorganic fluorides and hydrogen fluoride
Formaldehyde (December 1976; revised in CIB April 1981; revised May 1986 as part of NIOSH testimony at OSHA hearing)	3 ppm, 8-hr TWA; 5 ppm acceptable ceiling; 10 ppm maximum ceiling (30 min)	Ca 0.1 ppm ceiling (15 min); represents the lowest reliably quantifiable concentration	Potential for cancer in humans; produced tumors of the nasal cavity in animals	Medical surveillance; skin protection
Foundries (September 1985)	Many aspects covered under the numerous OSHA regulations for general industry (29 CFR 1910)	Various recommendations emphasizing good work practices, engineering controls, and medical surveillance	Cancer; respiratory disease; heat-induced illness; noise-induced hearing loss; vibration-induced disorders; eye injuries; traumatic and ergonomic injuries	Recommendations limited to foundries that pour molten metal into sand molds
Furfuryl alcohol (March 1979)	50 ppm (200 mg/m^3), 8-hr TWA	50 ppm (200 mg/m^3) TWA	Respiratory effects	None

*Date recommendation was published or testimony presented is in parentheses.

†NIOSH TWA recommendations are based on exposures up to 10 hours unless otherwise noted.

§Unless otherwise noted health effects cited are for humans.

Potential Hazard*	OSHA PEL'S/ Standard	NIOSH REL'S[†]/Other Recommendations	NIOSH Health Effect(s) Considered[§]	NIOSH Comments
Glycidyl ethers (June 1978; revised in CIB October 1978)	All values in ppm (mg/m^3): allyl glycidyl ether (AGE): 10 (45) ceiling; n-butyl glycidyl ether (BGE): 50 (270), 8-hr TWA; di-2,3-epoxypropyl ether (DGE): 0.5 (2.8), 8-hr TWA; isopropyl glycidyl ether (IGE): 50 (240), 8-hr TWA; phenyl glycidyl ether (PGE): 10 (60), 8-hr TWA	All are ceiling values (15 min) in ppm (mg/m^3): AGE: 9.6 (45); BGE: 4.4 (30); DGE: 0.2 (1) Ca; IGE: 50 (240); PGE: 1 (5)	DGE: Potential for cancer in humans; produced skin tumors in animals; DGE and other glycidyl ethers: skin and mucous membrane effects; sensitization potential; possible hematopoietic and reproductive system effects	Possible additive effects with mixtures; medical surveillance
Glycol ethers (CIB May 1983)	2-Methoxyethanol: 25 ppm (80 mg/m^3), 8-hr TWA (Skin); 2-ethoxyethanol: 200 ppm (740 mg/m^3), 8-hr TWA (Skin)	Reduce exposure to lowest feasible level	Reproductive effects; teratogenicity	Prevent skin contact
Grain elevators and feed mills, occupational safety in (Technical Guideline September 1983; reaffirmed June 1984 as part of NIOSH testimony at OSHA hearing)	Many general aspects (e.g., protective equipment, dust control, etc.) covered under the numerous OSHA regulations for general industry (29 CFR 1910)	Various recommendations for control of combustible dusts and ignition sources, machine guarding, isolation and lockouts, bin entry, training, and personal protective equipment	Injury and death	Health hazards from exposure to fumigants, pesticides, and grain dust
Hexachloroethane (CIB August 1978)	1 ppm (10 mg/m^3), 8-hr TWA (Skin)	Ca Reduce exposure to lowest feasible level	Potential for cancer in humans; produced liver tumors in animals	None

Hot environments (June 1972; revised April 1986)	None	Sliding scale limits based upon environmental and metabolic heat loads	Heat-induced illnesses	Recommendations include acclimatization, strict work practices, protective equipment, and medical surveillance
Hydrazines (June 1978)	All values in ppm (mg/m^3): hydrazine: 1 (1.3), 8-hr TWA; 1,1-dimethyl-hydrazine: 0.5 (1.0), 8-hr TWA; phenylhydrazine: 5 (22), 8-hr TWA; methylhydrazine: 0.2 (0.35) ceiling	Ca All are ceiling values (120 min) in ppm (mg/m^3): hydrazine: 0.03 (0.04); 1,1-dimethylhydrazine: 0.06 (0.15); phenylhydrazine: 0.14 (0.6); methylhydrazine: 0.04 (0.08)	Potential for cancer in humans; produced tumors of the lung, liver, blood vessels, and intestines in animals; blood, liver, and skin effects	Blood and urine monitoring and chest x ray required; bowel examination for workers over age 40
Hydrogen cyanide and cyanide salts (October 1976)	Hydrogen cyanide: 10 ppm (11 mg/m^3), 8-hr TWA (Skin); cyanide: 5 mg CN/m^3, 8-hr TWA (Skin)	4.7 ppm (5 mg CN/m^3) ceiling (10 min)	Thyroid, blood, respiratory system effects	Concurrent measurement required for HCN when measuring for cyanide salt; trained first-aid personnel and first-aid kits to be available during use; prevent skin and eye contact
Hydrogen fluoride (March 1976)	3 ppm, 8-hr TWA	3 ppm (2.5 mg F/m^3) TWA; 6 ppm (5.0 mg F/m^3) ceiling (15 min)	Skin, eye, and airway irritation; bone effects	Pelvic x ray (male workers only) and urine testing required
Hydrogen sulfide (May 1977)	20 ppm acceptable ceiling; 50 ppm maximum ceiling (10 min)	10 ppm (15 mg/m^3) ceiling (10 min)	Irritation; severe acute effects involving nervous and respiratory systems	Continuous monitoring required if potential exists for exposure to $\geq$ 70 mg/m^3 (47 ppm); evacuation required at this level

*Date recommendation was published or testimony presented is in parentheses.
†NIOSH TWA recommendations are based on exposures up to 10 hours unless otherwise noted.
§Uniess otherwise noted health effects cited are for humans.

Potential Hazard*	OSHA PEL'S/ Standard	NIOSH REL'S[†]/Other Recommendations	NIOSH Health Effect(s) Considered[§]	NIOSH Comments
Hydroquinone (April 1978)	2 mg/m^3, 8-hr TWA	0.44 ppm (2 mg/m^3) ceiling (15 min)	Eye and skin effects	Special provisions for darkroom use
Identification system for occupationally hazardous materials (December 1974)	Sections of Hazard Communication (29 CFR 1910.1200) and carcinogen standards may be applicable	Complete designation system for occupationally hazardous materials	None	Includes definition, safety data sheets, alert symbols, and label statements
Isopropyl alcohol (March 1976)	400 ppm (980 mg/m^3), 8-hr TWA	400 ppm (984 mg/m^3) TWA; 800 ppm (1,968 mg/m^3) ceiling (15 min)	Mucous membrane irritation; possible cancer threat in manufacturing process	Stringent work practices and medical surveillance for manufacturing workers required
Kepone (January 1976)	None	Ca 1 μg/m^3 TWA	Liver cancer; nervous system effects	Liver function testing required
Ketones (June 1978)	All are 8-hr TWA values in ppm (mg/m^3): acetone: 1,000 (2,400); methyl ethyl ketone: 200 (590); methyl n-propyl ketone: 200 (700); methyl n-butyl ketone: 100 (410); methyl n-amyl ketone: 100 (465); methyl isobutyl ketone: 100 (410); methyl isoamyl ketone: none; diisobutyl ketone: 50 (290); cyclohexanone: 50 (200); mesityl oxide: 25 (100); diacetone alcohol: 50 (240); isophorone: 25 (140)	All are TWA values in ppm (mg/m^3): acetone: 250 (590); methyl ethyl ketone: 200 (590); methyl n-propyl ketone: 150 (530); methyl n-butyl ketone: 1 (4); methyl n-amyl ketone: 100 (465); methyl isobutyl ketone: 50 (200); methyl isoamyl ketone: 50 (230); diisobutyl ketone: 25 (140); cyclohexanone: 25 (100); mesityl oxide: 10 (40); diacetone alcohol: 50 (240); isophorone: 4 (23)	Irritation; liver, kidney, and nervous system effects	Urinalysis required; workers exposed to methyl n-butyl ketone to be warned of nervous system effects

Land-based oil and gas well drilling, comprehensive safety recommendations for (Technical Guideline September 1983; reaffirmed March 1984 as part of NIOSH testimony at OSHA hearing)	Many aspects covered under the numerous OSHA regulations for general industry (29 CFR 1910)	Various recommendations for safe work practices and technologic improvements	Injury and death	Many tasks, types of equipment, and conditions are not covered by existing regulations
Lead, inorganic (January 1973; revised May 1978)	50 μg Pb/m^3, 8-hr TWA; over 8-hr exposure to be determined by formula 29 CFR 1910.1025	<100 μg Pb/m^3 TWA; air level to be maintained so that worker blood lead remains ≤ 60 $\mu g/100g$	Kidney, blood, and nervous system effects	Blood monitoring required
Lockout/tagout, guidelines for controlling hazardous energy during maintenance and servicing (Technical Guideline September 1983)	Many aspects covered under OSHA regulations for general industry (29 CFR 1910) and construction standards (29 CFR 1926)	Work-practice recommendations for controlling hazardous energy during maintenance and servicing activities	Injury and death	"Energy" defined in this document as kinetic energy, potential energy, electrical energy, and thermal energy
Logging from felling to first haul (July 1976)	None	Extensive work-practice and personal protection recommendations	Primarily trauma and falls	Tetanus toxoid inoculations and first-aid programs to be instituted
Malathion (June 1976)	15 mg/m^3, 8-hr TWA	15 mg/m^3 TWA	Nervous system effects	Prevent skin contact; blood monitoring required

*Date recommendation was published or testimony presented is in parentheses.

†NIOSH TWA recommendations are based on exposures up to 10 hours unless otherwise noted.

§Unless otherwise noted health effects cited are for humans.

Potential Hazard*	OSHA PEL'S/ Standard	NIOSH		
		REL'S[†]/Other Recommendations	Health Effect(s) Considered[§]	Comments
Mercury, inorganic (August 1973)	0.1 mg/m^3 acceptable ceiling	0.05 mg Hg/m^3, 8-hr TWA	Central nervous system and mental effects	Work practices, sanitation, monitoring, and medical surveillance emphasized
2-Methoxy-ethanol (see Glycol ethers)				
Methyl alcohol (March 1976)	200 ppm (260 mg/m^3), 8-hr TWA	200 ppm (262 mg/m^3) TWA; 800 ppm (1,048 mg/m^3) ceiling (15 min)	Blindness; metabolic acidosis	None
Methyl chloro-methyl ether (NIOSH testimony at OSHA hearing September 1973)	No PEL Cancer-suspect agent Stringent workplace controls, recordkeeping, and medical surveillance required 29 CFR 1910.1006	Ca Use 29 CFR 1910.1006	Lung cancer	None
4,4'-Methylenebis-(2-chloroaniline) (MOCA) (Special Hazard Review September 1978)	Standard formally revoked by OSHA, August 1975	Ca 3 $\mu g/m^3$ TWA (lowest detectable limit)	Potential for cancer in humans; produced liver and lung tumors in animals	Chest x ray; blood and urine testing required
Methylene chloride (March 1976; revised April 1986 in CIB)	500 ppm, 8-hr TWA; 1,000 ppm acceptable ceiling; 2,000 ppm acceptable maximum peak for 5 min in any 2-hr period above the acceptable ceiling for an 8-hr shift	Ca Reduce exposure to lowest feasible limit	Potential for cancer in humans; produced tumors of the lung, liver, salivary, and mammary glands in animals	None

4,4′-Methylene-dianiline (CIB July 1986)	None	Ca Reduce exposure to lowest feasible limit	Bladder cancer; skin and liver effects	Prevent skin contact
Methyl parathion (September 1976)	None	0.2 mg/m³ TWA	Central nervous system effects	Prevent skin contact; blood monitoring required
Monohalo-methanes (CIB September 1984)	Methyl chloride: 100 ppm, 8-hr TWA; 200 ppm ceiling; 300 ppm acceptable maximum peak for 5 min in any 3-hr period above the acceptable ceiling for an 8-hr shift; methyl bromide: 20 ppm (80 mg/m³) ceiling (Skin); methyl iodide: 5 ppm (28 mg/m³), 8-hr TWA (Skin)	Ca Exposure to methyl chloride, methyl bromide, and methyl iodide should be reduced to lowest feasible level	Potential for cancer in humans; produced tumors of the kidney, forestomach, and lung in animals; methyl chloride should also be considered a potential teratogen	None
alpha-Naphthyl-amine (NIOSH testimony at OSHA hearing September 1973)	No PEL Cancer-suspect agent Stringent workplace controls, recordkeeping, and medical surveillance required 29 CFR 1910.1004	Ca Use 29 CFR 1910.1004	Bladder cancer	None
beta-Naphthyl-amine (NIOSH testimony at OSHA hearing September 1973)	No PEL Cancer-suspect agent Stringent workplace controls, recordkeeping, and medical surveillance required 29 CFR 1910.1009	Ca Use 29 CFR 1910.1009	Bladder cancer	None

*Date recommendation was published or testimony presented is in parentheses.

†NIOSH TWA recommendations are based on exposures up to 10 hours unless otherwise noted.

§Unless otherwise noted health effects cited are for humans.

Potential Hazard*	OSHA PEL'S/ Standard	NIOSH REL'S†/Other Recommendations	NIOSH Health Effect(s) Considered§	NIOSH Comments
Niax® Catalyst ESN (Joint NIOSH/ OSHA CIB May 1978)	OSHA and NIOSH recommend that exposure to Niax® Catalyst ESN and its components, dimethylaminopropionitrile and bis [2-(dimethylamino)ethyl] ether, be minimized		Urological disorders; nervous system effects	Work-practice and engineering controls to reduce exposure
Nickel carbonyl (Special Hazard Review May 1977)	1 ppb (7 μg/m^3), 8-hr TWA	Ca 1 ppb (7 μg/m^3) TWA (lowest detectable level)	Lung and nasal cancer	Chest x ray, pulmonary function testing, and urine monitoring required
Nickel, inorganic compounds (May 1977)	1 mg Ni/m^3, 8-hr TWA	Ca 15 μg Ni/m^3 TWA	Lung and nasal cancer; skin effects	Chest x ray and pulmonary function testing required
Nitric acid (March 1976)	2 ppm (5 mg/m^3), 8-hr TWA	2 ppm (5 mg/m^3) TWA	Dental erosion; nasal/lung irritation	Prevent skin and eye contact; chest x ray required
Nitriles (September 1978)	Acetonitrile: 40 ppm (70 mg/m^3), 8-hr TWA; tetramethyl succinonitrile: 0.5 ppm (3 mg/m^3), 8-hr TWA (Skin)	All are TWA values in ppm (mg/m^3): acetonitrile: 20 (34); n-butyronitrile: 8 (22); isobutyronitrile: 8 (22); propionitrile: 6 (14); malononitrile: 3 (8); adiponitrile: 4 (18); succinonitrile: 6 (20). All ceiling values (15 min) in ppm (mg/m^3): acetone cyanohydrin: 1 (4); glycolonitrile: 2 (5); tetramethyl succinonitrile: 1 (6). When present as mixtures or with other sources of cyanide, exposure to be considered additive and environmental limit to be calculated.	Hepatic, renal, respiratory, cardiovascular, gastrointestinal, and nervous system effects	Chest x ray and pulmonary function testing required; trained personnel and first-aid kits to be available during use; prevent skin and eye contact

4-Nitrobiphenyl (NIOSH testimony at OSHA hearing September 1973)	No PEL Cancer-suspect agent Stringent workplace controls, recordkeeping, and medical surveillance required 29 CFR 1910.1003	Ca Use 29 CFR 1910.1003	Potential for cancer in humans; produced bladder tumors in animals	None
Nitrogen, oxides of (March 1976)	NO_2: 5 ppm ($9\ mg/m^3$) ceiling; NO: 25 ppm ($30\ mg/m^3$), 8-hr TWA	NO_2: 1 ppm ($1.8\ mg/m^3$) ceiling (15 min); NO: 25 ppm ($30\ mg/m^3$) TWA	Respiratory and blood effects	Pulmonary function testing required
Nitroglycerin and ethylene glycol dinitrate (EGDN) (June 1978)	Nitroglycerin: $2\ mg/m^3$ (0.2 ppm) ceiling (Skin); EGDN: $1\ mg/m^3$ (0.2 ppm) ceiling (Skin)	$0.1\ mg/m^3$ ceiling (20 min) recommended limit for either substance alone or mixtures	Circulatory system effects	Prevent skin contact
2-Nitro-naphthalene (CIB December 1976)	None	Ca Reduce exposure to lowest feasible level	Bladder cancer	Compound metabolizes to beta-naphthylamine, a known carcinogen
2-Nitropropane (CIB April 1977; revised October 1980 in Joint OSHA/NIOSH Health Hazard Alert)	25 ppm ($90\ mg/m^3$), 8-hr TWA	Ca Reduce exposure to lowest feasible level	Potential for cancer in humans; produced liver tumors in rats	Medical monitoring with specific emphasis on liver function tests
N-Nitroso-dimethylamine (NIOSH testimony at OSHA hearing September 1973)	No PEL Cancer-suspect agent Stringent workplace controls, recordkeeping, and medical surveillance required 29 CFR 1910.1016	Ca Use 29 CFR 1910.1016	Potential for cancer in humans; produced tumors of the liver, kidney, lung, and nasal cavity in animals	None

*Date recommendation was published or testimony presented is in parentheses.

†NIOSH TWA recommendations are based on exposures up to 10 hours unless otherwise noted.

§Unless otherwise noted health effects cited are for humans.

Potential Hazard*	OSHA PEL'S/ Standard	NIOSH REL'S†/Other Recommendations	NIOSH Health Effect(s) Considered§	NIOSH Comments
Noise (August 1972)	90 dBA, 8-hr TWA	85 dBA TWA; 115 dBA ceiling	Hearing damage	None
Organotin compounds (November 1976)	0.1 mg/m^3, 8-hr TWA	0.1 mg Sn/m^3 TWA	Eye, skin, liver, nervous system, and cardiovascular effects	Chest x ray, blood and urine monitoring, eye tests, heart examination, and nervous system testing required; prevent skin and eye contact
Paint and allied coating products, manufacture of (September 1984)	Many aspects covered under the numerous OSHA regulations for general industry (29 CFR 1910)	Various recommendations for the handling of raw materials and finished products; dispersion of pigment or resin particles; thinning, tinting, and shading; filling; and laboratory functions	Injury and a wide range of toxicities considered	Paint and allied coating products include paints, varnishes, lacquers, stains, putties, and paint and varnish removers
Parathion (June 1976)	0.1 mg/m^3, 8-hr TWA (Skin)	0.05 mg/m^3 TWA	Nervous system effects	Prevent skin contact; blood monitoring required
Pentachloroethane (CIB August 1978)	None	To be handled in the workplace with caution	Central nervous system effects; possible liver and kidney effects	Exposures should be minimized due to the structural similiarity to the carcinogenic chloroethanes
Pesticides, manufacture and formulation of (July 1978)	Current OSHA PEL's or previous NIOSH REL's to be followed; stringent work-practice and medical surveillance requirements to be instituted. Pesticides considered in groups based on toxicity.		Wide range of toxicities considered; cancer; nervous and reproductive system effects	Blood monitoring required for some groups; workers to be warned of reproductive effects for some compounds; prevent skin contact

Phenol (July 1976)	5 ppm (19 mg/m^3), 8-hr TWA (Skin)	5.2 ppm (20 mg/m^3) TWA; 15.6 ppm (60 mg/m^3) ceiling (15 min)	Skin, eye, central nervous system, liver, and kidney effects	Prevent skin and eye contact
Phenyl-beta-naphthylamine (CIB December 1976)	None	Ca Reduce exposure to lowest feasible level	Bladder cancer	Compound metabolizes to beta-naphthylamine, a known carcinogen
Phosgene (February 1976)	0.1 ppm (0.4 mg/m^3), 8-hr TWA	0.1 ppm (0.4 mg/m^3) TWA; 0.2 ppm (0.8 mg/m^3) ceiling (15 min)	Respiratory effects	Pulmonary function testing and chest x ray required
Polychlorinated biphenyls (September 1977)	42% chlorine: 1 mg/m^3, 8-hr TWA; 54% chlorine: 0.5 mg/m^3, 8-hr TWA	Ca 1 μg/m^3 TWA (the minimum reliably detectable concentration using the recommended sampling and analytical methods)	Potential for cancer in humans; produced tumors of the liver, pituitary gland and leukemias in animals; skin, liver, and reproductive system effects	Blood testing required; female workers of child-bearing age and nursing mothers to be warned of potential adverse effects
Polychlorinated biphenyls (PCB's), potential health hazards from electrical equipment fires or failures (CIB February 1986)	42% chlorine: 1 mg/m^3, 8-hr TWA; 54% chlorine: 0.5 mg/m^3, 8-hr TWA	Ca Reduce exposure to lowest feasible limit	Potential for cancer in humans; produced tumors of the liver, pituitary gland and leukemias in animals; skin, liver, and reproductive system effects	Fire-related incidents involving PCB's have resulted in widespread contamination of buildings with PCB's and, in some cases, with PCDF's and PCDD's including TCDD. Emergency response personnel, maintenance or cleanup workers, or building occupants may be exposed to these compounds.

*Date recommendation was published or testimony presented is in parentheses.

†NIOSH TWA recommendations are based on exposures up to 10 hours unless otherwise noted.

§Unless otherwise noted health effects cited are for humans.

Potential Hazard*	OSHA PEL'S/ Standard	NIOSH		
		REL'S[†]/Other Recommendations	Health Effect(s) Considered[§]	Comments
Precast concrete products industry, comprehensive safety recommendations for (Technical Guideline June 1984)	Many aspects covered under numerous OSHA regulations for general industry (29 CFR 1910)	Various recommendations for safe work practices and worker training	Injury and death	Equipment, conditions, and many of the tasks specific to the industry are not covered under the existing regulations
beta-Propiolactone (NIOSH testimony at OSHA hearing September 1973)	No PEL Cancer-suspect agent Stringent workplace controls, recordkeeping, and medical surveillance required 29 CFR 1910.1013	Ca Use 29 CFR 1910.1013	Potential for cancer in humans; produced tumors of the liver, skin, and stomach in animals	None
Refined petroleum solvents (July 1977)	2,900 mg/m^3 (500 ppm), 8-hr TWA (Stoddard solvent)	Kerosene: 100 mg/m^3 TWA; all other solvents: 350 mg/m^3 TWA; 1,800 mg/m^3 ceiling (15 min)	Eye, nose, and throat irritation; dermatitis; nervous system effects	Blood and urine monitoring required; action level for petroleum ether, rubber solvent, naphtha: 200 mg/m^3 TWA; action level for mineral spirits and Stoddard solvent: 350 mg/m^3 TWA; action level for kerosene: 100 mg/m^3 TWA; prevent skin contact
Silica, crystalline (November 1974)	250/%SiO_2+5 in mppcf, or 10 mg/m^3/%SiO_2+2 (respirable quartz)	50 μg/m^3 TWA, respirable free silica	Chronic lung disease (silicosis)	Chest x ray, pulmonary function testing required
Sodium hydroxide (September 1975)	2 mg/m^3, 8-hr TWA	2 mg/m^3 ceiling (15 min)	Respiratory irritation	Prevent skin and eye contact

Styrene (September 1983)	100 ppm, 8-hr TWA; 200 ppm acceptable ceiling; 600 ppm maximum ceiling (5 min in 3 hr)	50 ppm (213 mg/m^3) TWA; 100 ppm (426 mg/m^3) ceiling	Nervous system effects; eye and respiratory system irritation; reproductive system effects	Prevent skin contact; workers to be warned of possible adverse reproductive effects
Sulfur dioxide (February 1974; revised May 1977 as part of NIOSH testimony at OSHA hearing)	5 ppm (13 mg/m^3), 8-hr TWA	0.5 ppm (1.3 mg/m^3) TWA	Respiratory effects	Pulmonary function testing required
Sulfuric acid (June 1974)	1 mg/m^3, 8-hr TWA	1 mg/m^3 TWA	Pulmonary irritation	Prevent skin and eye contact
2,3,7,8-Tetra-chlorodibenzo-*p*-dioxin (TCDD) (CIB January 1984)	None	Ca Reduce exposure to lowest feasible level	Potential for cancer in humans; produced tumors at many sites in animals; chloracne	None
1,1,1,2-Tetra-chloroethane (CIB August 1978)	None	To be handled in the workplace with caution	Central nervous system effects; possible liver and kidney effects	Exposures should be minimized due to the structural similiarity to the carcinogenic chloroethanes
1,1,2,2-Tetra-chloroethane (December 1976; revised in CIB August 1978)	5 ppm (35 mg/m^3), 8-hr TWA (Skin)	Ca Reduce exposure to lowest feasible level	Potential for cancer in humans; produced tumors of the liver in animals; liver, gastrointestinal, and nervous system effects	Prevent skin contact; blood monitoring required

*Date recommendation was published or testimony presented is in parentheses.

†NIOSH TWA recommendations are based on exposures up to 10 hours unless otherwise noted.

§Unless otherwise noted health effects cited are for humans.

Potential Hazard*	OSHA PEL'S/ Standard	NIOSH		
		REL'S[†]/Other Recommendations	Health Effect(s) Considered[§]	Comments
Tetrachloro-ethylene (July 1976; revised January 1978 in CIB)	100 ppm, 8-hr TWA; 200 ppm acceptable ceiling; 300 ppm maximum ceiling (5 min in 3 hr)	Ca Minimize workplace exposure levels; limit number of workers exposed	Potential for cancer in humans; produced tumors of the liver in animals	None
Thiols: n-alkane mono thiols, cyclohexanethiol, and benzenethiol (September 1978)	Butylmercaptan: (1-butanethiol) 10 ppm (35 mg/m^3), 8-hr TWA; ethylmercaptan (1-ethanethiol): 10 ppm (25 mg/m^3) ceiling; methylmercaptan (1-methanethiol): 10 ppm (20 mg/m^3) ceiling	All values in ppm (mg/m^3) ceilings (15 min): 1-methanethiol: 0.5 (1.0); 1-ethanethiol: 0.5 (1.3); 1-propanethiol: 0.5 (1.6); 1-butanethiol: 0.5 (1.8); 1-pentanethiol: 0.5 (2.1); 1-hexanethiol: 0.5 (2.4); 1-heptanethiol: 0.5 (2.7); 1-octanethiol: 0.5 (3.0); 1-nonanethiol: 0.5 (3.3); 1-decanethiol: 0.5 (3.6); 1-undecanethiol: 0.5 (3.9); 1-dodecanethiol: 0.5 (4.1); 1-hexadecanethiol: 0.5 (5.3); 1-octadecanethiol: 0.5 (5.9); cyclohexanethiol: 0.5 (2.4); benzenethiol: 0.1 (0.5); Mixtures of thiols to be controlled by calculation of equivalent concentrations	Irritation; eye, skin, blood, and nervous system effects	Blood and urine monitoring required; prevent skin contact

o-Tolidine (August 1978)	None	Ca 20 $\mu g/m^3$ ceiling (60 min)	Bladder cancer; nasal irritation	Urine testing required; quarterly urine monitoring recommended; prevent skin contact
o-Tolidine-based dyes (Joint NIOSH/OSHA Health Hazard Alert December 1980)	None	Ca Should be handled in the workplace with caution; minimize exposures	Bladder cancer	Substitute less toxic dyes wherever possible
Toluene (January 1974)	200 ppm, 8-hr TWA; 300 ppm acceptable ceiling; 500 ppm maximum ceiling (10 min)	100 ppm (375 mg/m^3), 8-hr TWA; 200 ppm (750 mg/m^3) ceiling (10 min)	Central nervous system depressant	None
Toluene diisocyanate (July 1973; revised—see Diisocyanates September 1978)	0.02 ppm (0.14 mg/m^3) ceiling	0.005 ppm (0.036 mg/m^3) TWA; 0.02 ppm (0.14 mg/m^3) ceiling (20 min)	Respiratory effects	Chest x ray, blood tests, pulmonary function testing required
1,1,1-Trichloroethane (July 1976)	350 ppm (1,900 mg/m^3), 8-hr TWA	350 ppm (1,910 mg/m^3) ceiling (15 min); action level set at 200 ppm (1,091 mg/m^3) TWA	Central nervous system, liver, and cardiovascular effects	Medical warning of possible congenital abnormalities required
1,1,2-Trichloroethane (CIB August 1978)	10 ppm (45 mg/m^3), 8-hr TWA (Skin)	Ca Reduce exposure to lowest feasible level	Potential for cancer in humans; produced liver tumors in animals; central nervous system effects	None

*Date recommendation was published or testimony presented is in parentheses.

†NIOSH TWA recommendations are based on exposures up to 10 hours unless otherwise noted.

§Unless otherwise noted health effects cited are for humans.

Potential Hazard*	OSHA PEL'S/ Standard	NIOSH REL'S†/Other Recommendations	NIOSH Health Effect(s) Considered§	NIOSH Comments
Trichloroethylene (July 1973; revised in Special Hazard Review January 1978)	100 ppm, 8-hr TWA; 200 ppm acceptable ceiling; 300 ppm maximum ceiling (5 min in 2 hr)	Ca 25 ppm TWA	Potential for cancer in humans; produced liver tumors in animals; central nervous system effects	Workers to be warned of hazards; 25 ppm level can be achieved by use of existing engineering control technology
Trimellitic anhydride (CIB February 1978)	None	Should be handled in the workplace as an extremely toxic substance	Pulmonary edema; immunologic sensitization; irritation of pulmonary tract, eyes, nose, and skin	Limit exposure to as few workers as possible while minimizing workplace levels
Tungsten and cemented tungsten carbide (September 1977)	None	Insoluble tungsten: 5 mg W/m^3 TWA; soluble tungsten: 1 mg W/m^3 TWA; dust of cemented tungsten carbide (containing > 2% cobalt): 0.1 mg Co/m^3 TWA; dust of cemented tungsten carbide (containing > 0.3% nickel): 15 μg Ni/m^3 TWA	Lung and skin effects	Pulmonary function testing and chest x ray required
Ultraviolet radiation (December 1972)	None	For spectral region of 315-400 nm: 1.0 mW/cm^2 for periods >1,000 sec; for periods ≤1,000 sec, 1,000 mW · sec/cm^2 (1.0 J/cm^2); for spectral region 200-315 nm: consult criteria document	Skin and eye effects	None
Vanadium (August 1977)	Vanadium pentoxide (dust): 0.5 mg/m^3 ceiling; (fume): 0.1 mg/m^3 ceiling; ferrovanadium: 1 mg/m^3, 8-hr TWA	Vanadium compounds: 0.05 mg V/m^3 ceiling (15 min); metallic vanadium and vanadium carbide: 1 mg V/m^3 TWA	Eye, skin, and lung effects	Pulmonary function testing and chest x ray required

Vibration syndrome (CIB March 1983)	None	Jobs should be redesigned to minimize the use of vibrating handtools; powered handtools should be redesigned to minimize vibration	Vibration syndrome; adverse circulatory and neural effects in the fingers	None
Vinyl acetate (September 1978)	None	4 ppm (15 mg/m^3) ceiling (15 min)	Irritation	None
Vinyl chloride (March 1974; reaffirmed June 1974 as part of NIOSH testimony at OSHA hearing)	1 ppm, 8-hr TWA; 5 ppm ceiling (15 min) 29 CFR 1910.1017	Ca Lowest reliably detectable level	Liver cancer	Liver function testing required
Vinyl halides (September 1978)	None except for vinyl chloride	Ca Vinyl halides to be controlled as specified for vinyl chloride in 29 CFR 1910.1017 with eventual goal of zero exposure	Liver cancer for vinyl chloride; potential for cancer from the other vinyl halides that have produced liver and kidney tumors in animals	Vinyl halides include vinyl chloride, vinylidene chloride, vinyl bromide, vinyl fluoride, and vinylidene fluoride monomers
Waste anesthetic gases and vapors (March 1977)	None for substances when used as anesthetic agents	Halogenated anesthetic agents: 2 ppm ceiling (1 hr); nitrous oxide: 25 ppm TWA during periods of use	Reproductive system effects and audiovisual performance decrements	Workers to be advised of potential effects; abnormal outcome of pregnancies of workers and spouses to be documented
Xylene (May 1975)	100 ppm (435 mg/m^3), 8-hr TWA	100 ppm (434 mg/m^3) TWA; 200 ppm (868 mg/m^3) ceiling (10 min)	Central nervous system depressant; respiratory irritation	None
Zinc oxide (October 1975)	5 mg ZnO/m^3, 8-hr TWA	5 mg ZnO/m^3 TWA; 15 mg ZnO/m^3 ceiling (15 min)	Metal fume fever	None

*Date recommendation was published or testimony presented is in parentheses.

†NIOSH TWA recommendations are based on exposures up to 10 hours unless otherwise noted.

§Unless otherwise noted health effects cited are for humans.

1,1,1-TRICHLOROETHANE

CAS: 71-55-6

CH_3CCl_3 1988 TLV-350 ppm

Synonyms: Methylchloroform; α-trichloroethane

Physical Form. Colorless liquid

Uses. Solvent and cleaning agent

Exposure. Inhalation; moderate skin absorption

Toxicology. 1,1,1-trichloroethane causes central nervous system depression.

A number of human fatalities related to industrial exposure in closed spaces have been reported, some of which may have been "sudden deaths" due to sensitization of the myocardium to epinephrine.[1]

Based on effects caused in monkeys and rats, the following effects are expected in humans: 20,000 ppm for 60 minutes, coma and possibly death; 10,000 ppm for 30 minutes, marked incoordination; 2000 ppm for 5 minutes, disturbance of equilibrium.[2] Human subjects exposed to 900 to 1000 ppm for 20 minutes experienced lightheadedness, incoordination, and impaired equilibrium; transient eye irritation has also been reported at similar concentrations.[1] Impairments in psychomotor task performance have been demonstrated at levels around 350 ppm.[3,4] Other studies at similar exposure levels have failed to show any impairment, but the type of task chosen to test behavioral effects and the times at which behavioral measures were sampled during the course of exposure may explain the variations from study to study.[3]

An epidemiologic study of 151 matched pairs of exposed textile workers revealed no evidence of cardiovascular, hepatic, renal, or other effects as a function of exposure; for some workers, exposures exceeded 200 ppm and duration of exposure ranged from several months to 6 years.[5]

A few scattered reports have indicated mild kidney and liver injury in humans from severe exposure; animal experiments have confirmed the potential for liver, but not for kidney, injury.[1] Skin irritation has occurred from experimental skin exposure to the liquid and from occupational use. The liquid can be absorbed through the skin, but this is not a significant route of toxic exposure.

In dogs, myocardial sensitization to epinephrine occurred at concentrations of 5,000 to 10,000 ppm.[6] In a carcinogenicity study, rats and mice were given the liquid orally at two different dose levels, 5 days a week for 78 weeks.[7] Both female and male test animals exhibited early mortality compared with untreated controls, and a variety of neoplasms were found in both treated animals and controls. Although rats of both sexes demonstrated a positive dose-related trend, no relationship was established between the dosage groups and the species, sex, type of neoplasm, or sites of occurrence. The IARC concluded that an evaluation of the carcinogenicity of 1,1,1-trichloroethane could not be made.[8]

In a more recent study, rats exposed at 1500 ppm for 6 hr/day 5 days/week for 2 years showed no oncogenic effects.[9]

Inhalation exposure of female rats before mating and during pregnancy to 2100 ppm caused an increased incidence of skeletal and soft tissue variation in the offspring indicative of developmental delay; no persistent detrimental effects were found in the offspring at 12 months of age.[10]

The odor threshold has been described by various investigators as ranging from 16 to 400 ppm.[1]

REFERENCES

1. National Institute for Occupational Safety and Health: Criteria for a Recommended Standard. Occupational Exposure to 1,1,1-Trichloroethane (Methyl Chloroform). DHEW (NIOSH) Pub No 76-184, pp 16–96. Washington, DC, US Government Printing Office, 1976
2. 1,1,1-Trichloroethane—emergency exposure limits. Am Ind Hyg Assoc J 25:585, 1964
3. Mackay CJ et al: Behavioral changes during exposure to 1,1,1-trichloroethane: Time-course and relationship to blood solvent levels. Am J Ind Med 11:223–239, 1987
4. Gamberale F, Hultengren M: Methylchloroform exposure. II. Psychophysiological functions, work environ. Health 10:82–92, 1973
5. Kramer C et al: Health of workers exposed to 1,1,1-trichloroethane: A matched-pair study. Arch Environ Health 33:331–342, 1978
6. Reinhardt CF, Mullin LS, Maxfield ME: Epinephrine-induced cardiac arrhythmia potential of some common industrial solvents. J Occup Med 15:953, 1973
7. National Cancer Institute: Bioassay of 1,1,1-Trichloroethane for Possible Carcinogenicity, Technical Report Series No. 3. DHEW (NIH) Pub No 77-803. Washington, DC, US Government Printing Office, 1977
8. IARC Monographs on the Evaluation of the Carcinogenic Risk of Chemicals to Humans, Vol 20, Some Halogenated Hydrocarbons. pp 515–531. Lyon, International Agency for Research on Cancer, 1979
9. Quast JF, Calhoun LL: Chlorothene VG*: A Chronic Inhalation Toxicity and Oncogenicity Study in Rats and Mice. Part II. Results in Rats. Final Report, Feb 5, 1986, pp 1–165. Midland, Michigan, Mammalian and Environmental Toxicology Research Laboratory, Dow Chemical, 1986
10. York RG et al: Evaluation of teratogenicity and neurotoxicity with maternal inhalation exposures to methyl chloroform. J Toxicol Environ Health 9:251–266, 1982

Index of Compounds and Synonyms

Principal compounds are capitalized. Synonymous names appear in lowercase unless they are trade names.

CAS Number Index